Biodesign

The Process of Innovating Medical Technologies

Where do you begin as a medical technology innovator? What lessons can you learn from experienced inventors? How can you improve your chances of success?

Learn to innovate, recognize market opportunities, apply the design process, and develop business acumen with this "hands-on" guide to medical technology innovation. The biodesign innovation process begins with careful identification of a clinical need and moves in a stepwise approach through inventing and planning the implementation of a marketable solution. The process is based on the combined experience of literally hundreds of medtech innovators who are featured in the book through quotations, vignettes, and case studies.

- Master the three-phase biodesign process for innovating medical technologies – *identify → invent → implement*
- Understand the complete picture of medtech innovation through medical, engineering, and business perspectives
- Take action using the step-by-step instructions and supporting resources outlined in the *Getting Started* section for each chapter
- Access thousands of active links and additional information via the online companion to the book – *ebiodesign.org*

"Everything you ever wanted to know about medical device entrepreneurship and more. [The authors] have led an A-class team of experienced device company builders to produce a reference document to guide aspiring device entrepreneurs through all the challenges of getting an idea to market. These are tough times. Whether you're a physician with an idea, an engineer or a businessman, this is a unique and powerful resource."

John Abele, *Founder/Chairman Boston Scientific*

"I don't know of any other text that has the wealth of practical and usable information on the entrepreneurial process as *Biodesign*. This is a much needed 'how-to' book written by people who actually have done it many times themselves. No thirty-thousand foot views necessary or appropriate here. Each chapter has a 'Getting Started' section that will help guide the budding entrepreneur through the necessary steps. This book should be required reading for anyone wanting to develop a new medical device or to start a new company in the medical field. "

William Brody, *President of the Salk Institute and Former President of Johns Hopkins University*

"The chapters are thoughtfully organized. With an excellent blending of scientific information, clinical problems, and examples of solutions, including case studies, the book has succeeded in accomplishing its goal of being very practical... *Biodesign* will be the standard in this very important field. It will be of great value in the education of undergraduate and graduate students in biomedical engineering and related fields, as well as for industrial scientists and university faculty who educate/train young bioengineers or want to pursue the process of innovating new medical technologies themselves."

Shu Chien, *Professor of Bioengineering, University of Califonia, San Diego*

"*Biodesign: The Process of Innovating Medical Technologies* is a wonderful guide with lucent case studies that illustrate the critical steps necessary for the translation of ideas into commercial solutions. It is the *Grey's Anatomy* of device innovation."

William Hawkins, *Chairman and CEO of Medtronic*

"*Biodesign: The Process of Innovating Medical Technologies* is direct, clear, and simultaneously sophisticated yet practical as it unravels the many issues related to successfully navigating the entire biodesign path from concept to final product launch. I highly recommend that anyone seriously interested in developing an entrepreneurial venture in the medical products field read this book. It is likely to spare budding entrepreneurs a lot of trial-and-error and painful on-the-job training."

Dean Kamen, *Inventor and Founder/President of DEKA Research and Development*

"In *Biodesign*, the Stanford team has assembled a treasure trove of methods for medical device innovation. The book is certain to become an invaluable reference for students, instructors, and practitioners alike."

Karl T. Ulrich, *CIBC Professor of Entrepreneurship and eCommerce, The Wharton School*

"This comprehensive text provides clear guidance through every step of the biodesign process, from identification of market need to successful entrée into a complex, competitive marketplace. The authors of this book – faculty in Stanford's Biodesign Program – have done innovators a great service in shaping the study of biodesign and training students to put this knowledge into practice. Their expertise is self-evident, and, with this book, is now accessible to anyone serious about succeeding in biotechnology."

Miles White, *Chairman and Chief Executive Officer, Abbott*

Biodesign
The Process of Innovating Medical Technologies

Senior editors
Stefanos Zenios
Josh Makower
Paul Yock

Associate editors
Todd J. Brinton
Uday N. Kumar

Principal writer
Lyn Denend

Specialty editor
Thomas M. Krummel

Web editor
Christine Kurihara
(ebiodesign.org)

CAMBRIDGE UNIVERSITY PRESS
Cambridge, New York, Melbourne, Madrid, Cape Town, Singapore, São Paulo, Delhi

Cambridge University Press
The Edinburgh Building, Cambridge CB2 8RU, UK

Published in the United States of America by Cambridge University Press, New York

www.cambridge.org
Information on this title: www.cambridge.org/9780521517423

Copyright © 2010 by the Board of Trustees of
the Leland Stanford Junior University. All rights
reserved. Used with permission from the
Stanford University Graduate School of Business.

This publication is in copyright. Subject to statutory exception
and to the provisions of relevant collective licensing agreements,
no reproduction of any part may take place without
the written permission of Cambridge University Press.

First published 2010. Reprinted 2010.

Printed in the United States at Edwards Brothers Inc., Ann Arbor, MI.

A catalogue record for this publication is available from the British Library

ISBN 978-0-521-51742-3 hardback

Additional resources for this publication at www.cambridge.org/9780521517423 and ebiodesign.org

Cambridge University Press has no responsibility for the persistence or
accuracy of URLs for external or third-party internet websites referred to
in this publication, and does not guarantee that any content on such
websites is, or will remain, accurate or appropriate.

*To innovators – past, present, and future
– and the patients who inspire them.*

Contents

Foreword	page viii
Preface	ix
The Biodesign Community	xiv
Biographies	xix
Glossary	xxi

IDENTIFY

Stage 1 Needs Finding — 1
1.1 Strategic Focus — 4
1.2 Observation and Problem Identification — 20
1.3 Need Statement Development — 37
Case Study: Stage 1 — 51

Stage 2 Needs Screening — 57
2.1 Disease State Fundamentals — 60
2.2 Treatment Options — 74
2.3 Stakeholder Analysis — 95
2.4 Market Analysis — 117
2.5 Needs Filtering — 143
Case Study: Stage 2 — 165

INVENT

Stage 3 Concept Generation — 173
3.1 Ideation and Brainstorming — 176
3.2 Concept Screening — 193
Case Study: Stage 3 — 205

Stage 4 Concept Selection — 207
4.1 Intellectual Property Basics — 210
4.2 Regulatory Basics — 273
4.3 Reimbursement Basics — 299
4.4 Business Models — 319
4.5 Prototyping — 340
4.6 Final Concept Selection — 367
Case Study: Stage 4 — 378

IMPLEMENT

Stage 5 Development Strategy and Planning — 385
5.1 Intellectual Property Strategy — 388
5.2 Research and Development Strategy — 407
5.3 Clinical Strategy — 425
5.4 Regulatory Strategy — 458
5.5 Quality and Process Management — 473
5.6 Reimbursement Strategy — 503
5.7 Marketing and Stakeholder Strategy — 536
5.8 Sales and Distribution Strategy — 556
5.9 Competitive Advantage and Business Strategy — 580
Case Study: Stage 5 — 596

Stage 6 Integration — 609
6.1 Operating Plan and Financial Model — 612
6.2 Business Plan Development — 657
6.3 Funding Sources — 676
6.4 Licensing and Alternate Pathways — 708
Case Study: Stage 6 — 727

Image Credits	734
Index	735

See **ebiodesign.org** for active web links to the resources listed in each chapter, additional references, content updates, video FAQs, and other relevant information.

Foreword

As you begin ... a note from Tom Fogarty

Over the years I have spent developing new technologies, and watching innovators succeed or fail, I have identified some basic principles that are critical to success, and those that cause failure. The most important principle is that we innovate to improve the lives of patients. Commitments to ourselves, the institution we serve, and others are secondary. Distractions along the way are multiple. The love of money, the lure of technology, personal advancement, and recognition by our peers are only a few. Even with these distractions and institutional encumbrances, innovators are here to serve our patients first and foremost. If this is done well, benefits to the innovator will follow.

I have always thought that innovation is something you learn by doing. However, I do believe that certain individuals are born with a capacity to innovate that is significantly greater than that of others. It is much like the field of sports; some are innately more capable. Regardless of where one lies in this spectrum, listening to your mentors is probably the most critical component of your success. Persistence is the second most important factor (knowing when to hold 'em and when to fold 'em). Before you give up, reference anybody knowledgeable in the field, including your mentors, friends, and enemies. Yes, enemies – they often have insights and offer perspectives that friends will ignore or not articulate. Seek the truth, no matter where it lies.

An idea, by itself, has no importance whatsoever; it is the implementation of that idea and its acceptance by others that brings benefit to our patients. In this day and age, it is extremely difficult to successfully bring a concept to reality without the help of a myriad of others from different disciplines. The importance of their contributions should never be underestimated. The concept of value allocation becomes very important. Innovators often handle this badly. If there is no implementation of the concept or idea, there might as well be no concept or idea.

How to go about implementation is not intuitively obvious – and this is an area where the *Biodesign* text is useful. There is practical material in these chapters that can make the path to implementation clearer, particularly for the physician or engineer who may have seen only parts of this process before. It is also important that the first third of this book focuses on how to get the clinical need right. There is nothing more critical in the innovation process than starting with a truly significant patient need.

One final thought: the path to successful innovation is very often lonely and frustrating. Innovation by its very definition means something different than what exists. Basically we are defying standards and sometimes basic concepts. Be prepared to be criticized, ostracized, called crazy, inappropriate, outlandish, stupid, intolerable, and bound to fail. I myself have been called all of these names and many more that I can't remember or mention. Take solace from the fact that these challenges can be a useful part of the process of innovation. Overcoming obstacles that you recognize (and those that you don't) will occur. Ultimately, your ability to prevail through these challenges will benefit patients, caregivers, and institutions.

Thomas Fogarty, MD, is a cardiovascular surgeon and one of the most prolific medical device inventors in history, with many of his technologies in active use across a wide spectrum of patient care. He has founded or co-founded over 30 companies and was inducted into the National Inventors Hall of Fame in 2001.

Preface

If you have the desire to develop new medical technologies, there is a world of opportunity available to you. Health and quality of life are central issues for every human being on the planet. Through advances in science and technology, the complexities of the human body are being revealed, creating new ways to solve clinical needs that no one imagined previously. Medicine and surgery are more open for innovation than at any time in history.

Despite this promise, however, medtech innovators face significant hurdles. If not managed skillfully, patents, regulatory approval, reimbursement, market dynamics, business models, competition, financing, clinical trials, technical feasibility, and team dynamics (just to name a few of the many challenges) can all prevent even the best idea from reaching patient care. So, where should you begin as an innovator? What process can you use to improve your chances of success? What lessons can you learn from the inventors, engineers, physicians, and entrepreneurs who have succeeded and failed in this endeavor before? This book has been developed to provide practical answers to these important questions.

The text is based on a simple premise: that innovation is both a process and a skill that can be learned. While some may have more natural ability than others, everyone can be an innovator. The biodesign innovation process, as we call it, is described here in a way that is specific to the development of medical technologies, but the same general approach is followed by successful innovators in many fields.

This process is intended to provide you with a starting point. Each phase, stage, and core activity detailed within the book includes information to help you effectively capitalize on important opportunities and overcome common obstacles. Yet, as an innovator, you should adapt and modify this approach to reflect your own style and personal emphasis. It is our hope that by executing your own version of the biodesign innovation process, you will be able to navigate confidently the many twists and turns that lie ahead.

Genesis of the book

The idea for the book is the result of our experience in developing the biodesign innovation and fellowship programs at Stanford over the past eight years. It began as a collaboration between Josh Makower and Paul Yock, triggered by a chance conversation at breakfast at Il Fornaio in Palo Alto. Makower had previously created a medical devices innovation training program at Pfizer called "Pfreshtech" before launching his career as a serial medtech founder and entrepreneur. Yock, a professor of bioengineering and medicine, was interested in developing a graduate program in medical technology innovation that could leverage the deep medtech expertise and inventive culture of the Silicon Valley. The two agreed to work together to create a training initiative as a part of the Stanford University Program in Biodesign, which Yock directs. Stefanos Zenios, a professor of operations, information, and technology, and an expert in health systems from the Graduate School of Business (GSB), joined the biodesign faculty group and provided the conceptual organization for the biodesign process that is presented here. Todd Brinton, an alum of the fellowship program, current fellowship director, and a medtech company founder, served as an associate editor and contributed his insights. Uday Kumar, also an associate editor and alum of the fellowship, contributed to text from his experience as a cardiac electrophysiologist and founder and chief medical officer of a medtech company. Tom Krummel, chair of surgery, joined as codirector of biodesign and further

Preface

Figure P1 The team. Back row: Todd J. Brinton, Thomas M. Krummel, Uday N. Kumar; middle row: Josh Makower, Paul Yock, Christine Kurihara, Stefanos Zenios; front row: Lyn Denend.

refined the innovation process with a focus on clinical needs finding and validation. Principal writer Lyn Denend, from the Stanford GSB, initially joined us to help develop a series of notes to support the teaching syllabus for the course. Over time, the team worked together to significantly expand and enhance this material for the book. Chris Kurihara developed the web companion to the text, ebiodesign.org (see Figure P1).

In providing the material here, we have drawn from the talks of more than 200 industry speakers who have presented in the class and also, more significantly, from our experience advising more than one hundred project teams over the past eight years. Already ten of these projects have been converted to venture-backed companies, and 20,000 patients have been treated with devices from these organizations. To validate the principles described in the book we have also performed extensive field-based research. Being located in Silicon Valley with hundreds of medtech start-ups within a 50-mile radius of Stanford, we are fortunate to have unparalleled access to seasoned practitioners who have shared their insights with us. We have interviewed dozens of innovators and captured their experience as brief vignettes and more extensive "From the field" case studies to help demonstrate how many of the key issues we highlight manifest themselves in real-world situations.

Organization of the book and its supporting website

Biodesign: The Process of Innovating Medical Technologies divides the biodesign innovation process into three distinct phases.

- **Identify**: How do you identify an important unmet medical need where there is good clinical, scientific, and market knowledge to suggest that a solution to the need will be feasible and will have a reasonable likelihood of commercial viability?
- **Invent**: How do you next develop a solution to this need, taking advantage of the creative group process and the power of prototyping?
- **Implement**: How do you then transform an idea and a prototype into a product that can be used at the bedside to treat patients?

These three phases are further subdivided into a total of six stages and 29 core activities (with a chapter on each one). The diagram shown in Figure P2 summarizes the overall process and illustrates the interaction among the phases, stages, and activities. To help you navigate the content if you are new to innovation, we have organized the book in a linear fashion that parallels the course we teach and the process followed by many of the innovators we have interviewed. The fact that, in practice, many of these activities require a parallel and iterative approach

IDENTIFY

STAGE 1 NEEDS FINDING
1.1 Strategic Focus
1.2 Observation and Problem Identification
1.3 Need Statement Development

STAGE 2 NEEDS SCREENING
2.1 Disease State Fundamentals
2.2 Treatment Options
2.3 Stakeholder Analysis
2.4 Market Analysis
2.5 Needs Filltering

INVENT

STAGE 3 CONCEPT GENERATION
3.1 Ideation and Brainstorming
3.2 Concept Screening

STAGE 4 CONCEPT SELECTION
4.1 Intellectual Property Basics
4.2 Regulatory Basics
4.3 Reimbursement Basics
4.4 Business Models
4.5 Prototyping
4.6 Final Concept Selection

IMPLEMENT

STAGE 5 DEVELOPMENT STRATEGY AND PLANNING
5.1 Intellectual Property Strategy
5.2 Research and Development Strategy
5.3 Clinical Strategy
5.4 Regulatory Strategy
5.5 Quality and Process Management
5.6 Reimbursement Strategy
5.7 Marketing and Stakeholder Strategy
5.8 Sales and Distribution Strategy
5.9 Competitive Advantage and Business Strategy

STAGE 6 INTEGRATION
6.1 Operating Plan and Financial Model
6.2 Business Plan Development
6.3 Funding Sources
6.4 Licensing and Alternate Pathways

Figure P2 The biodesign innovation process.

Preface

is addressed in the chapters where it is most essential. If you are a more experienced reader, we have attempted to make individual chapters as complete and self-contained as possible so you can refer directly to the chapter most relevant to the challenges you are currently facing, without necessarily having to read those that precede it.

At the end of each chapter is a Getting Started section that outlines a practical, action-oriented roadmap that you can follow to execute the steps in the biodesign innovation process when working on an actual project. The roadmaps are supported by lists of resources and references to provide you with additional information, and they are mirrored on the website **ebiodesign.org** with active web links for each step.

Who will use the book?

Initially, this material was developed to support project-based classes in medical technology innovation. Over time, however, we have used the content in a variety of settings and with different audiences, both inside and outside the university, and have found it to be valuable for a much broader cross-section of readers. Certain parts of the text are particularly appropriate for these different groups.

Undergraduates will benefit most from the 16 chapters (1.1–4.6) in the first two phases (Identify and Invent). Students in capstone biomedical engineering design classes can use this book as a primary resource, coupled with an engineering text from the relevant discipline (mechanical, electrical, or biomedical engineering). For classes in which the clinical need is provided up-front, we recommend beginning with Chapter 2.1 in the Needs screening stage.

Graduate students in medicine, business, or engineering can use the book to learn a process for inventing and commercializing medical technologies. The chapters on implementation (5.1–6.4) deal with more advanced, strategic topics that innovators will encounter as they move toward commercialization of their concepts.

Students in business plan courses with a medical product idea will benefit by using the book as a medical-specific template for organizing their business plan development, with the chapters from the Needs screening stage (2.1–2.5) and from the Implement phase (5.1–6.4) being the most directly relevant.

Faculty interested in translational research may follow the steps in this book to develop a research and implementation plan for a technology or an idea with a potential clinical application. Chapters 1.1 to 4.6 will be the most directly useful.

Emerging entrepreneurs and inventors can leverage the book from beginning to end, using it as a roadmap for all steps in their journey – from evaluating a potential area of focus for their venture, to developing an execution plan, raising funds, and beyond.

Investors can draw information from the text to support a detailed due diligence checklist for evaluating opportunities and business plans in the medical device field.

Last, but not least, ***industry executives*** will discover that this book provides an innovation template and nomenclature that they can adopt within their own organization.

Medtech versus biotech and pharma

The book has an intentional focus on medical technologies, which we define as medical devices, diagnostics (including imaging and molecular diagnostics), and drug delivery. Its content is not as relevant to biotechnology and pharmaceuticals, primarily because some of the distinctive features of medical technology innovation do not translate directly to these other sectors. Although much has been made about the ultimate convergence of medtech and biotech/pharma, we believe that, for the foreseeable future, the innovation process for these areas will continue to have fundamental differences.

If you work in the pharmaceutical and biotechnology sector, you will still find that several portions of this book are relevant (e.g., 2.1 Disease State Fundamentals, 2.2 Treatment Options, 2.4 Market Analysis, 5.3 Clinical Strategy, and 5.6 Reimbursement Strategy) but other stages in the process (such as Stage 1 Needs Finding or Stage 3 Concept Generation) do not apply directly. In particular, in medtech there is a distinct emphasis on clinical need identification as the *initial step* in the innovation process. In contrast, most recent innovations in biotechnology or pharmaceuticals start with a breakthrough in the understanding of basic biological mechanisms at the bench, not at the bedside. The ideation phase is also unique in medtech, both because it starts with an explicit clinical need and because it is characterized by a close, cyclical interaction between prototyping and idea generation. The implementation stage has superficial similarities between medtech and biotech/pharma (such as FDA approval), but the processes and strategies used to overcome these hurdles are significantly different. This applies also to the business characteristics of medtech innovation – a new medtech product can reach the market in less than half the time a drug takes, and is likely to cost a small fraction in development expenses. All these differences mean that a biodesign innovation process can be applied to the pharmaceutical and biotechnology sectors, but it will have fundamental differences from the process described here.

Geographic focus

This material has a primary focus on the United States for two main reasons. The United States continues to be the world's largest medical device market, and our location in Silicon Valley provides us with unique insights from the epicenter of medtech start-ups. However, the overall process is global and can be readily applied by innovators targeting other markets. Of course, there are differences across markets that are driven by regulatory, reimbursement, and clinical policy variations. To address this, the book highlights where such differences exist, provides directional guidance for some important global markets, and gives you resources and ideas for how to further investigate markets outside the United States.

Ethics

As a prospective medical device innovator, your endeavors will involve patients' lives. If you are successful, your inventions may prolong life and alleviate pain, but the process of developing and testing these devices may expose patients to risks. Well-articulated ethical principles should guide the conflicts of interests that have the potential to arise throughout the biodesign innovation process. For this reason, rather than addressing ethics in a single, dedicated chapter, a discussion of ethics is embedded in the chapters where conflicts and ethical issues are most likely to arise. Guiding principles for effectively managing these ethical concerns are also provided to ensure that patients' best interests always come first in your journey.

Web resources: ebiodesign.org

Given the dynamic, fast-paced nature of the medtech industry, we have created *ebiodesign.org* as a companion to this text. Important updates and information about relevant industry changes will be posted here, along with video commentary from experts and frequently asked questions for each chapter. Additionally, ebiodesign.org provides an up-to-date list of active references that support each chapter of the book. We intend this to be a valuable resource and welcome your suggestions regarding useful material to include on the site.

Launching the biodesign innovation process

As the many innovators who have contributed to this book will tell you, biodesign is an exhilarating journey: you have in front of you the opportunity to deliver ideas and technologies that will transform healthcare for generations to come. We hope this book will help you to move more effectively toward that goal.

The Biodesign Community

This book carries the fingerprints of literally hundreds of contributors. One key set of experts who helped shape the material is the Leadership Group for the Program in Biodesign. We particularly wish to thank Richard Popp who heads our ethics and policy section; Craig Milroy who directs the biodesign prototyping collaboratory; Tom Andriacchi who advises on educational programs; Mike Gertner, Geoff Gurtner, and Paul Wang who are members of the core faculty; and Chris Shen, Julian Gorodsky, Jack Linehan, and Peter Fitzgerald who mentor the fellows. Our international focus has expanded recently through a new program called Stanford-India Biodesign, led by Executive Directors Raj Doshi and Balram Bhargava. The Biodesign fellowship program is generously supported by prominent medtech innovators who have also contributed to this text, including Tom Fogarty, Eberhard Grube, Julio Palmaz, John Simpson, and Simon Stertzer. A number of other key individuals and firms provide advice and support to the program, as outlined on the Stanford Biodesign website.

Biodesign is a unit of Stanford's innovative life sciences initiative called Bio-X. We are grateful to the leaders (Matt Scott, Carla Schatz, and Heideh Fattaey) who have provided encouragement and support as biodesign has grown up. The innovation class on which the text is based is hosted in the Department of Bioengineering and the Graduate School of Business (GSB). We have had the great benefit of advice and guidance from founding chair of the department, Scott Delp, as well as the ongoing support of the subsequent chair Russ Altman. Through the department, our experience with the Wallace H. Coulter Translational Research Partnership program has provided valuable experience in university technology transfer in the medtech space.

Our approach to biodesign draws heavily from our colleagues in the design initiatives at Stanford (the Hasso Plattner Institute of Design), as well as their colleagues at IDEO, Inc. We want to particularly acknowledge David Kelley, Tom Kelley, Dennis Boyle, Tad Simmons, and George Kembel for their considerable input into the program and this project. We are also grateful for the support of the Stanford Technology Ventures Program, especially Tom Byers and Tina Seelig.

The development of the biodesign program would not have been possible without the explicit support of Dean Philip Pizzo and Senior Associate Dean Harry Greenberg from the School of Medicine, Dean James Plummer from the School of Engineering, and Dean Robert Joss as well as Associate Deans Dave Kreps and Mary Barth from the GSB. Their camaraderie, willingness to experiment with an unusual interdisciplinary program, and ongoing support were critical to our success.

The text grew out of the biodesign fellowship and class. One of the first fellows in the program, Asha Nayak, developed a manual for the fellowship that contained practical information on needs finding, inventing, and developing ideas. The manual served as a motivation and guide for developing an expanded teaching syllabus and, ultimately, this text. One of the first business school students in biodesign, Darin Buxbaum, played a crucial role in building on Asha's manual to develop a prototype for several of the early chapters and Getting Started sections. Trena Depel and several others helped to develop and refine specific content. Their contributions are individually acknowledged at the end of the relevant chapters. The organization of the weblinks in ebiodesign.org, the online companion to the text, was coordinated by Abigail Garner and was supported by grants from the Kauffman and Argosy Foundations.

Subsequent generations of biodesign fellows and students have been "test subjects" for the material in this book. We are grateful for their input and proud of what they are accomplishing in their careers as innovators.

We wish to thank the staff of the biodesign program for the extensive efforts required to keep the various educational aspects of the program running smoothly. We are particularly grateful to Roula El-Asmar, Andrea Daniel, Mary Gorman, and Dawn Wojick, as well as alumni staff, including managing director Sandy Miller, educational coordinator Teresa Robinson, along with Quynchi Nguyen, Tracy Okamoto, Rebecca Huang, and Laura Dyball. From the GSB, Margot Sutherland of the Case Writing Office and Kim Simmons from Jackson Library were especially supportive of our efforts to develop a comprehensive set of teaching notes for our course, which led to this book. Diana Reynolds Roome and Malisa Young also provided key support in finalizing the manuscript. Michelle Carey, our primary contact at Cambridge University Press, provided invaluable assistance in helping us navigate the publishing process.

Finally, this book has been shaped by input from hundreds of medtech experts who have participated in the biodesign program as lecturers, speakers, mentors, coaches, and advisors. These experts have helped us to frame the biodesign process and hone the teaching material that has evolved into this text. We would like to thank sincerely the members of the community who are listed here – and those in the updated index of contributors found at **ebiodesign.org**.

John Abele
David Adams
John Adler
Tom Afzal
Todd Alamin
Cliff Alferness
Russ Altman
Evan Anderson
Roger Anderson
Thomas Andriacchi
Aimee Angel
Patrick Arensdorf
Paul Auerbach
David Auth
Kityee Au-Yeung
Michael Baker
Juliet Bakker
Lonnie Barish
David Barlow
Mary Barth
Shubhayu Basu
Amir Belson
Ian Bennett
Michael Berman
Balram Bhargava
Annette Bianchi

Michael Billig
Gary Binyamin
Howard Birndorf
Jeffrey Bleich
Nikolas Blevins
Dan Bloch
Mark Blumenkranz
Karen Boezi
Leslie Bottorff
David Boudreault
Kathryn Bowsher
Dennis Boyle
Corinne Bright
Earl "Eb" Bright
Sal Brogna
Bruce Buckingham
Edmund Bujalski
Darin Buxbaum
Robert Buyan
Brook Byers
Thomas Byers
Colin Cahill
Matthew Callaghan
John Capek
Michelle Carey
Dennis Carter

Michael Carusi
David Cassak
John Cavallaro
Kathryn Cavanaugh
Venita Chandra
John Chang
Kevin Chao
Henry Chen
Robert Chess
Kyeongjae Cho
Michael Chobotov
Tony Chou
Douglas Chutorian
Thomas Ciotti
Jessica Connor
Kevin Connors
Christos Constantinou
Brent Constantz
Craig Coombs
Jim Corbett
Benedict James Costello
Jack Costello
Robert Croce
Gary Curtis
Robert Curtis
Mark Cutkosky

Karen Daitch
Michael Dake
Ronald Dalman
Andrea Daniel
Reinhold Dauskardt
Liz Davila
Alison de Bord
Mark Deem
Jeani Delagardelle
Scott Delp
Trena Depel
Carey deRafael
Parvati Dev
Ronald Dollens
Ricardo Dolmetsch
Dennis Donohoe
Rajiv Doshi
David Douglass
Maurice Druzin
Laura Dyball
Debra Echt
Zachery Edmonds
Stephen Eichmann
Roula El-Asmar
Erik Engelson
William Enquist

The Biodesign Community

Laura Wilkes-Evans
Christian Eversull
William Facteau
Brian Fahey
Steve Fair
James Fann
Heideh Fattaey
Joelle Faulkner
William Fearon
David Feigal
Jeffrey Feinstein
Nina Fernandes
Richard Ferrari
John Ferrell
Louie Fielding
Frank and Jeanne Fischer
Robert Fisher
Harvey Fishman
Peter Fitzgerald
Thomas Fogarty
David Forster
George Foster
Stuart Foster
Daniel Francis
Matthew Frinzi
Jan Garfinkle
Robert Garland
Abigail Garner
Michi Garrison
James Geriak
Gary Gershony
Michael Gertner
Hanson Gifford
Jack Gill
Nicholas Giori
Benjamin Glenn
Gary Glover
Paul Goeld
Tom Goff
Garry Gold
Jeffrey Gold
Charlene Golding
Rex Golding
Joel Goldsmith
Richard Gonzalez

Stuart Goodman
Jared Goor
Joy Goor
Judy Gordon
Mary Gorman
Julian Gorodsky
Ginger Graham
Linda Grais
Ralph Greco
Joshua Green
Harry Greenberg
Jessica Grossman
Eberhard Grube
Deborah Gruenfeld
Fabio Guarnieri
Linda Guidice
Geoff Gurtner
Lee Guterman
Gary Guthart
Ken Haas
Sami Hamade
Ron Hanson
Basil Hantash
John Harris
Ali Hassan
Bernard Hausen
William Hawkins
Mike Helmus
Michael Henricksen
James Heslin
Bonnie Hiatt
Judith Hickey
Doug Hiemstra
Rick Hillstead
Tomoaki Hinohara
Russell Hirsch
Edwin Hlavka
David Hoffmeister
Janice Hogan
Linda Hogle
Charles Holloway
Howard Holstein
Phil Hopper
Michael Horzewski
Syed Hossainy

John Howard
Thomas Hsu
Bob Hu
Rebecca Huang
Tom Hutton
Wende Hutton
Karl Im
Mir Imran
Frank Ingle
Ronald Jabba
Robert Jackler
Paul Jackson
Chris Jacobs
Jamey Jacobs
Wilfred Jaeger
Ross Jaffe
Matthew Jenusaitis
Jeremy Johnson
Marie Johnson
Robert Joss
Ron Jou
James Joye
Steven Jwanouskas
Alan Kaganov
Vera Kallmeyer
Dean Kamen
Aaron Kaplan
Michael Kaplan
David Kelley
Ken Kelley
Tom Kelley
David Kelso
George Kembel
Jim Kermode
Fred Khosravi
Gilbert Kilman
Deborah Kilpatrick
Daniel Kim
Ted Kim
Paul King
Gil Kliman
Laura Knapp
Joseph Knight
Dorothea Koh
Thomas Kohler

Ellen Koskinas
Gregory Kovacs
David Kreps
Katharine Ku
Michael Lackman
Joseph Lacob
Nandan Lad
Greg Lambrecht
Ted Lamson
Jack Lasersohn
Sue Ann Latterman
Anne Lawler
Kenneth Lawler
Guy Lebeau
David Lee
Donald Lee
Tracy Lefteroff
Larry Leifer
Michael Lesh
David Liang
Bryant Lin
Richard Lin
Jack Linehan
Jacques Littlefield
Jeannik Littlefield
Sandy Littlefield
Frank Litvack
Rich Lotti
David Lowsky
Angela Macfarlane
Sean Mackey
John MacMahon
Steve MacMillan
Swaminatha Mahadevan
Anurag Mairal
Zachary Malchano
William Maloney
Joe Mandato
John Maroney
Chris Martin
Ken Martin
David Mauney
Calvin Maurer
Allan May
Mika Mayer

Milton McColl
Michael McConnell
Casey McGlynn
Dana Mead
Vinod Menon
Carlos Mery
Lachman Michael
Maria Millan
David Miller
Eric Miller
Sandy Miller
Timothy Mills
David Milne
Craig Milroy
Oscar Miranda-Dominiguez
William Mobley
Fred Moll
Kevin Montgomery
John Morton
Susan Moser
Nicholas Mourlas
Michael Mussallem
Michael Nash
Asha Nayak
John Nehr
Charles Nelson
Drew Nelson
William New
Bob Newell
Quynchi Nguyen
Gunter Niemeyer
Julian Nikolchev
Guy Nohra
Gordie Nye
Santiago Ocejo-Torres
Stephen Oesterle
Tracy Okamoto
John Onopchenko
William Overall
Michelle Paganini
Julio Palmaz
Olin Palmer
Bhairivi Parikh
T. Kim Parnell

Jay Pasricha
Ron Pearl
Donna Peehl
Rodney Perkins
Timothy Petersen
David Piacquad
Jan Pietzsch
Peter Pinsky
Moshe Pinto
Philip Pizzo
Hank Plain
Ben Pless
Sylvia Plevritis
Todd Pope
Richard Popp
Stuart Portnoy
Friedrich Prinz
Mary Beth Privitera
Michael Raab
Geetha Rao
Andrew Rasdal
Alok Ray
Mahmood Razavi
Michael Regan
Robert Reiss
Mehrdad Rezaee
Kelly Richardson
Jeff Rideout
Dan Riskin
Robert Robbins
Gregory Robertson
Teresa Robinson
William Robinson
Douglas Roeder
Campbell Rogers
Erica Rogers
Diana Reynolds Roome
John Avi Roop
Susan Rowinski
Geoffrey Rubin
Vahid Saadat
Eric Sabelman
Maria Sainz
Amr Salahieh
Bijan Salenhizadeh

Stephen Salmon
Will Samson
Terence Sanger
Alan Schaer
Carla Schatz
Stephen Schendel
Jeffrey Schox
Bob Schultz
David Schurman
Matt Scott
Randy Scott
Tina Seelig
Matthew Selmon
Bilal Shafi
Ramin Shahidi
James Shapiro
Adam Sharkawy
Hugh Sharkey
James Shay
Chris Shen
Jay Shukert
Kevin Sidow
Kim Simmons
Tad Simmons
Chuck Simonton
Carl Simpson
John Simpson
Baird Smith
R. Lane Smith
Yuen So
Roy Soetikno
Sarah Sorrel
Dan Spielman
George Springer
Sakti Srivastava
Fred St. Goar
Richard Stack
Neil Starksen
Bill Starling
Brett Stern
Simon Stertzer
John Stevens
Jackson Streeter
Mitchell Sugarman
Margot Sutherland

Robert Sutton
Judith Swain
Jim Swick
Daniel Sze
Katie Szyman
Raymond Tabibiazar
Karen Talmadge
Beverly Tang
Larry Tannenbaum
Tatum Tarin
Charles Taylor
Hira Thapliyal
Stephen Thau
Patty Thayer
Robert Thomas
Troy Thorton
James Tobin
Ravi Tolwani
Sara Toyloy
Julie Tracy
Alexandre Tsoukalis
Sean Tunis
Sara Little Turnbull
Ted Tussing
P. J. Utz
J. Sonja Uy
Brad Vale
Sigrid Van Bladel
Jacques Van Dam
Machiel Van Der Loos
Jamie van Hoften
Vance Vanier
Richard Vecchiotti
Ross Venook
Claude Vidal
Kenneth Waldron
Amrish Walke
Jeff Walker
James Wall
Mark Wan
Paul Wang
Sharon Lam Wang
Tom Wang
Kevin Wasserstein
Jay Watkins

The Biodesign Community

Steven D. Weinstein
Eric Weiss
John White
Ken Widder
Bernard Widrow
Allan Will

Parker Willis
Jim Wilson
Dawn Wojick
Scott Wolf
Timothy Wollaeger
Russell Woo

Kenneth Wu
Walter Wu
Alan Yeung
Malisa Young
Philip Young
Reza Zadno

Christopher Zarins
Mark Zdeblick
Robert Zider

Names of other members of the biodesign community can be found at **ebiodesign.org**.

Biographies

Stefanos Zenios is the Charles A. Holloway Professor at the Graduate School of Business, Stanford University. His pioneering work on maximizing the benefits of medical technology to patients when resources are limited has influenced policies in the United States and Europe. His research was featured in the *Financial Times* and Times.com. At Stanford University, he was the first to introduce courses on the interface between medicine, engineering, and management in the MBA curriculum. Dr. Zenios advises medical device and biopharmaceutical companies on health economics and outcomes studies for marketing and reimbursement strategies. He is also a co-founder of Culmini Inc., a company funded by the National Institutes of Health. It develops web-tools for patients and their families. He has published more than 30 papers and received numerous research grants and awards from professional Societies. He holds a Ph.D. in operations research from MIT and a B.A. in mathematics from Cambridge University.

Josh Makower is the founder and chief executive officer of ExploraMed, a medical device incubator. He is also a venture partner with New Enterprise Associates, a consulting associate professor at Stanford University Medical School, and a co-founder of Stanford's Biodesign Innovation Program. Dr. Makower has founded several medical device businesses including Moximed, Vibrynt, NeoTract, Acclarent, TransVascular, and EndoMatrix. Up until 1995, he was founder and manager of Pfizer's Strategic Innovation Group. He holds over 61 patents in various fields of medicine and surgery, an MBA from Columbia University, an M.D. from NYU, and an S.B. in mechanical engineering from MIT.

Paul Yock is the director of the Stanford Biodesign Program and the founding co-chair of the Department of Bioengineering at Stanford University. He is known internationally for his work in inventing, developing, and testing new medical devices, including the Rapid Exchange™ balloon angioplasty and stenting system, which is now the principal system in use worldwide. He also authored the fundamental patents for mechanical intravascular ultrasound imaging and founded Cardiovascular Imaging Systems. In addition, he invented a Doppler-guided access system known as the Smart Needle™ and PD-Access™. Dr. Yock holds 55 US patents and has authored over 300 papers, mainly in the area of catheter-based interventions and technologies. He has been elected to membership in the National Academy of Engineering and has received several prestigious awards, including the American College of Cardiology Distinguished Scientist Award.

Todd J. Brinton is a clinical assistant professor of medicine (Cardiovascular) and lecturer in Bioengineering at Stanford University. He is an interventional cardiologist at Stanford University Medical Center and investigator in interventional-based therapies for coronary disease and heart failure. He is also the fellowship director for the Biodesign Program, and co-director of the graduate class in Biodesign Innovation at Stanford University. Dr. Brinton completed his medicine, cardiology, and interventional training at Stanford University. He holds an M.D. from the Chicago Medical School and B.S. in bioengineering from the University of California, San Diego. He is cofounder of BioParadox, Inc., a venture-backed medical device company and serves on the advisory board for a number of early-stage medical device companies. Prior to medical school he was the clinical research director for Pulse Metric, Inc., a medical device start-up company.

Biographies

Uday N. Kumar is the founder and chief medical officer of iRhythm Technologies, Inc., a venture-backed medical device company focused on developing new devices and systems for the detection of cardiac rhythm disorders. He is also the associate director for Curriculum of Stanford-India Biodesign and a lecturer in Bioengineering, and has served as an adjunct clinical instructor of cardiovascular medicine, all at Stanford University. In these capacities, he mentors, advises, and teaches students and fellows about the biodesign process. Dr. Kumar completed a Biodesign Innovation fellowship at Stanford, cardiology and cardiac electrophysiology fellowships at the University of California, San Francisco (UCSF), an internal medicine residency at Columbia University, and his medical and undergraduate education at Harvard University. He was also chief medical officer and vice-president of Biomedical Modeling Inc., a medical start-up company.

Lyn Denend is a research associate at Stanford University's Graduate School of Business, where she has authored numerous case studies for use in graduate-level and executive education programs in areas such as strategic management, international business, supply chain management, healthcare, and biodesign innovation. Previously, Ms. Denend was a senior manager in Cap Gemini Ernst & Young's management consulting practice and vice-president of Operations for a start-up providing human resource services. She has an MBA from Duke University's Fuqua School of Business and a BA in Communications from the University of California, Santa Barbara.

Thomas M. Krummel is Emile Holman Professor and chair in the Department of Surgery, and co-director of the Stanford Biodesign Program at Stanford University. He has been a pioneer and consistent innovator throughout his career, and has served in leadership positions in many of the important surgical societies including the American College of Surgeons, the American Pediatric Surgical Association, the American Surgical Association, the American Board of Surgery, the American Board of Pediatric Surgery, and the American Board of Plastic Surgery. Over the last 14 years, Dr. Krummel has pioneered the application of technology to simulation-based surgical training and surgical robotics. For his work in this area, and developing a collaborative simulation-based surgical training system, he has received two Smithsonian Information Technology Innovation Awards.

Christine Kurihara is manager of special projects, Biodesign Program, Stanford University, where she oversees the development of new projects. She is currently developing the online companion to the biodesign textbook. Ms. Kurihara joined the Biodesign Program after an 11-year career with Stanford in media services. In her previous role she spearheaded media development efforts for an on-campus service unit, where her teams produced websites, online courseware, and video and broadcast products. Prior to running Media Solutions, she developed the first official Stanford University website and served as managing editor. In 1997, Ms. Kurihara co-chaired the Sixth International World Wide Web Conference.

Glossary

483 — A form letter issued by the FDA if actionable problems are uncovered during an FDA audit.

510 (k) — One of several pathways for medical devices through the regulatory process at the FDA. This pathway is used when similar devices are already in use.

Acquisition — A transaction in which the seller of the property (technology, IP, company) completely relinquishes control of the property to the acquirer.

Administrative detention — A temporary "cease and desist" order from the FDA.

AdvaMed — Advanced Medical Technology Association. The advocacy group for medical device companies.

AHA — American Hospital Association. An association that represents hospitals, healthcare networks, and their patients and communities.

AHRQ — Agency for Healthcare Research and Quality. A US government agency responsible for collecting evidence-based data on healthcare outcomes. Longitudinal data are available through its website.

AIMDD — Active Implantable Medical Device Directive 90/385/EEC. One of the key regulatory approval directives used in the European Union.

AMA — American Medical Association. The primary association of physicians in the United States. The AMA controls the issuance of new CPT codes.

Angel investor — Experienced individual investor who uses his or her own wealth to fund start-up companies. Angel investors may be organized in groups.

ANSI — American National Standards Institute. The US standards organization that is representative to ISO.

APC — Ambulatory Payment Classification. Codes for classifying hospital outpatient procedures.

Arm — Any of the treatment groups in a randomized trial. Most randomized trials have two arms, but some have three or even more (see Randomized trial).

ASIC — Application specific integrated circuit. One potential component of the electrical circuitry of a device.

ASQ — American Society of Quality.

BATNA — Best alternative to a negotiated agreement. The course of action that will be taken if a negotiation fails to lead to an agreement.

BCBS — Blue Cross Blue Shield. Health plans that operate in various regions in the United States. There are 39 BCBS plans and the BCBS Association is a trade group that, among other things, helps establish guidelines for reimbursement.

Bench testing — Testing prototypes (materials, methods, functionality) in a controlled laboratory environment (not in animals or humans).

Glossary

Beneficence — A basic principle of bioethics that all medical work is for the good of the patient; contrast to maleficence.

Bias — When a point of view prevents impartial judgment on issues relating to the object of that point of view. In clinical studies, blinding and randomization control bias.

Biocompatibility — The property of a material that indicates that it is suitable to be placed in humans.

Blind trial — A trial in which neither the members of the patient group nor any participating doctors, nurses, or data analysts, are aware of which treatment or control group the patients are in.

Blue-sky need — A large-scale need that would require major new medical or scientific breakthroughs and/or significant changes in practice.

Bottom-up model — A market model that uses a series of detailed sales factors, including sales cycle, adoption curve, hiring effort, commercial effort, etc. to predict future sales.

Breadboard — A board that can be used to assemble electronic components and connect them for use in prototyping devices with computer parts.

Bridge loan — An interim debt financing option available to individuals and companies that can be arranged relatively quickly and span the period of time before additional financing can be obtained.

Budget impact model — A model for demonstrating product value that examines the cost and treatable population within a health plan, as well as the expected annual cost to the plan for covering a device.

Bundled pricing — Setting a single price for a combination of products and/or services.

CAB — Conformity Assessment Body. The body that determines compliance to ISO 13485.

CAC — Carrier Advisory Committee. The committee that performs a review of all local coverage decisions through Medicare.

CAF — Contracting administration fee. The fee that a global purchasing organization will charge for managing the purchasing contracts for many end users, paid by the manufacturer.

CAGR — Compound annual growth rate. The annual growth rate for an investment.

CAPA — Corrective and preventive actions. One subsystem of a quality management system. The system to implement corrections upon and to avoid future problems in quality control.

Capability-based advantages — An advantage over competitors that is driven by a company's capabilities. This type of advantage is based on the ability to do something better or less expensively than the competition and/or customers.

Cash flow statement — An accounting statement that shows the cash that flows in to the company in each period (typically quarter) minus the cash that flows out in the same period.

CBER — Center for Biologics Evaluation & Research. The part of the FDA that approves biologics.

CDER — Center for Drug Evaluation & Research. The part of the FDA that approves drugs.

CDRH — The center within the FDA responsible for medical device regulation.

CE mark — Resulting "mark" that is given to a device in the EU to indicate regulatory approval.

Term	Definition
Citation	A formal warning to a company from the FDA. Prosecution will follow if changes are not made.
Civil penalties	Monetary penalties imposed on a company after a hearing for violations.
Class I	Classification of a medical device by the FDA that indicates low risk to a person.
Class II	Classification of a medical device by the FDA that indicates intermediate risk to a person. Class II devices are typically more complex than class I devices but are usually non-invasive.
Class III	Classification of a medical device by the FDA that indicates the highest risk to a person. Class III devices are typically invasive or life sustaining.
Clinical investigator	A medical researcher in charge of carrying out a clinical trial protocol.
Clinical protocol	A study plan on which all clinical trials are based. The plan is carefully designed to safeguard the health of the participants, as well as to answer specific research questions. A protocol describes what types of people may participate in the trial; the schedule of tests, procedures, medications, and dosages; and the length of the study.
Clinical trial	A research study performed to answer specific questions about diagnoses or therapies, including devices, or new ways of using known treatments. Clinical trials are used to determine whether new treatments are both safe and effective.
CME	Continuing medical education. Additional training required to maintain a license for physicians and others in healthcare-related fields.
CMS	Centers for Medicare and Medicaid Services. The primary government payer of healthcare charges for the elderly and disabled in the United States.
Coding	The process of assigning a specific, identifiable code to a medical procedure or process.
COGS	Cost of goods sold. Raw materials costs for a product.
Common stock	Equity in a company that confers on shareholders' voting and pre-emptive rights (the right to keep a proportionate ownership of the company by buying additional shares when new stock is issued).
Comparables analysis	Evaluating the pricing strategies (and associated reimbursement status) of similar offerings in the field.
Conditions precedent	Section of a term sheet that outlines what steps must be taken before the financing deal proposed in the term sheet can be finalized.
Controlled trial	A trial that uses two groups: one that receives treatment, and a second, control group, that does not, in order to compare outcomes.
Conversion, automatic conversion	Section of a term sheet that describes how preferred shares will convert to common shares.
Convertible bonds	A hybrid debt-equity alternative to companies seeking financing. A type of bond that can be converted into shares of stock of the issuing company, usually at some preannounced ratio.
Core laboratory	Laboratories that analyze data from a clinical trial; these laboratories often have specialized equipment and expertise.
Corporate investment	When corporations invest in new companies by: (1) the purchase of equity in support of a research and development or a licensing agreement, or (2) traditional venture investments.

Glossary

Correction
Repair or modification of a distributed product while it is still under the control of the manufacturer.

Cost-effectiveness model
A model for determining product value where cost is expressed per unit of meaningful efficacy, usually used comparatively across interventions.

Cost-utility model
A model for determining product value where cost is assigned for quality of life and years lived. It is based on clinical outcome measures related to quality of life and/or disability and mortality.

Cost/benefit analysis
A model for determining product value that demonstrates that the money spent on the device is lower than the total cost of the outcomes of the disease or of current standard therapy.

CPT codes
Common Procedural Terminology codes. Codes used to classify medical procedures in a standard way so that the same procedure is reimbursed in the same way across all facilities. Also known as HCPCS Level 1 codes.

CRO
Contract (or Clinical) Research Organization. An independent organization that provides management services for clinical trials.

Cycle of care
A description of how a patient interacts with the medical system.

Debt funding
Funding that is repaid with interest. A loan.

Design controls
One subsystem of a quality management system. Controls that ensure the device being designed will perform as intended when produced for commercial distribution.

Design creep
Ongoing, minor changes in developing a device that can lead to significant delays and issues with intellectual property and regulatory clearance.

Design validation
Ensuring that a design does what it is intended to do.

Design verification
Ensuring that a design meets product specifications.

Determination meeting
A formal meeting with the FDA to request approval of the design for a clinical study.

Differential pricing
Pricing the same product or service differently for different customer segments, e.g., discounts for large buyers.

Dilution
Section of a term sheet that stipulates how conversion prices will be calculated if future rounds of financing are dilutive to preferred shareholders' holdings (i.e., they reduce the total value of the shareholders' ownership stake in a company).

Direct sales model
Hiring a sales force within a company to sell to customers directly.

Discounted cash flow analysis
An analysis that uses cash flows discounted back to present at a discount rate that reflects the returns the shareholders expect from their investment in the company. The higher the risk in the investment, the higher the discount rate investors will use in this analysis.

Distribution play
To focus on product breadth and channel relationships rather than on product superiority.

Dividend provisions
Section of a term sheet that describes the conditions for which dividends will be paid. A dividend is a payment to the shareholder that is proportional to the shareholder's ownership of a company.

Divisional
A type of patent application that claims a distinct or independent

	invention based upon pertinent parts carved out of the specification in the original patent.		interventions in the interest of public health.
DRG	Diagnosis related group. A set of codes that are grouped together by diagnosis; used specifically for coding hospital related billing for patient encounters. Replaced recently by MS-DRG (Medical Severity DRGs) that include adjustments based on comorbidities and complications.	EPO	European Patent Office. The office that provides unified patent filing for 38 European countries.
		Equity funding	Funding in which the investor provides a cash infusion to a company and in exchange obtains equity in the company.
		Ethnographic research	Understanding a particular culture or way of life by studying the members of that culture or group.
DSMB	Data Safety and Monitoring Board. An independent body that reviews results of clinical trials.	Evergreening	The process of introducing modifications to existing inventions and then applying for new patents to protect the original device beyond its original 20-year term.
DSP	Digital signal processor. One potential component of the electrical circuitry of a device.
DTC	Direct-to-consumer. A type of marketing that targets the end user of a product, as opposed to the physician or other medical professional.	Evidence-based	Treatments, guidelines, and processes based on the results and outcomes generated from experiments and observation, which use specific evidence of outcomes and suggest treatment or processes based on such evidence.
DTP	Direct-to-patient. A type of marketing that targets patients directly, as opposed to physicians or other healthcare professionals.
Due diligence	An iterative process of discovery, digging into detail about the various elements of a start-up company's business plan or licensing opportunities.	Exclusion criteria	Characteristics or contraindications that eliminate subjects from participating in a clinical study.
		Exclusive rights	The rights of the inventor or group of inventors, who has/ve been issued a patent on an invention, to be the sole person(s) creating and marketing that invention.
Earn-out	An acquisition in which additional payments are made to the seller after the sale day if the acquired company reaches prespecified milestones.
		Exclusive license	A license that grants only the licensee (and not even the licensor) the right to use a technology.
Efficacy endpoints	A result during an animal or clinical study that demonstrates efficacy (i.e., a therapeutic effect). Endpoints are what a study is designed to prove.	Exit	When a company is either acquired or has an IPO.
		Expansion funding	Funding required to ensure completion of clinical trials, initiation of additional trials or initial product launch. Such funding is often acquired through VCs or corporate investment.
Epidemiology	Study of factors affecting the health and illness of a population that are used as the basis of making

Glossary

Facility and equipment controls	One subsystem of a quality management system. It ensures, in part, that standard operating procedures have been designed and implemented for all equipment and facilities.
Fast follower	A company that leverages its own corporate advantages to quickly capture market share from the first mover.
Field of use	A licensing option that allows an existing patented device to be used within a restricted domain, such as one clinical area.
FPGA	Field-programmable gate array. One potential component of the electrical circuitry of a device.
FIM	First-in-man. The first time a device or technology is used in a human subject.
Financial model	A detailed numerical articulation of a company's costs and revenue over time. It tracks both the cost of developing the innovation and bringing it to the market as well as market revenue, and it follows these costs and revenue over a period of five to seven years.
First mover strategy	An attempt by a company to be first to market with any innovation.
Flow of money	Analysis, aimed at identifying key stakeholders, that is focused on payments to providers of healthcare services.
Freedom to operate	The ability to commercialize a product, without infringing on the intellectual property rights of others.
Fully burdened cost	The total cost of an employee, including salary, benefits, associated overhead, and fees.
Gainsharing	When hospitals negotiate reduced prices with certain manufacturers in exchange for increased volume.
GCP	Good Clinical Practices. Guidelines from the FDA that outline specific standards for holding clinical trials.
GLP	Good Laboratory Practices. A system of management controls for laboratories that assures consistent and reliable results.
GMP	Good Manufacturing Practices. Formerly used by the FDA to promote quality; replaced by Quality Systems Regulation (QSR).
GPO	Global purchasing organization. An organization that brings together multiple hospital groups, large clinics, and medical practices into buying cooperatives.
HCPCS Level II	Health Care Financing Administration's Common Procedure Coding System Level II. Coding for supplies and services obtained outside the physician's office that are not covered by a CPT or APC code.
HDE	Humanitarian Device Exemption. An exemption to the normal regulatory pathways for a medical device that is intended to benefit patients in the treatment or diagnosis of a disease or condition that affects or is manifested in fewer than 4,000 individuals in the United States per year.
HIPAA	Health Information Portability and Accountability Act. Ensures comprehensive protection of patient health information (PHI).
HCUP net	Healthcare Cost and Utilization Project. A website with data about healthcare cost and utilization statistics in the United States (e.g., hospital stays at the national, regional, and state levels).
Hypothesis	A supposition or assumption advanced as a basis for reasoning or argument, or as a guide to experimental investigation.

Term	Definition
IACUC	Institutional Animal Care and Use Committee. A committee that institutions must establish in order to oversee and evaluate animals used for trials.
ICD-9-CM codes	International Classification of Diseases, 9th edition, Clinical Modification codes. Codes for classifying morbidity data and describing patient diagnoses and procedures; variation on ICD-9 used by the United States.
ICD-9 codes	International Classification of Diseases, 9th edition codes. Codes for classifying patient diagnoses.
IDE	Investigational Device Exemption. An exemption to a hospital or doctor from the FDA that allows the hospital or doctor to use a device prior to its regulatory approval, usually as part of a trial.
IDN	Integrated delivery network. An organization that aggregates hospitals, physicians, allied health professionals, clinics, outpatient facilities, home care providers, managed care, and suppliers into a single, closed network.
IFU	Indications for use. Instructions on how to use a device. Mandated by the FDA. Typically a package insert.
Illiquid	Not liquid (e.g., stock or other property that is not easily sold or converted to cash).
Inclusion criteria	Characteristics or indications that subjects must have in order to participate in a clinical study.
Incubator	Small companies that specifically serve to develop a need or concept at the early stages. An incubator may incubate multiple device concepts for a significant period of time. Successful products may result in the spin-out of a company from the incubator into a stand-alone entity.
Indirect sales model	A sales and distribution agreement with an existing distributor, or forming a third-party partnership with another manufacturer.
Information rights	Section of a term sheet that defines what and how much information about the company is shared with investors.
Informed consent	Consent by a research subject that indicates they are fully aware of all aspects of the trial prior to participating, including both the risks and potential benefits.
Injunction	An order issued by the courts that requires a medical device company to refrain from some action (manufacturing, selling, etc.).
Innovation notebook	A notebook in which an innovator documents each aspect of the invention. This notebook may be used in infringement trials to prove inventorship.
Interference proceeding	A proceeding held by the USPTO to determine who was first to invent a claimed invention.
IRB	Institutional Review Board. A committee that monitors clinical trials to ensure the safety of human subjects.
Intrapreneur	A person within a company who is tasked to develop new products or business models – an internal entrepreneur.
IPO	Initial public offering. The first offering of a company's stock for public sale in a stock exchange such as the New York or London stock exchanges.
ISA	International Searching Authority. The organization that performs patent searches as part of an international patent filing.

Glossary

ISO International Organization for Standardization. A non-governmental network of national standards institutes that establishes standards of quality. The name ISO is not an acronym but rather based on the Greek word *isos* meaning equal.

ISO 13485 International Standards Organization Certification 13485. The European Union's Quality System (compare to QSR).

ISO 9001 International Standards Organization Certification 9001. A quality certification in use around the world between the 1980s and 1990s.

IVMDD In Vitro Diagnostic Medical Device Directive 98/79/EEC. One of three regulatory approval directives used in the EU.

KOL Key opinion leader. Physicians and others in the medical device arena who are often consulted when new devices are readying for the market.

LCD Local coverage determination. One of two types of reimbursement determinations made by Medicare that provides guidance on national reimbursement coverage. Typically applies to payments for outpatient services. LCDs are decisions made by one of 28 Medicare contractors and apply only to the contractors' area of coverage.

Lexicographer An inventor may use his/her own language and definitions in a patent application, thus becoming a lexicographer.

Licensing One option in getting a technology to market by transferring the rights to the technology from the innovator to a licensee in exchange for ongoing royalties and/or other payments.

Liquidity event The transaction that enables an investor to receive cash in exchange for its equity stake in a company. Also referred to as exit events.

LLC Limited Liability Company. A type of corporation that establishes a board and limits liability to the owners of the company.

Longitudinal data Data collected in studies that take place over several years, often decades or more.

Loss leader An item sold at a lower cost (often below the cost to the manufacturer) in order to stimulate additional sales of profitable items.

Management controls One subsystem of a quality management system. Controls that ensure adequate management support and participation in quality systems.

Manufacturing costs Costs for material (COGS), manufacturing labor, facilities, and equipment.

Market segmentation Using specific parameters to partition the market into identifiable, homogeneous segments in order to understand sales and marketing needs.

Market withdrawal A response to a minor violation that is not caused by legal action by the FDA.

Marquee physicians High-profile practitioners who are influential with their colleagues.

Material controls One subsystem of a quality management system; controls that ensure material quality and consistency.

MAUDE The FDA database of all significant adverse events due to medical devices.

MDD Medical Device Directives 93/42/EEC. One of three regulatory approval directives used in the European Union.

MDR	Medical Device Reporting. The reporting vehicle through which the FDA receives information about significant medical device adverse events that was established by the Safe Medical Devices Act.		decreased quality of life, extended hospital stay, physical impairment, etc.
MDUFMA	Medical Device User Fee and Modernization Act. The federal act that established user fees in the medtech industry.	**NAI**	No Action Indicated. A classification for an FDA audit that indicates no further action is required by the inventor or company in order to seek approval for a device.
Me-too products	Products that are relatively undifferentiated from products already on the market.	**NCD**	National coverage determination. One of two types of reimbursement determinations made by Medicare that provides guidance on national reimbursement coverage. Typically applies to payments for inpatient services.
Mechanism of action	The specific biochemical or biomechanical interaction through which a drug or device produces its effect.		
MEDLINE	Medical Literature Analysis and Retrieval System Online. A literature database of biomedical research papers.	**NCHS**	National Center for Health Statistics. The US-based principal health statistics agency; they compile statistical information to guide actions and policies to improve health.
Medtech	Medical device technology. A short form to allow comparisons to Biotech, for instance.	**NDA**	Non-disclosure agreement. An agreement between two parties such that the party receiving confidential information from another party will not disclose the information to anyone for a fixed period of time.
MEPS	Medical Expenditures Panel Survey. The longitudinal data on health expenditures of 30,000 US households provided by AHRQ.		
Mezzanine funding	Funding that is required when some of the most significant risks have been resolved but the company has yet to generate sufficient revenue to be self-sustaining.	**Niche strategy**	A strategy whereby a company seeks to own the customer relationships in a specific, focused area of medicine.
		Non-exclusive license	A license that allows the licensee rights of use within a given field and within whatever other limitations are provided by the license, but allows the licensor to grant similar rights to other parties.
MHRA	Medicines and Healthcare Products Regulatory Agency. The organization that approves devices and drugs for Europe (including the UK).		
Mixed need	A need with features that are easily achievable (more incremental to existing approaches) and other elements that introduce significant technical or clinical risk.	**NGO**	Non-governmental organization. Non-profit organizations working for a cause. These organizations provide resources and assistance to parties when the governments will not or cannot provide them.
Morbidity	When a human is harmed in some way (short of death) by infection,	**Notice of Allowance**	The notice from the USPTO to indicate the patent has been accepted.

NSE	Not Substantially Equivalent. A determination by the FDA that a new device is not equivalent enough to a predicate device and therefore cannot use the 510k pathway.	**OpEx**	Operating expenses. Costs considered not to be manufacturing costs, including R&D, sales staff, general and administrative functions, and non-production facilities costs.
OAI	Official Action Indicated. A classification for an FDA audit indicating that action is required by the inventor or company in order to seek approval for a device.	**Opportunity cost**	The opportunity forgone by choosing a different opportunity.
		OPRR	Office for Protection from Research Risks. Now called the Office for Human Research Protections.
Observational studies	Studies that make conclusions about the efficacy of a treatment or device on a group of subjects where the assignment of subjects into the treated versus control groups is outside the control of the investigator.	**Option pool**	The total number of stock options available for a company to grant, typically to employees.
		OTC	Over the counter. Drugs or devices that are sold directly to the end consumer.
OCP	Office of Combination Products. The section of the FDA that reviews medical technology comprising a combination of drugs/device or drugs/biologics to determine which center of the FDA will regulate it.	**OTL**	Office of Technology Licensing. The office within a university that manages its IP assets.
		OUS	Outside United States. Refers to clinical trials (or other activities) that are performed outside the United States. Often used in reference to obtaining regulatory approval.
OEM strategy	Original equipment manufacturer strategy. When a company provides technology and/or components to another company that then assembles and sells the finished product.	**P&PC**	Production and process controls. One subsystem of a quality management system. Requires that production processes be controlled and monitored to ensure product conforms to specifications.
Off-label use	The use of a treatment for conditions other than those approved by the FDA.		
Office action	A document issued by the USPTO that outlines objections or necessary changes to an application or claim due to finding prior art.	**Partnering strategy**	One option in getting an idea to market – joining with another company to help develop a device.
OHRP	Office for Human Research Protections. A federal agency that helps assure the protection of humans participating in clinical research.	**Pass-through code**	Also called a c-code. A code that is issued to cover the cost of a device that is incremental to the services provided under an existing APC code, or set of codes. The cost of the device may be bundled into this transitional APC code, or may still be billed separately under a temporary pass-through code.
OIPE	Office of Initial Patent Examination. The first agency that examines patent applications for completeness.		
Operating profit	The difference between income and the expense incurred during operations.	**Pathophysiology**	Study of the change of the normal mechanical, physical, and biochemical functions of a human

	due to disease or other interruption to normal function.	PIC	Programmable intelligent computer. One potential component of the electrical circuitry of a device.
Patient towers	Segments of patient populations based on epidemiological factors, market size, and other important factors to determine the most favorable target for a device.	Pilot trial	Early clinical trial, usually conducted as a registry.
		Pivotal trial	Typically larger, controlled studies designed to test specific hypotheses when significant clinical data are necessary. Are often used to support the submission of a PMA application for a new device.
Pay-for-performance	Setting prices contingent on the realization of specific results from a product or device. A new, if somewhat controversial, approach to healthcare fee structuring.		
Payer advisory board	A board consisting of key opinion leaders and medical directors from select payers in a target payer segment. They advise start-up companies in issues of reimbursement.	PMA	Premarket approval. The most stringent pathway through the FDA regulatory process for a device. A PMA is necessary when a new device is not substantially equivalent to any existing devices that were approved before 1976.
PCT	Patent Cooperation Treaty. A treaty that establishes unified patent filing for foreign countries. Issued by the World Intellectual Property Organization (WIPO).	POC	Point of care is the place in which the actual surgery, medical intervention, or other medical procedure is done.
PDP	Product development protocol. A contract that describes agreed-upon details of design and development activities, the outputs of these activities, and the acceptance criteria for these outputs.	Porter's five forces	Five forces, identified by Michael Porter (Harvard), that influence competition within an industry: rivalry, new entrants, substitute products, suppliers' bargaining power, and buyers' bargaining power.
Peer review	Review of a clinical trial or study by experts. These experts review the trials for scientific merit, participant safety, and ethical considerations. Peer-reviewed literature refers to scientific articles published in credible academic journals where a panel of physicians has reviewed the trial report.	Postmarketing trial	Trials performed following the commercial approval of a device in order to gain acceptance of the device in the field.
		Post-money valuation	A company's value after it obtains outside funds.
		Power	Statistical power refers to the probability of detecting a meaningful difference, or effect, if one were to occur. Ideally, studies should have power levels of 0.80 or higher – an 80 percent or greater chance of finding an effect, if one exists.
PHI	Patient health information. A patient's personal health information. Protected by HIPAA.		
PI	Principal investigator. The person responsible for conducting a clinical trial. May also be someone who manages a grant or contract for a particular research project.	Power calculation	A calculation used in clinical trials to calculation determine the optimum number of patients to include in order to demonstrate efficacy.

Term	Definition
Pre-money valuation	A company's value before it receives subsequent outside round of financing.
Preferred stock	Equity in a company that gives shareholders a liquidation preference, meaning that preferred stock holders are paid before common stock holders if a company is sold or its assets are liquidated. Holders of preferred stock are also given priority regarding the payment of dividends. Investors, such as venture capitalists, often are granted preferred stock in return for funding.
Primary endpoints	Criteria that the company, investigators and FDA agree are required to be met to prove device efficacy during a clinical trial. For the device to receive approval, these endpoints must be met. The number of patients needed in a clinical trial is based on finding a statistically meaningful difference in these endpoints.
Primary prevention patients	Patients who are at risk of experiencing a particular episode of a disease state.
Prior art	Any subject matter that is previously published or known generally in the field prior to an invention.
Priority date	The first filing date of a patent application. This date is used to help determine the novelty of an invention compared to prior art.
Protective provisions	Section of a term sheet that outlines the percentage of preferred shareholders whose approvals are needed for a company to take certain actions.
Provisional patent	A patent filing that is less costly and rigorous to file but that only ensures a filing date prior to filing for a regular patent. Provisionals expire in 12 months if the provisional patent is not converted to a regular patent during that time.
Proxy companies	Companies whose operations resemble what is required to commercialize an innovation. Also known as comparable companies. Used to validate financial models.
Proxy material	Material that may be used in prototyping due to its similar characteristics to the material that needs to be tested.
Public disclosure	The moment an inventor describes an invention in a public setting; discussions with a single individual may be considered public settings.
PubMed	Published Medicine, a government-funded search engine that provides access to biomedical research papers that have been published in professional journals.
QA	Quality assurance. The process that ensures that the product will operate according to its specifications.
QC	Quality control. The activities designed to catch defective products in the manufacturing before they are released to the customer.
QMS	Quality management system. A system through which quality assurance and quality control are implemented.
QSIT	Quality Systems Inspection Technique. A guide describing the way in which FDA investigators conduct an inspection of quality systems.
QSR	Quality System Regulation. Regulations regarding the manufacturing, design, material handling, and product testing of devices.
Quality of life	The degree of well-being felt by an individual.

Term	Definition
Randomization	A method based on chance by which study participants are assigned to a treatment group. Randomization minimizes the differences among groups by equally distributing people with particular characteristics among all the trial arms.
Randomized trial	A trial that uses random assignment of patients to the treatment and control groups.
Recall	A method of removing all of a particular product from the market in an effort to minimize the risk to patients.
Records, documents & change controls	One subsystem of a quality management system. Controls for managing documentation and records for all aspects of the quality systems in place.
Redemption	Section of a term sheet used to allow investors to redeem all or a portion of their investment if liquidity has not been achieved within a finite number of years.
Reduction to practice	Taking an invention beyond the concept stage. Showing that the invention actually works, as opposed to merely outlining its theoretical workings.
Reference pricing	Using the prices of comparable products used in healthcare system of one country to make decisions on payments within another, without necessarily considering any compensation for economic and regulatory differences.
Registration rights	Section of a term sheet that outlines the rights of preferred investors to force the company to register its stock for public offering under SEC rules.
Registry	A collection of cases that have been performed in real-world settings outside the scope of a formal comparative protocol; also called observational studies.
Reimbursement	The act of paying for a medical device, procedure, visit, or other element of patient care by a third-party payer.
Reimbursement dossier	A set of documents that describes all aspects of a device (including basic information, studies' results, modeling reports, etc.). Used by companies in introducing devices to payers.
Revenue ramp	The degree to which a company expects to grow its revenue over time.
Right of first refusal	The right that one first party may obtain via an investment or other strategic transaction wherein the second party must offer the terms of any deal with respect to the sale of any shares (preferred or common) or other proposal to the first party before it can offer those same terms to an outside party.
ROI	Return on investment. The amount of return made on a given investment.
Royalties	A percentage of commercial sales paid by the licensee/partner to the patent or rights holder.
Royalty anti-stacking provisions	Provisions in a licensing deal that provide a mechanism for the licensee to prevent the licensor from reducing the royalty when other royalties are required for the same product.
RUC	AMA/Specialty Society Relative Value Scale Update Committee. The committee within the AMA that assigns relative value units to a new code.
RVU	Relative value unit. Units that measure the resources required for a new procedure; used to determine the payment for a new CPT code.

Glossary

SE
Substantial equivalence. Demonstration that a new device is nearly the same as a predicate device (in terms of safety and effectiveness) in order to use the 510k pathway.

Secondary endpoint
Additional criteria that may be met during a clinical trial, but that are not required to obtain a successful positive clinical trial result. The FDA rarely uses secondary endpoints to gain approval or clearance.

Secondary prevention patients
Patients who have experienced a particular episode of a disease state and are candidates for preventing a recurrence of that episode.

Seed funding
Early-stage funding that supports company creation through prototypes and proof of concept. Typically, this round is often funded by friends, families, and/or angels due to low capital requirements.

Seizure
When the FDA takes action against a specific device, taking control over inventory or materials.

SG&A
Selling and general administrative expenses. Expenses that include salaries, commissions, and travel expenses for executives and salespeople, advertising costs, and payroll expenses.

Significant risk
A term used by the FDA to indicate that a device is: (1) an implant and presents potential for serious risk to the patient; (2) is used to support or sustain life; (3) has a significant use for diagnosing or treating a disease; or (4) otherwise presents serious risk.

SMDA
Safe Medical Devices Act. A federal regulation that established reporting mechanisms for mortality and morbidity due to medical devices.

Stakeholders
All parties with some interest in the delivery and financing of medical care for patients with a specific medical need.

Standard of care
A treatment process that is well supported by evidence and that a doctor or medical facility should follow for a particular type of patient, disease, or procedure.

Standard treatment
A treatment that is currently in wide use and approved by the FDA and considered to be effective in the treatment of a specific disease or condition.

Start-up funding
Funding that is required to make substantial investments in a company. This type of funding often requires millions of dollars and comes from angels or VCs.

Statistical significance
The probability that an event or difference occurred (or did not occur) by chance alone. In clinical trials, the level of statistical significance depends on the number of participants studied and the observations made, as well as the magnitude of differences observed.

Subject
A human who participates in an investigation, either as an individual on whom, or on whose specimen, an investigational device is used or who participates as a control. A subject may be in normal health or may have a medical condition or disease.

Surrogate endpoint
Substitute criteria that are used to prove efficacy in lieu of actual endpoints that may take too long to prove.

Switching costs
The costs of a hospital or medical provider to change from one piece of capital equipment to another or from one treatment process to another; these costs are often a deterrent to trying a new vendor's equipment or new procedures.

SWOT analysis	An analysis, identified by Albert Humphrey (Stanford), of factors affecting a company: strengths, weaknesses, opportunities, and threats.	**USPTO**	United States Patent and Trademark Office. The government organization that issues patents and trademarks.
Syndicate	When more than one investor is involved in funding a company.	**Utility patent**	A patent. A grant from the government to the inventor for exclusive rights to make, use, sell, or import an invention. A utility patent is the most common type of patent for medical devices and describes an invention in sufficient detail to determine that it is novel, useful, and unobvious.
TEC	Technology Evaluation Committee. A committee, established by a payer (such as Blue Cross/Blue Shield) that evaluates a medical device or procedure to assess whether to pay for the device or procedure and how much.		
		VAI	Voluntary Action Indicated. A classification from an FDA audit that indicates objectionable conditions or practices were found with a device (or procedure), but the agency is not prepared to take or recommend any administrative or regulatory action.
Term sheet	A document that outlines the terms for a deal and serves as a letter of intent between the investor providing funds and the company receiving them. Terms may include percentage ownership in the company, intellectual property division, and other financial and legal concerns.		
		Valuation	The worth assigned to the business.
		Value proposition	The sum total of benefits that one party promises to a second party in exchange for payment (or other value transfer).
Third-party payer	Insurance companies (both public and private) who pay on behalf of the patient.		
Time to profitability	The amount of time it takes a company to break even, and then turn a consistent operating profit.	**Voting rights**	Section of a term sheet that spells out how voting will be orchestrated when shareholder approvals are required.
Top-down model	A market model that forecasts projected yearly revenue by multiplying the number of customers by the price of the innovation.	**Warrants**	A form of equity that gives the holder the right to purchase a number of shares of a company's stock at a predetermined price per share, sometimes subject to other conditions as well.
Trade secret	Information, processes, techniques, or other knowledge that is not made public but provides the innovator with a competitive advantage.		
		WHO	World Health Organization. The direct coordinating authority for health within the United Nations system.
Treatment gap	Reviewing methods of treatment for a particular disease to uncover areas where treatment is not used, where it is not effective, or where it does not exist. Treatment-gap analysis is performed in order to find areas for innovation.		
		WIPO	World Intellectual Property Organization. The organization responsible for the Patent Cooperation Treaty.

IDENTIFY Needs Finding

STAGE 1 NEEDS FINDING
1.1 Strategic Focus
1.2 Observation and Problem Identification
1.3 Need Statement Development
Case Study: Stage 1

STAGE 2 NEEDS SCREENING
2.1 Disease State Fundamentals
2.2 Treatment Options
2.3 Stakeholder Analysis
2.4 Market Analysis
2.5 Needs Filtering
Case Study: Stage 2

STAGE 3 CONCEPT GENERATION
3.1 Ideation and Brainstorming
3.2 Concept Screening
Case Study: Stage 3

STAGE 4 CONCEPT SELECTION
4.1 Intellectual Property Basics
4.2 Regulatory Basics
4.3 Reimbursement Basics
4.4 Business Models
4.5 Prototyping
4.6 Final Concept Selection
Case Study: Stage 4

STAGE 5 DEVELOPMENT STRATEGY AND PLANNING
5.1 Intellectual Property Strategy
5.2 Research and Development Strategy
5.3 Clinical Strategy
5.4 Regulatory strategy
5.5 Quality and Process Management
5.6 Reimbursement Strategy
5.7 Marketing and Stakeholder Strategy
5.8 Sales and Distribution Strategy
5.9 Competitive Advantage and Business Strategy
Case Study: Stage 5

STAGE 6 INTEGRATION
6.1 Operating Plan and Financial Model
6.2 Business Plan Development
6.3 Funding Sources
6.4 Licensing and Alternate Pathways
Case Study: Stage 6

If you want to have good ideas you must have many ideas.
Linus Pauling[1]

If I had asked my customers what they wanted, they would have said a faster horse.
Henry Ford[2]

Needs Finding

Both Pauling and Ford offer great insights into this *most important* starting point. Identifying a compelling clinical need may seem simple and obvious, but it is not. Get it right and you have a chance, get it wrong and all further effort is likely to be wasted. The process of identifying needs involves first a broad screening survey, which we call "Needs finding." The follow-on process, "Needs screening," is covered in Stage 2. By way of analogy, needs finding is akin to snorkeling; needs screening is more like a deep dive.

Needs finding is a simple and yet profound process. The diagnostic and therapeutic workings of the healthcare system offer fertile ground to search for unsolved problems. From the back of an ambulance, to the operating room (OR), or the outpatient clinic, real problems abound. The principle is to observe real people and real life situations in order to fully understand clinical procedures and techniques, *as they are currently practiced*. The observer should then look for difficulties that healthcare providers or patients encounter, and major obstacles or technical barriers that may be modified. Look for what might be missing (Henry Ford). The essential task is to identify the real clinical challenges and problems that impose a significant medical burden.

This is neither an armchair exercise nor an isolated epiphany. Rather, thoughtful observation of clinical encounters with "fresh eyes" is most likely to identify substantial unsolved problems. It may be a spoken need, such as a surgeon asking for a "third hand"; it may be the unspoken need, only appreciated when clinical troubles or complications are the expectation of the treating team. When an untoward clinical outcome or complication is met with the retort, "Oh, we see this ..." – **pay attention**. This is a great stimulus to ask: "*Why* do you see this?" "Should you see this?" "Is this inevitable?"

This sequential and iterative process from *observation* to *problem* to *need* produces real clarity. For example, a chance observation – that an elderly woman was admitted to a nursing home because of urinary incontinence – sparked the interest of a team. Subsequent inquiries unearthed the fact that more money is spent on adult diapers than on infant diapers and that urinary incontinence is the leading cause of admission to a nursing home. Thus a compelling clinical need was identified.

Notes
1. As quoted by Francis Crick in his presentation "The Impact of Linus Pauling on Molecular Biology," 1995.
2. Unsourced quotation widely attributed to Henry Ford.

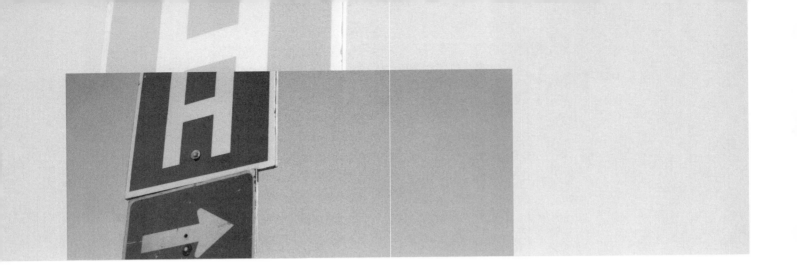

1.1 Strategic Focus

Introduction

An engineer with a needle-phobic mother decides to design an alternate method for administering the daily insulin she takes to control her diabetes. A spinal surgeon, frustrated with the limitations of the implants she uses to treat vertebral compression fractures, starts working on improvements to the device. A business student observing a birth at a hospital in Africa is struck by the need for a technology to prevent blood spray during the process to protect healthcare workers when the mother is infected with HIV. A resident studying oncology becomes passionate about understanding the disease and commits himself to cancer research and the pursuit of a cure. While all of these paths are worthwhile, they are not universally appealing. The course that excites one innovator may be uninteresting or overwhelming to another. But, the one thing that these paths have in common is that they are compelling to the people undertaking them. Their commitment to these unique focus areas will drive them forward through the many challenges that await them as they begin the innovation process.

One of the first, most important steps in the biodesign innovation process is for innovators to discover and explicitly commit themselves to the strategic focus area that stimulates their personal enthusiasm. To make an effective, meaningful decision about a strategic focus area – which could be represented by a medical practice area, a specialty, or a specific need – innovators must ask themselves questions about why they want to pursue this path, what they hope to accomplish, and how their strengths and weaknesses may affect their efforts. Additionally, a high-level assessment of the characteristics of the medical area should be taken into account relative to these goals. Ultimately, the most rewarding and successful biodesign projects are those that achieve a high degree of alignment between the values and competencies of the innovators and the defining characteristics of the strategic focus area that is chosen.

OBJECTIVES

- Understand that innovators must explicitly choose their strategic focus.

- Appreciate the importance of achieving alignment between the strategic focus area that is chosen and the mission and strengths/weaknesses of the individual or team.

- Recognize the steps involved in choosing a strategic focus.

Strategic focus fundamentals

As Mir Imran, CEO of InCube Labs and founder of more than 20 medical device companies, said:[1]

> I knew once I found a problem, I could solve it. The biggest challenge for me was which problem to solve.

Choosing a strategic focus area is an essential decision that launches the biodesign innovation process. If innovators think of this process as a journey – from discovering medical needs to developing new medical technologies that solve those needs – then the selection of a strategic focus is analogous to charting a course. The myth that innovators spontaneously create new ideas and inventions in a sudden stroke of genius, and that the process of innovation has no structure or predictability, could not be further from the truth. For most medical technology (**medtech**) innovators, ideas do not just happen – they are the result of an intentional decision to go out and make observations in a specific area, study multiple aspects of the healthcare landscape, identify opportunities where poor (or no) solutions exist, and then generate new solutions that address the gaps that have been discovered.

By explicitly deciding in what areas to focus, innovators accept different risks, challenges, and potential rewards (e.g., working on heart problems is much different from working on knee problems). As the stories and case examples in this book reflect, the choices made by individual innovators early in their journey have a direct and meaningful effect on the obstacles and opportunities they encounter on their path. As a result, deciding on a strategic focus is one of the most significant and directionally important decisions that innovators will make, and one that can have a major impact on the ultimate outcome of their efforts.

Steps toward developing a strategic focus

As one of the first steps in choosing a strategic focus, it is helpful to conduct a personal inventory. Importantly, the inventory should be performed before the innovator begins thinking about any particular practice area, specialty, or specific need. The purpose of the inventory is to identify the mission of the individual or team, as well as their strengths and weaknesses. It should also result in the definition of project "acceptance criteria." These

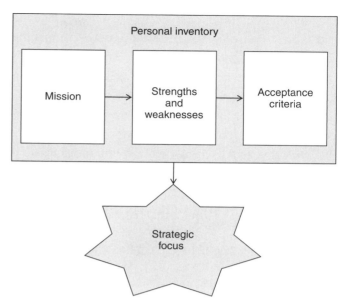

FIGURE 1.1.1
Taking the time to perform a comprehensive personal inventory can lead the innovator to the identification of an appropriate and exciting strategic focus.

criteria will be used to evaluate and decide on an area of strategic focus later in the process when the innovators begin scanning the external environment for needs and opportunities. See Figure 1.1.1.

Performing a personal inventory is equally important for individual innovators, academics/researchers, small teams, young companies, and large corporations, in that it helps ensure a good fit between the chosen strategic focus and the person (or people) undertaking the innovation process. The issues and priorities that emerge as a result of the inventory will be different based on the constituency performing it; however, the value of the exercise will be the same.

Determine a mission

Innovators need to be explicit about their mission. A mission is a broad, directional aspiration that defines what an individual or group wants to accomplish. Articulating a mission sets a desired destination for an innovation project and provides clarity about the ultimate goal the individual or group hopes to achieve.

To define a mission, individuals and groups should think about their priorities, beginning with questions about what is most important to them (or, conversely, what is not important to them). For example, a priority for someone pursuing a career in research or academia might be to engage in an exceptionally compelling research project that, if successful, would have a dramatic impact on healthcare worldwide. While such a long-term mission might take an entire career to achieve, the magnitude of the potential outcome would be large enough to make that commitment worthwhile to someone with this goal. Getting involved in a project with a less significant outcome might take less time and effort to achieve, but would be less interesting to the individual due to the misalignment with his/her mission.

In companies and other established organizations, the mission sometimes takes the form of what is commonly known as a mission statement. The Medtronic example below illustrates how a corporate mission statement might look.

FROM THE FIELD: MEDTRONIC

Defining a meaningful mission statement

Medtronic was founded in 1949 by Earl Bakken and his brother-in-law Palmer Hermundslieco as a medical equipment repair shop. The fledgling company quickly expanded into services and then into device design, development, and manufacturing.[2]

During the early years, Bakken was moved by the emotional response patients had to the company's products. Many were overjoyed to regain mobility, feel better, and sometimes even to be alive as a result of Medtronic's work (see Figure 1.1.2).[3] Inspired by their stories and the desire to make this type of human benefit the purpose of the organization's efforts, he and the board of directors created the Medtronic Mission, which remains an integral part of the company's culture and the driving force behind every project that it undertakes. This Mission guides the company's day-to-day work and keeps employees focused on the goal of changing the face of chronic disease for millions of people around the world.

Medtronic's Mission is to:[4]

- Contribute to human welfare by application of biomedical engineering in the research, design, manufacture, and sale of instruments or appliances that alleviate pain, restore health, and extend life.
- Direct our growth in the areas of biomedical engineering where we display maximum strength and ability; to gather people and facilities that tend to augment these areas; to continuously build on these areas through education and knowledge assimilation; to avoid participation in areas where we cannot make unique and worthy contributions.
- Strive without reserve for the greatest possible reliability and quality in our products; to be the unsurpassed standard of comparison and to be recognized as a company of dedication, honesty, integrity, and service.
- Make a fair profit on current operations to meet our obligations, sustain our growth, and reach our goals.
- Recognize the personal worth of employees by providing an employment framework that allows personal satisfaction in work accomplished, security, advancement opportunity, and means to share in the company's success.
- Maintain good citizenship as a company.

FIGURE 1.1.2
Earl Bakken with a young Medtronic patient (courtesy of Medtronic).

As William Hawkins, CEO of Medtronic, explained, "The Mission is our moral compass. It is the glue that binds all of our businesses together. It underpins everything we do. In good times and tough times, the one constant in our business model is our core values. We use the Mission to ensure that we work on the right things *and* that we strive to do things right."[5]

Large corporations may also choose to define specific missions for their divisions or groups. At this level, other priorities may surface as they approach the innovation process. With established portfolios of products to leverage (and protect), a division might not always be interested in finding the biggest near-term innovation. Instead, it may focus on driving incremental improvements in existing product lines that enable it to stay ahead of the competition. Or, with more extensive resources at its disposal, a company might be willing to make slightly larger, longer term investments with the intent of leapfrogging competitors over time.

The missions of aspiring entrepreneurs or young start-up companies may be different still. First and foremost, these individuals and teams do not necessarily need to create mission statements that are as formal or expansive as those of a large company. As long as the mission is clearly articulated, it can be significantly more informal (although it is still advisable to put it in writing). Second, the mission might be somewhat more practical or applied. For example, without the resources to support a vast, long-term research program, two innovators working together on a shoestring budget might decide that one important aspect of their mission is to identify a solution that is readily achievable (within one to two years) and compelling enough from a business perspective to raise financial support. Unlike the researcher or aspiring academic, these innovators would be more focused on near-term opportunities that are sizable, but not too expensive to pursue.

Identify strengths and weaknesses

In addition to thinking about a mission, individual innovators, academics/researchers, small teams, young companies, and large corporations will all benefit from assessing their strengths and weaknesses. Specifically, they should evaluate what they do well, and how they can capitalize on these strengths. They should also consider in what areas they are less experienced, competent, or confident, and how they can compensate for these relative weaknesses.

Some people can be successful in leading the innovation process (especially in its early stages) on their own. However, many individuals *and* groups recognize after they assess their strengths and weaknesses that they will benefit by collaborating with others who offer different, complementary skill sets. For example, if an innovator is a strong clinician, but not an engineer, it might be a helpful to partner with an engineer if the mission is to develop a device technology. Or, if that same innovator is interested in developing a business plan to pursue a concept, s/he might want to consider collaborating with someone with business training or experience to help construct and execute that plan. Wildly creative types are best paired with grounded, detail-oriented types, and so on. Fundamentally, the most important objective of this step is to identify where certain competency gaps and opportunities exist so that the innovator can address them when the time is right. It is rare for one person to embody all the talents necessary to identify, invent, develop, and commercialize a technology all on his/her own. However, if the innovator is aware of areas where help may be necessary, s/he can begin building a team with the strengths that complement known weaknesses, and can make sure that team expands as the need for more diverse skills increases.

Define acceptance criteria

The identification of a mission and the evaluation of strengths and weaknesses are direct inputs to the definition of project "acceptance criteria." At their most basic, acceptance criteria are parameters that must be met to make an innovation project attractive to the innovator. These criteria are used to choose an area of strategic focus, as well as to evaluate the specific opportunities that are discovered in the early stages of the biodesign innovation process. Common examples of factors to consider in defining acceptance criteria are found in Figure 1.1.3.

For example, suppose that a large corporation has a mission to develop a product that expands its portfolio into a new clinical area within the next two to three years to drive increased growth within the company. Before defining its acceptance criteria, the company would have to think about what strengths and weaknesses it has that would enable it to achieve this goal.

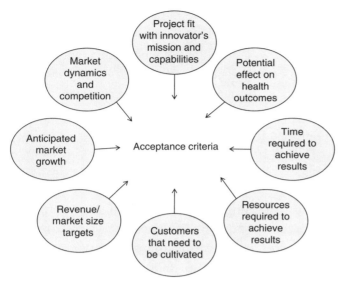

FIGURE 1.1.3
These and many other factors help shape an innovator's acceptance criteria, which can then be used to help define a strategic focus.

The availability of resources (staff, funding, and time) could certainly be a strength. However, the way in which the company's existing sales force is deployed (i.e., which types of doctors it already calls on) could be a strength or a weakness, depending on the specific area of focus that is chosen. After performing an assessment, the corporation might decide to engage in a project only if it meets the following acceptance criteria.

- The clinical practice area is new to the company and is growing at a minimum of ten percent per year and/or can generate a minimum of $100 million in revenue per year.
- Technologies in this space have a relatively simple regulatory pathway and straightforward **clinical trials** requirements so they can be brought to market quickly.
- The company's established sales force already calls on these same customers, so the commercial fit is very good.

For example, in the mid-1990s, American Medical Systems (AMS) had two primary products: an implantable urinary sphincter and a penile prosthetic line. The company had a mission of becoming a well-rounded urology company by broadening its focus to include other urological products. As it began to think about its acceptance criteria for new opportunities, the list included the following: (1) technologies that could be sold to the same customer or at the same "call point," (2) technologies that were more mechanical in function than biological, and (3) opportunities/areas that could grow at greater than 20 percent per year to add to the company's revenue growth.[6] Under different circumstances (for instance, if the company had saturated its existing customer base), the corporation might have eliminated the criterion to stay within the same customer group. While this would have made a wider cross-section of potential projects attractive to the company, it might not have allowed the company to achieve certain economies of scale by offering the same customers a wider line of products through the existing sales force. In this respect, the acceptance criteria defined by the company appropriately reflected the priorities of AMS at the time and capitalized on the perceived strength of its established sales arm.

Without any limitations imposed by a pre-existing business, an innovator or young company might define acceptance criteria around the magnitude of the impact its innovations can have on peoples' lives. In this scenario, with a mission to improve important outcomes for patients on a major scale, the acceptance criteria might require a project that:

- Has a total potential market of $1 billion or more.
- Will be attractive to investors (so it gets adequate financial support).
- Results in an innovation that has a significant impact on patients' **quality of life** (as opposed to an innovation that makes a device cheaper, faster, or easier to use).
- Has platform potential so that the benefits from one medical specialty can be rapidly leveraged to affect patients in other practice areas.
- Is focused on a patient segment where head-to-head competition can be avoided, especially if the company is concerned about its ability to compete with entrenched firms.

The acceptance criteria above are similar to those used by medtech **incubators** such as ExploraMed, The

Foundry, or The Innovation Factory. Such criteria enable these organizations continually to deliver powerful innovations in a number of diverse fields.

Fundamentally, acceptance criteria are the mechanism through which a mission, priorities, strengths, and weaknesses are woven together into a list of requirements that an innovation project must meet. There is no single set of acceptance criteria that works for every individual or team. However, whether they are driven by charitable motives, purely academic or scientific interest, or entrepreneurial drive, setting these criteria early will help ensure that their goals are ultimately achieved.

Articulating a strategic focus

Once specific acceptance criteria have been defined, the innovator can start exploring different medical specialties and practice areas for a good fit. Innovators are encouraged to look at a broad range of areas, keeping in mind that deep expertise in a field is not necessarily required. All too often, people who are deeply immersed in a field fail to see the opportunities and needs that surround them because they have been indoctrinated into a certain way of doing things. Individuals and teams that bring diverse experiences and different backgrounds to a field can sometimes be more successful in identifying needs and opportunities because they are more willing to question the status quo.

While a sweeping investigation of opportunities across the healthcare landscape is useful for some innovators, others have defined acceptance criteria that point them to a specific field based purely on a personal interest or passion for a practice area. For instance, someone might be committed to addressing needs in the breast cancer field after losing a loved one to the disease. While this is certainly a valid approach, such individuals are encouraged to get even more specific about their strategic focus. For instance, would it be a better fit to embark on a long-term research-based path to cure the disease, or to pioneer near-term improvements in the effectiveness of breast cancer treatment? The innovator can use his/her other acceptance criteria to define a focus within the desired field that is most likely to lead to a fulfilling experience and outcome.

As exploration of the healthcare landscape begins, certain choices can be immediately eliminated. For example, an innovator who is determined to have a major impact on treating or curing chronic illness can quickly set aside the investigation of any acute conditions. One who has defined acceptance criteria around the treatment of heart disease has no need to evaluate opportunities in other practice areas. If speed to market is a priority, areas that would require long regulatory or clinical processes are best avoided. All of these decisions, if identified early, can shape the strategic focus and have a powerful impact on the outcome.

One way an innovator can begin the process of screening focus areas against his/her acceptance criteria is to examine high-level data related to a practice area (note that more in-depth research will be performed in subsequent steps of the biodesign innovation process). Statistics to consider include the number of people affected by a disease state, the clinical impact of the disease or the outcomes of existing treatments, the profitability of existing treatments, and the rate at which spending is growing. See Table 1.1.1. Innovators can also glean insights from the total revenue realized each year in a particular medical field. See Figure 1.1.5.

The more rigorous this evaluation process, the better. However, even a cursory evaluation of different treatment areas (and their sub-specialties) will potentially help to narrow one's focus. For example, an innovator or company seeking a large business opportunity might review certain statistics and other data and immediately become interested in the cardiovascular field. Yet, the fact that this is a relatively well-established, mature field may conflict with some of the other acceptance criteria that the innovator has defined. If s/he is committed to new opportunities and needs that have not yet been defined or where innovation has not occurred for quite some time, another field outside of cardiology might be a better fit (e.g., respiratory medicine or urology). In an area with a well-defined market opportunity, there may be intense competition and a great deal of pressure to be first to market with technology that could set the new **standard of care**. In less popular

Table 1.1.1. Data such as the percentage of total change in healthcare spending accounted for by the 15 most costly medical conditions, as shown in the table, can be an interesting source of ideas regarding areas that meet an innovator's acceptance criteria (copyrighted and published by Project HOPE/Health Affairs as Kenneth E. Thorpe, Curtis S. Florence, and Peter Joski, "Which Medical Conditions Account for the Rise in Health Care Spending?" Health Affairs, web exclusive, August 25, 2004; the published article is archived and available online at www.healthaffairs.org).

	Treated prevalence per 100,000		Spending (millions of dollars)		Approximate percentage change in total healthcare spending
Condition	1987	2000	1987	2000	(1987–2000)
Heart disease	6,189	6,226	30,450.1	56,678.6	8.06
Pulmonary conditions	10,389	15,526	11,684.5	36,476.5	5.63
Mental disorders	4,373	8,575	9,935.8	34,439.1	7.40
Cancer	2,862	3,348	21,167.5	38,901.8	5.36
Hypertension	9,734	11,382	8,008.6	23,394.5	4.24
Trauma	17,866	12,338	26,527.6	41,124.2	4.64
Cerebrovascular disease	410	854	3,859.8	14,938.8	3.52
Arthritis	5,479	6,966	7,403.5	17,686.3	3.27
Diabetes	2,961	4,260	8,661.1	18,287.9	2.37
Back problems	3,400	5,092	7,964.6	17,451.0	2.99
Skin disorders	6,754	7,990	4,758.0	12,044.5	2.26
Pneumonia	1,537	1,370	5,437.6	12,641.3	2.29
Infectious diseases	6,588	5,841	3,658.0	9,849.5	1.35
Endocrine	5,515	7,322	5,247.8	10,276.9	1.18
Kidney	675	908	4,938.1	8,169.5	1.03

areas, the advantages of weaker competition are balanced by greater uncertainties. Both issues impact the ability to attract investment and motivate behavior change among physicians who are entrenched in the old ways of treating patients. This is where the innovator's acceptance criteria (and prioritization) can help to resolve inherent conflicts and facilitate effective trade-offs, which become clearer when evaluating these different risks and rewards.

Once preliminary data about the defining characteristics of various practice areas have been considered against the acceptance criteria, a strategic focus or a few acceptable focus areas should begin to emerge.

While the focus area will be different for every individual or group, the key is to ensure that it is aligned with the innovator's mission, strengths, weaknesses, and acceptance criteria. For example, one innovator might choose to pursue opportunities related to chronic obstructive pulmonary disease, while another decides to go after opportunities associated with retinal detachment in the eye. In either scenario, a strong sense of "the right fit" is essential to anyone embarking on the biodesign innovation journey.

The following story from ExploraMed describes how one innovator worked through the process of choosing a strategic focus.

FROM THE FIELD | EXPLORAMED

Applying acceptance criteria in evaluating a strategic focus

Making an explicit decision about the strategic focus to be pursued is an essential exercise for individual innovators, teams, companies, and company incubators alike. According to Josh Makower, founder and CEO of medical device incubator ExploraMed, "Choosing what is *not* a fit is as important as determining what is."[7] ExploraMed, which was started in 1995, has embedded this step in its process for identifying, creating, and developing new medical device businesses. When Makower initiates a new business, he and his team spend time assessing their relative strengths and weaknesses and articulating the acceptance criteria against which they will screen potential opportunities.

ExploraMed's defined mission is to "focus on clinical needs where there is an opportunity to dramatically improve outcomes and build freestanding businesses." As Makower explained, "I get excited about working on things that are going to have a major impact on medicine. We want to work on projects that make a substantial contribution, can potentially change the direction of healthcare, and affect outcomes for thousands or millions of patients. If a large number of people are affected by a problem and currently have poor outcomes from the existing set of treatments, it could be a hot area for us to investigate." Recognizing their own strengths and weaknesses, Makower and team further constrained their efforts to medical device opportunities, leaving drugs, diagnostics, and other healthcare technologies to a different set of innovators and entrepreneurs. Finally, they specifically decided that they liked the idea of being "contrarians." "We like to go where others haven't gone and where people believe there aren't reasonable opportunities. You can create a competitive advantage for yourself by being the first to go in another direction. The other thing that we're trying to do is create big enterprises. To do this, we almost always have to be willing to go into a space where there aren't a lot of other players. A little fish can grow to be pretty big if he finds himself in a large pond all by himself. I like that a lot better than trying to establish a foothold in an already crowded market."

These defined acceptance criteria are routinely used by ExploraMed to evaluate which opportunities to pursue, as the following example demonstrates. Early in the company's history, when the team was actively investigating new projects, Makower's elderly aunt fell and broke her hip. "Before the accident, she was energetic, vibrant, and active. After she fell, her life changed dramatically. She had trouble with her daily activities, as well as doing the things she loved like seeing her children and grandchildren. Suddenly, she was an old lady, when that wasn't how she lived before." With a new passion to address patient needs in this area, Makower and Ted Lamson, an ExploraMed project creator at the time (see Figure 1.1.4), began to investigate the space. What they quickly learned was that hip fractures represented a sizable problem. Some 350,000 elderly people broke a hip each year in the United States alone. Only 50 to 60 percent of those individuals recovered fully; 20 percent never walked again; and 40 percent ended up in a nursing home. Furthermore, within a year, 37 percent of Americans suffering a hip fracture were dead.[8] "It's a shocking mortality rate," noted Makower. "We speculated that there was a need for a less invasive alternative to hip replacement, and that the size of the incision or **morbidity** from the operation itself was the key to the problem. Eventually, we discovered that this guess was wrong and that the real need was not in the surgery, but in post-surgery recovery. If we hadn't been following a defined process for identifying and understanding needs, we easily could have become biased towards a solution early on that would have sent us in the wrong direction," he emphasized.

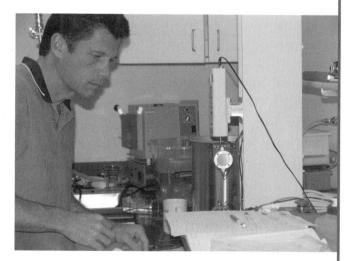

FIGURE 1.1.4
Lamson, of ExploraMed, working on device research (courtesy of Explora Med).

FROM THE FIELD

EXPLORAMED

Continued

ExploraMed's acceptance criterion related to the size and severity of the problem was the first screen the team applied to the problem, and hip fractures appeared to be a promising market. Makower and Lamson conducted further preliminary research to understand what companies and innovations were active in the space. It turned out that numerous advancements had been made in hip surgery and the devices used to support it, but that few new technologies existed to improve post-operative care and recovery. "If you get the patient up immediately post-procedure, and you effectively manage their pain locally so they can walk around and never waste any of their muscles, then their outcomes are fantastic. But if they stay in bed more than they should because their pain is not managed well, they do terribly. What happens is that they lose muscle mass, they get sick or become depressed, and then they die of pneumonia or some other complicating condition." With few individuals or companies working to address the non-surgical issues associated with hip fractures, the field appeared to be wide open to ExploraMed.

Unfortunately, when it came to ExploraMed's desire to focus on medical devices, this is where the project took a different turn. "We discovered that the real need was for something to improve local pain management to help patients ambulate more quickly," recalled Makower. "It was a big opportunity in an open market, but we realized it could probably be addressed best by a drug. We didn't have the right technology, skills, or resources to take on a drug project. We really wanted to figure it out, but we realized that we weren't the right guys to do it. Regardless of your passion for an area, you have to be honest with yourself about you and your team's strengths and weaknesses."

The team confirmed its finding through additional research, observations, needs finding, and consultation with experts in the field (as described in 1.2 Observation and Problem Identification and 1.3 Need Statement Development). "Upon further research we actually discovered there already existed systems to do exactly what we wanted to do to address this need, but they were not being utilized because of healthcare management constraints or cost. This was very discouraging ... the answer was there and doctors were actually aware of it, but they were not using it for one reason or another," Makower commented. Eventually, the team decided to reject the project, and continue their search elsewhere. "You have to be willing to accept a lot of failure," he said, reflecting on the experience. "But you've got to keep on trying – and failing if necessary – in order to understand the parameters that will make you successful and ultimately enable you to choose the right path." Later, Makower and Lamson redirected their focus to an entirely different clinical area and, after several months of investigation, found a compelling opportunity that met all their criteria and became a company called NeoTract, Inc. (see 5.2 R&D Strategy for more information about NeoTract).

Global considerations in choosing a strategic focus

In a 2008 article, worldwide medical device sales were estimated at $200 billion, of which the United States accounted for 45 percent, Europe 30 percent, and Japan 10 percent.[9] The remaining 15 percent represented the rest of the world, including the large but still developing markets of China and India. While the vast majority of medtech innovation is presently centered in the West, this was expected to change dramatically over the next two decades. As a result, more and more innovators would begin choosing strategic focus areas outside the United States and Europe.

An important issue for innovators to recognize when seeking opportunities outside these well-developed markets is that their acceptance criteria will likely be markedly different. Particularly in the United States, there tends to be a heavy emphasis on practice areas with target populations that can support cutting-edge products with high profit margins. In emerging markets, such as China, India, and Africa, there are vast groups of patients that may only require a simple solution, but who have a limited ability to pay for new technologies. As the appetite for devices and the ability to pay for them develops in these countries, the innovators who focus on these markets will be drawn to them

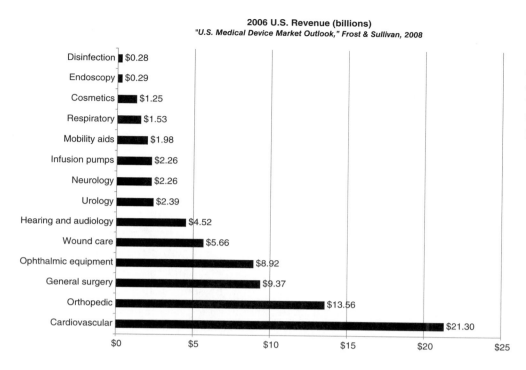

FIGURE 1.1.5
Information about medical device revenues by major medical segment can be helpful in choosing a strategic focus (from "U.S. Medical Device Market Outlook," Frost & Sullivan, 2008; reprinted with permission).

by their passion, commitment to helping others, and their desire to have a major impact on large numbers of prospective patients, rather than by an interest in optimizing their financial return. See Figure 1.1.6.

Beyond financial considerations, innovators exploring strategic focus areas in emerging markets also face an increased level of risk. For instance, in places like China, intellectual property (IP) enforcement and regulatory processes are still largely underdeveloped. This additional uncertainty prevents some companies and innovators from moving into these markets and creates additional challenges for those who do. Many of these types of risks will almost certainly be reduced or resolved over time, but the efforts of many motivated and committed innovators will be required to make this happen.

Ethics in the biodesign innovation process

Choosing a strategic focus is among the first of many steps in the biodesign innovation process where innovators may face ethical dilemmas. The potential for ethical conflicts exist at nearly every stage of an innovator's journey. Ethics focus on the intentional choices that people make and the basic moral principles that are used to guide these decisions. Ethics do not provide a specific value system for making choices, but rather a set of basic principles that can be followed to guide decision making.[10] Stated another way, ethics provide the rules or standards that guide (but do not determine) the conduct of a person or the members of a profession. In developing and bringing new medical technologies to market, the need to maintain the highest ethical standards extends to everyone involved in the process.

At the heart of most ethical issues are conflicts of interest, which arise when one person's interests are at odds with another's. For example, confidentiality, or the practice of discerning what is privileged information and rigorously protecting it, is an important principle in the medical field. If one party has an incentive to disclose confidential information about another party, a conflict of interest may arise. Because there are so many individuals and groups in the development and commercialization of any medical innovation, conflicts of interest are inevitable. Realistically, the objective of any innovator should not be to avoid such conflicts, but to ethically address and resolve them when they arise. In particular, in any scenario where conflicts of interests involve patients and the care they receive, innovators have a special obligation to act ethically

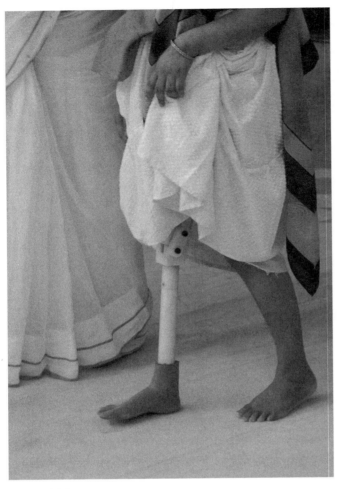

FIGURE 1.1.6
A Jaipur artificial limb (as shown above) costs $35–40 to manufacture in India, while an artificial limb in the United States can cost anywhere from $6,000 to $35,000. Disparities such as these have sizable implications for the innovator in choosing a strategic focus (courtesy of the Stanford-Jaipur Knee team from Professor Tom Andriacchi's mechanical engineering course: L. Ayo Roberts, Joel Sadler, Angelo Szychowski, and Eric Thorsell).

because of the potential to both improve and harm human lives.

Ethics in the medical field have a difficult past, with trials, such as the Tuskegee syphilis experiment, creating issues of fear and distrust between medical providers and the patients they are meant to serve. In this particular case, the US Public Health Service ran an experiment on 399 black men from 1932 to 1972. These patients, who were mostly poor and illiterate, had late-stage syphilis but were not informed from what disease they were suffering. The doctors involved in the experiment had no intention of curing the men – instead, their objective was to collect scientific data from their autopsies. Over years, the patients experienced tumors, heart disease, paralysis, blindness, insanity, and, eventually, death.[11] Since then, significant strides have been made in enforcing a strong code of ethics across the medical community. Moreover, every member of the medical community, medtech innovators included, has an important role to play in promoting and adhering to ethical behavior.

Early in the biodesign innovation process, for instance when choosing a strategic focus, innovators often struggle with the tension between altruistically addressing important medical needs and the imperative to do so in such a way that the solution has a viable chance of reaching the market for which it is intended. An inspirational new therapy cannot, in most cases, reach patients without the necessary capital to develop it; yet capital will only be provided by commercial investors if they feel that a reasonable profit can be obtained. It can be frustrating to innovators to identify important medical needs only to discover that the market or profit potential for a solution is too small or risky to attract funding. Although capital can be obtained from government grants, non-governmental organizations (**NGOs**), or beneficent donors, there are inherent limitations associated with this type of funding that can prevent an innovation from achieving its full potential.

To develop a truly sustainable solution, most innovators have to strike a balance between satisfying the needs of the target audience and satisfying the interests of investors. Protecting investors is known as a fiduciary duty. A fiduciary is any individual or group that has the legal responsibility for managing somebody else's money.[12] As a fiduciary, the innovator has an obligation to carry out the responsibility of managing others' funds with the utmost degree of "good faith, honesty, integrity, loyalty, and undivided service of the beneficiary's interest."[13] At the most basic level, this means that the innovator has a duty not to favor anyone else's interests (including his/her own) over those of the beneficiary.[14] If the fiduciary violates this responsibility, s/he may be subject to legal liability, which is another reason why ethical behavior is so important throughout the innovation process.

Table 1.1.2. *An example of a medical ethics (from the American Medical Association's "Principles of Medical Ethics;" reprinted with permission).*

American Medical Association's principles of medical ethics
A physician shall be dedicated to providing competent medical care, with compassion and respect for human dignity and rights.
A physician shall uphold the standards of professionalism, be honest in all professional interactions, and strive to report physicians deficient in character or competence, or engaging in fraud or deception, to appropriate entities.
A physician shall respect the law and also recognize a responsibility to seek changes in those requirements which are contrary to the best interests of the patient.
A physician shall respect the rights of patients, colleagues, and other health professionals, and shall safeguard patient confidences and privacy within the constraints of the law.
A physician shall continue to study, apply, and advance scientific knowledge, maintain a commitment to medical education, make relevant information available to patients, colleagues, and the public, obtain consultation, and use the talents of other health professionals when indicated.
A physician shall, in the provision of appropriate patient care, except in emergencies, be free to choose whom to serve, with whom to associate, and the environment in which to provide medical care.
A physician shall recognize a responsibility to participate in activities contributing to the improvement of the community and the betterment of public health.
A physician shall, while caring for a patient, regard responsibility to the patient as paramount.
A physician shall support access to medical care for all people.

Striking an appropriate balance can be difficult when conflicting interests arise. The important thing to remember is that the right solution may vary for each individual innovator based on his/her ethical compass. By openly acknowledging the fact that "gray areas" exist and taking time for self-reflection, innovators can more readily determine the approach that is most closely aligned with their values.

Regarding other ethical conflicts in the innovation process, innovators are generally advised to maintain a primary focus on the needs of patients in resolving issues. Seeking input and advice from objective third parties can be an invaluable resource for resolving conflicts. However, more often than not, innovators must rely on their own codes of personal and professional ethics. The following four principles are widely accepted as ethical standards in the medical field.[15]

Respect for autonomy
Respect for autonomy refers to others' rights to make their own choices. This means, for example, that all parties with an interest in a new innovation need to be informed about its risks and benefits, any potential conflicts of interests among those involved in its development and delivery, and about any other factors that could conceivably affect their choice.

Beneficence
Beneficence is the practice of doing good. In the medical field, this mandate extends to maximizing benefits while seeking to minimize potential harm.

Non-maleficence
The mandate of non-maleficence is also captured by the phrase "First do no harm." Often beneficence and non-maleficence cannot be separated. In the process of providing a medical benefit, healthcare providers may also expose patients to risk. For instance, in clinical trials, patients are exposed to risks for the sake of others, by making it possible for life-saving devices to reach the market. The Hippocratic oath taken by each physician essentially combines the principles of beneficence with non-maleficence, by stating that the obligations of healthcare professionals is to provide the greatest net medical benefit at minimal risk.[16]

Justice or fairness

All those in the medical field have an obligation to fairly decide among competing concerns and interests. At a minimum, this requires recognizing potential conflicts of interest and objectively determining, sometimes with third-party assistance, how they should be resolved.

Because so many interactions in the biodesign innovation process involve clinicians, innovators should become familiar with the specific ethical codes developed by relevant medical professional societies. For instance, it may be helpful to familiarize oneself with the World Medical Association's Physician's Oath as defined in the Declaration of Geneva in 1948.[17] Table 1.1.2 summarizes the American Medical Association's *Principles of Medical Ethics*, the foundation of the **AMA** *Code of Medical Ethics*, one of the most well-known and widely practiced codes of ethics in the medical field. The complete code, consisting of these *Principles* and the *Opinions of Council on Ethical and Judicial Affairs*, is available through the council's web page.[18]

When conflicts must be resolved or difficult decisions must be made, a strong code of ethics can be used by innovators as an essential guide.

GETTING STARTED

Choosing a strategic focus is an exciting first step in the biodesign innovation journey. Use the following steps to help chart a course.

See **ebiodesign.org** for active web links to the resources listed below, additional references, content updates, video FAQs, and other relevant information.

1. Take inventory

1.1 **What to cover** – Before thinking about a specific practice area to pursue, perform a personal inventory. Define a mission that reflects the defined purpose, priorities, and specific goals in launching the biodesign innovation process. Next, assess key strengths and weaknesses, and begin thinking about how to complement the innovator's skills with those of different team members as the process progresses. Then, identify the key acceptance criteria that will be used to ensure the most appropriate strategic focus area. Prioritize these acceptance criteria in case trade-offs become necessary in the evaluation of specific opportunities.

1.2 **Where to look**
- **Personal reflection** – Spend time alone reflecting on the questions and issues outlined in the Fundamentals section of this chapter. Complete an honest assessment of personal motivators, capabilities, and the characteristics that any project must possess to be fulfilling.
- **Facilitated session(s)** – If working in a group, consider holding one or more facilitated session to build consensus regarding the team's mission, strengths/weaknesses, and acceptance criteria.
- **Advice from respected advisors** – Sometimes others can help an innovator identify his/her strengths and weaknesses. Through the process of getting to know the innovator, advisors can offer important insights that can help one develop a vision.

2. Articulate a strategic focus

2.1 **What to cover** – Begin by performing research to identify a series of different strategic focus areas that might perform well against the defined acceptance criteria. Evaluate opportunities carefully and objectively against each criterion. Remain cognizant of the potential for conflicts of interest and use the highest ethical standards in choosing a strategic focus. Continue narrowing down the list of strategic focus areas until the best fit can be identified. Be as specific as possible (e.g., the desire to explore improvements in early-stage breast cancer treatment) without jumping ahead to medical needs or solutions (which are addressed in the next two chapters).

2.2 **Where to look**
- **PubMed** – PubMed is a database of the US National Library of Medicine that includes more than 16 million citations from **MEDLINE** and other life science journals dating back to the 1950s.
- **Agency for Healthcare Research and Quality (AHRQ)** – Sponsors and conducts research that provides **evidence-based** information on healthcare outcomes, quality, cost, use, and access. From its website, one can gain access to **longitudinal data** regarding patient interactions with the healthcare system, including all transactions and their codes (indicating procedures and diagnosis), as well as location of service, which can be used to help develop market segments. Important databases accessible via the site include:
  - **HCUPnet** – A free, online query system based on data from the Healthcare Cost and Utilization Project (**HCUPnet**). It provides access to health statistics and information on hospital stays (inpatient encounters) at the national, regional, and state levels.
  - **MEPS Data** – The Agency for Healthcare Research and Quality provides longitudinal data on the health expenditures of 30,000 US households via the Medical Expenditures Panel Survey (**MEPS**). This data is publicly available for primary analysis. It is useful for more detailed analyses of market segments (and sizing), but working with the data can be labor intensive. This source is probably most helpful if the innovator or company needs to support a need specification with actual publications as part of a marketing strategy. Data is available for conditions with a 1 percent prevalence rate or greater.
- **US Census Bureau** – Provides online access to the latest US census data.
- **World Health Organization** – WHO is the directing and coordinating authority for health within the United Nations system. Information on global health trends, research, policies, and standards can be found through this group.
- **Professional societies** – These associations can provide a wealth of information to help innovators understand issues and opportunities within a potential focus area. See Appendix 1.1.1 for a table that includes a sample of the professional societies that exist within fields associated with the 15 most costly medical conditions.
- **Industry-specific news resources** – Online and offline publications in the medtech field, such as *Medtech Insight*, *InVivo*, and *Start-Up Magazine* are also useful sources of relevant information.

Credits

The editors would like to thank William Hawkins and Richard L. Popp for their contributions to this chapter.

Appendix 1.1.1

Professional associations for select medical conditions

Medical condition	Related professional societies	Websites
Heart disease	American Heart Association American College of Cardiology	www.americanheart.org www.acc.org
Pulmonary conditions	American Lung Association	www.lungsusa.org
Mental disorders	American Mental Health Counselors Association	www.amhca.org
Cancer	American Cancer Society American Association for Cancer Research	www.cancer.org www.aacr.org
Hypertension	National Hypertension Association Pulmonary Hypertension Association	www.nathypertension.org www.phassociation.org
Trauma	Orthopaedic Trauma Association American Trauma Society American Association for the Surgery of Trauma	www.ota.org www.amtrauma.org www.asst.org
Cerebrovascular disease	American Stroke Association National Stroke Association American Association of Neurological Surgeons (AANS)/Congress of Neurological Surgeons (CNS) Cerebrovascular Section	www.strokeassociation.org www.stroke.org www.neurosurgery.org
Arthritis	Arthritis Foundation	www.arthritis.org
Diabetes	American Diabetes Association International Diabetes Federation	www.diabetes.org www.idf.org
Back problems	American Spinal Injury Association North American Spine Society National Spinal Cord Injury Association	www.asia-spinalinjury.org www.spine.org www.spinalcord.org
Skin disorders	American Skin Association American Dermatological Association	www.americanskin.org www.amer-derm-assn.org
Pneumonia	American Lung Association	www.lungsusa.org
Infectious disease	Infectious Disease Society of America	www.idsociety.org
Endocrine disease	American Association of Clinical Endocrinologists American Association of Endocrine Surgeons International Association of Endocrine Surgeons The Endocrine Society	www.aace.com www.endocrinesurgery.org www.iaes-endocrine-surgeons.com www.endo-society.org
Kidney	National Kidney Foundation Renal Association	www.kidney.org www.renal.org

Notes

1. From remarks made by Mir Imran as part of the "From the Innovator's Workbench" speaker series hosted by Stanford's Program in Biodesign, April 28, 2004, http://biodesign.stanford.edu/bdn/networking/pastinnovators.jsp. Reprinted with permission.
2. "The Medtronic Story," Medtronic.com, http://www.medtronic.com/about-medtronic/our-story/garage-years/index.htm (September 18, 2008).
3. "Our Mission," Medtronic.com, http://www.medtronic.com/about-medtronic/our-mission/index.htm (September 18, 2008). Reprinted with permission.
4. Ibid.
5. From an exchange with William Hawkins, CEO of Medtronic, Fall 2008. Reprinted with permission.
6. From an exchange with Thom Gunderson, Medical Device Analyst for Piper Jaffray, Fall 2008. Reprinted with permission.
7. All quotations are from interviews conducted by the authors, unless otherwise cited.
8. Muriel R. Gillick, "Break a Leg?" Perspectives on Aging, March 14, 2006, http://www.drmurielgillick.com/2006_03_01_/archive.html (November 1, 2007).
9. "Japan Device Maker Terumo To Boost Catheter Production In U.S., Europe, Vietnam," *Medical Devices Today*, July 7, 2008, http://www.medicaldevicestoday.com/2008/07/japan-device-ma.html (September 18, 2008).
10. R. J. Devettere, *Practical Decision Making in Health Care Ethics* (Washington D.C.: Georgetown University Press, 2000).
11. "The Tuskegee Syphilis Experiment," Infoplease.com, http://www.infoplease.com/ipa/A0762136.html (September 18, 2008).
12. Jerry Sais Jr. and Melissa W. Sais, "Meeting Your Fiduciary Responsibility," Investopedia, http://www.investopedia.com/articles/08/fiduciary-responsibility.asp (September 18, 2008).
13. Errold F. Moody, Jr., "Fiduciary Responsibility," EFMoody.com, http://www.efmoody.com/arbitration/fiduciary.html (September 18, 2008).
14. Ibid.
15. T.L. Beaucham and J.F. Childress, *Principles of Biomedical Ethics* (New York: Oxford University Press, 1989).
16. R. Gillon, "Medical Ethics: Four Principles Plus Attention to Scope," *British Medical Journal*, July 16, 1994, p. 184.
17. See The World Medical Association's Physician's Oath, Declaration of Geneva, 1948, http://www.cirp.org/library/ethics/geneva/ (September 18, 2008).
18. See *"Council on Ethical and Judicial Affairs,"* American Medical Association, http://www.ama-assn.org/ama/pub/category/4325.html (December 9, 2008).

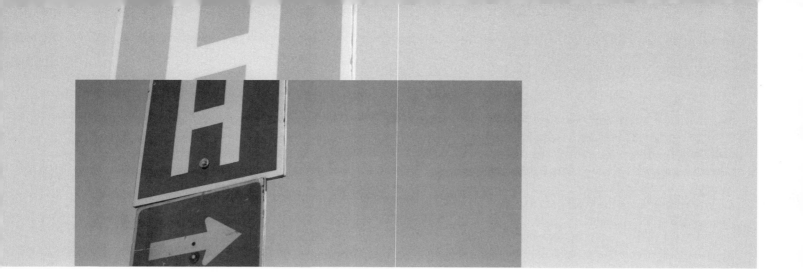

1.2 Observation and Problem Identification

Introduction

Two aspiring innovators are looking for opportunities to improve sternotomy[1] procedures, and they both contact a leading clinician. One asks the physician for an interview. The other requests permission to watch a procedure and spend time in post-operative recovery. Both may glean important insights from their investigations, but the innovator who actually sees the sternotomy performed, watches for inefficiencies, malfunctions, and risks in the process, and observes points in the procedure and post-operative care where stress, pain, or uncertainty occur, will learn dramatically more about the real opportunities for improvement. Rarely can an innovator fully appreciate the circumstances and characteristics of a particular opportunity simply by hearing or reading about it.

Before the development of new solutions can actually take place, innovators must first identify and understand the clinical problems that are associated with their chosen strategic focus area. The process of identifying important clinical problems requires innovators to utilize observation skills and find new ways of looking at processes, procedures, and events. The well-observed problem, in turn, provides the basis for describing a need – the fundamental building block of the biodesign innovation process.

OBJECTIVES

- Differentiate between observations, problems, and needs.
- Understand how to perform effective, meaningful observations.
- Identify the types of problems that are likely to result in significant clinical needs.

Observation and problem identification fundamentals

Medical needs exist in abundance. The key is to target needs that arise from genuine and important clinical problems. Problems are significant issues that have been identified in a diversity of real-world healthcare settings. While some problems are obvious, others have not yet been recognized, even by those closest to them. For this reason, the process of identifying a clinically relevant project must begin with objective, direct observation.

Clinical observations are not problems themselves, but rather a component of the methodology for identifying them. This methodology includes three important steps: (1) observing a specific clinical situation, (2) identifying the problem inherent in that situation, and (3) reshaping one's understanding of the problem into a need. Consider the following example.

> ***Observation:*** A medical resident in training struggles to intubate a patient (place a breathing tube into a patient's trachea) in the emergency room, leading to a drop in the patient's oxygen levels.
>
> ***Problem:*** For the unskilled practitioner, the time required to place an endotracheal breathing tube in an emergency setting can, at least in some cases, be extensive and can dramatically impact the outcome for the patient.
>
> ***Need:*** A way to reduce the time required for unskilled medical practitioners to place endotracheal tubes in an emergency setting.

An observation statement (in the context of the biodesign innovation process) is a description of a singular event that the innovator witnessed during a clinical observation. This differs substantially from a problem statement, which describes a recurring situation in which doubt, uncertainty, or difficulty is met in the process of providing clinical care. Ultimately, these will form the foundation for the need statement, which identifies the change in outcome that is required to address a given problem.

The language used to express each type of statement is critical and worthy of substantial thought. For instance, in the examples above, the innovator should ask whether the identified problem might be a concern for a larger population than just residents. The existence of a more widespread problem is possible, but that determination can only be made with further observations and data (e.g., the innovator may need to observe intubations performed by experienced physicians, paramedics, etc.). Further, all assumptions made in translating an observation to a problem must be validated and tested. For instance, in the example, the problem noted is based on the assumption that the resident's lack of skill led to the requirement of extra time to place the tube. If this proves to be incorrect (i.e., the problem is instead caused by certain types of patients with challenging anatomy and may not change with improved skill), the innovator could potentially invest time, effort, and money in pursuing a need that does not exist. Notice also that, in this case, reduced time is defined as the outcome goal. However, time may not be as important to patient care as a more specific clinical endpoint, such as minimizing oxygen saturation changes during the procedure. For this reason, in the observation process, the innovator needs to continually validate the variables surrounding an issue before naming the problem and subsequently translating it into a need.

The information outlined in this chapter serves as a guide for performing observations and developing problem statements. The next chapter, 1.3 Need Statement Development, describes how to transform problem statements into need statements.

Uncovering problems through observations

Direct observation of the healthcare environment ensures that the innovator will have the opportunity to identify clinically important problems in the field.

Even when innovators have specific ideas regarding needs that should be addressed in their areas of focus, observations are essential to confirm that these problems are "real." This is particularly relevant when the innovator is a clinician working in his/her field of expertise and may have strong preconceptions about the important problems in that area, or when s/he has gathered ideas and information from those entrenched within the specialty.

Without direct verification, the innovator may end up developing devices that are technically interesting

Figure 1.2.1
Direct observations help ensure that important clinical problems (and the associated needs) are not overlooked.

but ill-adapted to address the actual problems (and resulting needs) experienced by the target audience. Or, the innovator may miss the identification of certain problems that have not yet been recognized (or are routinely accepted) by those involved see Figure 1.2.1. Furthermore, observations make it easier to verify objectively that the observed problems are sufficiently important to guarantee that a new solution would be well received by the affected **stakeholders**.

Observational perspectives

The best way to begin problem identification is to observe the process of patient care within a given strategic focus area. In conducting these observations, the innovator must keep in mind the perspective of the patient, but also all of the other people involved in delivering care – patients, physicians, nurses, and other representatives of the healthcare system. Problems often are uncovered when the inadequacies or limitations of current approaches are identified. To be sure that nothing is missed, it is important to observe each component of the care process and ask the most basic questions at every stage and of every participant. These questions should be framed from multiple perspectives to ensure a broad understanding of issues and opportunities across a focus area.

For example, the questions in Figure 1.2.2 could be used to guide observations that provide multiple perspectives for a patient undergoing a hospital procedure.

The key is not only to observe what procedure was performed (or what device was used), but what the patient, provider, or system *experienced* as a result of the process.

It is also essential to make observations that span the entire **cycle of care**. It is not enough to watch a few minutes of a surgery when a specific device is used. Observers need to understand what is involved in the preparation, procedure, and post-operative care to truly understand potential problems (and corresponding needs) in a focus area. The importance of understanding different perspectives and evaluating the full cycle of care is exemplified by a patient preparing for and then undergoing colonoscopy to screen for colon cancer. According to most patients, one of the biggest problems with colonoscopy procedures occurs during "bowel preparation," the day prior to treatment. In this case, the patients are asked to drink a medication that often causes severe diarrhea, bloating, flatulence, and other forms of discomfort. In contrast, when physicians are asked about the most difficult aspects of the procedure, they might comment on the technical challenges of reaching certain regions of the bowel with the colonoscope.

What people *say* about who, what, when, where, how, and why they do something can also be somewhat misleading. As Thomas Fogarty, inventor of the embolectomy balloon catheter, as well as dozens of other medtech devices, said:[2]

> Innovators tend to go out and ask doctors what they want rather than observe what they need. When you talk to physicians, as well as others involved in the delivery of care, you've got to learn the difference between what they say, what they want, what they'll pay for, and what they actually do.

For this reason, it is far more valuable to *see* how they act (and react) to a situation and *watch* the problems that arise.

In addition, an unbiased party may be able to observe problems that those involved in daily care are unaware of or have begun to overlook. For instance, a physician skilled at a procedure may be so used to performing it that s/he is unconscious of the fact that the

The patient

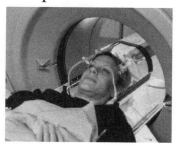

- What did the patient have to undergo in terms of pre-operative tests, appointments, etc., prior to the procedure?
- What time did the patient have to get up to prepare for the procedure?
- Was s/he allowed to eat the night before?
- What sort of preparation was required?
- Did the preparation have any negative or unintended side effects?
- What did the patient experience when s/he arrived at the hospital?
- How long did s/he have to wait?
- Was the patient taken to the operating room in a wheelchair or on a gurney?
- How long did the procedure take?
- What were the steps of the procedure and how long did each one take?
- Did the procedure require a general anesthetic?
- How much pain (or discomfort) did the patient experience during the procedure? Post-operatively? After discharge?
- What was involved in the post-operative process?
- What sort of bandage did the patient receive?
- Did the wound require dressing changes or drains?
- How often was the bandage changed/wound drained?
- Was a urinary catheter required?
- Was intravenous (IV) access required?
- Were there any complications that resulted from these procedures?
- How long was it before the patient could discontinue the drain, catheter, or IV?
- Are there any variations in the ways patients are prepared for, treated during, or cared for after a procedure, depending on the environment?
- Did the patient need to stay in the hospital overnight? For how many nights?
- Did the patient need any assistance after hospital discharge?
- What was the time required before the patient could resume normal activities?

The provider
(physician, nurse, physician's assistant, etc.)

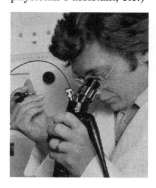

- Who prepares the patient for the procedure?
- How many people are present in the operating room?
- What are their various roles?
- Does the same person perform the procedure from start to finish?
- Are practitioner staffing levels and roles the same across different environments?
- Why is work allocated across practitioners in this way?
- How long has this been the standard of care?
- How was the procedure performed before the current standard?
- What are the accepted primary limitations or difficulties associated with the current procedure?
- Do the devices (or other tools used in the procedure) perform as the providers want/need them to?
- How does the provider use the device?
- Does the provider appear confident using the device? Did the provider have difficulties using the device? Operating? Implanting it? How many hands were required to operate/implant/use the device properly (i.e., did the provider need assistance operating the device)?
- Did the provider make any errors while using the device?
- How much follow-up is required of the surgical provider(s) following the procedure?
- What are the most common complications associated with the procedure?
- Who treats the complications?
- How (and where) are they treated?

Figure 1.2.2
An approach for probing multiple perspectives about all aspects of patient care (photos from the National Institutes of Health image bank).

| Others in the healthcare system (facility, payer, etc.) | · How much does the procedure cost?
· At what rate is the procedure reimbursed?
· Is the procedure profitable?
· What factors are most likely to drive up (or down) costs?
· How long does the procedure take to perform?
· What aspect(s) of the procedure take the longest to complete?
· How many resources are tied up as the procedure is being performed?
· What facilities (e.g., rooms) are tied up as a result of the procedure?
· Is the procedure performed in only one setting (e.g., operating room) or can it be performed in other venues (e.g., outpatient procedure or radiology lab)?
· What devices, equipment, or supplies are required to support the procedure?
· How much do the devices, equipment, and supplies cost?
· To what extent do they affect the profitability of a procedure?
· What risks do complications from the procedure present to the system?
· If there are complications to the procedure, who bears the cost? |

Figure 1.2.2
Continued

process may be suboptimal in some way. Returning to the colonoscopy example, even when asked directly about patient discomfort, physicians often describe the colonoscopy procedure as relatively easy to tolerate. From their point of view, they observe only limited patient discomfort when the scope is in the bowel and may forget about or inadvertently minimize the pain that some patients experience when preparing at home since they are not present to witness it directly. A shrewd observer has the ability to assess the end-to-end process so well that s/he recognizes these types of disconnects. S/he may also discover problems and needs that have not yet been recognized, let alone addressed.

How questions are framed is also vitally important since the way something is asked can influence or **bias** the answer. When asking questions in the observation process, they must be presented in such a way that the respondents can see situations and their inherent problems in an objective fashion. For example, when balloon angioplasty was in its early stages of development, many cardiac surgeons were simply unable to envision the potential benefits of a procedure that did not require stopping the heart, cardiopulmonary bypass, and sternotomy. Other surgeons conceptualized the problem differently, and saw the potential for angioplasty to be useful in patients who could not tolerate surgery. Thus framed, the question elicited a more objective response from the cardiac surgeons, and a greater willingness to acknowledge the need for an alternative to coronary artery bypass graft (CABG) surgery.

Preparing for observations

As Seneca, a mid-first century Roman philosopher once said, "Luck is what happens when preparation meets opportunity." While one may or may not agree that luck is involved in identifying significant clinical needs, there is little argument that preparation is essential.

A large number of medtech inventions have come from physician inventors. One main reason is that their work allows them to observe relevant problems directly. Creating the opportunity to perform observations for non-clinician innovators can be challenging. In most cases, these individuals must leverage their personal networks (and often their extended networks, i.e., friends of friends, distant family members, and introductions gained through casual acquaintances) to make the necessary connections. Another strategy is to add a physician to the innovation team. However, even without a physician directly on the team, many non-clinicians have been successful at defining needs and bringing new solutions to the market.

Innovators should prepare carefully for observations in the field they are investigating. This preparation involves performing background research. Understanding medical terminology and basic facts related to the clinical situation to be observed is essential to ensure that key communications and interactions among the people being observed are not lost or misunderstood. While delving deeply into understanding a disease and its treatments will be covered as part of the needs screening process (see 2.1 Disease State Fundamentals and 2.2 Treatment Options), it may be useful to review those chapters and perform some basic research before making observations.

Conducting observations

Once access to the clinical environment has been granted, an innovator must then immerse him/herself in the clinical situation of interest. The process of observation advocated within this chapter is linked to an approach called **ethnographic research**. Ethnography is concerned with the study of the "ways of life of living human beings."[3] The basic ethnographic research method involves the researcher becoming immersed in the activities of the group that s/he wants to study with the goal of gaining the in-depth perspectives of that group. In the biodesign innovation context, this means trying as much as possible to become an integral part of the group being studied to understand the perspective of the "insiders." This does not mean becoming part of the treatment team (i.e., delivering care). Being immersed in the team without performing its work provides first-hand insight to the problems and conflicts members face while still allowing the innovator to maintain the role of an observer. This is in contrast to an in-depth interview, for example, where probing questions are used in isolation to learn about a process. In this case, the interviewer is not part of the team, but instead is inquiring as an outsider about team members and their activities.

The primary steps in ethnographic research include those outlined in Figure 1.2.3. There are three common techniques for performing effective observation that complement the ethnographic approach: (1) become part of the team being observed, (2) expect the unexpected, and (3) observe the same process in different settings.

Become part of the team

As described in the context of ethnographic research, becoming part of the team is essential. Immersion enables the innovator to collect more accurate and complete information from his/her observations. To accomplish this, the innovator literally needs to shadow the activities of the person or group being observed. S/he needs to get into the mindset of the people delivering (or receiving) treatment to thoroughly understand the circumstances within which problems may exist. For example, when using a device, how many other things are coming at the physician at once? How much time does s/he have to focus on the use of the device and the issues it is designed to address? As noted, complementary observations should also be made from the patient's perspective, as well as the perspectives of other healthcare representatives involved in the process.

To be considered part of the team, the observer needs to be willing to commit substantial time and energy to

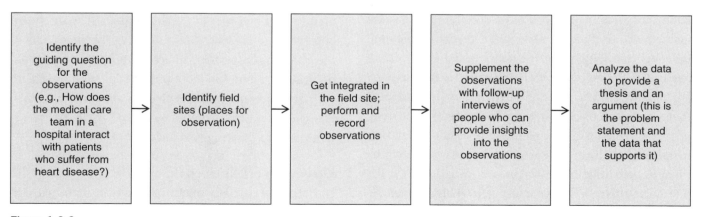

Figure 1.2.3
Ethnography is a simple but powerful approach to observation and problem identification (based on information from the Penn Anthropology website of the University of Pennsylvania; reprinted with permission).

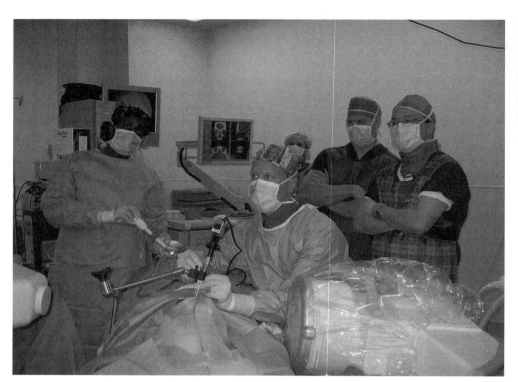

Figure 1.2.4
Insights gleaned from directly observing activities within the physician's office, laboratory, or the operating suite can be invaluable (courtesy of Josh Makower).

conducting observations. For example, if one is seeking to identify the problems associated with the use of a device in a certain surgical procedure, s/he needs to arrive at the hospital when the surgeon does, watch several unrelated cases, and then observe the entire procedure and the post-operative routine. See Figure 1.2.4.

If, instead, the innovator arrives just before the device is used and leaves the room as soon as it is put away, s/he is almost certain to miss important problems and opportunities. Furthermore, the surgeon, nurses, and other assisting professionals are unlikely to share insights or observations that might reveal problems if they feel that the observer is not invested in the entire process of patient care. If the observer stands as long as the team stands, rests only when they do, and joins them in the cafeteria, people are far more likely to open up. This behavior is just as likely to result in relevant insights regarding problems and needs revealed by them as watching the device in use. Usually this is not due to any *formal* statements but to insights gleaned from watching members work and participating in their *informal* discussions. As noted, having some understanding of the medical situation being observed (through basic research and a review of the medical literature) will also be viewed favorably and will help an innovator frame questions in an educated manner.

One other issue to consider when striving to become part of the team for observations is known as the "observer effect." The observer effect refers to changes in the phenomenon being observed that are caused by the mere act of observation. For instance, physicians and other providers may perform tasks or respond to problems differently when they are being watched. Patients, too, may modify their behavior (e.g., their response to pain) when observers are present. As the observer becomes part of the team, the observation effect can often be diminished because members feel more comfortable with the observer's presence and less conspicuous about their actions. However, it is still essential to make repeated observations prior to drawing conclusions.

Expect the unexpected

Discovery is partially about being in the right place at the right time. However, it is also about being receptive to new ideas and opportunities when they arise. Expecting the unexpected requires an innovator to go into the observation process with an open mind and an outlook that opportunities will reveal themselves in interesting and unexpected ways. If an innovator goes into an observation thinking, "I've already seen this before," it is unlikely that s/he will be in the state of mind necessary to pick up on the subtle insights that often lead to great ideas.

All of the observer's senses should be focused on identifying issues or problems that have never before been seen (or noticed). For example, an off-handed remark might stimulate an idea. Or a rare complication might reveal something interesting about physiology that had not yet been considered. Because one never knows when these types of triggers may occur, multiple levels of observation are essential. Observing the same procedure or event multiple times is also important, since this may allow an innovator to pick up on subtle differences; for example, the slight variations in the way different physicians perform the same step of a procedure may lead to an interesting insight.

Perform observations in different healthcare settings

While common clinical problems exist throughout the healthcare system, there is tremendous variability in how similar problems are handled in different healthcare environments. It is often necessary to perform observations of the same clinical problem or process in different settings since this can potentially reveal unique insights about these problems. For instance, if an innovator's area of focus involves procedures performed in a hospital setting, observations performed in private or community hospitals may result in the identification of problems that are distinct from those observed in academic medical centers affiliated with major universities.

To illustrate this point, consider the process of closing a skin flap after plastic surgery, a practice that requires suturing by hand and can often take as long as three hours to complete. In a university hospital, the surgeon would complete a procedure and then turn the patient over to a resident to close the flap. With many residents on staff, all eager for experience, this time-consuming, labor-intensive process is not viewed as a problem. In a community hospital, however, there are no "extra" resources to complete these kinds of tasks. Surgeons must close their own flaps, which ties up their time, potentially limits the number of cases they can manage on any given day, affects their ability to deliver other forms of patient care, and may lessen the overall amount of money that the physician and the facility can earn. Without this type of differential insight, an innovator could potentially miss a problem and an important driver of the related need.

Clues for identifying problems that may lead to clinical needs

During the observation process, certain types of events and behaviors can serve as "clues" that may be indicative of a significant problem. These clues are often specific to one particular perspective. When clues are observed, they should be investigated further as they often lead to the identification of important clinical needs.

From the patient perspective, the innovator should look for:
- **Pain** – If a procedure is painful to a patient, seek to uncover what problems currently prevent it from being delivered in a less painful manner.
- **Death** – Anytime that a procedure results in death, some problem has occurred. Carefully evaluate these events for preventable issues and complications.
- **Stress** – Stress refers to physical, mental, or emotional strain or tension.[4] Anytime that a physician, patient, or other stakeholder experiences visible stress, seek to understand what problem(s) might be causing it.

From the provider perspective, watch out for:
- **Risk** – Risk is exposure to the chance of injury or loss.[5] Generally, in their quest to "do no harm," physicians seek to minimize risks when delivering care. If a physician (or other provider) advocates a treatment alternative with higher perceived risk, understand what problems have necessitated the riskier approach.
- **Malfunction** – Whenever a device or other piece of equipment malfunctions, look closely at what caused the problem.
- **Uncertainty** – Watch for instances in which a provider is unsure or indecisive about how to proceed. These occurrences may point to problems that have yet to be solved.

- **Dogma** – Dogma refers to settled or established opinions, principles, or beliefs[6] that may or may not represent optimal behavior. If an observer asks why a procedure is performed in a certain way and the provider says, "Because that's how I was trained to do it," or "This is always how it is done," this may be indicative of a dogma problem (or at least a practice area or procedure that no one has evaluated critically in quite some time).

From the system perspective, consider:

- **Cost** – Any aspect of a procedure or treatment regimen that significantly increases cost may be indicative of a problem (or a problem in need of a different solution).
- **Inefficiency** – Evaluate the treatment process from the perspectives of the patient, the provider, and the system when seeking to identify problems of inefficiency. For example, in what instances must patients be held overnight while they await test results? Or when is additional staff required to perform only a small part of a procedure?

Although each type of clue is linked to a different perspective, too often the physician's perspective is given top priority when there is much to be gained from taking another point of view. For example, instead of considering how to find a faster way to cut during surgery, the patient's perspective would advocate the elimination of cutting altogether. It can be equally helpful to consider problems by making observations from the **third-party payer's** perspective (i.e., public and/or private health insurance providers) and the facility's point of view (hospitals, outpatient clinics, etc.).

A note on ethics and observations

When scheduling and performing observations, it is essential to remain professional at all times and respectful of the approach/limitations of key contacts. People seek medical care due to illness and, therefore, the medical environment is fraught with fear of the unknown and the possibility of impairment or death. Patients and families are fragile during these periods, and providers are ethically obligated to provide them with a safe, respectful environment when delivering care. For these reasons, innovators must remain sensitive to privacy-related issues while working in the medical environment and also cognizant of the boundaries that providers and healthcare facilities may put into place to protect their patients.

For instance, under the Privacy Rule set forth by the Health Insurance Portability and Accountability Act (**HIPAA**), patients are provided with comprehensive federal protection for the privacy of their personal health information.[7] This rule, which took effect in April 2003, establishes regulations for the use and disclosure of an individual's protected health information (**PHI**) and has caused many healthcare providers to become increasingly conservative and risk averse in terms of sharing patient information, including granting admission to observers in the clinical environment. Any individual seeking to perform observations that involve patients and/or patient data should have a thorough understanding of HIPAA regulations and demonstrate sensitivity to the healthcare provider's constraints and limitations under the law. Most facilities and providers require an observer to become HIPAA certified, which is accomplished through a several-hour long training session. Others may request that the observer get written patient consent. As a rule, innovators should be responsive and resourceful in responding to these requests in an effort to increase their likelihood of gaining permission to conduct observations. When in doubt, they should always seek guidance before entering a patient care environment.

Remember that it is a privilege for an innovator to gain access to a healthcare team and patients to conduct observations. The people on the healthcare team are providing real medical care to patients in need during observations. As a result, the innovator's purpose or agenda must always be secondary to allowing the normal pace and manner of healthcare delivery to occur. The innovator must gauge where and when it is appropriate to be present and ask questions. This can be determined by talking with the healthcare team during more informal, less critical periods to gain a better understanding of the team's expectations of the innovator during observations.

Documenting observations in an innovation notebook

Thoroughly documenting observations in an **innovation notebook** is essential. The data captured as part of the observation process is central to an innovator's follow-up analysis and the development of an effective need statement. Moreover, although the innovators are not recording information about any inventions during the early stages of the biodesign innovation process when observations are initiated, they are establishing a pattern of documentation that may be useful when they eventually seek to protect their work (see 4.1 Intellectual Property Basics). Capturing observations in the early pages of an innovation notebook helps an innovator tell a holistic story of how his/her idea came to fruition, which can be helpful if the invention is ever contested.

When documenting anything in an innovation notebook, the innovator should following these specific guidelines.

- **Format** – Choose a *bound* notebook with *numbered* pages. Never tear out or add pages.
- **Process** – Date and sign each page. Never backdate any entry. Even if a problem was observed (but not noted) on the previous day, always date each entry with the actual day it was written. An explanation for the delay in making an entry can be made, but do not falsify any information as this can constitute fraud if an issue ever goes to court.
- **Authentication** – Have a non-innovator act as a witness, signing and dating each page. This can be done every few weeks – it does not need to be done daily.
- **Additions** – Any added material that is pasted in must be "signed over" (this means pasting in the page and then signing/dating the addition in such a way that the signature is partially on the original page and partially on the addition).
- **Blank space** – Cross out blank spaces to ensure that no retroactive entries can be made.
- **Deletions** – Never white-out anything. Use a single line to strike through any errors and initial the corrections.

When documenting observations, record only what is seen. For example:

> The patient was laid flat on the table. The physician's assistant sterilized the groin area. Then, the doctor tried to gain vascular access through the groin. This took multiple attempts. The doctor mentioned that this was because the vessels were deep and non-palpable. The patient seemed to experience pain each time the needle was inserted and the physician became increasingly frustrated.

Avoid the temptation to editorialize. Do not begin filtering or classifying information at this point. And do not risk trying to interpret information before adequate data is collected. Stay focused on capturing raw data for analysis and interpretation at a later date (see 1.3 Need Statement Development).

The following types of information should be routinely recorded as part of the observation process.[8]

- Date, time, and place of observation.
- Specific facts, numbers, details of what happens at the site.
- Sensory impressions: sights, sounds, textures, and smells.
- Personal responses to the fact of recording fieldnotes (i.e., did someone comment when this particular effect was noted?).
- Specific words, phrases, summaries of conversations, and insider language.
- Timing of various steps of a process, procedure, or interaction; often good to have a stopwatch available.
- Questions about people or their behaviors to be investigated later.

One of the challenges related to documentation deals with inventorship. Remember that framing the problem is important, but that this is only a small portion of the total biodesign innovation process. Inventorship will be addressed in more detail in 4.1 Intellectual Property Basics. However, innovators are advised to document the discussions they have with others that help to frame a problem, even if they fall short of direct invention. Such notes may be useful later in the process.

Knowing when to stop and transition to the next step

The process of identifying important clinical problems is inherently unpredictable and inefficient. Observers may have to watch dozens (or even hundreds) of procedures before any significant issues are revealed. In some cases, even that might not be enough. There are certainly instances of smart people exploring interesting focus areas without uncovering any meaningful needs. This potentially means that there is a mismatch in the fit between the focus area and the innovator. Rather than pursuing one strategy indefinitely, there may be times when one is better served by going back and reevaluating the chosen focus area (see 1.1 Strategic Focus).

In general, a dedicated team or innovator should expect to spend at least two weeks ramping up on the observation process (e.g., learning the basics of a procedure, getting acclimated to the stakeholders, and becoming informed enough to identify complications and unexpected issues). Then, they might spend another two to three weeks performing intense, embedded observation activities. After that, in most cases, the idea flow tends to taper off. A sign of this may be that there are fewer and fewer *new* observations but a few others keep coming up repeatedly. If the team or innovator has identified viable clinical problems that seemingly now require translation into compelling needs, it may be time to begin harvesting those ideas (see 1.3 Need Statement Development). If not, it might be time to consider shifting focus areas.

When preparing to move on to the next step, be certain to maintain good relationships with the patients, providers, and representatives of the system who have been observed. Once a need statement has been developed and additional research performed, it will be necessary to return to the clinical environment to validate the need before concept generation begins. Having these relationships to leverage in the validation process is extremely helpful.

The following story about a multidisciplinary innovation team at the University of Cincinnati provides an example of how observations and problem identification can effectively be performed.

FROM THE FIELD — UNIVERSITY OF CINCINNATI MEDICAL DEVICE TEAMS

Observing problems as part of the needs finding process

Mary Beth Privitera, an assistant professor of biomedical engineering, is always on the lookout for problems in the medical field. As the codeveloper of the University of Cincinnati's Medical Device & Entrepreneurship Program, she is responsible for bringing together students in their senior year from biomedical engineering, industrial design, and the business honors program, dividing them into multidisciplinary teams, and assigning them real-world medical issues to investigate as part of a year-long innovation process. Each academic year, these projects are sponsored by companies and/or physician researchers in the medical device field and guided by experienced faculty from the colleges of design, art, architecture and planning, engineering, medicine, and business.

In the fall of 2006, Privitera was approached by Respironics, a medical device company with a focus on sleep and respiratory solutions, to identify the problems and needs of sleep apnea patients. Obstructive sleep apnea (OSA) is a condition that causes an individual to stop breathing repeatedly during sleep because the

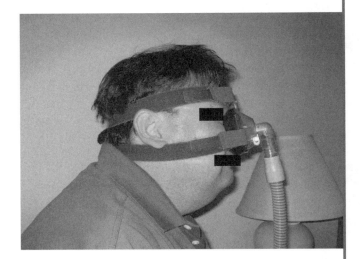

Figure 1.2.5
A patient with obstructive sleep apnea demonstrates how he wears the CPAP device (courtesy of Mary Beth Privitera and the University of Cincinnati; note: the patient is disguised to protect his privacy).

airway collapses. The most common symptoms of OSA are loud snoring and restless sleep, but it can also cause headaches, forgetfulness, depression, and anxiety, as well as other mood changes. In severe cases, sleep apnea causes pressure on the heart that can ultimately lead to heart failure or stroke. Approximately one in five Americans suffers at least minor sleep apnea.[9] Although some patients undergo surgical procedures to modify the airway mechanics, most patients opt for a non-invasive solution. Continuous positive airway pressure (CPAP) is a mask-like device worn during sleep, which supplies a constant stream of pressurized air to prevent the airway from collapsing. See Figure 1.2.5.

While CPAP was generally effective in preventing the symptoms of sleep apnea, many patients find the device uncomfortable and difficult to use. As a result, "Respironics was particularly concerned with patient compliance and promoting a more positive patient experience during device use," said Privitera.[10]

Privitera assembled a team of biomedical engineers and industrial designers to better understand the problem. Despite the availability of sleep clinics that provide an environment in which a patient's sleep patterns can be carefully monitored, Privitera, along with Respironics and a team of faculty, guided the students to interview and observe patients in their homes. "I'm a firm believer that the problems people have don't happen in a lab," she said. While carefully controlled experiments have their place in observing problems, Privitera felt that the student observers would learn more by being able to see how patients interact with their CPAP devices in the environment where they use them. "I always try to send students to the location where the problems really occur," she said, emphasizing that this approach is consistent with the focus on ethnographic research that she advocates in her courses.[11]

To gain a clinical perspective on the problems associated with non-invasive treatment of sleep apnea, the faculty team suggested that students should also meet with the specialists treating sleep apnea patients in the sleep centers as part of the observation process. To prepare for these interactions, the students researched sleep apnea and acquainted themselves with the current CPAP devices. Privitera highlighted the importance of prior research before beginning to interview doctors: "We don't expect a physician to talk in laymen's terms. I want the student to speak in the language of who they're interviewing, and who they're observing."

The students also developed extensive interview protocols to ensure consistency in the observation and data collection processes. "With multiple people on the team, we needed to do everything we could to achieve consistency across interviews," Privitera said. "The research protocols outlined specific questions they would ask, and then specific activities that they would observe. For example, in patient interviews the team would ask what they liked and disliked about the device, what improvements they would recommend, and how they used the apparatus. They would also observe the patient using the equipment, cleaning it, and performing other common behaviors.

When this background work was done, the next challenge was to identify patients and physicians willing to participate in the observation process. "I'm a firm believer that students need to learn how to make these contacts themselves," explained Privitera. Working with nothing more than a list of sleep centers in the area, the students made "cold calls" to the clinics to schedule appointments and gain access to lists of patients who might be willing to participate.

Reflecting on common pitfalls associated with the observation process, Privitera noted that knowing when to stop observing can be a challenge. "Typically you have enough information when you start to recognize repeat patterns of behavior," she said. "It might be after 12 hours, or it might be after 30 hours, but you have to stop at the point when the same situation repeats itself a couple of times." She also cautioned observers to take their time and let the patients [or physicians] do the talking. In her experience, observers are often eager to volunteer information to show their knowledge. They can also have a tendency to anticipate people's answers or misinterpret a response if it differs from what they are expecting.

As with every observation process, the students on the sleep apnea project identified problems that they did not initially foresee. Through careful observation, it became clear that sleep apnea patients were actually quite diverse in the issues they faced and the extent to which these problems affected their compliance with their recommended treatment regimen. This led the team to develop a series of different "personas" to help differentiate patients and their needs. Privitera explained, "When we put forth our plans for the quarter, we didn't expect to develop the personas, but it happened. We saw different interrelations that led us to these six categories of users, which did not necessarily reflect one

FROM THE FIELD: UNIVERSITY OF CINCINNATI MEDICAL DEVICE TEAMS

Continued

FIGURE 1.2.6
The six personas defined by the team included the Nomad, Hipster, Metro, Dude, Gramps, and Trucker (courtesy of Mary Beth Privitera and the University of Cincinnati).

person, but were combinations of people that had some of the same sensitivities and were like-minded." See Figure 1.2.6.

These personas helped the team define much more detailed patient needs that varied by market segment. For example, the Hipster persona was young, single, socially motivated, and concerned about appearances. To address the specific needs of this group, a solution would have to be quiet (so as not to disturb roommates), compact and easily camouflaged for communal apartment and/or dorm room living, non-institutional in appearance, and customizable in its fit and style. In contrast, the Metro persona was health conscious, spiritual, concerned with personal fulfillment, and interested in enhanced experiences. The needs of this segment would be driven by the desire for a serene user experience and could include criteria such as a built-in sleep mask (to block out light) and integrated audio (for "white noise" or other soothing sounds). The Dude persona had completely different needs. Members of this segment were relatively unhygienic, unconcerned with their appearance, and motivated by convenience above all else. This led to the need for a disposable contact interface, a rugged and durable device, self-cleaning functionality (or some mechanism that provided automatic feedback when it was time to clean or replace it), and an exceptionally easy user experience.

As the biodesign innovation process progressed, the team used this information to develop individual need requirements for each persona. They presented this information to the company, giving Respironics the opportunity to potentially develop unique solutions for the segments of greatest interest.

When initiating observations to uncover important clinical problems, use the following process as a guide. Note that this process ties closely to the ethnographic research method presented in Figure 1.2.3. It is summarized in Figure 1.2.7, along with an example.

See **ebiodesign.org** for active web links to the resources listed below, additional references, content updates, video FAQs, and other relevant information.

1. Set up observations

1.1 **What to cover** – Observations can be performed with patients, physicians, other healthcare providers (nurses, and physician assistants), third-party payers, facility administrators, and other stakeholders involved in the delivery and/or management of care. Innovators should determine which ones are the most relevant and then tap into their personal networks to gain permission to conduct observations. Refer back to the defined strategic focus to make sure that the chosen contacts and fieldsites will provide information aligned with the area of interest. If possible, try to find various settings in which to conduct observations. Many hospitals are often part of larger integrated healthcare networks, which may be a way to access other clinical settings.

1.2 **Where to look** – Innovators should look for contacts within the colleges and universities with which they are affiliated, particularly if they include a medical school. Personal or family physicians can be contacted for leads. Friends, family, and neighbors may be able to make introductions. Business associates with investments in healthcare-related companies and products may also be able to provide contacts. Cold calling, while sometimes frustrating, can be effective, especially when no other contacts have been identified through networking.

2. Prepare for observations

2.1 **What to cover** – Understand the environment in which the observations will be conducted. Assess the level of conservatism in the location's policies and be sensitive to concerns related to HIPAA and other relevant regulations. Get smart about the practice area in which the observations will be made so that there are few surprises when observations begin. Prepare a list of guiding questions to ask, be prepared to engage in a meaningful, educated discussion about what is observed. Always keep an open mind.

2.2 **Where to look**
- **Primary research** – Individuals within the field can be an excellent source of information, even if they do not grant the innovator access to perform observations. Talk with these people about their work in an effort to prepare for observations. Just be careful not to let their perspectives and opinions bias the observation process. It can also be helpful to talk with others who have performed observations in the past regarding their experiences, how they handled themselves onsite, and the lessons that they learned.
- **Literature review of clinical problem** – Review information about the chosen strategic focus area through online sources such as:
  - **Harrison's online** – Contains the complete contents of *Harrison's Principles of Internal Medicine*.
  - **Medical references** – *Guyton's Textbook of Medical Physiology* or another standard medical textbook will provide valuable, credible, and detailed information.
  - **PubMed** – A database of the US National Library of Medicine that includes more than 16 million citations from MEDLINE and other life science journals back to the 1950s. General reviews will be particularly helpful in understanding **pathophysiology** (locate Reviews by selecting this type of article under the "Limits" tab from the main page of the site before beginning a search).
  - **UpToDate** – A database of evidence-based clinical information.

- **eMedicine** – Clinical knowledgebase with more than 10,000 physician authors and editors.

3. Conduct observations

3.1 What to cover – Work hard to become embedded within the team. Perform rounds with physicians and shadow them "on call" in an effort to better understand the culture of what they do and the environment in which they do it. Pay careful attention to the clues that may signal problems with the delivery of care. View problems through different perspectives to uncover new ways of thinking about the core issues. Also, during the observation process, remember to:
- Ask questions, but do not be constrained by what individuals *say* their problems are. Do not hesitate to question dogma.
- Watch for latent problems. Many times, the greatest issues are those that are not yet recognized.
- Keep in mind that failures are an important source of information. Investigate what is most likely to go wrong and find out what can be learned from these failures.
- Be on the look-out for competing problems (e.g., accuracy of procedure versus procedure time) and seek to understand what drives these tensions.

3.2 Where to look – Observers should exercise their best judgment regarding when and how to embed themselves within the team and the approach they take to conducting observation. In addition to watching for clues that can signal problems, observers should be aware of and respond appropriately to verbal and non-verbal signals from those being observed regarding the observer's presence. Maintain the utmost professionalism and discretion at all times.

4. Document observations

4.1 What to cover – Create an innovation notebook and take detailed and copious notes regarding what is observed. Do not filter or editorialize when recording what has been seen. Record the date, time, and place of all observations; specific facts, numbers, details of what happens at the site; sensory impressions; personal responses to the fact of recording fieldnotes; specific words, phrases, and insider language; summaries of specific conversations; and questions about people or behaviors at the site for future investigation. Based on HIPAA rules, do not record patient identifying information (e.g., medical record number, social security number, etc.).

4.2 Where to look – Refer back to the information provided in the chapter for important guidelines about maintaining an innovation notebook.

5. Name the problem

5.1 What to cover – As the idea flow begins to wane, begin prioritizing and evaluating the problems that have been observed. The critical output from this step is a statement of the primary problem to be pursued. Precisely how the problem is framed or stated can have sizable implications on the need, so try stating it multiple ways before deciding on the best approach.

5.2 Where to look – Seek input from advisors, as needed.

6. Test and refine problem statement

6.1 What to cover – Conduct additional research and observations to test and triangulate the problem statement. The intent of this step is to ensure that the problem statement is supported by multiple data points. Ask questions such as:
- Why does the problem occur? What are the possible explanations and causes for the problem?
- What are the medical implications of the problem (anatomy, physiology, **epidemiology**, etc.)?
- Which constituencies are affected by the problem – Patients? With what specific condition(s)? Providers? What type, and in

what specialties? The overall healthcare system? In what ways?
- In what ways are they negatively affected (clinical outcomes, safety/risk, inconvenience, recovery, ease-of-use, productivity, cost, etc.)?
- How severe is the effect of the problem?
- In what setting does the problem occur – during procedure in physician's office, operating room, etc.? During inpatient/outpatient recovery? Anywhere (without notice)?

These questions can also be asked of stakeholders in the environment, but be careful to recognize potential biases in their responses. Summarize the most important data gathered and refine the problem statements as needed. Remember, it is still too soon to begin thinking about solutions. While this can be frustrating to entrepreneurs and innovators, they must resist the tendency to allow "solution biases" to taint their understanding of the problem (for more information, see 1.3 Need Statement Development).

6.2 **Where to look** – Refer back to the sources listed above to perform additional, focused literature searches. Talk with stakeholders in the environment where observations have been conducted. Schedule additional observations with new or existing contacts.

Process	Description	Example
Set up observations	Target patients, providers, payers, and representatives from the facility.	Make arrangements to spend a week performing observations in two different orthopedic surgery centers. Plan to observe full cycle of care across stakeholders.
Prepare for observations	Perform background research on procedure/treatment, as well as the facility and its policies. Be prepared to ask educated questions.	Research hip implant procedure and standard of care. Investigate track records, policies, and attitudes of the facilities and physicians to be observed. Refresh understanding of HIPAA policies. Make note of questions to ask and things to watch for during observations.
Conduct observations	Become embedded in the team. Maintain an open mind. Watch for clues that signal a potential problem.	Arrive early. Shadow the team. Be helpful (when appropriate). Stay out of the way (when appropriate). Identify clues and watch for their recurrence. Talk with stakeholders, but be careful to recognize opinions, dogma, etc.
Document observations	Create an innovation notebook. Record only what is seen – do not editorialize.	Observation = Elderly patient with degenerative hip disease who had received elective hip replacement surgery needed hip implant removed due to infection.
Name the problem	Summarize critical data. Prioritize what has been learned. Articulate the most compelling problem.	Problem statement = Elective hip implants to treat degenerative hip disease in the elderly can become infected and require removal.
Test and refine problem statement	Conduct additional observations. Perform secondary research of the literature in the field. Ask who, what, when, where, and why? Refine problem statement, as needed.	Find out the answers to important questions, e.g., How often does infection occur in elderly versus the general population? What causes infection? Where does infection occur? Is this problem unique to hip implants? Etc.

FIGURE 1.2.7
The biodesign process for performing observations and identifying problems.

Credits

The editors would like to acknowledge Steve Fair and Asha Nayak for their help in developing this chapter. Many thanks also go to Mary Beth Privitera, Respironics, and the University of Cincinnati student teams that worked on the project described in this chapter's case example: Laurie Burck (Biomedical Engineering Leader), Nate Giraitis (Industrial Design Leader), Celina Castaneda, Christina Droira, Adam Feist, Tom Franke, Christine Louie, Bryan Porter, Nicole Reinert, and Rebecca Robbins.

Notes

1. In a sternotomy, a vertical incision is made along the sternum and it is cracked open to access the heart or lungs during surgery.
2. From remarks made by Thomas Fogarty as part of the "From the Innovator's Workbench" speaker series hosted by Stanford's Program in Biodesign, January 27, 2003, http://biodesign.stanford.edu/bdn/networking/pastinnovators.jsp. Reprinted with permission.
3. "What is Ethnography?" Penn Anthropology, University of Pennsylvania, http://www.sas.upenn.edu/anthro/anthro/whatisethnography (November 17, 2008).
4. "Stress," www.dictionary.com, http://dictionary.reference.com/browse/stress (August 30, 2007).
5. "Risk," www.dictionary.com, http://dictionary.reference.com/browse/risk (August 30, 2007).
6. "Dogma," www.dictionary.com, http://dictionary.reference.com/browse/dogma (August 30, 2007).
7. See United States Department of Health and Human Services, "Medical Privacy – National Standards to Protect the Privacy of Personal Health Information," http://www.hhs.gov/ocr/hipaa/ (November 17, 2008).
8. "Fieldnotes," Penn Anthropology, University of Pennsylvania, http://www.sas.upenn.edu/anthro/anthro/fieldnotes (November 17, 2008).
9. A.S. Shamsuzzaman, B.J. Gersh, V.K. Somers, "Obstructive Sleep Apnea: Implications for Cardiac and Vascular Disease," *Journal of the American Medical Association* (October 2003): 1906–14.
10. All quotations are from interviews conducted by the authors, unless otherwise cited. Reprinted with permission.
11. According to Privitera, ethnography is a research method completed through in-depth user interviews and directed observations in the context of people and tasks targeted with the design problems. Its primary advantages are that the approach: (1) helps uncover the differences between what people say and what they do; and (2) enables the researcher to describe what a device needs to do in context.

1.3 Need Statement Development

Introduction

Innovators usually grasp the nature of observed clinical problems quickly. The greater challenge is in understanding the associated clinical need and in translating problems into a meaningful clinical need statement. For example, after observing the difficulties some physicians have when transecting the sternum for thoracic surgery, a well-intentioned innovator may define a "need for a more effective cutting device to perform a sternotomy." This need will lead him to investigate multiple solutions for cutting the skin and bone during thoracic procedures. However, if someone else is simultaneously exploring the "need for a way to access the chest to perform procedures on organs of the thorax," the options for innovative solutions will be much broader. In fact, the second need statement would likely lead to minimally invasive thoracotomy, which the first need statement would not. Ultimately, the first innovator may find that he faces a significantly diminished demand for his new cutting tool if the broader need is met by the second innovator's solution.

OBJECTIVES

- Learn how to translate a problem into a clinical need statement that is accurate, appropriate in scope, and solution independent.
- Understand the importance of targeting a specific outcome in a need statement.
- Understand some of the pitfalls associated with developing a poor need statement.
- Understand the different categories of need statements and how these may relate to solution risks and benefits.

Needs, which represent the change in outcome or practice that is required to address a defined clinical problem, may be thought of as the bridge between problems and solutions. Far too often, clever innovations fail because they have not been developed to address "real" customer and/or market needs. Once a need is clearly articulated, the most important parameters or criteria that will guide the design and development of the solution can be defined. Concepts and potential solutions can then be evaluated against these criteria to ensure that they effectively meet the clinical need.

Stage 1 Needs Finding

Need fundamentals

Once a problem statement has been developed and validated (as outlined in 1.2 Observation and Problem Identification), an innovator's next challenge is to translate that problem into a need statement. Need statements articulate the problem and the change in outcome that is required to satisfy the clinical dilemma. Importantly, a need should not address *how* the change in outcome will be accomplished – this is specified by the solution later in the biodesign innovation process. Rather, the need should be focused solely on defining *what* change in outcome is required to resolve the stated problem.

About the need statement

Shaping a need statement has been described by seasoned innovators as something of an art form. It is a challenging, but essential, exercise that is likely to require a significant amount of trial and error. However, getting the need statement right is imperative because it defines the scope of the need specification (see 2.5 Needs filtering) and, ultimately, the parameters that any solution must satisfy.

In developing a need statement, it is necessary to carefully choose every single word that is used. Innovators are encouraged to experiment with different variations of the need statement as the specific wording can potentially lead to dramatically different solutions. For example, consider the differences among the following simplified need statements.

- *A method to prevent hip dislocation in high-risk patients.*
- *A method to prevent recurrent hip dislocations in high-risk patients.*
- *A method to prevent recurrent hip dislocations in patients after surgical treatment of a first hip dislocation.*

All three statements address the same general clinical issue (hip dislocation in high-risk patients). Yet, some identify certain existing conditions (past dislocation or previous surgery for past dislocation) that would target different patient populations and that could potentially send an innovator in an entirely different direction in terms of what s/he attempts to accomplish.

A slightly more elaborate example further illustrates this point. Imagine that a new innovator has observed a problem with long-term urinary catheters causing infections in patients in the intensive care unit (ICU), and is discussing the need statement with his/her chosen mentor.

INNOVATOR: "I think I've finally defined a need statement for *a catheter that will not track infection.*"

MENTOR: "Are you talking about all catheters or just urinary catheters?"

INNOVATOR: "Just urinary catheters."

MENTOR: "OK. Do you intend to address all urinary catheters?"

INNOVATOR: "No, just long-term catheters that are used for more than two weeks."

MENTOR: "Great. But is catheterization the only possible solution to this problem?"

INNOVATOR: "Well ... I suppose we might be able to develop something that allows for the evacuation of urine without keeping a catheter in place at all times .... Or maybe we could come up with a subcutaneous implant that releases localized antibiotics in the area of the urethra ..."

MENTOR: "Good. So maybe the need you're trying to address is really more appropriately defined as the need *for a way to reduce the incidence of urinary tract infections in ICU patients.* Right?"

While both need statements address the same basic problem, the final statement expresses with much greater clarity the true nature of the problem and the metric necessary to achieve the desired outcome. It is also more focused on the target audience. If the innovator were to adopt the first need statement, a vast array of potential (non-catheter based) solutions would never be considered. Without a specific, clearly identified target user, potential solutions could have been developed for users to whom the need was not truly applicable. The example also highlights the importance of defining an appropriate scope for the need. The goal is to establish the need as broadly as possible while keeping it linked to a specific, verifiable problem.

1.3 Need Statement Development

Common pitfalls in need statement development

Developing effective need statements is highly experiential – many innovators master this skill "the hard way" (by making mistakes and learning from them). This can be a costly process, since a poorly defined need statement usually is not discovered until the solution for that need statement misses the mark much later in the biodesign innovation process, after significant time, money, and effort have been invested. While there is no specific approach that can guarantee the crafting of successful need statements, there are at least two common mistakes that can be recognized and avoided: (1) embedding a solution within the need, and (2) inappropriate definition of the scope. By maintaining a keen awareness of these pitfalls and proactively steering clear of them, innovators can increase the likelihood of developing an effective need statement that serves them well throughout the entire biodesign innovation process.

Embedding a solution within the need

Embedding a solution within a need is technically one way that innovators can define the scope of a need too narrowly. However, it is such a prevalent problem in the biodesign innovation process that it warrants special attention.

As noted, a need statement should address *what* change in outcome is required to resolve a stated problem, not *how* the problem will be addressed. It is a common pitfall for innovators to incorporate elements of a solution into their need statements because they quickly envision ideas to solve the problems they observe. This is especially true when they are bombarded with potential solutions from respected senior physicians who tell them how they would solve the needs in question. Sometimes this occurs blatantly, sometimes subtly. In either scenario, embedding a solution into a need statement seriously reduces the range of possible opportunities that are explored, constrains the creativity of the team, and places unnecessary boundaries on the potential market. More importantly, it can lead to a biased need statement that may not truly represent the actual clinical problem and, thus, may lead to solutions that do not effectively address the need.

One young company, for example, observed that although stents (a mesh-like tubular scaffold that can be deployed in blood vessels to expand a narrowed region) are beneficial in holding open arteries, they can cause a shower of emboli (clots that become dislodged, travel through the bloodstream, and potentially become lodged in other smaller blood vessels, creating blockages) during deployment. This company presumably defined a need *to develop a stent that did not result in the sequelae of emboli during placement*. Based on this need, they began designing a solution. The company surmised that holes in the mesh of the stent could cause fragments of atherosclerotic plaque or thrombus to break free, resulting in distal embolization. They decided to develop a stent incorporating a material that would stretch over the holes and prevent the emboli from dislodging. However, after some research and development, they discovered that the covering prevented the natural blood vessel surface from reforming around the stent after the procedure (which could create other serious complications, including inducing subsequent clots and emboli to form). After obtaining initial funds to pursue the concept, the company failed to deliver a product to the market due to the challenges created by the material and was eventually shut down.

Another company, however, took a completely different approach to solving the problem of emboli after stenting. This company presumably defined a broader need: *a way to prevent the sequelae of emboli secondary to an interventional procedure*. Their solution was to develop a basket that would catch the emboli downstream if they were dislodged during a procedure. This company was dramatically more successful, and was acquired by one of the major medical device companies within a few years of demonstrating the clinical success of the approach. By defining the need independent of any particular solution, it avoided the inherent limitations of a stent-based approach and opened up a more sizable range of possibilities. Both companies were staffed with talented and creative engineers. The first, however, was at a disadvantage because of the solution-bias embedded within its need statement. Ultimately, taking a stent-based approach imposed artificial constraints on the team and prevented it from considering more feasible and effective solutions. As shown in

Stage 1 Needs Finding

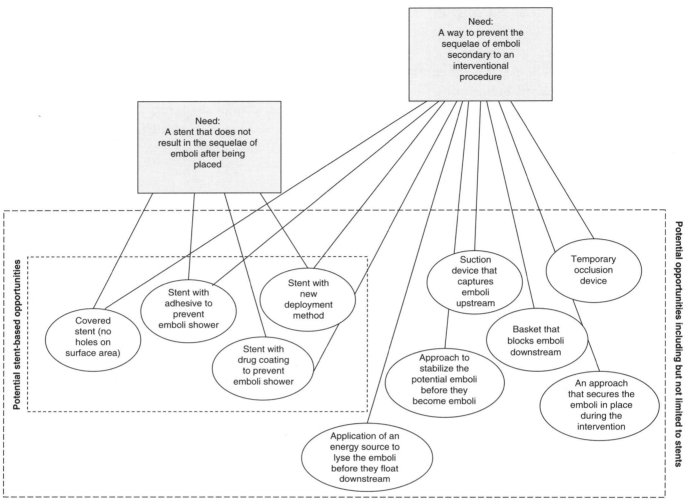

FIGURE 1.3.1
Embedding a solution within a need can dramatically limit the number and types of solutions that an innovator eventually explores.

Figure 1.3.1, the range of possible solutions considered by a team can be affected by embedding versus not embedding a solution in the need statement.

Inappropriate definition of the scope

Defining the scope of a need either too narrowly or too broadly can be equally problematic. As described above, one way that innovators can define the scope too narrowly is to embed a solution within the need. This also occurs when one is unnecessarily specific about what s/he hopes to accomplish. Consider the sample need: *a way to prevent or decrease the incidence of infections associated with hip implants in the elderly.* While there is no explicit solution-bias in the need, there may be some other words included in the statement that impose unnecessary boundaries. For example, even though this particular need has been observed in the elderly, the innovator should ask if it might be generalizable to a broader segment of the population. Through research and additional observations, the innovator may determine that the need to decrease the incidence of infections associated with hip implants actually applies to all recipients, not just to those over a certain age. A similar finding may result from investigating whether the need applies to other types of joint implants (e.g., artificial knees). The result could be that the potential target market is significantly larger and, thus, more compelling than originally estimated. Upon closer examination of the need statement, the innovator may also want to ask whether there is a superior or superseding need that warrants consideration. In the example, infections associated with hip implants may be a need worth addressing. However, it is also part of a larger need to *find a better way to treat osteoarthritis*. The innovator should at least consider whether s/he would be better served to work on this "higher level" need (see section on Needs and Risk later in this chapter for more information about superseding needs).

Just as innovators often make the mistake of defining a need too narrowly, they must also be careful to avoid defining a need too broadly. In an effort to avoid unnecessary constraints, innovators sometimes overgeneralize a need by making the assumption that it applies to a broader population when, in fact, it does not. For instance, in revisiting the example above, "*a way to prevent or decrease the incidence of infections associated with hip implants in the elderly,*" an innovator may assume that the risk of infection associated with hip implants in elderly patients is similar for knee implants. This would lead to a much larger market opportunity. However, ultimately, the risk of infection may not be similar due to unique differences in susceptibility and **mechanisms of action**. Similarly, the need for "*a way to decrease the incidence of infections associated with joint implants in the elderly*" is much broader, but may not appropriately address the need for a solution to the unique infection problems observed in elderly hip patients.

Innovators are cautioned not to broaden the scope of a need statement beyond what has been observed without careful, exhaustive research and additional observations. Every word that is (and is not) included in the need statement must be defensible with hard, quantitative data and real-world observations. Need statements that are broadened based on assumption or conjecture will almost certainly lead to serious problems later in the innovation process since the words that are included in (and omitted from) the need statement are integral to the path taken to develop solutions.

Ongoing research and information gathering

While information about the target audience's desires will have been gleaned through the observation process, more data can be gathered at this stage by talking directly with potential users, customers, and other stakeholders about the problem. This is important since the change outlined by a need *must* be driven by what the target audience wants and/or requires. If an innovator seeks to solve a problem that is not important to the target user, the innovation may not be widely adopted. If the problem identified through observation is an issue about which the target audience is not readily aware, it can be more challenging for the innovator to confirm the need. For example, in approximately 5 to 20 percent of colonoscopy procedures, the cecum (the pouch at the base of the ascending colon) is never reached due to the difficulty in navigating the endoscope through the entirety of the colon.[1] As a result, some instances of colon cancer that exist deep within the colon are not detected. Using current technology, it is difficult for a physician to know with certainty whether or not the cecum has been reached. However, despite the fact that published colonoscopy completion rates vary substantially, most physicians believe that this is not a problem that they personally experience. In cases such as this, the innovator may be successful in initiating a productive dialog if s/he identifies the problem in a generalized manner, based on research, without asking if a physician has personally ever experienced the problem. With this approach, physicians can acknowledge and discuss the need for more reliable colonoscopy completion results, whether or not they feel comfortable admitting any personal familiarity with the problem. Moreover, if a solution is introduced that makes the procedure more failsafe for all physicians, with no additional cost or risk, most would be likely to adopt it.

In talking with target users, it is essential to ask them exactly *what* results they would want, not *how* to achieve them to keep the discussion focused on the need and not potential solutions. Deconstruct the problem, breaking it down to each component to ensure that it is understood at every level. Seek to understand any possible interactions between the various components of the problem and develop hypotheses for the root causes. This is important since these hypotheses can be validated or refuted by the target audience. Then, with the input of target users, try to identify the key elements that an ideal solution would have to include to satisfy

Table 1.3.1 *Common changes in outcomes and how they are measured in need statements.*

Necessary change	As measured by …
Improved clinical efficacy	Treatment success rates in clinical trials
Increased patient safety	Rate of adverse events in clinical trials
Reduced cost	Total cost of procedure relative to available alternatives
Improved physician/facility productivity	Time and resources required to perform procedure
Improved physician ease of use	Decrease in the number of complex workarounds and/or the simplification of workflow
Improved patient convenience	Frequency and occurrence of required treatment, change in treatment venue (inpatient versus outpatient, physician's office versus home), etc.
Accelerated patient recovery	Length of hospital stay, recovery period, and/or days away from work

them (ideally, these elements should be linked back to the root causes they are likely to address). Target users can be asked to not only identify these elements, but also prioritize them in order of importance. Keep in mind that experts may have different requirements than "common users," but that common users often represent the greater market. They may also have different biases based on their own experience and perspectives that should be considered when gathering feedback from them about an observed clinical problem.

Incorporating outcome measures into the need statement

Once the parameters of a specific change have been identified based on observations and target user input, measures can be chosen that translate into direct benefits to the members of that target user group. An effective need statement incorporates a metric or indicator of how the desired change in outcome will be measured once the need is addressed. For example, consider the following need statement: *a way to reduce the time required for unskilled medical practitioners to place endotracheal tubes in an emergency setting*. In this statement, the specified outcome measure is the reduction in time associated with the procedure. In the need statement, *a way to prevent or decrease the incidence of infections associated with hip implants in the elderly*, the chosen indicator is improved clinical efficacy as measured by a reduction in the incidence of infection. Including an outcome measure in a need statement is essential so that the innovator knows, in no uncertain terms, when the need has effectively been addressed.

It is worth noting that a need may ultimately be solved in such a way that it results in multiple benefits (or a number of improved outcomes). For example, the development of a surgical procedure that can be performed in the physician's office as opposed to the operating room will likely result in measurably lower costs (primary benefit), but may also be significantly more convenient for the patients and physicians (secondary benefits). However, a need statement is typically crafted to include just the primary outcome measure, rather than listing all outcomes that potentially may be positively affected. This approach can be less cumbersome to the innovator and keep him/her focused on the most important result. It also helps eliminate the perception that a need has been successfully addressed only if it performs well against *all* outcome measures (e.g., improved efficacy and safety, reduced cost, etc.) – a difficult and often unattainable challenge.

One way to think about how to incorporate measures into a need statement is to consider what change is necessary to eliminate or improve the undesirable outcome (and then how it can be measured). Table 1.3.1 summarizes some of the key outcomes that are important in clinical care and the metrics that are commonly associated with them.

Once the innovator has identified the most appropriate change and outcome measure to incorporate into the need, s/he can craft the actual need statement, being careful to avoid the pitfalls noted earlier in the chapter. More information about the process for developing a need statement is provided in the Getting Started section.

Need criteria

Once the need statement has been crafted, the detailed information that supported its creation can be used to define the need criteria that the solution must meet. Need criteria should be based on information gathered in interviews and discussions with providers, patients, and other stakeholders, as well as their observed behaviors. For instance, for the need *to reduce the incidence of urinary tract infections in ICU patients*, the need criteria might include the following.

- Whatever the solution may be, it needs to be deployable by personnel that are already available within the ICU.
- It has to last for at least two weeks (since this is how long the average patient spends in the ICU). And if it is limited to two weeks, it needs to be repeatable, if necessary, with no adverse consequences.
- It must have a similar safety profile to existing treatment (traditional urinary catheters) so as not to introduce the risk of a consequence more dangerous than a urinary tract infection.
- The cost of the solution should be comparable to the cost of a traditional catheter (or only slightly more, based on the incidence of urinary tract infection and the costs associated with its treatment in those who become infected) in order for it to be commercially viable.

While the most important need criteria are integrated into the need statement itself, many of the supporting principles are of value and will be used later in developing the need specification (see 2.5 Needs Filtering). The innovator might also think about defining other need criteria that address "softer" solution attributes, such as ease of use, speed, patient convenience, etc. Fundamentally, these criteria provide an innovator with additional "boundary conditions" that will be used for further investigation and initiation of solution development. Note that if the definition of need criteria necessitates modifications to the need statement, it is essential to reconfirm (with data and additional observations) that the revised need statement is still valid.

FROM THE FIELD **NORTHWESTERN UNIVERSITY BIOMEDICAL ENGINEERING TEAMS**

Navigating the challenges of needs finding

In 2004, the leaders of Northwestern University's capstone course in biomedical engineering, David Kelso and Matt Glucksberg, became interested in design issues associated with global health problems. "The challenge with the equipment and devices used to address health issues in developing countries was not that they were poorly designed, but that they were not designed for the environment in which they would be used," Kelso said.[2] "If people began designing devices specifically for resource-poor settings, they could come up with much better solutions."

Motivated to make a difference, Kelso and Glucksberg initiated a program that gave senior students the opportunity to design solutions that specifically addressed medical needs in developing parts of the world. While their goal was to have students work on "real" projects for "real" end users, they initially launched the program targeting health-related issues identified by the World Health Organization (WHO) or other universities around the world. In one case, they read about a project initiated by engineers at the Massachusetts Institute of Technology (MIT) to develop an incubator that would help address the high rate of infant mortality in developing nations. In countries such as Bangladesh, the area where Kelso and a team of five students decided to focus, as many as 30 percent of all births were premature, a figure that translated into approximately 3,500 premature babies a day.

The team committed to developing *a better incubator that would be designed for local conditions to help reduce infant mortality*. Early in the process, they also defined specific need criteria for the solution, including

Stage 1 Needs Finding

FROM THE FIELD — NORTHWESTERN UNIVERSITY BIOMEDICAL ENGINEERING TEAMS

Continued

the capacity of the system to operate without electricity, maintain a baby's temperature at a constant 37 degrees Celsius, help protect the infant from infections, contain a high percentage of local material, and have low manufacturing and operating costs. Over the course of the ten-week academic quarter, the team networked extensively, seeking input about incubator design from contacts with experience in healthcare delivery in South Asia as well as those with prenatal education and neonatal baby care. After developing an initial prototype in a plastic laundry basket (see Figure 1.3.2), the team decided to make the container from jute so it could be sourced and manufactured at low cost in Bangladesh.

The phase-change material they used to control temperature in the new model worked just as well as electrically powered devices. "We were really excited about this," noted Kelso, who immediately began seeking ways to take the project beyond prototype and into production.

Tapping into a Northwestern study abroad program, Kelso formed a second team of students located in Capetown, South Africa and began working with this group to make the incubator relevant for the South African market. Targeting the most prominent neonatal intensive care unit in the area, they scheduled a meeting with the head of that department at Karl Bremer hospital. Kelso and team brought with them photographs and storyboards that described their incubator project. "But as we got off the elevator, we saw incubators piled up in the corner," he recalled. "They were not at all interested in our incubator solution, but invited us to come in and see how they care for premature babies. There were 30 mothers in the neonatal intensive care unit, all caring for their newborns with something called Kangaroo Mother Care."

Kangaroo Mother Care (KMC) had been pioneered in 1979 in Bogotá, Colombia to overcome the inadequacies of neonatal care in developing countries. The basic idea is to place the infant (without clothes, except for diapers, a cap, and booties) upright between the mother's breasts.[3] The baby is held inside the mother's blouse by a pouch made from a large piece of fabric. The method promotes breastfeeding on demand, thermal maintenance through skin-to-skin contact, and maternal–infant bonding.[4] While few large-scale studies have been conducted or published in mainstream medical journals, KMC is believed to help babies stabilize faster and to provide more protection from infection (from the antibodies gained through frequent breastfeeding) compared to babies isolated in incubators.[5] Many believe it also leads to reduced mortality rates among premature and low birth weight infants, although these results are still being studied. "Basically, it provides superior results at no cost," summarized Kelso.

"During our early discussions, we heard about Kangaroo Mother Care as a method they were trying to teach in Bangladesh," he remembered. "But the concept didn't affect the team's design." Kelso continued, "The right way to specify a design challenge is to do it in solution-independent form. By saying we would develop an incubator, we had over-constrained the need. Otherwise, almost all of the need criteria on our list were spot-on." It just so happened that KMC also met these need criteria, while offering other benefits to the infants as well as the mothers who preferred not to be separated from their babies.

Recognizing that the incubator solution was no longer appropriate for this environment, Kelso and his team quickly revisited the needs finding process. One of the associated needs they uncovered was a way to identify apnea in neonates (a problem that could have tragic consequences). Ultimately, they developed an innovative monitor that was appropriate for babies and could be used in conjunction with KMC.

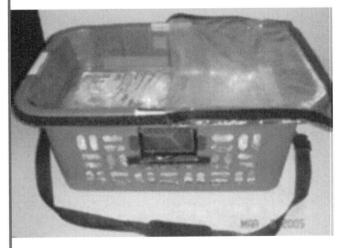

FIGURE 1.3.2
A photograph of the team's early prototype (courtesy of David M. Kelso).

1.3 Need Statement Development

Needs and risk

While all needs should be associated with positive changes in outcome, innovators should recognize that they are also associated with certain risks. The risk inherent to a need is partially related to what change in outcome or practice the need seeks to address:

- If the goal is to directly improve a significant clinical outcome (e.g., morbidity or mortality), the need will tend to have:
  - A relatively high likelihood that many solutions may fail.
  - A high risk of intense competition.
  - Somewhat unpredictable biology.
  - The need for randomized, **controlled trials**.
- If the goal is to expand the market by adding new features or benefits to an existing approach, the need will have a lower relative risk.
- If the goal is to increase speed or ease of existing procedures, the risk will be even lower because the solution will not attempt to affect significant clinical outcomes.
- If the goal is to improve safety, the risk of failure drops lower still. No randomized, controlled trials will typically be required since there is no associated departure from conventional methods.
- If the goal is to reduce cost, the need may have the lowest possible risk, but also may provide the least clinical benefit. Clinical studies may be needed to support the claim of cost reduction and there may be other significant benefits if, for example, reducing cost may expand access to care. Such benefits may also need to be substantiated through clinical studies.

As noted, another source of risk that innovators face deals with the concept of superseding needs. The closer a need is to addressing the fundamental aspects of a disease state, the less likely it is that the need will be displaced by an alternate needs and its solution. One analogy that is often used to describe the concept of superseding needs is that of a tree. For instance, take the case of atrial fibrillation, a disease in which the irregular beating of the heart causes clots to form that can potentially dislodge and travel to the brain, causing a stroke. The maintenance of normal (or

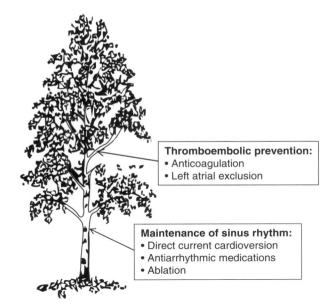

FIGURE 1.3.3
An example of superseding needs and related solutions.

sinus) rhythm by various means might be the trunk of the tree, while one branch coming off the trunk would be treatment with medications to prevent clot formation (thromboembolic prevention). See Figure 1.3.3. The further away the innovator works from the trunk of the tree on related needs, the more likely it is that the branch where his/her innovation exists could be cut off (or superseded by another invention). Of course, there is also the risk that the entire tree could be chopped down – in this example, if someone found a cure for atrial fibrillation. As Mir Imran, serial inventor, entrepreneur, and founder of InCube Labs, summarized:[6]

> One of the things that device company executives worry about most is that a technology they've worked very hard on for many years will do everything they want it to do and solve exactly the clinical problem that they figured, but by the time they actually get it to the market, it's been passed by another technology, or the clinical problem has been solved in some other way.

The point is that there is a cascade of events that create a need, and each event within the series may

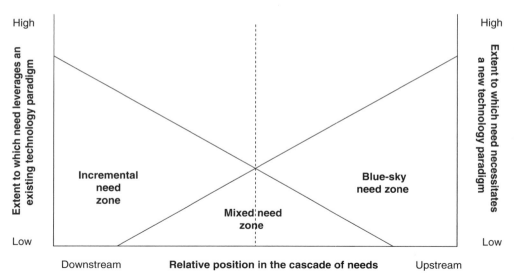

FIGURE 1.3.4
Different types of needs carry with them different benefits and risks.

be associated with its own unique need. This creates a hierarchy of related needs that directly affects the risk profile associated with the issue the innovator is seeking to address. In general, broad needs (e.g., that seek to cure, eliminate, or prevent a disease) often have the potential to supersede other needs. In contrast, needs focused on changes to existing treatments are often at risk of being superseded. Despite these general guidelines, however, this is not to say that focusing on broadest possible need is always the best course of action. In fact, while broad needs may have the lowest likelihood of being superseded, they are often the hardest and riskiest to solve due to an insufficient understanding of the disease mechanisms and other complicating factors. Recognizing where a need exists relative to the cascade of events that creates it can help an innovator identify one more source of risk associated with the target need.

Categorizing needs

Once defined, needs can be organized into three general categories: *incremental*, *blue-sky*, and *mixed*. Organizing them in this way can help innovators appreciate the potential impact that their solution might have on medicine, and also give them a gauge for the relative difficulty of addressing them. Categorization may also help innovators eliminate entire clusters of needs, depending on their strategic focus (see 1.1 Strategic Focus), thus improving their efficiency and likelihood of success.

The three primary need categories exist upon a continuum (with incremental needs on one end, blue-sky needs on the other, and mixed need in between) based on the extent to which they operate within existing treatment paradigms. See Figure 1.3.4.

Incremental

An incremental need is focused on addressing issues with or making modifications to an existing solution, such as the function of a device or other technology. For example, *a way to prevent clogging of the machinery used to remove tumors during neurosurgery* would be considered an incremental need. It is important to understand that incremental needs typically assume that underlying treatment paradigms or technologies will continue to be used and applied. This is not the same as saying that solutions are built into the need, but rather that solutions are constrained by being far downstream in the cascade of needs. As noted, incremental needs are generally approachable, but run the risk of being superseded when new technology paradigms are introduced.

Blue-sky

Blue-sky needs, on the other hand, require solutions that represent a major departure from currently available alternatives and address needs that are further upstream in the cascade of needs. As a result, they may be difficult to define. Blue-sky needs are often focused on curing, eliminating, or preventing various disease states and, therefore, are more focused on physiology and mechanisms of action than existing treatments or solutions. For example, rather than focusing on improvements to existing machinery or procedures, a blue-sky need might be something like *a way to prevent the spread of cancer*. Blue-sky needs, if solved, will often supersede most other related needs within a treatment area. A delicate balance exists in determining whether a blue-sky need is approachable at the current point in time or if it requires more study. For a blue-sky need to be entertained as something that could be solved, it is important that at least some of the underlying disease mechanisms are understood in the medical community. This differentiates a blue-sky need from a "science experiment" – or an exercise for which the fundamental building blocks of a solution are not yet known and, thus, are unlikely to be solved.

Mixed

A mixed need can be thought of as existing somewhere in between an incremental and blue-sky need on the continuum. With a mixed need, most of the problem may be defined, yet the solution requires expansive thinking. *A better way to surgically remove cancer at the time of surgery to ensure all of it has been eliminated from the site* is an example of a mixed need.

Importantly, the scope of the problem or type of need does not necessarily correspond to the size of the business opportunity or potential market. Small, incremental needs can be solved by incremental solutions that can result in sizable business opportunities if undertaken at the right time. Conversely, blue-sky needs may result in solutions that are direction changing in the industry but may not necessarily translate into significant commercial opportunities.

GETTING STARTED

When it is time to translate a problem into a need statement and need criteria, the steps below can help make this challenging process more approachable. This process is summarized in Figure 1.3.5, along with an example.

See **ebiodesign.org** for active web links to the resources listed below, additional references, content updates, video FAQs, and other relevant information.

1. Translate problem into a need statement

1.1 What to cover – Begin the process of translating problems into needs by asking a series of probing questions to reduce each problem to a simple, causal factor that results in an undesirable outcome. Next, evaluate what change in outcome or practice the problem calls for from the target user's perspective and determine how it can be measured. Then, capture the essence of the need in a one-sentence statement that defines the specific issue that requires a solution with a focus on the goal or desired endpoint. Writing an effective need statement may take multiple iterations. Make every word count.

1.2 Where to look
- **Primary research** – Perform more primary research with target users to validate desired outcomes (and the changes they necessitate).
- **Business publications** – Review business publications in the field such as "Turn Customer Input into Innovation" by Anthony Ulwick (*Harvard Business Review*, January 2002, pp. 91–97) and "Customers as Innovators: A New Way to Create Value" by Stefan Thomke and Eric von Hippel (*Harvard Business Review*, April 2002, pp. 74–81) for help thinking about the customer's perspective.

- **Networking** – Network with other innovators and review their need statements.

2. Verify the accuracy of the need statement against the problem

2.1 What to cover – Carefully evaluate every word in the proposed need statement to ensure that it has been validated by observations and supporting research. Make sure each element of the need statement can be traced back to the critical problem that was observed and requires solving. Innovators should also challenge themselves to be certain that no assumptions, inferences, or other forms of judgment have found their way into the need statement. Try to trace the cascade of events leading to the need statement to ensure that a good understanding of all related needs at levels above and below the need are identified. Mitigate the risk of superseding needs by determining whether there are superior needs that may be more appropriate to pursue. If a superior need is subsequently chosen, follow the same process to verify the accuracy of the new need statement.

2.2 Where to look – Revisit the innovation notebook and review notes on all observations to be sure the need statement captures the critical aspects of the problem(s) observed. Clarify the language of the need with the patients, physicians, and other providers where it was identified. Confirm the need by getting other opinions from individuals at different centers or geographic locations.

3. Confirm that the need is solution independent

3.1 What to cover – Evaluate the need statement to confirm that there is no solution embedded within it, unless it is a fundamental component of an incremental need. This includes references to current solutions, as well as emerging possibilities. Any reference (no matter how subtle) to a specific solution or treatment path can introduce artificial constraints into future thinking. Be aware that any time a specific device is mentioned in a need statement (e.g., catheter, stent, or scalpel), it is likely that the need has at least some portion of a preconceived solution embedded within it. Again, the only time this may be appropriate is for certain incremental needs but, even in these cases, the innovator should be careful not to unnecessarily constrain the need statement.

3.2 Where to look – There are few external resources to assist in this exercise. An innovator must rely on his/her own critical analysis skills to perform this assessment.

4. Validate that the scope of the need is appropriate

4.1 What to cover – Evaluate the need statement on a word-by-word basis to ensure that every one is necessary. Too often, superfluous words included within the need statement end up making a need too narrow and limiting the range of possible solutions (and the market opportunity). Next, evaluate the need statement on a word-by-word basis to ensure that it has not been defined too broadly. A need that has been overgeneralized can result in a solution that does not effectively address the needs of the true target audience. It can also cause innovators to overestimate the size of the market opportunity later in the biodesign innovation process. Carefully confirm each word in the need statement to ensure that it can be verified with data and that nothing has been assumed.

4.2 Where to look – Innovators should refer back to observations and secondary research to perform this step. If necessary, additional observations can be scheduled.

5. Define the need criteria and classify the need

5.1 What to cover – Use the detailed information collected to define criteria that will meet the requirements of the people most likely to use a solution to the need. Consider issues such

1.3 Need Statement Development

as where, when, and by whom the solution is most likely to be used, as well as ease of use, time of use, duration of use, cost, and other factors that could potentially affect its adoption. These assessments should be made based on the behaviors directly observed through the observation process. The need can also be classified as incremental, blue-sky, or mixed to help raise the innovator's awareness of some of the characteristics of the need s/he is seeking to address. Depending on the innovator's strategic focus, there may be other criteria that can be used to help bring organizational clarity and direction in seeking to address the identified needs.

5.2 **Where to look** – Innovators should refer back to observations and secondary research for need criteria. The need statement must be directly evaluated to classify the type of need being addressed.

Process	Description	Example
Translate problem into need statement	Reduce the problem into a simple causal factor that results in an undesirable outcome.	Need statement = Need a way to prevent or decrease the incidence of infections associated with hip implants in the elderly.
Verify accuracy of need statement against problem	Double check to be sure the need statement accurately embodies the problem that has been observed.	Types of questions to ask = Does the need statement accurately capture the critical problem that was observed? Has any element of the need statement been inferred or assumed? Can each element be backed by direct observations?
Confirm that need is solution independent	Evaluate the need statement to be certain it does not unnecessarily constrain or limit the solution to any particular technology or approach.	Types of questions to ask = Does the need statement specify any particular technology or approach that must be used to address the issue of infections? Does the need statement place constraints on the solutions that can be considered?
Validate that scope of the need is appropriate	Evaluate need statement word by word to ensure that every word is necessary and does not unnecessarily constrain the need. Be equally certain that every word is validated by data and/or observations and that the need has not been inappropriately generalized.	Types of questions to ask = Is infection issue only related to hip implants? Does similar issue exist for other orthopedic implants? Implants outside orthopedics? Orthopedic procedures that do not require an implant? Is it only relevant to the elderly or can it be generalized to a larger population? Does it affect all elderly or only a subset with certain disease characteristics? Is there a specific type of infection that should be addressed or should need address any/all infection resulting from hip implant?
Define need criteria and classify need	Use the detailed information that has been collected to define what criteria the solution to the need must meet to be successful. Classify the need and related business opportunity.	Types of question to ask = Is there a target practitioner for the solution? Target location? Specific treatment parameters that must be met? Is this an incremental, blue-sky, or mixed need?

FIGURE 1.3.5
The biodesign process for developing a need statement and need criteria.

Credits

The editors would like to acknowledge Asha Nayak for her help in developing this chapter. Many thanks also go to David M. Kelso and the student team at Northwestern University for sharing their story.

Notes

1. Jane Neff Rollins, "Many New Colonoscopic Devices are in Pipeline," *Internal Medicine News*, August 1, 2006, http://www.articlearchives.com/medicine-health/diseases-disorders-cancer-colorectal/664881–1.html (November 17, 2008).
2. All quotations are from interviews conducted by authors, unless otherwise cited. Reprinted with permission.
3. "Kangaroo Mother Care," Population Council, http://www.popcouncil.org/rh/kanga.html (November 20, 2007).
4. Ibid.
5. "What is KMC?" Kangaroo Mother Care, http://www.kangaroomothercare.com/whatis03.htm (November 20, 2007).
6. From remarks made by Mir Imran as part of the "From the Innovator's Workbench" speaker series hosted by Stanford's Program in Biodesign, April 28, 2004, http://biodesign.stanford.edu/bdn/networking/pastinnovators.jsp. Reprinted with permission.

Acclarent Case Study

Although the biodesign innovation process may appear to be relatively straightforward and linear, in reality it is complex, iterative, and dynamic. The innovator faces a continual stream of interconnected challenges and opportunities as s/he moves from needs finding to integration, and then on to commercial launch. New information becomes available at every stage of the process, which can require the innovator to revisit previous decisions, address new risks, and consider complicated trade-offs. Moreover, the biodesign innovation process takes place within the competitive medtech field, against the backdrop of the increasingly demanding and complicated domestic and global healthcare environment.

Nothing demonstrates the difficult, ever-changing, yet potentially rewarding nature of this process better than a real-world example. The following case study, which is presented in six parts, tells the story of a company called Acclarent, Inc. as it moves through each stage of the biodesign innovation process toward the commercial launch of its innovative new technology.

Stage 1 Needs finding

After completing the sale of his most recent company, TransVascular, to Medtronic, Josh Makower was at a crossroads. With Makower's leadership, TransVascular had pioneered the development of a proprietary catheter-based platform to facilitate existing and emerging intravascular procedures. The new technology could be used to bypass occluded vessels in the coronaries and peripheral vasculature, rescue failed attempts to navigate total occlusions, and deliver therapeutic agents (e.g., cells, genes, and drugs) to precise locations within the vascular architecture.[1] One potential application for the system was to repair the damaged heart tissue that resulted from the more than 1.5 million heart attacks suffered annually.[2] In September 2003, Medtronic acquired substantially all of TransVascular's assets for a deal valued up to $90 million, leaving Makower in a position to decide what he wanted to do next. While the sale to Medtronic was a positive financial outcome for the investors and employees, it fell far short of the expectations the TransVascular team had for the business when its members set out to "pioneer the vascular highway" in 1996, the year in which the company was founded.

1.1 Defining a strategic focus

"Anytime you have an opportunity to stop, step back, and reassess, it's just a great time to check in on your priorities and the things that you want to try to accomplish in life to make sure that you're heading in the right direction," Makower said.[3] Recognizing the chance to define a fresh strategic focus in his career, he initiated a personal inventory and started by thinking about his mission. "For me, it was about trying to make sure that I learned from the mistakes that I had made in the past. And I also wanted to stay true to what I initially set out to do, which was to work on medical problems of a magnitude that, if solved, would result in a significant improvement of quality of life for thousands, if not millions, of patients," he recalled.

In terms of his strengths, Makower, who holds a SB in mechanical engineering from MIT, an MD from the New York University School of Medicine, and an MBA from Columbia University, had now delivered successful **liquidity events** for investors from his first two companies. This gave him not only a valuable educational background, but the battle-scars of real-life experience on which to depend. In addition, "I felt that a skill I could rely upon was the ability to sift out from the weeds important things to work on, and to create projects that were compelling enough to draw extremely talented people together. I also knew that I had selling skills that would help me raise money and communicate enough enthusiasm to others that they would see the vision, commit themselves to a project, and join me in giving it all we have," he commented. Makower further recognized that he had an advantage in his close relationship with venture capital firm NEA, which had invested in TransVascular and his prior start-up, EndoMatrix.[4] His ties with this entity bolstered his belief that he would be able to secure enough funding to get his next idea off the ground. Based on his past experiences (with start-ups as well as Pfizer's Strategic Innovation Group), his education and, most importantly, his mastery of the biodesign innovation process, Makower also felt confident in his ability "to venture with a blank sheet of paper into any clinical field and come up with something that was meaningful to improve the lives of patients."

As far as assessing his weaknesses, Makower tried to be brutally honest with himself. "One weakness," he recalled, "was that I knew that I did not want to be a CEO again, at least not for a long period of time. While I understood the skills required for management, appreciated its value, and probably could do it, it just wasn't fun for me. I knew managing large groups of people was not for me and it was not until I brought a new CEO into TransVascular, Wick Goodspeed, that I started enjoying my role again. I liked being a part of finding the solution, being a problem-solver, and providing a vision of the future. But I didn't enjoy having to manage people by milestones, conduct performance reviews, and all the other things that good managers do to manage and lead a company." This brought him to the conclusion that he preferred working with small teams during their start-up phase and growing the business to the point where he could reasonably hire a CEO to lead the dozens or even hundreds of employees that might come afterwards.

Moreover, "I realized that while I deeply enjoy pushing the edge of medicine and exploring completely new concepts in medical areas that are not well understood, basic research was not a good place to operate a venture-backed company," he said. Due to the highly theoretical nature of this kind of work and the extreme levels of uncertainty that innovators face, Makower recalled, "I had strong feelings about not wanting to get people – employees and investors – on board with a vision and then have them be disappointed because our theory was wrong after so much good effort and hard work. We had been there before with TransVascular, and I just didn't want to do that to myself and the people that I worked with again. I wanted to look for opportunities that were much more concrete and could be realized commercially in a reasonable time frame."

After evaluating his strengths and weaknesses, Makower thought about specific project acceptance criteria that would make a new project attractive to him. "I viewed EndoMatrix and TransVascular as good learning experiences, but not tremendously successful. I wanted to have the opportunity to deliver on a project that was very successful. I felt like it was time to take what I had learned and really apply it." This led him to focus on opportunities with a reasonably high chance of success. "No more science experiments," became a mantra of sorts as he and his eventual team began evaluating possibilities.

Another key acceptance criterion was Makower's desire to work on problems that affected a large number of people. Finding a compelling market – opportunities with the potential to reach millions of people and achieve $1 billion in revenue – was another important factor. "We knew we needed to create a company that within a ten-year timeframe could have $100 million to $200+ million in annual revenues with a reasonable growth rate to achieve our investor's return expectations," he recalled.

Finally, he decided to commit himself to projects that would not involve any patient deaths. According to Makower, "At TransVascular, we worked on a technology targeted at critically ill patients. Our interventions risked patient lives in an effort to try to save them. This kind of project requires a certain level of intestinal fortitude and a willingness to accept dire consequences for miscalculating the unknowns. I respect it. I've done it. But I didn't want to do it again, at least not as my next big thing. It's just too much emotion and stress, worrying about the patients."

With these (and a handful of other) acceptance criteria defined, Makower set out to identify one or more specific strategic focus areas that would meet his requirements. To help accomplish this, he restarted medical device incubator ExploraMed. ExploraMed I, which was originally founded by Makower in 1995, spawned EndoMatrix and TransVascular. In its new form, ExploraMed II (as it would be called) was intended to become a platform for launching two to three new medical device businesses.

Makower's first move was to secure trusted team members in key roles. Karen Nguyen signed on to oversee the finances and Maria Marshall agreed to continue on this new ExploraMed venture as his executive assistant. His first technical hire was John Chang, a seasoned R&D veteran who had been a core part of the engineering team at TransVascular. See Figure C1.1. In fact, Makower accelerated his plans to restart ExploraMed in an effort to help prevent Chang from accepting another job. "I think one of the important parts of my model is an emphasis on people," he said. "At the end of the day, the value of a business is in the people. Ideas are great, but the people who make it all work are the reason why you're successful. I wanted to work with John again because he's just the most positive, energetic, happy, hard working, smart, dedicated, loyal, and trustworthy guy anyone would ever want to have on a team. So I restarted ExploraMed sooner than I wanted to, so we could have a chance to work together again."

Together, Makower and Chang decided to explore four key areas. Two of these areas were orthopedics and respiratory disease. Another was focused on trying to find a niche within the congestive heart failure (CHF) arena that would meet Makower's acceptance criteria. This field was interesting to the two men because of their prior experience at TransVascular, but they needed to define a scope and focus that would be more applied.

FIGURE C1.1
Makower and a skull model join Chang on his first day of work with ExploraMed (courtesy of ExploraMed).

"The thought was, 'Is there anything that we can do that would be simpler than what we were working on at TransVascular – something that's not going to require us to create new science?'" Makower recalled. They started to focus on pulmonary edema associated with CHF (the effect that causes patients to become starved of oxygen during the night and unable to sleep), and noticed it was an important side effect that dramatically affected patients' quality of life, yet seemed to have some interesting mechanical implications.

The fourth potential focus area was in the ear, nose, and throat (ENT) specialty – a space in which Makower had already performed some preliminary research and generated some ideas. As someone who suffered from chronic sinusitis, a condition involving the recurring inflammation of the cavities behind the eyes and nose commonly caused by bacterial or viral infections,[5] Makower was intimately familiar with the inadequacies of existing treatment alternatives for this condition. Patients usually were treated with over-the-counter and prescription medications, including antibiotics when severe infections occurred. In fact, sinusitis was the fifth most common condition for which antibiotics were prescribed in the United States.[6] Steroids were another type of therapy that was used when these other treatments failed to produce or sustain results. In a relatively small number of the worst cases, chronic sinusitis patients qualified for certain surgical procedures, the most common of which was functional endoscopic sinus surgery (FESS). Frustrated with the efficacy and nature of his treatment alternatives, Makower had informally begun exploring new approaches in this area. When Chang joined ExploraMed, the time was right to more formally evaluate the strength of potential opportunities in ENT. "We set aside my previous work for the time being to try to treat this like any of the other projects," said Makower. "We needed to start with the basic clinical problem, develop a deep understanding of the real clinical needs, and do our due diligence to see if we would arrive at the same general conclusions and ideas that I had going in."

1.2 Problem identification and observation

According to Chang, they developed a plan to spend their first few weeks collecting general information about what was going on in the areas of orthopedics, respiratory disease, CHF, and ENT. "I scheduled a number of meetings with physicians I knew, and I attended several conferences," he said. In these meetings, Chang remembered, "We said, 'So, tell us about what you do. Tell us about some of the patients you see. Tell us about the challenges you face as a physician. What are some of your greatest needs?'" Sometimes the physicians had specific ideas to share but, more often than not, they simply talked about their experiences and discussed whatever frustrations were giving them problems. The team knew that the real insights would come from observing physicians in the operating room (OR), with their patients in the clinic, and at clinical meetings as they debated and discussed current therapy.

For this reason, the next step was to schedule first-hand observations. Makower explained: "We needed to find clinicians that would allow us to see a large volume of cases, get a lot of patient experience, and quickly come up to speed on the space." Makower had already begun teaching at Stanford University's Program in Biodesign and, as a result, had a rich network at the University's medical center to tap into for contacts in certain specialties. However, in some fields, such as ENT, he and Chang had to "cold call" the hospital. "We called and asked, 'Who's the rhinologist at this hospital?'" Makower remembered.

Once they made the appropriate connections, a formal observation process was launched. Again, in the ENT space, "We spent a couple of days following a surgeon around in clinic, getting an appreciation for his day-to-day routine – what patients were coming in, and why. Was it an initial visit, a post-operative follow-up, or some other type of appointment?" said Chang. "We also spent time in the OR asking 'dumb' questions like, 'Hey, I notice you did that four times. Is there a reason why you have to do that?' The idea was to keep the eyes and the mind open."

"We went to clinic, we watched cases, we talked to patients, and we observed surgeries," Makower recapitulated. "And basically, the more we heard and saw, the more we began to feel that we really had something here." Functional endoscopic sinus surgery to treat severe cases of chronic sinusitis became the dominant form of sinus surgery in the mid-1980s, led by the introduction of the endoscope to the field. This approach involved the insertion of a glass-rod optic, called an endoscope, into the nose for a direct visual examination of the openings into the sinuses. Then, under direct visualization, several cutting and grasping instruments were used to remove abnormal and obstructive tissues in an effort to open the sinus drainage pathways. In the majority of cases, the surgical procedure was performed entirely through the nostrils (rather than through incisions in the patient's face, mouth, or scalp, as was previously necessary).[7] "Conceptually, the specialty made this huge leap forward 20 years ago from large open incisions to FESS, which was considered to be minimally invasive and atraumatic," commented Chang. "But when observing a FESS procedure, the video image coming off the endoscopic camera was often a sea of red. As an outsider, you just think, 'Gosh, maybe that's better than it used to be, but that doesn't seem so atraumatic to me. Ouch!'"

According to Makower, one of the reasons the process was so bloody was because a significant amount of bone and tissue was being removed in every procedure. As an example, he explained:

> The uncinate process is a bone that sits at the edge of the maxillary sinus. It is never diseased, yet it is completely removed in almost every conventional sinus surgery. To me, it was amazing that they were removing a structure solely because it was in the way. I had to ask the question two or three times: 'So, the only reason why you're taking it out is because you can't see around it?' 'Yes.' 'But it's not diseased?' It was kind of incredible. And that was the beginning of how we came to the realization that there was a lot more cutting involved in the procedure than ideally needed to be done. We used to joke that it was analogous to someone deciding that they needed to make their bedroom door a little bit bigger, and choosing the method of driving a bulldozer through the entry doorway, through the living room, and demolishing the kitchen along the way just to get to the bedroom door, because all those things were just 'in the way.'

Other major problems identified through their observations included post-operative scarring, mostly related to the trauma imparted during surgery that often led to suboptimal outcomes and a need for repeated procedures solely to address the recurrence of their scars. Additionally, "There were other potential complications, like cerebrospinal fluid (CSF) leaks, as well as a high level of complexity of the procedure, not to mention the significant post-operative pain and bleeding," Makower added.

1.3 Need statement development

Using the information gleaned from their initial observations, Makower and Chang defined a need statement and a preliminary list of need criteria. Summing up their take-aways, Makower said, "We saw a need for a minimally invasive approach to treating chronic sinusitis that had less bleeding, less pain, less bone and tissue removal, less risk of scarring, and that was faster, easier, and safer to perform. We tried hard to put aside the ideas that we had come up with already and stay true to the process, but our excitement about the opportunity was clearly building."

Notes

1. "Medtronic Completes Transaction with TransVascular, Inc.," *Business Wire*, September 24, 2003, http://findarticles.com/p/articles/mi_m0EIN/is_2003_Sept_24/ai_108088865 (August 27, 2008).
2. "Medtronic Agrees to Acquire Assets of TransVascular, Inc., Maker of Next-Generation Vascular Devices," EuroPCROnline.com, August 11, 2003, http://www.europcronline.com/fo/exchange/news/press_releases.php?news_id=417 (August 27, 2008).
3. All quotations are from interview conducted by the authors, unless otherwise cited.
4. EndoMatrix was a medical device company focused on the treatment of incontinence and gastro-esophageal reflux. It was acquired by C.R. Bard in July, 1997.
5. "Balloon Therapy," *Forbes*, May 22, 2006, p. 82.
6. Carol Sorgen, "Sinus Management Innovation Leads to an Evolution in Practice Patterns," *MD News*, May/June 2007.
7. "Fact Sheet: Sinus Surgery," American Society of Otolaryngology – Head and Neck Surgery, http://www.entnet.org/HealthInformation/SinusSurgery.cfm (August 28, 2008).

IDENTIFY Needs Screening

STAGE 1 NEEDS FINDING
1.1 Strategic Focus
1.2 Observation and Problem Identification
1.3 Need Statement Development
Case Study: Stage 1

STAGE 2 NEEDS SCREENING
2.1 Disease State Fundamentals
2.2 Treatment Options
2.3 Stakeholder Analysis
2.4 Market Analysis
2.5 Needs Filtering
Case Study: Stage 2

STAGE 3 CONCEPT GENERATION
3.1 Ideation and Brainstorming
3.2 Concept Screening
Case Study: Stage 3

STAGE 4 CONCEPT SELECTION
4.1 Intellectual Property Basics
4.2 Regulatory Basics
4.3 Reimbursement Basics
4.4 Business Models
4.5 Prototyping
4.6 Final Concept Selection
Case Study: Stage 4

STAGE 5 DEVELOPMENT STRATEGY AND PLANNING
5.1 Intellectual Property Strategy
5.2 Research and Development Strategy
5.3 Clinical Strategy
5.4 Regulatory Strategy
5.5 Quality and Process Management
5.6 Reimbursement Strategy
5.7 Marketing and Stakeholder Strategy
5.8 Sales and Distribution Strategy
5.9 Competitive Advantage and Business Strategy
Case Study: Stage 5

STAGE 6 INTEGRATION
6.1 Operating Plan and Financial Model
6.2 Business Plan Development
6.3 Funding Sources
6.4 Licensing and Alternate Pathways
Case Study: Stage 6

Successful entrepreneurs do not wait until 'the muse kisses them' and gives them 'a bright idea': they go to work ... Those entrepreneurs who start out with the idea that they'll make it big – and in a hurry – can be guaranteed failure.

Peter Drucker[1]

I find out what the world needs. Then, I go ahead and invent it.

Thomas Edison[2]

Needs Screening

After the winnowing of many needs, a rigorous follow-on process of screening and specification is required before you begin inventing – the deep dive. This is not necessarily an intuitive skill for most; typically, bright people will encounter a clinical need and proceed directly to devising solutions without first validating their chosen focus.

In fact, careful scrutiny of all facets of the need is essential. While serial innovators may do this intuitively, a formal process is recommended for those with less experience. The iterative process of "walking around the problem" with a concept map may be enhanced by an occasional cooling-off period. There is a perfectly natural human tendency to fall in love with a problem statement and remain anchored to it. While an early formulation of the concept may be directionally correct (but imperfect), dispassionate review and reflection can prevent missteps.

By the end of this deep dive, the innovation team should have become absolutely expert on the problem, with a detailed specification of the need including anatomy, physiology, pathophysiology, epidemiology and market dynamics, competitors and their current solutions, and customer requirements – both "must haves" and "desirables." At the heart of the need specification is a need statement, which is a single sentence focusing on the goal or the endpoint, not the problem. We call this the genetic code of the eventual solution.

Using the urinary incontinence observations from Stage 1, one of our innovation teams developed the following need: *A simple, safe mechanism to reduce the frequency of stress urinary incontinence in elderly women.*

In crafting the needs statement, the innovation team revisited and revalidated the early work of needs finding. Needs are in fact relatively easy to find, and all can look appealing ... at least initially. It is only through the rigorous deep-dive process that the real importance of a need can be determined. *Only* then is it worth investing intellectual capital and the other resources needed to pursue a solution.

Notes
1. Peter Drucker, *Innovation and Entrepreneurship: Practice and Principles* (HarperCollins, 2006).
2. Robert A. Wilson and Stanley Marcus, *American Greats* (PublicAffairs, 1999).

2.1 Disease State Fundamentals

Introduction

In the excitement of having identified one or more compelling needs, an innovator's instinct may be to quickly jump ahead and begin inventing. However, establishing a detailed knowledge of the relevant disease state, with a particular focus on its mechanism of action, is fundamental to validating any need and understanding how it can best be addressed. Disciplined disease state research is an essential part of the biodesign innovation process and an invaluable activity for clinician and non-clinician innovators alike.

Understanding disease state fundamentals involves researching the anatomy and physiology, pathophysiology, symptoms, outcomes, epidemiology, and economic impact of a disease. This information is pertinent to the process of finding a clinical need or in validating a need that has already been established. The process also provides the innovator with a critical level of knowledge about a condition so s/he can be credible when speaking to external stakeholders, such as physicians and other experts in the field.

OBJECTIVES

- Understand the importance of disease state analysis.
- Know what factors to investigate as part of this research.
- Appreciate how to effectively search for and summarize this information into a useful format to aid the needs screening process.

Disease state fundamentals

Disease state research may be performed in conjunction with the creation of a new need statement or in the process of validating a need statement that has already been defined (e.g., in a project-based biodesign program where students are assigned specific needs to investigate). Disease state research also serves a critical role in forming the basis for screening multiple needs against one another later in the biodesign innovation process (see 2.5 Needs filtering).

Disease state research is first performed using general scientific resources such as medical textbooks or medical information websites, then transitions over time into a more comprehensive, in-depth review of historical and current medical literature. This approach allows the innovator to begin by developing a general understanding of a disease and then become increasingly knowledgeable about aspects of the condition that are most relevant to the need. Obtaining an understanding of a condition's mechanism of action – or the science behind how the disease works from a biologic or physiologic perspective – is especially important. Some disease states are well understood and, therefore, needs in the field are more readily approachable. Disease states in which the mechanism of action is unclear may pose a significant challenge.

Because disease state research can be tedious, innovators may be tempted to skip this step. Those with a medical background may figure that they already know enough to understand the disease state associated with a need. In contrast, innovators from business or engineering backgrounds may have a tendency to shortcut the research in their enthusiasm to evaluate the market or other factors that will help determine if an opportunity is promising. However, under-investing in this area is almost always short-sighted. Disease research not only provides a foundation for understanding the underlying disease state, but also lends valuable knowledge that aids in the investigation of existing treatments, the current market, and important stakeholders. Later in the biodesign innovation process, this information can be used again to assess the clinical, technical, and commercial feasibility of any solution concept that will eventually be developed.

The following example, which references one of the great medtech success stories of the 1990s, illustrates that even the most experienced innovators and companies, regardless of their prior experiences and training, can realize significant value from disease state research and should regard the analysis as indispensable.

FROM THE FIELD | **JOHNSON & JOHNSON**

Understanding disease state fundamentals as part of the needs screening process

Johnson & Johnson (J&J), through its subsidiary Cordis, was an early pioneer in the market for bare metal stents, small mesh-like tubular scaffolds that can be used to open narrowed heart arteries. The company dominated the treatment space after the introduction of its Palmaz-Schatz® coronary stent in 1994. Johnson & Johnson held a firm leadership position until 1997 when competition from other medical device manufacturers began to intensify, particularly with the launch of Guidant's Multi-Link® bare metal stent. Seeking a way to regain the company's leadership position while further reducing the need for repeat procedures in patients with coronary artery disease, Bob Croce, J&J's group chairman of Cordis Corporation at the time, went back to the drawing board with his team to reexamine disease state fundamentals as part of the need screening process.

Coronary artery disease occurs when plaque, a mixture of cholesterol and other substances, accumulates over time within the arterial wall, through a process called atherosclerosis. This causes a reduction in the available area for blood flow. See Figure 2.1.1. The restriction of blood flow to the heart can result in angina (chest pain) or lead to a myocardial infarction (heart attack), depending on

FROM THE FIELD — JOHNSON & JOHNSON

Continued

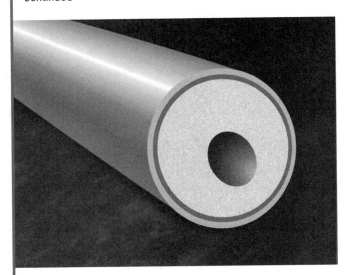

FIGURE 2.1.1
A blood vessel narrowed by atherosclerosis (developed by Yasuhiro Honda; reprinted with permission).

the severity of the narrowing. According to the American Heart Association, more than 6.5 million people in the United States suffer from angina, and 1.2 million people experience new or recurrent heart attacks each year. Approximately 40 percent of these heart attacks are fatal, making cardiovascular disease the leading cause of death in the United States.[1] The treatment of coronary artery disease is a major contributor to the roughly $21 billion market for cardiovascular devices.[2]

Angioplasty, an interventional procedure in which a physician inserts a balloon-tipped catheter into a narrowed artery to increase blood flow, revolutionized the treatment of coronary artery disease in 1977 by providing a less invasive, lower risk alternative to coronary artery bypass surgery.[3] However, there was often a recoiling effect of the arterial wall, which meant that the artery remained only partially open after the balloon catheter was removed. In addition, there was scarring within the artery as a response to the injury from the balloon – called restenosis – that occurred in 30 to 40 percent of patients within six months of the angioplasty as the body sought to heal the artery.[4] For these reasons, many patients required repeat angioplasty procedures or bypass surgery, resulting in increased risk for the patient and added cost for the healthcare system.

Bare metal stents were incorporated into the balloon angioplasty procedure to address these issues. When the balloon was inflated at the site of the blockage, a stent – a small mesh-like tubular scaffold – was expanded and locked into the wall of the artery. See Figure 2.1.2.

The stent physically held the artery open and prevented it from recoiling once the balloon was extracted. As a result, the number of repeat procedures declined and patient restenosis rates dropped to approximately 20 to 25 percent. While bare metal stents were widely considered to be a major breakthrough, "The statistics weren't that great," said Croce.[5] "The stents corrected one problem, the retracting of the arterial wall, and they improved outcomes compared to using the balloon alone. Unfortunately, they also caused the closing of the arteries through neointimal growth." Neointimal growth was the formation of scar tissue within the stent as a result of the trauma involved with the insertion of the stent and the body's reaction to the stent. Thus, through a different mechanism, the arteries could still eventually become narrowed.

To better understand the need criteria that any new solution would have to satisfy, Croce and his team spent significant time revisiting the physiology of the coronary arteries and the pathophysiology associated with neointimal growth. One of the most significant realizations that came from this research was the recognition that the original disease state had shifted. The need was not just to address atherosclerosis, the build-up of plaque in the vessel wall, but, ultimately, the new disease state of restenosis caused by neointimal growth. The fundamentals of the disease state were generally known, so "It wasn't like the cycle of neointimal growth in the arteries was a brand new discovery," recalled Croce. However, "Many smart people in the area had not been trained in neointimal growth for a long time and in some cases they never did understand it since it wasn't important to them in their practice," he continued. Through a thorough study of the disease state, Croce and team increased their understanding, as well as their confidence that no opportunities would be overlooked. "No matter how experienced you are, you can't go into this process assuming that you know everything. It's essential to stay open-minded and force yourself to go through the analysis from all different aspects of the disease," he said.

2.1 Disease State Fundamentals

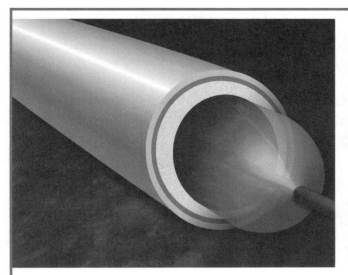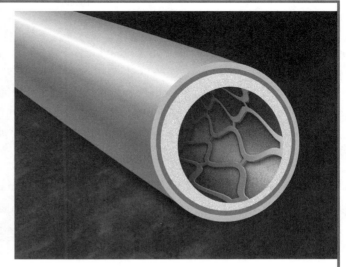

FIGURE 2.1.2
A balloon is used to deploy a stent within the arterial wall (developed by Yasuhiro Honda; reprinted with permission).

An in-depth understanding of the pathophysiology of the disease was particularly important in this case because of the solution that was eventually chosen – a combination of drug and device that eventually became the drug eluting stent. Only through revisiting the underlying need and studying neointimal growth, a fundamental aspect of the disease state, did the team determine that certain drugs, which were known to inhibit neointimal growth, could be used to prevent the neointimal growth due to stent placement. However, the marriage of medical device developers with pharmaceutical scientists was not an easy one. Medical device companies had a bias toward engineering a better stent that would not scar the arteries, while pharmaceutical companies were predisposed toward small molecule and biotech solutions and did not necessarily want or know how to consider device development. "If you look before drug eluting stents, there were no major drug–device combination products. So, there was a lot of hesitation on all sides to this project," remembered Croce. In addition, physicians were skeptical that such a novel concept could produce results. Nonetheless, the science behind the disease provided a key factor around which all parties could converge, and therefore served as a critical building block for the effort.

After years of development, J&J's drug eluting stent moved into clinical trials. In its first-in-man studies, the Cypher stent demonstrated in-stent restenosis rates of 0 to 3 percent and in-segment or vessel restenosis rates of up to only 9 percent, compared to 33 percent in the bare-metal stent arm.[6] When Cypher received FDA approval in April 2003, J&J decisively regained its leadership position in the treatment of coronary artery disease.

Approach to disease state analysis

Disease state research is best approached in a systematic manner, particularly if the need is related to a single, specific disease area. Six key areas, outlined in Table 2.1.1, should be addressed to ensure a thorough understanding of a disease.

For needs that cross more than one disease area, the innovator must establish a clear understanding of each related condition. Realistically, however, it may be necessary to take a somewhat broader perspective, paying close attention to those aspects of the various disease areas that are most directly related to the need. In these cases, the anatomy, physiology, and even pathophysiology that are studied may not be for a specific organ or system, but instead for a fundamental biologic process that is shared across the multiple disease areas.

In the remainder of this chapter, an example focused on atrial fibrillation (AF), a disease in which the heart has an abnormal rhythm, is used to further illustrate the

Table 2.1.1 *Six key areas of disease state analysis.*

Focus area	Description
Anatomy and physiology	Describes the normal anatomy and/or function of the organ system, which may include various organs or areas of the body affected by the need.
Pathophysiology	Describes the disturbance of normal anatomy and physiology caused by a disease or other underlying physical, mechanical, electrical, or biochemical abnormality.
Clinical presentation	Profiles the patient state and clinical status associated with a disease. These include the signs of the disease (what one might find on a clinical exam or with lab testing), as well as symptoms (what the patient feels and suffers from).
Clinical outcomes	Profiles the most common outcomes experienced by patients as a result of having the disease.
Epidemiology	Describes the causes, distribution, and control of disease in the population.[34]
Economic impact	Outlines the cost of the disease to the healthcare system.

type of disease state analysis that the innovator should perform at this stage of the biodesign innovation process. Importantly, the evaluation of disease state fundamentals is distinct from understanding the treatment options used to address the disease. An in-depth review of treatment options is covered in a separate chapter (see 2.2 Treatment Options). However, the analysis of treatment options may lead to a refined understanding of disease state fundamentals and vice versa.

Anatomy and physiology

Obtaining a basic working knowledge of the normal anatomy and physiology of the organ, system, or structure of the body that is affected by a need is important because it establishes a baseline against which abnormalities are understood. While some diseases affect a specific organ or system, other disease states affect multiple organs or systems within the body. Through research of the normal anatomy and physiology, the innovator should quickly be able to determine whether narrowing his/her focus to one organ or area is appropriate. This research also provides the innovator with an understanding of important vocabulary and context as s/he delves into further research.

The disease will be much easier to comprehend if the anatomy of the affected organ or organ system is clearly understood and can be visualized. For example, an innovator with an engineering mindset will likely be assisted in understanding the disease by knowing the position, size, and proximity of the affected organ or system in relation to other systems. In addition, both gross and cellular anatomy must be evaluated. Gross anatomy refers to the study of the anatomy at a macroscopic level (through dissection, endoscopy, X-ray, etc.). Cellular anatomy, also called histology, refers to the study of the body using a microscope. It is often most effective to start with gross anatomy, as this is generally easier for most innovators to grasp. Then, with knowledge of the gross anatomy serving as context, an innovator can more effectively tackle the topic of cellular anatomy.

Physiology, or the way in which biologic tissues function, is often much better understood after the innovator establishes a working knowledge of the normal anatomy. As with anatomy, physiology should be investigated at both the gross and cellular levels. Once the innovator learns about normal patterns of function within an affected area, s/he has a basis for understanding how the disease functions (as described in the next section). Biologic and physiologic processes can be evaluated in terms of their mechanical, electrical, and chemical mechanisms. Later in the biodesign innovation process, this information can serve as a basis for brainstorming new concepts that act on these mechanisms of action.

Using the AF example, an innovator would begin by determining that AF is a disease of the heart, which is part of the cardiovascular system. As the heart is the primarily affected organ, the innovator could then focus on investigating the basic gross anatomy of the heart and its normal function. Understanding the heart's size, location, and position in relation to other structures quickly establishes a baseline context for investigating more complex concepts and interactions, such as how the electrical system of the heart establishes a rhythm that affects the organ's ability to mechanically contract. While the appropriate level of detail to capture varies significantly with each specific need and its associated disease state, the working example below is representative of the detail that is appropriate for a *preliminary* disease state assessment.

Working example: Normal anatomy and physiology of the heart

The pumping action of the heart depends on precise electrical coordination between the upper loading chambers (atria) and lower pumping chambers (ventricles) as shown in Figure 2.1.3.

The contraction of the atria and ventricles is regulated by electrical signals. During normal sinus rhythm[7] the sinoatrial (SA) node (often referred to as the "pacemaker"), which is located in the high right atrium, releases an electrical discharge that causes the atria to contract. The electrical signal then propagates through the atria to the atrioventricular (AV) node, which is located between the atria and the ventricles to help regulate the conduction of electrical activity to the ventricles. Electrical signals are conducted from the AV node through Purkinje fibers[8] into the ventricles, causing the ventricles to contract.

The rate of the electrical impulses discharged by the SA node determines the heart rate. At rest, the frequency of discharges is low, and the heart typically beats at a rate of 60 to 80 beats per minute. During periods of exercise or excitement, an increase in the heart rate is mediated by the input of the central nervous system onto the SA node, which subsequently discharges more rapidly.

The heart's mechanical and electrical coupling is the result of the organ's fundamental cardiac cellular physiology. While the heart is composed primarily of connective tissue, cardiac muscle tissue is responsible for the electromechanical coupling of electrical signals and mechanical pumping. Muscle contraction is essentially the result of changes in the voltage of a cell due to the movement of charged ions across the cell's surface. This initial voltage change is the result of ions flowing from cell to cell, and is usually initiated by pacemaker cells, such as

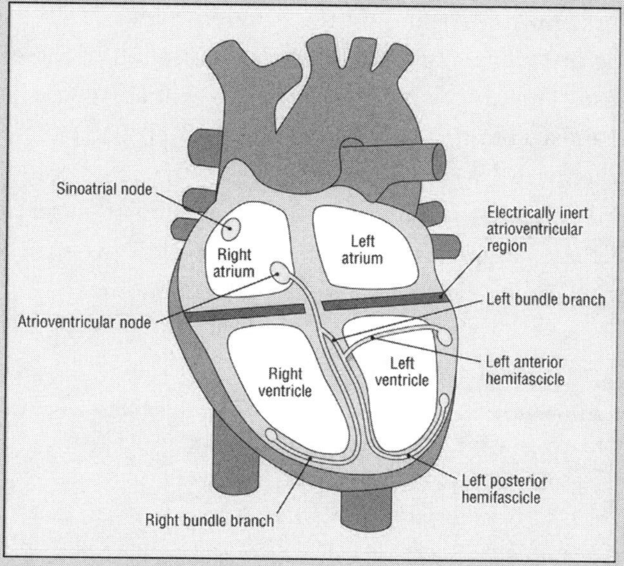

FIGURE 2.1.3
The heart's electrical system is one aspect of normal anatomy and physiology that an innovator must understand when initiating an investigation of AF (reproduced from "ABC of Clinical Electrocardiography: Introduction," Steve Meek and Francis Morris, 324, 415–418, 2002 with permission from BMJ Publishing Group Ltd.).

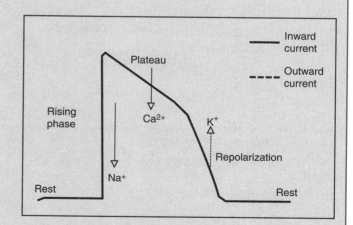

FIGURE 2.1.4
Depolarization is the result of the inward flux of sodium ions and the outward flux of potassium ions. Changes in the baseline voltage due to depolarization result in the inward flow and intracellular release of calcium and muscle contraction.

> **Working example:** *Continued*
>
> the cells of the SA node, which intrinsically cycle through voltage changes. This voltage change then triggers the movement of other ions within the cardiac muscle cell to cause changes in mechanical structures that result in contraction. The majority of cardiac muscle contracts due to depolarization, which is a change in voltage caused by the influx of sodium ions and the outflux of potassium ions.
>
> This flow of ions results in changes in the heart's baseline voltage, which causes both the influx of calcium ions and the release of internal calcium stores. Calcium, in turn, results in the interaction of various cellular components, bringing about a contraction in the mechanical filaments of the muscle cell. See Figure 2.1.4.
>
> This process propagates throughout the heart from cell to cell, such that the result of all the cellular filaments contracting is the temporally and spatially coordinated contraction of the heart muscle. Once contraction is complete, various cellular components are activated to reset the filament structure and ionic balance so that the process can begin again.

Pathophysiology

Once an understanding of anatomy and physiology of the relevant organ in a healthy individual is established, then the innovator can examine how the disease disturbs the normal structure and function. It is critically important to take into account the fact that most diseases are not homogeneous. Different subtypes of a disease often exist and the heterogeneous nature of patient populations can result in a broad range of effects for any given disease state.

When investigating pathophysiology, the first step is to better understand how the disease works from a biologic and physiologic perspective, and then how this affects the normal function of the organ or system. The second step is to identify the risk factors and causal associations (e.g., genetics, age, associated diseases, and lifestyle) that characterize the disease. Finally, the innovator can seek to understand the disease progression.

Disease progression examines the rate (e.g., days, weeks, or years) at which the disease leads to abnormal function. This includes the peak age of the effect and the types of changes that occur at each stage of the disease.

In the AF example, an innovator would explore how the heart might be structurally altered, leading to abnormal function, and whether or not the condition can lead to structural changes in the organ. S/he should also look at the common causes of AF, the primary risk factors, and how AF progresses. Time should be spent understanding the different types of AF and the unique characteristics of each variation of the disease. This might include looking at which type of AF is most common among different groups of patients, whether all AF patients progress in the same way (or if progression is more directly affected by other factors such as coexisting conditions), and how likely patients are to progress from one type of AF to another.

> **Working example:** Pathophysiology of atrial fibrillation
>
> ### Disease function
>
> In a normally functioning heart (as described in the Anatomy and physiology section), the rates of contraction for the atria and ventricles are typically equal and result in a regular heartbeat. However, during AF, ventricular and atrial contractions become irregular and unsynchronized. Instead of electrical discharges being regularly generated solely by the SA node, rapid and irregular discharges come from other areas in the atria. Since these other areas are discharging so fast, the SA node's slower, more regular rate is suppressed. There are several "trigger points" for this electrical activity, which create a pattern of rapid, chaotic electrical activity that is characteristic of AF. The majority of these focal sources (approximately 94 percent) are located in areas around the four pulmonary veins, which are connected to the left atrium. Other less common areas include the superior vena cava, right and

left atrium, and the coronary sinus.[9] Though not fully understood, causal factors (see below) may result in inflammation and injury to the heart, causing alterations in cell structure and predisposing it to abnormal electrical discharges that can initiate and maintain AF.

As a result of these irregular discharges, the atria contract between 300 and 600 times per minute.[10] However, the atria do not actually contract as a whole – the rapid contractions of parts of the atria may be better thought of as a quiver, or fibrillation, rather than regular beating. This results in improper filling and ejection of blood, as well as a decreased efficiency of the heart's pumping process. Since all electrical activity from the atria can typically only get to the ventricle via the AV node, the AV node is able to filter many of the irregular electrical discharges associated with AF, preventing the rapid rate of the atrial beat from being conducted into the ventricles. However, not all of the signals are blocked and AF is often accompanied by irregular ventricular beating, at 50 to 150 per minute.[11]

Causal factors

The most common causes of AF are advanced age, abnormalities in the heart's structure, uncontrolled hypertension (e.g., high blood pressure), thyroid disease (e.g., an overactive thyroid or other metabolic imbalance), and acute exposure to heart stimulants (e.g., alcohol).

Disease progression

According to the American Heart Association, AF can be classified into three clinical subtypes; paroxysmal, persistent, and permanent. In the case of AF, the subtypes parallel disease progression with one subtype transitioning to the next over several years in a large majority of patients. These subtypes are defined by the ease with which episodes of AF terminate. Paroxysmal AF refers to recurrent or lone[12] episodes that spontaneously self-terminate after a relatively short period of time. Persistent AF requires pharmacological or electrical cardioversion[13] (e.g., giving medicines or an electric shock) to restore regular sinus rhythm. In patients with permanent AF, regular sinus rhythm cannot be restored and the irregular heartbeat becomes the accepted rhythm.[14] These subtypes stand in stark contrast to AF associated with reversible causes (e.g., thyrotoxicosis,[15] electrolyte abnormalities) and the occurrence of AF secondary to acute myocardial infarction, cardiac surgery, or acute pulmonary disease. These conditions are considered separately since the AF is unlikely to recur once the precipitating condition has been treated.

Clinical presentation

Research of clinical presentation focuses on the impact of the disease on the patient. It emphasizes the symptoms (what the patient says s/he experiences) and the signs (what the astute healthcare provider identifies or observes during the patient examination) of a disorder or disease. Gaining an understanding of clinical presentation is important because it is often the target for improved care and the development of new therapies that address identified needs. When evaluating clinical presentation, it is important to describe what patients complain about when they see a clinician and how they feel. Note that patients with the same disease may present differently based on a number of factors, such as age, gender, ethnicity, and coexisting conditions. Since every individual is different, each is likely to experience symptoms slightly differently. Ultimately, clinical presentation may manifest itself in the signs/symptoms that result from the primary effect of the disease or from the long-term consequences of having and managing the disease over time.

When researching AF, an innovator would seek to understand the most common symptoms for patients with the disease, how they feel with AF, and the signs most commonly observed by physicians in patients with the disease. S/he should also consider whether all AF patients are affected by the same symptoms and what factors have the greatest impact on symptoms presented (e.g., age, coexisting conditions). For example, young patients are much more likely to report symptoms of palpitations with AF than older ones. This may directly impact the goal of therapy for different age groups.

One strategy that may be helpful in evaluating clinical presentation is to take the perspective of the healthcare delivery system (e.g., insurance company or hospital). From a provider's perspective, what symptoms or comorbidities bring patients in for clinical care? From an insurance company's perspective, what types of bills are submitted from the providers who first see a patient with the disease and what is the frequency of care?

> **Working example: Clinical presentation of atrial fibrillation**
>
> While some patients do not experience noticeable symptoms due to AF, others have fatigue, weakness, lightheadedness, shortness of breath, or chest pain. Palpitations – sensations of a racing, uncomfortable, or irregular heartbeat – are also quite common. Symptomatic AF is widely recognized as leading to reduced patient quality of life, functional status, and cardiac performance.[16]
>
> Importantly, one of the most common presentations for AF is stroke. In fact, AF is the heart condition that most commonly causes stroke.[17] Because the atria are fibrillating and not contracting, the flow of blood in the atria can become sluggish, especially in certain parts of the left atrium. This blood can coagulate leading to the formation of a clot. If a clot is dislodged and pumped out of the heart to the brain, it can cause a stroke. As a result, many patients with AF are treated by physicians for stroke using medicines that prevent the blood from clotting easily. Since these medicines can lead to the side effect of bleeding, a consequence is that physicians sometimes occasionally need to treat patients for a side effect of the stroke treatment itself.
>
> In general, younger patients tend to have more "palpitations symptoms," which cause them to seek medical care. Older patients tend to have few (or no) symptoms of palpitations, but may be more compromised by fatigue. Patients with pre-existing cardiac disease, such as heart failure, in which the heart does not function well at baseline, can become severely ill if they develop AF, sometimes resulting in the need for acute hospital care.

Clinical outcomes

Importantly, clinical outcomes are different from symptoms. Outcomes generally refer to hard data points associated with a disease that can be measured. The two most important types of clinical outcomes to consider are morbidity and mortality. Morbidity refers to the severity of the disease and its associated complications. Measures of morbidity may be evaluated using quality of life questionnaires, or they can be assessed by more specific endpoints such as distance walked in six minutes, hospital admissions, or a clinical event that does not cause immediate death (e.g., stroke, heart attack). Mortality refers to the death rate associated with a disease. Clinical outcomes are particularly important as they often serve as endpoints for clinical trials since they can be assessed more easily and objectively than symptoms and have a direct impact on cost.

In the AF example, key clinical outcomes to address are the morbidities associated with AF, their likelihood of occurrence, and what factors have the greatest impact on morbidities (e.g., age).

> **Working example: Clinical outcomes associated with atrial fibrillation**
>
> **Morbidities**
>
> In addition to its lifestyle implications, AF can lead to severe complications in patients, including a four- to five-fold increase in the risk of stroke.[18] The risk of stroke due to AF increases with age, rising from 1.5 percent for patients in their 50s to 23.5 percent for those in their 80s.[19] Overall, the annual risk of stroke in patients with AF ranges from 3 to 8 percent per year, depending on associated stroke risk factors[20] – a rate that is from two to seven times higher than the rate of stroke in patients without AF. Of all the strokes in the United States each year, more than 15 percent occur in people with AF.[21]
>
> Beyond the risk of stroke, AF is widely believed to reduce the heart's pumping capability by as much as 20 to 30 percent. As a result, AF (combined with a rapid heart rate over a sustained period of time) can lead to congestive heart failure (CHF). More directly, patients with existing heart failure often decompensate when they develop AF, requiring acute hospitalization.
>
> **Mortality**
>
> AF is also associated with an increased risk of death. According to the Framingham Heart Study, AF leads to a doubling of mortality in both sexes. After making adjustments for comorbidities, the risk remains 1.5 times higher in patients with AF. This increased rate of mortality is mainly due to strokes, progressive ventricular dysfunction and heart failure, and increased mortality from coronary events.[22]

Epidemiology

Effective epidemiology data are detailed and specific. Among other uses, they serve as background information for market analysis later in the biodesign innovation process (see 2.4 Market Analysis). Innovators should include data for the disease as a whole, as well as the most relevant patient subsegments. They should also try to find information about disease dynamics, such as growth rate, to illustrate how the disease will impact society in the future.

A thorough assessment of epidemiology will address the *incidence* of a disease, which is the rate at which it occurs (i.e., number of new cases diagnosed per year). It will also include *prevalence* data, or a measurement of all people afflicted with the disease at a given point in time.

For the AF example, one should start by capturing incidence and prevalence data for the overall disease state and its most meaningful subgroups (e.g., paroxysmal AF). It would also be important to understand how incidence and prevalence rates are changing.

Working example: Epidemiology of atrial fibrillation

Incidence

AF is the most common chronic cardiac dysrhythmia. More than 160,000 new cases of AF are diagnosed each year.[23] According to a recent study, the lifetime risk of developing AF is one in four for people over 40 years of age.[24]

Prevalence

In a 2001 study, the prevalence of AF was 0.01 percent among adults younger than 55 years of age and 9.0 percent in persons aged 80 years or older. Among individuals 60 years or older, 3.8 percent had AF. Overall, AF was more common in men than in women, in every age category. Among patients aged 50 years or older, the prevalence of AF was also higher in whites than in blacks.[25] Some 2.3 million individuals in North America and 4.5 million people in the European Union have paroxysmal or persistent AF.[26]

The prevalence of AF has been increasing. This is due, in part, to the aging of the general population.

Additionally, improvements in medical care are leading to increased longevity in patients with coronary artery disease, hypertension, and heart failure, which are all chronic cardiac conditions that increase an individual's risk for AF.[27] By 2050, the prevalence of AF is estimated to increase 2.5-fold, to 5.6 million people in the United States, with more than 50 percent of the affected individuals over 80 years of age.[28]

Economic impact

At this stage of the biodesign innovation process, the focus of economic research should be on the overall cost of the disease on the system at large, including the annual cost of treatment, hospitalization, and lost productivity due to absenteeism from work. Consider these costs at the system level, not necessarily for individual treatment alternatives. More detailed economic analysis will be performed as part of 2.2 Treatment Options, 2.4 Market Analysis, and 4.3 Reimbursement Basics.

Be diligent in trying to understand the distribution of costs. Is the primary expenditure for medication or for device treatment? Is it a result of hospitalization for symptoms, or is care mainly provided in the outpatient setting (which can be more cost effective)? These issues may point to opportunities as the innovator moves through the needs screening research process, since needs that reduce cost in some way often represent the greatest opportunities.

For AF, one might look at the aggregate, system-level cost of AF on an annual basis, the treatment-related annual cost of AF, the annual cost of hospitalization, and the annual cost of lost productivity from absenteeism due to AF.

Working example: Economic cost of atrial fibrillation

According to a 2005 article, approximately 350,000 hospitalizations, 7 million office visits, and 542,000 emergency department visits in the United States were attributable to AF annually (based on a retrospective analysis of three federally funded US databases).[29]

Patients with AF tend to have high healthcare resource utilization rates. For example, approximately

> **Working example:** *Continued*
>
> two-thirds of all emergency department visits with a primary diagnosis of AF result in hospital admissions, and 35 percent of all hospital admissions for arrhythmia are attributable to AF.[30] As the demographic composition of the United States becomes older, the costs associated with AF are expected to increase significantly.
>
> A 2005 study of privately insured patients in the United States under 65 years of age sought to quantify the direct and indirect costs of AF, from the perspectives of both the private third-party payer and the employer. Costs for individuals over 65 were excluded because Medicare claims data were unavailable (see 4.3 Reimbursement Basics for a description of the Medicare program and how Medicare claims data can support economic studies). Direct costs included items such as medical care and prescription drugs. Indirect costs included issues such as lost productivity due to time away from work. For private third-party payers, the average annual direct cost of a patient with AF was estimated to be $15,553 – more than five times the cost of a non-AF patient ($2,792). Of the $15,553 average direct medical cost of AF, $13,749 was attributable to medical services costs. Inpatient hospital costs accounted for the largest portion of the medical service costs (61.7 percent or $8,486). Outpatient costs accounted for 33.6 percent ($4,622) of medical services costs. In addition to medical services costs, $1,804 was spent on drugs. After controlling for coexisting conditions and other patient characteristics, AF was found to impose an average direct cost burden of $12,349 annually per patient. The excess annual total cost from the employer's perspective after adjusting for coexisting conditions and demographic factors was $14,875.[31]

Tips for performing disease state research

When performing disease state research, there are innumerable resources. To increase the credibility of the end results, however, it is best to give priority to professional medical resources (textbooks, **peer-reviewed** medical journals, and websites targeted toward physicians and backed by accredited medical institutions). Peer-reviewed medical journals are usually the most up-to-date resource; however, not all medical journals are equal – just because an article is published in a medical journal does not make it fact. The higher the quality of the medical journal, the more likely the research design, process, and conclusions are accurate. One method to evaluate the credibility of a journal is to review its "impact factor." The impact factor, a measure of the number of times a journal is cited in other accredited publications within a year, is often used as an indicator of the importance of the journal to its field.[32] A "citation impact" can be used to measure the significance of an individual work or author.[33]

In terms of approaching the research itself, it can be valuable to start with a series of general searches (e.g., on "atrial fibrillation") using the sources outlined in the Getting Started section of this chapter. Then, specific gaps in available information can be addressed through additional, more specific data searches (e.g., annual cost of hospitalization for atrial fibrillation) until the assessment is complete. In addition to moving from general to more specific inquiries, another helpful approach for completing disease state research is to look up references cited in some of the most informative documents and find and review the listed papers, especially the peer-reviewed ones. Lastly, analyst reports can be invaluable for understanding hard-to-find economic impact data, which is usually important to understanding the economic impact of a disease.

Summarizing the data

When summarizing what has been learned about a disease state, the innovator should strive to keep the target audience in mind. Write the overview in an appropriate manner (i.e., not too technical if intended for potential investors, but adequately scientific if targeted to clinicians). Additionally, be sure to cite the sources of all statistics, study results, and clinical outcomes, as well as the source of interviews with physicians or other experts. Unless this information is cited, the credibility of the research is subject to question. If conflicting information is uncovered during the research process, give priority to data from peer-reviewed medical journals or other similar resources, as noted above.

Regardless of whether or not the innovator has any sort of medical background, an effective assessment of disease state fundamentals can be developed for any given disease state following the steps outlined below.

See **ebiodesign.org** for active web links to the resources listed below, additional references, content updates, video FAQs, and other relevant information.

1. Assess anatomy and physiology

1.1 What to cover – Describe the normal anatomy and physiology of the affected organ(s) and/or system(s).

1.2 Where to look

- **Harrison's online** – Contains the complete contents of *Harrison's Principles of Internal Medicine.*
- **Medical references** – Guyton's *Textbook of Medical Physiology* or another standard medical textbook will provide valuable, credible, and detailed information regarding physiology. References such as Netter's *Atlas of Human Anatomy* or Gray's *Anatomy of the Human Body* are important anatomy references.
- **eMedicine** – Clinical knowledgebase with more than 10,000 physician authors and editors.

2. Understand the pathophysiology of the disease

2.1 What to cover – Address disease function, causal factors, and disease progression.

2.2 Where to look – In addition to the previously listed references, use:

- **Harrison's online**
- **PubMed** – A database of the US National Library of Medicine that includes more than 16 million citations from MEDLINE and other life science journals back to the 1950s. General reviews will be particularly helpful in understanding pathophysiology (locate Reviews by selecting this type of article under the "Limits" tab from the main page of the site before beginning a search).
- **Medical references** – Robbins and Cotran's *Pathologic Basis for Disease* is a useful resource for understanding pathology and pathophysiology (it is available online for a fee).
- **UpToDate** – A database of evidence-based clinical information.

3. Understand clinical presentation

3.1 What to cover – Profile the patient state associated with a disease.

3.2 Where to look – The *New England Journal of Medicine* (NEJM) and the *Journal of the American Medical Association* (JAMA) often have clinically relevant review articles that may provide an up-to-date summary of a disease state and its clinical presentation. Physician and/or patient interviews are another excellent source of information. More basic information can be found on patient advocacy group websites and healthcare company websites.

4. Assess clinical outcomes

4.1 What to cover – Elaborate on the morbidity and mortality rates associated with the disease.

4.2 Where to look

- **Harrison's online**
- **UpToDate**
- **eMedicine**
- **PubMed** – Clinical trial outcomes will be helpful (locate Randomized Controlled Trials [RCTs] by selecting this type of article under the "Limits" tab from the main page of the site before beginning the search).

5. Gather epidemiology data

5.1 What to cover – Outline the incidence and prevalence of the disease, as well as dynamics in the area.

GETTING STARTED

5.2 Where to look
- **PubMed** – See reviews and/or other journal articles.
- **Harrison's online**
- **US Government Resources** – Centers for Disease Control and Prevention – Part of the Department of Health and Human Services (HHS); National Center for Health Statistics – The principal health statistics agency in the US
- **Wrong Diagnosis.com** – A disease and symptom research center.
- **Other** – The websites of various associations (e.g., the American Heart Association, Heart Rhythm Society, American Stroke Association) provide general information.

6. Evaluate the economic impact

6.1 What to cover – Determine the overall cost of the disease on the system at large.

6.2 Where to look
- **PubMed** (Reviews and/or other journal articles).
- **Medscape** – Database of clinical information targeted at physicians.
- **MEPS Data** – The Agency for Healthcare Research and Quality provides data on the health expenditures of 18,000 US households via the Medical Expenditures Panel Survey (MEPS). This data is publicly available for primary analysis.

7. Assess and summarize the information

7.1 What to cover – Capture the most important information gathered through the disease state research and compile it into an overview suitable for the target audience.

7.2 Where to look – Refer back to the resources listed above if gaps are identified or questions arise when developing the summary.

Credits

The editors would like to acknowledge Darin Buxbaum and Steve Fair for their help in developing this chapter. Many thanks also go to Robert Croce for his assistance with the case example.

Notes

1. "Heart Attack and Angina Statistics," American Heart Association, http://www.americanheart.org/presenter.jhtml?identifier=4591 (January 4, 2006).
2. "U.S. Medical Device Market Outlook," Frost & Sullivan, 2008.
3. In coronary artery bypass surgery, a patient's chest is opened and an artery or vein from another part of the body is attached to the diseased artery such that it allows blood to flow around, or bypass, the narrowing or blockage.
4. Matthew Dodds, Efren Kamen, Jit Soon Lim, "DES Outlook: Adding the Wild Cards to the Mix," *Citigroup*, September 27, 2005, p. 4.
5. All quotations are from interviews conducted by the authors, unless otherwise cited. Reprinted with permission.
6. Robert J. Applegate, "Drug-Eluting Stents: The Final Answer to Restenosis?" Wake Forest University Medical Center, http://www1.wfubmc.edu/articles/CME+Drug-eluting+Stents (January 4, 2006).
7. Sinus rhythm is the term used to refer to the normal beating of the heart.
8. Purkinje fibers form a network in the ventricular walls and rapidly conduct electric impulses to allow the synchronized contraction of the ventricles.
9. Michel Haissaguerre, MD, Pierre Jais, MD, Dipen Shaw, MD, Atsushi Takahashi, MD, *et al.*, "Spontaneous Initiation of Atrial Fibrillation by Ectopic Beats Originating in the Pulmonary Veins," *New England Journal of Medicine* (September 3, 1998): 659.
10. "What is Atrial Fibrillation?" Cleveland Clinic: Heart and Vascular Institute, http://www.clevelandclinic.org/heartcenter/pub/atrial_fibrillation/afib.htm (October 2, 2006).

11. Ibid.
12. In a "lone" episode, AF occurs in a heart that seems to be otherwise structurally and functionally normal.
13. Cardioversion is a method to restore a rapid heart beat back to normal.
14. Maurits A. Allessie, MD, PhD, *et al.*, "Pathophysiology and Prevention of Atrial Fibrillation," *Circulation* (2001): 769.
15. Thyrotoxicosis, or hyperthyroidism, is a condition in which the thyroid gland produces excess thyroid hormone (thyroxine), which affects the whole body.
16. Alan S. Go, Elaine M. Hylek, Kathleen A. Phillips, *et al.*, "Prevalence of Diagnosed Atrial Fibrillation in Adults," *JAMA*, Vol 285, No. 18 (May 9, 2001): 2370.
17. "Advanced Imaging Can ID More Causes of Stroke Before They Strike," *Science Daily*, March 22, 2007, http://www.sciencedaily.com/releases/2007/03/070320084738.htm (December 18, 2007).
18. Gregory Y. H. Yip, Hung Fat-Tse, "Management of Atrial Fibrillation," *The Lancet* (2007): 612.
19. "ACC/AHA/ESC 2006 Guidelines for the Management of Patients with Atrial Fibrillation – Executive Summary," American College of Cardiology Foundation, the American Heart Association, and the European Society of Cardiology, 2006, p. 866.
20. Ibid.
21. "Atrial Fibrillation," American Heart Association, http://www.americanheart.org/presenter.jhtml?identifier=4451 (January 3, 2008).
22. Lloyd-Jones, Wang, Leip, *et al.*, "Lifetime Risk for Development of Atrial Fibrillation," *Circulation* (2004): 1042–1046.
23. J. Crayton Pruitt, Robert R. Lazzara, Gary H. Dworkin, Vinay Badhwar, Carol Kuma, George Ebra, "Totally Endoscopic Ablation of Lone Atrial Fibrillation: Initial Clinical Experience," *Annals of Thoracic Surgery* (2006): 1325–1331.
24. Lloyd-Jones, Wang, Leip, loc. cit.
25. Go, Hylek, Phillips, op. cit., pp. 2370–2373.
26. "ACC/AHA/ESC 2006 Guidelines for the Management of Patients with Atrial Fibrillation – Executive Summary," op. cit., p. 865.
27. J. S. Steinberg, "Atrial Fibrillation: An Emerging Epidemic?" *Heart* (2004): 239.
28. Go, Hylek, Phillips, loc. cit.
29. Peter Zimetbaum, MD, "An Argument for Maintenance of Sinus Rhythm in Patients with Atrial Fibrillation," *Circulation* (2005): 3146.
30. Eric Q. Wu, Howard Birnbaum, Milena Mareva, *et al.*, "Economic Burden and Co-Morbidities of Atrial Fibrillation in a Privately Insured Population," *Current Medical Research and Opinion*, October 2005, pp. 1693–1699. Abstract available via PubMed at http://www.ncbi.nlm.nih.gov/pubmed/16238910 (November 19, 2008).
31. Ibid.
32. "Impact Factor," Wikipedia.org, http://en.wikipedia.org/wiki/Impact_factor (November 5, 2008).
33. "Citation Impact," Wikipedia.org, http://en.wikipedia.org/wiki/Citation_impact (November 5, 2008).
34. "Epidemiology," Dictionary.com, http://dictionary.reference.com/search?r=2&q=epidemiology (December 11, 2007).

2.2 Treatment Options

Introduction

The innovator establishes a thorough understanding of the disease state and then dives into amassing detailed knowledge about existing treatment options in the field. An important gap in the treatment landscape is uncovered and the development of a solution ensues. Everything is going well until the innovator gets wind of a new treatment that has just begun testing in humans and has the potential to render the new solution obsolete. If only the innovator had included emerging treatments as part of the completed treatment research, s/he could have potentially modified the course of action to prevent a conflict.

The goal of any treatment is to improve outcomes in those patients with a disease or disorder. Treatment analysis involves detailed research to understand what established and emerging therapies exist, how and when they are used, how and why they work, their effectiveness, and their economics. Through the process of investigating treatments and creating a comprehensive profile of how a condition is typically addressed, areas for improvement or gaps in available therapies may become apparent. This analysis also helps provide the innovator with an understanding of the clinical and patient-related requirements that any new treatment must meet to be equivalent or superior to existing alternatives. It further establishes a baseline of knowledge against which the uniqueness and other merits of new solutions can be evaluated.

OBJECTIVES

- Appreciate the value of understanding treatment options for a given disease state.
- Know how to effectively search for and summarize treatment option information into a useful format.
- Understand how to perform a gap analysis that can lead to the identification of opportunities within the treatment landscape.

Treatment option fundamentals

Treatment options can only be evaluated after gaining a working knowledge of disease state fundamentals (see Chapter 2.1). As with disease state analysis, research of treatment options can be performed when identifying a clinical need or as part of validating an established need.

The primary goal of treatment research is to gain an understanding of potential gaps that may exist between a compelling, true clinical need and existing or emerging solutions. Large gaps that exist in fields where current treatments are sorely inadequate can lead to the development of unique solutions and innovative new approaches. In contrast, smaller **treatment gaps** are more likely to lead to modifications and incremental improvements to established tools and practices. In either scenario, treatment research provides the platform for better understanding patient-related requirements that will be a vital component of the need specification.

Types of treatments to consider

It is critical for an innovator to perform a comprehensive search of treatments and not overlook any therapies that may be outside his or her technological expertise or area of interest. For example, while one's focus may be on opportunities that involve devices, treatment research must include all therapies, irrespective of their technology platform. It is also imperative for the innovator to evaluate existing technologies, but also to be resourceful in gaining information about emerging technologies. A full range of treatments should be considered in treatment analysis as shown in Table 2.2.1.

Other treatment considerations

In addition to being exhaustive about the types of treatment options associated with a given disease state, an innovator should carefully consider what type of provider delivers each therapy. The mapping of different treatments to their practitioners will be a useful input to stakeholder analysis (see 2.3 Stakeholder Analysis). It also provides a construct to help one think about the capabilities of each physician type. The skills necessitated by each treatment can then validate (or call into question) some of the need criteria that have been defined. For instance, colonoscopy requires dexterity and finesse on the part of the physician to avoid patient discomfort. Gastroenterologists who routinely perform this procedure are more likely to have the skills required for other delicate or sensitive procedures using this approach, while other types of physicians might be less suited to deliver new treatments of this nature.

The skills or requirements that each therapy places on patients should also be taken into account. For example, orthopedic surgical procedures are usually associated with rehabilitation in order to optimize the surgical result. Clearly, compliance varies among different patient populations based on age, functional status, and even economic background. These issues are often tied to the likelihood of procedural success or follow-up complications. As a result, the innovator must think about the patient population indicated by

Table 2.2.1 *Be careful not to overlook any treatment types during treatment research.*

Type of treatment	Description
Behavioral and lifestyle modifications	Treatment alternatives such as diet and exercise.
Pharmacologic or biologic therapy	Small molecules or biologic agents; usually injected or orally delivered.
Energy-based therapies	Radiation or ultrasound therapies; may be used in conjunction with a particular delivery method.
Percutaneous treatments	Therapies administered through a catheter.
Minimally invasive treatments	Procedures performed using small incisions (e.g., laparoscopic procedures to access the abdomen or a joint; pacemaker placement).
Open surgery	Procedures that require the surgeon to cut larger areas of skin and tissues to gain direct access to the structures or organs involved.

Table 2.2.2 *Several different types of gaps can exist in the treatment landscape, each leading to different types of opportunities.*

Treatment gaps	Description
Gap between the desired treatment outcomes and the outcomes achieved by existing or emerging treatments	As treatment options are researched and evaluated, think about ways in which they fail to meet the desired outcomes defined in the need criteria and what may be the cause of their shortcomings. For instance, gaps may be caused by applying an inappropriate treatment to a clinical need (as often happens when developed technologies are adapted to unrelated problems). Or, they may stem from solutions that embody the correct approach, but need further development or specific refinements to achieve improved results.
Gap between a particular stage (or subtype) of a disease and existing/emerging solutions	Diseases and disorders often have different manifestations depending on the stage or severity of the condition. Particular subgroups, such as the elderly or patients with certain comorbidities (e.g., diabetes) may also respond differently to differing treatments. Accordingly, treatment gaps may exist for specific patient subtypes and/or stages of a disease.
Gap between existing technology platforms applied to treat patients with this disease state and existing/emerging platforms that have been applied elsewhere but not to this disease	This type of gap may be exemplified by thinking of a pharmacotherapy that could be introduced to address a disease whose treatment is currently dominated by open or laparoscopic surgery. If the innovator can identify this "white space" gap and conceptualize ways to align an unrelated technology with the need, the technological know-how that exists in one field can be used to create opportunities in another field.

the need and how the various treatments in the disease area are likely to align with their interests, motivations, and capabilities.

Finding gaps to identify treatment opportunities

The desired outcome of treatment analysis is to find a gap (or gaps) in the treatment landscape that represents an opportunity to address the stated need. There are at least three types of gaps to consider, as shown in Table 2.2.2.

When performing a treatment gap analysis, it is important to think about how the treatment landscape will look several years in the future, not just how it appears today. Treatments represent a moving target. If it takes three to seven years to design, develop, and commercialize a medical device (and more than ten years to bring a new drug to market), the innovator must take into account the incremental improvements to existing technologies that will occur within this time frame, as well as the implications of new breakthroughs, and then evaluate these projections against the defined need criteria. Without considering this temporal aspect to the improvement of existing treatments and the development of new treatment options, an innovator may focus on the gap, not realizing that it was already diminishing or even on the path to closing altogether.

Just as looking forward is important, a great deal can be learned from looking backward. If there appears to be a major gap in the treatment landscape, an innovator would be well served to do some research on therapies that sought to address that gap but failed. Studying the unsuccessful experiments of other innovators can highlight important pitfalls, risks to be avoided, and fundamental learnings, which can be leveraged to accelerate efforts moving forward. The example below, focused on The Foundry and Emphasys Medical, provides one example of this approach.

FROM THE FIELD: THE FOUNDRY AND EMPHASYS MEDICAL

Understanding treatment options as part of the needs screening process

The Foundry, a medical device incubator that helps inventors "rapidly transform their concepts into companies,"[1] has learned that early in the biodesign innovation process a thorough examination of all available options for treating a disease can spark ideas that others may have missed.

In the late 1990s, Hanson Gifford, CEO of The Foundry, and Mark Deem, its chief technology officer, were interested in pursuing opportunities related to the treatment of emphysema. As part of their early research, they performed a detailed assessment of established and emerging treatment alternatives in the space. At the time, Gifford pointed out, "The only real treatment for emphysema was lung volume reduction surgery (LVRS)."[2] Lung volume reduction surgery is a highly invasive procedure in which the surgeon opens a patient's chest and removes roughly 30 percent of the diseased tissue in order to increase the flow of oxygen to the remainder of the lungs. The operation was difficult to perform and extremely painful for the patient. Moreover, the associated mortality rate ranged from 6 to 10 percent, among the highest for any elective procedure.[3] "We spent a lot of time trying to understand how we might be able to do the surgery better. We did extensive literature research, talked with many surgeons and pulmonologists, and really explored the disease state, including its natural history, the cellular degradation of the lungs, and so on," explained Gifford. Besides disease state analysis, they also researched existing and emerging treatment options by cold-calling inventors, companies, and experts working in the field and networking with other entrepreneurs. Through this process, they realized that emerging technologies were predominantly focused on incrementally improving LVRS, and almost no research was being done to look for an alternative, non-surgical procedure.

Recognizing that there was a huge gap in the treatment landscape for emphysema (and driven by a desire to develop a less painful and invasive option for patients), Gifford and Deem decided to refocus their treatment research on non-surgical alternatives. "One of our major breakthroughs came when we decided that no matter what kind of a solution we came up with, what we really needed to do was make this a non-surgical procedure," recalled Gifford. "Everybody was looking at surgery, which became just this little corner of the treatment landscape for us," added Deem. "There was this huge area outside of surgery where nobody was working that was full of potential opportunities."

The absence of other innovators or companies pursuing non-surgical treatments for emphysema left the field wide open, but it also created a host of difficult challenges that they would potentially face in pioneering a new procedure and/or device. Their research had shown that pulmonologists (physicians specializing in the lungs) saw emphysema patients on a regular basis, but it was surgeons who performed LVRS. In order for a solution to be non-surgical, pulmonologists would have to be able to deliver treatment. At the time, "Pulmonologists didn't really perform therapeutic procedures," said Deem. "In some ways, we would have to develop a new field of medicine, or at least expand a traditional specialty to make a non-surgical treatment for emphysema possible." To evaluate the feasibility of such a shift, they invested significant effort in examining the current referral patterns among doctors, how equipment was procured and funded in hospitals, and whether pulmonologists had the right skills, resources, and physical space to perform therapeutic procedures.

They also evaluated economics in the treatment area. "We took a fairly broad brush toward the financials at that point," said Gifford. "We looked at the overall cost to the doctor, the hospital, and the healthcare system for comparable procedures because – right or wrong – if these entities are already used to paying 'X' dollars for the treatment of a disease, you're more likely to be able to get that same amount." He added, "LVRS cost in the range of $20,000 to $30,000. If we could come up with a therapeutic procedure that only costs a few thousand dollars, then there would be room for a reasonably well-priced device."

Despite the many challenges in the field, Gifford and Deem saw the potential for a breakthrough product. There are 3.1 million patients diagnosed with emphysema in the United States alone.[4] While a relatively small number of LVRS procedures are performed each year, the market for a less invasive treatment alternative would be significant, particularly given the mortality rates associated with the current LVRS procedure and the pain and suffering to patients associated with both the surgery and the disease. According to Deem, "The standard line that physicians would hear from emphysema patients was, 'Cure me or

> **FROM THE FIELD**
>
> **THE FOUNDRY AND EMPHASYS MEDICAL**
>
> *Continued*
>
> kill me. I don't care which one, but I can't stand being perpetually short of breath.'"
>
> Ultimately, The Foundry decided to move forward with this project, collaborating with John McCutcheon and Tony Fields to found Emphasys Medical in 2000. Emphasys developed a minimally invasive procedure utilizing a removable valve that could be inserted into a patient in as little as 20 minutes. See Figure 2.2.1. The valves could control air flow in and out of the diseased portions of the lungs to help healthier portions function normally. The concept was to collapse the diseased portions of the lungs without having to remove the tissue surgically. Clinical trials have now confirmed safety and efficacy of the valves in over 700 patients.
>
>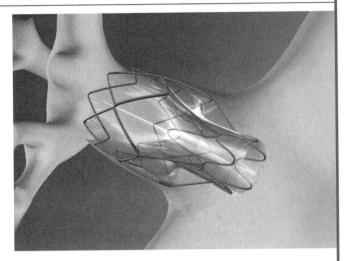
>
> FIGURE 2.2.1
> The Emphasys device (courtesy of Emphasys Medical).

Approach to treatment analysis

A complete analysis of treatment options should include information in the core areas outlined below.

- **Overview of treatment options** – Provide a high-level description of relevant treatments in the field, including a summary of each treatment, the common clinical approach, and the skills required by the user. This provides a foundation for both gap analysis and the refinement of needs criteria.
- **Clinical treatment profile** – Describe the clinical rationale for why and when each treatment is used. In particular, outline the clinical mechanism of action, clinical evidence on safety and effectiveness, indications, and patient segments. This information can be useful for performing gap analysis and refining needs criteria, but ultimately provides evidence of technical feasibility for new solutions that are similar to existing and emerging ones.
- **Economic treatment profile** – Outline the cost associated with each treatment, including both direct and ancillary costs and who or what entity is incurring them. This information is useful in performing a market analysis of the treatment space (see 2.4 Market Analysis).
- **Utilization treatment profile** – Describe how each treatment is used in clinical practice, by whom, and where. Utilization is another important component of the market analysis.
- **Emerging treatment profile** – Capture new products, procedures, and other treatments likely to affect the treatment landscape within the next three to ten years. Diligent research of possible emerging technologies helps prepare the gap analysis, refine the needs criteria, and provide support for the technical and clinical feasibility of potentially new solutions.
- **Summary of the treatment landscape** – Develop an assessment of potential opportunities within the current and emerging treatment environment, focusing on the gap between current/emerging treatments and the defined need criteria.

The remainder of this chapter examines each of these areas, using atrial fibrillation (AF) as an example. Detailed analysis is provided for one existing therapy (pulmonary vein isolation) and one emerging therapy (left atrial appendage occlusion). To better understand the example, refer back to the information given in 2.1 Disease State Fundamentals as it provides an overview of AF, which is relevant to understanding treatment options. The focus on only two sample treatments is for pedagogical purposes only. In reality, the innovator must examine all available treatment options within his/her area of interest.

As with disease state analysis, the appropriate level of detail for assessing treatment options varies significantly based on the number of treatment options that exist within a particular field of study, as well as the number of different needs being evaluated. Remember to take an iterative approach, exploring each treatment in progressively more depth as one's direction becomes increasingly clear. This analysis should also be revisited as additional information is gathered in other steps of the biodesign innovation process (such as those obtained as part of 4.3 Reimbursement Basics). The example below has been prepared at a relatively high level, as it is for illustrative purposes only.

Overview of treatment options

The first step in performing treatment analysis is to investigate and summarize the established treatments for the disease. Developing an overview of the treatment landscape may be best accomplished by categorizing the options based on common features. These common features are usually unique to the particular clinical need area, but may include patient populations, technology platforms, or even mechanism of action. Remember not to limit this analysis to any specific solution, but to include everything from lifestyle modification to open surgical therapies. For each one, explain the objective(s) of the treatment, the common clinical approach (e.g., are there progressive steps to treatment?), the person delivering treatment, and the skills required by the user to provide this treatment. Organize the information in such a way that it provides a complete sense of the treatment alternatives that are currently used in practice. It is also important to elaborate on the relative strengths and weaknesses of each approach.

Working example: Overview of treatment options for atrial fibrillation

Atrial fibrillation is a disease in which an irregular, typically rapid, and chaotic heart rhythm replaces sinus rhythm (normal heart rhythm) leading to a variety of symptoms as well as blood clot formation that can result in a stroke. Available treatments for AF typically seek to accomplish one of three different objectives: (1) the restoration and maintenance of normal sinus rhythm, (2) the control of the ventricular rate, or (3) the reduction of the risk of forming and dislodging a clot, or thromboembolic risk (to be explained in more detail below). Therefore, it is natural to classify the different treatments into one of three categories according to their objective. Within each of these general categories, a variety of pharmacological, surgical, and device therapies are utilized. Typically, a pharmacological approach is used first for treating AF, along with lifestyle changes such as limiting the intake of alcohol and caffeine. When medications do not work, are not tolerated, or lose their effectiveness over time, surgical and/or device therapy may be required.[5] These treatment alternatives may also be made available to patients when medicines that prevent blood clot formation, or anticoagulants, cannot be taken or when blood clots, which can cause strokes, do occur.[6]

Restoration and maintenance of sinus rhythm

One treatment strategy for patients with AF, termed rhythm control, is to reestablish and maintain normal sinus rhythm since this can improve symptoms, correct atrial function and structure, reduce the risk of blood clot formation (thereby reducing stroke risk), and potentially reduce the need for long-term treatment with anticoagulants. Furthermore, regardless of whether underlying heart disease is present in a patient, restoring sinus rhythm is associated with improved oxygen utilization, lifestyle improvements, and increased exercise capacity.[7]

The primary advantages of rhythm control treatments are improved cardiac function and reduced symptoms in some patients. While a rhythm control strategy has a greater probability of reducing the frequency of AF, it may not eliminate it altogether. As a result, most patients using rhythm control treatments are still required to take anticoagulants on a long-term basis to reduce their risk of blood clots and strokes.

Cardioversion and pharmacologic rhythm maintenance

The most common way to return the heart to sinus rhythm is through direct current cardioversion – the restoration of the heartbeat to normal function by electrical countershock. Another approach is chemical

Working example: *Continued*

cardioversion, which involves the use of antiarrhythmic drugs to convert the heart rhythm to normal.

Regardless of whether an electrical or pharmacological approach to cardioversion is used, patients are generally required to take anticoagulants for some period after the cardioversion to prevent blood clots from forming while the atria recover from the stunning of the cardioversion procedure. Some patients are also required to take such medication before the cardioversion to reduce the risk of stroke due to the cardioversion itself. This is separate from the need for long-term treatment with anticoagulants due to the possibility of AF recurrence.

The relatively high rate of AF recurrence using a cardioversion and antiarrhythmic medication strategy for rhythm control is one of the key disadvantages of this treatment approach. After successfully being returned to a normal sinus rhythm, only 20 to 30 percent of patients remain in sinus rhythm after one year. Although this percentage can be increased to 40 to 80 percent through the sustained use of antiarrhythmic drugs,[8] medications that affect the electrical properties of cells in the heart to help prevent the occurrence of AF, the overall risk of recurrence remains significant. Antiarrhythmic medications also have serious potential side effects, including the development of new, abnormal heart rhythms.[9]

Catheter-based pulmonary vein isolation

Catheter ablation is a minimally invasive procedure used to terminate AF by eliminating and/or "disconnecting" the pathways supporting the initiation and maintenance of AF. Catheters introduced into the heart direct energy to destroy tissue in specific areas (mostly located in or around the pulmonary veins) that are the source of AF, and to prevent it from initiating or conducting any type of electrical impulse to the rest of the heart, thereby allowing normal sinus rhythm to continue. While there are several variations of catheter ablation procedures currently in use to address AF, they all seek to either electrically isolate or eliminate the pulmonary vein triggers of AF and/or eliminate any other areas in the left or right atrium capable of initiating or maintaining AF. For the sake of simplicity, all of these varied procedures will be categorized together under the term "pulmonary vein isolation" in this example.

Cox-Maze surgery

The primary surgical approach to restoring sinus rhythm in patients with AF is the Cox-Maze procedure, during which a series of precise incisions in the right and left atria are made to interrupt the conduction of abnormal electrical impulses and to direct normal sinus impulses to the atrioventricular (AV) node, as in normal heart function.[10] Because this procedure is very invasive and complex, it is usually offered only to patients with a high risk of stroke and who are already undergoing another form of cardiac surgery.

Implantable atrial defibrillators

Another method of restoring sinus rhythm is through an implantable atrial defibrillator. This device delivers small electrical shocks via leads placed in the heart to convert abnormal rhythms to sinus rhythm. Patients can turn them on and off to treat AF when episodes occur, or they can be set to operate automatically. The device is inserted by a cardiologist in a cardiac catheterization laboratory using X-ray guidance. The primary limitations of atrial defibrillators are their relatively large size and, more importantly, the fact that the shocks they deliver can be quite painful. As such, they are not currently used to treat AF on a widespread basis.

Ventricular rate control

In patients who receive ventricular rate control treatments, no attempt is made to cure or eliminate AF. Rather, these treatment alternatives are focused on slowing the conduction of electrical impulses through the AV node, the part of the heart's conduction system through which all impulses from the atria typically need to pass before activating the ventricles. By controlling the rate at this junction point, the ventricular heart rate can be brought back into the normal range and thereby potentially mitigate symptoms due to the rapid pumping of the heart.[11] Because AF continues, anticoagulants are recommended to prevent blood clot formation and strokes, typically on an indefinite basis. Another disadvantage associated with these treatment options is the fact that it can be difficult to adequately control the heart rate and relieve the symptoms on a long-term basis.[12]

Pharmacological rate control

Rate therapy using various medications leverages the gatekeeper properties of the AV node to reduce the ventricular rate to 60 to 90 beats per minute during AF.

Such a change does not eliminate the irregular heartbeat but, by slowing the heart rate, reduces the workload of the heart and potentially the symptoms that a patient may be experiencing.

Catheter ablation of the AV node

Catheter ablation of the AV node is typically reserved as a last resort for AF patients who have failed other treatment options.[13] In this procedure, a catheter is inserted into the heart and energy is delivered to destroy the AV node, thereby disconnecting the electrical pathway between the atria and the ventricles. Without an atrial source to drive the contraction of the ventricles, which are responsible for pumping blood to the lungs and body, a permanent pacemaker needs to be implanted at the time of the procedure to restore and sustain regular ventricular contractions. The pacemaker is a device that sends electrical pulses to the heart muscle, causing it to contract at a regular rate. Even though the atria continue to fibrillate, the symptoms of AF are reduced in many patients.

Reduction of thromboembolic risk

Preventive treatment to reduce the risk of blood clots and strokes (thromboembolic risk) in patients with AF is an important consideration in any AF treatment regime. Anticoagulant or antiplatelet therapy medications are commonly used on both a short-term (e.g., before and after electrical cardioversion) and long-term (e.g., in conjunction with ventricular rate control treatments) basis. While anticoagulant and antiplatelet therapies carry a bleeding risk, their use is warranted in patients where the risk of thromboembolic events is greater than the risk of bleeding complications. Key risk factors for stroke in AF patients include previous history of ischemic attack or stroke, hypertension, age, diabetes, rheumatic, structural, or other heart disease, and ventricular dysfunction.[14]

Pharmacological therapy

The most common drugs used to treat the risk of thromboembolic events are aspirin and warfarin. Aspirin is an over-the-counter medication that is an antiplatelet therapy; platelets are blood constituents that play an important role in the clotting cascade. By affecting the adherence of circulating platelets to one another, aspirin can reduce the likelihood of clot formation. However, aspirin is usually only effective in AF patients who are young and who do not have any significant structural heart disease. Warfarin (Coumadin) is an anticoagulant used to prevent blood clots, strokes, and heart attacks. Warfarin is a vitamin K antagonist, which reduces the rate at which several blood clotting factors are produced. The metabolism and activity of warfarin can be affected by various other medications, foods, and physiologic states. As such, the dosage of warfarin needs to be monitored closely and usually necessitates frequent blood tests to check the degree to which it is anticoagulating a patient's blood. Over-anticoagulation with warfarin can lead to a significantly higher bleeding risk.

Clinical treatment profile

With the universe of existing treatments defined, the next step is to assess the clinical rationale for why and when each one is used. This includes researching the following areas.

- **Mechanism of action** – Review what has been learned about the disease in terms of the steps or sequence of events that occurs – also called its mechanism(s) of action (see 2.1 Disease State Fundamentals). Then identify which disease mechanisms are targeted by a given treatment option and how each treatment seeks to affect the disease.
- **Indications** – Identify the patient populations for which each treatment is indicated or contraindicated. For drugs and devices, consider the specific indications approved by the FDA or other regulatory bodies outside the United States. For any type of invasive procedure, determine whether there are specific patient segments for which the treatment is recommended and understand why.
- **Efficacy** – Define the benefits for each treatment. Ideally, these should be measurable benefits (e.g., reduction in mortality) which are best demonstrated by clinical trials.
- **Safety** – Describe the risks of treatment, including precautions and adverse reactions.

Working example: Clinical AF treatment profile: pulmonary vein isolation

Pulmonary vein isolation seeks to prevent abnormal electrical impulses that initiate and maintain AF from reaching the atria. The procedure is focused on destroying, or ablating, the abnormal "triggers" that originate in and around the pulmonary veins, or creating lesions to effectively isolate them such that they can no longer electrically communicate with the rest of the heart.

On the morning of the procedure, patients are admitted to the hospital and receive mild sedation and the injection of local anesthetic to the groin where catheters are inserted (there is little or no patient discomfort). The procedure is performed by a cardiac electrophysiologist and can take between three to six hours, depending on the number of areas treated (although estimates range from as few as two to as many as ten hours). Careful monitoring and a series of tests are completed following the procedure. If there are no complications, patients can be discharged from the hospital after approximately 24 hours. Patients generally resume their normal activities after two or three days.

Mechanism of action

During the pulmonary vein isolation procedure, multiple catheters are advanced through the blood vessels and positioned in various locations of the heart chambers. The catheters are used to electrically stimulate the heart and intentionally trigger AF, after which the catheters record and/or map the heart's electrical activity in and around the pulmonary veins and atria. Using this data, the tissue responsible for the abnormal electrical impulses causing AF can usually be identified. A different type of catheter then can be used to apply energy to destroy, or ablate, this tissue. Radio-frequency energy can be used to heat the tissue to create the lesions, or cryothermy can be used to create the lesions through freezing. Ultrasound and laser techniques are also under development.

Ablation lesions are placed at the interface between the atrial tissue and pulmonary veins to effectively create continuous encircling lesions that electrically isolate the pulmonary veins so that any abnormal impulses originating in them cannot reach the rest of the heart and initiate AF. Ablating too deep in the pulmonary veins can cause narrowing, which will cause long-term complications for the patient. The lesions heal within four to eight weeks, forming scars around the pulmonary veins. The use of this technique can "cure" AF in many patients. See Figure 2.2.2.

Indications

Various types of catheters and mapping systems approved for other indications are used by physicians to perform pulmonary vein isolation. However, because this treatment is a percutaneous procedure that is not enabled by a specific device, it is not directly subject to US Food & Drug Administration (FDA) approval. As a result, the use of different catheters to perform the procedure is technically **off-label**,[15] with no ablation catheter or device approved by the FDA specifically for the ablation of AF. Guidelines released by the American College of Cardiology, the American Heart Association, and the European Cardiology Society in August 2006 indicate that the procedure is appropriate for highly symptomatic patients with recurrent paroxysmal AF, or AF in which recurrent episodes spontaneously self-terminate after a relatively short period of time. These individuals should have failed to benefit from at least one antiarrhythmic drug and also should not have left atrial enlargement. Typically, these patients are younger (<70 years old) and do not have significant structural

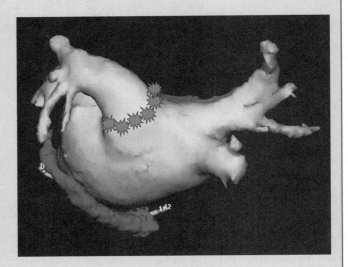

FIGURE 2.2.2
A three-dimensional image of the left atrium and pulmonary veins prior to pulmonary vein isolation (obtained using NavX™ navigation and visualization technology by St. Jude Medical). The star-shaped marks indicate the targeted region for pulmonary vein isolation; the structure beneath the left atrium illustrates the position of the coronary sinus, which wraps around the atria (courtesy of Amin Al-Ahmad and Paul Wang, Stanford University).

heart disease, which is the profile of many patients with recurrent paroxysmal AF.[16]

Based on the approach, pulmonary vein isolation is most successful for patients with AF originating in the pulmonary vein(s), although continued variations in the procedure have additionally incorporated the identification and ablation of non-pulmonary vein triggers, such as those in the atria. However, since it is nearly impossible for clinicians to determine the origin of the arrhythmia without intracardiac mapping, which is too invasive and resource intensive to use on all patients to determine whether they need subsequent pulmonary vein ablation, clinicians typically need to rely on the clinical pattern and patient profile to determine whether a patient will be a good candidate for the procedure.[17]

In terms of contraindications, more study is needed before patients with impaired left ventricular function, previous cardiac surgery, valvular heart disease, advanced age, or previous left atrial scarring are widely considered candidates for the procedure.[18] The procedure is also not yet recommended for patients who are minimally symptomatic. There is also some question about what to do with patients who have chronic or permanent AF, in which sinus rhythm cannot be restored long term. As of the 2006 guidelines, catheter ablation was not listed as one of the therapeutic options for individuals with permanent AF. The concern is that ablation techniques would be unlikely to "cure" these forms of AF for more than a year or two, given that patients with this type of AF may have more significant structural factors contributing to their AF versus mainly pulmonary vein triggers.[19]

Efficacy

There have been numerous non-controlled studies that demonstrate the efficacy of varied types of pulmonary vein isolation in patients primarily with paroxysmal AF, and there are several randomized, controlled trials comparing pulmonary vein isolation with **standard treatments**. While these studies have limitations, they all report differences in favor of an ablation strategy in terms of relevant outcomes.

Wazni et al. (2005) conducted a small, randomized, unblinded trial that compared the procedure with a rhythm-control strategy. In this study, the magnitude of benefit in reducing AF recurrence was large at one year (13 percent recurrence in pulmonary vein isolation versus 63 percent recurrence in the medical group). There was also an improvement in quality of life that was considered clinically significant.[20]

Pappone et al. (2006) conducted a larger randomized trial to compare pulmonary vein isolation with antiarrhythmic drug therapy in patients with paroxysmal AF. This study concluded that pulmonary vein isolation is more successful than drug therapy with few complications. Eighty-six percent of patients receiving pulmonary vein isolation were free from recurrent atrial tachyarrythmias versus 22 percent of patients receiving antiarrhythmic drug therapy.[21]

Stabile et al. (2006) conducted a multicenter controlled, randomized trial to investigate the adjunctive role of ablation therapy to antiarrhythmic drug therapy in preventing AF relapses in patients with paroxysmal or persistent AF in whom antiarrhythmic drug therapy had already failed. After 12 months of follow-up, 91.3 percent of control group patients had at least one AF recurrence, whereas 44.1 percent of ablation group patients had atrial arrhythmia recurrence (4 patients had atrial flutter; 26 had AF).[22]

Of the many non-randomized studies, the study by Pappone et al. (2003) was particularly important. It was a large, non-randomized study (1,171 patients) that compared pulmonary vein ablation to drug therapy and reported on a wide range of outcomes. The results showed significant benefits in survival, quality of life, and maintenance of sinus rhythm for the pulmonary vein isolation group. Specifically, the patients who had ablation had lower mortality due a lower rate of cardiovascular deaths, primarily heart failure, stroke, and sudden death. Recurrence of AF was also lower among patients receiving ablation therapy (20 percent versus 58 percent). Quality of life returned to normal within six months in patients who underwent ablation.[23]

Overall, while these studies are considered promising, more data is needed since various study limitations, differences in approach, and variations in patient populations, follow-up, and endpoints affect the generalizability of the findings. Larger trials incorporating more centers and practitioners are needed to determine if greater consistency and generalizability is achievable. They should also include measures of important clinical outcomes (survival, cardiovascular events,

> **Working example:** *Continued*
>
> complications from treatment) and evaluate adverse events. Nonetheless, given the accumulated positive data about the use of pulmonary vein isolation, the use of this technique was included as a reasonable alternative treatment for patients with symptomatic recurrent paroxysmal AF in guidelines published in 2006.
>
> Because the pulmonary vein isolation procedure is relatively new, there are also questions about whether the efficacy data will hold up long term. For example, in one promising study, pulmonary vein isolation showed an 86 percent success rate with only 7 percent of patients requiring a follow-up procedure at a mean of six months. Yet, at a mean of eight months the recurrence rate had increased to 25 percent.[24] However, another study showed a success rate of 87 percent after nearly a year of follow-up. Clearly, more data are needed to conclusively assess the efficacy of the procedure.[25]
>
> **Safety**
>
> While pulmonary vein isolation is generally considered safe when performed by experienced doctors, there are serious risks to consider, especially in patients in whom the likelihood of success may be low due to various factors such as structural heart disease. Complications from the procedure include those shown in Table 2.2.3 (based on pooled data from six studies involving more than 1,000 patients). The procedure also can result in valvular injury, esophageal injury, and proarrhythmia. Complications appear to be declining with modifications to the procedure, new technology, and greater clinician experience. Advocates for the procedure assert that it is safe and may reduce morbidity and mortality associated with medical therapy.[26] Critics acknowledge that the short-term safety of newer ablation procedures has improved, but maintain that serious life-threatening complications do exist and that long-term safety is relatively unknown.[27]
>
> Table 2.2.3 *Common complication rates for pulmonary vein isolation (from Atul Verma, Amdrea Natale, Benzy J. Padanilam, Eric N. Prystowsky, "Why Atrial Fibrillation Should be Considered for First Line Therapy,"* Circulation *(2005): 1220).*
>
> | Complication | Rate of occurrence |
> |---|---|
> | Transient ischemic stroke | 0.4 percent |
> | Permanent stroke | 0.1 percent |
> | Severe PV stenosis (greater than 70 percent, symptomatic) | 0.3 percent |
> | Moderate PV stenosis (40–70 percent, asymptomatic) | 1.3 percent |
> | Tamponade/perforation | 0.5 percent |
> | Severe vascular access complication | 0.3 percent |

Economic treatment profile

For each defined treatment, it is important to understand the financial impact at the individual level, as well as at the level of the healthcare system. The innovator should seek to identify the costs of providing the treatment (drug cost, procedural, hospital stay, etc.), as well as potential cost savings for using one particular treatment versus another. Remember to determine the total cost of a therapy, including costs for hospitalization, the surgical procedure, and the physician's fee. A long-term treatment, such as medical therapy, should be evaluated for its long-term beneficial potential for the estimated remaining lifetime of a patient versus the benefits of one-time surgical or device therapy. It is also important to take into account how cost-effective each treatment is perceived to be by the various stakeholders in the healthcare system. **Reimbursement** is another critical consideration; however, it will be evaluated in more detail later in the biodesign innovation process (see 4.3 Reimbursement Basics).

> **Working example: Economic treatment profile for pulmonary vein isolation**
>
> The cost of pulmonary vein isolation procedures varies and limited data is available. However, in a Canadian study designed to compare the cost of medical therapy to catheter ablation in patients with paroxysmal AF, the cost of the procedure was found to range from $16,278 to $21,294, with a median estimate of $18,151. Follow-up costs ranged from $1,597 to $2,132 per year.[28]
>
> In one study, Weerasooriya et al. found that the initially high cost associated with pulmonary vein isolation would offset the ongoing costs of antiarrhythmic medication in year three, assuming a 72 percent ablation cure rate after 1.5 procedures. After four years, ablation becomes more cost-effective than treatment with antiarrhythmic medication.[29]

Utilization treatment profile

Next, an innovator should seek to provide an overview of how each treatment is currently used in clinical practice. This may include when (and why) certain treatments are used and the frequency of procedures. Physician organizations disseminate best practices for treatment regimens via guideline documents. However, physicians may deviate from or modify these guidelines (as well as FDA regulations) in certain medically appropriate circumstances (e.g., prescribing drugs for off-label usage). This is particularly relevant in the area of pulmonary vein isolation devices, since there is currently no device approved for this application, yet tens of thousands of procedures are performed annually. This is accomplished by physicians using existing ablation devices approved for other forms of arrhythmia treatment on an off-label basis for the treatment of atrial fibrillation. This type of usage highlights the importance of gaining insight into how drugs, surgical procedures, and devices are used by the majority of physicians in everyday practice, not just what indications they are cleared for in the market.

> **Working example: Utilization profile for pulmonary vein isolation**
>
> According to survey estimates, the number of catheter ablation procedures performed each year is relatively large (10,199), with a sizable portion of those patients undergoing pulmonary vein isolation (6,600) instead of other procedures.[30]
>
> Traditionally, pulmonary vein isolation has been used as a therapy after a patient has failed at least one (if not two or more) antiarrhythmic drugs or has an intolerance or contraindication to antiarrhythmic therapy. As of 2006, the Centers for Medicare and Medicaid Services (**CMS**), as well as many private insurance plans (AETNA, Wellpoint), covered pulmonary vein isolation as a second-line therapy for specific patient groups. However, other sizable programs had yet to grant coverage (Blue Cross/Blue Shield). (See 2.3 Stakeholder Analysis for more information about the role of public and private insurance companies – or payers – in the medtech field.)
>
> Looking forward, a debate exists as to whether or not pulmonary vein isolation should be recommended to some patients as first-line therapy. Several small, randomized studies suggest that such a move could be reasonable, although more data are being called for by many clinicians. In the meantime, clinicians are making decisions about the medical appropriateness and necessity of the procedure as a first-line therapy based on their professional judgment and experience. In general, cardiologists are more likely to prescribe combinations of antiarrhythmic and anticoagulant drugs to patients before considering the procedure, whereas cardiac electrophysiologists might be more inclined to recommend the procedure earlier in the treatment process.

Emerging treatment profile

Clinical treatment options change over time as companies develop new products and physicians create new techniques. For some disease states, new treatment alternatives are introduced at a rapid rate and have the potential to dramatically change the treatment landscape. Researching emerging treatments provides the innovator with a working understanding of new solutions potentially on the horizon, the areas they are targeting, and their timeline for their development and/or entry into the market. Although available information may be limited, try to cover as many of the topics outlined under the clinical treatment profiles section as possible.

When investigating emerging treatments, look for leads by exploring what new treatments are being tested in clinical trials and/or being funded by investors. For each emerging treatment, understand which mechanism or symptom of the disease it addresses and how it works. Evaluate the hypothesized efficacy and safety of the treatment, as well as its anticipated time to market.

Working example: Emerging treatment profile for left atrial appendage occluder

As much as 90 percent of the embolisms, or clots that dislodge, associated with AF originate from the left atrial appendage (LAA), a small pouch-like sac attached to the left atrium.[31] Traditionally, chronic anticoagulation therapy has been used to manage this risk; however, these drugs can present safety and tolerability problems, particularly in patients older than 75 years of age (the group accounting for approximately half of AF-related stroke patients). Unfortunately, antiplatelet medications, such as aspirin, are less effective compared to drugs such as warfarin. Long-term warfarin therapy has additional downsides in that it requires costly and inconvenient patient monitoring, can have unpredictable drug and dietary interactions, and can be difficult to administer due to the frequent dosage adjustments needed to keep the risks of clotting and bleeding events appropriately balanced.

Since most clots in AF form in the LAA, one option for preventing clot formation would be to occlude or eliminate the LAA through surgical means. However, the Left Atrial Appendage Occlusion Study (LAAOS), which evaluated LAA occlusion performed at the time of coronary artery bypass grafting, showed that complete occlusion was achieved in only 45 percent of cases using sutures and in 72 percent of cases using a stapler.[32]

A new device-based approach seeks to address the shortcomings of surgery and traditional drug therapy by controlling embolisms that form in the LAA to reduce the risk of stroke. These devices generally would work by placing a filter at the mouth of the LAA where it meets the left atrium, thereby isolating or occluding the LAA from blood flow, which would prevent harmful-sized emboli that may form in this location from exiting into the blood stream and traveling to the brain, resulting in a stroke.[33]

Though several companies have been working on LAA occluders in the United States, none are currently available on the market. Appriva Medical (later acquired by ev3, Inc.) received CE Mark approval in Europe for its nitinol-based PLAATO™ (Percutaneous Left Atrial Appendage Transcatheter Occlusion) system in 2002 (see 4.2 Regulatory Basics for more information about CE marking and other forms of regulatory clearance).[34] This device is delivered directly to the LAA during a transseptal[35] catheter-based procedure. The nitinol wire cage is held in place by small hooks and its polymeric (ePTFE) covering promotes endothelialization on the atrial side such that approximately four months following implantation, the device becomes part of the atrial wall. Studies completed by the company demonstrated that, as a result, blood and thrombus did not attach to the device on the atrial side.[36] In these studies, the device was used only in patients with contraindications to warfarin and who had at least one (if not more) increased risk factors for stroke.

In 2005, however, ev3, Inc. decided to discontinue the development and commercialization of PLAATO. For three years following the device's CE Mark certification, ev3, Inc. explored regulatory pathways and trial designs for establishing the PLAATO technology in the United States, finally determining that the cost and risk of enrollment and the trial design ultimately mandated by the FDA was not commercially feasible for the company. ev3, Inc. also planned to phase out the availability of the device in Europe by the end of the year.[37]

Another device under development is the Watchman® filter system. As with the PLAATO system, the Watchman is placed using a standard transseptal approach just within the LAA opening. With the device, the cumulative risk of stroke for patients treated dropped 59 percent to approximately 1.7 percent per year (compared to the expected 4.2 percent annual rate for this population).[38] Patients receiving the Watchman are initially placed on warfarin for 45 days after the implant. After that period, 90 percent of the 60 patients treated as of 2006 have been removed from warfarin therapy.[39]

Despite the promise of these new technologies to reduce the stroke risk associated with AF, they still remain largely limited to clinical trials and warfarin continues to be the standard of care. One of the main reasons that they remain in trials is that, to be approved, the trials would require neurological endpoints, such as the occurrence of strokes. Since AF-related strokes are still relatively rare, especially if patients are taking warfarin, the overall time and number of patients required to demonstrate a difference in this clinically significant endpoint (by having enough strokes actually occur) between the device and standard therapy could be impractically high and enormously expensive. Such a requirement (imposed by the FDA in the United States) is an obstacle for many companies interested in this treatment space.

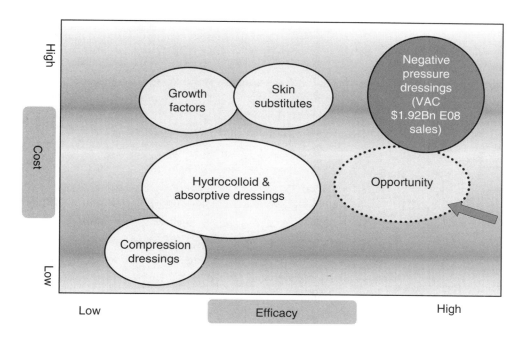

FIGURE 2.2.3
One example of a treatment cost/benefit analysis (courtesy of Spiracur, Inc.; reprinted with permission).

Treatment landscape

At this point, the innovator should synthesize all of the findings from the previous treatment research into a comprehensive framework that summarizes the treatment landscape. Such a summary should include an overview of what is known about a disease's mechanisms of action, which mechanisms are targeted by what treatment options, and which are not. This information can be used to highlight gaps in available treatment options, as well as the areas of greatest innovation opportunity. Ideas for effectively summarizing this information are outlined below.

Costs versus benefit

Cost/benefit analysis describes the trade-offs of the various treatment options relative to one another. This approach is illustrated by work of Spiracur, a medical device start-up targeting treatment of venous ulcers of the lower extremities. While not related to AF, the work of this team provides a strong example of how cost/benefit analysis can be used to summarize a treatment landscape. During its early work, the team evaluated solution efficacy versus cost, as shown in Figure 2.2.3.

This analysis led to the identification of a gap in the market for solutions with a moderate cost and efficacy equivalent to leading solutions.

Causes versus consequence

Certain treatments attack the cause of the disease while others ameliorate symptoms. With atrial fibrillation, rhythm control works to target the disease mechanism and rate control targets the improvement of symptoms associated with rapid ventricular rates. It might be helpful to draw a tree diagram, with the causes shown nearer to the trunk and symptoms branching out from them. Place treatments above the location in the diagram that they target.

Stage 2 Needs Screening

Mechanism of action frequency plot

Typically, a number of treatments may target a single mechanism of action. Based on the clinical need area, multiple mechanisms may be targets for therapy. Count the treatments in each area to see which spaces are more crowded than others. Atrial fibrillation therapies may also be illustrated with this strategy. In this case, the mechanism of action could be the treatment of primary arrhythmia versus the control of ventricular rate as a result of the arrhythmia.

Patient segment frequency plot

Different treatments are indicated and utilized for different patient populations. Plotting how many options are available to each segment may uncover underserved populations. This is particularly important for AF, in which elderly patients may have better outcomes with a rate control strategy as opposed to a rhythm control strategy, but this may not be the same for younger patients.

Working example: Summary of the treatment landscape for atrial fibrillation

Atrial fibrillation is the most common sustained cardiac arrhythmia, affecting 2.2 million people in the United States each year, yet there is not a clearly defined and agreed-upon strategy for the treatment of the disease.[40] Treatments to address the three categories of sinus rhythm maintenance, rate control, and thromboembolic risk are often administered according to the general guidelines that exist in the field. Typically, clinicians seek to achieve rhythm control through the use of cardioversion and antiarrhythmic drugs. If that approach fails, the next step is to try ventricular rate control therapy or AV node ablation with the insertion of a pacemaker. Throughout the treatment process, thromboembolic risk is treated on an as-needed basis, typically with anticoagulants. A visual representation of how to organize and think about these treatment options is shown in Figure 2.2.4.

Without delving into all the various treatment options considered throughout this chapter, as an example of how to compare treatment options, consider the debate regarding rhythm and rate control. Until recently, there was little evidence that supported the traditional, somewhat sequential approach that favored rhythm control over rate control as a first-line therapy. In the AFFIRM trial, which included 4,060 patients, researchers compared the effects of long-term rhythm management and rate management treatment strategies to determine whether one approach offered significant advantages over the other in terms of benefits and risks to patients. Over a mean follow-up time of 3.5 years, the results of the study showed that there were no clear advantages to rhythm versus rate control. In fact, there was a trend toward increased mortality in the rhythm control group. The patients in the rhythm control group were also significantly more likely to be

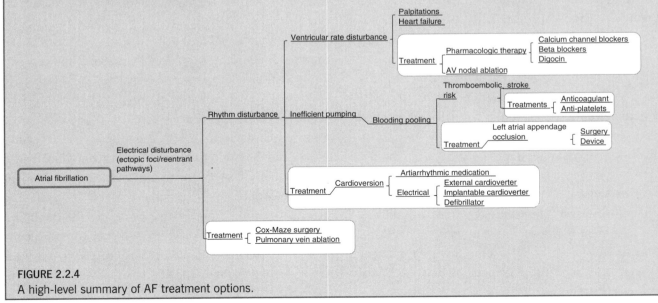

FIGURE 2.2.4
A high-level summary of AF treatment options.

hospitalized and have adverse drug effects than those in the rate control group.[41] The conclusion of the researchers was that rate control therapy should be considered a primary approach to the treatment of AF and that rhythm control should be abandoned early if it is not fully satisfactory.[42] However, this study focused on an older patient population so the generalizability of the study needs to be approached cautiously. As well, this trial did not include catheter ablation strategies for achieving rhythm control.

This highlights the fact that while this study is an important first step in starting to evaluate one type of treatment option against another in the management of AF, more research is needed before conclusive guidelines that address all patient variations are in place. Furthermore, a wider variety of treatment alternatives need to be taken into account (e.g., pulmonary vein isolation) in the development of these guidelines.

At a high level, the benefits and risks of the primary treatment alternatives for AF are summarized in Appendix 2.2.1. A gap analysis is provided in Appendix 2.2.2. As can be seen, despite the significance and consequences of AF, many areas remain open for further research and development. Some examples include higher efficacy ablation procedures, improved medications for rhythm or rate control, better devices for the treatment of the thromboembolic complications of AF, and safer and less invasive surgical procedures. By understanding the entire treatment landscape, gaps and inadequacies in current treatment methods can be used to help identify where promising clinical needs may exist that require innovative solutions.

GETTING STARTED

Treatment option research is a sizable job. However, the steps outlined below can help the innovator complete a thorough assessment of the treatment landscape for any given disease state.

See **ebiodesign.org** for active web links to the resources listed below, additional references, content updates, video FAQs, and other relevant information.

1. Develop an overview of treatment options

1.1 What to cover – Create comprehensive descriptions of all relevant treatments in the field.

1.2 Where to look
- **Medical references** – Refer to *Harrison's Principles of Internal Medicine*, *Guyton's Textbook of Medical Physiology*, or *Schwartz's Principles of Surgery* for credible information regarding established treatments.
- **PubMed** – A database of the US National Library of Medicine that includes more than 16 million citations from MEDLINE and other life science journals back to the 1950s. General reviews will be particularly helpful in understanding pathophysiology (locate "Reviews" by selecting this type of article under the "Limits" tab from the main page of the site before beginning a search).
- **eMedicine** – Clinical knowledgebase with more than 10,000 physician authors and editors.
- **UpToDate** – A database of evidence-based clinical information.

2. Evaluate clinical treatment profiles

2.1 What to cover – For each treatment option, assess the clinical rationale for why and when each one is used, including the mechanism of action, indications, efficacy, and safety.

2.2 Where to look
- Medical references
- PubMed (Reviews)
- UpToDate
- **FDA website** – The FDA posts the label or approval summary for new medical technologies.
- **BlueCross/BlueShield (BCBS) TEC Assessment** – Offers an evidence-based review of clinical treatment options.

3. Analyze economic treatment profiles

3.1 What to cover – Consider the financial impact associated with each treatment, including the costs of providing the treatment and the potential cost savings for using one particular treatment versus another available treatment.

3.2 Where to look
- **PubMed** (Reviews)
- **Medscape** – Database of clinical information targeted to physicians.
- **MEPS data** – The Agency for Healthcare Research and Quality provides data on the health expenditures of 18,000 US households via the Medical Expenditures Panel Survey (MEPS). This data is publicly available for primary analysis.
- **Reimbursement analysis** (read ahead to 4.3 Reimbursement Basics).

4. Utilization treatment profiles

4.2 What to cover – Provide an overview of how each treatment is currently used in clinical practice. Be sure to capture when/why each treatment is advocated and the frequency of procedures performed.

4.3 Where to look
- **PubMed** (Reviews and Guidelines documents)
- **UpToDate**
- **BCBS TEC Assessment**
- **Primary research** – Interviews or direct observations of clinical practitioners.
- **Reimbursement analysis**

5. Investigate emerging treatment profiles

5.1 What to cover – Understand emerging treatments that target the given disease state to identify what new areas are being focused on and how quickly new treatments will come to market.

5.2 Where to look
- **PubMed** – In addition to searching for relevant reviews, watch for information about early trials (e.g., animal trials) and papers presented at conferences to stay current with the most recent developments in a given treatment area.
- **ClinicalTrials.gov** – A government database that tracks clinical trials with medical technology.
- **Controlled-Trials.com** – A site for Europe that is comparable to ClinicalTrials.gov in the United States.
- **US Patent Office** – A government database that contains all issued and **provisional patents** and patent applications (see 4.1 Intellectual Property Basics for more information, as well as other sources).
- **VentureXpert** – Provides data on mergers and **acquisitions**, **IPOs**, and venture capital (VC) funding based on data from Securities Data Corporation. A useful source to get information on emerging companies,
- **Other** – Additional venture capital firm websites.

6. Summarize the treatment landscape

6.1 What to cover – Synthesize all of the findings from the first five topics into a comprehensive framework that summarizes the treatment landscape. At a minimum, this should include a summary of treatment options, risks and benefits, and a gap analysis with respect to need criteria and patient indications. Validate the need criteria with respect to the outcomes of existing therapies.

6.2 Where to look – Synthesize the information from steps 1 to 5. Bring in information from 2.1 Disease State Fundamentals for the patient segment gap analysis and from 1.3 Need Statement Development for the need criteria.

Credits

The editors would like to acknowledge Darin Buxbaum and Steve Fair for their help in developing this chapter. Many thanks also go to Hanson Gifford and Mark Deem of The Foundry for their assistance with the case example.

Appendix 2.2.1

Risks and benefits of AF treatments

Treatment category	Treatment option	Benefits	Risks
Rhythm control	Electrical cardioversion	• Immediate relief of symptoms • Rapid return to sinus rhythm	• Patient must be anesthetized • Possibility of AF recurrence • Risk of thrombosis before and after procedure so patient requires anticoagulation
	Pharmacological cardioversion	• Relief of symptoms • Return to sinus rhythm	• Patient may require hospitalization and monitoring when drugs are first administered • Drugs may lose effectiveness over time, with possibility of AF recurrence • Patient can develop new abnormal heart rhythms • Patient likely still needs anticoagulation
	Catheter ablation	• Minimally invasive • Alternative for patients who have resisted or failed other treatments	• Practitioner experience important • Likelihood of repeat procedures • Lower success rates in patients with chronic AF and/or structural heart disease
	Cox-Maze surgery	• Alternative for patients with intolerable symptoms or those that fail other treatments	• Highly invasive; requires general anesthetic • Usually only done in patients undergoing other cardiac surgery • Comparatively high rates of comorbidities and/or mortality
	Implantable atrial defibrillators	• Potential long-term solution for maintaining sinus rhythm • Can operate automatically or be controlled by the patient when AF occurs	• Relatively large size (inconvenient to patients) • Shocks are usually painful • Patient may still need anticoagulation
Rate control	Pharmacological rate control	• Reduces workload of the heart • May be associated with lower rates of hospitalization, adverse effects, and mortality than rhythm control drugs	• Patient required to take anticoagulants indefinitely • May not be as effective long term
	Catheter ablation of the AV node	• Long-term solution for managing ventricular rate • More effective than medication in reducing palpitations, controlling dyspnea, and improving quality of life	• No improvement in cardiac performance • Patient required to take anticoagulants indefinitely • Pacemaker requires life-long management
Thromboembolic risk	Pharmacological therapy	• Reduces the risk of stroke and mortality	• Patients may experience severe bleeding complications • Some therapies need frequent blood tests

Appendix 2.2.2

Gap analysis matrix for AF

		Return to sinus rhythm/ relief of symptoms	Reduction in recurrence of AF	Improvement in cardiac performance	Reduction in strokes and associated mortality	Limited side effects	Minimally invasive	Non-invasive	Convenient for the patient	Economics
Rhythm control	Electrical cardioversion	++	-	+	+/-	+	+	-	-	+
	Pharmacological cardioversion	++	-	+	+/-	-	+	++	-	-
	Catheter ablation	++	+	+	+	-	+	-	- -	-
	Cox-Maze surgery	+	+	+	+	- -	- -	- - -	- - -	- -
	Implantable defibrillators	+	-	+		- -	-	- -	- -	- -
Rate control	Pharmacological rate control	-		+		-	+	+	+	+
	Catheter ablation of the AV node	-		+		-	+	-	-	- -
Thromboembolic risk	Pharmacological therapy				++	- -		++	-	+

Legend:
+ shows a positive effect.
– shows a negative effect.
A blank shows a neutral effect.

Notes

1. The Foundry, http://www.the-foundry.com/ (November 19, 2007).
2. All quotations are from interviews conducted by the authors, unless otherwise cited. Reprinted with permission.
3. Lung Volume Reduction Surgery, Thoracic Surgery Division, University of Maryland Medical Center, http://www.umm.edu/thoracic/lvrs1.htm (November 16, 2007).
4. "How Serious is Emphysema?" American Lung Association, http://www.lungusa.org/site/apps/s/content.asp?c=dvLUK9O0E&b=34706&ct=67284, (November 16, 2007).
5. "What is Atrial Fibrillation?" Cleveland Clinic: Heart and Vascular Institute, http://www.clevelandclinic.org/heartcenter/pub/atrial_fibrillation/afib.htm (October 2, 2006).
6. Ibid.
7. Nicholas S. Peters, Richard J Schilling, Prapa Kanagaratnam, Vias Markides, "Atrial Fibrillation: Strategies to Control, Combat, and Cure," *The Lancet* (2002): 596.
8. "Patient Information: Atrial Fibrillation," Up-to-Date, http://patients.uptodate.com/topic.asp?file=hrt_dis/4882 (October 2, 2006).
9. Ibid.
10. "What is Atrial Fibrillation?" op. cit.
11. "Patient Information: Atrial Fibrillation," op. cit.
12. Ibid.
13. Peters, Schilling, Kanagaratnam, Markides, loc. cit.
14. Ibid., p. 600.
15. If a device is used in a way that deviates from the indication on the FDA-approved label, it is called "off-label" usage. While physicians may elect to engage in off-label use, the manufacturer of the device is not allowed to promote any methods of use other than those cleared by the FDA.
16. "ACC/AHA/ESC 2006 Guidelines for the Management of Patients with Atrial Fibrillation – Executive Summary," American College of Cardiology Foundation, the American Heart Association, and the European Society of Cardiology, 2006, p. 862.
17. Ibid.
18. Atul Verma, Amdrea Natale, Benzy J. Padanilam, Eric N. Prystowsky, "Why Atrial Fibrillation Should be Considered for First-Line Therapy," *Circulation* (2005): 1220.
19. "New ACC/AHA/ESC Guidelines for the Management of Atrial Fibrillation," op. cit.
20. Wazni *et al.*, "Radiofrequency Ablation versus Antiarrhythmic Drugs as First-Line Treatment of Symptomatic Atrial Fibrillation," *JAMA* (2005): 2634–2640.
21. Pappone *et al.*, "A Randomized Trial of Circumferential Pulmonary Vein Ablation Versus Drug Therapy in Paroxysmal Atrial Fibrillation," *Journal of the American College of Cardiology* (2006): 2340–2347.
22. Stabile *et al.*, "Catheter Ablation Treatment in Patients with Drug-Refractory Atrial Fibrillation," *European Heart Journal* (2006): 216–221.
23. Kenneth A. Ellenbogen and Mark A. Wood, "Ablation of Atrial Fibrillation: Awaiting the New Paradigm," *Journal of the American College of Cardiology* (2003): 198–200.
24. Melvin M. Scheinman and Fred Morady, "Nonpharmacological Approaches to Atrial Fibrillation," *Circulation* (2001): 2122.
25. Verma, Natale, Padanilam, Prystowsky, op. cit., p. 1218.
26. Ibid., p. 1219.
27. Benzy J. Pandanilam, MD and Eric N. Prystowsky, MD, "Should Ablation Be First-Line Therapy and for Whom: The Antagonist Position," *Circulation* (August 23, 2005): 1227.
28. Khaykin *et al.*, "Cost Comparison of Catheter Ablation and Medical Therapy in Atrial Fibrillation," *Journal of Cardiovascular Electrophysiology* (September 2007): 909.
29. Verma, Natale, Padanilam, Prystowsky, op. cit., p. 1219.
30. "Transcatheter Radiofrequency Ablation of Arrhythmogenic Foci in the Pulmonary Veins as a Treatment for Atrial Fibrillation," WellPoint, December 1, 2005, p. 2.
31. B. Meier, I. Palacios, S. Windecker, M. Rotter, Q.L. Cao, D. Keane, C.E. Ruiz, and Z.M. Hijazi, "Transcatheter Left Atrial Appendage Occlusion with Amplatzer Devices to Obviate Anticoagulation in Patients with Atrial Fibrillation," Catheter Cardiovasc Interv, November 1, 2003; 60(3): 417–422. http://www.summerinseattle.com/syl_pdf/22/plaato.pdf (October 17, 2006).
32. Jeff S. Healey, Eugene Crystal, Andre Lamy, Kevin Teoh, Lloyd Semelhago, Stefan H. Hohnloser, Irene Cybulsky, Labib Abouzahr, Corey Sawchuck, Sandra Carroll, Carlos Morillo, Peter Kleine, Victor Chu, Eva Lonn, Stuart J. Connolly, "Left Atrial Appendage Occlusion Study (LAAOS): Results of a Randomized Controlled Pilot Study of Left Atrial Appendage Occlusion During Coronary Bypass Surgery in Patients at Risk for Stroke," MedScape, September 9, 2005, http://www.medscape.com/viewarticle/510570 (January 2, 2007).
33. Ibid.
34. "APS: Appriva Medical Receives Approval to Commercialize PLAATO," ANP Pers Support, http://www.perssupport.anp.nl/Home/Persberichten/Actueel?itemId=40776&show=true (October 17, 2006).

35. A procedure that is performed by entering the left atrium via the right atrium across the septum, which is the wall that separates them.
36. "Early Results Show *PLAATO* Device Safe and Feasible for Stroke Prevention in Patients with Atrial Fibrillation," Medscape Today, April 11, 2003, http://www.medscape.com/viewarticle/452162 (October 17, 2006).
37. "Form 8-K for EV3, Inc." Yahoo! Finance, http://biz.yahoo.com/e/050927/evvv8-k.html (October 17, 2006).
38. "Left Atrial Appendage Closure," Rapid New Summaries, American College of Cardiology Annual Scientific Session 2006, March 2006, http://www.cardiosource.com/rapidnewssummaries/index.asp?EID=22&DoW=Sun&SumID=151 (October 17, 2006).
39. Ibid.
40. The AFFIRM Writing Group, "A Comparison of Rate Control and Rhythm Control in Patients with Atrial Fibrillation," *New England Journal of Medicine* (December 5, 2002): 1825–1833.
41. Ibid.
42. Ibid.

2.3 Stakeholder Analysis

Introduction

A clinical need often begins with a patient, his/her symptoms, and the underlying medical problem. But that is just the tip of the iceberg. Think about the physician and the nurses involved in the patient's care. Also, somewhere in the back office, there is a facility manager crunching numbers to decide whether or not to invest in the necessary equipment and infrastructure to support the patient's treatment. And then, perhaps a thousand miles away, is an insurance company administrator, who decides whether or not to pay for the care that has been delivered. All of these are stakeholders – individuals and groups who are touched by the need and have a stake in how it is ultimately addressed.

In stakeholder analysis, the innovator systematically examines the direct and indirect interactions of all parties involved in financing and delivering care to the patient. The purpose of this analysis is to understand how these entities are affected by the need and to determine their requirements (or their stake) in how it is addressed. Stakeholders have different perspectives – for instance, some will benefit if the need is addressed, but others may be adversely affected. Uncovering these perspectives and any potential conflicts is critical to shaping and refining the need statement and need criteria. It also allows the innovator to anticipate resistance, as well as to define and prioritize the requirements that will shape the eventual solution to maximize its chance of adoption among the most important and influential stakeholders. For these reasons, stakeholder analysis should start early in the biodesign innovation process, while needs are being identified and assessed. It can then be repeated as more information becomes known and progress is made.

OBJECTIVES

- Learn to identify important stakeholders.

- Understand each stakeholder's perception of the medical need (initially) and the proposed solution concept (eventually).

- Recognize which stakeholders are in conflict and/or alignment with one another and for what reasons.

Stage 2 Needs Screening

This is the first of two chapters focused on stakeholders. The output from the basic stakeholder analysis described here informs 5.7 Marketing and Stakeholder Strategy. The chapter is also closely linked to 2.4 Market Analysis. Among other topics, market analysis focuses on the assessment of competitors (i.e., businesses offering competing products) and other suppliers of products that address a given need. Competitors obviously have an important stake in the need and could, as a result, technically be considered stakeholders. However, because they will always resist new solutions proposed by competing innovators, they are excluded from traditional stakeholder analysis and considered among the other market forces that can create barriers to the adoption of a new idea.

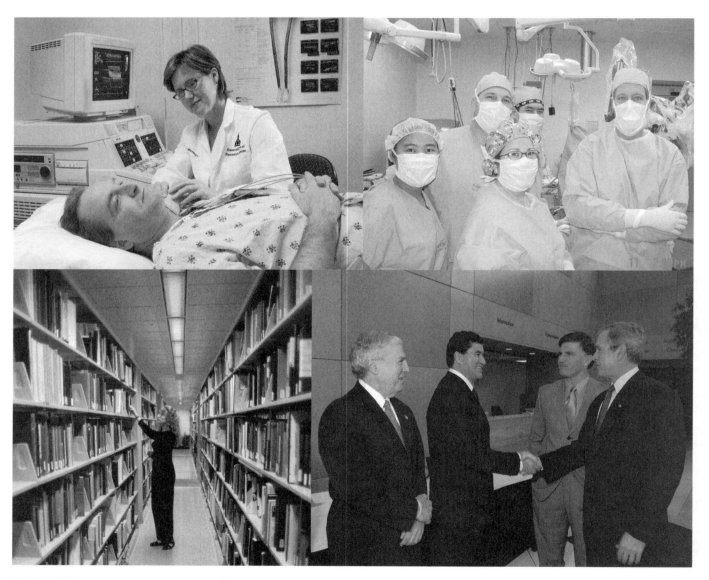

FIGURE 2.3.1
Stakeholders include (clockwise from left) patients and physicians; surgeons, anesthesiologists, and operating room nurses; elected and career government officials, facility administrators, and others. Anyone with an interest in the need and how it is addressed is considered a stakeholder.[15]

Stakeholder fundamentals

The need for stakeholder analysis is caused by the fact that in the medical field multiple groups and individuals, not just a single customer, drive the adoption of the product. Richard Stack, physician, inventor, and investor, succinctly summarized the multifaceted nature of medical innovation and reinforced the need for in-depth stakeholder analysis:[1]

> You have to know who your customer is. Certainly you have to know what the patient wants … But the person actually buying the product, you have to know that psychology very, very well.

One of the early steps in performing a stakeholder analysis is to identify the many different parties involved in delivering and financing care related to the need. See Figure 2.3.1.

There are two primary ways to identify stakeholders in the medical field. The first focuses on stakeholders involved in the cycle of care – patient diagnosis and the delivery of treatment. With this approach, the innovator studies how the patient moves through his/her healthcare experience, making note of all of the different players, their roles, and interests. The second method is concerned with stakeholders involved in financing patient care. In this analysis, the innovator follows the **flow of money** from one entity to the next as charges and payments are made. The results from these two methods can be triangulated by referring back to the observations made during the problem identification process (see 1.2 Observation and Problem Identification), as well as the data collected as part of the disease state and treatment options analysis (see 2.1 Disease State Fundamentals and 2.2 Treatment Options). Figure 2.3.2 provides a representation of the many different stakeholders with a

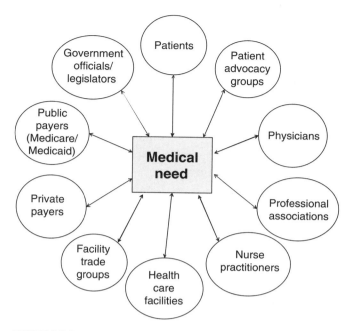

FIGURE 2.3.2
All stakeholders have the capacity to embrace or resist new medical technologies. While some exert more influence than others, all should be considered, particularly in a preliminary stakeholder analysis.

potential interest in a new medical technology to address a defined need.

Cycle of care analysis

Cycle of care analysis is based on how patients interact with the medical system. The focus of such an assessment is specifically on understanding the patient's diagnosis and treatment (not payments). The innovator should investigate who diagnoses the condition, who provides preliminary treatment, who provides next level treatment if the condition progresses, and what parties are involved in the ongoing management of the disease. It is especially important to pay attention to which different medical specialties are involved in this cycle and the referral patterns that exist between specialties.

Working example: The cycle of care for end-stage renal disease

Consider, for example, a patient with end-stage renal disease (ESRD). End-stage renal disease is characterized by chronic failure of the kidneys. The cycle of care for this condition often resembles the flow outlined in Figure 2.3.3. It starts with the patient developing certain symptoms that may trigger a visit to a primary care physician or, in extreme cases, to the emergency room (ER). In both situations, a series of laboratory tests are performed that will be used by the attending physician to make a diagnosis. When a diagnosis of ESRD is confirmed, the patient is referred to a nephrologist (specialist in kidney disease). The nephrologist evaluates the patient and determines whether dialysis is needed. If so, s/he refers the patient to both a dialysis clinic and a vascular surgeon, who prepares the patient's access for dialysis. (In dialysis, a patient's blood stream is accessed with a needle through a permanent access point established by a vascular surgeon. The blood flow is diverted through a filter in the dialysis machine to clear the toxins that the failing kidneys cannot eliminate. Dialysis involves three lengthy treatments per week.) Simultaneously, the nephrologist determines whether the patient is a good candidate for a kidney transplant (a surgical procedure in which an ESRD patient receives a new organ from a donor). If so, s/he refers the patient to a transplant center and a transplant surgeon. Once the patient receives a transplant, the transplant center is initially involved in follow-up care with the involvement of a transplant nephrologist. Eventually, the patient may be referred back to the original nephrologist for long-term follow-up care.

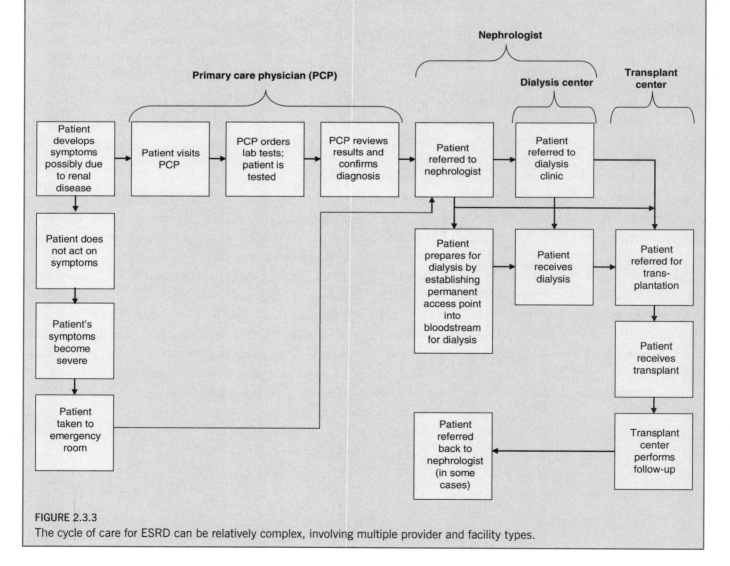

FIGURE 2.3.3
The cycle of care for ESRD can be relatively complex, involving multiple provider and facility types.

Once the cycle of care has been mapped, as in the ESRD example, the innovator's objective is to examine who interacts with the patient, the nature of their relationships with the patient, the duration and timing of the interactions, and the financial incentives linked with those interactions (even though the flow of actual payments should not yet be taken into account). All of the individuals and groups in the cycle (including the patient) should be considered stakeholders in the process, each with unique needs, requirements, and interests. Stakeholders involved in the diagnosis and treatment of ESRD include multiple clinical specialists (primary care physician, emergency care physician, nephrologist, transplant nephrologist, vascular surgeon, and transplant surgeon), different nursing specialties, and numerous facility types (doctor's office, ER, dialysis clinic, and transplant hospital). Not all of these stakeholders will be intimately involved with every need considered in the ESRD cycle of care. However, by using a method for identifying everyone with even a remote stake in a need (and how it is solved) innovators ensure that no one is overlooked.

In conducting this analysis, it is important to be aware that referral patterns in the cycle of care can be a source of potential conflicts between stakeholders, especially in cases where multiple specialties are involved. In the ESRD example, nephrologists lose their patients at the point when transplantation occurs, with only some patients referred back to them for long-term care (the remaining patients are cared for by the transplant center). A similar conflict exists between dialysis centers and transplant centers due to the loss of revenues to dialysis providers after patients receive transplants. So, it follows that if a new breakthrough became available that allowed more patients to receive transplants, nephrologists and dialysis clinics might mount some resistance since the breakthrough could potentially lead to a substantial loss in their patient-care revenue.

This particular point of view is not entirely speculative. A study published in the *New England Journal of Medicine* confirmed that the likelihood of being placed on the waiting list for a renal transplant was lower for patients treated at for-profit dialysis centers than non-profit ones, which led to increased mortality in this patient group.[2] While this study and the previous discussion appears to suggest that healthcare providers and physicians may not always have the best interests of their patients in mind, the real message is more nuanced. In the medical field, choosing the proper course of treatment for each patient requires the careful balance of the risks and benefits associated with the treatment options. Yet, this equation is ambiguous and varies from provider to provider. Transplantation, for example, may be perceived by some providers to be too risky for certain patients (based on their age, coexisting conditions, or other factors). The equation is further complicated when a provider has a financial incentive that makes one course of treatment preferable to another (dialysis versus transplantation). These situations create conflicts of interest that can affect the likelihood of adoption of a particular treatment for certain high-risk patients. Such resistance can sometime be overcome by understanding and addressing the motivations and concerns of the involved stakeholders. For instance, if ESRD patients are explicitly referred back to their initial nephrologists for ongoing care after transplantation (even though the transplant centers are capable of administering such services), then this may reduce resistance to high-risk transplantation among nephrologists and stimulate more transplant referrals.

Flow of money analysis

A flow of money analysis identifies stakeholders acting behind the scenes to finance the cycle of care. It focuses on who pays for the services and procedures performed in diagnosis and treatment of patients. In the process, it highlights all of the entities with a direct financial stake related to the need.

In most developed countries, healthcare payments are managed under a third-party payer system – the patient does not directly pay the medical provider for his/her medical treatment; instead, payments are made on the patient's behalf through a public or private insurer (see 4.3 Reimbursement Basics for more information). The National Health Service (NHS) in the U.K. and Medicare in the United States [administered by the Centers for Medicare and Medicaid Services (CMS)] are both examples of public health insurance programs financed by taxpayer money. Examples of private health insurers in the United States include companies such as Blue Cross/Blue Shield, UnitedHealthcare, and Kaiser. Such companies provide insurance on behalf of their subscribers, who pay a premium in exchange for health insurance coverage.

> **Working example:** Overview of the US healthcare financing system
>
> The US healthcare financing system is undeniably complex, involving both public and private payers. A simplified view of the stakeholders involved in financing the US healthcare system and their interactions with the stakeholders delivering and receiving care is depicted in Figure 2.3.4. The United States has two public insurance programs (Medicare for the elderly and disabled, and Medicaid for the poor), as well as a multitude of private insurers. Medicare and Medicaid are funded by individual and corporate taxpayer money and administered by the Centers for Medicare and Medicaid Services, with significant state involvement for Medicaid. In many instances, both Medicare and Medicaid subcontract with private insurers for the administration of the benefits they cover. Private insurers collect premiums from individual subscribers or from employers who provide health insurance benefits to their employees (historically, most of the employed non-elderly obtain health insurance through their employer, while some may purchase individual insurance). Insurers (both private and public) then pay healthcare providers (facilities and physicians) for the services they provide to the individuals they insure. In many instances, the payments made by the insurers do not cover all of the charges made, such that the individuals receiving treatment must pay the balance (called a co-payment). Individuals without any insurance coverage have to pay for all healthcare services on their own (out-of-pocket).
>
>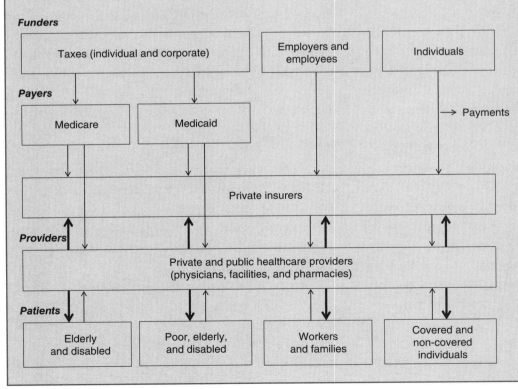
>
> FIGURE 2.3.4
> There are many interrelated entities involved in the flow of money in the US healthcare system (based on N. Sekhri, *Bulletin of World Health Organization*, 2000, 78 (6) 832; reprinted with permission).

Despite the involvement of multiple players and the many handoffs between participants, the flow of money in many treatment areas follows a standard path. For example, in the United States, most patients who receive an implantable cardiac defibrillator for the treatment of heart rhythm disorders are covered by private insurance until they turn 65 years of age (or become disabled), at which time they are covered by Medicare. However, because there are always idiosyncrasies and variations within the system, innovators should be cautioned about making assumptions regarding the flow of money. For instance, ESRD patients without private insurance are covered by Medicare three months after they begin dialysis and continue to be covered until three years after transplantation, regardless of their age. End-stage renal disease patients with private insurance are covered under their private health insurance policy for the first three years of treatment, but then convert to Medicare (again regardless of age). In this scenario, it would be imperative for an innovator working in the ESRD space to understand the role of Medicare in the flow of money for the treatment area and then focus on this group as a primary stakeholder, even if the innovator's solution targets a subset of the population under the age of 65. Flow of money analysis is especially useful for validating one's understanding of the potential financial incentives that may affect the behavior of providers (as uncovered in the cycle of care analysis), as well as determining the financial incentives of insurers.

Stakeholder interests

Once all key stakeholders in a clinical area have been identified, comprehensive analysis can be performed to further evaluate the barriers that might cause a stakeholder to resist the adoption of a new innovation, as well as the benefits that may drive their adoption. An effective stakeholder analysis analyzes the costs and benefits to each stakeholder identified. Direct costs and benefits should be considered (e.g., loss/gain in revenue, decrease/increase in profitability, decrease/increase in time away from work). However, non-monetary costs and benefits that drive stakeholder behavior should also be evaluated (e.g., impacts on reputation, ease of use, and **opportunity costs**, which are defined as the cost and benefits of giving up one alternative to pursue another).[3] Indirect factors are becoming increasingly important to stakeholders and can be a valuable source of leverage for innovators. Understanding the drivers of stakeholder behavior is essential to being able to influence stakeholder actions. The following sections articulate some of the common drivers of stakeholder behavior for patients, providers (physicians, nurses), facilities, and payers.

Patients

Patients are the ultimate decision makers on whether or not to receive a specific treatment. Traditionally, these decisions have been made based on information and advice primarily received from physicians. However, pharmaceutical companies, medical device manufacturers, and payers are increasingly trying to influence patient decisions through financial and non-financial means, such as direct-to-consumer advertising, co-payment requirements, and other mechanisms. Patients are also seeking medical information from other resources (e.g., Internet advertisements, online blogs, bulletin boards, and support groups) and using this data to help make their decisions (even though not all available data may be credible). All inputs to a patient's decision, as well as the patient's needs, must be understood to determine how likely s/he is to consider a new treatment alternative.

In addition to evaluating individual patients, the innovator should evaluate relevant patient advocacy groups. These entities have the ability to influence patient opinions. They are also frequently sponsored by major corporations whose interests may be served or threatened by the new innovation. Consider the sponsors' interests alongside the interests of the advocacy groups and the patients they are intended to serve.

With respect to accepting a certain treatments, patient behavior is often driven by the following *direct* factors.

Clinical outcomes

Patients are interested in the treatment that will best resolve their primary problem. However, they are also concerned with the elimination of symptoms and the avoidance of unintended consequences from the treatment. The order of importance of these factors is likely to be different for each patient. For instance, when considering a patient's reaction to a new treatment alternative that may have clear benefits, it is still necessary to think about the amount of pain the patient will experience, whether or not the patient's appearance will be altered, and/or other potential side effects that might be associated with one treatment but not another. Also, the innovator should evaluate the benefit of living longer if a treatment helps delay mortality.

Safety
While procedures and their associated risks may be considered routine from a physician's perspective, the idea of undergoing certain treatments can be traumatic for a patient. Patients must consider whether the "cost" of living with a disease is higher than the risk of being treated, based on the safety profile of the treatment and a patient's own individual preferences.

Economic impact
In the case of new technologies, patients are often required to cover a larger portion of the total treatment cost. Or, in some cases, they are required to cover all treatment expenses. Determine what out-of-pocket expenses a patient should expect to incur relative to other treatment alternatives and evaluate this cost against the anticipated change in clinical outcome.

Convenience
The impact that a new treatment may have on a patient's life can vary from inconvenient to life changing (in a positive or negative way). In deciding on a treatment, patients often think about whether the treatment is available nearby, how easy it is to schedule, what impact it will have on days off work, and the long-term implications on their quality of life.

Other *indirect* factors that can influence patient behavior include the following.

Opportunity cost
An innovator should consider what a patient could do with his or her time, money, and energy if s/he elected to have one treatment over another. Think about this question on both a near-term and long-term basis. Also, remember to include a patient's choice to do nothing about the problem as one potential alternative.

Perceived risk
Sometimes perceived risk can be a major factor in a patient's decision-making process, even if the actual risk associated with the treatment is relatively low. Particularly for experimental treatments, consider how the perceived risk is likely to affect a patient's behavior. It is also important to take into account the psychological effect that the new treatment is likely to have on the patient relative to established treatments.

Importantly, not all patients facing the same medical need will perceive it the same way. Just as patients respond to pain in different ways, their reaction to medical needs will vary dramatically. In evaluating the stake patients have in a need, it can be useful for the innovator to differentiate between patient types by developing a series of patient profiles. Within these profiles, patients will view a need similarly and have comparable reactions to different treatment alternatives. However, across profiles, patient perceptions of the need and how it is treated will be distinct. To understand how these profiles can be developed, refer back to the case example in 1.2 Observation and Problem Identification that described how a biodesign team from the University of Cincinnati created different profiles for patients to better understand needs in sleep apnea. Identifying these types of profiles will also prove helpful in 2.4 Market Analysis where one of the goals will be to define clear segments of patients with uniform perceptions toward the need.

Physicians

Because physicians are the primary individuals recommending patient treatment, they are critical stakeholders for almost every medical treatment option. While physicians are first and foremost driven by the desire to provide patients with the best possible treatment, they also face the need to earn a living. Innovations that make new procedures possible can often help physicians achieve both of these desired outcomes. Yet, from time to time, conflicts can arise. For instance, if a technology shifts patients from one specialty to another, this may cause "turf wars" between physicians and create serious obstacles to the technology's adoption. As physicians become increasingly specialized within a single field, these kinds of conflicts are becoming more common. Guy Lebeau, a physician and businessman who led the growth of Cordis Corporation's cardiology, endovascular, neurovascular, and electrophysiology businesses, commented on the benefits of this trend, using the field of cardiology as an example:[4]

> I think the fact that we are no longer going to have one cardiologist with one set of knowledge, but probably ten different types of cardiologists who are going to focus their energy on treating one type or one part of the disease, is excellent because this creates a situation where the learning and the competency of physicians is going to be higher.

The downside is that, as new technologies disrupt referral patterns, they create "winners" and "losers," particularly among physicians within these narrowly defined sub-specialties. When considering physicians as stakeholders, the innovator must be on the look-out not just for the relevant specialties, but the potential sub-specialties that exist within them. Then, the interests of all such stakeholders need to be taken into account.

The role of professional societies is another important factor to consider. For most well-established medical fields, there is an association that addresses the area within which a need exists. Physicians involved in these groups are often considered thought leaders in their respective fields and are likely to have strong opinions on the benefits and costs of any new developments in the field.

Physician behavior with respect to the potential utilization of a new treatment is generally driven by the following *direct* factors.

Agency
The physician's mandate is to represent the best interests of his/her patients. As an agent of the patient, a physician's treatment recommendations must carefully balance the risks and benefits to patients and take into account patient preferences. Medical ethics underlie this relationship (see the section of this chapter entitled Ethical considerations in stakeholder analysis).

Clinical outcomes
Clinical outcomes refer to the manner and degree to which physicians will be able to improve medical outcomes through the use of a new treatment. In combination with agency, this is one of the most persuasive factors for getting physicians to adopt a new technology.

Economic impact
The financial impact on the physician of adopting a new treatment is an important consideration. This includes how (and how much) a physician might be reimbursed for the treatment and the number of treatments a physician would perform annually. These calculations must be understood relative to existing treatment options. Major equipment that must be purchased and any expensive equipment that may be rendered obsolete by the innovation should also be taken into account (because this may be a potential source of resistance).

Risks
The risks to the physician of adopting a new treatment primarily focus on issues related to patient safety. They also cover malpractice liability and/or the liability of not complying with evidence-based guidelines, as well as the impact of the change on the physician's malpractice insurance. In some cases, new innovations can reduce physician risks and malpractice liability.

Physician behavior with respect to the utilization of a new treatment is also driven by a series of *indirect* factors.

Opportunity cost
In determining how a new treatment might fit into a physician's current practice, an innovator should analyze how the physician currently uses his/her time. Understand how long current treatment options take to administer, how many providers are involved, and how many procedures are typically performed within a given period of time. This information can then be used as a benchmark as more becomes known about the need and the solution that will eventually address it. Specifically, when a physician adopts a new solution,

the revenue s/he would earn per unit of time should be at least the same as the revenue currently generated from the same unit of time. Otherwise, the opportunity cost associated with adopting the new treatment alternative may be perceived as being too high to make its adoption appealing. In these cases, the clinical benefit would have to be extremely compelling to make the desired change in physician behavior feasible.

Workflow

While all new treatments may not carry with them significant capital expenditures (e.g., investments in new equipment), almost all will require a change in workflow (or the process by which they are used). Consider how disruptive a new treatment may be to established physician practices, or whether it can be integrated relatively seamlessly into common processes and, if so, how much this may cost. It is important to remember that it is much more difficult to get physicians to adopt a new treatment if it requires a significant change in their accustomed workflow, compared to one that can be integrated easily into their existing daily routine.

Ease of use

Evaluate whether or not a new innovation is likely to make the physician's job easier or harder to perform. While workflow takes into account the process by which a device is used, ease of use considers how difficult or easy the device is to use at each stage of that process. In terms of costs, the innovator should consider what training might be required to perform a new treatment and any new or special skills that physicians may need to develop. On the benefit side, identify the ways in which an innovation might make it easier for physicians to provide effective treatment. Ease of use can be a potent advantage for a new technology and a factor that can rapidly drive acceptance. This was demonstrated when Guidant Corporation introduced a new bare metal stent to treat coronary artery disease. Because it was so much easier to use (and did not require nearly the same amount of physician experience and skill to effectively place within a patient's artery), it quickly displaced the market leader, a bare metal stent marketed by Cordis Corporation, despite the fact that Cordis was first to market with its product.

Reputation

Consider whether the adoption of a new treatment or innovation might be perceived positively or negatively by patients and, in turn, what its effect might be on a physician's standing in the physician community. If a physician is known for being a leader in his/her field, evaluate the indirect benefits (e.g., visibility) to his/her practice of adopting the innovation. Conversely, if a physician is risk averse and takes pride in providing proven treatments, consider the reputational cost of adopting an exploratory treatment. There are significant differences in perceptions of new technologies and their impact on physician reputation among both physicians and specialties. For example, while interventional cardiologists pride themselves on being quick adopters of new technologies, cardiac surgeons are more conservative in their approach. This may explain why the adoption rates of new technologies vary dramatically between these two specialties.

As noted, nurse practitioners are another important group of stakeholders with a significant interest in the adoption of new technologies for addressing identified needs. Many of their interests parallel the benefits and concerns facing physicians. However, there are several important distinctions. First, nurse practitioners are far more likely than physicians to be salaried and, therefore, they are less likely to be financially affected, in the near-term, if referral patterns are disrupted. Secondly, they may pay more attention to how a new device fits their daily workflow and have a better perspective about the fit of a technology in the broader healthcare environment. This is because they tend not to be as specialized in the care they administer and may interact more with patients than the physicians do. This potentially gives them greater visibility across the patient's cycle of care.

Facilities

The interests of facilities, such as hospitals, surgical centers, laboratories, and other centers where care is delivered, are largely cost driven. Innovations that increase procurement costs are most likely to meet resistance from this stakeholder group, while those that reduce procurement costs and are assured of third-party reimbursement are most likely to be accepted. Facilities are also sensitive to innovations that shift the location where a treatment, procedure, or test is delivered since the revenue they receive is often adjusted based on the location. As an example, consider point-of-care (**POC**) testing for hemoglobin A1c (HbA1c), a variant of hemoglobin (an oxygen-carrying molecule of blood) that can be used as a marker of glucose control in diabetics. Testing for HbA1c typically requires a patient to go to an outpatient lab. Point-of-care testing would change the venue of testing from the lab to the clinic (or the doctor's office). For such tests, the stakeholders include not only the physicians who would perform the POC tests, but also the laboratories that previously provided this service. Following approval by the FDA of one such POC test (Metrika InView), a series of studies were performed (not managed or influenced by the manufacturer) which compared the results from the POC tests to those from tests performed in the lab, with lab tests considered the "gold standard." The studies showed that the correlation between the results was high, but not high enough to make the POC test a substitute for the test in the lab.[5] An editorial accompanying one of these studies, written by a professor of pathology, stated that there are numerous issues to consider when evaluating a new method for HbA1c POC testing, including whether or not the method is NGSP certified,[6] how well it performs in a field setting, and if it is free from common interferences.[7]

Such a response demonstrates that pathologists working in labs have a stake in the adoption of POC testing. When their views appear in medical journals, they have the potential to hinder or catalyze the adoption of a new test. Innovators should try to anticipate such viewpoints and develop a strategy to pre-empt them, for example, by securing appropriate certifications and by carefully designing clinical studies that may go beyond FDA requirements.

While facilities would hope for all of their procedures to be profitable, it is not uncommon for some to be designated as **loss leaders** – procedures billed for less than they cost because they generate business (e.g., patient traffic, additional revenue) in other areas of the facility. For example, at many dialysis centers in the US, the delivery of treatment is performed at a loss since it allows profits to be generated from other services, such as the administration of epogen, a drug necessary to stimulate red blood cell production and help control anemia, which is typically administered while patients receive dialysis. If an innovation will not be profitable for the facility where it is administered or utilized, the innovator should think creatively about related products/services that can be bundled with it, modified, or eliminated such that the innovation results in a net benefit to the facility.

Many groups shape the treatment decisions of a facility, including physicians, facility management, and purchasing professionals. When evaluating facility stakeholders, however, the primary focus should be placed on understanding how management and purchasing will respond to the new innovation.

Direct factors driving facility stakeholder behaviors for adoption of a new treatment include the following.

Economic impact
Depending on how the costs of a new treatment will be covered, innovators should begin thinking about whether a potential change may increase, decrease, or hold constant the overall cost of treating a given disease state. Since facility payments for treatments typically do not adjust higher for increased costs that may be incurred (i.e., facilities typically receive fixed payments for a treatment), a new innovation must reduce ancillary costs associated with the treatment to decrease overall cost or have a neutral effect on a facility's budget. For example, reducing a patient's length of stay in the hospital following surgery can provide large financial incentives for a facility to adopt a new treatment if the facility's payment for the surgery is fixed and does not increase with a longer stay (see 4.3 Reimbursement Basics). Also, think carefully about innovations that may change the location where treatment is administered (i.e., takes business away from a facility), as the POC example illustrated. Finally, consider other factors, such as whether a facility is for-profit or non-profit, an academic or community institution, and if it is located in an affluent or underserved area. The drivers of facility behavior can be quite different across these various settings.

Risk
Consider the effect of the new treatment on a facility's risk profile. Some procedures may significantly reduce

facility risk while others may increase it. An increase in risk can carry with it direct financial costs by affecting liability and insurance.

Indirect factors influencing the behavior of facility representatives regarding the adoption of new treatments include the following.

Opportunity costs
Facilities have limited resources in terms of their providers, support staff, and physical space in which to provide care. If a new treatment will change the number of procedures performed each year, it may create or consume procedural time in the operating room or other settings. This is time that could be used on other procedures. Therefore, the potential profit that could be generated per unit of procedure time should exceed the profit generated by existing procedures.

Reputation
Being seen as a leader in a certain field can attract patients to a facility. If an innovation serves as a magnet for a facility to draw additional patients, the facility may be willing to make trade-offs in other areas to achieve the benefit of additional patient traffic, especially if the additional patient traffic results in the need for additional ancillary services, such as testing. The DaVinci robot, an innovative surgical robot that can be used to gain improved surgical results and make procedures less invasive, provides a good example of a technology that is used by hospitals to enhance their reputation. For example, in August 2008, the hospital at the University of California at San Diego had the top sponsored link on a Google search of "Da Vinci Robot." This entry stated that "UC San Diego's Surgeons Are National Leaders in Da Vinci Surgery."[8]

Payers

If payers grant adequate payment (referred to as reimbursement) for a medical innovation, it is a powerful force in stimulating adoption. On the other hand, if payers deny, delay, or restrict reimbursement, it can be extremely detrimental to the success of a new treatment.

To payers, new treatment innovations often tend to be synonymous with increased costs. The reason that payers continue to fund new treatments is the promise of better outcomes, especially if this is coupled with the possibility of lower long-term costs for a given patient (e.g., fewer hospitalizations, surgeries, or other expensive forms of care). However, if the cost burden becomes too great or the perceived clinical benefits are not significant enough, both public and private payers may deny coverage and/or limit the number of patients eligible for a new treatment by dividing the patient population into sub-groups and restricting reimbursement to a specific sub-group. Another potential scenario that payers use is implementing step-therapy guidelines, which force physicians to try alternative therapies before utilizing the new treatment.

In general, payers are most likely to cover new and potentially expensive medical technology if it is proven (through robust clinical trials) to improve hard clinical endpoints (mortality and morbidity). Softer endpoints, such as patient convenience, physician convenience, or quality of life, are less likely to gain reimbursement unless the improvements are shown to be medically necessary (and medical necessity is a rather ambiguous concept that can often be shaped by the innovators as part of their reimbursement and marketing efforts – see 5.6 Reimbursement Strategy and 5.7 Marketing and Stakeholder Strategy). Payers need to be convinced that any additional cost incurred will improve clinical outcomes without added risk. Increasingly, innovators and medical device companies are discovering that the data required for regulatory approval is sometimes not enough to make a compelling case to payers (4.3 Reimbursement Basics and 5.6 Reimbursement Strategy provide more details on payers, their reimbursement decisions, and how innovators can influence them).

Genomic Health, a company that developed a high-end genetic test to help determine if women with early-stage breast cancer will benefit from chemotherapy (a commonly prescribed treatment that is effective in only a small percentage of patients), provides a compelling example of the role of payers in the adoption of a new medical technology. Although the effectiveness of the diagnostic is backed by strong clinical evidence, its value to payers is realized only if women with a negative test result choose *not* to receive chemotherapy (thereby saving money in administering ineffective treatment). However, according to one payer in the US, UnitedHealthcare, too many women are still receiving chemotherapy even if the test suggests they do not need it. For this reason, United entered into a conditional agreement with Genomic Health under which it agreed to cover the cost of the test for an 18-month trial period while the outcomes are monitored. If enough women with low scores on the diagnostic do not abstain from chemotherapy, then United will seek to negotiate a lower price with Genomic Health on the grounds that the test is not having the intended impact on actual medical practice. According to Dr. Lee N. Newcomer, senior vice-president for oncology at UnitedHealthcare, this arrangement was designed to make the manufacturer more responsible for how its product is used in the medical marketplace.[9]

Direct factors driving payer behavior regarding adoption of a new treatment can be outlined as follows.

Clinical outcomes
Innovators should consider both the near-term medical benefits, as well as the longer term effects of ongoing treatment in improving outcomes relative to any existing treatment alternatives. The elimination of symptoms and side effects that often require separate treatment can also be significant from a payer's perspective.

Economic impact
Be prepared to evaluate the total cost of any new treatment relative to existing treatment alternatives. When more is known about a potential solution, the innovator can start with the payment per treatment (how much the payer would be willing to reimburse for the new procedure) and then multiply this by the anticipated number of treatments per year. Compare this to data for alternative treatments to calculate by how much the new treatment will increase payer costs. In some cases, an innovator may be able to evaluate whether the innovation can decrease near-term or long-term costs to payers by reducing other services requiring reimbursement, such as hospitalization or additional testing (e.g., blood test or X-rays). Another way to think about the financial impact to payers is the incremental increase or decrease to its cost per member per month.

It is also important to consider if the new treatment will expand the market for treatment in such a way that significantly more patients will seek treatment, which can represent a sizable cost increase to payers.

Indirect factors that can also influence payers regarding adoption of a new treatment are as follows.

Competition
Payers often move as a group. In the United States, Medicare makes reimbursement decisions first and private payers usually follow closely behind. Consider competitive dynamics among payers in making reimbursement decisions (particularly in the private sector) and think about the benefits (e.g., in terms of market share) and costs of being the first (or last) payer to cover a new treatment.

Reputation
It can also be helpful for the innovator to think about the effect on the payer's reputation of offering the new treatment. If any new treatment is widely perceived as being ground breaking, the payer will have a more difficult time justifying a decision not to cover it. Conversely, if a new treatment is marginally effective yet costly relative to available alternatives, a payer will have little incentive to justify reimbursement in a cost-conscious environment where its own internal and external stakeholders would be critical of such a move.

Understanding global stakeholders
As markets outside the large US medical device market continue to evolve and grow, innovators should not assume that a US-based stakeholder analysis will necessarily address the needs of other international markets. If they are serious about addressing needs outside the United States, additional analysis in the countries of interest must be performed to ensure that subtle (and not so subtle) distinctions in stakeholder attitudes, preferences, and perceptions are understood. The principles and approach to the analysis will be the same as those described within this chapter, but the output will likely be different.

High level differences in the financing and organization of healthcare systems across different countries have profound implications on the stakeholder landscape. For instance, the cycle of care for a group of patients would be remarkably different in China than in the United States, Europe, or other Western nations. Services that are routinely part of healthcare delivery in one country may be considered non-health-related in another. A heart failure patient, for example, might be prescribed ACE inhibitors in the United States (to control high blood pressure), but not in some Southern European nation where such treatments are not as widely used. In China, treatments could differ even further, with a patient receiving traditional Chinese medicines in lieu of Western medicine. Significant payer differences exist as well. In most European countries there is taxpayer-funded universal health coverage with reimbursement decisions made by a single entity in each country, sometimes after referring to the decisions of their counterparts in other countries. In China, the central government routinely underfunds hospitals and healthcare providers making many hospital services unaffordable for most patients and inducing providers to over-deliver those services that patients find affordable (such as prescription drugs) and are willing to pay for out-of-pocket. Understanding such cross-country variations for the key markets addressed by a need is an important part of stakeholder analysis.

Linkages between stakeholder groups

In addition to considering stakeholders outside the United States, relationships between key stakeholder groups should be explored. In the medical environment, no single stakeholder group operates in isolation from the others. For example, facilities have their own issues, priorities, and considerations but also must satisfy the needs and demands of their associated physicians. Importantly, they may or may not be in a position to directly influence physician decisions. Patients typically follow the instructions of their physicians, but are increasingly exercising greater control over medical decision making, including treatment alternatives and locations. Physicians and patients may seek to embrace a new innovation but have their adoption hindered by the decision-making process of the payer system. As a result of these types of interconnected issues, the prioritization of stakeholder interests is critical because it can determine the order in which these interests are addressed.

As the forces that drive stakeholder behavior are understood, stakeholders can be classified based on their unique characteristics, motivations, and level of potential impact. Generally, the more central stakeholders are to the success of an innovation, the more time, effort, and resources should be devoted to understanding and managing their involvement and commitment.[10] One simple strategy for classifying stakeholders is to identify which ones are critical to driving the adoption of an innovation (or who have the power to derail it) versus those without such an important voice.

The following story of InnerPulse, Inc. demonstrates how stakeholder analysis works in practice and highlights some of the linkages between different stakeholder groups.

FROM THE FIELD — INNERPULSE, INC.

Anticipating and managing stakeholder reactions in the innovation process

Founded in 2003, InnerPulse, Inc. (formerly Interventional Rhythm Management) is an early-stage medical device company developing PICDs™ (percutaneous implantable cardioverter defibrillators), or miniaturized ICDs that can be placed via a catheter-based approach. Implantable cardioverter defibrillators are used to prevent sudden cardiac death by issuing a lifesaving jolt to the heart when a patient suffers sudden cardiac arrest. The company's product is made of a chain of pencil-thin components measuring 56 centimeters in total length. See Figure 2.3.5. The device can be placed in the vascular system percutaneously (across the skin), using a standard catheter-based approach, in under ten minutes. Percutaneous implantable cardioverter defibrillators are differentiated from conventional ICDs in two ways: their size and delivery method. Conventional ICDs are roughly the size of a hockey puck, and are usually surgically implanted in the upper chest of a patient. Because of the complexity of ICD devices and the accompanying procedure, implantation is usually only performed by a small group of heart rhythm specialists known as cardiac electrophysiologists (EPs). In contrast, the PICD, due to its less invasive delivery method, can be implanted by EPs and interventional cardiologists (ICs), who typically use catheters to treat blockages in blood vessels. As a consequence, it can be made accessible to a much larger group of defibrillator candidates who would potentially benefit from ICDs.

There are two primary groups of patients who benefit from ICDs: (1) **secondary prevention patients** – patients with a prior episode of sudden cardiac arrest; and (2) **primary prevention patients** – patients at high risk of sudden cardiac arrest who have a weakened heart manifested by a left ventricular ejection fraction <35 percent (normal ejection fraction is typically ≥55 percent). The primary prevention market dramatically increased as a result of two recent clinical trials [MADIT-II (2001) and SCD-HeFT (2004)], which showed that ICDs dramatically reduced the mortality rates of these patients. Based on these results, the indication for ICD implantation extended to more than 1 million similar individuals at risk for sudden cardiac death. However, the large number of primary prevention patients seeking ICDs is currently overwhelming the 1,800 EPs nationwide. The consequence of this problem is that only 10 to 20 percent of all patients who would potentially benefit from an ICD have received the device. Percutaneous implantable cardioverter defibrillators have the potential to unlock this large, previously untapped market. This under-penetration of the market due to the limited number of EPs available to initiate defibrillator therapy was an important factor in helping understand the role PICDs could play.

Because PICDs can be administered by ICs, instead of just EPs, they raise some important stakeholder issues. When InnerPulse was initially pursuing the *need to develop a less invasive way to prevent sudden cardiac death*, it had to consider the effect a non-surgical implant would have on stakeholders in the field. There is an unwritten conventional rule in medical devices that if a new device "steals" business from one specialty to benefit another specialty, turf wars between the specialties may ensue, creating obstacles to the adoption of the new technology. For example, the conflict between cardiac surgeons and ICs that followed the introduction of balloon

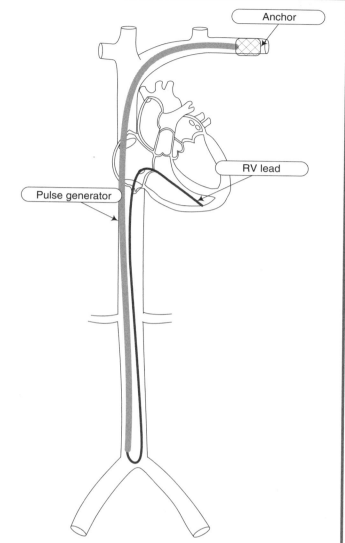

FIGURE 2.3.5
A schematic of the InnerPulse PICD (courtesy of InnerPulse, Inc.).

angioplasty as a minimally invasive alternative to invasive coronary artery bypass surgery was legendary. The team at InnerPulse had to ask itself if the introduction of a less invasive technology implanted by non-EPs would create similar tensions with EPs, who controlled the traditional ICDs market.

Bill Starling, chairman of the InnerPulse board, company cofounder, and an early investor in the venture, believed that turf wars could be avoided in this case if the need for a less invasive solution was focused on the primary (as opposed to the secondary) prevention market. He explained: "The great opportunity here is the primary prevention market. Electrophysiologists have traditionally

FROM THE FIELD: INNERPULSE, INC.

Continued

ignored this market because they do not see these patients. EPs only see the secondary prevention patients who have developed an arrhythmia. Primary prevention patients are seen by the interventionists [ICs]. Primary prevention patients usually have coronary artery disease. The cardiologists put in stents, give them some drugs, and send them home because there is nothing more they can do. With a less invasive solution, the interventionist [IC] would now be able to do something more for the patient. And remember, these are patients that the EPs would not see anyway."[11] Thus, the fact that a patient would typically see a cardiologist or IC for some initial treatment or evaluation prior to being referred to an EP was key in understanding the stakeholder relationships. Without referrals from colleagues in other areas of cardiology, an EP would not have patients requiring ICD implants.

According to InnerPulse, the argument in favor of pursuing the need that eventually led to the PICD was even stronger than this. The company predicted that addressing this need would not only create new business for ICs, but it would expand the market for EPs as well. Using an analogy shared by the company, PICDs are like an airbag: they prevent death when an accident (sudden cardiac arrest) occurs, but they also lead to additional service and maintenance work. When a patient with a PICD experiences an electric shock, he will visit his IC. Chances are he will then be referred to an EP, since this is what ICs do for their patients with an arrhythmia. Coming back to the airbag analogy, the IC installs the airbag (PICD), but when the airbag inflates (sudden cardiac arrest), the patient – now a secondary prevention patient – is sent to the EP for treatment/repair, potentially with a more complex, traditional ICD.

Despite their early hypothesis that both ICs and EPs would respond favorably to the need for a less invasive solution, Starling and cardiologist Richard Stack (the other company cofounder – see Figure 2.3.6) did not leave anything to chance. Toward the beginning of the biodesign innovation process, they assembled an advisory board of internationally recognized scientific thought leaders. They used this group of five ICs and five EPs to test their basic assumptions. Through this process, they confirmed that both the ICs and the EPs could see that a solution like the PICD would expand the practices and markets for both specialties. Also, it

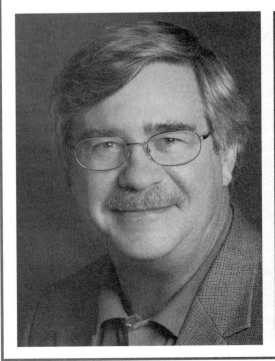

FIGURE 2.3.6
Bill Starling and Richard Stack (courtesy of InnerPulse, Inc.).

became clear to them that ICs would want to use these devices. Since EPs would be central to the early adoption of the technology, the clinical trials would be largely managed by EPs. This was partly a pragmatic business decision (EPs have deep experience running trials with ICDs, so the trial design would be more readily accepted by the FDA). However, it was also a savvy business strategy that would create a greater sense of buy-in for the device among EPs (by positioning the PICD as a device for them, yet one that could also be used by ICs in patients that the EPs did not normally treat), thereby helping to minimize any potential turf wars.

Beyond that, Starling and Stack were careful in considering all other possible stakeholders in deciding to move forward. A summary of the key stakeholders, their concerns, and how the company ultimately believed each stakeholder would respond to the PICD is shown in Table 2.3.1.

As of November 2008, one stakeholder "wild card" that emerged was the reaction of payers to the new technology. While reimbursement codes were in place that would cover PICDs, there was always the risk that once PICDs become widely adopted, major payers would seek to revise the payment levels downwards to physicians and facilities, in order to reflect the lower duration and lower complexity of using and implanting PICDs compared to ICDs. In fact, following the release of MADIT-II, there was widespread informal agreement among major payers that if the technology achieved 100 percent penetration it would have a significant negative effect on their income statements. InnerPulse wondered if payer reactions to the introduction of PICDs could create obstacles to their vision of expanding the market for ICDs.

Starling was convinced that any payer-related issues would be minimal and could be managed by the company. Implantable cardioverter defibrillators addressed such an important medical need that efforts by payers to reduce reimbursement or restrict coverage would backfire, he asserted. Recent history seemed to support his belief: Medicare analyzed the results from the MADIT-II and SCD-HeFT trials to argue that not all primary prevention candidates should get ICDs. However, clinician opposition was so strong that Medicare had to backtrack shortly afterwards. Starling maintained that the same dynamics that overcame payer resistance in the past would play to InnerPulse's advantage in its quest to expand this market.

Output from stakeholder analysis

After a stakeholder analysis is complete, an innovator should compile the information into a comprehensive summary of the stakeholder environment. As noted, Table 2.3.1 provides a sample of that output for InnerPulse. It summarizes the key stakeholders, outlines the primary benefits and costs associated with a new solution to address the defined need, and provides a subjective assessment of the overall net impact of a new technology on each stakeholder. Importantly, this example deals with a stakeholder analysis for which the solution is already known. In most cases, at this stage in the biodesign innovation process, the innovator should not yet have defined a solution. As a result, the stakeholder summary would capture the *potential* factors that might stimulate adoption or resistance. As more becomes known about the eventual solution, the stakeholder summary can be made more specific until it resembles the one shown. At this point, it can serve as the foundation for developing specific stakeholder management strategies and to facilitate decisions about how much time and energy should be invested in winning over each stakeholder group. It can also be leveraged to help forge important relationships and develop key messages appropriately targeted at various stakeholder groups (see 5.7 Marketing and Stakeholder Strategy).

Ethical considerations in stakeholder analysis

As described in 1.1 Strategic Focus, ethics focus on the intentional choices people make and the basic moral principles they use to guide their decisions. They do not provide a specific value system for making choices, but rather outline a set of basic principles to follow in decision making and in ethically managing the conflicts of interest that these choices may create.[12]

Stakeholder analysis informs the innovator about the types of interests that could be affected by a given decision, and can help identify potential conflicts that may arise. Personal interests (what is best for the individual making the decision), social interests (what is

Table 2.3.1 *The sample stakeholder analysis for InnerPulse demonstrates how the key take-aways from a stakeholder analysis can be summarized in a concise, actionable format.*

Stakeholders	Primary benefits	Primary costs	Assessment of net impact
Patients	Reduced invasiveness compared to traditional ICDs. More convenient, shorter recovery time.	Need for re-implantation after a documented arrhythmia.	**Positive**: Reduced invasiveness expected to increase patient's comfort with the procedure.
Physicians: interventional cardiologists (ICs)	Expanded practice and market, additional revenue. Allows for retention of primary prevention patients without need for EP referral.	Learning of new procedure.	**Positive**: ICs are typically quick to embrace new technologies that expand their market, especially if combined with an attractive reimbursement.
Physicians: electrophysiologists (EPs)	Expanded overall referrals from ICs of primary prevention patients who develop arrhythmias and thus become secondary prevention patients.	Possible loss of primary prevention patient referrals (but these may represent only a small component of an EP's practice).	**Positive**: EPs seemingly are interested in using PICDs and, by nature, prefer to focus on complex arrhythmia cases not seen by ICs.
Facilities: EP labs and IC catheterization labs	Increase device implantation volume for EPs and ICs.	Overall costs, including expensive components, for EPs and ICs become similar.	**Positive**: As long as reimbursement for PICDs remains the same as for ICDs.
Payers	Expansion of life-saving technology to many patients who could benefit from it.	Increased costs.	**Negative**: Total cost of delivering defibrillator therapy will go up. Will they try to reduce reimbursement for ICDs and/or PICDs, given that PICDs can be implanted more quickly?

good for a community or society at large), and professional interests (what is good for the company or the patient/client) are examples that usually need to be considered when making almost any decision. At times, these interests are aligned and the level of conflict is low. At other times, they directly conflict and the "right" answer may not be obvious.

As a general rule, innovators must make every effort to avoid putting stakeholders in a position that might potentially compromise their ethics. This can be accomplished if the innovator systematically uses a code of well-established ethical principles in interacting with each stakeholder group and making choices available to them (this involves treatment options offered, but also treatment options withheld). As profiled in 1.1 Strategic Focus, key ethical principles for medical device innovators include truthfulness, fairness, beneficence, non-maleficence, respect, and confidentiality.

An appreciation of these principles by innovators not only helps them think through their own choices, but can explain why stakeholders might potentially resist the adoption of a new innovation if it presents an ethical dilemma according to these guidelines (e.g., trying a new treatment if there are questions about its efficacy or safety as compared to other options). For example, this issue was seen in the case of left ventricular assist devices (LVADs), which are mini-pumps implanted into the chest to help a patient's failing heart pump blood. Left ventricular assist devices were initially approved as a temporary "bridge" to help maintain the functioning of the heart while patients with severe heart disease awaited heart transplantation. The FDA eventually approved LVADs for permanent use in terminally ill patients who were not eligible for a heart transplant due to comorbidities or age.[13] However, despite the fact that the LVAD manufacturer had completed FDA approval, achieved relatively strong clinical data that showed improved survival rates compared to drug therapy, and launched an aggressive marketing campaign, physicians resisted widespread adoption of the treatment based, in part, on poor economics (the procedure was reimbursed at a fraction of its actual cost).[14] More importantly, physicians raised concerns about LVAD-related infections, which often were quite serious. Without effective clinical evidence to verify the magnitude of the infection risk, physicians did not perceive the potential benefits to their patients as adequate to justify the increased safety risk under the guideline of beneficence, or the obligation to do no harm. As a result, they did not adopt the innovation in large numbers.

While this response alone created significant resistance to the technology, the interpretation of the LVAD example is even more nuanced. With healthcare budgets continually being squeezed, physicians increasingly recognize that expensive technologies strain patient access to care. In the case of LVADs, one could reasonably argue that certain physicians resisted their adoption because they considered it unjust in the face of severe budget constraints. Some questioned the fairness of spending upwards of $200,000 on a device with a questionable benefit and serious safety issues while other patients had difficulties obtaining access to more basic care. Even though the principles of medical ethics discourage individuals from using their own values to determine what is just and unjust in making such decisions, it is inevitable that individual physician judgments will play a role. The innovator should try to anticipate how such individual value judgments may affect the adoption of a new innovation.

As noted previously, the risk/benefit ratio associated with a new technology ultimately drives its adoption. Medical ethics dictate that physicians carefully balance and discuss the risks and benefits of any treatment with their patients. Regulatory authorities make decisions to approve devices by evaluating evidence on their safety and risk at an overall system level. However, when physicians evaluate the same data one-on-one with a patient, the risk/benefit ratio might look significantly different. Physicians in the field are empowered (and obligated) to make their own decisions, together with their patients. Their own personal experiences in successfully and (perhaps more importantly) unsuccessfully treating a disease may often be a stronger force in their decision making than clinically validated evidence, especially with regard to issues of safety. When the risks are grave, they may not recommend a treatment to a patient even if, for example, it has been shown statistically to prolong the life of a terminally ill patient, as the LVAD example illustrates.

GETTING STARTED

Understanding stakeholders can be a fascinating exercise that provides the innovator with an even better understanding of a medical need and the optimal way to address it. The following steps can be used to initiate the process of identifying key stakeholders and isolating their perceptions about a defined need.

See **ebiodesign.org** for active web links to the resources listed below, additional references, content updates, video FAQs, and other relevant information.

1. Identify stakeholders

1.1 What to cover – Scan the landscape to identify all stakeholders with an interest in the innovation. Start with a cycle of care analysis to identify stakeholders in the clinical environment. Then, conduct a flow of money analysis to identify additional stakeholders in other areas (see 4.3 Reimbursement Basics for more information about following the flow of money in the US healthcare system). Do not overlook stakeholders that, while peripheral, may pose risks for the need (e.g., suppliers of basic components to competitors).

1.2 Where to look
- **Primary research** – There is no better source of information than data gathered directly from stakeholders themselves. Whenever possible, seek to conduct interviews and/or focus groups with a wide variety of target

stakeholders. While such primary research, when performed by a young enterprise, does not typically include a large enough sample to be considered truly representative, it can certainly provide invaluable directional information that can then be supported with secondary research on a larger scale. When performing primary research, actively seek input from members of the stakeholder group who are likely to be supportive of a new innovation, as well as those who are not. Seek the advice of individual professionals and patients, as well as input from representatives of professional associations. In addition to targeting physicians and patients, speak with case managers at payer organizations, facility nurses, purchasing managers, facility managers, and any/all other relevant contacts.

- **Professional societies** – Professional societies can provide a wealth of information to help innovators understand the cycle of care and make contact with appropriate stakeholders. Refer back to Appendix 1.1.1 in 1.1 Strategic Focus for a table of select medical conditions and some of the professional societies in those fields.
- **Innovation notebook** – Refer back to notes made during observations and as a result of research conducted during the development of the preliminary need statement (see 1.2 Observation and Problem Identification and 1.3 Need Statement Development).
- **Disease and treatment research** (see 2.1 Disease State Fundamentals and 2.2 Treatment Options).

2. Outline benefits and costs for each stakeholder group

2.1 What to cover – Identify the specific benefits and costs that an innovation might create for each unique stakeholder group. Carefully consider direct benefits and costs, as well as those that are indirect. Be exhaustive in considering each stakeholder's perspective.

2.2 Where to look
- **Primary research** – Refer back to interviews with target stakeholders or conduct additional interviews.
- **Other chapters** – Review 2.1 Disease State Fundamentals, 2.2 Treatment Options, 4.3 Reimbursement Basics, and 5.6 Reimbursement Strategy, as well as the information and resources collected as a result of following the processes outlined in these chapters.
- **Professional societies** – These groups can be helpful for understanding physician and/or facility costs and benefits.
- **Patient advocacy and support groups** – These groups can be helpful for understanding patient costs and benefits.
- **Online patient blogs** – Even though opinions expressed in web blogs may not be representative, they can provide awareness of issues, concerns, and opportunities for further study.
- **Private insurance websites** – Sites of private insurers such as The Regence Group (TRG) Medical Policy or Wellmark Blue Cross/Blue Shield Medical Policies outline under what circumstances medical services may be eligible for coverage under these private insurance programs and can provide insight into payer behaviors. The Blue Cross/Blue Shield Technology Evaluation Center (TEC) is another applicable resource.
- **Public insurance websites** – Sites such as Medicare's Coverage Center or the UK's National Institute for Health and Clinical Excellence may provide relevant information.
- **Facility-related resources** – To locate facility-related perspectives, try conducting a Medline search using keywords that reflect the need and author affiliations/departments related to the people working in the facility (e.g., hospital management, pathology). Trade group websites, such as the **American Hospital Association** and the Medical Group Management Association, are

other possibilities (although the extent of the information provided may be limited).

3. Summarize net impact and key issues for each stakeholder group

3.1 What to cover – After a list of costs and benefits has been developed for each stakeholder group, make a determination regarding the overall impact of the innovation on each group. This will be a somewhat subjective determination – the important thing is to understand directionally whether a stakeholder group is likely to be supportive or resistant to the introduction of the innovation (positive, negative, or neutral). Then, summarize the most important issues that are likely to drive stakeholder behavior so it is clear what opportunities/challenges need to be addressed when preparing to interact with each group.

3.2 Where to look
- **Interviews** – Go back to any stakeholders already interviewed to validate the assessment of net impact and the most important issues to be addressed.
- **Treatment research** – Review the gap analysis performed as part of 2.2 Treatment Options.

4. Classify stakeholders and assess trade-offs

4.1 What to cover – Further evaluate the stakeholder groups and classify them based on their unique characteristics, key issues, and level of power/control. Which stakeholder groups are imperative to win over? Which stakeholders are not central to the adoption of the innovation, though one would hope to influence them? Next, understand what relationships exist between stakeholders. Examine the trade-offs required to address the interests of the most powerful/important stakeholders. If their needs are met, how does that affect the needs/interests of other parties? Will it change the likelihood of other groups supporting the innovation? Again, this will be a relatively subjective evaluation. However, in developing a comprehensive understanding of the stakeholder environment, it is imperative to consider how meeting the needs of one group affects the needs of other groups.

4.2 Where to look – Refer back to the resources listed above, as needed. Review case studies and articles on other companies to see how they have prioritized stakeholders and managed trade-offs among them in bringing medical devices to market.

Credits

The editors would like to acknowledge Darin Buxbaum, Donald K. Lee, and Richard L. Popp for their help in developing this chapter. Many thanks also go to William N. Starling for sharing the InnerPulse story.

Notes

1. From remarks made by Richard Stack as part of the "From the Innovator's Workbench" speaker series hosted by Stanford's Program in Biodesign, March 6, 2006, http://biodesign.stanford.edu/bdn/networking/pastinnovators.jsp. Reprinted with permission.
2. Pushkal P. Garg, Kevin D. Frick, Marie Diener-West, and Neil R. Powe, "Effect of the Ownership of Dialysis Facilities on Patients' Survival and Referral for Transplantation," *New England Journal of Medicine*, November 25, 1999, http://content.nejm.org/cgi/content/abstract/341/22/1653 (July 26, 2007).
3. "Opportunity Cost," Dictionary.com, http://dictionary.reference.com/browse/opportunity%20cost (September 22, 2008).
4. From remarks made by Guy Lebeau as part of the "From the Innovator's Workbench" speaker series hosted by Stanford's Program in Biodesign, April 5, 2005, http://biodesign.stanford.edu/bdn/networking/pastinnovators.jsp. Reprinted with permission.
5. Laurence Kennedy and William H. Herman, "Glycated Hemoglobin Assessment in Clinical Practice: Comparison of the A1cNow™ Point-of-Care Device with

Central Laboratory Testing (GOAL A1C Study)," *Diabetes Technology & Therapeutics*, 2005, 7(6): 907–912.

6. NGSP (National Glycohemoglobin Standardization Program) is a certification method beyond what was required for FDA approval.

7. Kennedy and Herman, loc. cit.

8. Per a Google search on "Da Vinci Robot" performed in August 2008.

9. Andrew Pollack, "Pricing Pills and the Results," *The New York Times*, July 14, 2007.

10. "Stakeholder Analysis," NHS Institute for Innovation and Improvement, http://www.institute.nhs.uk/NoDelaysAchiever/ServiceImprovement/Tools/IT145_stakeholder.htm (November 13, 2006).

11. All quotations are from interviews conducted by the authors, unless otherwise cited. Reprinted with permission.

12. R. J. Devettere, *Practical Decision Making in Health Care Ethics* (Washington D.C.: Georgetown University Press, 2000).

13. Susan Conova, "FDA Approves Heart Pump for Terminally Ill Patients," *In Vivo*, December 4, 2002, http://www.cumc.columbia.edu/news/in-vivo/Vol1_Iss20_dec04_02/index.html (July 26, 2007).

14. Muriel R. Gillick, "The Technological Imperative and the Battle for the Hearts of America," *Perspectives in Biology and Medicine*, 2007, http://muse.jhu.edu/journals/perspectives_in_biology_and_medicine/v050/50.2gillick.html (July 26, 2007).

15. Images of doctor/patient and government officials from the National Institutes of Health image bank. Surgery image courtesy of Thomas Krummel. Administrator image courtesy of the Stanford Graduate School of Business.

2.4 Market Analysis

Introduction

Not all needs are created equal. Without a compelling, accessible market to support them, even the seemingly most important needs are unlikely to be solved. For better or worse, the number and characteristics of the people in the target market, their ability and willingness to pay, and the presence and nature of competition in the field all directly affect the financial viability of pursuing a particular need and the innovator's ability to successfully commercialize a solution.

Market analysis focuses on the systematic examination of these issues. By estimating the total potential market size associated with the need (measured in terms of revenue), an innovator can determine how many resources can realistically be dedicated toward addressing the need and whether or not it can support a sustainable business. Market analysis also allows the innovator to understand the dynamics of the market – whether it is growing or shrinking, becoming more or less competitive – to determine whether opportunities in the market will increase or diminish over time. Finally, it provides a mechanism for identifying competitors, assessing their products, and understanding their positioning in the field, in order to detect gaps in the existing offerings. These gaps can be used to help define and confirm key stakeholder requirements that ultimately can lead to product differentiation and success in the target market. Market analysis should start early in the biodesign innovation process, while needs are being identified and assessed. It can then be repeated and performed in more depth as progress is made.

The information gathered through 2.4 Market Analysis is important throughout the biodesign innovation process, but plays a central role in 5.7 Marketing and Stakeholder Strategy and 6.1 Operating Plan and Financial Model.

OBJECTIVES

- Learn to divide the market into homogeneous market segments defined by similar perceptions of and/or responses to a medical need.

- Appreciate how to define the market size and competitive dynamics in each market segment.

- Recognize how well customers' needs are currently being addressed in each segment and their willingness to pay for alternate solutions.

- Know how to identify which key market segment(s) to target.

- Understand how to package this information into a reasonable, fact-based assessment of a market opportunity.

Market analysis fundamentals

Market analysis is performed early in the biodesign innovation process to confirm whether or not the need being considered is associated with a commercially viable market. Importantly, each innovator (and investor) must determine for him/herself what constitutes a viable market. As the costs associated with developing and commercializing medical devices continue to escalate, many innovators believe that having a large market is one key to success. As Richard Stack, who serves as the president of Synecor Inc., a Silicon Valley-based medical device incubator, summarized:[1]

> The need has to have a very large market ...
> There's only so much time in the day, and it's really just as easy to develop a solution for a large market as it is for a small market.

In contrast, other innovators prefer a different approach: starting with a smaller market and expanding over time. John Abele, cofounder of Boston Scientific, provided this perspective:[2]

> I'm a big fan of niches – of not trying to take on an entire market at once – because there's less resistance to innovation when you do it on a smaller scale.

While different innovators may favor different market approaches, there is widespread agreement regarding the importance of performing market analysis early in the innovation process and using the results as one of the filters for determining whether or not a need is worth pursuing. The outcome also enables the innovator to appropriately refine the need statement and develop an expanded need specification (see 2.5 Needs Filtering) that focuses on the unique needs, dynamics, and characteristics of the chosen target market.

Market analysis is a multistep process that leads the innovator through an investigation that becomes increasingly specific, as described in Table 2.4.1. The remainder of this chapter explores each of these steps in more detail. However, keep in mind that, while each of these steps is important, an innovator with dozens of needs to evaluate cannot realistically perform such an in-depth market assessment for each one. Rather, it makes sense to start with a higher-level approach (focused on broad customer segmentation, market sizing, and competitive analysis), and then work toward increasing depth and detail as the list of needs is narrowed down through the iterative needs filtering process.

Step 1: Market segmentation

Market segments divide patients, payers, and providers (in this context, physicians and facilities) into distinct groups that share similar perceptions about a need. The purpose of this exercise is to recognize and understand the subgroups that comprise a total market. It is rare that any new treatment designed to address a need will meet the requirements of *all* customers in a broad medical field (e.g., all patients with the heart rhythm disorder known as atrial fibrillation); it is far more likely for it to address the needs of a subset of the total population (e.g., patients with paroxysmal atrial fibrillation originating in the pulmonary veins, over the age of 65, treated by an electrophysiologist, and covered by Medicare).

An effective **market segmentation** strategy identifies subsets of the population with the greatest similarity while maintaining maximum differences between the groups. Market segments should be measurable, accessible, durable (so as not to change too quickly), and substantial enough to be profitable.[3] The market segment(s) that demonstrate(s) a need most directly aligned with the need an innovator is seeking to address will likely become the target market (see below for more information about target markets).

Because of the complex interactions among the multiple stakeholders that exist within medical markets, market segmentation should be thought of as an iterative, progressive activity. By beginning with a relatively simple patient-based analysis and then layering in increasingly complex factors (e.g., payers and providers), an innovator can eventually account for the many different stakeholders and interactions that need to be considered in the assessment of the market. In most cases, four steps can be used to navigate the market segmentation process:

1. Segment patients based on symptoms and risk factors (using approaches outlined in 1.2 Observation and Problem Identification and 2.1 Disease State Fundamentals).

2.4 Market Analysis

Table 2.4.1 *There are six key steps for performing a market analysis.*

Step	Topic	Questions to investigate
1	Market segmentation	What are the key factors that can be used to divide potential customers into distinct market segments in which the population shares common needs and perceptions (e.g., patient characteristics, treatment options, provider attributes, and payer mix)?
2	Market size	What is the size of the market opportunity in each market segment and its potential for growth and expansion?
3	Market dynamics	What are the competitive dynamics in each market segment that a new entrant would face? Are new companies created and are they successful? What is the nature of their competitive relationship with existing companies? Are companies acquired in this space?
4	Market needs	How well are the needs of customers within each segment addressed by existing solutions? How closely aligned are they with the need the innovator is seeking to address?
5	Willingness to pay	How willing is each market segment to pay for a new solution and, if applicable, what do customers pay for existing solutions?
6	Target market	Which market segment(s) is (are) most likely to embrace a solution?

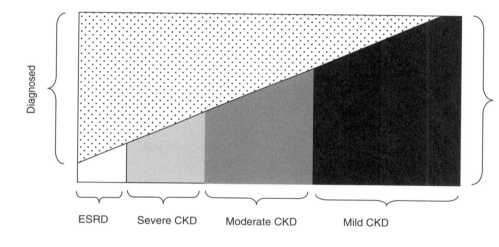

FIGURE 2.4.1
Segmentation of CKD patients based on diagnosis and disease stage.

2. Segment patient subgroups based on treatment (using approaches outlined in Chapters 1.2, 2.2 Treatment Options, and 2.3 Stakeholder Analysis).
3. Segment patient/treatment subgroups based on providers (using approaches outlined in Chapters 1.2, 2.2, and 2.3).
4. Segment patient/treatment/provider subgroups based on payers (using approaches outlined in Chapters 1.2, 2.3, and 4.3 Reimbursement Basics).

The high-level example that follows, which is focused on patients with chronic kidney disease (CKD), describes in more detail how this approach to market segmentation works.

Segment patients based on symptoms and risk factors

First, to segment patients based on symptoms and risk factors, an innovator can divide patients into diagnosed and undiagnosed categories, then further refine these groups based on the stage of their disease [stages 1–2 = mild, 3 = moderate, 4 = severe, and 5 = end-stage renal disease (ESRD)]. Most stage 1 patients are undiagnosed, while most stage 5 are diagnosed. Figure 2.4.1 provides a representation of the resulting segments. In this and the other segmentation figures that follow, the area of each subgroup reflects the size of the segment in terms of the number of patients it includes.

Stage 2 Needs Screening

Segment patient subgroups based on treatment

The next step is to further subdivide patients based on the treatment they receive. Common treatment categories for CKD might include self-care management (e.g., dietary and fluid restrictions), drug treatment (glucose control medication for diabetics; blood pressure control therapy for those suffering from high blood pressure; or other treatments for patients with anemia or bone disease); hemodialysis (HD); peritoneal dialysis (PD); or transplantation (Tx). Figure 2.4.2 shows the subdivision of the diagnosed ESRD patient subgroup into six segments based on treatment type.

Segment patient/treatment subgroups based on providers

These six segments can be refined further based on the type of provider delivering the treatment to the patient. Typically, in this market, there is a one-to-one correspondence between providers and the associated treatments. For example, for PD and HD, treatment is delivered by a nephrologist in combination with a dialysis clinic. For transplantation, treatment is administered by a surgeon in collaboration with a transplant program, as shown in Figure 2.4.3.

Segment patient/treatment/provider subgroups based on payers

The final segmentation step is to evaluate payer type relative to the subgroups. To prevent the example from becoming too complex, Figure 2.4.4 seeks to differentiate only between patients covered by private insurance versus public insurance (Medicare in the United States).

This simplified CKD example demonstrates steps for performing progressive patient, provider, and payer

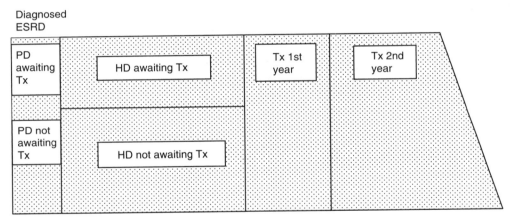

FIGURE 2.4.2
Further segmentation of patients diagnosed with ESRD based on their treatment.

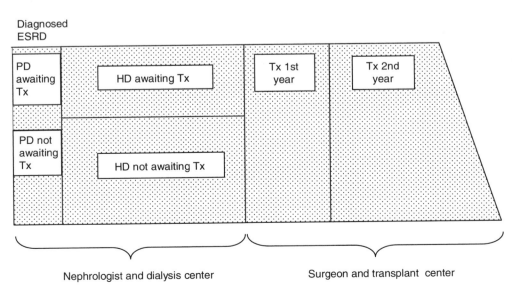

FIGURE 2.4.3
Further segmentation of diagnosed ESRD patients based on the provider administering care.

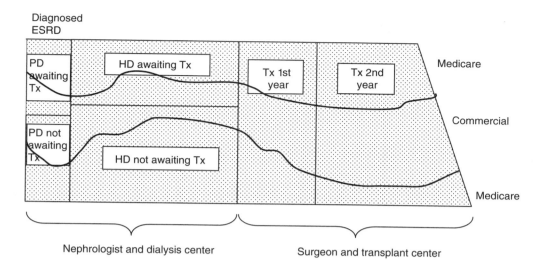

FIGURE 2.4.4
Additional segmentation of diagnosed ESRD patients based on payer type. The center section, bounded by the lines, indicates the percentage of private or commercially insured patients, while the remaining areas, top and bottom, represent those covered by Medicare.

segmentation using a sequence of Venn-like diagrams. However, there are other ways of illustrating this kind of hierarchical process, including the use of decision trees. While the process illustrated by Figures 2.4.1 to 2.4.4 may appear to be complex and potentially tedious, it provides a structured way for relatively inexperienced innovators to think about market segmentation in the early stages of the biodesign innovation process. More experienced innovators also perform these steps, although they may do so in a less structured, more instinctive fashion.

At each level of a progressive market segmentation approach, the innovator can use one or more attributes to divide larger groups into smaller ones. Figure 2.4.5 outlines many more of the different dimensions that can be used to differentiate between market segments.

When conducting an actual market analysis, recognize that meaningful differences between market segments may not exist in all of these dimensions. However, they all should be considered to be certain that no important sources of similarities and differences between subgroups are overlooked. Furthermore, keep in mind that there may be additional factors to consider based on the unique and distinctive characteristics of the market being assessed, and that some market segments may be combined.

Step 2: Market size

After the various market segments have been defined, the innovator should size the market opportunity associated with each one. This involves answering the following four categories of questions:

1. How many patients are there in each market segment?
2. What treatment options are currently being used within each segment? What companies provide these solutions? What is their relative market share?
3. What is the total dollar value of each market segment (i.e., total cost of treatment or total medical expenditures)?
4. What is the epidemiologic growth pattern in the segment? Is the segment contracting or expanding? What do growth projections look like in the near, medium, and long term?

Innovators can take either a "**top-down**" or a "**bottom-up**" approach to calculating the dollar value of each segment. A top-down approach begins with the overall spending on the disease state, and then divides total spending into categories based on the percentage of customers in each segment. The problem with the top-down approach is that it essentially assumes the same spending per patient across the various segments. However, its advantage is that it does not demand as much data, which can be helpful when an innovator is still evaluating multiple needs. A bottom-up approach derives the total market size by multiplying the number of customers within each market segment by the associated costs of their treatment. It is more precise than

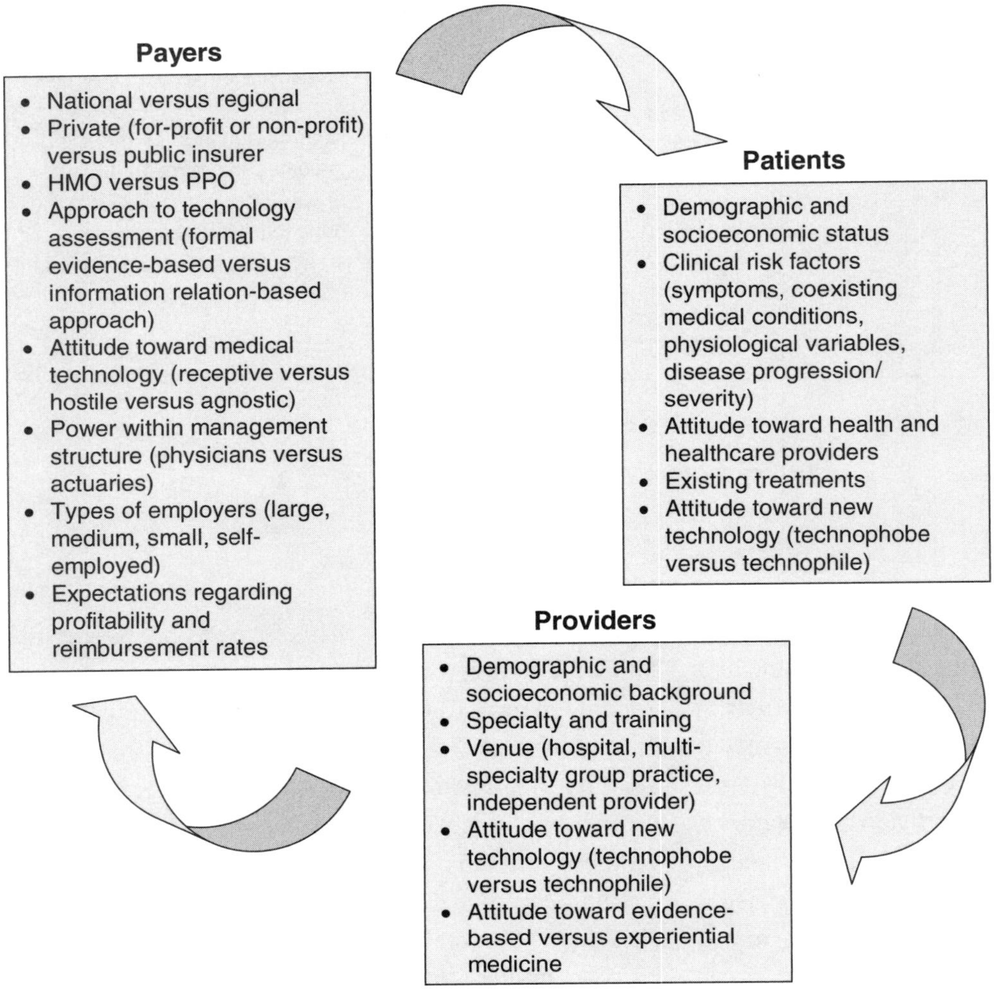

Stage 2 Needs Screening

Payers
- National versus regional
- Private (for-profit or non-profit) versus public insurer
- HMO versus PPO
- Approach to technology assessment (formal evidence-based versus information relation-based approach)
- Attitude toward medical technology (receptive versus hostile versus agnostic)
- Power within management structure (physicians versus actuaries)
- Types of employers (large, medium, small, self-employed)
- Expectations regarding profitability and reimbursement rates

Patients
- Demographic and socioeconomic status
- Clinical risk factors (symptoms, coexisting medical conditions, physiological variables, disease progression/severity)
- Attitude toward health and healthcare providers
- Existing treatments
- Attitude toward new technology (technophobe versus technophile)

Providers
- Demographic and socioeconomic background
- Specialty and training
- Venue (hospital, multi-specialty group practice, independent provider)
- Attitude toward new technology (technophobe versus technophile)
- Attitude toward evidence-based versus experiential medicine

FIGURE 2.4.5
There are many attributes that can be used to differentiate between market segments. All should be considered by the innovator when performing market analysis.

the top-down approach but may require data that are not readily available at this point in the biodesign innovation process. Initially, either of these approaches can be used to determine a high-level estimate of market size. However, it may be useful to perform both types of analysis and compare/validate the resulting data. (An example of a bottom-up analysis is provided later in this chapter in the section entitled Bringing it all together. More information about top-down and bottom-up market estimates can also be found in 6.1 Operating Plan and Financial Model).

An example of top-down market sizing is found in the story of a team of innovators who eventually founded Spiracur. When Moshe Pinto, Dean Hu, and Kenton Fong were students in Stanford University's Program in Biodesign, they began working on the need *to promote the healing of chronic wounds*. As part of the team's need screening research, Pinto and his teammates discovered that the wound care market is immense (accounting for more than $20 billion in annual healthcare costs in the United States alone and $566 million in wound care products),[4] as well as diverse, with at least five major segments: diabetic ulcers and resulting amputations; post-trauma wounds; venous ulcers; pressure sores; and burns. After examining the market, they decided to focus on venous ulcers, primarily because they felt that the gap between existing treatments and desired outcomes was so great – most ulcers would remain open after 12 weeks of treatment (an unacceptably long period).

Once the team identified the venous ulcer segment, it proceeded with more formal market analysis. To determine the size of the segment, the group estimated that between 0.5 and 1.5 percent of the US population suffers from venous ulcers based on extrapolations of available epidemiologic statistics. Competitive analysis revealed that the aggregate daily cost of the leading product for venous ulcers was $150.[5] To obtain the potential segment value, the team multiplied this amount by the duration of the therapy and by the prevalence of venous ulcers, and then incorporated growth estimates into the calculations to reach a total potential market size of $12 billion by 2009[6] (the total market size included all patients with venous ulcers). At the time of the analysis, that market segment had a penetration rate of approximately 30 percent (i.e., only 30 percent of the patients with a venous ulcer received some form of treatment). Even without any increased penetration in the market, Pinto and his team faced a total potential market opportunity worth $3.6 billion by 2009.[7]

Step 3: Market dynamics

After all market segments have been sized, it is important to spend time understanding the competitive dynamics within them. Who are the potential competitors? How do they compete with each other? What are their strengths and weaknesses? This analysis will help an innovator anticipate key market opportunities, uncover possible hurdles, and begin to get a sense for how much market share s/he can realistically hope to capture in the early years following a product's launch. It is important to note that while the previous steps in the market analysis process were primarily product-agnostic (i.e., they were focused on the need), an assessment of market dynamics involves the direct examination of competitors and their offerings so it is no longer product agnostic.

At least two well-established frameworks can be applied to help assess the competition in a market. The first is **Porter's five forces**, which can be used to evaluate the overall competitive landscape in a field. The second is **SWOT analysis**, which is focused on understanding the dynamics surrounding individual competitors. This section describes how both of these approaches can be used and provides an example of their application in the medtech field. It also revisits the InnerPulse story (introduced in 2.3 Stakeholder Analysis) to demonstrate the importance of considering market dynamics.

Porter's five forces

Porter's five forces framework, developed by Michael Porter of Harvard Business School, identifies the five primary forces that drive competition within an industry:[8]

- The threat of new competitors to the market (and the barriers that prevent them from entering).
- The bargaining power of suppliers (as measured by the number of suppliers in a market and the costs of switching from one to another).
- The bargaining power of buyers (in terms of the extent to which buyers can directly affect profitable sales).
- Pressure from substitute products or services (which necessitates differentiation).
- The intensity of rivalry among existing competitors.

These five forces interact with one another to shape the profit potential of an industry (and the primary firms within it).[9] By investigating each factor, and how they are interrelated, an innovator can objectively characterize the market, the potential for profitability (which is directly related to the market opportunity), and the critical issues that must be managed. Figure 2.4.6 outlines the type of issues to consider when investigating each of these competitive forces.

The application of this framework to the medical device industry illustrates the value of the approach. Of course, evaluating the medtech field as a whole would be far too broad to be meaningful in this context – a manufacturer of cardiac pacemakers does not compete directly (or even indirectly) with a manufacturer of orthopedic implants. For optimal results, the five forces framework should be applied within a single specialty or therapeutic field, such as drug eluting stents (DES).

Threat of new entrants

The DES market, although highly competitive, was controlled by two companies for roughly four years. Cordis, a unit of Johnson & Johnson, introduced the first drug eluting stent in April 2003. Boston Scientific followed suit with its own product in March 2004. These two companies had the only two DES in the market until Medtronic made a product launch in February 2008, followed by Abbott in July 2008. One factor accounting for the absence of new entrants was the immense

Stage 2 Needs Screening

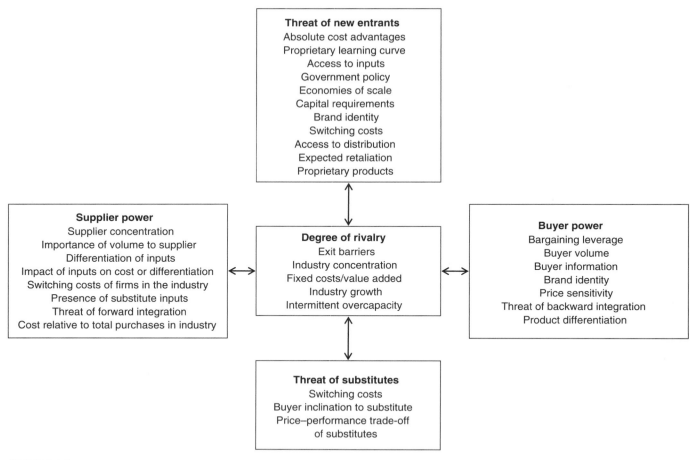

FIGURE 2.4.6
Porter's five forces (from an overview of the framework developed by QuickMBA.com; reprinted with permission).

capital requirements associated with developing, testing, and launching a DES. The level of risk is also daunting. Furthermore, the incumbents worked diligently to create additional barriers to entry. For example, both companies constructed formidable patent portfolios, which they enforced aggressively. Evidence of this can be found in Cordis, which had sued Abbott five times for infringing on its patent before Abbott's product was even released.[10]

Threat of substitutes

Substitutes for DES include any drugs that may prevent coronary artery disease and the need for angioplasty. For instance, a patient could take cholesterol-reducing drugs in lieu of receiving angioplasty (which necessitates the use of stents). However, such drugs need to be used regularly and preventively. Once coronary artery disease is diagnosed, the patient may not have access to any comparable substitutes. Bare metal stents, which preceded DES, do not perform as effectively (in terms of restenosis rates), and coronary artery bypass grafting (CABG) is a far more invasive procedure for which many patients do not qualify.

Supplier power

Suppliers of the raw material used to make DES do have some power, especially those companies that provide the drugs used in the stents. Evidence of this power can be found when suppliers exert themselves against the stent makers. For example, in October 2006, Wyeth, the supplier of the drug used in the Cordis stent, sued Cordis for breach of contract.[11] The two companies eventually settled.[12] Such actions demonstrate that suppliers in the DES market do have power that must be managed to prevent them from negatively affecting industry/firm profits.

Buyer power

Buyers of stents include hospitals (the direct purchasers) and payers (indirect buyers through third-party reimbursement). Because of the strength of the clinical evidence favoring DES, and the lack of substitutes, buyers have limited power in this field. However, as buyers (especially insurers) consolidate in the United States, their purchasing power is likely to increase. This change may eventually lead to pricing pressures on firms in the industry.

Degree of rivalry

The lack of sufficient buyer power and substitutes in the field, combined with significant barriers to entry, creates an environment that supports high profitability for the established firms in this industry. These desirable conditions, however, have led to an environment characterized by extreme levels of rivalry in the field as the incumbents jockey for position. This intense competition and the presence of some supplier power in the field are among the few factors keeping DES manufacturers in check.

SWOT analysis

SWOT analysis, which is credited to Albert Humphrey of Stanford University, focuses on the internal and external factors affecting a project or company (including strengths, weaknesses, opportunities, and threats).[13] Using SWOT analysis to understand what is occurring within the competitive environment requires an innovator to identify the primary competitors in a market and assess each one individually. In doing so, the first step is to identify each competitor's primary objective. Then, the following factors can be considered to determine the relative power of their competitive position:[14]

- **Strengths** – *Internal* attributes of the organization that are helpful in achieving the objective. Identify the company's competitive advantage and what differentiates its products/services.
- **Weaknesses** – *Internal* attributes of the organization that are harmful in achieving the objective. Consider what barriers or hurdles the company is facing.
- **Opportunities** – *External* conditions that are helpful in achieving the objective. In the medtech field, these might include intellectual property (IP) and regulatory issues, partnerships, stakeholder satisfaction, and other economic, social, and technological factors.
- **Threats** – *External* conditions that are harmful in achieving the objective. Explore the same factors listed above, in addition to the activities of the company's direct and indirect competitors.

An assessment of competitors' strengths and weaknesses helps an innovator understand the capabilities and limitations of other players in the market. Strengths represent capabilities that must be overcome (e.g., barriers to entry and/or attributes against which a company must differentiate itself). Weaknesses represent limitations that have the potential to derail a competitor's success. The innovator must be careful to avoid or mitigate the same risks within his/her own company (and may be able to exploit the weaknesses of competitors by turning them into his/her own opportunities).

Opportunities and threats can be shared by all players within a market. For example, opportunities facing established companies may have the potential to be better or more quickly addressed by a new entrant. Similarly, threats to an established competitor may translate into opportunities for an innovator. In general, SWOT analysis is easier to perform for large, established firms about which more information is publicly available. However, in some cases, it may be strategically unwise for a small, emerging company to enter into direct competition with a major, existing company (at least initially, while it is getting started). In these cases, understanding what threatens the established player can be a shrewd defensive move. Offensively, it can also help shed light on what companies may be suitable partners or acquirers.

Table 2.4.2 provides an example of SWOT analysis for Abbott Vascular. Abbot Vascular was the latest entrant into the DES market in 2008. The company offers an array of medical devices used to treat vascular conditions. It is a division of Abbott, a multinational diversified healthcare products firm with 68,000 employees and approximately $25 billion in annual sales.[15]

It is worth noting that the strengths and weaknesses presented above tend to be similar for many established medtech firms. The opportunity reflects the expectation

Table 2.4.2 *A high-level SWOT analysis for Abbott Vascular.*

Strengths	Weaknesses
• Extensive R&D capabilities • Strong relations with key clinicians and opinion leaders • Strong distribution channels • Deep pockets for making large investments • Extensive product portfolio of complementary products • Relations with other Abbott divisions, including pharmaceutical division with drug expertise	• Most significant growth potential comes from acquisitions, not organically developed products (e.g., the company's DES was based on platform acquired from Guidant) • Bureaucratic decision processes that slow down innovation
Opportunities	Threats
• Several other conditions can involve local delivery of drugs to the affected area of the human body, leveraging platform potential of DES	• Adverse safety risk of DES (e.g., risk of stroke) • Pricing pressures

that the model of combining drugs with localized delivery mechanisms has applications beyond those in DES, while the primary threat reflects the growing awareness that combination products of drugs and devices pose safety risks that are not common to pure device innovations. Overall, the SWOT analysis reveals that there are potential opportunities for a nimble new entrant.

After a SWOT and/or five forces analysis has been performed on a handful of the most important competitors within the market segments, the resulting data can be summarized across companies and the innovator can look for trends and issues across organizations. At a minimum, such a summary should include the following information:

- Number of established and emerging companies working within the market segment.
- Specific products or services offered.
- Maturity of products or services offered.
- Market penetration.
- Estimated market share.
- Pricing for each product or service.
- Revenue per product or service.
- Profit per product or service.

InnerPulse revisited

Porter's five forces can be used to examine the decision made by InnerPulse (introduced in 2.3 Stakeholder Analysis) to develop a less invasive implantable cardioverter defibrillator (ICD) for patients at risk of experiencing sudden cardiac death (SCD). As of 2007, the existing competitors in the ICD space were Boston Scientific, Medtronic, and St. Jude Medical (Boston Scientific entered the ICD market after its acquisition of Guidant in 2006, following a bidding war with Johnson & Johnson). These firms traditionally enjoyed a comfortable oligopoly because of significant barriers to entry, which include the high capital requirements to develop and get approval for new ICD devices, as well as their strong IP positions. Suppliers of components for the industry used to have some power, but it was diminishing as some manufacturers were now independently developing components that they previously would have had to purchase from suppliers. Buyers (payers, such as private insurance companies and Medicare) had significant power, but it had been weakened through clinical and patient advocacy efforts supporting the need for ICDs. There was also some evidence of competitive rivalry (e.g., litigation around IP and the bidding war between Boston Scientific and Johnson & Johnson for the acquisition of Guidant). However, this rivalry could primarily be observed among new entrants, rather than between established firms. Further, alternative or substitute treatment options were not available to treat patients who had suffered from SCD (secondary prevention) or were at risk of SCD (primary prevention) (i.e., no drugs had been shown to be as effective at reducing the risk of SCD as an ICD). Importantly for InnerPulse, unlike the smaller secondary market, which was well penetrated, significant growth opportunities existed in the larger primary prevention market since the overall market penetration was low (approximately 20 percent). Stated another way, only one in five patients at risk for SCD received an ICD. This was primarily due to the inability of primary prevention patients to be seen by the physicians who implanted ICDs.

This quick analysis suggests that the incumbents are highly profitable (in fact, gross profit margins for ICDs are in excess of 80 percent) and that they will protect their market aggressively. InnerPulse's strategy to focus on the underpenetrated primary prevention market (in which it could fill the major void in the total ICD market as opposed to trying to "steal" market share from incumbents) shows that it recognizes the challenges of direct head-to-head competition with the incumbents. By taking this approach, InnerPulse can potentially maximize its likelihood of differentiating its own product from products of the established companies as it enters the market. Because the incumbents are undoubtedly looking into the primary prevention market as a growth opportunity, InnerPulse may eventually become an acquisition target (see 6.3 Funding Sources).

The primary lesson from this five forces analysis is that, whenever possible, innovators should focus on market segments where they will avoid direct head-to-head competition with strong entrenched players. If they can focus instead on untapped or unpenetrated market segments with a significant market size and need, they can more effectively differentiate their solutions from the products of the incumbents.

Step 4: Market needs

In addition to understanding the size and dynamics of each market segment, it is important to determine how the needs of each subgroup may vary slightly from one another. The innovator should refer back to his/her notes from observations and other data to see if they shed light on any unique problems or needs that exist within the segment (see 1.2 Observation and Problem Identification, 1.3 Need Statement Development, and 2.2 Treatment Options). If possible, observations, problems, and needs identified early in the biodesign innovation process should be mapped to specific market segments. The idea is to systematically verify the unique needs of each segment and to be sure that any segments that may have been overlooked during the observation process have the same need(s). In some cases, it may be necessary to perform additional observations.

It is equally important to determine how well the needs of customers within each subgroup are currently being met. To accomplish this, an innovator can begin by evaluating outcomes for patients within each segment. For example, what is the prognosis for patients over the age of 65 with paroxysmal atrial fibrillation originating in the pulmonary veins? What outcomes should they expect if their condition goes untreated?

Once this is understood, the next step is to evaluate how expected outcomes change as a result of the existing treatment options that represent the standard of care within the market segment. Using the same example, to what extent does the prognosis change when patients over the age of 65 with paroxysmal atrial fibrillation originating in the pulmonary veins are treated by a cardiologist using a minimally invasive procedure such as pulmonary vein ablation (or another relevant treatment)? Be sure to take note of the improvement in outcome, as well as any complications or new risk factors introduced by the treatment, as these must also be considered (see 2.1 Disease State Fundamentals and 2.2 Treatment Options).

Determining how well existing treatment options meet patient needs requires some qualitative analysis, as there are no "rules" that define how great an improvement in outcome a treatment must provide to satisfy a patient's needs. Ideally, each treatment would provide a long-term "cure" or prevent the consequences of the condition it was developed to address. However, in the absence of a complete cure, the innovator must apply his/her judgment to evaluate how effective existing treatments are in improving patient outcomes *and* the extent to which new complications and risks associated with each treatment are considered an acceptable trade-off.

These factors, in combination, determine how satisfied patients and physicians are likely to be with the available treatment options. By extension, their level of satisfaction with existing treatments typically corresponds to the level of perceived need for new treatment alternatives within a market segment. If the level of satisfaction is relatively high, it does not necessarily mean that a new treatment is not needed. It does, however, mean that any new treatment option introduced by the market will have to perform significantly better than the available treatment (in terms of improved outcomes and diminished risks/complications) in order to be adopted. See Figure 2.4.7.

In addition to looking at individual treatment alternatives, an innovator should review the SWOT analysis that s/he performed when exploring the dynamics within each market segment. The results of a SWOT analysis can help an innovator identify broad unmet market needs and gaps in competitor offerings that should be taken into account when evaluating overall market needs.

Step 5: Willingness to pay

Another important factor to consider is how much any given segment is potentially willing to pay for a new innovation. This is not the same as asking how much the segment is willing to pay for an existing treatment, since new treatments may afford stakeholders new benefits that could result in a greater willingness to pay. Instead, the question is tightly linked to the information uncovered via 2.3 Stakeholder Analysis and 4.3 Reimbursement Basics. The answer depends to a large degree on which stakeholder(s) will be making the payments and how the new innovation will affect him or her. Table 2.4.3 outlines the key stakeholders who are likely to determine a segment's willingness to pay and the issues to consider when evaluating their response.

To evaluate and understand the answers to these questions, the innovator can use one of the following three approaches. Ideally, more than one of these approaches will be used to allow the innovator to triangulate the information collected.

1. **Comparables** – Evaluate the prices for similar devices (across medical fields) and existing treatment options (within the same medical field). These comparable prices provide a benchmark for how high a price the market segment will bear.
2. **Value analysis** – Assess the anticipated financial gain to the user (provider, patient, or payer) for adopting the new technology (e.g., increased revenue, reduced cost). The total expected financial gain per patient is an upper bound on how much the user is willing to pay.
3. **Willingness-to-pay survey** – Collect information directly from stakeholders regarding the maximum payment they would be willing to make for different degrees of innovation. The survey should provide a description of the benefits a potential solution might provide and a range of prices. Users can then be asked whether or not they would be willing to use the device for each of the prices presented.

Step 6: Target market

The target market represents the market segment that is most likely to respond favorably to a new solution related to the medical need in which the innovator is interested. Choosing a target market takes into account information uncovered throughout the market analysis process. Conceptually, the innovator puts all potential

FIGURE 2.4.7
Market analysis often provides the innovator with conflicting answers to important questions. But a rigorous assessment approach will almost certainly result in less ambiguous information than shown above.[23]

Table 2.4.3 *Providers, patients, and payers can all affect willingness to pay for a new medical innovation.*

Stakeholders	Common issues affecting their willingness to pay
Providers (physicians and facilities)	• Will physicians pay for the device from the reimbursement they receive for the procedure? • Is the reimbursement adequate for the device? • Does it afford physician opportunities to expand the number of patients they treat? • What is the total cost of using the device and how does it compare to the cost of existing treatment(s)? • What is the clinical benefit that the device can offer relative to existing treatment options? How much have physicians paid for devices that offered similar benefits? • Will the payment be part of the costs of supplies incurred by the facility or can it be reimbursed separately? • Does the new device provide market benefits for the facility such as the opportunity to grow its market penetration? • What is the total cost of supplies for the procedure that would utilize this device and what is the profitability of the procedure with the new device? • How does it compare to the profitability of comparable procedures?
Patients	• Will patients pay out-of-pocket (this is especially relevant for cosmetic procedures)? • Historically, how much have patients been willing to pay for similar procedures?
Payers	• Is the device covered by existing reimbursement codes or is there a need for a new code? • What is the total cost of care for patients in each segment with existing treatment options and how will the new device affect the total cost of care?

market segments through a funnel that screens them against the desired characteristics for market size, market dynamics, and market needs. The target market is the segment that scores best against these criteria. Figure 2.4.8 summarizes many of the factors discussed to this point in the chapter. However, a few additional factors are worth consideration when deciding on a target market.

Investor funding

Ideally, a target market will have sufficient investor interest that the innovator feels comfortable about raising necessary funds as the biodesign innovation process progresses. If so, the challenge for the innovator will not be to persuade the investors that the market has commercially attractive needs (which can be a difficult challenge). Instead, it will be convincing them that the need identified and the concept chosen to address the need are unique and can be differentiated from existing or future competitors. Signs that investor funding is available include the creation of investor-funded new businesses in the space, and merger and acquisition activity (which provides liquidity for investors and drives new investments – see 6.3 Funding Sources). The presence or absence of these signs will be revealed as part of the market dynamics steps.

Accessibility of the market segment

Some market segments are easier to reach than others. For example, undiagnosed patients are significantly more difficult to target and attract since they do not yet understand that they are potential customers. Generally, a strong target market will be made up of patients who are already part of the healthcare system (so they can easily be reached), who recognize their need (so they are potentially receptive to new, improved solutions), and who are seen by a small number of physicians who can easily be targeted by a sales force. This means that specialties are typically more easily targeted than the primary care market. Think about market accessibility when performing the preliminary market segmentation step.

Likelihood of segment to adopt new technology

In some medical areas, new treatments are being introduced all the time; in others, the same treatment

Stage 2 Needs Screening

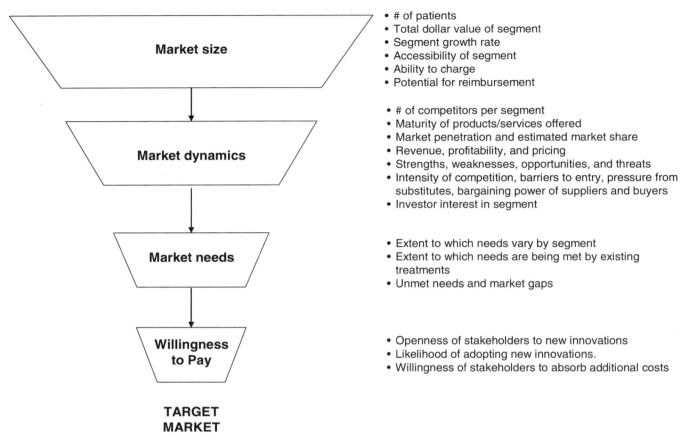

FIGURE 2.4.8
The target market should be the market segment that appears the most promising in terms of size, competitive dynamics, market need, and willingness to pay.

paradigms have been in use for decades. Similarly, certain patient, provider, and payer groups have characteristics that make them more or less willing to consider new solutions. To a large degree, this has to do with the general personality type of the physicians comprising a particular specialty (e.g., aggressive versus conservative), which is influenced by the type of medical care they provide and the environment in which they work. Do not underestimate the importance of this willingness when selecting a target market (and consider it along with willingness to pay). For example, interventional cardiologists have a culture of embracing the latest innovation, while gynecologists tend to be more conservative because of high malpractice liability insurance. Such biases must be understood and taken into account during market analysis.

These additional factors can potentially make one market segment more attractive than another and should be used to help evaluate the results of any preliminary analysis. An example from cardiology illustrates the impact that these factors can have. Following the successful development and adoption of percutaneous transluminal angioplasty by interventional cardiologists, there was a flurry of start-up activity focused on the development of devices that would enable minimally invasive cardiac surgery. The driver behind this activity was the idea that, just as interventional cardiologists had rapidly adopted angioplasty as a minimally invasive treatment option for patients with coronary artery disease, cardiac surgeons would quickly embrace minimally invasive techniques as well. Heartport, a company profiled in 5.4 Regulatory

Table 2.4.4 *These criteria can be used as indicators to identify favorable markets.*

Criteria	Desirable attributes
Market size expressed as total number of treatments multiplied by the price per treatment	Large
Market growth in total market size and in number of procedures	Growing
Existence of a need	Clear and compelling
Ability of existing products/service to meet the market need	Clear gap
Pricing and profitability of existing products	High

Strategy, was one of the early pioneers in this space. Founded in 1991, Heartport struggled for several years to achieve adoption for its device that enabled minimally invasive surgery through a keyhole cut in the patient's chest. By the late 1990s, Heartport had trained more than 500 surgeons, but half of them had abandoned the procedure.[16] The company was ultimately acquired by Johnson & Johnson in 2001 for $81 million.

A key lesson from the Heartport experience is that a tremendous difference in adoption patterns exists among clinical specialties that, on the surface, may seem interrelated. Interventional cardiologists have a maverick reputation and they always want to try the latest and greatest new gadgets. Cardiac surgeons tend to be more conservative, preferring to rely more on their skill as opposed to gadgets, partly due to the hierarchical nature of their training. As a consequence of the Heartport experience, the interventional cardiology space remains a fruitful space for new innovation and a highly attractive (although competitive) market segment. In contrast, little medical device innovation takes place in minimally invasive cardiac surgery, even though the need is great, due to the adoption patterns of the surgeons in the field.

While no market segment will rank positively on all of the dimensions that innovators are encouraged to consider, the overall goal of market analysis is to identify and target the strongest possible market segment across the greatest number of these dimensions, particularly with respect to size, growth potential, and the absence of negative characteristics. Importantly, there are no clear rules for evaluating one market segment against another. Some innovators take a purely qualitative approach to this exercise. Others create quantitative ratings systems to help them score and prioritize segments. In either scenario, the innovator must exercise judgment in making the final determination about the most attractive target market to pursue. More often than not, s/he must also be willing to make a decision with limited information and be prepared to defend the decision. The approach outlined in this chapter is designed to help the innovator develop a compelling argument to support his/her selection of a target market.

Bringing it all together

The process of performing detailed market analysis and identifying a target market involves collecting an immense amount of information and synthesizing it into a coherent story that supports the chosen target market. This data is also essential to determining whether or not an adequate market opportunity exists to support the development of an innovation that addresses the need(s) within that market. Table 2.4.4 summarizes the key criteria that should be included in the output of any market analysis and the attributes of desirable markets for each criterion.

The results of a preliminary market analysis enable the innovator to refine the need statement (and develop a need specification) to appropriately address the needs of the target market. In fact, the need statement may be customized slightly by market segment for maximum impact (see 2.5 Needs Filtering). Eventually, this output can also be used as the foundation of a detailed sales and marketing plan (see 5.8 Sales and Distribution Strategy), as well as to support the company's financial model and funding strategy (see 6.1 Operating Plan and Financial Model, as well as 6.3 Funding Sources).

Working example: Market analysis for chronic kidney disease

The following example provides an overview of market analysis performed in the area of chronic kidney disease. It demonstrates how publicly available databases can provide important information to support market analysis and how the results can be presented in using so-called **patient towers**.

Patients with CKD are often diagnosed late in the disease's progression and are frequently hospitalized. However, early, effective diagnosis of CKD reduces the incidence of hospitalization and also diminishes the rate of disease progression. As a result, there is a need for a device, method, or system that would successfully diagnose patients with CKD earlier in the cycle of care to help prevent disease progression.

When performing market analysis for this need, the innovator can begin by doing patient-based market segmentation, using markers of disease progression to identify basic similarities and differences among patient populations. Recall that for CKD, the marker is a measure of kidney function that can be used to divide patients into five CKD stages: 1–2 = mild, 3 = moderate, 4 = severe, and 5 = ESRD (end-stage renal disease). (Note that provider and payer considerations will be taken into account later in this example.)

Once these basic segments are defined, primary and secondary research can be conducted to identify the size of each segment (number of patients), extent of the medical need in each segment (mortality and morbidity rates, expenses per patient), and segment growth rates. Literature searches can be used as a starting point, but beyond that, public databases can provide relevant information. Two such databases that are widely used are the National Health and Nutrition Examination Survey (NHANES)[17] and the Medical Expenditure Panel Survey (MEPS).[18] These surveys provide longitudinal data for large samples of patients regarding medical condition(s) and medical expenditures for families and individuals, and they provide a wealth of information that can be used for any preliminary market analysis. Using data from sources such as these, a bottom-up analysis of the market size for the various segments can be performed.

The NHANES database provides information that allows the innovator to calculate the prevalence of each stage of CKD, as well as mortality rates, morbidity rates, and

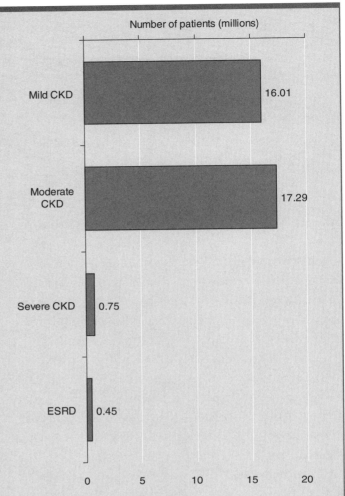

FIGURE 2.4.9 (a)
Patient towers for CKD: number of patients by segment.

hospitalization expenditures. Figure 2.4.9(a) presents the number of patients in four segments. Figure 2.4.9(b) shows the corresponding rates of hospitalization.

These data are presented in a patient tower format in which the total patient population is divided into segments of diminishing size, but increasing medical needs. Reading from the top of the diagram (mild CKD) to the bottom (ESRD), the number of patients typically decreases but the severity of the condition increases. While other visual methods can be used to present these data, the patient tower approach has the advantage of visually contrasting the trade-off between the number of patients in each segment and the severity of the medical issues affecting each segment.

To identify a target market, the innovator also needs to get a sense for the cost associated with each stage of the disease. Sources such as NHANES do not provide

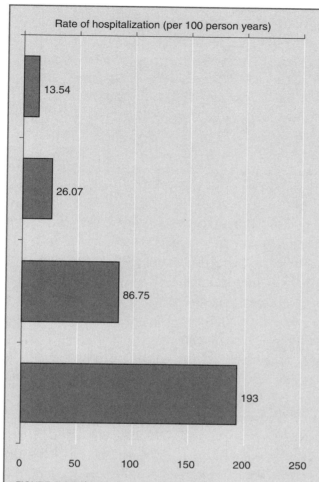

FIGURE 2.4.9 (b)
Patient towers for CKD: hospitalization rate by segment.

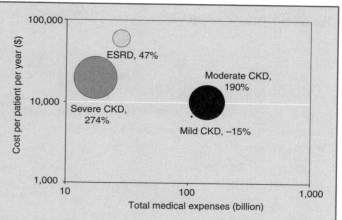

FIGURE 2.4.10
Market size and growth analysis for CKD.

cost information (even though they provide information on the utilization of medical services). Therefore, some data extrapolation using numbers obtained from a literature search may be required. Another potential strategy is to use data from MEPS, which typically includes expenditure information or data from disease-specific data sources. Figure 2.4.10 summarizes the data (the horizontal axis gives the total medical expenses for each segment, the vertical axis gives the cost per patient in each segment, and the size of the bubble indicates the expected growth rate). The data in the figure were based on expenditure data provided in the 2007 US Renal Data System report (the annual medical cost per prevalent case of severe CKD was $20,784 in 2003 and for ESRD patients it was $59,412).[19] To obtain cost estimates for the other two segments (where no data were reported), an assumption can be made. In this case, the CKD cost for earlier stages was scaled by taking the cost for stage 4 and multiplying the ratio of the hospitalization rate in each stage, divided by the hospitalization rate for beyond stage 4 (cost for stage 1 (or 2) = cost of stage 4 × hospitalization rate in stage 1 (or 2) / hospitalization in stage 4). Using this approach, the total annual medical cost for each stage was estimated; then the growth in the prevalent population (estimated from NHANES) was used to obtain the growth in total costs for each stage.

With this information, the innovator can define the characteristics of the ideal market segment: large total medical expense (all segments satisfy this), high expenses per patient (which makes the total number of patients relatively small and easier to access), and a high growth rate. Using these characteristics as a filter, the ESRD and severe CKD segments emerge as most attractive.

However, before the analysis is complete, the innovator needs to study whether all patients who have the condition are actually aware of it (if not, they are unlikely to seek treatment, which will negatively affect the total potential market size). In the example, data provided via NHANES included a response to a survey question that asked patients whether or not they were aware of diminished kidney function. This survey showed that only 25 percent of the patients with serious CKD were aware of it, and all those who were aware of it had other symptoms that helped make the diagnosis (e.g., elevated albumin to creatinine ratio – ACR). This information suggests that patients who are correctly diagnosed are often patients with high ACR. As a result, the target market could be defined in even more specific terms: patients with severe CKD and high ACR.

> **Working example:** *Continued*
>
> To some, this definition of the target market may appear to be surprising (after all, these are the patients who are already diagnosed). However, what makes this patient segment attractive is that their disease is already recognized by the healthcare system and, therefore, this segment can be accessed by the innovator. For this reason, addressing the need for better diagnosis and/or control of complications for these patients is more likely to be clinically feasible than targeting patients who are not currently diagnosed. The latter segment (of undiagnosed patients) could potentially provide an innovator with an opportunity for expansion, once the preliminary target market has been addressed, but may not be as easy to target.

Common pitfalls

When performing market analysis, it is not unusual for innovators to make certain common mistakes. One of the most frequent pitfalls is the tendency to make estimates that are too high. The following three errors lead to overly optimistic market estimates:

1. The innovator calculates market size based on the total market, not the market segments most likely to adopt an innovation. This error can be driven by failure to understand the needs of specific segments and their different adoption patterns.
2. The innovator fails to recognize that not all people with a need will take steps to address it (e.g., seek treatment). This could be the consequence of relying on "sanitized" data obtained from well-controlled studies. When obtaining estimates for the number of patients, it is important to confirm that the estimates reflect community practices and not well-controlled academic environments.
3. The innovator overestimates the amount of market share that a new entrant can capture in an established market or the rate at which a new innovation will be adopted in an emerging market. This can be partly due to inexperience and partly due to underestimating the effect of competition and the capital requirements needed to sell a product. In most cases, innovators should not expect to gain more than 1 percent market share in the first year, increasing to a maximum of 15 to 30 percent based on market conditions.

Overestimating a market opportunity can give innovators unrealistic expectations and result in a business plan that is unsustainable. It can also hurt their credibility with stakeholders (e.g., physicians and investors) if they present an idea that is based on overly aggressive or impractical assumptions.

Just as innovators may overestimate the market potential, they may also define a market opportunity too narrowly, or fail to recognize that an innovation has the potential to address similar needs in markets beyond the immediate target. For example, EndoNav, a company that developed a disposable scope that could be inserted over a standard colonoscope to reduce the difficulty of complex colonoscopies and minimize adverse effects associated with such procedures, fell victim to this problem. In its early development work, the company failed to recognize that the need for improved navigation within the body existed in multiple practice areas, not just in colonoscopy. As a result, it defined its market too narrowly. In contrast, NeoGuide, a competitor to EndoNav, recognized the ability of its technology to serve as a platform across practice areas. Based on this approach, the company was able to make a more compelling case to investors and received funding much earlier. By the end of 2007, EndoNav was acquired by a larger company that recognized the platform potential while NeoGuide was refocusing its business strategy on expanding the platform potential of its technology.[20]

Remember that market analysis can be complicated and difficult, potentially requiring a certain amount of trial and error. An iterative, increasingly detailed approach to market analysis is recommended. Innovators should be sure to revisit and adjust their preliminary market analysis as more becomes known about the need, as well as potential solutions. The story of Genomic Health demonstrates how one company tackled the challenge of market analysis.

FROM THE FIELD — GENOMIC HEALTH INC.

Market analysis for a revolutionary product

Identifying the target market segment and sizing the overall market can be especially challenging for an innovator considering a need with ground-breaking potential, but no proven market. Genomic Health Inc., founded by Randy Scott, Joffre Baker, and Steve Shak in August 2000, is one company that faced this challenge.

In his role as the cofounder of Incyte Corporation, Scott and his team had made massive databases of genomic information available to major pharmaceutical companies to aid their research and development efforts in identifying targeted therapies at the molecular level. However, as the cost of genomic information came down (due to what Scott called a "Moore's Law effect" in biotechnology), he recognized that it would become possible to analyze genomic information on a patient-by-patient basis and develop truly personalized regimes to treat disease. Passionate about helping make this happen, he set out to pursue *the need for high-value, information-rich diagnostics based on patient-level (gene expression) genomic testing to enable more personalized treatment decisions*.

Scott initially pitched the idea to Incyte, but the need was met with mixed reviews. "One of my more controversial views with Incyte at the time," recalled Scott, "was that drugs would ultimately be commodities, and that greater value would come from genomic information about disease. I believed that there was more power in the information than in the solution because what's more important than understanding exactly, precisely what the molecular cause of a disease is? Once you have that information, there will be ten potential therapies developed to treat it. These drugs will ultimately become generic, but the value in the diagnosis of that information will hold."[21] Scott positioned this as a paradigm shift in the healthcare industry that would occur gradually over the next 30 years. When Incyte decided it was not interested in taking this direction, Scott decided to pursue the idea himself, recruiting Baker and Shak to work with him.

In terms of analyzing the market opportunity related to the need, Scott remembered that they had an initial interest in focusing on cancer patients. However, with so many other life-threatening conditions affecting relatively sophisticated patients who would understand the benefits of personalized, genomic-based medicine, more in-depth research of the potential market segments was required. To narrow the scope of their segmentation analysis, they identified several fields (such as oncology, inflammation, cardiovascular, infertility) that would potentially benefit from personalized medicine and ranked them according to several clinical and market criteria, including market size; potential for genomic information to predict disease progression; drug development pipelines (and whether the percentage of responders for the new drug would be high); patient involvement in treatment choices; and physician willingness to adopt new treatment paradigms. After performing this analysis, "The bottom line," said Scott, "was that we just kept coming back to cancer." Unlike other diseases, such as cardiovascular, where lifestyle factors played a big role in disease progression, cancer is mostly a genome-based disease. Further, response to existing drugs (or drugs under development) was considered to be variable and uncertain, and a strong history of patient advocacy showed that patients would drive the adoption of new technologies. Beyond that, the total market for oncology drugs was in the multibillions of dollars, with most drug manufacturers developing several new therapeutics.

While identifying cancer as a focus area was a good first step, it was only the beginning of their segmentation analysis. The next challenge was to determine which type of cancer to target. Ultimately, they decided to focus on breast cancer for five primary reasons: (1) *Prior experience*: Shack led Genentech's clinical development program for Herceptin® (a novel, genomic-based cancer therapeutic[22]) and had a vast network of contacts, as well as a strong reputation in that clinical area; (2) *Top four*: Breast cancer is one of four most prevalent cancers (the others being colon, prostate, and lung cancer) and the team wanted to "do something big"; (3) *Market accessibility*: Breast cancer is characterized by a large system of patient advocacy groups, readily accessible education channels that make it easy for a new company to reach the market segment, and patients that tend to be highly involved in their treatment decisions; (4) *Likelihood of adopting new technology*: Both physicians and patients are likely to embrace new treatments in this clinical area; (5) *Clinical knowledge*: There is a deep and broad body of knowledge on the genetic basis of breast cancer and a vast library of breast cancer tissue samples that make it technically feasible to develop and clinically validate any new test.

FROM THE FIELD: GENOMIC HEALTH INC.

Continued

Within breast cancer, the founders focused on further refining the target market segment. The team examined the various drugs available to breast cancer patients, and their effectiveness or ineffectiveness at various stages of the disease. What they learned is that for late-stage cancer patients, treatment choices are unambiguous and highly aggressive. Patients and physicians do not want to "give up hope" and they will try everything to stop or slow the disease. In contrast, decisions about what treatments to pursue (beyond surgery) for early-stage patients were relatively subjective and varied significantly from physician to physician, making it difficult for patients to determine the appropriate course of treatment to prevent recurrence. Most early-stage patients do well with only surgery and hormone therapy. However, chemotherapy is commonly prescribed to minimize a woman's chance of repeat tumors even though only a small fraction of early-stage breast cancer patients benefit from it (approximately 4 percent). This led to an "aha" moment for the team: predicting distant recurrence of breast cancer for early-stage patients would likely help them make better decisions about what treatment to pursue. For example, patients with a low risk of distant recurrence could potentially be counseled to forego chemotherapy, a physically stressful and disruptive therapy, since their cancer was unlikely to return. Taking the process one step further, based on what was known about the disease, the team defined its preliminary target market segment as patients with early stage, node-negative (N–), estrogen receptor positive (ER+) breast cancer. They also further refined their need statement to address requirements for *high-value, information-rich diagnostics based on patient-level (gene expression) genomic testing to predict the recurrence of early stage, N–, ER+ breast cancer and enable personalized treatment decisions*.

With the market segment identified and need refined, the next challenge was to determine the total market size in order to determine if the market potential would justify the high anticipated cost of clinical development and the anticipated risk to prospective investors. According to the company's estimates at the time, there were approximately 200,000 new breast cancer cases diagnosed in the United States each year, of which roughly 50 percent were early stage, N–, ER+ cases. Quantifying the total market size (in dollars) was challenging, though, because of the lack of comparable products available in any clinical field. Any solution that would address Genomic Health's defined need would be classified as a diagnostic test. However, Scott and team determined that diagnostic companies traditionally charge between $25 and $50 for their tests, commanding margins of just 5 to 10 percent. Using a $50 price for the new test, the total potential market size (assuming full penetration) would be just $5 million per year. This estimate clearly illustrated that the success of the venture would necessitate a completely different pricing paradigm to support the heavy investment in research and development required to make genomic testing practical.

Two types of analysis suggested that a price in the range of $1,000 to $7,000 per test could be viable. The first was *comparables analysis*. Kim Popovits, COO of Genomic Health, recalled, "There was another diagnostic in the marketplace at that time, a genetic test that looked at the mutation of the BRAC-1 and -2 genes to assess a woman's hereditary risk of breast cancer." This test was priced around $3,000 and was on its way to being reimbursed on a relatively broad scale. The second was value-based pricing analysis, which linked back to Scott's original **value proposition**. Over time, Genomic Health's test had the potential to save money for the overall healthcare system and could, thus, shift the pricing power from therapeutics to diagnostics. Specifically, the total cost of chemotherapy for early-stage breast cancer patients was conservatively $15,000. If the test cut the number of patients undergoing chemotherapy by 50 percent (by predicting low recurrence risk), then the total savings to the healthcare system would be roughly $7,000 per patient. This meant that Genomic Health could command a price of up to $7,000 per test, making the total potential market $700 million.

In fact, this estimate brought Scott and team back to an earlier market estimate they developed when they raised their preliminary funding. Back in 2001, Genomic Health's business plan was not as precise about the market segment and size (highlighting the iterative nature of market analysis as part of the biodesign innovation process). The original vision was that consumers would pay for genomic test results out-of-pocket. As a result, the team felt that a reasonable price would be somewhere in the four figures range (primarily because consumers pay analogous amounts for cosmetic surgery and other

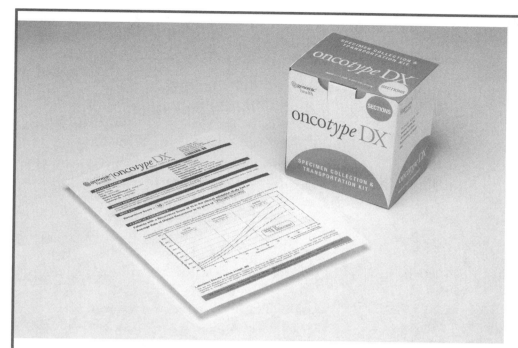

FIGURE 2.4.11
The Oncotype DX specimen collection and transportation kit, along with a results report (courtesy of Genomic Health).

elective medical procedures funded out-of-pocket). Using similar projections regarding the total number of consumers willing to pay for the test, the founders reached a total market estimate in the billion dollar range and raised more than $30 million in investments. About a year later, they focused their market segment and further developed the company's value-based pricing model.

In summary, through the careful analysis of the needs of different patient segments, the founders of Genomic Health decided to focus on the development of a test that would predict the risk of distant recurrence for early-stage breast cancer patients. They realized that success of their venture required a value-based pricing strategy, with a target price in the $1,000 to $7,000 per test range. Their product and clinical development strategy would be focused on addressing their stated need and providing the evidence that would support a value-based pricing strategy. By the end of 2007, Genomic Health's first product, Oncotype DX®, had been used by more than 40,000 patients (see Figure 2.4.11).

Several clinical studies established that the test not only predicted the likelihood of recurrence, but also a patient's response to chemotherapy. Priced at $3,650, the test was covered for more than 60 percent of insured patients. It also was included in the treatment guidelines set forth by the American Society of Clinical Oncology and the National Comprehensive Cancer Network.

A note on global markets

The overall market opportunity for a clinical need is often global in nature. While the approach and examples outlined in this chapter are US-based, they can be used to help assess opportunities in different markets. When performing a global market analysis, innovators should be aware of important differences across countries that may affect their results. At a high level, many key differences are driven by how healthcare is financed and delivered in other countries, as well as by variations in standards of care. Important variations can also be seen between competitors based in the United States versus other countries, such that understanding the nature of local competition becomes important. Given the complexity of global medtech markets, innovators are advised to focus on a single market at this stage of the biodesign innovation process to avoid spreading themselves too thin.

Stage 2 Needs Screening

Validating the market opportunity associated with a need is of critical importance, but can also be an exciting and rewarding experience. Use the following steps to tackle this challenge.

See **ebiodesign.org** for active web links to the resources listed below, additional references, content updates, video FAQs, and other relevant information.

1. Perform market segmentation

1.1 What to cover – Take an iterative, progressive approach to defining market segments. To address the complex interactions with multiple stakeholders in medical markets, start by segmenting patients based on symptoms and risk factors. Then, segment these patient subgroups based on treatments. Next, segment the patient/treatment subgroups based on their providers. Finally, segment the patient/provider subgroups based on payer attributes. Consider whether there are other unique characteristics of patients, providers, or payers in the particular area of need that should be "layered" in to the segmentation analysis to ensure a complete understanding of the similarities and differences among customers in this market.

1.2 Where to look
- **Business plans** – Reviewing the business plans of other innovators and entrepreneurs may help stimulate ideas regarding the development of segmentation frameworks.
- **Other chapters** – The analysis described in 2.1 Disease State Fundamentals may be useful for performing risk-based segmentation. The information in 2.2. Treatment Options may be helpful for provider segmentation and segmentation based on treatment options. 2.3 Stakeholder Analysis and 4.3 Reimbursement Basics may be helpful with provider/payer segmentation.
- **UpToDate** – A database of evidence-based clinical information.
- **PubMed** – A database of the US National Library of Medicine that includes more than 16 million citations from MEDLINE and other life science journals back to the 1950s.
- **Wrong Diagnosis** – A disease and symptom research center.
- **Centers for Disease Control and Prevention** – Part of the Department of Health and Human Services (HHS).
- **Clinical trials** (see clinicalTrials.gov or published clinical trials at PubMed) – Enrollment criteria for clinical trial designs can provide insights into how competitors segment the market.
- **Product labels for other treatments** – Located on the FDA website, the labels for approved treatments can provide insights into different market segmentation approaches.
- **Marketing consultants** – Professional market research analysts and marketing consultants can help to complete the process.

2. Size the market

2.1 What to cover – The goal of this activity is to calculate the size and value of each market segment that has been defined. Perform both top-down and bottom-up evaluations and rationalize/reconcile the results. By combining the top-down and bottom-up approaches, confidence intervals also can be derived for the size of the market. Other factors to take into account include the potential cost of the new treatment relative to existing treatment, the likelihood and rate of reimbursement, and the growth rate within the segment. All of these factors will influence the total value of each market segment.

2.2 Where to look
- **Agency for Healthcare Research and Quality** – AHRQ sponsors and conducts research that provides evidence-based information on healthcare outcomes; quality; and cost, use, and access. From its website,

one can gain access to longitudinal data regarding patient interactions with the healthcare system, including all transactions and their codes (indicating procedures and diagnosis), as well as location of service, which can be used to help develop market segments. Important databases accessible via the site include:

- **HCUPnet** – A free, online query system based on data from the Healthcare Cost and Utilization Project (HCUP). It provides access to health statistics and information on hospital stays (inpatient encounters) at the national, regional, and state levels.
- **MEPS Data** – The Agency for Healthcare Research and Quality provides longitudinal data on the health expenditures of 30,000 US households via the Medical Expenditures Panel Survey (MEPS). This data is publicly available for primary analysis. It is useful for more detailed analysis of market segments (and sizing), but working with the data can be labor intensive. This source is probably most helpful if the innovator or company needs to support a need specification with actual publications as part of a marketing strategy. Data are available for conditions with a 1 percent prevalence rate or greater.

- **Nutrition Examination Survey** – Provides longitudinal data for large samples of patients regarding medical condition(s) and medical expenditures for families and individuals.
- **Professional societies** – Professional societies often provide statistics and data that can be useful in sizing a market.
- **Analyst reports** – When investment analysts begin coverage for a company, they typically write an initiating coverage report that gives an overview of the company's products and the market for those products. These reports are available using a database such as Reuter's Research On Demand, OneSource's US Business Browser, and Business Insight. Typically, it is easier to find reports focused on established companies and treatment areas than for emerging companies and treatment areas.
- **Market research reports** – Depending on the topic, MarketLine, MarketResearch.com Academic, and Mintel may be useful sources (all three offer healthcare selections). Market research reports can also be accessed on the Internet (e.g., through Google). Sites such as Datamonitor and Marketresearch.com have healthcare sections that provide detailed reports that sometimes include primary research. Unfortunately, these reports often cost several thousand dollars to purchase. However, if a market research budget happens to be available, this may be a lead worth pursuing.
- **Press releases** – Market research firms put out press releases when they publish new research reports. If the report itself is too expensive, these one-page "teasers" usually provide overall dollar amounts spent on treating a given disease. Business Wire is one source for this information.
- **Disease and treatment research** (see 2.1 Disease State Fundamentals and 2.2 Treatment Options for basic epidemiology and economics).
- **Annual reports and SEC filings** – Large firms traded in public markets such as the New York Stock Exchange issue annual reports with detailed financial information and they also file annual financial reports, called 10-Ks, with the Securities and Exchange Commission (SEC).

3. Understand market dynamics

3.1 What to cover – Create profiles of the products and companies associated with the most dominant treatments. Identify the product costs and market power of each player. Also

Stage 2 Needs Screening

include emerging companies that may not have a product on the market yet. Perform a SWOT analysis for each of the most powerful market players. Use Porter's five forces to evaluate the market as a whole. Then, summarize the results.

3.2 Where to look
- **Treatment research** (see 2.2 Treatment Options to identify competitors and results from gap analysis).
- **Analyst reports**
- **Market research reports**
- **Annual reports of publicly traded companies**
- **Company websites**
- **Press releases**

4. Assess market needs

4.1 What to cover – Evaluate how market needs may vary by segment by referring back to observations and research performed early in the needs finding process. Assess how well needs within each segment are being addressed by current treatment alternatives. Start by evaluating outcomes for the disease state without treatment. Then, analyze how these outcomes change when existing treatments are utilized. Make a qualitative judgment regarding the extent to which patient needs are being satisfied relative to improved outcomes and new/different complications and risks introduced by the treatment options. Revisit the SWOT analysis to ensure that no broad unmet needs or market gaps are overlooked.

4.2 Where to look
- **Innovation notebook** – Refer back to notes made during observations and as a result of research conducted during the development of the preliminary need statement (see 1.2 Observation and Problem Identification and 1.3 Need Statement Development).
- **Disease and treatment research** (see 2.1 Disease State Fundamentals and 2.2 Treatment Options).
- **SWOT analysis** – Look back at the materials used to conduct the SWOT analysis and review the summary of market/competitor dynamics.

5. Assess willingness to pay

5.1 What to cover – Determine the price per procedure that the market will bear. Consider which stakeholder would be making the payment and each stakeholder's willingness to pay by evaluating comparable devices, performing financial analysis related to the cost and benefits of the device, or surveying a sample of stakeholders to ascertain this information.

5.2 Where to look
- **2.3 Stakeholder Analysis**
- **4.3 Reimbursement Basics**
- **Market research reports**
- **Company websites**

6. Define the target market

6.1 What to cover – Once various market segments have been analyzed, a target market must be chosen. Evaluate each segment according to the factors and criteria outlined above to determine the most favorable target. Remember to take into account other market enablers, such as the availability of investor funding, the accessibility of the market, and the receptiveness to new technology.

6.2 Where to look
- **Quarterly Venture Capital Activity Reports** – Provides quarterly listings of venture capital investment activity in the United States based on data from PricewaterhouseCoopers, Thomson Venture Economics, and the National Venture Capital Association.

- **VentureXpert** – Provides data on mergers and acquisitions, IPOs, and venture capital funding based on data from Securities Data Corporation.
- **Venture capital firm websites**
- **Analyst reports**

7. Bring it all together

7.1 What to cover – Validate the target market through the development of patient towers that summarize the size of the patient population, the extent of the medical need, and the growth rate for each of the primary market segments considered. These charts can be used not only to confirm that the right target market has been chosen, but also to convince stakeholders and potential investors that a sizable opportunity exists.

7.2 Where to look – Refer back to references used to segment the market and determine the size of its subgroups. Discuss the findings with key advisors and team members.

Credits

The editors would like to acknowledge Darin Buxbaum and Donald K. Lee for their help in developing this chapter. Many thanks also go to Randy Scott and Kim Popovitz for sharing the Genomic Health story, Amir Belson for discussing NeoGuide, and Ross Jaffe of Versant Venture for sharing his perspective on market analysis.

Notes

1. From remarks made by Richard Stack as part of the "From the Innovator's Workbench" speaker series hosted by Stanford's Program in Biodesign, March 6, 2006, http://biodesign.stanford.edu/bdn/networking/pastinnovators.jsp. Reprinted with permission.
2. From remarks made by John Abele as part of the "From the Innovator's Workbench" speaker series hosted by Stanford's Program in Biodesign, February 2, 2004, http://biodesign.stanford.edu/bdn/networking/pastinnovators.jsp. Reprinted with permission.
3. "Market Segmentation," www.QuickMBA.com, http://www.quickmba.com/marketing/market-segmentation/ (August 21, 2007).
4. "U.S. Medical Device Market Outlook," Frost & Sullivan, 2008.
5. KCI 10-K form for FY2006 and analyst reports revealed that the VAC device is rented at a monthly rate of $2,000. Adding to that the cost of the disposable elements of the therapy (screens, dressings, etc.) and the cost of the healthcare professional changing the dressings, the total cost was estimated at $150 per day.
6. It should be noted that ulcers larger than 5 cm^2 take longer to heal (68 days) but, to be conservative, it was assumed that the average healing time is 52 days, as it is for smaller venous ulcers.
7. Estimates from Spiracur analysis.
8. Michael E. Porter, "How Competitive Forces Shape Strategy," *Harvard Business Review*, March-April 1979.
9. Ibid.
10. Manuel Baigorri, "Abbott Fights Fifth Patent Accusation," *Medill Reports*, January 16, 2008, http://news.medill.northwestern.edu/chicago/news.aspx?id=74855&print=1 (September 25, 2008).
11. "J&J/Cordis' Cypher DES Threatened by Supplier Dispute," *Medical Design Technology*, November 2006, http://findarticles.com/p/articles/mi_hb4875/is_200611/ai_n17921724 (September 25, 2008).
12. "Wyeth, Cordis Drop Stent Lawsuit," *Cardiovascular Business News*, January 3, 2008, http://www.cardiovascularbusiness.com/index.php?option=com_articles&view=article&id=12685 (December 4, 2008).
13. "SWOT Analysis," www.wikipedia.org, http://en.wikipedia.org/wiki/Swot_analysis (August 22, 2007).
14. Ibid.
15. "Fast Facts," Abbott.com, http://www.abbott.com/global/url/content/en_US/10.17:17/general_content/General_Content_00054.htm (September 25, 2008).
16. Ralph T. King, Jr., "New Keyhole Heart Surgery Arrived with Fanfare, but Was It Premature?" *Wall Street Journal*, May 5, 1999, p. A1.
17. National Center for Health Statistics, Center for Disease Control and Prevention, http://www.cdc.gov/nchs/nhanes.htm (September 25, 2008).

18. Medical Expenditure Panel Survey (MEPS), Agency for Healthcare Research and Policy, http://www.ahrq.gov/data/mepsix.htm (September 25, 2008).
19. "Cost of CKD and ESRD," 2007 U.S. Renal Data System, 2007, http://www.usrds.org/2007/pdf/11_econ_07.pdf (September 25, 2008).
20. See "EndoNav," GSB No. E-214, 2006.
21. All quotations are from interviews conducted by the authors, unless otherwise cited. Reprinted with permission.
22. Only those patients who over-expressed the genetic alteration for HER2 responded to the drug Herceptin. By screening patients with a genetic diagnostic for this marker, physicians could make more personalized, effective treatment decisions and prescribe Herceptin to only those patients with the greatest probability of benefiting from its effects.
23. Cartoons created by Josh Makower, unless otherwise cited.

2.5 Needs Filtering

Introduction

To aspiring biodesign innovators, the process of filtering needs can feel a lot like comparing apples and oranges when they like both kinds of fruit. One unmet need that applies to millions of patients might offer the opportunity to do little more than alleviate certain symptoms of their underlying condition. Another unmet need involving fundamental mechanics of curing a disease might require years of research to affect just a hundred-thousand sufferers. Yet another could address a vital need common to a vast population, but in a geographic market where patients have little or no ability to pay for the remedy. With all of these trade-offs and challenges swirling in their minds, innovators require an objective process for deciding which needs to pursue.

Needs filtering uses all of the data already collected through the biodesign innovation process to identify a smaller set of needs, or sometimes a single need, that warrants further research and investigation. While the process of needs filtering is inherently subjective, which can be frustrating, one of its key purposes is to incorporate those issues and priorities that are most important to the innovators, as defined by their strategic focus. By taking a rigorous, structured approach to this exercise, innovators increase the likelihood of focusing on those needs that are best aligned with their interests and strengths. The process further ensures that data about important factors, such as the commercial viability of each need, are considered, which also increases the innovator's eventual chance of success. The output of needs filtering is the development of a need specification – a document that synthesizes all of the important data gathered through observations and research, and also outlines the criteria that any solution must meet in order to satisfy the need. This information is then used as the starting point for the generation of preliminary solution concepts.

OBJECTIVES

- Understand how to develop a needs ranking system in which:
  - Key factors to consider, including the innovator's goals, are identified.
  - An objective rating is assigned to each factor for each need.
  - The values assigned to each factor are combined into a score for each need.
  - The needs are filtered and prioritized using these scores.
- Learn how to create a need specification for the highest priority needs.

Stage 2 Needs Screening

Needs filtering fundamentals

Needs filtering requires the innovator to compare one need statement against another to identify the most promising opportunities before moving into concept generation. Since most innovators transition from needs finding into needs screening with an abundance of needs and no clear sense of which ones hold the greatest potential, an objective process to filter needs helps ensure that sound, unbiased decisions are made using the data gathered from 2.1 Disease State Fundamentals through 2.4 Market Analysis. Through this process, the innovator can narrow the list of opportunities s/he is pursuing to between one and ten top priority needs.

The process of needs filtering can be performed in many different ways, but generally involves four essential steps:

1. Select screening criteria or the factors to consider in making a decision.
2. Assign ratings for each factor for each need.
3. Combine values to produce a score that can be used to prioritize the needs.
4. Filter the needs to produce a small set for further investigation.

Each of these steps is described in more detail in the sections that follow. Once the list of needs under consideration has been refined, a need specification can be developed for each one to guide subsequent steps in the biodesign innovation process.

Some innovators are quite systematic about their approach to needs filtering, whereas others may work almost entirely from "gut" feel. Similarly, certain innovators move quickly from many needs to just a few, while others spend more time gradually sharpening their focus. Figure 2.5.1 provides a summary of how one team worked through their needs filtering exercise.

Since the biodesign innovation process fundamentally leads the innovator to identify dozens (or even hundreds) of needs, it is unrealistic to think that s/he can spend weeks investigating each one. Research, as described in Chapters 2.1 to 2.4, should initially be kept at a relatively high level. As needs are filtered through an iterative process and the list of those being considered becomes progressively smaller, the innovator can perform more detailed and thorough research. Keep in mind that preparing a need specification may also lead the innovator to come across information relevant to the needs filtering process. Thus, throughout the needs filtering process, an innovator must strike a delicate balance in deciding how much time and energy to invest in investigating each need before moving on to concept generation.

Step 1: Select screening criteria

The first step in the needs filtering process is to clearly identify which factors to evaluate in comparing the needs under consideration. Some of the most common factors that innovators use to evaluate one need against another are outlined in Table 2.5.1. The research

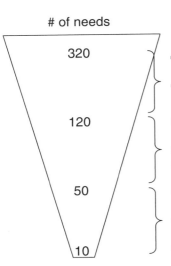

FIGURE 2.5.1
An example of the needs filtering approach used by a multidisciplinary team in Stanford University's Program in Biodesign. The team started with a set of 320 needs, which it whittled to 120, then to 50, and finally to 10 target needs. Increasingly detailed criteria (developed through more thorough research) were used at each step in the filtering process (with permission from Uday N. Kumar; developed with John White, Kityee Au-Yeung, and Joseph Knight).

Table 2.5.1 *Some combination of these factors can be chosen to structure the needs filtering process.*

Factor	Question(s)	Issues/implications
Disease state mechanisms (see 2.1 Disease State Fundamentals)	Are the mechanisms of the disease state associated with the need well understood?	• Needs that are focused on poorly understood disease states (e.g., autoimmune deficiencies) may be more challenging to solve, especially with devices. • Needs focused on multifactorial disease states may not yield simple/direct solutions (e.g., diabetes). • Needs focused on disease states with expanding numbers of affected people are favorable since information can be discovered and learned quickly (e.g., gastroesophageal reflux disease).
Treatment landscape (see 2.2 Treatment Options)	To what extent is the need currently being addressed? Is there a gap between the need and existing treatments (i.e., a "white space" characterized by no solutions)?	• If treatment options currently exist that address the need (or could be "tweaked" to provide a solution), there may be significant competition in the market. • If established treatments exist but the need is still unmet, understand their limitations or shortfalls as a source of information. • Pay particular attention if no treatment options exist to address the need. This can signal both opportunities and risks.
Stakeholder impact (see 2.3 Stakeholder Analysis)	How significant would the benefits to patients be if the need is addressed? How would physicians and facilities be affected? How likely are payers to provide reimbursement for a solution that addresses the need?	• Any perceived or real ethical issues relating to solving a need can have an impact on how that need is received by key stakeholders. • Needs to improve patient outcomes are generally considered particularly important. • In addition to the benefits realized by some facilities and/or physicians from addressing the need, keep in mind that a solution might adversely affect others, e.g., by taking away business or shifting care from one specialty or facility to another. • Payers are more likely to provide reimbursement if the need is related to improved safety or efficacy rather than increased convenience or a better user experience.
Market size and competitive dynamics factors (see 2.4 Market Analysis)	How large is the market for a solution to the need? Which players (and how many of them) offer or are developing therapies in the need space? What magnitude of funding will be required to develop and commercialize a solution in the need area? Is the market large enough to justify this requirement?	• Market size helps quantify the potential impact that an innovator can have by addressing the need. • Market size is also an important factor in determining the likelihood that funding can be raised. • Blue-sky needs tend to have greater financing requirements so they require a larger market to support them. • Incremental needs may have lower funding needs, thus, can be feasible with a somewhat smaller market. • The level and nature of competition in an area of need has significant implications for how easy (or difficult) it will be to penetrate the market.

performed for each need should enable the innovator to answer these kinds of questions.

Another important factor to consider is the type of need that is being evaluated – blue-sky, mixed, or incremental (see 1.3 Need Statement Development). Some innovators specifically want to evaluate a diversity of need types, while others prefer only to investigate those in a particular category. The decision of which type (or types) of needs to pursue is ultimately linked to the risk–reward equation that the innovator desires. Blue-sky needs typically involve larger risks, but can lead to significantly larger rewards; the opposite is usually true for incremental needs. For this reason, it may be helpful to categorize all needs under consideration by type and then determine how many of each type to pursue based on the innovator's preferences.

Innovator preferences should also be taken into account via a series of other subjective factors. Many of the subjective criteria used in needs filtering link back to the strategic focus decided upon at the outset of the biodesign innovation process (see 1.1 Strategic Focus). As described in that chapter, acceptance criteria are the mechanism through which an innovator's mission, strengths, and weaknesses are woven together into a list of requirements that an innovation project must meet. The innovator should revisit his/her defined acceptance criteria and use them to define additional factors to consider. This helps ensure that the projects emerging from the needs filtering process are, in fact, aligned with the innovator's interests. Again, there is no "correct" set of factors to use and they will likely vary based on the types of needs being assessed and the interests of the innovator and his/her team. However, choosing several objective factors and one or two key subjective factors (e.g., the team's level of enthusiasm for the need or the extent to which the need is life-saving) should be sufficient to understand and compare all of the needs on a level playing field. For teams of innovators working together, a personal interest rating provided by each team member can be used to collect and incorporate everyone's feedback on the most important subjective factors to evaluate.

The screening criteria that an innovator decides to consider in the needs filtering process can be tracked in many different ways. Some experienced innovators do little more than keep a mental list. Others keep track of them on paper or include them in their innovation notebooks (see 1.2 Observation and Problem Identification). However, for most innovators, especially those who are just starting out, it can be extremely helpful to construct a spreadsheet or database to perform this tracking function. An electronic tool provides great flexibility in organizing, manipulating, and sorting information, particularly when there are many needs to track. Figure 2.5.2 provides an example of the screening criteria chosen by one team, and the method used to compile and manage this information.

Step 2: Assign ratings for each factor

Once screening criteria have been selected, the innovator can evaluate each need and assign a rating for every factor. When assigning ratings, it is helpful to develop a scale and apply it consistently across factors. For example, one might decide to use a scale of 1 to 5, where 5 is the best possible score and 1 is the lowest. For each factor, the rating scale must be further defined in meaningful terms for the factor. For instance, for an objective factor related to *patient impact*, the innovator might define the scale shown in Table 2.5.2. For an objective factor related to *treatment landscape*, the scale shown in Table 2.5.3 might be used.

Ratings should also be assigned to the subjective factors that the innovator has defined. For example, if one of the acceptance criteria set forth in the innovator's strategic focus is to work on projects that can be life-saving, s/he may define a factor (and a rating scale) that evaluates this issue as shown in Table 2.5.4.

The most difficult part of assigning ratings is that innovators must do so based only on what is known about the *need*. Because the innovator should still be thinking in solution-independent terms, it is essential not to take into account information or biases related to any particular solution. Innovators have a tendency to jump ahead – for example, by rating the market size associated with a new type of catheter, rather than the size of the market associated with a better way to perform interventional procedures. The effect of this bias is to skew the rating process based on preconceived notions (e.g., "Catheter-based solutions are really hot right now, so I'm going to rate the need high," or "Medication pumps are ubiquitous commodities, so I'm going to rate the need low"). The best way to prevent this type of bias is to stay focused on rating the impact of solving the need rather than the impact of addressing it with a particular type of solution (see 1.3 Need Statement Development for more information about the pitfalls associated with embedding solutions within needs).

As ratings are assigned to each factor for a given need, they should be captured in the database or spreadsheet used to manage the filtering process, as shown in Figure 2.5.3.

After ratings have been assigned to all factors for every need, it is useful to look across needs, factor by factor, to be sure that the values have been assigned consistently, fairly, and without bias.

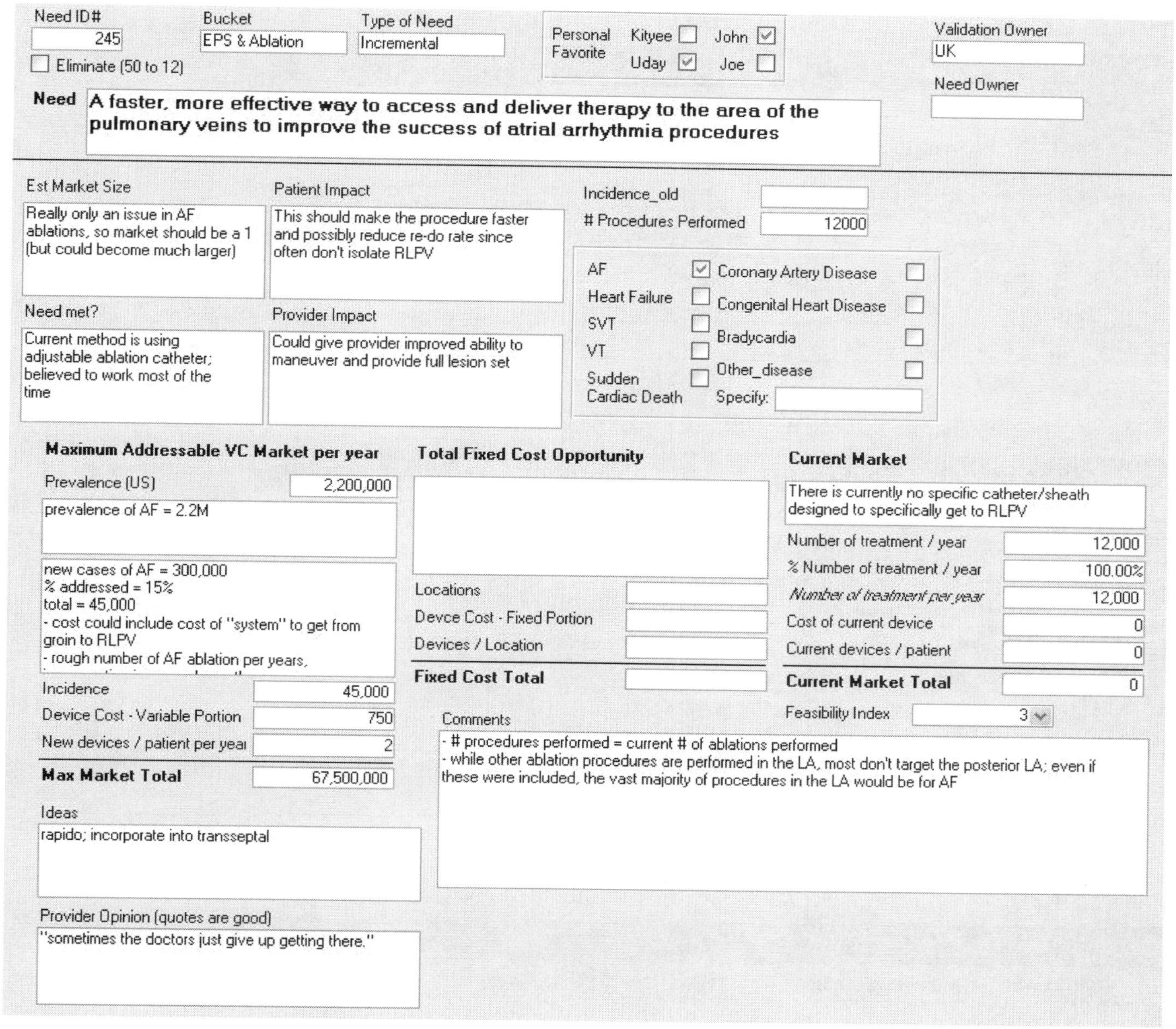
Need ID# 245
Eliminate (50 to 12)
Bucket: EPS & Ablation
Type of Need: Incremental
Personal Favorite: Kityee ☐ John ☑ Uday ☑ Joe ☐
Validation Owner: UK
Need Owner:
Need A faster, more effective way to access and deliver therapy to the area of the pulmonary veins to improve the success of atrial arrhythmia procedures

Est Market Size: Really only an issue in AF ablations, so market should be a 1 (but could become much larger)
Patient Impact: This should make the procedure faster and possibly reduce re-do rate since often don't isolate RLPV
Incidence_old:
Procedures Performed: 12000
AF ☑ Coronary Artery Disease ☐
Heart Failure ☐ Congenital Heart Disease ☐
SVT ☐ Bradycardia ☐
VT ☐ Other_disease ☐
Sudden Cardiac Death ☐ Specify:
Need met?: Current method is using adjustable ablation catheter; believed to work most of the time
Provider Impact: Could give provider improved ability to maneuver and provide full lesion set

Maximum Addressable VC Market per year
Prevalence (US) 2,200,000
prevalence of AF = 2.2M
new cases of AF = 300,000
% addressed = 15%
total = 45,000
- cost could include cost of "system" to get from groin to RLPV
- rough number of AF ablation per years,
Incidence 45,000
Device Cost - Variable Portion 750
New devices / patient per year 2
Max Market Total 67,500,000

Total Fixed Cost Opportunity
Locations
Devce Cost - Fixed Portion
Devices / Location
Fixed Cost Total

Current Market
There is currently no specific catheter/sheath designed to specifically get to RLPV
Number of treatment / year 12,000
% Number of treatment / year 100.00%
Number of treatment per year 12,000
Cost of current device 0
Current devices / patient 0
Current Market Total 0
Feasibility Index 3

Ideas: rapido; incorporate into transseptal
Comments:
- # procedures performed = current # of ablations performed
- while other ablation procedures are performed in the LA, most don't target the posterior LA; even if these were included, the vast majority of procedures in the LA would be for AF
Provider Opinion (quotes are good): "sometimes the doctors just give up getting there."

FIGURE 2.5.2
A screen shot from a database developed by a team in Stanford University's Program in Biodesign to support the needs filtering process (with permission from Uday N. Kumar; developed with John White, Kityee Au-Yeung, and Joseph Knight).

Step 3: Combine values to produce a score

The next step is to figure out how to combine ratings in such a way that each need ends up with a single numeric score. These scores should reflect the relative strengths and weaknesses of each need to allow for their direct comparison and prioritization.

One approach is to simply add or average the ratings assigned to each factor to come up with a score for each need. If the right mix of objective and subjective factors has been developed, this kind of simple approach to calculating an overall score may be appropriate. Another approach is to assign different weights to the various factors and then calculate a total (or average) score. Weighting the factors is important if the innovator believes that certain criteria are significantly more essential than others. For example, if an innovator is

Table 2.5.2 *A sample rating scale for the objective factor: patient impact.*

Rating	Description
5	Addressing the need would be life-saving to patients.
4	Addressing the need would reduce morbidity and/or eliminate the risk of serious complications.
3	Addressing the need would reduce or eliminate undesirable symptoms of the disease, but not have an impact on morbidity or quality of life.
2	Addressing the need would be convenient for the patient, but not have an impact on symptoms, morbidity, or quality of life.
1	Addressing the need would not have an impact on patients.

Table 2.5.3 *A sample rating scale for the objective factor: treatment landscape.*

Rating	Description
5	There are no existing treatments available to address the need and the field is *not* littered with inventors/companies who have tried to address the need and failed.
4	There are no existing treatments available to address the need but numerous inventors/companies have tried to address the need and failed.
3	Treatments exist to address the need but have *serious* deficiencies that must be overcome.
2	Treatments exist to address the need but have *minor* deficiencies that must be overcome.
1	Treatments exist to address the need that are generally well accepted by the target user population and address the need well.

Table 2.5.4 *A sample rating scale for the subjective factor: need must be life-saving.*

Rating	Description
5	Without addressing the need, patients will face likelihood of death.
4	Without addressing the need, patients will face a *serious* risk of increased mortality.
3	Without addressing the need, patients will face a *moderate* risk of increased mortality.
2	Without addressing the need, patients will face a *minor* risk of increased mortality.
1	Without addressing the need, patients will not face an increased risk of death.

driven to build a thriving business more than anything else, it might be appropriate to assign a greater weight to a factor such as market size. On the other hand, if an innovator is motivated above all else to improve outcomes for patients, factors such as patient impact might warrant greater weight. An example of the weightings and factors considered in calculating a score used by the team mentioned in Figure 2.5.1 is shown in Figure 2.5.4. Note also that, as suggested earlier, the team classified the needs based on their type.

The process of weighting factors is inherently subjective – there really is no right way to do it. As a result, innovators must trust their instincts in determining the best possible approach. For example, blue-sky needs often "gain the most points" simply because they are so broad, but may not be the correct needs

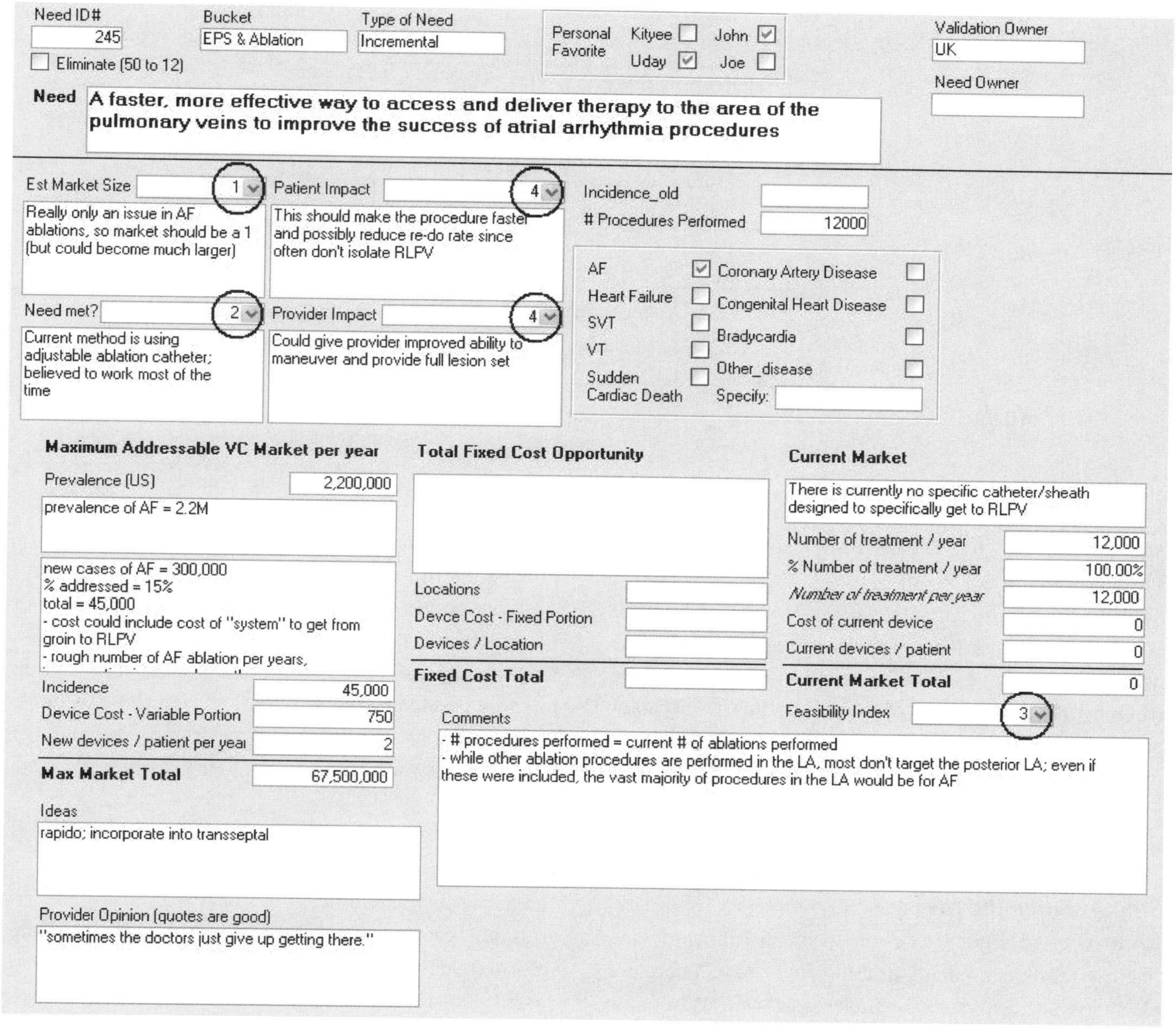

FIGURE 2.5.3

A screen shot from a database developed by a team in Stanford's Program in Biodesign showing assigned ratings (with permission from Uday N. Kumar; developed with John White, Kityee Au-Yeung, and Joseph Knight).

to pursue if an innovator indicates a strong desire to bring a tangible product to market in a relatively short period of time. After calculating the overall scores for each need, do a "gut check" on a sample of the final scores to make sure that they accurately reflect the innovator's priorities and interests, as well as the more objective factors that characterize a strong opportunity. Furthermore, keep in mind that the scores represent estimates of the strength of each need, but are not perfectly precise. This means that a need with a slightly lower score than another need (e.g., 27 versus 28) should not be quickly discarded without first evaluating the underlying reasons why it scored lower.

A database or spreadsheet can be used to capture the approach to weighting certain factors (if appropriate)

Stage 2 Needs Screening

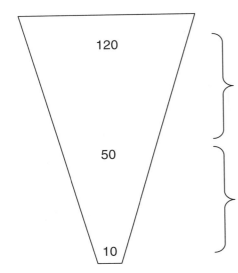

FIGURE 2.5.4
To get from 120 to 50 and then to 10 needs, the team categorized the needs as blue-sky, mixed, or incremental and then assigned weightings (the numbers in the table) to its chosen filtering criteria. These weightings were multiplied by the ratings assigned to each criterion to calculate a score for each need.

and to calculate the overall scores for each need as shown in Figure 2.5.5.

In this case, the spreadsheet was organized by need type – blue-sky, mixed, and incremental – to allow different weightings to be assigned to each need type (thereby allowing needs of a similar type to be compared against one another). Working with different weightings is important to adjust for the inherent bias towards blue-sky needs, due to their typically greater rewards.

Step 4: Filter the needs

With scores assigned to each need, an innovator can begin directly comparing and filtering them. There are many ways to approach the filtering process. While one method is not necessarily better than another, it is helpful to select and apply a defined approach to ensure consistency in the process. One way to do this is to determine how many needs statements to consider in the "first pass" through the filtering process, list the needs in order from highest to lowest score, and then select the chosen number of needs with the highest scores. Another approach would be to start eliminating any needs with a score that falls below a certain value.

Remember that the filtering process should be iterative. The first time through the process, an innovator might cut the number of possible need statements in half, go back to do some additional research, adjust certain ratings and scores, and then work through this process again to reduce the list by another 50 percent. As shown in the example in Figure 2.5.1, the team went from 320 initial needs to 120 needs in its first pass. Such an approach results in more time spent researching only the needs that progressively continue to "make the cut," which can be important from a time management standpoint when an innovator starts with many needs.

Typically, the quantitative scores developed through this process provide an excellent way to make the "easy" decisions about which needs to eliminate. For example, some needs clearly will not be as compelling as others in terms of the potential opportunity they represent, and their scores will cause them to be quickly set aside. However, most innovators find that, as the list of needs gets smaller and smaller, the quantitative scores become less helpful and they must rely on more qualitative decision-making criteria to reduce the list of need statements to ten or less. This is a natural part of the process, demonstrating that there is fundamentally no substitute for an innovator's professional judgment. Again, as long as the process is followed consistently, this will bring some level of objectivity to the exercise.

An alternative approach to filtering needs

Recognizing that needs filtering is an inherently subjective process, some innovators choose to bypass a

2.5 Needs Filtering

Need ID.	Need	Est Market Size	Patient Impact	Provider Impact	Type of Need	Degree to Which the Need is Met	Feasibility Index	Personal Favorite - Kit	Personal Favorite - Uday	Personal Favorite - Joe	Personal Favorite - John	Total	Rank
INCREMENTAL		3	1	2	0	0	2	0	0	0	0		
116	A method to ablate more tissue, over a larger area, with each application of energy	2	2	2	1	3	3	0	0	0	0	18	1
84	A way to change the RF energy output location along the length of the catheter without moving the catheter at all	2	1	2	1	4	3	0	0	0	0	17	3
123	A cheap, simple method with which to accurately and easily enter the arterial and venous systems	3	1	1	1	2	3	0	0	0	0	18	1
51	A better way to stabilize device lead in coronary sinus	2	1	2	1	2	3	0	0	0	0	17	3
152	A way to quickly map and ablate within a 3-D structure	2	2	2	1	3	2	0	0	0	1	16	5
245	A better way to access the right lower pulmonary vein. (changed better "catheter design" to better way)	1	2	2	1	2	3	0	1	0	1	15	9
159	A way for one cath lab operator to perform all necessary equipment movements for ablation and EPS procedures	3	1	2	1	4	1	0	0	0	0	16	5
70	Lighter x-ray shielding outfits	1	1	3	1	3	2	0	0	0	0	14	11
21	A way to avoid esophageal injury during ablation	1	1	2	1	2	3	0	0	0	0	14	11
56	A way to increase the battery life in the average implanted device	3	3	1	1	3	1	0	0	0	0	16	5
52	A more effective way to close an implanted device pocket	1	1	2	1	2	3	0	0	0	0	14	11
321	A way to determine the settings of an implanted device without a programmer available	3	1	2	1	4	1	0	0	0	0	16	5
40	A way to prevent electrophysiology catheter cables from building up torque outside of the patient	1	1	2	1	4	3	0	0	0	0	14	11
191	A way to prevent thrombosis/vegetation on device leads	3	2	1	1	2	1	0	0	0	0	15	9
24	A cost effective way to remotely monitor many inpatients	3	1	1	1	2	1	0	0	0	0	14	11
22	A better handle for ablation catheters	2	1	2	1	2	0	0	0	0	0	11	16
MIXED		2	1.5	1.5	0	1	2	0	0	0	0		
300	A way to remove the left atrial appendage from within the left heart	3	3	2	2	4	1	0	0	0	0	19.5	1
119	A method to assess the depth of the lesion created during ablation	2	2	3	2	4	2	1	1	0	0	19.5	1
325	A way to access the pericardial space	2	2	2	3	3	3	0	0	0	0	19	3
76	A better, more reliable way to determine if an arrhythmia that was present is now gone	2	2	3	2	3	2	0	0	0	0	18.5	4
149	A better way to determine/estimate ablation site prior to the actual procedure	2	2	3	2	3	2	0	0	0	0	18.5	4
144	A way to improve and angioplasty balloon	3	2	2	2	2	2	1	0	0	0	18	6
57	A way to recharge an implanted device without removing the entire generator	3	3	1	2	4	1	0	0	0	0	18	6
226	A way to anchor the wire / lead / catheter once you are in a vessel you are interested in.	2	1	2	2	3	3	0	0	0	0	17.5	8
118	A method or a device/suite of devices that would allow one person to perform all the functions necessary to perform a complete ablation	2	1	3	2	3	2	0	0	0	0	17	9
155	A way to determine the appropriate ablation "dosage" to account for tissue variability	2	2	2	2	3	2	1	0	0	0	17	9
279	A better way to more effectively deliver energy during defibrillation	3	2	1	2	2	2	0	0	0	0	16.5	11
284	A way to change the stiffness of a balloon (complaint to non-compliant)	3	1	2	2	2	2	0	0	0	0	16.5	11
180	A better way to non-invasively monitor arrhythmias long-term	3	2	1	3	2	2	0	0	0	0	16.5	11
248	A way to treat vessel perforation during a procedure	1	3	1	2	4	2	0	0	0	0	16	14
73	A better way to keep an ablation catheter against the wall of the heart	2	1	2	2	3	2	0	0	0	0	15.5	15
298	A way to make an ablation catheter assume any shape	2	1	2	2	3	2	0	0	0	0	15.5	15
47	A better way to access the coronary sinus	2	1	2	2	2	2	0	0	0	0	14.5	17
195	A way for an EP recording system to interpret signals, not just display them.	2	1	2	2	2	2	0	0	0	0	14.5	17
314	A way to guide a catheter to the previously stored location	2	1	2	2	2	2	0	0	0	0	14.5	17
307	A way to turn different parts of a catheter (REMOVED " without having to torque at the handle")	2	1	2	2	2	2	0	0	0	0	14.5	17
131	A way to alter the properties (stiffness, softness, slickness) of the wire/stylet without exchanging it	2	1	2	2	1	2	0	0	0	0	13.5	21
86	A way to include depth perception on a fluoro image	2	1	2	2	3	1	0	0	0	0	13.5	21
99	A way to eliminate putting 12 individual ECG patches on the patient	2	1	1	2	0	3	0	0	0	0	13	23
299	A way to place permanent leads in the left heart without an increased risk of thrombus	2	1	1	2	2	2	0	0	0	0	13	23
67	A way to prevent bleeding locally in the device pocket.	1	1	2	2	2	2	0	0	0	0	12.5	25
143	A way to change the visibility of equipment on the imaging system	2	1	2	2	2	1	0	0	0	0	12.5	25
305	A way to outline the borders of the endocardium for the duration of a procedure	1	1	2	2	3	1	0	0	0	0	11.5	27
301	A way to image the left atrial appendage without using a transesophageal approach	1	1	2	2	1	1	0	0	0	0	10.5	28
163	An easier way to compare/characterize/determine the direction of waveform propagation and its origin	1	1	2	2	2	1	0	0	0	0	10.5	28
BLUE SKY		2	1.25	1	0	1	2	0	0	0	0		
326	A way to improve the contractility of the heart	3	3	2	3	4	1	0	0	0	0	17.75	1
122	An alternative to fluoroscopy that will allow visualization of internal structures in real-time which would also allow real-time procedures to be performed	3	2	3	3	4	1	0	0	0	0	17.5	2
107	A method to non-invasively image coronary and peripheral arterial plaques susceptible to rupture	3	3	2	3	3	1	0	0	0	0	16.75	3
14	A better way to treat a stenosis	3	3	1	3	2	2	0	0	0	0	16.75	3
8	A method to identify the likelihood of malignant arrhythmias in patients who are currently not symptomatic from the arrhythmia	3	3	1	3	4	1	0	0	0	0	16.75	3
112	A non-invasive method to ablate internal cardiac tissue	2	3	3	3	4	1	0	0	0	0	16.75	3
210	A way to cause scar (not iatrogenic fibrosis e.g. from a lead placement) tissue to conduct again	3	3	1	3	4	1	0	0	0	1	16.75	3
324	A better way to treat AF	3	3	1	3	3	1	0	0	0	0	15.75	8
124	A method with which to assess electrical and mechanical ventricular dyssynchrony	3	2	2	3	3	1	0	0	0	0	15.5	9
322	A way to prevent arrhythmias in post-MI patients with low EF	2	3	1	3	4	1	0	0	0	0	14.75	10

FIGURE 2.5.5

A screen shot from a spreadsheet developed by a team in Stanford University's Program in Biodesign shows the assigned ratings, weightings, and total scores for numerous needs. Note: the highlighted numbers indicate the weighting used for each factor (with permission from Uday N. Kumar; developed with John White, Kityee Au-Yeung, and Joseph Knight).

scoring-based approach in favor of a slightly simplified process that seeks to: (1) align funding requirements with market size, and then (2) create a balanced portfolio of needs to pursue. Given the intrinsic differences between blue-sky, mixed, and incremental type needs and their funding requirements, some innovators start the filtering process by separating all the needs into different groups based on these categories. After the needs have been separated, the innovator can:

- Determine which blue-sky needs have a market of $500 million to $1 billion. Such a market would be required to justify the fact that it often takes approximately $100 million to develop a solution to a blue-sky need.
- Set a threshold for mixed needs. Usually mixed needs necessitate a market of roughly $250 million (since $25 million to $50 million in funding is often required).
- Establish a market threshold for incremental needs. Generally, a market of $50 million to $100 million is required to justify the typical investment of $5 million to $10 million in funding.

Note that these market thresholds are "back of the envelope" estimations that are meant to take into consideration the opportunity relative to the investment required.

With these thresholds in place, the innovator may choose to select a number of needs within each group to create a balanced portfolio of needs (e.g., three blue-sky, three mixed, and four incremental). The needs within each group would then be weighted and scored, using the other identified factors. While the use of a funding estimate is in itself a filtering step, there is a potential danger in simply selecting a few needs based on this alone. For instance, if among the list of needs, there are seven blue-sky needs out of 20 that meet the funding estimate, but an innovator only wants to choose three for the portfolio, it would be wise to score all of the seven against the other factors and then pick the top three, rather than simply selecting three of the seven purely subjectively.

Again, while there is no correct way to utilize, assign, weight, or apply scores, it can be useful to try different approaches (i.e., modifying the order of these steps) as the results vary significantly.

The following case example describes an approach used by one medical device incubator, The Foundry. It also highlights lessons that can be learned when performing needs screening and filtering.

FROM THE FIELD — THE FOUNDRY

An incubator's approach to filtering needs

As noted in 2.2 Treatment Options, The Foundry is a medical device incubator that helps innovators "rapidly transform their concepts into companies."[1] Because of the large number of needs the company evaluates each year, Hanson Gifford, CEO, and Mark Deem, CTO, are constantly in the needs filtering process as they decide which projects to pursue and which opportunities will translate into the most promising solutions.

Gifford and Deem are inundated with new ideas every day. Some of these needs arise organically from related projects they are working on, while others are brought to them by entrepreneurs and inventors. With limited resources and even more limited time, Gifford and Deem are forced to make tough and relatively quick decisions about which needs should proceed to concept generation and which should not. While Gifford and Deem appreciate the merits of a well-defined, clearly structured needs filtering process, they admit that their approach is only partially systematic. As Deem commented, "I'd like to say our project analysis is completely calculated, but each opportunity is different. At the end of the day, you have to take the facts around each of them and make a judgment call. Sometimes, as we move along, one need starts to make more sense. We've got better ideas and information in one area than we do in another and it becomes the lead horse," he explained.[2]

Despite their informal characterization of The Foundry's process for filtering needs, further discussion revealed that Gifford and Deem do, in fact, evaluate many of the primary objective and subjective factors described within this chapter. While they may not formally assign quantitative ratings and explicitly weight the factors via rigid analysis to come up with an overall score, they do much more than depend simply on their gut instincts. When pressed for information on how The Foundry makes decisions regarding which needs to pursue, Gifford highlighted the main factors that they consider: "The market opportunity, the clinical benefit, and the overall investment landscape – is the need attractive to investors? Are we going to be able to get there within our lifetime and with less cost than the national debt? We also carefully consider how likely something is to really succeed in the market." These criteria, consistently applied across needs, provide a basis of comparison that lends objectivity to the filtering process. If there is only a small market for a product, or if the intellectual property is already substantially owned by others, The Foundry may be inclined to pass on the opportunity. Likewise, Gifford and Deem may steer away from a need if the referral and reimbursement patterns are unduly complex, or if prior research suggests clinical infeasibility in the treatment space.

Continuing to describe the factors they take into account in filtering needs, they emphasized the importance of stakeholder analysis, abiding by the saying that "A new therapy needs to be attractive to patients, physicians, and payers." Typically, the Foundry investigates the value proposition to all three groups before making a decision.

Gifford explained that it is also important to take the innovator's interests and motivations into account in the needs filtering process: "This isn't just a one-time decision about whether projects A, B, or C are interesting to work on. It's a decision followed by several years of blood, sweat, and tears to make it happen. We have to take a good, hard look at ourselves and ask which opportunity we want to commit a portion of our lives to." That is not to say that they always stick with what they know and like best – as Deem pointed out, The Foundry has taken on highly diverse projects over the course of its history. Rather, he said, they must be excited about any need they decide to pursue. Over time, they have learned that a project only succeeds if the people involved want to work on it and not simply because it looks financially promising. If they are not enthusiastic about and truly interested in addressing the need, Gifford and Deem are more than willing to shelve it and move on. They recognize that it is this sense of passion that has kept them excited about coming to work every day through ten successful start-ups since 1998.

One of The Foundry's companies, called Cierra, Inc., provides an example of how the group's needs filtering criteria are applied to new opportunities. Cierra was founded to develop a novel approach for the treatment of patent foramen ovale (PFO). This condition results from the incomplete closure of the septal wall between the right and left atria (upper chambers) of the heart, an opening that exists before birth to allow oxygenated blood to circulate throughout the fetus without having to pass through its lungs. In 75 percent of those with the condition, the FO closes naturally after birth. In the remaining 25 percent, it does not seal completely, allowing blood to flow directly, under certain conditions, from the right atrium to the left atrium.[3] As a result of the existence of a PFO, the natural filtration that the lungs provide is partially bypassed. Consequently, blood clots and other agents in the blood can go directly into the arterial system, potentially causing paradoxical embolism, cryptogenic stroke, and right to left nitrogen embolism in severe decompression illness.

Early in the innovation process, when Gifford and Deem first began looking at PFO, they did so at the urging of cardiologists in their network who anticipated that this was going to become an increasingly important need. At the time, there was no clearly defined standard of care. Many physicians believed there was no reason to treat for PFO unless a patient experienced cryptogenic stroke (a stroke with unknown causes). In these cases, patients often were prescribed chronic anti-coagulation therapy. The condition could also be addressed through open surgery or a transcatheter intervention that used an implantable closure device to address the problem. Two transcatheter devices had received approval from the FDA for the treatment of PFO under a humanitarian device exemption (**HDE**), a category of FDA clearance that applies to devices designed to treat a population of less than 4,000 patients.[4]

After initially studying the disease and evaluating stroke-related PFO opportunities against their filtering criteria, Gifford and Deem were somewhat less than enthusiastic about the need. The market opportunity was relatively small, the attractiveness to investors was dubious, and the clinical benefits were not compelling relative to the amount of effort and investment required to conduct a stroke trial (which were notorious for being costly, lengthy, and difficult to enroll). "We just couldn't get excited about signing up for a project where we were going to have do

FROM THE FIELD: THE FOUNDRY

Continued

a one-thousand patient clinical trial to try to show an improvement in stroke rates from one single digit number to another single digit number," recalled Deem. "We were about to shelve it," he continued, "when we came across some of the first articles that were being published on the migraine–PFO association." As Erik Engelson, who later became CEO of Cierra, described, "There were observations and some single-arm studies that were published. In deep-sea divers who were having PFO treated for decompression illness and in stroke patients who also had migraines and had their PFO closed, the observation was that the frequency of migraine decreased after the procedure." "That caught our interest, and we started studying it a little bit more," continued Deem. "It was still a relatively unproven association, but we felt like there was probably something there that could transform this into a huge opportunity." Although there was still a fair amount of uncertainty surrounding the linkage between migraines and PFO closure, as well as the amount of time and cost required to develop a solution, Gifford and Deem became passionate about pursuing the need.

When Gifford and Deem reapplied their filtering criteria to the migraine-related opportunity, it was more appealing, in part, because it seemed that the market could potentially be more easily and quickly accessed than the stroke market. While the symptoms of this larger population of migraine sufferers were not life threatening, they could be severely debilitating, so The Foundry perceived the clinical benefit of a solution to be significant in terms of improving their quality of life. Moreover, Gifford and Deem anticipated being able to more effectively get investors interested due, in part, to recent activity in the field. "All of a sudden this market was hot," recalled Engelson. "There was this public company called NMT Medical, the market cap of which doubled on the buzz of upcoming completion of their UK-based migraine study [the MIST-I study]."

The apparent attractiveness of the opportunity was also enhanced by the novel concept of closing PFO with a non-implant solution (rather than leaving behind an implant in the heart). As a result, The Foundry team refined its need criteria to include this design requirement. "An implant is all well and good if you're a 75-year-old patient who has already had one or two strokes and you're trying to prevent another one," Deem said. "But if you're a 25-year-old migraine sufferer who is otherwise reasonably healthy, having a metal implant in your heart for the rest of your life is a completely different value proposition due to known [and yet-to-be known] complications associated with some of the PFO implant devices. The notion of a non-implant solution played positively to cardiologists, patients, neurologists, and investors. So, we decided that we needed a solution that left nothing behind in the heart." It would have been a lot easier and faster to simply pursue a better implant. But, as Deem summarized, "We felt like the benefit of leaving absolutely nothing behind versus having the next best clip warranted the decision to go forward there."

When Engelson was recruited to head Cierra, the company had 12 employees. The team had evaluated clips, snaps, staples, sutures, and patches to build a broad IP position, but ultimately decided on an implant-free system that closed the PFO using suction and "tissue welding" performed percutaneously using radiofrequency (RF) energy. See Figure 2.5.6.

Under Engelson's leadership, Cierra refined its product, completed animal testing, treated eight human patients in Germany (with an initial 75 percent PFO closure rate), and raised $21 million in next-round funding (incremental to the previous $8 million). The team also had a series of promising meetings with the FDA. In order for the PFO–migraine solution to be successful, the company needed to be able to design a clinical study that would address the agency's safety concerns related to applying RF energy to the heart. "The FDA really seemed to like the non-implant solution. They gave us a verbal thumbs-up on our safety data," Engelson recalled.

Shortly thereafter, however, Cierra faced a reversal of fortune. "When we formally submitted our clinical protocol, the written feedback we received seemed to differ significantly from the verbal feedback on the conceptual study design we had previously discussed

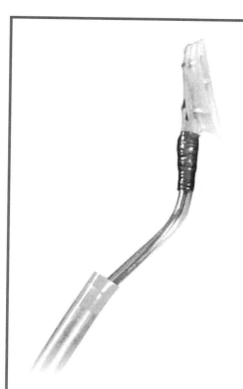

FIGURE 2.5.6
The Cierra device (courtesy of Cierra, Inc.).

with the FDA," said Engelson. The FDA's written feedback (questions and suggestions) directed the company to conform its study design to that of the other migraine studies, which were not enrolling. Around the same time, the company found itself achieving mixed technical results. "We had learned to segment by size and we were getting good PFO closure in smaller PFOs, but still had work to do in the larger ones. Closing the large PFOs was becoming an insurmountable challenge with the RF technology," noted Engelson. Additionally, when NMT released partial results of the MIST-I trial in spring 2006, the buzz around PFO–migraine opportunities diminished. In that study of 147 patients, there was no difference in headache cure between migraine sufferers receiving PFO closure and those receiving the sham procedure.[5] Despite these challenges, the Cierra team continued to press forward. "We believed in this," said Engelson. "We had signed up and we were fighting this war together. So we carried on." The company's efforts continued until late 2007, when it finally discontinued its operations. Interestingly, NMT Medical announced that it was halting its MIST-II (US-based migraine) trial one month later, citing difficulty in enrolling patients in the study and the need to redirect the millions of dollars necessary to sustain their stroke investigation.[6]

Reflecting on the Cierra experience, Engelson commented that some of the challenges faced by the company could be traced back to steps in the needs screening/filtering process. While the team's market analysis demonstrated that migraine was, in fact, a bigger and potentially more easily-accessible market segment, it later became clear that neurologists, not cardiologists, were the key stakeholder group since they would be the ones to control study enrollment. And, unfortunately, they were generally not supportive of PFO–migraine studies. "It wasn't a cardiology deal," said Engelson. "It was a neurology deal. Compared to cardiologists, neurologists tend to be less aggressive – they were not enthusiastic to try new, interventional technologies. They're used to using drugs to manage their patients," he noted. While initiating an interventional procedure to address a quality-of-life issue might have been broadly accepted in cardiology, the concept was viewed skeptically in neurology.

The definition of a need criterion to develop a non-implantable solution also may have unnecessarily constrained the company's efforts and eventually contributed to its difficulties in overcoming the technical challenges associated with closing large PFOs. According to Engelson, "We spent a lot of time and money developing a totally non-implantable solution when, in hindsight, I'm not sure that this was as critical as initially thought. We may have been better served by developing a succession of products for PFO closure: first a very small implant such as a clip, which would be much smaller than the existing implants, followed by the non-implant technology. This technology–product trade-off decision (or product portfolio planning) process might have been more effective had we had a visionary and inventive physician as part of our internal thinking process."

Overall, summarized Engelson, "It was the FDA's high study-design hurdle, the lack of support by migraine neurologists, and the remaining technical challenge in closing larger PFOs without an implant that collectively led to our shut-down decision." Similar issues had proven to be challenging for numerous other device-based companies working in the field.

Stage 2 Needs Screening

Need specification

Once a small set of needs has emerged from the needs filtering process, a need specification should be created for each one. A need specification is a detailed, but succinct, stand-alone document that summarizes the data gathered through the needs screening process (see Chapters 2.1 to 2.4). Importantly, it also outlines the key criteria that any solution must meet in order to satisfy the need. Together this information serves as guiding principles for the next stage of the biodesign innovation process – concept generation. For this reason, it is essential that each specification is prepared thoroughly. However, this should not be viewed as an exhaustive task. Roughly the same level of effort and rigor should be applied to creating a need specification as one would apply to the development of a research paper, although a need specification can typically be implemented in more of an outline form.

Sometimes innovators may try to short-cut this process by preparing one background document for multiple needs that overlap, particularly when they all focus on a given disease state. However, in doing this, the innovator runs the risk of creating confusion about the key drivers of each need, especially if those are not clearly separated. Additionally, since a need specification provides a key input into upcoming brainstorming sessions, it would be undesirable for information about competing or alternate needs to distract participants or bias their thinking if it was included in a combined need specification. Needs will also continue to fall off the list of opportunities that the innovator is pursuing as more information becomes known. If each need has its own specification, these can easily be set aside. To avoid these pitfalls, it is essential to create a self-contained need specification for each need.

In terms of content, a need specification provides a several-page summary of the most important and relevant information gathered in support of the need, including the information outlined below.

The need statement

The need statement can be used as written (see 1.3 Need Statement Development) or refined based on what has been learned through the needs screening process.

The problem

The summary of the problem should address affected anatomy and physiology, disease mechanisms of action, disease progression, past and current approaches to the problem, outcomes, complication rates, and a gap analysis of existing and emerging treatments (see 2.1 Disease State Fundamentals and 2.2 Treatment Options).

Market description

The market overview should cover high-level market-related information from the needs screening process, including data about the target market, procedure volume, treatment penetration, competitive landscape, value (revenue per procedure), and burden on the healthcare system (see 2.3 Stakeholder Analysis and 2.4 Market Analysis).

Need criteria

As noted, need criteria are the key elements required and/or desired by the customer (e.g., efficacy rates, compatibility with other devices, ease of use) that a solution must and/or should satisfy in order to address the need. A preliminary set of these requirements is usually developed as the need statement is defined (see 1.3 Need statement development). However, unlike these criteria, which are primarily based on observations, the criteria in the need specification are significantly influenced by the data gathered from research about the problem and market (see the two sections above). This data allows an innovator to more deeply understand the key requirements that must be met for a need to truly be addressed. For instance, with the need *to* **reduce the incidence of urinary tract infections in ICU (intensive care unit)** patients, one initial need criterion was that any solution had to be *deployable* by personnel that are already available within the ICU. This criterion could have been derived from preliminary observations. However, with additional research performed as part of a market analysis, it may be learned that many of these catheters are removed during a patient's stay in the ICU. As a result, the need criteria may then be refined to indicate that a solution needs to be *deployable and removable* by personnel that are already available in the ICU. Continuing with this example, the innovator may also uncover the fact that these catheters are only purchased from hospital distributors in bulk. So another more specific requirement might be that a solution needs to be attractive from a pricing and inventory standpoint for hospital distributors to offer it.

When thinking about need criteria, additional data can be used to determine which requirements are absolute or "must-haves" and which are desirable or "nice-to-haves." In the above example, requiring that ICU personnel be able to deploy and remove the catheter might be an absolute necessity, whereas the criterion about hospital distribution might be desirable, but not necessary, if other distribution methods are possible. This distinction is important when innovators begin developing solutions and evaluating their relative strengths and weaknesses.

While there is no "right" way to develop need criteria, it is important that the information gathered through needs screening be used to help refine and specify the criteria in such a way that absolute requirements can be distinguished from those that are desirable but not on the critical path. However, recognize that creating too many absolute criteria (more than eight to ten) will place too many constraints on concept generation and screening (see 3.2 Concept Screening). Stay focused on the few criteria that are absolutely essential to address the need.

Finally, while some need criteria may specify product attributes (e.g., small, able to be placed via a blood vessel), these more specific attributes should still be kept at a high level. The point is to avoid the implication of a solution while reflecting the accepted constraints for the need (e.g., developing a solution for a problem accepting that fact that it needs to leverage a catheter-based platform). Need criteria should also continue to evolve and become increasingly concrete as an innovator iterates the need specification.

References
Include in the need specification physician testimonials, quotes, articles, market surveys, and other such information to help validate the information.

Keep in mind that the need specification does not have to follow the outline above exactly. Innovators have the freedom to experiment with different headings and ways of organizing the information, as long as all of the most relevant and important information is included within the specification. See Appendix 2.5.1 for two real-world examples of need specifications that have quite different forms, but cover most of the same fundamental information.

In the process of creating and pulling together all the elements of a need specification, an innovator's perspective may change, throwing the needs filtering process into a new light. For example, when compiling more detailed data about various treatment options, s/he may realize that one treatment option, previously thought to be unimportant, is actually of significance. If elements about competitive landscape were included in the filtering process, this information could suggest that the weighting scheme used, for instance, may need to be adjusted. Although minor adjustments are not likely to have a significant impact on the overall ranking of needs (especially if the scores have been viewed more as general estimates), it is important to confirm this by repeating the filtering process. Major adjustments in the ranking will typically only occur if important information was somehow missed in the initial screening process or if the research performed at each stage of filtering was kept relatively superficial, even as the number of needs was being diminished. This interaction of the needs filtering process with the creation of the need specification once again highlights the iterative nature of the biodesign innovation process.

Similarly, innovators may revisit the needs filtering process and, in particular, the need specification if they have difficulty screening concepts in a meaningful way. This may be an indication that the need criteria are too broad (see 3.2 Concept Screening).

GETTING STARTED

Even though needs filtering is somewhat subjective and highly experiential, the steps here can be followed to initiate the process. Remember to experiment with different screening criteria and rating methods, and always take an iterative approach to ensure that nothing is missed.

See **ebiodesign.org** for active web links to the resources listed below, additional references, content updates, video FAQs, and other relevant information.

1. Select screening criteria

1.1 What to cover – Define the most important criteria that will be used to filter the needs. Consider several objective factors that accurately identify strong opportunities of interest to investors, physicians, and prospective employees. Next, consider subjective factors that reflect the interests and priorities of the innovator. Capture these factors in a database or spreadsheet.

1.2 Where to look – Refer to information gathered through research as part of the needs screening process (see Chapters 2.1 to 2.4) for ideas on what factors to consider. An innovator should also revisit his/her strategic focus for help identifying the more subjective criteria (1.1 Strategic Focus). Network with others in the field and ask for their input on the most important criteria before filtering needs.

2. Assign ratings

2.1 What to cover – Develop an appropriate rating scale and rate each need against the chosen factors. Remember to evaluate the benefit that results from addressing or solving the need to ensure that the assessment remains solution independent. Document all ratings in the database or spreadsheet.

2.2 Where to look – Refer back to 1.3 Need Statement Development for information about keeping needs solution-independent.

3. Calculate scores

3.1 What to cover – Decide on an approach for combining all ratings to come up with a single score for each need. Assign weights to each factor, as appropriate. Calculate scores and do a "gut check" of the approach being used. Make required adjustments and record the scores in the database or spreadsheet.

3.2 Where to look – There are few external resources to assist in this exercise. An innovator must rely on his/her own critical analysis skills to perform this step in the process.

4. Filter needs

4.1 What to cover – Decide on an approach for filtering the needs with respect to a particular number of desired needs versus a threshold score. Directly compare scores across needs and eliminate some according to the chosen approach. Perform additional research, adjust ratings and scores (as needed), and go through the filtering process again until the list of needs has been refined to a manageable number.

4.2 Where to look – As with the previous step, there are few external resources to assist in this exercise.

5. Create need specification(s)

5.1 What to cover – The final step in the needs finding process is to create a need specification for each of the remaining needs. Leverage common information across needs where it makes sense, but make sure that a unique specification (especially the need criteria) is created for each need. Where appropriate, revisit the needs filtering process if information is gleaned that may affect the prioritization process.

5.2 Where to look – Network with other innovators and ask to review need specifications they may have written in the past. Review and consider the outputs from Chapters 2.1 to 2.4 independently and collectively in creating the specification.

Credits

The editors would like to acknowledge Steve Fair for his help in developing this chapter. Many thanks also go to Hanson Gifford and Mark Deem of The Foundry and Erik Engelson, formerly of Cierra, for their assistance with the case example.

Appendix 2.5.1

Sample need specifications

As noted, need specifications can take many different forms. The two that follow are actual examples from teams in Stanford University's Program in Biodesign. Their purpose is to provide the innovator with different perspectives on the final form that need specifications can take.

Example 1[7]
The need
An improved, long-term way to defibrillate the ventricle.

The problem
- Defibrillation is the use of energy to "reset" normal rhythm.
  - Mainly used to terminate ventricular fibrillation (VF), a potentially fatal arrhythmia.
- Only long-term way to defibrillate is via an implantable cardioverter defibrillator (ICD).
- There are numerous limitations to the use of an ICD:
  - Amount of energy required to defibrillate with an ICD usually causes pain.
  - ICD batteries last <3-7 years so ICD generators need replacement.
  - Though small, ICDs are placed in the chest resulting in cosmetic effects.
  - Getting "shocked" (defibrillated) can be psychologically traumatic.

The burden of VF
- ICD statistics can serve as a surrogate for burden of potential and real VF.
  - >260,000 patients with ICD indications, including those at risk (prophylactic).
  - >150,000 ICDs implanted annually in the United States.
  - Cost >$3 billion worldwide currently.
  - Medications have been shown not to be as effective as ICDs.
- Patient types:
  - Higher risk of VF in patients with low ejection fraction (EF), usually <30-35 percent.
    - Candidates for primary prevention ICD implantation (prophylactic).
  - Patients who have had documented VF or survived cardiac arrest.
    - Candidates for secondary prevention ICD implantation.

The mechanisms of VF
- VF is theorized to occur and persist via a four-stage mechanism:
  - Stage I (tachysystolic stage): Premature ventricular beats can induce initial wavebreak with large reentrant wave; usually lasts 1-2 seconds.
  - Stage II (convulsive incoordination): Multiple wavelets and reentry over smaller regions of myocardium; lasts 15-40 seconds.
  - Stage III (tremulous incoordination): Wavelets and independent contraction in even smaller areas; lasts 2-4 minutes.
  - Stage IV (atonic fibrillation): Slow passage of wavelets over small distances with loss of all visible contractility.
- VF most often seen in the setting of structural heart disease or ischemia.
  - These can promote reentry and VF stages III and IV.
- ICDs deliver defibrillation energy in stage II.

The mechanism of defibrillation
- The exact mechanism of how administration of a pulse of energy "resets" heart is unknown.
- Hypothesized to be due to cessation of most of the multiple wavelets and reentrant areas in VF, thereby permitting resumption of coordinated excitation and contraction.

- Requires rapid induction of changes in transmembrane potential of myocytes.
  - Prolonging of refractoriness of action potential may be important.
- Requires a critical mass of myocardium (>75 percent) to be involved.
- Must not reinitiate fibrillation:
  - If only part of the heart affected, heterogeneity of refractoriness can occur leading to reentry and eventual fibrillation.
  - Shock that halts VF must not reinitiate VF through same mechanism by which a shock of same strength during vulnerable period can initiate VF (upper limit of vulnerability).
- Exact ionic mechanisms of changes in transmembrane potential not known.
- Defibrillation can be achieved using various waveforms:
  - Biphasic waveforms require less current and energy than monophasic waveforms to defibrillate and are usually more successful.
  - Waveform duration, waveform shape, electrode impedance, and defibrillator capacitance all important in success of defibrillation.
- Energy required to defibrillate is based on minimum potential gradient that needs to be created in a sufficient amount of myocardium.
  - Location of defibrillation electrodes play significant role:
    - From body surface, usually need 200–360 joules for defibrillation – only 4–20 percent of current reaches heart, due to loss through tissue.
    - With leads in heart, usually only need 20–34 joules.
  - Distance between electrodes, amount of heart "within" circuit, size of electrodes, respiration, etc., also play a role.
    - Large epicardial patches require less energy for defibrillation.
  - Mathematical formulations model how a pulse of energy in one area can affect transmembrane potential at a distance away from electrode.
    - The cell membrane and gap junctions are viewed as high-resistance barriers while the intra- and extracellular spaces are low-resistance spaces.
    - Current is forced to "go around" these high-resistance barriers:
      - This exit and reentry of current causes various degrees of depolarization or hyperpolarization in different parts of the intracellular space.
      - Since the barriers function as discontinuities, they can be thought of as secondary sources of action potentials.
      - This causes perpetuation of the current at a distance from the electrode.

Methods of defibrillation

- ICD – only current long-term solution.
  - Circuit is from coil on lead in ventricle (usually right) to a second coil more proximal on lead, to ICD generator, or to subcutaneous array on patient's side/back.
  - May also be attached to epicardial patches, but uncommon.
- LifeVest – for short-term use outside hospital. Usually if temporary risk of VF or as bridge to ICD.
- External pads or paddles – for emergency or while in hospital. Circuit formed by paddles or two patches.
- Internal paddles placed on surface of heart – usually only during open heart surgery.

Patient issues with ICDs

- Pain
  - Defibrillation feels like being "kicked in the chest." Likely due to activation of skeletal muscles and nerves in the chest and abdomen.
  - Patient can feel even 1–2 joules of energy.
  - Changes in sensing require retesting of defibrillation threshold. Also requires induction of VF, with potential risk of death and sensation of "shock."
- Generator implantation and replacement
  - Internal leads require permanent implantation in venous system and fixation to heart.
    - Can become infected requiring removal, which carries significant risk.

- Leads may fail, requiring additional leads, which can reduce lumen of vessel.
- Current ICD batteries last <3–7 years leading to need for generator changes.
  - Generator changes expose patient to risks of additional procedure and infection.
  - Requires retesting of defibrillation threshold.
- Cosmetic
  - Though ICDs small, implantation in the chest results in small scar.
  - In thin patients, may often see outline of device.
- Psychological
  - Sensation of being "shocked" can be traumatic.
    - Increases thoughts about potentially lethal events and possible death.
    - Increases fear of performing activities that may lead to VF and "shock."
- Other
  - Devices may be affected by magnets and may limit patient's suitability for MRI.
  - Devices can only be interrogated with programmer.
    - Sometimes difficult to ascertain device settings in emergency situation.

New approaches being developed to treat arrhythmias
- Subcutaneous, leadless ICDs only for defibrillation
  - Sensing may be an issue requiring methods beyond sensing only rate.
  - May require very high energy to successfully defibrillate.
- Intravascular ICD
  - Could be low-cost and implanted in many patients. Possible first-line therapy prior to permanent ICD.
  - Few shocks only.
  - May be thrombogenic.
- Cell therapy
  - May be able to reconstitute damaged, non-functioning tissue, thereby preventing VF.

Need criteria
- Absolute requirements:
  - Must improve at least one of the four main limitations of current ICDs without leading to new limitations or worsening of current benefits or limitations of ICDs:
    - Pain from defibrillation.
    - Need for replacement every <3–7 years.
    - Cosmetic effects.
    - Psychological effects.
  - If affects procedure, should not be any more invasive than current implantation or replacement procedures.
  - Should have no effect on current medical therapy and should not be affected by medications.
- Desirable requirements:
  - No effect on cardiac function.
  - Easy to use/learn for average electrophysiologist.
  - Patient friendly.

Example 2[8]

The need
A way to reduce incidence of central venous catheter infections in ICU patients.

Problem description
The most common source of central venous catheter (CVC) related infections is colonization of the intracutaneous and intravascular portions of the catheter by microorganisms from the patient's skin and occasionally the hands of healthcare workers. Microorganisms gain access to the catheter wound and migrate along the catheter's subcutaneous tract into the fibrin sheath that surrounds the intravascular catheter. Scanning electron micrographs reveal that both the external and internal surfaces of catheters can become colonized with microorganisms.

Market description
Because CMS will no longer pay for line infections, all hospitals with intensive care units (ICUs) are potential customers. However, because the incidence of infection is low (1–3 percent) we will target high volume ICUs. High volume is defined as any center with greater than 1,000 ICU catheter days per year or with greater than 10 ICU beds.[9]

The anti-microbial CVC market is estimated at $98.7 million in 2002 representing a growth rate of 18.5 percent over the previous year. The market is forecast to reach $374.0 million in 2009 growing at a compound annual growth rate (**CAGR**) of 21.0 percent over the forecast period of 2002 to 2009.[10]

Fifteen million CVC days are logged in ICUs in the United States per year. On average, 5 percent of the 3 million short-term, unimpregnated CVCs that are inserted annually in the United States lead to bloodstream infection, resulting in 150,000 cases of catheter-related bloodstream infection per year.[11] Approximately 50,000 deaths per year are attributed to hospital-acquired bloodstream infections, resulting in a healthcare burden of $2.3 billion per year.[12]

Arrow International has about 84.1 percent share of the antimicrobial central venous catheter market. The antimicrobial central venous catheters are fast penetrating the conventional CVC market. As of 2002, the total market for conventional CVCs is expected to be around $300 million. Currently, approximately 20.4 percent of the market consists of the antimicrobial CVCs. By the end of the forecast period, it is expected that about 68.7 percent of the CVC market will comprise antimicrobial CVCs.[13]

In 1990, Arrow International launched Arrowgard Blue®, the first antimicrobial CVC in the US market. The technology involved impregnation of catheter walls by two agents, chlorhexidine and silver sulfadiazine, bonded together. The catheter was preferred over catheters that used antibiotics by most hospitals due to concern over bacterial resistance from overuse of antibiotics. Supported by the existing market access to the US CVC market, the products enjoys an overwhelming share of more than 80 percent of the market. In addition to Arrow, Edwards Critical Care, Cook Critical Care, and Johnson & Johnson (J&J) have commercially available products. The Edwards Vantex™ antimicrobial CVC, with Oligon agent, is the first central venous antimicrobial catheter made of a biocompatible antimicrobial material. The Cook Spectrum® Glide™ catheters are impregnated with minocycline and rifampin, which have been shown to minimize the risk of bacterial colonization of the catheter and catheter-related bacteremia during use. The J&J BioPatch® is the only dressing clinically proven for use in skin antisepsis. The BioPatch is uniquely designed to continually release CHG over seven days, providing 360° protection.

Need criteria

Indication: Patients requiring central line access for greater than three days.

Location: ICU physicians and nurses.

Procedure time: Equivalent time to place as the existing CVC or not additive of substantial time to line placement, upper bound is 30 minutes.

Procedure frequency: If the solution is integral to the catheter, then the effect must last a minimum of four days. It is preferable that said solution reduces the frequency of catheter change; therefore, target effect is eight to ten days. If the solution is not integral to the catheter, then increased frequency is tolerated.

Follow-up: If the solution is integral to the catheter, it must not change the standard practice of catheter inspection every other day. Minimum equivalent to BioPatch.

Efficacy: Solution must provide statistically significant improvement to current baseline of 5.3 infections per 1,000 catheter days.[14] Target is a 50 percent reduction in infection meaning, 1 percent or 2.6 infections per 1,000 catheter days.

Complication rate: Solution shall not introduce new complications.

Value: Less than $100 if catheter based, and less than $10 if adjunctive.

Comparative pricing: The BioPatch is currently sold for $3.75 per patch and antimicrobial central venous catheters sell for approximately $50 per catheter.

Human factors analysis: Solution must easily integrate into the existing ICU cycle of care. If the solution is integral to the CVC, then it shall not prohibit any standard functionality or significantly affect the ease of placement or inspection. If the solution is not integral to the CVC, then it shall be no more cumbersome than the J&J BioPatch.

Attributes: May not affect catheter function: flow rate, injection pressure or diameter (inner and outer).

Notes

1. The Foundry, http://www.the-foundry.com/ (November 19, 2007).
2. All quotations are from interviews conducted by authors, unless otherwise cited. Reprinted with permission.
3. "Congenital Heart Disease: Septal Defects," Revolution Health, March 29, 2007, http://www.revolutionhealth.com/conditions/heart/congenital-heart-disease/types/septal-defects (December 4, 2008).
4. "Surgery Section: Closure Devices for Patent Foramen Ovale and Atrial Septal Defects," Priority Health, April 11, 2007, http://www.priorityhealth.com/pdfs/medical-policies/91528.pdf (December 4, 2008).
5. Shelley Wood, "Mixed Results for PFO Closure in Migraine Cloud Interpretation of MIST," TheHeart.org, March 13, 2006, http://www.theheart.org/article/840151.do? (July 15, 2008).
6. Shelley Wood, "NMT Announces Termination of Its MIST II Trial of PFO Closure for Migraine," HeartWire, January 24, 2008, http://www.medscape.com/viewarticle/569169 (July 15, 2008).
7. Developed by Uday N. Kumar, John White, Kityee Au-Yeung, and Joseph Knight as part of Stanford's Program in Biodesign. Reprinted with permission.
8. Developed by John Avi Roop, Matthew John Callaghan, Joelle Abra Faulkner, Kevin Zi Jun Chao as part of Stanford's Program in Biodesign. Reprinted with permission.
9. 15 million catheter days in the ICU per year/5,000 hospitals = 3,000 catheter days per hospital per year. 3,000 days/365 days = ~8 catheters in parallel per year. Because not every hospital will have an ICU, then assume that average hospital has 10 ICU beds.
10. "U.S. Antimicrobial Devices Markets: Challenges and Strategies," Frost & Sullivan, April 22, 2003.
11. R. O. Darouiche, I. I. Raad, S. O. Heard, et al., "A Comparison of Two Antimicrobial-Impregnated Central Venous Catheters," New England Journal of Medicine (January 7, 1999): 1–8.
12. R. P. Wenzel, M. B. Edmond New England Journal of Medicine (January 7, 1999): 48–50.
13. Frost & Sullivan, op. cit.
14. According to the Centers for Disease Control.

Acclarent Case Study

Stage 2 Needs screening and specification

At the same time that Makower and Chang were investigating the chronic sinusitis need, they continued to press forward with investigating needs in orthopedics, respiratory disease, and congestive heart failure (CHF). However, delays in gathering information and gaining access to perform observations allowed chronic sinusitis to become somewhat of a "lead horse" in their efforts to identify which project to work on.

Importantly, they developed more than one need in this area. For instance, in addition to finding a potential alternative to FESS surgery, they also saw the need for improved, more dynamic diagnostics that would allow ENT specialists to more accurately assess a patient's condition. Chang explained: "In ENT surgery, they use endoscopy, they use computer tomography (CT) scans, and they assess symptoms. But the results from these measures don't necessarily correlate. Computer tomography scans are static – they provide a snapshot in time, just like endoscopy. So, you can have a picture-perfect CT scan from a week ago, but the patient feels terrible today, and vice versa. So it's pretty obvious that the surgeons are working virtually in the dark and going on their best clinical judgment rather than a single conclusive diagnostic test."

With an understanding of the clinical problems in the area and a working hypothesis for addressing more than one important need, the next step in the process was to perform more detailed research regarding the disease state, treatment options, stakeholders, and market. Once this was completed, they would be in a much better position to screen the needs and decide which one should become their chief focus.

2.1 Disease state fundamentals

To understand the disease state, Makower and Chang threw themselves into a thorough literature search. They sought to understand the anatomy and physiology of the sinuses by looking at surgical textbooks, cadaver dissection books, and hundreds of CT scans (see Figure C2.1).

In terms of understanding the pathophysiology of chronic sinusitis, they did more book research and combed the peer-reviewed clinical articles, paying particular attention to the disease's mechanism of action. However, as Sharon Lam Wang, a consultant to the ExploraMed II team, pointed out, "Chronic sinusitis is not a given. It's not like we understand what this is. Even the physicians in the field don't agree on the definition. Is it a syndrome? Is it a disease? How do allergies play into it? How does the anatomic makeup of the patient affect the condition? It's a multifactorial disease that seems to have a life of its own."

For this reason, Makower and Chang felt that secondary research would not be enough to ensure an adequate understanding of the disease. Their primary research strategy would involve direct physician input, as well as

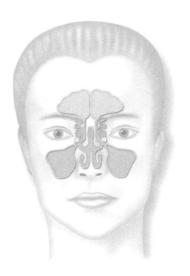
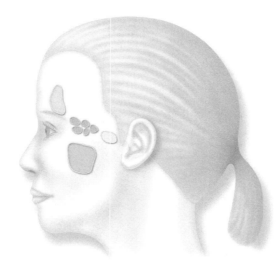

FIGURE C2.1
The sinuses are four bilateral sets of air-filled cavities with ostia that connect to the nose. The frontal sinuses are in the forehead, the ethmoid sinuses are groups of small air pockets located between the eyes, the sphenoid sinuses are behind the eyes, and the maxillary sinuses are in the cheek area. Each sinus has an opening through which mucus drains. The drainage of mucus is a normal process that keeps the sinuses healthy. Mucus moistens the nasal lining and protects the inside of the nose from impurities such as dust and bacteria (provided by Acclarent).

cadaver dissections. However, they wanted to be highly selective about interfacing with ENT specialists so early in their process. "We needed somebody who we really felt was a safe person, who understood this space very well, but ideally was someone who did not work directly as a rhinologist," explained Makower. "This kind of person would be able to provide great clinical feedback and training for us, but we wouldn't have burned a bridge or potentially mismanaged a relationship with anyone in our target market. We needed someone without a vested interest in what we were doing." The team found such an individual in Dr. Dave Kim, who was an ENT physician, but a specialist in facial plastics for cosmetic and post-traumatic applications. The referral to Dr. Kim came through one of ExploraMed's board members. According to Makower, "We went to him and basically started from scratch. What's a sinus? Where are they located? Why do they remove that structure? We revisited all of the basic physiology, how it functions, and so on. And then we went into the anatomy lab and did a detailed dissection with him where he pointed out all the structures, which was very helpful." In terms of a key take-away from the investigation, Makower stated, "For most people it seemed that when you have good flow out of your sinuses, you have healthy sinuses. That seemed to be the bottom line" (see Figure C2.2). Other important information gleaned from the team's disease research is shown in Table C2.1.[1,2,3]

2.2 Treatment options

In parallel with learning about the disease state of chronic sinusitis, Makower and Chang studied existing and emerging treatment options in the field. "This is where being a patient myself came back into play," said Makower, "because I had been very frustrated with my choices." He elaborated on the progression of clinical treatment:

First, there are a variety of over-the-counter drugs, like Sudafed and other allergy medications, to reduce mucus and relieve minor symptoms. At the next level, under doctor's prescription are antibiotics and sprayable steroids, such as Flonase. These always scared me – I never liked the idea of spraying steroids into my nose. Then, before someone would qualify to get surgery, a lot of doctors prescribed high doses of oral steroids, which can introduce significant side effects. Finally, in the worst cases, functional endoscopic surgery was performed. Over the years, I had seen multiple ENTs regarding the possibility of surgery. But each time, they would look at my CT scans and say, 'You know, you just don't have bad enough disease in your sinuses for us to do anything.' And that never really made sense to me, because I was really suffering. But, after watching these surgeries, I finally understood

Table C2.1 *Chronic sinusitis is a serious medical condition that affects a significant number of patients.*

Sinusitis facts
Sinusitis affects approximately 39 million people each year (or 14 percent of the adult US population), making it one of the most common health problems in the United States.
It is more prevalent than heart disease and asthma and has a greater impact on quality of life than chronic back pain, diabetes, or congestive heart failure.
Common symptoms, which can affect patients physically, functionally, and emotionally, include facial pain, pressure and congestion, discharge and drainage, loss of one's sense of smell, headache, bad breath, and fatigue.
Direct healthcare expenditures due to sinusitis total more than $8.1 billion per year.
Patients with chronic sinusitis have symptoms that last more than 12 weeks (patients with acute sinusitis usually have their symptoms resolved in less than four weeks).
Approximately 1.4 million patients are medically managed for chronic sinusitis each year.
Patients with chronic sinusitis make 18–22 million annual office visits per year and have twice as many visits to their primary care doctors and five times as many pharmacy fills as those who do not.

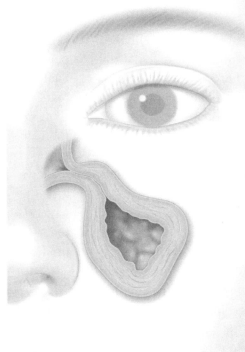

FIGURE C2.2
Sinusitis is an inflammation of the sinus lining most commonly caused by bacterial, viral, and/or microbial infections, as well as structural issues such as blockage of the sinus opening (ostium). If the ostium becomes swollen shut, normal mucus drainage may not occur. This condition may lead to infection and inflammation of the sinuses (provided by Acclarent).

why they were steering patients away. They were protecting me from the potential complications of surgery, which are severe when they happen. The surgical tools they use to perform FESS (see Figure C2.3) can sometimes cause scarring, which actually can create a whole new level of disease itself. For a recurrent sinusitis sufferer like me, without surgery, I might have had months of symptoms from swollen sinus that would eventually subside. But after FESS, it would be possible to end up with a permanently scarred ostium that would have had little or no chance of draining, with the possibility of even worse symptoms than I had before.

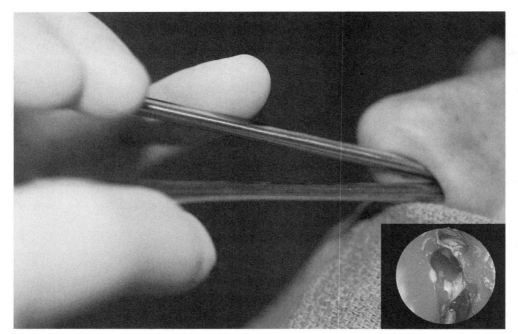

FIGURE C2.3
Functional endoscopic surgery uses straight, rigid tools in the tortuous sinus anatomy. As a result, surgeons are often forced to remove healthy tissue and bone simply to reach the infected sinus(es) (courtesy of Acclarent).

Of the 1.4 million people who are medically managed for chronic sinusitis each year, approximately 970,000 are considered candidates for FESS surgery. The goal of this procedure is to remove bone and tissue to enlarge the sinus opening and restore drainage. Yet, a surprisingly small number of the qualified surgical candidates – just 500,000 per year – undergo the FESS procedure.[4] Some of the most common reasons patients declined the surgery are shown in Table C2.2.[5]

After investigating these risk factors, Makower noted, "It was clear that this procedure had not been completely fixed yet." Chang added, "Patients described feeling terrible for a week or two post-operatively, as if they were recovering from being hit in the face with a baseball bat. This was often accompanied by swelling and the occasional black eye. But, even worse is the fact that there can be a significant revision rate. That's horrible! How do you ask patients to go through such a painful surgery and recovery with an almost one-in-four chance that they will need to go through it again?"

2.3 Stakeholder analysis

Understanding the risks involved with FESS helped provide Makower and team with insights into how the patients felt about the procedure. As a result, they spent the majority of their time during their preliminary stakeholder analysis evaluating the perspectives of physicians. Research was also performed on payers and facilities, but this work will be described in more detail in 4.3 Reimbursement Basics.

Lam Wang performed much of this analysis, joining the team to determine if the group could realistically make a business out of the idea. She had worked with Makower and Chang at TransVascular and then played a role in starting-up device company Kyphon®. When Lam Wang got involved in this project, she believed that ENT specialists would respond positively to new innovations in the area of chronic sinusitis:

> I thought that they were open to new technology, primarily because of the FESS revolution that had occurred in the 1980s. Before then, sinus surgery was done open through cuts on the face or through the gums to get to your sinuses. They would literally peel away the skin on your face to get to your sinuses. With the advent of the endoscope, physicians switched over to accessing the sinus through the nose. It was almost like an overnight switch, and they adopted the new approach with no clinical data indicating that it worked better. Even when we were researching the problem, there was very little data out there.

Table C2.2 *Issues the team identified with FESS.*

About FESS
FESS is successful in approximately 76 to 98 percent of all procedures[6] (with revision rates up to 24 percent).
Post-operative visits to remove debris, clots, and scars from the surgical site are often needed and can be painful.
Rigid steel surgical instruments are used to perform the procedure through the patient's nostrils, which creates the need to remove bone and tissue so that the physician can reach target sinus and open the ostium to facilitate drainage.
The procedure is conducted under general anesthesia.
Given the proximity of the sinuses to the skull and eyes, there is a small chance with every surgery (one in 200) that the surgeon will inadvertently remove a piece of thin bone between the sinus and the brain, which can cause a spinal fluid leak and can introduce the risk of meningitis.
Blindness is another possible complication from the procedure if the wall of the sinus against the eye socket is breached.
Due to the highly vascular nature of the nose and sinuses, the nasal cavity can sometimes require packing with gauze to absorb bleeding from the procedure, and its subsequent removal can be painful.
Although the procedure is usually performed on an outpatient basis, it is often described by patients as one of the worst experiences of their lives.
Following surgery, it can take weeks for the ancillary effects of the surgery (e.g., swelling) to subside in order to determine whether the procedure was successful in opening the ostium.

And so that told me these folks are willing to be practical and open-minded to new ways of doing things.

To confirm this point of view, she conducted interviews with four or five well-respected doctors in the ENT specialty – those physicians who routinely performed FESS. "In these initial discussions, nothing led me to believe that they would be resistant," she recalled.

From Chang's perspective, however, there were some mixed signals coming from their physician stakeholders. He explained:

On the one hand, we'd talk to them and they'd claim to be 'gadget guys.' They'd tell us how much they loved new toys, like powered instrumentation and surgical navigation. This gave us the feeling they might be excited to adopt new technology that did things a little better. On the other hand, as we peeled back the onion, we recognized that most of what they were talking about were incremental innovations taken from other specialties – powered shavers from orthopedics and surgical navigation for neurosurgery. Procedurally, they were still cutting and removing bone, where the shaver and navigation system allowed them to do it more efficiently and safely. These were certainly improvements, but not huge leaps forward in their approach to surgery. They were also very proud of keeping the same instrument set for the last 12 years. They would handle these steel instruments with the utmost care because they didn't want to break them and subsequently have to pay for a new set. This did concern us.

Despite the uncertainty this created, the team felt reasonably comfortable pressing forward with their investigation of the needs in this space.

2.4 Market analysis

The next step was to perform a more detailed assessment of the market. The team believed that to get a truly accurate estimate of the market, it would not be enough to rely on the broad-brush estimates published in the available literature. So, they started from scratch. To determine the incidence of people affected by chronic sinusitis each year, Lam Wang researched the clinical literature and found a study that used ICD-9 diagnosis codes to estimate the true number of adults diagnosed with the condition.

Stage 2 Needs Screening

Estimated Market Potential

US Market Model for 2004

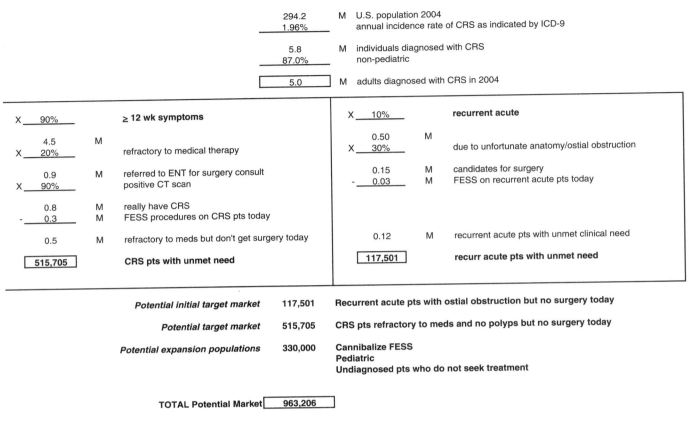

FIGURE C2.4
The market analysis for chronic sinusitis pointed to a large unmet medical need (provided by Acclarent).

Next, she worked through a cycle of care analysis to identify the "low hanging fruit" – people who do not benefit from drug therapy, are considered candidates for surgery, but elect not to undergo the FESS procedure. After diving into all the numbers, validating them with practicing physicians in the field, and testing all the assumptions in the model through an iterative process, she had a preliminary answer. "It turns out it was a phenomenally high number, like 900,000 patients per year, who might be candidates for an alternate procedure," Lam Wang recalled. "So that spoke to a huge unmet need." Importantly, this figure included the more than 300,000 patients per year currently undergoing FESS (assuming they might eventually move toward an alternate procedure). It also took into account a certain type of chronic sinusitis called recurrent acute sinusitis, which met the definition of sinusitis but was "considered a different animal." According to Lam Wang, "These are folks who have frequent episodes, possibly because of their anatomy – a narrowing in their passageway. Getting at these numbers required a lot of iteration with the doctors." See Figure C2.4 for the basic model.

The fact that there were "tons and tons of possible patients, but not a lot of procedures being performed," indicated to Lam Wang, Makower, and Chang that the current and future opportunity was highly attractive.

2.5 Needs filtering

By the time the team had determined that the total potential market for an alternative technology for FESS surgery could reach nearly a million patients per year, they had all the information necessary to confidently commit to this opportunity in its ongoing innovation efforts. Enough was known about the disease state to make new solutions feasible, the treatment landscape was ripe for innovation, patients were eager for an alternative technology for FESS, and physicians seemed to be generally open to considering new ideas. It was also a strong fit with the defined strategic focus and Makower's related acceptance criteria.

While the team was still investigating needs related to orthopedics, respiratory disease, and CHF (as well as other chronic sinusitis needs), the need for ***a minimally invasive approach to treating chronic sinusitis that had less bleeding, less pain, less bone and tissue removal, less risk of scarring, and that was faster, easier, and safer to perform*** had strongly overtaken any of the other opportunities being considered. As Makower explained, rather than performing all of their needs filtering activities at one point in time, the team had been filtering its needs on more of a rolling basis as new information became available:

> We did our filtering along the way. When we teach the biodesign innovation process, we focus on isolating the fundamentals of how people innovate to make each step clear, but in doing so, it can make the process seem like all of these activities occur in sequence. In practice, they can happen almost simultaneously. There's an analogy that I like to use with my project architects: getting to the finish line with a winning project is like a horse race. You can put the need areas or ideas (horses) in at any time during the race, and they can get knocked out at any time, too. Some horses start at the beginning of the race and go very far before being knocked out by other screening criteria. Others don't make it very far at all. Eventually there's one horse that keeps on going and gets out ahead of the others, clearing each screening hurdle with ease. Once you see one getting far ahead of the others, you solely focus on that one horse and you try to ride it to the finish line.

As a seasoned innovator and entrepreneur, Makower was comfortable leaving some opportunities behind when others seemed more promising. "You can only win by eventually focusing on one thing at a time. If you spread your efforts too thin, you'll be too distracted to execute well on any one need, and you'll never get there."

Notes

1. "Sinusitis Overview, Acclarent, http://www.acclarent.com/patients/introduction.html (August 28, 2008).
2. Sorgen, op. cit.
3. Stephen Levin, "Acclarent: Can Balloons Open Sinuses and the ENT Device Market?" *In Vivo*, January 2006.
4. Ibid.
5. Ibid.
6. R. S. Jiang and C. Y. Hsu, "Revision Functional Endoscopic Sinus Surgery," *Annals of Otology, Rhinology and Laryngology*, February 2002, pp. 55–59.

INVENT
Concept Generation

IDENTIFY

STAGE 1 NEEDS FINDING
1.1 Strategic Focus
1.2 Observation and Problem Identification
1.3 Need Statement Development
Case Study: Stage 1

STAGE 2 NEEDS SCREENING
2.1 Disease State Fundamentals
2.2 Treatment Options
2.3 Stakeholder analysis
2.4 Market Analysis
2.5 Needs Filtering
Case Study: Stage 2

INVENT

STAGE 3 CONCEPT GENERATION
3.1 Ideation and Brainstorming
3.2 Concept Screening
Case Study: Stage 3

STAGE 4 CONCEPT SELECTION
4.1 Intellectual Property Basics
4.2 Regulatory Basics
4.3 Reimbursement Basics
4.4 Business Models
4.5 Prototyping
4.6 Final Concept Selection
Case Study: Stage 4

IMPLEMENT

STAGE 5 DEVELOPMENT STRATEGY AND PLANNING
5.1 Intellectual Property Strategy
5.2 Research and Development Strategy
5.3 Clinical Strategy
5.4 Regulatory Strategy
5.5 Quality and Process Management
5.6 Reimbursement Strategy
5.7 Marketing and Stakeholder Strategy
5.8 Sales and Distribution Strategy
5.9 Competitive Advantage and Business Strategy
Case Study: Stage 5

STAGE 6 INTEGRATION
6.1 Operating Plan and Financial Model
6.2 Business Plan Development
6.3 Funding Sources
6.4 Licensing and Alternate Pathways
Case Study: Stage 6

Innovation is now recognized as the single most important ingredient in any modern economy.

The Economist[1]

The devil's advocate may be the biggest innovation killer ...

Tom Kelley[2]

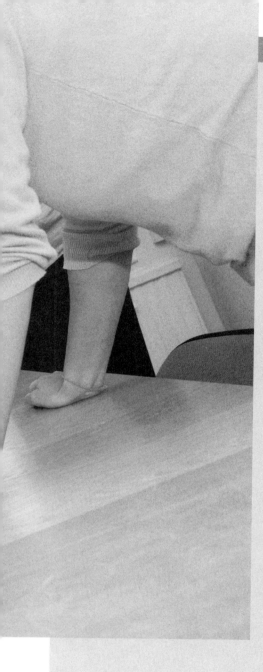

Concept Generation

If invention/innovation is so important, why are there so many devil's advocates? Can we banish them ... at least early on?

Once an important clinical need is clearly identified, it's time to have some fun. It's time to invent. The recurring theme of constant iteration developed in Phase 1 of the biodesign innovation process (Identify) continues as an essential component, during both the Concept Generation and Concept Selection stages.

Concept generation, getting the ideas, begins with ideation and brainstorming. This approach originated half a century ago in Alex Osborn's *Applied Imagination*, which launched the study of creativity in business development. Its premise is clear. There are three things to work with – facts, ideas, and solutions; *each* deserves quality time. The natural tendency is to leap from facts to solutions, skipping over the play and exploration that is at the heart of finding new ideas. Most of us are experienced with fact finding, it's a consequence of contemporary education's preoccupation with facts. We're also familiar with solutions; most of us like to solve problems and move on. Idea finding may seem childlike (and it should be) but at its heart is the exploration of possibilities, *free from as many constraints as possible*.

Brainstorming is not new-age nonsense, rather it is a studied process and practiced art for inventing something that has value. If nothing revolutionary, weird, or goofy surfaces, this stage has failed. The vibe should be upbeat – a chance to try things out, to free associate, and to challenge the wisdom of the present.

Brainstorming is enormously enabled by participants with diverse backgrounds, points of view, and expertise. In a perfect world it might go like this. An engineer will know the properties of lasers. A chemist will understand hydrogels, and perhaps their polymerization with laser light. An orthopedic surgeon might add awareness of the clinical need for fracture stabilization. Strategic insights into markets and product development might be added by an experienced entrepreneur – voila, a laser light polymerized hydrogel, not only to aid in fracture stabilization but to serve as a drug delivery system to enhance fracture repair – and an invention is born.

Notes
1. As cited in Tom Kelley, *The Ten Faces of Innovation* (Doubleday Business, 2005).
2. Ibid.

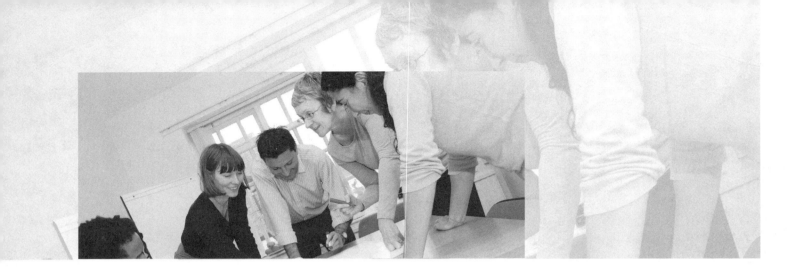

3.1 Ideation and Brainstorming

Introduction

One compelling medical need. Six knowledgeable participants with diverse backgrounds and experiences. Ninety minutes. And a goal to develop 100 new ideas. While this recipe may not be familiar to all medical device inventors, it represents a proven formula for developing creative solutions to a defined need.

"Ideation" refers to the process of creating new concepts or ideas – it comes into play at key steps in the biodesign innovation process when new solutions are required to address well-defined needs. "Brainstorming" has developed a specific meaning in the design tradition, as a set of methods for ideation that is based on the power of group creativity. This approach requires participants to temporarily suspend their instinct to criticize new ideas and open their minds to a rapid flow of new possibilities and connections.

OBJECTIVES

- Understand the role of ideation in the context of the biodesign innovation process.
- Learn the basic methods of brainstorming and how to plan and execute a session.
- Consider special brainstorming approaches and tips that are specific to biomedical technology inventing.

Ideation and brainstorming fundamentals

Since the time of Thomas Edison, the stereotypic picture of the process of invention has been the "aha" moment – the burst of inspiration from the brilliant mind of the inventor when the light bulb literally (or figuratively) flashes on. Edison himself, however, made it clear that inventing is a disciplined process that involves patience and hard work, saying, "None of my inventions came by accident. I see a worthwhile need to be met and I make trial after trial until it comes. What it boils down to is 1 percent inspiration and 99 percent perspiration."[1] Although he became an individual icon as an inventor, the key to Edison's productivity was that he developed a multidisciplinary *team* of innovators in his Menlo Park laboratory. He also knew that the *quantity* of ideas was important and that the *failure* of many potential solutions was inevitable – for example, the lab tried several thousand different filaments before finding a stable material for the first successful light bulb.[2]

Ideation in the medtech field

Invention of medical devices is, if anything, a more involved and complex process than invention in many other technology sectors. As the biodesign innovation process emphasizes, there are multiple factors that must be carefully considered, including the upstream issues associated with the medical need, gaps in the treatment landscape, stakeholder interests, and the market opportunity, as well as the downstream concerns of patenting, regulation, reimbursement, and deployment in the healthcare system. Against this backdrop, the "aha" model of ideation provides a particularly poor representation of how medical technologies are truly invented. John Simpson, medical device pioneer and founder of Advanced Cardiovascular Systems (ACS), Devices for Vascular Intervention (DVI), Perclose, and FoxHollow Technologies summarizes his perspective in this wry but insightful way:[3]

> I know that there have been people who have had these visions and suddenly the design of a catheter pops into their mind and they make it that afternoon and then the following day they achieve enormous wealth. Every morning, I get up, I check my mind to see if it's in there and if it's not, then I go to work. And so far, I haven't missed a day of work yet.

Specific strategies and methods for ideation are described in the sections below on brainstorming. However, there are a few general guidelines related to ideation in the medtech field that should be emphasized regardless of the approach being used. First, it is important to understand that ideation at this step in the biodesign innovation process requires a specific mindset that is different from what is required at any other stage. While critical filtering is necessary and important at other points in the innovation process, it can be counterproductive to a team's results when they first begin considering solutions. Inventors need to open their minds to a creative flow of ideas, set aside their preconceived notions, and look beyond the solutions that they may have been consciously or subconsciously forming during needs screening. They should also suspend their tendency to evaluate and make judgments about new concepts as they arise. This can be particularly difficult for bright engineers, business people, and physicians who have built their academic or professional careers by learning how to get the right answer quickly and consistently. The need to be perfect is the enemy of ideation.

A second main theme is the importance of cross-pollination in the ideation process. In medtech, there are abundant opportunities to look across specialties to adapt the technologies and approaches from one area to another. Acclarent, a company profiled in this text, is a case in point: it developed a new, minimally invasive approach to functional endoscopic sinus surgery to treat chronic sinusitis by adapting the equipment and procedures of coronary angioplasty for the heart. Medtech is rich with similar opportunities for cross-pollination: between different medical specialties; between physicians and engineers; and even between medical and non-medical technologies. Julio Palmaz, inventor of the coronary stent, thought of the metal lattice approach when studying masonry expanded metal.[4] Thomas Fogarty, one of today's most prolific medtech innovators, likes to walk the aisles of a hardware store to look at different materials and tools when he is struggling for new approaches to a medical technology.[5]

A third general point is that ideation approaches (and brainstorming, specifically) can be applied at many different stages in the biodesign innovation process. Brainstorming is particularly useful after the inventor has developed a need specification and is ready to begin thinking about different solution concepts to address the need. But there are many other circumstances that may be appropriate for brainstorming, including when an inventor or team is exploring a new or more specific direction for a solution (e.g., finding a delivery method to place a device in a particular spot) or when a team is refining its approach to an already accepted solution (e.g., exploring ways to modify the design of a device to improve its function). It can also be used to address different types of problems, including technical, clinical, or market-related challenges. For instance, brainstorming could be used to consider alternative business models or various market segmentation strategies. In fact, solutions to many of the strategic issues that will need to be addressed later in the biodesign innovation process can be developed using the brainstorming process. Brainstorming can also be helpful in addressing smaller, tactical challenges that arise along the way, such as how to overcome unanticipated hurdles or break through issues that might be slowing the team's progress.

Finally, it is important to realize that ideation and brainstorming are part of an iterative or cyclic approach. The team should never consider itself to be completely done with the brainstorming process. New information and new circumstances crop up at all stages of the process and may require the team to go back into brainstorming mode. There is an especially important feedback loop between prototyping (4.5 Prototyping) and brainstorming: good prototypes provide powerful stimuli for new ideas.

A guide to brainstorming

Most people would say they know about brainstorming and have some experience with the process. When it comes to the formal design of solutions, however, brainstorming refers to a highly specific set of methods and tools that distinguish this approach from other forms of ideation. In this context, brainstorming explicitly describes a group process, where participants with different backgrounds leverage their collective creative power to bring forward a larger number of more innovative ideas than an individual could generate.

The purpose of brainstorming is to push the biodesign innovation process forward toward the final goal of *one* good idea and *one* good development strategy to support the idea. However, it is important to recognize that, like needs finding, brainstorming may result in tens or hundreds of ideas that need to be screened, sorted, and then evaluated before any single idea is chosen. See Figure 3.1.1.

The process that follows the brainstorming meeting provides the mechanism for evaluating these ideas and identifying the best concepts to investigate further (see 3.2 Concept Screening through 4.6 Final Concept Selection).

Brainstorming rules

The seven most widely recognized "rules" for brainstorming come from the product design company

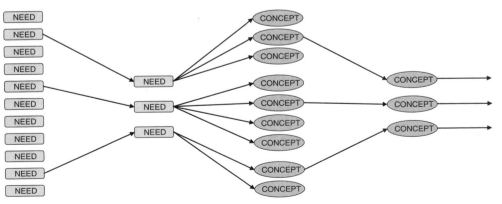

FIGURE 3.1.1
An overview of the ideation process: multiple needs are filtered into the most important few; then brainstorming produces many concepts that are screened down to the leading prospects.

IDEO. The company's brainstorming methods, which are based on years of research at universities and other experiences in the design community, are summarized below. A more detailed treatment can be found in *The Art of Innovation*, by Tom Kelley,[6] IDEO's general manager (other references are listed in the Getting Started section of this chapter).

1. Defer judgment

Deferring judgment is perhaps the most counter-intuitive and difficult to follow of all the brainstorming rules. The point is to suspend any critical thoughts or commentary until later in the innovation process (well after the brainstorming session). The purpose of brainstorming is to open up both individual creativity and the group's creative process. One good way to make this happen is to accept any new idea – even those suggestions that seem at first to be impractical or silly – and move on quickly to the next concept.

Learning to defer judgment can be challenging for many people. If some of the participants in a brainstorming session are new to the process, it is especially important to explain this first rule and to practice it with an informal warm-up (see below). Another effective way to make sure that this first rule is followed is to be careful when selecting participants for the brainstorm. There are certain people who are extremely valuable as critics and will be essential later in the biodesign innovation process (e.g., as concepts are screened). However, these individuals do not belong in an early-stage brainstorming session because they may negatively affect the flow of creative ideas. Conversely, there are other people who, at first glance, might not seem to have relevant backgrounds or expertise, but make great collaborators in that they bring out the best in others by generating and encouraging ideas not limited by the established biases of a given field.

2. Encourage wild ideas

When brainstorming, participants should do more than suspend their critical filters. They should actually practice thinking in new and different ways to generate ideas that are "outside the box." Because ideas will be filtered later in the process, it is acceptable to articulate wild ideas at this stage. Fun (or even silly) ideas serve an important purpose in that they can stimulate collective creativity by building a connection to something far removed from the current conversation. This, in turn, can lead to new inspirations. As one former member of the IDEO team shared informally, "Great ideas hide right behind the goofy ones." In an effective brainstorming session there is also a rhythm, or a speed, that builds up, and part of this is maintained by group members offering far-fetched ideas. Additionally, wild ideas help keep the energy in the session upbeat and thought provoking, which is essential to the creative process.

3. Build on the ideas of others

Building on the ideas of others means leveraging one idea as a foundation from which to make another suggestion. The power of this method is seen as one participant's idea stimulates other participants to come up with solution enhancements, novel connections, and even new ideas they would not have thought of otherwise.

Because building on the ideas of others is complementary to the concept of deferring judgment (rule 1), this rule can be just as unnatural and difficult for some inexperienced brainstormers to embrace. One technique for helping participants with this behavior is to encourage them to explicitly say the words, "Building on [this person's] idea, what if . . ." This approach acknowledges the contribution of the person with a preceding idea, while also offering a way to enhance or improve it without being critical. This works especially well in brainstorming sessions that include participants across different functional areas or disciplines.

4. Go for quantity

As noted, a successful brainstorming session builds a momentum or flow that breaks through the usual inhibitions of a group. One way to achieve this flow is to set a target goal for the group to create a large number of concepts without regard to how "good" the ideas are. A typical brainstorming session lasts for approximately 60 to 90 minutes (more than 90 minutes of intense brainstorming can be exhausting and unproductive). Within this timeframe, a team might expect at least 60 new ideas to be generated, with a "stretch" goal of

100 or more. Clearly, developing so many ideas within such a short period of time requires the group to move quickly from idea to idea rather than dwelling on any single suggestion.

5. One conversation at a time

The critical concept underlying this rule is that listening can be as important as talking during the creative process. Enforcing this rule usually falls to the facilitator (see below), although each participant should help keep the group focused on one discussion. As with all of the brainstorming rules, the need for one conversation at a time should be established as an expectation going into the session so that the facilitator has an accepted, non-threatening basis for holding people accountable to the rule if a problem arises. For example, the individual leading the session might use a "spoof punishment" to keep a group member on track without being too heavy-handed (e.g., "Do that again and I'm going to toss these M&Ms at you!"). It can also be helpful to have the brainstorming rules posted in the room, both as a reminder for everyone and to make it a little bit easier for someone with a concern to point to the rule and say, "Remember: one conversation at a time."

6. Stay focused on the topic

Even the most disciplined participants may have a tendency to let their conversations "wander" in a brainstorming session. While these digressions sometimes result in valuable ideas and information, they tend to have a negative effect on both the productivity and the flow of the meeting. To help a group stay focused, it is important to avoid distractions, side conversations, real-time analysis, and filtering of ideas (this comes later). One useful strategy to minimize the occurrence of these distractions is to have a special flip chart or section of a whiteboard where unrelated suggestions or questions can be tabled until another time. If the momentum of the meeting begins to slow down, the facilitator should try to jump to a new area of brainstorming within the general topic of the session. If the meeting is seriously stalling, the group should consider taking a short break or stopping altogether for the day.

Keep in mind that wild ideas are not digressions, but valuable components of the brainstorming process (even though they may seem somewhat off-topic at the time). However, there is no clear rule for distinguishing one from the other. As a result, the facilitator and participants should not get too concerned about policing this distinction. Remember that brainstorming is an iterative process. If a brainstorming area is important, it will be worth conducting more than one session. Through this approach, today's digression may turn into tomorrow's great idea.

7. Be visual

Being visual begins with the physical space in which brainstorming occurs. It is useful to have blackboards, whiteboards, large flip pads, sticky notes and/or other means of drawing and writing that encourage an open, fast, and unlimited flow of ideas. Everyone needs to be able to see what is being written. It is important to have a scribe – sometimes two, if the idea flow is rapid – to make sure that all ideas are captured and made visible to participants throughout the session. In capturing ideas for the group, use as few words as possible to describe the idea clearly. Catchy labels or headlines are helpful, as long as everyone will remember what they mean.

Cartoon-like doodles can increase the speed and economy of communication. The point is not to create great art, but to stimulate the flow of ideas. This is similar to the game "Pictionary" in which players must use drawings (not words) to help their team guess a word, saying, or concept. In this game, it is not the most skilled artist who wins, but rather the player who can most quickly use rudimentary drawings to communicate critical information.

It can also be helpful to cluster ideas into different "regions" of the room if they seem to be related (sticky notes can be particularly useful when it comes to clustering ideas). Showing a flow of ideas (i.e., one idea building on the other) with arrows is another way to increase the visual aspects of a brainstorming session. If flip pads are used, it is helpful to be able to tape them up around the room as the groups of ideas expand (rather than turning from one page to the next such that earlier notes can no longer be seen). A typical brainstorm can fill several whiteboards, so be prepared with plenty of space for capturing ideas. One physical

FIGURE 3.1.2
An immersion room for brainstorming – lots of space and the fun of writing all over the walls (by Anne Knudsen, courtesy of the Stanford Graduate School of Business, 2006).

configuration that supports the brainstorming process well is an "immersion" room, with floor-to-ceiling whiteboards on all four walls (see Figure 3.1.2).

However, it is not necessary to have a space specifically designed for brainstorming – productive sessions have been conducted in nearly every type of space using little more than large flip pads and tape. Once the session is underway, it is important to confirm whether or not the space, as well as the configuration of whiteboards, sticky notes, and other tools, is working for the group. If not, it may be worthwhile to interrupt the session and try a different approach.

Keeping track of ideas as the number reaches 100 or more can be challenging. Planning the spatial locations where ideas will be written down (or where sticky notes are placed) in the room can be an important way of staying oriented and organized. Give this careful thought prior to the session. Another simple, but useful, strategy is to number all ideas as they are generated so that they can be quickly and easily referred to when building on previous concepts.

One more strategy for making a brainstorming session more visual is to make various artifacts available (toys and props) since they can be useful in prompting new connections and inspiring insights. Simple props, such as blocks, pipe cleaners, clay, balls, tubes, or Legos, can be used to help create an interactive and fertile mood during the session. They also can be used to visually stimulate ideas and demonstrate three-dimensional (3-D) concepts. Tape and staplers help support the assembly of more complex (although still quick and crude) mock-ups. In one recent brainstorming session focused on a new multitool laparoscopic device, a three-color click pen became the key prop for stimulating ideation. Miming, role playing, or physically simulating the use of a device can also be a good way to express a 3-D concept to others, in order to provide rapid understanding and provoke new ideas.

The question of whether and how to use computers and the Internet during a brainstorm is becoming increasingly common. The answer depends on the participants and purpose of the session. Online connectivity can be useful to quickly pull up images, search through devices, and watch procedure videos, but the opportunity for distraction and breaking the rhythm of the session is obviously quite high. In general, it is probably best to avoid the use of computers and the Internet while brainstorming unless someone has been designated as the web searcher and has had the chance to prepare ahead of time. Some groups like to use computer-based mind maps (see 3.2 Concept Screening) as a way of tracking ideation in real time during a brainstorming session, but this, too, can be distracting if not carefully managed.

While these seven brainstorming rules may seem like common sense, it is easy to forget one or more in the "heat" of a brainstorming session. When some members of the team have not had practice with this approach to brainstorming, it is helpful to introduce them to the rules ahead of time and/or review them at the beginning of the session. Of course, this needs to be done in a way that fits with the overall spirit of brainstorming – namely, it should be upbeat and energetic.

Special considerations for medtech brainstorming

Before identifying participants and a facilitator for a brainstorming session, it is important for the individual or group hosting the session to clearly define the topic to be addressed. For medtech brainstorming, the need specification (see 2.5 Needs Filtering) generally should be used as a starting point for defining the scope of the session. The idea is to focus on the core problem while also leaving enough room for participants to be able to get creative and think beyond existing solutions. The topic should neither be too broad nor too restrictive.[7] In practice, it is not always easy to dial in the right "focal zone" in advance of the session, but it is possible to avoid major errors. For example, a 60-minute session devoted to "reducing the pain and disability of arthritis" is too general; a session on "slowing or reversing the progression of early cartilage deterioration in the knee" has a good chance at being productive. If the given problem is particularly large in scope, consider dissecting it into multiple topics that can be addressed through a series of brainstorming sessions. Good topic choices come with practice. Do not worry too much about the topic at first; remember, the group can always schedule another brainstorming session with a different focus or even make a change in the middle of a session when the focus becomes clearer.

Selecting participants

Choosing the team for medtech brainstorming can be a particular challenge, given that domain experts in the medical device field are typically physicians and engineers who, by training, can be reflexively critical when it comes to new ideas (including their own). Medical training is based on the maxim of "doing no harm" (this is, in fact, part of the Hippocratic Oath). The deeply ingrained value of avoiding damage to patients tends to make many experts in the field conservative when it comes to innovation or change. This factor is influential enough that an inventor or team may want to think carefully whether or not to invite experts in the problem area to participate in the brainstorming session, particularly early in the biodesign innovation process. Guidance from these physicians and engineers may be more appropriate later in the process, when screening and improving ideas become the focus.

A related challenge with involving experts who are currently active in the field is that they may be disproportionately focused on near-term improvements that can make their work safer or more efficacious. For example, some physicians will be full of good ideas for the next generation of a device (one to two years in the future). However, they may not have spent as much time thinking about breakthrough concepts or radical new ideas that will revolutionize a practice area in the long term (a 10- to 20-year vision). Similarly, executives, managers, and engineers within medical device firms may have a wealth of knowledge to share in the brainstorming process, but bring with them certain preferences or biases based on the products their companies are investing in and/or currently marketing.

Among all these cautions about the "expert problem" in brainstorming, it is important to mention that there are many exceptions to this profile – that is, there are many wonderfully innovative physicians, engineers, and executives who have no trouble whatsoever contributing to freewheeling brainstorms. Finding these individuals can supercharge a brainstorming session and lead to impressive results.

With or without domain experts, it can be helpful to include participants with little or no medical expertise, but who bring a unique perspective, or whose background can cross-pollinate the brainstorming session. For example, in addition to one or two physicians with relevant clinical experience in a practice area, it may be useful to invite specialists from other areas unrelated to the one in which the need exists. The contributions of these individuals are often based on technologies or procedures they know from their own fields. When applied to the medical area under consideration in the brainstorming session, these technologies and procedures may provide innovative solutions to the problem(s) at hand. Similarly, it is helpful to include different types of engineers in the process (e.g., mechanical, electrical, and chemical engineers). It can also be interesting to involve a "gizmologist" in the group – an expert at prototyping and making devices, as opposed to an expert in the in-depth engineering science.

In developing the list of participants to invite to a brainstorming session, consider all of the areas that potentially will come into play in designing and developing a medical device solution. This can be hard to do when brainstorming solutions to a need since, in theory, it is impossible to predict what the ultimate solution will be and, thus, what fields should be represented. Similarly, it is important to try to find people who understand the field of interest and existing technologies, but also have the ability to see past their own knowledge so as not to bias the group toward a particular type of solution. For example, if a team is brainstorming new solutions for visualizing the gastrointestinal (GI) tract, which is currently performed using an endoscope, it may want to enlist cross-functional representation from individuals who understand fluids, light, electronics, displays, optics, and the mechanics of scopes. In finding a person with an understanding of endoscope mechanics, the group should target someone with an open mind and the ability to think beyond the known mechanics of today's endoscopes to fruitfully brainstorm new solutions for visualizing the GI tract.

In the same example, it would also be helpful to include physicians from the clinical area in which the scope would be used, as well as experts who could represent how other body corridors are assessed (e.g., catheters for blood vessels). The team should not overlook the possibility of including "non-technical" individuals within the group, too. Business people from various functional backgrounds (finance, operations, marketing, sales) can all make strong contributions to the creative process, particularly since they may overlook (or overcome) conventional practices entrenched in the status quo. In a biodesign course several years ago, the device concept that went forward to form a start-up company was proposed by an MBA with a finance background as he brainstormed with two PhD engineers and a cardiologist.

Choosing a facilitator

The facilitator's job is to run the session, enforce the rules of brainstorming (as described previously), and make sure the process works smoothly. S/he needs to stimulate ideas, prevent lingering too long on any one idea, and move the group in fresh directions when idea generation lags. Often, it is best to have the facilitator refrain from playing a participant role so that s/he can be completely focused on managing the effectiveness of the session.

Especially for new inventors and teams, it may be helpful to avoid having an expert clinician or engineer as the session leader as these individuals may naturally wind up in an authority position within the group, which could stifle less experienced participants from sharing ideas. Depending on the dynamics of the team, the participants, and the topic, having a neutral third-party act as facilitator of the session can be effective. The facilitator must, of course, understand the area well enough to effectively navigate the participants' contributions and stimulate their thinking. However, it is equally important for the facilitator to have the ability to keep the session balanced and well paced, and serve as a referee regarding adherence to the brainstorming rules.

Although the model of one facilitator per brainstorm is a good starting point, the team may want to consider other alternatives in certain situations. It can be useful to switch facilitators during the course of a brainstorm to keep the session fresh, to use the talents of the facilitator as a participant, or to make other adjustments in response to a change in the group dynamic that becomes obvious during the session. The point is to stay alert regarding the effectiveness of the session and be flexible about the facilitator's role.

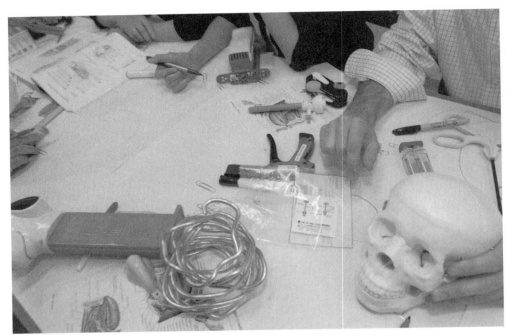

FIGURE 3.1.3
Medtech brainstorming tools – a creative jumble of anatomy aids, devices, toys, and doodles (by Anne Knudsen, courtesy of the Stanford Graduate School of Business, 2006).

In advance of the session

Once participants and a facilitator have been chosen, the individual or group hosting the brainstorming session should think about what background information can be provided in advance of the session to aid the idea generation process. For brainstorming sessions focused on creating solutions to needs, usually the need specification can be the foundation of this background information (see 2.5 Needs Filtering). A disease state fundamentals report (see 2.1 Disease State Fundamentals), an overview of existing treatment options (see 2.2 Treatment Options), or other additional data relevant to the creation of the need specification may also be provided. However, the aim is to achieve a balance between offering ample information for effective brainstorming without constraining creativity by restrictive framing of background information about the need itself. This is another reason to mix in participants who are not specific experts in the medical or technical areas under consideration.

Medical props

As mentioned, it is useful to make props available to participants in a brainstorming session. For medtech sessions, consider having a few real medical tools among the generic blocks and Legos on hand in the brainstorming room. Simple instruments – clamps, retractors, trocars, basic balloon catheters, etc. – can be mixed in with the other artifacts since they can be useful in quickly conveying a concept or inspiring a connection to a new idea (see Figure 3.1.3).

Anatomical drawings and models are other tools that can be useful in a medical-related brainstorming session, as long as the discussion does not get bogged down in detail. Many companies sell large anatomical charts of the different organ systems that are both helpful and visually stimulating. Simple pictures or drawings of a particular operation or procedure can also help motivate ideas. Even better are 3-D models of the human anatomy (or portions directly related to the brainstorming focus area, e.g., the GI tract, the heart, and the skeleton). These plastic models are typically expensive, but can be purchased online or borrowed from a medical school or specialty practice office. Brainstorming with actual tissue parts is another idea to consider, especially when the physical characteristics of the organ are particularly important to understanding the need and developing potential solutions (e.g., looking at a pig stomach when exploring GI needs – see case example). Just be sure to understand the biosafety regulations in the chosen setting if any actual tissues will be brought to the session.

Sugar, caffeine, and brainstorming endorphins

It is part of the brainstorming ritual to provide participants with candy, cookies, and cola to help ensure a high energy level. Whether or not a medtech group thinks a metabolic "buzz" is desirable, it can certainly help create a positive, engaging, and upbeat environment. The props and the set-up (and the candy and cookies as well), all contribute to the right tone and feeling for the session. It can also be a good idea for the facilitator to encourage participants at the outset of the meeting by pointing out that the attendees have been chosen for their talent and creativity, and conveying that it is an honor to have been invited. The positive reinforcement of ideas also goes a long way and can be helpful as the group is just getting going.

If all goes reasonably well, participants will generally get some real enjoyment and satisfaction from the process of coming up with new ideas. Experiencing both one's own creativity and the power of the group to innovate is intrinsically pleasurable for bright and motivated people, and it should provide an incentive for coming back next time.

Managing the session

It is essential to be clear with participants that the brainstorming session will require 60 to 90 minutes of uninterrupted attention. Physicians are especially vulnerable to being paged or phoned by the hospital. If possible, get them when they are not on call or are able to "sign out" to someone else during the session. It is appropriate to ask participants to turn off their cell phones and personal assistants (e.g., Blackberries) during the meeting. If an important interruption occurs, the participant should leave the session to take the call. The idea is to focus on the topic, avoid interruptions, and maintain the momentum that the group has built.

At the beginning of the session, it can be valuable to invest a few minutes in doing some kind of quick warm-up exercise. This is particularly true when dealing with participants who are new to the brainstorming process. It can be especially difficult for physicians to switch from the rapid-fire decision-making mode of medical care to a more whimsical, creative frame of mind. An unrelated ice breaker can help in making this transition. Finding an "easy" medical-related exercise to lead with can be effective. For example, the facilitator might ask the group to come up with as many suggestions as possible within two minutes for improving a hospital bed or nightstand. By making it clear that s/he is looking for creative and even wacky suggestions delivered at a rapid response rate, the facilitator can bring the group up to ideation speed without spending too much time on a warm-up. Another approach is to coax a warm-up need from the group. Ask a simple question like, "What was the most irritating technical problem you had today?" Make a quick list of responses on the whiteboard and then choose the one that seems to resonate most with the group. Some groups may appreciate a non-medical warm-up. For the right participants, these can be fanciful or even humorous: "How could you prove there is life on Mars?" "What's the best way for the facilitator to get into *People* magazine?" "What can the group do to make sure its favorite football team wins this weekend (without getting caught for illegal activities)?" Dedicate two or three minutes to coming up with solutions, with the facilitator encouraging the group to think of quick, interesting, and goofy responses. To get the most from a warm-up exercise, it is important to build a rhythm of rapid responses and to have some fun with this initial experience.

One of the most difficult issues for both the facilitator and the group is dealing with a problematic participant. Challenging participants come in different varieties. The most common type in both academic and corporate settings is the shy or reticent contributor. A non-threatening, easy warm-up exercise can help this person to get started. Participants can help further by taking every opportunity to build on this person's ideas. The facilitator needs to look for opportunities to engage him/her gently, without being heavy handed. In general, the facilitator should work to achieve a reasonable balance of contributions from all participants, accounting for differences in style, comfort, and facility with the topic. It is absolutely not the case that everyone should contribute equally – in fact, one of the best ways to kill a brainstorm is to "cold call" on participants to make sure everyone is having an equal say.

On the other end of the spectrum is the overbearing, controlling, or dominating participant. Here the facilitator, joined by any brave members of the group, needs to help direct this problem participant toward a more positive contribution. The brainstorming rules generally provide a useful framework for making these suggestions. If this is not successful, the facilitator should declare a break and speak to the participant privately.

Capturing the results

It is important to capture and preserve the results of every brainstorm with as much accuracy and richness as possible. If large paper or sticky notes have been used, these should be gathered up and saved. It is worth paying attention to the spatial order in collecting them, because remembering where a particular note was physically made in the room may aid more detailed recollection later on. If a follow-on brainstorm is conducted, it can be useful to place the sheets back in the position they were in at the end of the first session. When whiteboards are used, taking digital photographs of the boards is a good idea. The photos can be projected later for a second session and are also easy to archive electronically. It is essential that some method of recording is performed immediately at the end of the brainstorm, even if that requires the brute force method of simply copying down the output of the session. Again, it is most useful to do this in a way that is spatially close to the original organization. Remember to include the original numbering scheme. Further recommendations for collecting and organizing the results of a brainstorm are presented in 3.2 Concept Screening.

Intellectual property ownership

Brainstorming creates new ideas, and new ideas are the heart of new intellectual property (IP) (see 4.1 Intellectual property basics). The inventor or team needs to be clear with the members of a brainstorming group regarding the implications of their participation in the session with respect to IP. If the participants are all from the same company or university, the ownership of an important idea may stay within that organization (and, thus, ownership is uncomplicated). If the team employs a facilitator or other participants who are not part of the company or university, it may be reasonable to ask them to sign a non-disclosure agreement (see Chapter 4.1).

Even in a situation where all participants are from the same entity, it is important for each person to recognize that s/he may become an inventor as a result of ideas discussed in the session and understand, in general, that ownership will be decided in a fair manner. It is impossible to give specific guidance on how to do this, as each situation is unique. In practice, it will only be clear later in the biodesign innovation process whether an idea coming from a brainstorming session will move forward into a patent. Basically, as described in Chapter 4.1, an inventor is someone who materially contributes to one or more claims of a patent. In some cases it will be clear that a single individual from the group came up with the key concept that was patentable, while in other cases it will be some combination of individuals or the whole group. One good way of beginning to clarify this from the outset is to jot down the initials of the key contributor(s), alongside the idea, in real time as the concepts are generated. Ultimately, it is the responsibility of the team leader (the person or persons responsible for driving the project) to make a determination of IP ownership in consultation with the group's members. It is, of course, best to clarify who is an inventor, and what his/her relative contribution is, as early as possible in the patent filing process.

One more important point worth making about IP relates to the date of conception for an idea. Because date of conception is key to pursuing patent rights, it is important to have documentation of the initial concept(s) emerging from the brainstorming session in the innovation notebook that the inventor and/or team is maintaining (as described in 1.2 Observation and Problem Identification).

The following case example demonstrates how all of these considerations come together in planning and executing a medtech brainstorming session.

FROM THE FIELD

A TEAM IN STANFORD'S BIODESIGN INNOVATION PROGRAM

Brainstorming a concept that becomes a company

As part of a biodesign innovation course, Darin Buxbaum, a first-year Stanford MBA student, and his team collaborated to address the need for *a less invasive way to help morbidly obese people lose weight*. In addition to Buxbaum, who had a business background in the medical device field, the team included one postgraduate resident who had completed a rotation in general surgery and was working on a specialty in plastic surgery, two students pursuing master's degrees in bioengineering with a specialty in biomaterials, and a PhD student in bioengineering with a background in mechanical engineering (see Figure 3.1.4).

After investigating the disease state of morbid obesity, understanding the current treatment landscape, performing stakeholder and market analyses, and developing a need specification, the team was ready to begin the process of concept generation. "The need was so big to begin with," recalled Buxbaum, "that we agreed to focus our first brainstorming session on coming up with broad mechanisms for losing weight, like increasing energy expenditure or reducing calorific intake. And then we would devote a session to exploring the categories that seemed to be the most promising."[8]

All of the team members were relatively new to brainstorming, so they decided to bring in an outside facilitator to lead the first session. "When we were first getting together, there was no formal organization," noted Buxbaum. "Bringing in an outside facilitator helped impose some structure on the brainstorm – there was a designated facilitator and everyone else on the team was an equal contributor." With limited funds at their disposal during this early stage, the team called in a favor with a friend who worked at a design firm, asking him to lead the session. However, according to Buxbaum, it would have been worth paying a junior design consultant to facilitate based on the value they extracted from the role. Because the chosen facilitator was a personal friend, they did not ask the individual to sign a non-disclosure agreement. However, Buxbaum advised other teams to be more cautious. "You just never know who's going to be the one to come up with something really interesting in this kind of session," he said.

The facilitator started the session by leading a couple of short, fast-paced warm-up exercises to get the team going, for example asking the participants to name as many different animals as they could in one minute. "This is something we would have completely glossed over if we hadn't used a facilitator," commented Buxbaum. "But it

Figure 3.1.4
Buxbaum (standing by laptop) and his team at an early brainstorming session (by Anne Knudsen, courtesy of the Stanford Graduate School of Business, 2006).

Stage 3 Concept Generation

FROM THE FIELD — A TEAM IN STANFORD'S BIODESIGN INNOVATION PROGRAM

Continued

really helped people get used to the approach of throwing out ideas as fast as they could and building on the ideas of others."

Where the facilitator really demonstrated value, however, was in helping the group manage a team member who had difficulty allowing for wild and crazy ideas without passing judgment. According to Buxbaum, "Any time a really far-fetched idea would be thrown out like, 'Let's put a black hole in the stomach,' this person would say, 'That's not possible.' He complained about any idea that wasn't physically feasible. In the end, he shut down and stopped giving suggestions. It was really frustrating for the team, and must have been painfully frustrating for him. If it wasn't for the facilitator, we probably would have had a team argument about what we should and shouldn't be able to brainstorm." However, the facilitator was able to mediate the situation, enforce the rules, and keep the session going. Over time, the problematic team member grew increasingly comfortable with the brainstorming approach. "In subsequent sessions, he loosened up and eventually came up with some of our most creative ideas," noted Buxbaum. In later sessions, the team was able to facilitate the brainstorms on its own. "The norms became so ingrained in us that we were able to self-police," he said.

Because the team members came from diverse backgrounds, they decided that some advance preparation was required to get everyone up to speed before each session. "One of the things we worked incredibly hard on was looking at the mechanisms by which the human body works," said Buxbaum. "The physician on our team would create a presentation before each meeting and teach us everything about the relevant anatomy. For instance, if we wanted to talk about mechanical ways for making people eat less, he went over all the different parts of the stomach and how they interact. One time before a session, he presented the entire gastrointestinal system, how the chemical pathways work, and how the hormones are released to stimulate satiety. Having that depth of knowledge led to some really neat ideas. For example, one of our more interesting concepts was to put a stent in the intestines where it would elute fatty acids to activate a hormonal pathway to make people feel full and slow down the entire digestive tract." However, Buxbaum admitted that the concept of advance preparation could be taken too far. "One way that we may have overprepared was to look at IP early in the process. While it seemed like a good idea to understand what concepts were already being worked on, this may have actually clouded our judgment and put unnecessary constraints on our thinking," he said.

Another technique that proved to be invaluable to the team was using drawings, crude prototypes, and medical props during the sessions, including pig stomachs in a few sessions (see Figure 3.1.5). In fact, Buxbaum credited one of the team's major breakthroughs to its use of the stomachs in a brainstorming session: "We were playing with a pig stomach in one of our meetings, literally just manipulating it while we were thinking, and it just flopped into the right geometry. We never would have imagined that the stomach could take this particular shape unless we had a real piece of tissue in front of us." Acquiring the pig stomach was an adventure in itself. First the team tried a traditional butcher shop. "They made a really big deal about it," remembered Buxbaum. "They said they would have to call the place where they got their meat and convince them to stop the line, pull out the stomach, inspect it, and package it for us. They charged

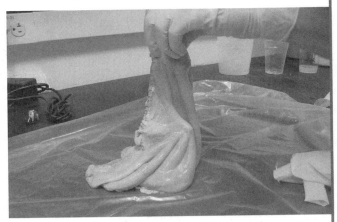

FIGURE 3.1.5
Having a pig stomach available in the brainstorm provided the key insight for the team's invention (courtesy of Jennifer Blundo, Darin Buxbaum, Charles Hsu, MD, Ivan Tzvetanov, and Fan Zhang).

us $30 and it took about two weeks for us to get it." He continued: "In the meantime, a couple of our team members happened to be at a Chinese grocery and saw a whole stack of pig stomachs for $1.99." These stomachs ended up working just as well, at much lower cost and greater convenience to the team.

In total, the team held approximately 12 brainstorming sessions that typically ranged from 60 to 120 minutes. "Usually, we wouldn't go longer than 90 minutes," said Buxbaum. "Our longest session was probably two hours but that was pretty draining. If we went over our time limit, we would table the process and pick it up in another session." In these sessions, the team's goal was to generate two to three ideas a minute, sometimes walking out with as many as 200 ideas from a single meeting.

Reflecting back on the process, he noted, "We have a lot of fond memories of brainstorming. It was a real coming together for the team." With the team's device still in development, Buxbaum indicated that he hoped to do more brainstorming, even though they were well past the concept generation stage in the biodesign innovation process. "We haven't really done much brainstorming recently," he admitted. "But whenever you're dealing with an undefined space, brainstorming can really help you fill it out. The team didn't brainstorm to figure out how to design the device. To our detriment, we sat around tables at coffee shops or in people's living rooms instead. The space wasn't as good and we didn't use all the rules. As a result, our ideas weren't as creative and we didn't get anywhere near the same quality. Now, we're looking at filling out our IP space and I think it would be really helpful to get together and brainstorm again."

Other versions of medtech ideation

While brainstorming works well in many situations, there are other effective ways to approach ideation. Sometimes the "expert problem" in brainstorming can be turned into an advantage. Medical device companies frequently use a format where they invite in a "key" physician (thought leader) to visit the company and lead a working session. While this type of facilitated discussion may draw on many of the same aspects of the brainstorming process described in this chapter (e.g., having different types of people from engineering, sales and marketing, etc. in attendance), it differs in one important way. The purpose of these working sessions is usually to "uncork" the expert's mind and stimulate interesting ideas. Unlike a brainstorming session in which all participants are encouraged to contribute more equally, in these situations the group is asymmetrical, with the expert's ideas being given the most attention. Despite this important distinction, the results of such sessions can be effective and the approach can be successfully employed by inventors, entrepreneurs, and established device companies.

Some medical device companies also conduct ideation sessions with groups of expert physicians. It can be challenging for these experts to switch into a brainstorming-like mode of building ideas together when they are used to competing with each other (although this is by no means true of all groups). This kind of session may require a particularly skillful facilitator.

A final note on brainstorming

The real power of brainstorming is that it leverages a group of cross-functional contributors, all with different perspectives, to solve a problem instead of looking to a single expert or group of experts for a solution. While not every inventor will feel the need to conduct formal brainstorming sessions, this technique has proved to be a valuable tool that is still under-utilized in the field of medtech innovation. What brainstorming teaches is that ideation is a valuable process in itself, requiring the consideration of different possibilities – and lots of them – precisely because it is impossible to pick the right idea immediately. Brainstorming is not about coming up with one brilliant idea, but about creating new directions to pursue – it is a process tool that opens doors to an inventor.

Stage 3 Concept Generation

GETTING STARTED

Brainstorming is one of the most stimulating and enjoyable stages of the biodesign innovation process. As a result, inventors and team members usually do not experience much hesitation as they initiate the ideation process. Use the steps below to get started in planning and carrying out a fruitful brainstorming session.

See **ebiodesign.org** for active web links to the resources listed below, additional references, content updates, video FAQs, and other relevant information.

1. Understand basic ideation and brainstorming concepts

1.1 What to cover – Be sure that all team members share a common understanding of key ideation and/or brainstorming processes and rules (see Figure 3.1.6 for a summary of the rules). Agree on the approach that will be used to address the need defined in the team's need specification(s).

1.2 Where to look
- Tom Kelley, *The Art of Innovation* (New York: Currency/Doubleday, 2001).
- Stefan Thornke and Ashok Ninigade, *IDEO Product Development* (Harvard Business School, 2000).

2. Define the topic

2.1 What to cover – Define a topic that clearly lays out the scope of the session. The topic should identify the core need or problem being investigated without giving the group any preconceived notions about what the "right" solution might be.

2.2 Where to look – Refer to the need specification defined as part of 2.5 Needs Filtering as a starting point.

3. Identify participants

3.1 What to cover – Identify individuals who are creative, can bring interesting, insightful viewpoints, are open-minded, and will represent a diversity of perspectives. Strike a balance between "experts" and more general but creative contributors, and be certain that different functional areas are represented (clinical, engineering, business).

3.2 Where to look – A group of four to seven people is ideal for brainstorming, not including the facilitator and scribe. The inventors hosting the session should leverage their personal networks to identify brainstorming participants.

4. Identify a facilitator

4.1 What to cover – Based on the topic being discussed and the participants who will attend, choose a facilitator with enough subject matter knowledge to command respect, but not so much that s/he will unduly bias or otherwise influence the group. Be sure the facilitator is well versed in the brainstorming rules and (whenever possible) has experience leading successful brainstorming sessions.

4.2 Where to look – The inventors hosting the session should also leverage their personal networks to identify an appropriate facilitator. Professional innovation and idea creation consulting firms, such as IDEO and Phillips+Co., can be hired to lead formal brainstorming sessions for inventors and companies with access to adequate funding.

5. Prepare for the session

5.1. What to cover – Meet with the facilitator to ensure his/her understanding of the session's topic and to agree on the desired outcome. Identify an appropriate location and decide on an approach for conducting the meeting (whiteboards, flip charts, sticky notes, etc.). Put together and distribute pre-reading materials for all participants to review before the session, including the brainstorming rules. Pull together appropriate "props" to stimulate thinking in the meeting and plan the warm-up exercise. Identify scribes to take notes and prepare them for this role.

5.2 Where to look

- **Location** – If available, book a brainstorming room. Keep in mind, however, that any area will do, as long as there is ample space, the right supplies, and an environment in which the group will not be distracted.
- **Widgets and medical props** – Bring modeling clay, Legos, pipe cleaners, and other materials to help stimulate creativity. Companies such as the Anatomical Chart Company, the Anatomy Warehouse, EduGraphics.net, or Anatomy Resources.com sell anatomical charts and 3-D anatomical models. Other medical-related supplies can be purchased from companies such as Surgical Medical Instruments Corporation. Animal organs and bones can typically be purchased inexpensively from a local butcher supply.

6. Conduct the session

6.1 What to cover – Execute the plan that has been developed, enforce the brainstorming rules, and stay within the allotted time (60 to 90 minutes). During and immediately after the session, record all ideas generated. Remember to capture the results by gathering the flip chart sheets and crude prototypes and/or taking digital photographs of the whiteboards, props, etc., beyond the notes taken by the scribe. Consider taking a photograph of the team during one of the brainstorming sessions to show later at the dinner celebrating the FDA approval of the device!

6.2 Where to look – Refer back to the brainstorming rules to help keep the session on track.

IDEO's seven rules for effective brainstorming[9]

1. **Defer judgment** – don't dismiss any ideas.
2. **Encourage wild ideas** – think "outside the box."
3. **Build on the ideas of others** – no "buts," only "ands."
4. **Go for quantity** – aim for 100 ideas in 60 minutes!
5. **One conversation at a time** – let people have their say.
6. **Stay focused on the topic** – keep the discussion on target.
7. **Be visual** – take advantage of your space; use objects and toys to stimulate ideas.

Figure 3.1.6
A relaxed and playful space for a company brainstorming session (courtesy of Cordis Corporation).

Credits

The content of this chapter is based, in part, on lectures in Stanford University's Program in Biodesign by David Kelley, Tom Kelley, Tad Simmons, and George Kembel. The editors would also like to acknowledge Darin Buxbaum and Asha Nayak for their help in developing this chapter, as well as Tad Simmons for editing the material. Additional thanks go to Darin Buxbaum, Jennifer Blundo, Charles Hsu, Ivan Tzvetanov, and Fan Zhang for contributing to the case example.

Notes

1. "Thomas Edison," Wikiquote.org, http://en.wikiquote.org/wiki/Thomas_Alva_Edison (July 31, 2008).
2. Paul Israel's, *Edison: A Life of Invention* (John Wiley & Sons, 1998).
3. From remarks made by John Simpson as part of the "From the Innovator's Workbench" speaker series hosted by Stanford's Program in Biodesign, March 3, 2003, http://biodesign.stanford.edu/bdn/networking/pastinnovators.jsp. Reprinted with permission.
4. From remarks made by Julio Palmaz as part of the "From the Innovator's Workbench" speaker series hosted by Stanford's Program in Biodesign, February 10, 2003, http://biodesign.stanford.edu/bdn/networking/pastinnovators.jsp. Reprinted with permission.
5. From remarks made by Thomas Fogarty as part of the "From the Innovator's Workbench" speaker series hosted by Stanford's Program in Biodesign, January 27, 2003, http://biodesign.stanford.edu/bdn/networking/pastinnovators.jsp. Reprinted with permission.
6. Tom Kelley, *The Art of Innovation* (New York: Currency/Doubleday, 2001).
7. Tom Kelley, *The Ten Faces of Innovation* (New York: Doubleday, 2005), pp. 151–152.
8. All quotations are from an interview conducted by the authors, unless otherwise cited. Reprinted with permission.
9. Based on a summary by Patrick Dominguez, Green Business Innovators, June 16, 2008, http://www.greenbusinessinnovators.com/7-rules-of-brainstorming-from-ideo (August 1, 2008). Reprinted with permission.

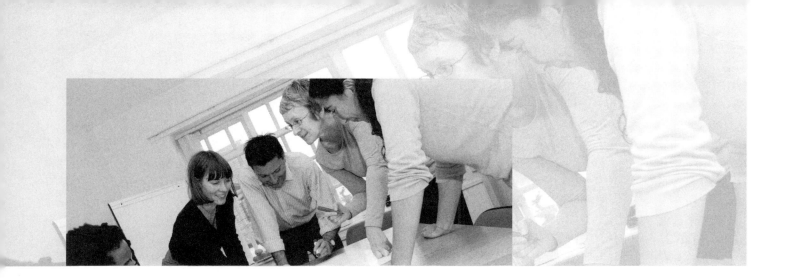

3.2 Concept Screening

Introduction

A "leap of faith" is defined as an act or instance of accepting or trusting in something that cannot readily be seen or proved.[1] Innovators are required to choose a handful of solution concepts to pursue from among the hundreds of ideas coming out of one or more brainstorming sessions, with little or no detailed information about each solution's ability to satisfy the criteria laid out in the need specification. To this extent, concept screening is a leap of faith. Is the solution technically feasible? Is enough understood about the mechanisms of the disease or critical interactions between different engineering concepts to make the solution viable? How much time, effort, and money will it take to determine the idea's practicality? How will doctors and other important stakeholders respond to the concept? In the face of these unknowns, innovators leap forward, setting the majority of their ideas aside to investigate the few that appear most promising.

Being selective about which concepts to pursue from the hundreds generated through ideation and brainstorming is of the utmost importance. Concept screening involves comparing all of these ideas against the defined need specification to evaluate how well they satisfy the need. This process also involves organizing the ideas into related groups to identify potential gaps or biases in the proposed solutions, as well as opportunities to combine ideas into unique, synergistic solutions that better address the need than any individual concept. If significant gaps are found or certain biases are discovered that unnecessarily constrain the solutions, additional brainstorming may be required. The final output of concept screening is a few concepts with which to begin the in-depth concept selection phase of the biodesign innovation process.

OBJECTIVES

- Understand how to cluster and organize the output of a brainstorming session so it can be presented and analyzed in a meaningful way.

- Learn to objectively compare solution concepts against a need specification to determine which concepts to pursue.

Concept screening fundamentals

As described in 3.1 Ideation and Brainstorming, one of the key brainstorming rules is to "go for quantity," encouraging as many ideas as possible. This is important to ensuring that a wide variety of solutions is considered before an innovator settles on any particular concept. However, generating a vast number of ideas can create a challenge in that it is not possible (or even advisable) for the innovator to research and develop them all. At the most basic level, the process of concept screening can be thought of as an objective, comprehensive method for organizing and evaluating the information generated through ideation and brainstorming in order to go from many concepts to a few.

Transitioning from ideation to concept screening

Before initiating concept screening, the innovator should first make certain that s/he has carefully reviewed the raw output from the brainstorm to make sure that all ideas are fully and accurately captured and clearly understandable (see Figure 3.2.1).

Ideally, this should be done immediately following the session so that the innovator can seek clarification from the participants if something is unclear. Misinterpretation can be a source of error and may inadvertently lead to the elimination of potentially good ideas. If each idea is associated with the contributing participants immediately following (or during) the brainstorm, as suggested in 3.1 Ideation and Brainstorming, it should be relatively easy for the innovator to follow up.

Remember that it is important to consider all of the notes, pictures, drawings, diagrams, and 3-D mock-ups that were created during the brainstorm, and not just the documentation produced by the scribe. The goal is to review all ideas, in their entirety, so that they can be adequately assessed.

Once an innovator feels comfortable that s/he has accurately identified and interpreted all of the raw data from the brainstorm, the information can be recorded in some meaningful and easily understandable way. One approach is to tag each idea with a short phrase or name to make it quickly identifiable. By standardizing the way ideas are referred to, an innovator can more efficiently compile and organize the concepts. For example, in the obesity space, the idea of converting white fat cells (which store energy) into brown fat

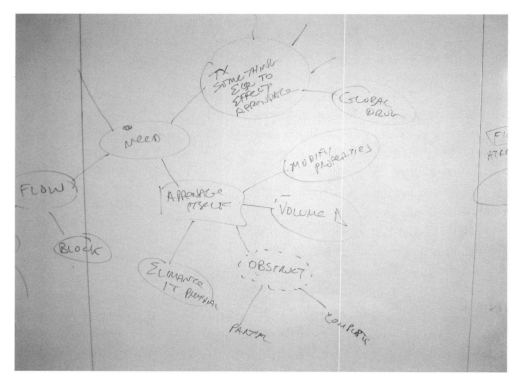

Figure 3.2.1
Example of the raw output from a brainstorming session (provided by Uday N. Kumar; developed with John White, Kityee Au-Yeung, Joseph Knight, and Josh Makower).

cells (which burn energy) might be given the label "reprogram fat cells." An idea to put an implant in the stomach that elutes a substance to simulate satiety, or the sensation of feeling full, might be named "drug eluting implant."

Grouping and organizing ideas

Once the raw data from a brainstorming session has been reviewed, "cleaned up," and labeled, the innovator can organize the concepts. As described in more detail below, this is typically done by grouping ideas according to some organizing principle, such as the mechanism of action or technical feasibility. Organizing concepts is important because it allows the innovator to identify gaps, biases, and synergies, as well as to differentiate between concepts and approaches.

Identifying gaps

Identifying gaps among groups of solutions is necessary so that additional brainstorming can be performed to address any opportunities that may have been missed. For example, if an innovator working on the need for **a way to prevent strokes due to clots coming from the heart** generated a preliminary set of solutions that did not include any concepts involving the blood vessels that convey clots to the brain, this would be an obvious gap to focus on during additional ideation.

Uncovering biases

The innovator should also watch out for biases in the types of solutions proposed. For example, if all the concepts initially generated to solve the stroke need (described above) were heavily focused on mechanical engineering solutions, individuals from other engineering disciplines might suggest different approaches and concepts (e.g., chemistry or drug-based solutions, such as locally administered clot dissolving agents).

Discovering synergies

An effective approach to organizing concepts will also allow the innovator to identify commonalities and complementarities between concepts so that they can be merged to create synergistic, combined concepts. For example, for the same stroke need, one group of solutions might focus on meshes to capture clots while another set focuses on medications to dissolve clots. By putting these clusters together, a new set of ideas, centered on a mesh that elutes a substance to dissolve a clot, may emerge. In many cases, the combination of concepts can result in a stronger overall solution.

Differentiating between concepts and approaches

Organizing concepts also helps an innovator determine whether a disproportionate number of *approaches*, rather than tangible solutions, have been generated. For instance, for the stroke need above, the idea to "capture a clot in a blood vessel" is representative of a general approach. In comparison, the idea to "use a stent with an integrated mesh to capture clot in carotid artery in neck," is far more concrete. If most of the concepts are general in nature or too high level to be actionable, additional brainstorming can be used to translate promising approaches into tangible concepts.

More about clustering ideas and concept mapping

Effectively organizing data prior to initiating concept screening requires the innovator to focus on two primary activities: (1) clustering the ideas, and (2) visually organizing them into a concept map. While the approaches outlined below are not the only methods for accomplishing this, they provide helpful guidelines for meaningfully conveying the output from a brainstorming session.

Clustering

The notion of arranging ideas into groups is simple in theory, but can be difficult to accomplish in practice (requiring more art than science). At its most basic, clustering requires the examination of ideas to identify common themes or similarities and group them based on their related characteristics. For instance, the roughly 100 ideas generated in a brainstorm may fit within 10 general clusters based on the most dominant traits they have in common.

The first step in the grouping process is to identify the primary organizing principle for creating the clusters. Some examples of common ways to group ideas in the medtech field are outlined in Table 3.2.1.

Table 3.2.1 *One or more organizing principles can be used to help cluster ideas during concept screening.*

Organizing principle	Description	Example
Anatomic location	Group ideas according to the part of the anatomy they pertain to and/or target. Differences between groupings might be small if all solutions are in a highly focused area (e.g., the vertebral discs). Alternatively, groupings might span entire regions/organ systems if the solutions focus on a need pertinent to a significant portion of the body (such as the various places that emboli dislodged from the heart can travel through the vasculature).	For solutions to address the problem of obesity, ideas might naturally cluster around the mouth, esophagus, stomach, pylorus, small intestines (duodenum, jejunum, and ileum), large intestines, and the various valves in the GI tract.
Mechanism of action	Group ideas according to how the solutions are intended to work.	Increasing energy expenditure, regulating food intake, reducing nutrient absorption, and reducing the motivation to take in energy are all different mechanisms of action for reducing weight.
Engineering or scientific area	Group ideas according to the type of engineering or scientific approach underlying the solution.	Solutions could be supported by three main types of engineering: *chemical* (pharmacological weight control), *electrical* (gastric pacing), and *mechanical* (laparoscopic banding, bariatric surgery, liposuction).
Technical feasibility	Group ideas according to their likelihood of coming to fruition. This is based on understanding what is feasible using current engineering and scientific methods, which implies some knowledge of the science behind the solution and/or the engineering development timeline required.	Solutions such as reprogramming fat cells might have low feasibility; a drug eluting implant might have moderate feasibility; a space-occupying stomach device could have high feasibility.
Funding required	Group ideas around the amount and/or source of funding required to develop them. While this may be difficult before researching the funding landscape, a "best guess" based on prior information (see Chapters 2.4 Market Analysis and 2.5 Needs Filtering) may suffice.	For obesity, a solution such as filling the stomach with a space-occupying device would likely require less money to develop than a drug eluting implant.
Affected stakeholder	Group ideas around the stakeholder most affected, typically the patient or healthcare provider. While this may result in rather general groupings, it can provide insights into which concepts are more likely to be adopted (see 2.3 Stakeholder Analysis). While payers will be affected by all concepts, a consideration of variations in cost-effectiveness will be useful.	For obesity, solutions focusing on medications might be more attractive to patients than those requiring surgeries. From a payer's perspective, solutions requiring a one-time payment may be more attractive than solutions requiring recurrent payments.

The challenge to the innovator is to find the most meaningful organizing principle for the solutions being evaluated. Factors like anatomical location and engineering area are simple to understand and provide easy comparisons, but they may not be relevant for some needs. Factors such as mechanism of action may be harder to define and apply, but more significant in terms of identifying the similarities and differences between ideas and for assessing the likelihood of success. Innovators are encouraged to experiment with several different organizing methods before choosing the one or two that make most sense.

Another approach is to create a hierarchy of organizing principles where, for example, concepts are first clustered according to one particular organizing principle (e.g., anatomical location) and then, within each of these clusters, arranged into subgroups according to a different organizing principle (e.g., mechanism of action). The innovator can then continue this process, incorporating additional organizing principles at deeper and deeper levels. For instance, in the example focused on preventing strokes due to clots from the heart, the first-level organizing principle might be anatomical (solutions could be broken down by location: heart, blood vessel, and brain). The concepts within each anatomical location could next be organized by engineering area (mechanical solutions, chemical solutions, or biological solutions). Alternatively, the innovator might start by organizing concepts according to engineering area and then create subgroups based on anatomical location. Again, the key is to try several different approaches, watching for patterns that illuminate important gaps or commonalities among the concepts.

Concept mapping

After one or more organizing principles have been chosen and the ideas placed into groups, they can be documented in a concept map (also called a mind map). A concept map visually illustrates how ideas relate to one another and to the main problem or need. As noted, they are meant to help the innovator recognize patterns and build connections between solution concepts, as well as between the concepts and the need. Concept maps are particularly useful for highlighting gaps in the solution set – for example, innovators who are well versed in the chosen organizing principle (e.g., all the anatomical locations relevant to the chosen need) can quickly spot what might be missing if it is not represented in the concept map.

When developing a concept map, the need is placed at the center, with the clusters of ideas spanning out in different directions. Figure 3.2.2 shows one concept map generated for a need focused on obesity.

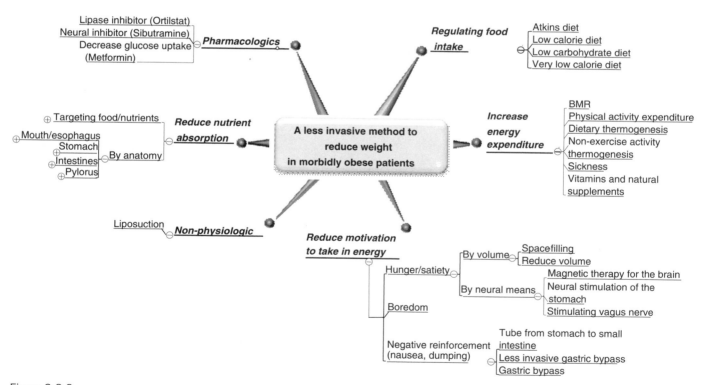

Figure 3.2.2
A sample concept map for the need "a less invasive method to reduce weight in morbidly obese patients," as developed by a team in Stanford University's Program in Biodesign (with permission from Darin Buxbaum; developed with Jennifer Blundo, Charles Hsu, Ivan Tzvetanov, and Fan Zhang).

Stage 3 Concept Generation

The primary clusters are broken down by various mechanisms of action: (1) regulating food intake, (2) increasing energy expenditure, (3) reducing the motivation to take in energy, (3) reducing nutrient absorption, (4) pharmacologics, and (5) non-physiologic solutions. These clusters are further broken down into subgroupings as they are laid out in the concept map. For example, the cluster focused on "ways to reduce the motivation to take in energy" is further organized into a subgroup called "ways to reduce hunger." This includes solutions that affect volume (space-filling, volume reduction) and reduce hunger through neural means (neural stimulation of the stomach, stimulating the vagus nerve). Using the chosen organizing principle, the actual solutions are then placed where they fit best within these clusters and subgroupings. In studying the ideas at the edges of the map, it is apparent that some are well-defined concepts while others are mostly approaches rather than actual concepts. At this stage, the innovator and his/her team could choose to focus only on the clusters with well-defined solutions. Alternatively, each of the less defined clusters and/or subgroups could serve as the topic for subsequent brainstorming sessions to fill out the concept map with more concrete solutions before determining which ones to pursue. Regardless of the path chosen, the exercise of concept mapping plays an invaluable role in helping a team identify the correct next step.

Figure 3.2.3 provides another example focused on the need for *an improved way to access the pulmonary veins to deliver ablative therapy*.

This figure shows that many solutions have been identified to support the cluster called "systems to aim at the PV" (pulmonary vein). However, when looking at the cluster "ablation systems," more brainstorming may be needed to flesh out the ideas since many of these solutions are much less well defined in scope and are not closely linked to the central need. To be effective, an innovator must strive to ensure that all of the clusters

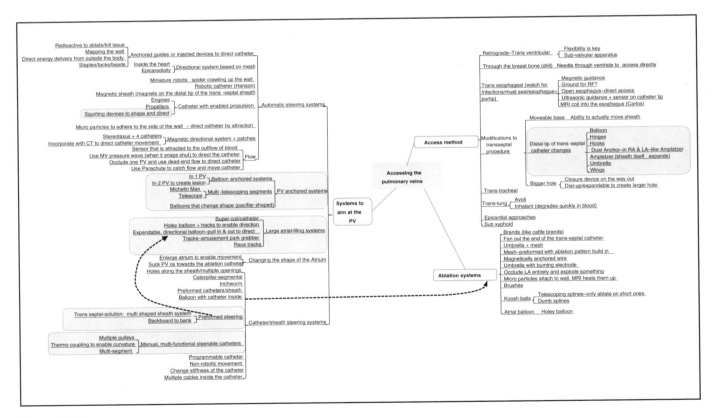

Figure 3.2.3
A sample concept map for the need "a better way of accessing the pulmonary veins to deliver ablative therapy," as developed by a team in Stanford University's Program in Biodesign (with permission from Uday N. Kumar; developed with John White, Kityee Au-Yeung, and Joseph Knight).

have an obvious relationship to the need. S/he should also check to be sure that all solutions fit comfortably underneath the chosen headings. If an idea appears "forced" to fit within a group, this may indicate that the map is not based on the optimal organizing principle or that another heading is missing. This sometimes occurs if an innovator has gone into the brainstorming session with a bias towards certain types of solutions (which places an artificial constraint on the brainstorming process). For example, if the need being explored is for *a way to prevent bleeding from an artery*, a group with a mechanical engineering bias may subconsciously focus on mechanical solutions, completely forgetting about chemical or electrical ideas. When such biases are discovered, the team should complete more brainstorming. As mentioned earlier and emphasized in the guidelines presented in 3.1 Ideation and Brainstorming, involving people with different backgrounds, expertise, and experiences can help keep a brainstorming session balanced.

These two examples show that no two concept maps are exactly alike. The basic layout is similar, but the innovator has a good deal of freedom to decide exactly how to visually present the concepts in the most meaningful and value-added way. Again, it can be helpful to experiment with multiple concept maps based on various organizing principles before choosing the one that best presents the data from the brainstorming session. Additionally, it is important to remember that by creating maps, different gaps in the solution set can be more readily identified. Furthermore, well-defined solutions assigned to different clusters on differently arranged concept maps may help the innovator better define each concept along a variety of parameters. Both of these outcomes can lead to a more effective assessment of each solution against the need.

There are no specific guidelines for determining at what point an innovator has achieved the optimal presentation of the data. Instead, s/he must rely on making an appropriate judgment call on a map that is logical, balanced, and comprehensive before moving to the next step of concept screening.

Comparing ideas against the need

With a solid understanding of the intent behind the various solutions and a concept map created, an innovator is ready to compare the proposed solutions against the need. The goal of the screening process is to filter the universe of ideas to the ones that best address the need. To accomplish this, the innovator must rigorously compare each solution against the need statement and the explicitly defined need criteria laid out in the need specification to see which concepts satisfy the requirements and which do not.

As this comparison is performed, it is essential for the innovator to apply the need criteria as originally defined. Modifying the need criteria or compromising in terms of how they are applied to different solutions can tarnish the integrity of the screening process and lead to poor choices, since they will no longer be driven by the pure requirements of the need. Importantly, while it can be tempting to let personal interest in a particular solution bias the process, this can ultimately be a costly mistake, leading to a solution that does not appropriately address the real need. This, in turn, can undermine the final success of any product coming out of the biodesign innovation process.

That said, the screening process is inherently subjective in that the innovator is using his/her best judgment to determine which solution ideas are best aligned with the need. While there is no fail-safe way to eliminate this subjectivity, the chance of successfully identifying the most promising solution concepts can be increased by considering each concept in a consistent manner. As mentioned above, the creation of different concept maps will naturally lead to a greater understanding of the different parameters along which each solution is aligned. This may also help in comparing these solutions with the need criteria. Involving more individuals in the screening process is another specific tactic that can be used to help minimize personal bias.

Many innovators choose to conduct a solution screening meeting. This is similar to a brainstorming session, in that a diversity of individuals should be invited and a clearly defined protocol used. A facilitator should preside over the meeting with the goal of driving the group to identify the most promising solutions. The facilitator will be in charge of selecting the process to filter the ideas, as well as ensuring that the selection committee follows the process. However, unlike the facilitator's role in the brainstorming process, s/he should actively participate in the screening process.

Because concept screening requires a different skill set from what is required for brainstorming, it makes sense to involve different types of individuals. For example, this is a stage where experts in the field can be invaluable. These experts might include physicians, nurses, patients, and technologists who can help evaluate the practicality or usefulness of the various solutions. Having multiple perspectives can enhance the decision-making process. However, for practical purposes, the innovator should seek to limit the number of people from which s/he solicits feedback. If too many people are involved, it can become impossible to rationalize each person's input, based on their individual preferences and biases.

In the meeting itself, participants provide input and may even be asked to vote to eliminate and prioritize ideas. For example, one approach is to give each person approximately five sticky notes and ask them to place one next to each of their top five ideas on the concept map (which can be drawn/copied on to a whiteboard or flipchart). After tallying the votes, the list of ideas can be reduced to roughly 10 leading concepts. Once the list of ideas is shortened to 10, the group can assess them in more detail relative to the need criteria, until a smaller set of concepts has been agreed upon.

At times, it may not be feasible to get the desired participants into a room at the same time and place to participate in the screening process. Another approach is to meet with the targeted individuals one-on-one to seek their input. The feedback gathered from the experts can then be used as input to the screening process when it is performed by the innovator and his/her team. Regardless of how an innovator collects input from experts in the field, this input is invaluable to concept screening, as illustrated in the Simpirica Spine story below.

FROM THE FIELD SIMPIRICA SPINE

Seeking expert advice to help screen concepts

Colin Cahill, Ian Bennett, and Louie Fielding, three student innovators in Stanford University's Program in Biodesign in 2004, recognized the importance of seeking expert input during the biodesign innovation process. While they were initially focused on disparate student projects, they ultimately joined together to work on the need for *a way to stabilize the spine without compromising mobility*, using expert input to help screen their concepts, refine the need, and develop a meaningful solution. The output from their collaboration with Todd Alamin, a Stanford spine surgeon, led to the founding of Simpirica Spine, based in Redwood City, California.

Early in the biodesign innovation process, Fielding was working with two other students, Pat Arensdorf and Ralph Crisostomo, to address the stated need. Their team held a series of 10 brainstorming sessions that generated hundreds of diverse solution ideas before they gradually narrowed down the list to the five most promising concepts. Describing the process, Fielding recalled, "We developed concept maps of the different solutions and then came up with a scorecard to help us evaluate them."[2] Along with the market need, surgical and mechanical feasibility were two important factors included on the scorecard. As a result, some ideas were easily eliminated. "We looked at certain concepts and saw right away that they wouldn't really be possible – they were voodoo," said Fielding. Other concepts required more in-depth evaluation or resources that were out of scope for a small team with limited capital. Acknowledging the presence of some personal bias in the exercise, Fielding admitted, "We tended to favor those ideas that were aligned with our expertise or were interesting engineering challenges. But, usually, those concepts scored pretty well against our criteria."

Once the team felt comfortable that it had identified the best potential solutions from among the ideas, the members prepared to seek expert input. "We put together a presentation that included the background on what we were trying to do, the different diseases we were potentially targeting, what treatments were currently available, where we were trying to fit in, and how we'd come up with our ideas." Additionally, they developed an overview of each of the five solution concepts. Tapping into their personal networks for contacts, they met with one spine surgeon who dismissed the concepts without serious consideration. However, before going back to the drawing board, they decided to meet with a second expert for a different point of view.

Taking advantage of another referral, they approached Todd Alamin, a spine surgeon at Stanford Hospital. Alamin, who completed a fellowship in spinal surgery at Stanford in 2000, was an active innovator and entrepreneur, as well as a practicing surgeon. "We pretty quickly went through the background material," said Fielding, "and started talking about the five different ideas." This time, they got a different reaction. "Todd was able to recognize that the ideas were still very conceptual and that this was not necessarily what they would look like in their final implementation. He was able to use his imagination to see through to something that could be realistic," explained Fielding. From the list, Alamin focused on one concept for a minimally invasive implant that clinically made sense to him. When evaluating concepts, Alamin said he routinely assesses if: (1) the device has a clear function, (2) it addresses a clear clinical need, and (3) it offers the simplest possible solution for solving the given problem. "Particularly with the spine," he noted, "anything that's too mechanically complicated isn't going to work because you have to worry about wear debris being created over time and individual pieces malfunctioning. Anything that is designed to repetitively take on axial load over the life of the patient is also a concern because the likelihood of ultimate failure is high."

The minimally invasive implant concept presented by Fielding and team appealed to Alamin because it had a clear mechanism of action. As for the clinical need, he saw the potential for the device to help relieve early-stage degenerative disc pain, an area with a vast unmet need. In addition, the concept was simple and would not necessitate large axial loads so it would therefore be unlikely to fail mechanically. In combination, these factors caused Alamin to recommend to the team that it further develop the idea.

At this point, Cahill and Bennett got involved (while Arensdorf and Crisostomo phased out their participation in the project). Over six months, the team worked to refine the need statement and significantly evolve the solution concept, periodically checking in with Alamin and using his input to validate and drive their progress. Shortly thereafter, Alamin formally joined the team and the four agreed to pursue further development of the concept together, a partnership that eventually led to the formation of Simpirica Spine in September 2006.

According to Alamin and the team, it makes sense to use an expert in the field to help narrow down the list of potential solution concepts because such physicians have an invaluable understanding of what will work and how it will be received. "I know exactly which clinical problems we're trying to treat and I've used all sorts of different implants for different spinal problems. I know what the size of the anatomy is, and I know from a surgeon's standpoint what will work and what won't. I can even provide input about almost silly things, like whether it will bug the surgeon to have part of the device a little off center or if it's too big," he said. However, he cautioned innovators to look for physicians that are willing to think "outside the box" and question the status quo. "You need someone who hasn't been in the field so long that they can't step back and take a different perspective. It also helps to find someone who's a bit of a cynic and doesn't believe that everything they do is perfect. A healthy cynicism balanced with a little optimism is what you need to make things better." Cahill concurred: "You need to find a doctor who thinks for him- or herself and isn't trying to come up with a 'right-sounding' answer."

Alamin also recommended that, at this stage in the biodesign innovation process, innovators "find one clinician who you think is smart and creative and knows a lot about his field, and stick with him. If you talk to too many people, you end up creating a cacophony of noise that doesn't make any sense, with some physicians recommending exactly the opposite course from what other physicians recommend. No matter what design you have, even if you think it's the best design in the world, you can always find someone who will tell you that you're the stupidest person they've ever heard of and that your idea is ridiculous. That's just the way it is. The risk in asking too many people for their advice is that you'll end up getting stuck doing nothing." Bennett, Fielding, and Cahill agreed, but acknowledged that there are times in the biodesign innovation process where more than one opinion is helpful. In these cases, Bennett advised learning about the physicians in advance of seeking their input. "You have to know a little bit about their backgrounds, what they like to do, and their natural tendencies so that you can frame their answers with these factors in mind," he said. "This provides a context for their answers and helps you understand them a little bit

> **FROM THE FIELD: SIMPIRICA SPINE**
>
> *Continued*
>
> better, rather than necessarily taking their input at face value."
>
> In choosing an expert to work with, the team offered additional advice. "If you're only seeing one or two surgeons, it's important to make sure that they're not out on the fringe somewhere," said Fielding. "Make sure that whoever you're working with is middle of the road, or at least not a total outlier." Cahill added, "Find out who you can trust." Using a variety of sources to understand an expert's reputation is important before committing to any sort of advisory relationship. Once the right contacts are identified, "Be prepared to meet them anytime, anywhere. Their schedules are crazy," noted Bennett. Finally, Cahill emphasized, "Self-educate as much as possible before the first interaction. If you're new to the field, you will not get everything totally right, but at least you'll show that you've done your homework."
>
> Simpirica Spine was in the process of finalizing its minimally invasive treatment in preparation for clinical testing. Using a customized process to organize, evaluate, and score their concepts early in the biodesign innovation process helped this team effectively identify the most promising solutions to address its need. Seeking expert advice during concept screening (and beyond) helped the group select, refine, and develop the concept that would eventually lead to an innovative device.

Keep in mind that some solutions may meet the need criteria, yet have to be eliminated from consideration because they are simply impractical or infeasible. Usually, ideas are determined to be infeasible due to technology constraints. However, other factors should also be taken into account, including potential patient, provider, and payer concerns. While more primary and/or secondary research may shed light on whether or not an idea is impractical, an innovator must usually rely on a "gut check" regarding what can realistically be accomplished within a practical timeframe. Another consideration that may cause an innovator and team to eliminate a solution relates to their mission, strengths and weaknesses, and project acceptance criteria (as defined in 1.1 Strategic Focus). For example, if a team has conceptualized an electrical engineering solution that meets the need criteria but the members are dedicated to working on mechanical solutions based on the interests they outlined in their acceptance criteria, they may choose to abandon the electrical idea. Another approach would be to modify their acceptance criteria. However, regardless of any modifications to acceptance criteria, the need criteria must remain the same.

As noted, in some cases, an innovator may end up with different general "approaches" that address the need, rather than specific solutions. Sometimes "approaches" may seem to meet the need criteria more effectively because of their lack of detail; concrete details are often the key factors in determining why a particular concept does not meet a need. For instance, in the example shown in Figure 3.2.3, which focuses on the need for ***a better way of accessing the pulmonary veins to deliver ablative therapy***, some of the need criteria were: (1) the ability to reach all configurations of PVs, (2) easy to use, (3) faster than current methods, (4) low/no morbidity, and (5) must enable at least radiofrequency energy delivery. Other factors taken into account included the team's interest in pursuing solutions in line with current catheter-based procedures, the feasibility of the solution, and the expertise of the team. Based on these criteria, this particular team chose the subgroupings and associated solutions in the outlined and shaded areas on the concept map. Even after these initial choices, the team had to complete additional brainstorming and research to generate more specific concepts from these approaches, as there were still more than could realistically be taken forward.

In contrast, although it is relatively rare, if screening yields too many genuine concepts rather than approaches that meet the need criteria, then the need criteria may be too broad. This may require the innovator to revisit the need specification to generate more specific criteria (see 2.5 Needs Filtering). This is not to be regarded as a failure of the need specification process, but rather as an inherent aspect of the iterative nature of the biodesign innovation process.

As with so many other elements of the biodesign innovation process, there is no exact way to know when enough solution ideas have been generated and when they have been appropriately screened. Similarly, there is no right number with respect to the final number of concepts to take forward. In most cases, innovators will focus on one to five of the concepts that best meet their need criteria. These are the concepts that undergo additional research and prototyping, in sequence or in parallel, in the next stage of the biodesign innovation process: concept selection.

GETTING STARTED

During initial concept screening, the goal is to narrow down many concepts into a manageable set for research, evaluation, and, eventually, concept selection. A more formalized process (called the Pugh Concept Selection method) will be introduced in 4.6 Final Concept Selection for the purpose of selecting the final concept after additional information has been gathered on the most promising solution ideas, including their intellectual property (IP), reimbursement, regulatory, and technical profiles. For concept screening, the following steps can serve as a guide.

See **ebiodesign.org** for active web links to the resources listed below, additional references, content updates, video FAQs, and other relevant information.

1. Review and document raw data

1.1 **What to cover** – Review all output from the brainstorming session. Associate ideas with the individual(s) who generated them and seek clarification, as needed. Assign a name or label. Summarize the results.
1.2 **Where to look** – Refer directly to the output from the brainstorming session as described in 3.1 Ideation and Brainstorming.

2. Cluster ideas

2.1 **What to cover** – The next step is to organize similar ideas into clusters. Assess the data to identify the most meaningful organizing principle for the groupings (e.g., anatomical location, mechanism of action, engineering area, feasibility). Experiment with different approaches, as needed. Consider using multiple organizing principles at different levels.
2.2 **Where to look** – Different ways of clustering ideas can be learned by reviewing other concept maps. Examples can be found by searching the Internet for sample concepts or mind maps.

3. Develop a concept map

3.1 **What to cover** – Visually document the clusters in the form of a concept map. Start with the need in the center and then place the groupings, subgroupings, and ideas. Again, experiment with different approaches, if necessary.
3.2 **Where to look** – Software can facilitate the concept mapping process. Available software packages (e.g., Mindjet MindManager and IHMC Cmap Tools) can be found on the Internet.

4. Assess the concept map

4.1 **What to cover** – Take a broad view of the concept map to determine if there are: (1) obvious gaps based on the organizing principle(s), (2) biases in the solution set in terms of the recurrence of specific approaches or ideas, (3) mostly general approaches or specific solutions represented among the ideas, and/or (4) commonalities or complementarities between concepts such that they can be combined into new and unique ideas. Perform this analysis for all the concept maps that have been created, using different approaches to see if any themes emerge.

GETTING STARTED

4.2 Where to look – Revisit the output from Chapters 2.1 to 2.4 to validate the range of possible solutions for a given need and identify gaps and biases. For example, if the primary organizing principle chosen for a concept map is focused on anatomy, reviewing one's disease state analysis could highlight whether all relevant anatomical areas are represented on the concept map. Involving outside help [perhaps one or more people who will help with concept screening (see below)] to quickly assess the concepts can also help in identifying biases or gaps, since their fresh perspectives may more readily detect missed opportunities and prejudices influencing the defined concepts.

5. Compare concepts against the need to complete concept screening

5.1 What to cover – To prepare for a concept-screening meeting, identify the participants, decide on a facilitator, define the process to be used in the session, and distribute pre-reading (the need specification and concept map should be shared with participants in advance). If a concept-screening meeting is impractical (or if expert feedback is desired as an input to the meeting), schedule one-on-one meetings with targeted contacts to share the solutions and collect their thoughts. In either type of meeting, evaluate each solution idea against the need criteria until a small subset of the most promising solutions has been identified. These solutions will be the ones that go forward into the solution refinement stage of the biodesign innovation process. If too few solutions meet the criteria, additional brainstorming may be needed. Assuming that the solutions represent true concepts and not approaches, if too many solutions satisfy the need criteria, the need specification and need criteria may need to be revisited (2.5 Needs Filtering).

5.2. Where to look – The innovator should leverage his/her personal network to identify appropriate participants for the solution screening process.

Credits

The editors would like to acknowledge Joy Goor for her help in developing this chapter. Many thanks also go to Todd Alamin, as well as Ian Bennett, Colin Cahill, and Louie Fielding of Simpirica Spine for their assistance with the case example.

Notes

1. "Leap of Faith," Dictionary.com, http://dictionary.reference.com/browse/leap%20of%20faith (July 8, 2008).
2. All quotations are from interviews conducted by the authors, unless otherwise cited. Reprinted with permission.

Acclarent Case Study

Stage 3 Concept generation

With a well-validated need for *a minimally invasive approach to treating chronic sinusitis that had less bleeding, less pain, less bone and tissue removal, less risk of scarring, and that was faster, easier, and safer to perform*, as well as a firm commitment to finding a less invasive way to treat chronic sinusitis, Makower and Chang began aggressively exploring potential solution concepts and finally returned to some of the concepts they had developed previously. While the idea of developing "interventional devices" was intriguing when Makower first conceived it, the approach made even more sense in the context of the team's recent disease state and treatment research. "Seeing the three-dimensional anatomy in cadavers and reviewing so many CT scans caused us to appreciate how tortuous the bones and mucosa were in the sinuses. It reinforced the observation that those drainage pathways resembled blood vessels," he commented. "So we started to imagine how we could implement cutting, dilating, stenting, energy delivery, and drug delivery across this bony network, utilizing whatever we had learned from our experiences in the cardiovascular space."

3.1 Ideation and brainstorming

To develop a more specific set of potential solution concepts, Makower and Chang began exploring different possibilities in earnest. Makower, who had a bias against disclosing his ideas with a wider audience too early, advocated for a scaled-down, hands-on version of the ideation process. While they did not assemble a cross-functional group of participants to hold formal brainstorming sessions, Chang noted, they still benefited from multidisciplinary input: "In a way, we did get multiple perspectives, because Josh is an engineer, a doctor, a scientist, and a business person, and I'm an engineer." He went on to describe their ideation approach.

> Josh and I spent a lot of time brainstorming around what we had learned about the disease and anatomy and from our cadaver work. We dug into how we could use flexible instruments to get around and preserve anatomical structures while gaining access to the sinuses. And we considered what we could do to make access easier. For example, one way we developed this understanding was to section a cadaver head at one of the labs. By sectioning the specimen, we had an unobstructed view of the anatomy with our own eyes, instead of using an endoscope. We laid down malleable wire and got an appreciation for the dimensions of the anatomy, relative locations of the structures, openings, and where instruments would need to travel. It was hands-on brainstorming while we were assessing the anatomy.

Stage 3 Concept Generation

Through this process, they developed multiple types of potential solutions. First they thought about ways to dilate the anatomy to gain access to the sinuses, using tools such as lasers, balloons, stents, and what Chang described as a "chomper." They also generated ideas to address a blockage area once it was reached. "We thought about a catheter that cuts, a spinning burr, different kinds of lasers, and balloons. Maybe we could freeze it, or maybe we could use microwave energy. We considered anything that would ablate or cut, that we could deliver on a flexible platform," Chang said. See Figure C3.1.

3.2 Concept screening

After they felt that they had more or less exhausted the possibilities for ways to address the need, they started to narrow down the list of concepts. Makower and Chang both had a desire to do something fast and had a preference to try one idea at a time. For these reasons, they saw some logic in starting with one of the concepts that was quickest to prototype. If it worked, they would press forward. If not, they would go back to the drawing board and choose another solution to explore. When they rationalized their list, they decided that the quickest idea to investigate was related to the use of balloons. "We said, 'Look, if we can make it work with a balloon, that'd be great,'" Makower remembered. "It doesn't involve complicated instruments. It doesn't involve bleeding. There's no cutting, no energy, and no hardware. Given our backgrounds, we could also make or obtain balloons to test easily." He continued, "If the balloon works, we're in great shape. If not, we can try energy next, or blades, or maybe even stenting.

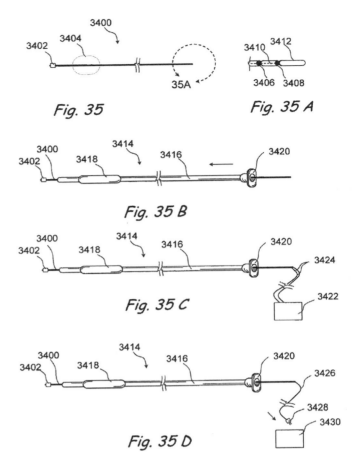

FIGURE C3.1
The team considered many methods of opening sinus pathways, including use of balloons, stents, drug delivery, and energy-based devices (excerpt from patent 7,462,175 with permission from Acclarent).

We knew our mission was to open the sinus drainage pathways in a flexible, less invasive way, but which technology was going to be required was really not clear yet."

INVENT | Concept Selection

STAGE 1 NEEDS FINDING
1.1 Strategic Focus
1.2 Observation and Problem Identification
1.3 Need Statement Development
Case Study: Stage 1

STAGE 2 NEEDS SCREENING
2.1 Disease State Fundamentals
2.2 Treatment Options
2.3 Stakeholder Analysis
2.4 Market Analysis
2.5 Needs Filtering
Case Study: Stage 2

STAGE 3 CONCEPT GENERATION
3.1 Ideation and Brainstorming
3.2 Concept Screening
Case Study: Stage 3

STAGE 4 CONCEPT SELECTION
4.1 Intellectual Property Basics
4.2 Regulatory Basics
4.3 Reimbursement Basics
4.4 Business Models
4.5 Prototyping
4.6 Final Concept Selection
Case Study: Stage 4

STAGE 5 DEVELOPMENT STRATEGY AND PLANNING
5.1 Intellectual Property Strategy
5.2 Research and Development Strategy
5.3 Clinical Strategy
5.4 Regulatory Strategy
5.5 Quality and Process Management
5.6 Reimbursement Strategy
5.7 Marketing and Stakeholder Strategy
5.8 Sales and Distribution Strategy
5.9 Competitive Advantage and Business Strategy
Case Study: Stage 5

STAGE 6 INTEGRATION
6.1 Operating Plan and Financial Model
6.2 Business Plan Development
6.3 Funding Sources
6.4 Licensing and Alternate Pathways
Case Study: Stage 6

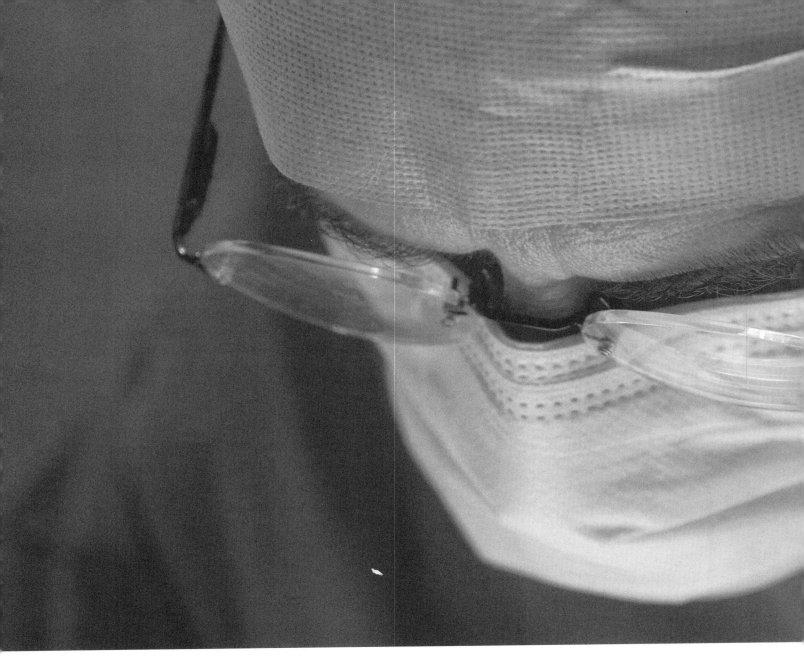

There's more than one way to skin a cat.

English proverb[1]

Junk can dance.

David Kelley[2]

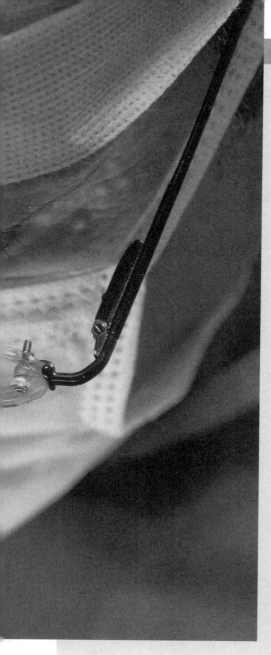

Concept Selection

After the careful screening of many needs, you have a clear statement of a focused problem worth solving. Brainstorming has generated hundreds of ideas, many whimsical or outlandish. In Phase 1 (Identify), many needs were winnowed down to a focused, clinical need statement. At this point in Phase 2 (Invent), it is time to commit. There is a frequent tendency to delay because of uncertainty. The plain reality is that no decision is a failure. It is impossible to consider a dozen concepts simultaneously.

The United States Marine Corps knows something about decision making in the middle of uncertainty. In the classic book *War Fighting* the message is clear – perfect certainty doesn't exist. It is far better to move expeditiously with the best information available at the moment and to react flexibly as the situation changes and as you learn on the fly.

So, too, choosing a concept that is directionally correct, and acting upon it, will frequently provide new insights and refine the direction. There is no shame in meeting a blind alley and even reversing if necessary. NovaCept, a radiofrequency endometrial ablation device, began with the concept of small joint arthroscopy; a good team backtracked to success.

Concept selection requires an understanding of the rules of the road for medtech innovation. The delicate and frequently conflicting interplay between intellectual property, reimbursement, regulatory, and business model options requires sophisticated judgment. There is seldom a single "right" way.

Experienced serial entrepreneurs will emphasize that time is money ... and more. A balance between thorough deliberation, and the opportunity cost of time lost to paralytic analysis is needed.

Just as concept selection is an iterative process, so too is prototyping. Prototyping involves several steps, the first of which is closely linked to brainstorming. We build to learn. Early on, crude mock-ups are constructed to serve as a "looks like" or "feels like" version of a device made of easily shaped and assembled materials. Access to junk makes this easy and fun. Foam core, plastic, cardboard, outdated surgical instruments, catheters, and endoscopes can be "cannibalized" for early prototyping. Duct tape rules. The goal is to fail early, fail cheaply, and ultimately to fail better.

Notes
1. Proverbs are short, pithy sayings that reflect the accumulated wisdom of a group of people.
2. As quoted in Tom Kelley, *The Ten Faces of Innovation* (Doubleday Business, 2005).

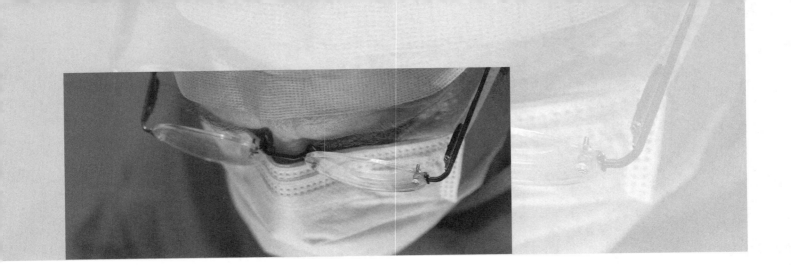

4.1 Intellectual Property Basics

Introduction

The first months of the project have sped by without any real roadblocks. The team's prototypes are working better than expected in the preliminary animal model. The 510(k) pathway appears straightforward. Existing reimbursement codes are in place to cover the new technology. And the project has begun to attract some investor interest. Then an e-mail arrives from the patent attorney: "See attached Patent Cooperation Treaty publication from Israel – just uncovered. Please call immediately."

As soon as a team has identified a promising new concept, it is time to start exploring intellectual property (IP) and developing an approach to patenting. A patent is a legal document that gives an inventor the right to exclude others from commercial use of the invention. The presence of existing patents in the field can complicate or even derail the innovator's ability to bring a related technology to market. Alternatively, a strong patent portfolio can maximize the value of the invention by creating a competitive barrier to others.

This is the first of two chapters on IP. It covers the fundamentals of medtech patents, focusing on the initial steps required to obtain a patent. The subsequent chapter, 5.1 Intellectual Property Strategy, describes how to build an effective medical device patent portfolio and addresses strategies, tactics, and methods for asserting the inventor's rights.

OBJECTIVES

- Understand the different types of patents, including the basic elements of provisional and utility patents.

- Recognize the requirements of patentability, including practical aspects of the filing process for medical devices.

- Develop familiarity with the patent search process.

- Understand the fundamentals of international patent coverage.

- Appreciate issues surrounding inventorship and ownership.

Intellectual property fundamentals

Intellectual property is defined as any product of the human mind or intellect (e.g., an idea, invention, expression, unique name, business method, or industrial process) that has some value in the marketplace and, ultimately, can be reduced to a tangible form, such as a device, software program, process, or other invention.[1] Intellectual property law governs how individuals and groups may capitalize on innovations by determining who owns the IP, when owners can exclude others from using the invention for commercial purposes, and the extent to which courts will enforce the patent holder's rights.[2]

There are several categories of intellectual property that are of major interest to the medical device innovator – patents, trademarks, copyrights, and **trade secrets**.[3] A *patent* is a grant of **exclusive rights** from the government to make, use, sell, or import an invention. Although it is commonly described as a "monopoly," it is more accurate to say that a patent is the right to *exclude others* from making, using, selling, or importing the technology. The government is willing to grant this right to the inventors for the life of the patent – generally 20 years – in exchange for the inventors disclosing the details of the invention and making best efforts to bring it into the marketplace. Patents benefit society because without this guarantee, innovators and companies would not have the incentive to invest the time and resources needed to bring a technology forward.

The main type of patent in medical devices is the **utility patent**. A utility patent describes an apparatus or method wherein the novel invention is *useful* for accomplishing a specific end result. In the medtech industry, utility patents cover devices themselves, as well as methods of use and how the devices are made and manufactured. It follows that utility patents are by far the most important category of medical device IP. There are also *design patents*, which cover the unique ornamental, visible shape, or design elements of a non-naturally occurring object. These are particularly important in the consumer products area (one example would be the design features of adhesive bandages). ***Provisional patents*** are preliminary filings that are generally submitted in order to establish a ***priority date*** for subsequent utility patents.

A *trademark* is a word or symbol that is consistently associated with a specific product and gives the holder exclusive rights to use a word or symbol as a brand name or logo. Trademarks come into play when medical devices are named or logos are created for companies or products. Trademark rights arise automatically from use of the name or logo in connection with the goods or services (that is, no formal application is required to obtain trademark protection). However, registering a trademark generally strengthens the holder's legal position for asserting the validity of the mark in court.[4]

A *copyright* grants authors and artists of written and graphical materials the right to prevent others from using their original works of expression without permission. In the medtech industry, copyrights are important for software (for example, the programs that run an ultrasound scanner) and for advertising and educational materials (written, web, audio, and video). Works are automatically copyrighted when they are placed in some tangible form. As with trademarks, federal registration of a copyright leads to additional procedural rights.

Trade secrets are information, processes, techniques, designs, or other knowledge not generally understood or made public, which provide the holder with a competitive advantage in the marketplace. With some medtech products, trade secrets may be more important than patents. For example, a knee joint device may involve a special manufacturing process that creates smoother contact surfaces. In this case, the company may choose not to pursue a patent for its manufacturing process in an effort to keep it a secret and avoid "tipping off" competitors about an innovation that could improve their products. Instead, the company may attempt to protect the process as a trade secret, putting in place various safeguards to keep other companies from finding out anything about this manufacturing know-how. Some strategic considerations regarding trade secrets versus patents are provided in 5.1 Intellectual Property Strategy.

The remainder of this chapter explores three basic IP considerations: criteria for obtaining a patent, **freedom to operate**, and **prior art** searching. It then provides an overview of provisional and utility patents, identifies important information related to international

Stage 4 Concept Selection

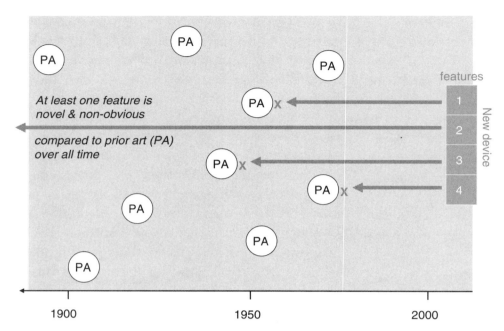

FIGURE 4.1.1
In order for a device to be patentable, at least one claimed feature of the device must be novel and non-obvious compared to any patents, publications, presentations, or other public disclosures (the prior art, PA) going back over all time, anywhere in the world. In this case of a device with four key features, a search going back in time reveals that features 1, 3, and 4 are blocked by prior art but feature 2 is novel and non-obvious, so a patent can be awarded.

patenting, and addresses a number of other fundamental IP concepts, such as confidentiality.

Criteria for obtaining a patent

There are three basic parameters for judging the *patentability* of an invention:[5]

1. **Utility** – The invention must do something useful.
2. **Novelty** – It must be new with respect to other patents, products, or publicly available descriptions anywhere in the world (collectively known as the "prior art").
3. **Obviousness** – The invention must not be obvious in light of the prior art to someone of average skill and knowledge working in the given field.

Only if an invention objectively meets these criteria is it considered to be patentable. In practice, the first criterion – utility – is rarely a problem for a new medical device invention. However, an inventor's ability to demonstrate novelty can be trickier. Prior art may exist in the field that anticipates an inventor's innovation but is not yet published (e.g., patent applications that are filed but not published). Beyond this is the world of medical device innovators who are constantly developing new technologies, testing them, and making **public disclosures** in some fashion (sometimes in fairly obscure places). It is not uncommon for an innovator to come up with a new device only to discover, for example, that a scientist working in some other country tested a similar concept in animals. In order for a patent to be granted, at least one feature of the device must be novel relative to any previous patents, publications, abstracts, speeches, or other presentations – in short, *any public disclosure – over all time* (see Figure 4.1.1).

The only way to uncover such information is diligently and patiently to search the literature. The advent of the Internet and powerful search engines (such as Google) has made this discovery process easier in recent years (see section below on patent searching).

The point that most often becomes the focus of medtech patent litigation is whether or not the invention is obvious. Obviousness can be complicated, and no inventor is ever completely secure on this point. The criterion that the invention is not obvious to someone "of average skill and knowledge" has a specific legal meaning. As a commonsense guideline, think of this hypothetical person as someone working in the field that uses the invention (say, gastrointestinal endoscopes) and who magically has complete knowledge of the prior art in the field of endoscopy. If the invention would be obvious to this hypothetical person, it is not patentable. In practice, the determination of obviousness becomes an issue that the patent examiner decides; later on, it may become a subject of litigation and be determined by the courts. In recent years the courts have tried to base decisions on objective measures of non-obviousness including commercial success of the invention, prior lack of a solution to a longstanding need, and failure of others to come up with the invention.[6] An experienced patent attorney is the inventor's best source of advice on the obviousness of an invention.

Beyond the three essential criteria for patentability, the law specifies that a patent must describe the invention in a clear, unambiguous, and definite way so that a person with knowledge in the field would be able to make and use it. Patent law also requires that the innovator describe the *best mode* for the invention, which generally means the "embodiment" that is, in the innovator's opinion, the most effective configuration or version of the invention. This may seem like a difficult requirement but, at its most basic, it simply asks the inventor to specify which approach s/he favors at the time of filing, regardless of whether a different approach proves to be better later. For example, if the inventor of a particular stent structure thought that the best mode at the time of filing was to use stainless steel in the construction of the stent – and in subsequent development and testing cobalt-chromium proved to be superior material – the patent would still be valid since the inventor discharged the obligation of disclosing best mode at the time of filing.

Reduction to practice

As described above, **reduction to practice** is another important factor for patentability because it shows that the idea is being diligently pursued. According to this requirement, the innovator must be able to demonstrate that s/he has made a good faith effort to bring an invention into practice. With medical devices this often means building a prototype and performing some kind of bench-top or animal testing. The model making and testing can occur after the patent filing, and there are no strict guidelines about timeliness of this reduction to practice other than what is reasonable for that type of technology. Again, an expert patent attorney can help clarify this issue on a case-by-case basis. As an alternative to making and testing a model, an invention can also be reduced to practice "constructively" by a careful description in the patent, with sufficient detail and in an adequately predictive manner that someone knowledgeable in the field could build the device. This is particularly useful when the development of a working model will be extremely costly, time intensive, or difficult to accomplish. Constructive reduction to practice via a patent application has the same legal effect as evidence of an actual reduction to practice.[7] Whether actual or constructive, early reduction to practice can help show diligence, and diligence can help prevent a first inventor from losing patent rights to a later inventor.

First to invent versus first to file

Under the US patent system, the first person to invent an innovation is awarded the patent, provided that the inventor was reasonably diligent in reducing the invention to practice. This means that if two inventors are competing for a similar patent, the person who can prove that s/he was working on the innovation earliest will be given the benefit of *first invention* (regardless of who filed the application first). If the US Patent and Trademark Office (**USPTO**) becomes aware of a dispute about the date of invention between two innovators, it will initiate an ***interference proceeding*** to make a determination of who was first to invent. Internationally, a different system is used whereby the first inventor to file a patent application is awarded the patent, regardless of when the work began. There is continuing discussion in the United States about whether to switch to a first to file system.

Freedom to operate: an introduction

In addition to patentability, the other key aspect of IP that determines the commercial usefulness of an invention is freedom to operate (FTO). Freedom to operate is often confusing for first-time inventors to understand. The key concept is that *receiving a patent on a device does not guarantee that the inventor is free and clear to make, sell, use, or import the technology (that is, has freedom to operate)*. There may be claims of existing patents that read on (i.e., are related to) important features of the device that incorporate the new invention. For example, an inventor could patent a new kind of intraocular lens that has a high resistance to infection – this would be a patentable feature that is not covered by any prior art. However, despite obtaining a patent

Stage 4 Concept Selection

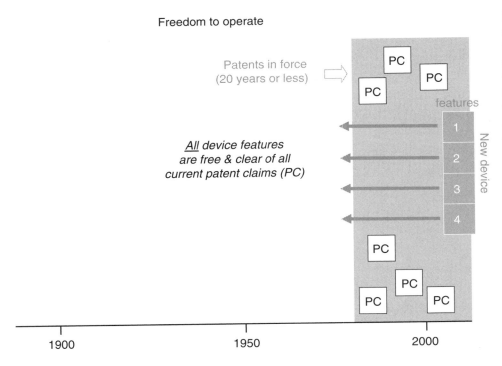

FIGURE 4.1.2
A new device has freedom to operate only if *all* of its features are free and clear of any patent claims (PC) that are currently valid or "in force." Generally, this means patents filed within the past 20 years. In this case, all four of the key features of the new device are free and clear of any current patent claims, so the device is free to be marketed and sold.

on the new lens, the inventor may not be free to commercialize this invention if there are current patents in force with claims that cover important features of the general structure and function of an intraocular lens – features that are also necessary parts of the new device. *A new device has FTO only if all of the features of the device are free and clear of valid claims from patents that are still in force* (generally patents that are 20 years old or less). This situation is illustrated in Figure 4.1.2, where all four of the key features of the new device are free and clear of existing claims going back 20 years.

Because it affects whether or not a device can actually be sold, determining FTO for a new invention is equally important as determining patentability and requires a diligent search of the prior patent art (see below). Strategies for dealing with FTO issues are outlined in 5.1 Intellectual Property Strategy.

Before filing: the prior art search

Understanding the prior art landscape for a new invention is one of the most critically important parts of the biodesign innovation process. Inventors naturally hope to discover that their idea is indeed completely new. Even if the search reveals that there is major prior art, however, the team benefits greatly from finding this information as quickly as possible. Such discoveries can potentially save an inventor huge amounts of wasted time and resources in pursuing the wrong idea. There is also the opportunity to modify the invention into an approach that is patentable and has freedom to operate.

Prior art searching is as much an art as a science, and at a later stage it is almost always wise to have a professional search conducted by an attorney or patent agent before the filing of a utility patent (see 5.1 Intellectual Property Strategy for considerations about hiring an attorney). These experts will usually locate prior art that the team has not uncovered. Still later in the process, it is not uncommon for the patent examiner to find art that the expert attorney has not discovered. Despite the fact that the initial search will not be perfect, it is essential for the inventor to conduct his/her own search early in the process. The inventor will learn a tremendous amount about the field, and this up-front diligence will make the work of the attorney or agent much more effective and economical.

For the inventor just launching a search, the vast scope of the territory for prior art may at first seem daunting. To search for patentability, the inventor is looking for any previous disclosure of the idea in a public manner – not only patents, but all written descriptions and public presentations at any time, anywhere in the world. In practice, searching patent databases is an effective way to start since most commercially significant technologies are patented (especially in the medtech field). One subtlety that is often overlooked is the fact that abandoned and expired patents are also public information and, thus, constitute important prior art.

Given the large scope of the search territory, the prior art search process is unavoidably time consuming. Depending on the invention, it is not excessive to allocate as much as 20 to 25 percent of the team's time to searching. Like other steps in the biodesign innovation process, searching should be performed at multiple points in the sequence. This is because there is a continuous flow of potential new prior art coming into public view, both in the form of patent applications and in public disclosures at conferences, in the literature, and in the press. Additionally, the patent search process is complex, involving many combinations of different keywords and expressions. Searching must be an iterative process – looking at patents and publications uncovers new keywords and expressions. This information should be used, in turn, to conduct additional searches.[8]

A preliminary patent search should be completed in relatively short order after concept screening (see 3.2 Concept Screening) to avoid the risk of getting locked into a particular way of thinking about a solution before the IP landscape is well understood. Patent searches are also valuable in assessing the depth and extent of the field of invention early in the biodesign innovation process. As a concept is developed, the patent search helps refine the idea and almost always leads to a stronger patent application.

In the United States, inventors have a legal requirement to disclose any relevant prior art they uncover to the patent office. Practically speaking, a careful cataloging of prior art can help the patent examiner more quickly assess the technology and review the application.[9] It is a foregone conclusion that any significant prior art will be uncovered at some point – either during the careful examination by the patent office or in litigation once the patent is issued – so it is in the inventor's best interests to find and disclose these materials as early as possible.

Search basics

Before starting a patent search, the inventor must have reached a precise understanding of the invention that will be claimed. This underscores the requirement for the idea to be developed in some detail before an in-depth patent search is performed. A concise summary should be written that captures the most fundamental elements of the invention, including the need or problem that the invention addresses, the structure of the invention, and its function. It can be useful to draft a few hypothetical claims that help clarify the scope of the invention in advance of the search. These do not need to be written in "legalese" – a plain language description of what characterizes the invention will suffice.

Several conceptual points are important to understand about the actual search process (see the Getting Started section of this chapter for more practical tips and guidelines). First, there are different types of searches that are conducted depending on the stage of development of the product and the purpose of the search. In the initial stages following an invention, there are two basic types of searches. A search for *patentability* focuses on novelty and obviousness in light of the prior art. This search is intended to locate any information within any of the parts of a patent (the specification, drawings, claims, or references) that would make the examiner conclude that the current concept is obvious or not novel. A *freedom to operate* search is directed toward finding *claims* of currently in-force patents that read on some feature of the new concept so that it is not possible to commercialize the device without **licensing** the existing patent (see 5.1 Intellectual Property Strategy). Claims are described more fully below but, basically, they are the numbered paragraphs at the end of the patent that stake out precisely what is protected in the invention. This is the reason that the claims are the focus of the freedom to operate search.

Second, there are two primary ways to approach searching. The most intuitive method is text searching, also called *keyword* searching. This is a familiar process

Stage 4 Concept Selection

in the Google era in which keywords are entered into a search engine that looks through a patent database (Google itself now maintains an excellent patent database – Google Patent Search – along with the USPTO and others listed at the end of this chapter). The text search usually starts with keywords that describe the invention, though it can also be productive later on to search for known inventor names in the field, assignees (companies that own an important patent), competitor companies, etc. The precision of a text search can be greatly improved by combining keywords using *Boolean logic operators* such as AND and OR. One other practical piece of advice is to be careful about the fact that there may be many versions of keywords for the same technology. For example, the guidewire used in angioplasty is variously described as "guidewire," "guide-wire," "guide wire," "guide element," "wire," and probably other terms. The search engines have a feature that helps with this problem called *wild card symbols* or "truncation limiters." These are symbols attached to the root of a keyword that allow the searcher to find any other words that have that same root. For example, by searching "guide*" (the exact wild card symbol depends on the search database), the searcher can look for all keywords with "guide-" as a common root.

The other general category of search is by *classification*. The USPTO and other international patent agencies have systems for grouping and coding patents that are based on the industry, the structure, and function of the invention and its intended use. In the United States, the database is called the patent classification (USPC) index. It can be useful to perform a search by the assigned classes and subclasses, since the keywords a prior inventor has chosen to describe his/her concept may not be standard or intuitively obvious (such that a text search using "reasonable" keywords may miss some important prior art). Some inventors like to start with a classification search as a quicker way to understand the landscape of their invention than a keyword search. The classification systems used by the USPTO and other international agencies are not themselves necessarily intuitive or consistent, so some patience and serendipity may be required. More search tips can be found in the following working example.

Working example: Tips for keyword searching

Start by evaluating three basic questions:[10]
1. What problem does the invention solve? In the context of the biodesign innovation process, what is the basic need?
2. What is the structure of the invention?
3. What is the function of the invention? What does it do?

In developing the answers, generate as many relevant terms as possible. These terms will provide the basis for the search. For instance, for a new angioplasty catheter, a sample response is shown in Table 4.1.1.[40]

Use these keywords to search either the USPTO or Google Patent website (other patent databases can be used in subsequent rounds). Make use of Boolean logic operators and the advanced search techniques available through the site to refine the searches. For example, search "atherosclerosis AND angioplasty" to find patents using both of these keywords; or "angina OR myocardial infarction" to find patents covering either of these two conditions. To make the search more efficient, use the wild card symbols or "truncation limiters" to reach all keywords that have a common root. Instructions on how to use these symbols can be found on the search site.

Table 4.1.1. *An effective way to approach keyword searches.*

Issue	Possible search terms
Need addressed by the invention	Atherosclerosis, arteriosclerosis, coronary artery disease, coronary stenosis, arterial blockage, arterial stenosis, arterial flow, coronary flow, myocardial infarction, heart attack, angina, chest pain, etc.
Structure of the invention	Catheter, balloon catheter, angioplasty catheter, atherectomy, tube, balloon, flexible member, etc.
Function of the invention	Dilate, expand, open, inflate, pressurize, remove blockage, compress, compact, etc.

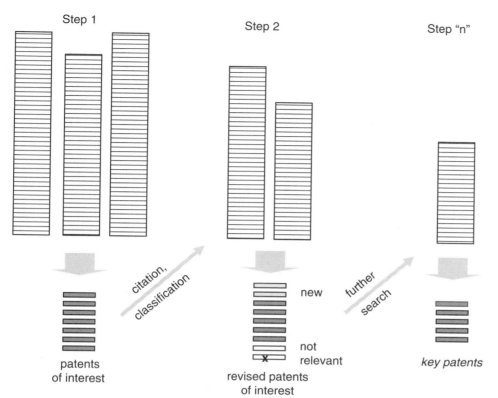

FIGURE 4.1.3
The cycle of patent searching: a repeated process of going out broadly into the patent literature, finding patents of interest, and using information from these patents to launch a new, more refined search.

Once a core set of interesting patents is identified by text and/or classification searching, it is helpful to use the citation search functions to look forward and backward in time to see what other patents are linked to these core examples. Search backwards in time using the citations (references) that are listed on the face sheets of the key patents that have been identified. For forward searches, start with patents where the drawing and specifications suggest that the patent represents a major advance in the field – search this patent forward in time to see what other, newer patents have referenced this invention.

In practice, combining these different methods leads to a search process that is cyclical in nature – a repeated process of looking out broadly into the patent art and then narrowing it down to progressively more relevant patents. (This iterative process is illustrated in Figure 4.1.3.) The first search may yield many hundreds, or even thousands, of patents. Through a thorough review of these, the innovator can filter the list down to perhaps 10 to 20 key patents. In step 2, these patents of interest are used to create another expanded search, triggered by citations in the patents of interest or classifications of the inventions (or both). This results in another large group of patents to search – but, typically, this group is smaller than the patents identified in Step 1. Careful review of the art uncovered in Step 2 produces a few new patents of interest (and probably also leads to the exclusion of some of the initial patents), resulting in a revised set of patents of interest. This process is repeated again, often several more times ("n" times in the figure), until there is a stable group of key patents – perhaps 5 to 15, depending on the area – that continue to be clearly relevant. This is the group of patents to focus on in moving forward.

In conducting searches, it is important to develop a system for capturing the output. This is essential because the volume of information is high, the search will likely go on in stages over a long period of time, and the information that is gathered is critically important. Some type of worksheet or spreadsheet is extremely helpful (see Figure 4.1.4 for one example). For the most relevant patents, this worksheet should include the patent number, title, assignee, key claims, inventor(s), publication and filing dates, and agency classification.

Stage 4 Concept Selection

Title	Assignee (Company)	Key claims	Search terms	Patent #	Inventor	Pub date	File date
Satiety – flow rate							
Gastrointestinal electrical stimulation	UT Austin	Retrograde feedback control of GI action by a stimulating electrode and a detection sensor. Method of external implantation	"gastric bypass"	6826428	Chen, Jiande; Pasricha, Punkaj Jay	30-Nov-04	11-Oct-00
Adjustable Sphincter System	Ethicon Endo-Surgery, Inc	Artificial sphincter that encircles a passageway that adjusts with fluid. Also discusses invasive and non-invasive manual adjustment methods.		20050272968	Byrum, Randal; Huitema, Thomas; Hassler, William	8-Dec-05	2-Jun-04
Gastric ablation followed by gastric pacing	N/A	Method of ablating pacemaker cells in stomach and then replacing them with electrical stimulation from a pacemaker		20050240239	Birinder, Boveja and Widhany, Angely	27-Oct-05	29-Jun-05
Satiety – space occupying							
Endoscopoic stomach insert for treating obesity and method for use	N/A	Upon release in the stomach, flexible blades for a dome shaped cage, applying pressure to stomach.	"gastric bypass"	5868141	Ellias, Yakub A	9-Feb-99	14-May-97
Endoscopic gastric baloon with transgastric feeding tube	N/A	A gastric baloon that also allows a transgastric feeding tube to go through it and nourishment to pass to the jejunum.		20060025799	Basu, Patrick	2-Feb-06	26-Jul-05
Satiety – by neural means							
Treatment of obesity by bilateral vagus nerve stimulation	Cyberonics	Prevent overeating by stimulating vagus nerves to condition the neural stimulus of the stomach.	"gastric bypass"	6587719	Burke, Barrett; Reddy, Ramish K; Roslin, Mitchell S.	1-Jul-03	1-Jul-99
Treatment of obesity by electrical pulses to sympathetic or vagal nerves with rechargeable pulse generator	N/A	Full system (pacemaker, programmer, leads) for stimulating or blocking sympathetic response		20050149146	Birinder, Boveja and Angely	7-Jul-05	31-Jan-05
Methods and devices for the surgical creation of satiety and biofeedback pathways	None	Sensors built into stomach restrictive devices to sense changes in the volume of the restrictive device; sensor then discharges a signal to induce satiety		2005/0267533	Gurtner, Michael	1-Dec-05	15-Jun-05

FIGURE 4.1.4
A sample format for capturing information from a prior art search related to the treatment of obesity as developed by a team in Stanford University's Program in Biodesign. Note that the team's analysis has been removed from this example for the purpose of confidentiality (courtesy of Jennifer Blundo, Darin Buxbaum, Charles Hsu, MD, Ivan Tzvetanov, and Fan Zhang).

A summary of the invention in the words of the searcher and a cut/paste of key claims and figures can also be useful. Additionally, it is a good idea to have a section for comments from the searcher about what is important about this invention and claims (though remember that this information is potentially discoverable in litigation, so it would be wise not to record a comment such as, "This looks identical to our idea..."). Beyond the key patents of interest, it is important to keep track of any interesting patent unearthed in a search, since that patent may emerge again in a later search and it is efficient not to have to restudy it. It is also worthwhile to keep track of the search terms used, and the number of "hits" generated by different combinations of terms.

Because the search process is often a complex and time-consuming undertaking, it is wise to plan the work flow ahead of time and divide tasks among the members of an innovator's team. For instance, one team member could start with a classification search while other members divide up the keyword list and begin these inquiries. It is very useful for the different team members to use the same central spreadsheet for recording search results – this can be web-based for convenience.

Remember that, in searching for patentability, any publicly disclosed information can serve as prior art. This includes scholarly papers in journals, popular articles, trade show presentations, conference abstracts, newspapers, analyst reports, and many other potential sources. Careful searching with Google, Google Scholar, PubMed, and other appropriate resources is essential. Talk to experts in the field for leads on where to look for public disclosures that may be relevant to the invention but are not yet published. Company booths at medical meetings can also be a great source of information.

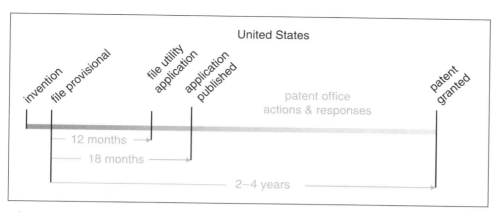

FIGURE 4.1.5
Typical timing associated with US patent filings.

A final general comment about searching is particularly important for the first-time inventor: it is essentially *inevitable* that the search process will uncover IP that appears to be similar to the new idea, particularly in a dynamic technology domain like medtech. Searching can be an emotional rollercoaster for this reason. Do not be immediately discouraged if something potentially problematic comes to light during a patent search – there are almost always ways to work through the issues, whether by modifying the invention or partnering with owners of existing IP. The important thing is to be thorough and accurate in the search in the first place, so that these issues surface as quickly as possible.

Provisional patents

Many times, medtech inventors file a provisional patent before pursuing a utility patent in the United States. Provisional patents are a relatively new mechanism, signed into law by President Clinton in 1994 as part of the General Agreement on Tariffs and Trade (GATT) accord. Basically, a provisional patent is an initial description of an invention that establishes a priority date that will ultimately be used by the patent examiner to help determine patentability, but only after the inventor files a utility patent on the same invention. As noted, provisional patent applications typically require less time and expertise to prepare than a utility patent, and are often written by inventors rather than attorneys (although it can be a good idea to have an attorney at least review a provisional patent application before filing). They are also relatively inexpensive to file with the USPTO, costing approximately $100 for an individual or small business entity. However, the provisional patent is *not itself examined by the USPTO* and can never become an issued patent with enforcement rights. In order to become an issued US patent, the inventor must file a non-provisional (utility) patent *within 12 months* of filing the provisional application (see Figure 4.1.5).

A second approach, which is much less commonly used, is to convert the provisional application to a non-provisional application by filing a "grantable petition" within 12 months of the provisional application.

It is important to note that the provisional patent is never published, so after filing it is possible to work on a new invention in complete secrecy. It is only after a US utility patent application is filed that the invention will ultimately be published (18 months after the provisional filing date).

In order for a utility patent application to claim the benefits of a provisional patent application, two criteria must be met. First, the written description and any drawing(s) of the provisional patent application must adequately support the subject matter claimed in the later-filed non-provisional patent application. Note that this does not mean that the written description and any drawings filed in a provisional patent application and a later-filed non-provisional patent application have to be identical. However, the non-provisional patent application is only entitled to benefit from the *common subject matter* disclosed in the corresponding provisional patent application. Second, the non-provisional patent application must have at least one inventor in common with the inventor(s) named in the provisional patent application to claim benefit of the provisional patent application filing date.

Many medtech inventors choose to write their own provisional patents. This is a reasonable approach, provided that the application is thorough and sufficiently detailed to communicate the basic invention and support the claims of a later utility patent. However, it bears repeating that it is wise to have a patent attorney review the application prior to filing, if at all possible.

There is no required format for the provisional application but it should include descriptions of:

1. The need or problem that the invention addresses.
2. Shortcomings of current solutions.
3. The motivation for the new invention.
4. The invention itself (in enough detail that someone skilled in the field would know how to make it).
5. Advantages of this solution to the problem, including why it is strategically valuable.

In addressing these different parts of the application, the inventor should make a compelling case for what is novel and non-obvious about the new device. In other words, the more that the provisional patent application looks like a utility application, the better the chance that the utility application will be able to successfully claim the benefit of the provisional filing date. Drawings can be very helpful in clarifying an invention and should certainly be included.

It is a common misconception that the provisional patent application process was developed to enable inventors to file "quick and dirty" applications. In fact, the process was put into place to give inventors in the United States the same rights as inventors overseas with respect to patent term. The 20-year patent term is based on the date a non-provisional patent application is filed. However, inventors outside the United States have one year from their ex-US filing date to file a patent application covering the same subject matter in the United States. In this way, they gain an early filing date, or showing of priority, without impacting the term of their US patent. Since, with the provisional patent approach, inventors in the United States now have exactly 12 months from the filing date of a provisional patent application to submit a non-provisional patent application, this process creates a US patent term essentially equivalent to the term afforded to those who originally filed first outside the United States.

Beyond preserving the term of the eventual utility patent, there are additional advantages to consider with provisional patents. One circumstance where a provisional patent is extremely useful is when an inventor is confronted with an unavoidable and urgent need for public disclosure – for example, a physician-inventor who is asked to give a talk or abstract presentation at a conference. As long as the provisional patent application is sent in prior to the presentation, the priority date is established and the ability to gain foreign as well as US patent rights is preserved. A more subtle advantage is that the provisional patent application constitutes a legal reduction to practice of the invention without actually building the device (provided there is sufficiently clear and detailed information about construction and use in the application). If there is litigation about the date of the invention later on, the provisional patent is a clear and unambiguous time point regarding both the invention and reduction to practice.

The main disadvantage of the provisional patent alternative is that it can provide a false sense of security. Inventors should be aware that filing a last-minute cocktail napkin sketch or a PowerPoint presentation that includes a few high-level ideas about an invention may not provide sufficient support to grant the ultimate utility patent. If the inventor does find him/herself in an urgent need-to-file situation, it is important to include as much detailed information as possible in the provisional patent. Many IP attorneys recommend investing nearly as much time in the development of a provisional patent application as a non-provisional patent application. (See Appendix 4.1.1 for a sample provisional patent application.)

Basic structure of utility patents

A utility patent has three primary parts: a specification, drawings, and claims.

Part 1: Specification

The specification is a description that explains the essential features of the invention. This is where the innovator must provide a detailed enough description to teach a person who is skilled in the field of the invention how to make and use the invention. For a medical device patent, the specification usually begins with a brief background section that clearly describes the need or basic medical problem (including a description of the disease or condition and a discussion of existing approaches to treatment – see 2.1 Disease State Fundamentals and 2.2 Treatment Options). Within the specification, uncommon medical terms should be explained so that anyone reading the patent, including the examiner, clearly understands the context for the invention, regardless of their level of medical background. The writing should be targeted to the "*Scientific American*" level – intelligible to a bright reader who is not necessarily versed in the field. The specification also must describe in detail what the new device does and how it is to be used (referring to diagrams with labels, as appropriate). This section can describe several different approaches for using the device but, as noted previously, the inventor has a legal obligation to describe the best mode of the invention, as well as sufficient detail to enable others in the field to be able to make and use the invention claimed in the patent.

Part 2: Drawings

Drawings refer to illustrations, typically with detailed labels, which convey the structure of the invention and how its parts work together. For medical device patents, it is often necessary to include anatomical figures to show the interaction of the device with the organ or system where it is intended to be used. Patent attorneys often employ illustrators who can help render complicated figures based on sketches provided by the inventor.[11] Figure 4.1.6 shows a sample patent drawing next to a photo of the commercial device that ultimately evolved from the invention.

Part 3: Claims

Claims are written statements that define the invention and the aspects of that invention that can be legally enforced. As such, they are the real "teeth" of the patent. It is the claims that provide the basis for comparison with the prior art to determine patentability and, later, are used to make a determination of infringement – a

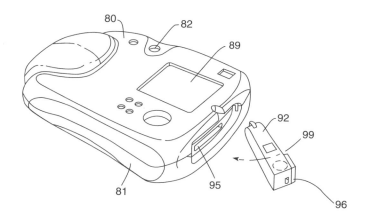

FIGURE 4.1.6
Patent drawing of an external cardiac defibrillator for public use, and the commercial product that was ultimately manufactured and sold (courtesy of Philips Intellectual Property & Standards; reprinted with permission).

ruling that another technology uses features covered by the patent, thereby making it illegal to make, use, sell, or import that device without a license. In this sense, the claims are like a deed to a piece of real estate – they specifically describe the boundaries of what is owned. Typically, claims are presented as a numbered list of points at the end of the patent and spell out, in precise terms, exactly how an invention or discovery differs from the prior art. Writing claims is such an arcane and expert exercise that inventors are advised to seek the assistance of an IP attorney. However, even when working with an IP expert, the inventor should have a basic familiarity with the art of claims writing in order to effectively understand and assist the work of the attorney.

United States utility patents can include two basic types of claims. *Independent claims*, as the name suggests, stand on their own. *Dependent claims* are claims that refer to one or more independent claims and generally express particular embodiments (typically they add additional elements or steps of carrying out the invention).[12] Dependent claims are used to clarify the language of an independent claim, so each dependent claim is more narrowly focused than the independent claim upon which it depends. Independent claims are typically written in broad terms to prevent competitors from circumventing the claim by modifying some aspect of the basic design. However, when a broad term is used, it may raise a question as to the scope of the term itself. For example, if there is any question about whether or not a "base" described in a patent application includes a set of legs, a dependent claim that included the phrase, "wherein said base comprises a set of legs," would clarify that it does.[13] While the independent claim broadly describes the invention, each dependent claim describes specific aspects of an embodiment (frequently, there is a string of dependent claims attached to an important independent claim). Strategically, dependent claims provide a safety margin in the case where the independent claim is invalidated by some prior art that comes to light after the filing (the dependent claim, which is more specific, may still be valid). Figure 4.1.7 shows how independent and dependent claims work together in the classic Palmaz patent for the intravascular stent.

Another way to think about claims is to categorize them based on what they cover. *Product or apparatus claims* cover a physical entity, such as a material or device. *Method claims* cover an activity, such as a process or method of use – for example, the way a medical device is deployed or used in the body. Note that certain method claims covering surgical or diagnostic procedures are not allowed in most countries outside the United States (see below for more about international patenting).

In addition to the major components in the body of the patent, the *face* or front page includes important

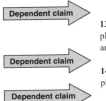

13. An expandable intraluminal vascular graft, comprising:
a tubular shaped member having first and second ends and a wall surface disposed between the first and second ends, the wall surface being formed by a plurality of intersecting elongate members, at least some of the elongate members intersecting with one another intermediate the first and second ends of the tubular shaped member;
the tubular shaped member having a first diameter which permits intraluminal delivery of the tubular shaped member into a body passageway having a lumen; and
the tubular shaped member having a second, expanded diameter, upon the application from the interior of the tubular shaped member of a radially, outwardly extended force, which second diameter is variable and controlled by the amount of force applied to the tubular shaped member, at least some of the elongate members being deformed by the radially, outwardly extended force, to retain the tubular shaped member with the second, expanded diameter, whereby the tubular shaped member may be expanded to expand the body passageway and remain therein.

14. The expandable intraluminal vascular graft of claim 13, wherein the plurality of elongate members are a plurality of wires, and the wires are fixedly secured to one another where the wires intersect with one another.

15. The expandable intraluminal vascular graft of claim 14, wherein the plurality of elongate members are a plurality of tantalum wires.

16. The expandable intraluminal vascular graft of claim 13 wherein the plurality of elongate members are a plurality of thin bars which are fixedly secured to one another where the bars intersect with one another.

FIGURE 4.1.7
Independent claims can stand alone, whereas dependent claims rely on a parent claim (from US utility patent 4733665, reprinted with permission of Julio C. Palmaz).

summary information that is very useful for searching (see Figure 4.1.8).

Key information to review includes:

- **Patent number** – Assigned by the USPTO.
- **Filing date** – The date the inventors submitted the application – and the start of the 20-year period of coverage.
- **Issue date** – The date in which the patent is officially granted.
- **Assignee** – The legal entity that has rights to the patent, often a company.
- **References cited** – The patents and literature cited by the inventor(s) and/or examiners in obtaining the patent.
- **Abstract** – A brief description of the invention.
- **Classes** – A list of the classes and subclasses assigned to the invention by the patent office.

United States Patent [19]	[11] Patent Number: 5,061,273
Yock	[45] Date of Patent: Oct. 29, 1991

[54] **ANGIOPLASTY APPARATUS FACILITATING RAPID EXCHANGES**

[76] Inventor: Paul G. Yock

[21] Appl. No.: 548,200

[22] Filed: Jul. 5, 1990

Related U.S. Application Data

[63] Continuation of Ser. No. 361,676, Jun. 1, 1989, abandoned, which is a continuation of Ser. No. 117,357, Oct. 27, 1987, abandoned, which is a continuation of Ser. No. 852,197, Apr. 15, 1986, abandoned.

[51] Int. Cl.^{5} A61M 25/00
[52] U.S. Cl. **606/194**; 604/96
[58] Field of Search 604/96, 101, 102; 606/192, 193, 194

[56] **References Cited**

U.S. PATENT DOCUMENTS

2,043,083	6/1936	Wappler	128/303.11
2,687,131	8/1954	Raiche	128/349
2,883,986	4/1959	de Luca et al.	128/351
2,936,760	5/1960	Gants	128/349
3,731,962	5/1973	Goodyear	128/351
3,769,981	11/1973	McWhorter	128/348
3,882,852	5/1975	Sinnreich	128/4
4,195,637	4/1980	Gruntzig et al.	128/348.1 X
4,198,981	4/1980	Sinnreich	128/344
4,289,128	9/1981	Rusch	128/207
4,299,226	11/1981	Banka	128/344
4,367,747	1/1983	Witzel	128/344
4,468,224	8/1984	Enzmann et al.	604/247
4,545,390	10/1985	Leary	604/95
4,569,347	2/1986	Frisbie	128/344
4,610,662	9/1986	Weikl et al.	128/348.1 X
4,616,653	10/1986	Samson et al.	128/344
4,619,263	10/1986	Frisbie et al.	128/344
4,652,258	3/1987	Drach	604/53

FOREIGN PATENT DOCUMENTS

591963 4/1925 France.
627828 10/1978 U.S.S.R..

OTHER PUBLICATIONS

A New PTCA System with Improved Steerability, Contrast Medium Application and Exchangeable Intracoronary Catheters, Tassilo Bonzel, PTCA Proc. Abstract, Course 3, Center for Cardiology, University Hospital, Geneva, Switzerland, Mar. 24–26, 1986.
Nordenstrom, ACTA Radiology, vol. 57, Nov. 1962, pp. 411–416.
Nordenstrom, Radiology, vol. 85, pp. 256–259 (1965).
Diseases of the Nose and Throat, at pp. 776–794–797 (S. Thomson) (6th Ed., 1955).
Achalasia of the Esophagus, at pp. 122–147 (F. Ellis, Jr. et al.) (1969) (vol. IX in the Series Major Problems in Clinical Surgery, J. Dunphy, M.D., Ed.)

Primary Examiner—Michael H. Thaler
Attorney, Agent, or Firm—Fulwider, Patton, Lee & Utecht

[57] **ABSTRACT**

Apparatus for introduction into the vessel of a patient comprising a guiding catheter adapted to be inserted into the vessel of the patient and a device adapted to be inserted into the guiding catheter. The device includes a flexible elongate member and a sleeve carried by the flexible elongate member near the distal extremity thereof and extending from a region near the distal extremity to a region spaced from the distal extremity of the flexible elongate element. The device also includes a guide wire adapted to extend through the sleeve so that the guide wire extends rearwardly of the sleeve extending alongside of and exteriorly of the flexible elongate element into a region near the proximal extremity of the flexible elongate element.

6 Claims, 3 Drawing Sheets

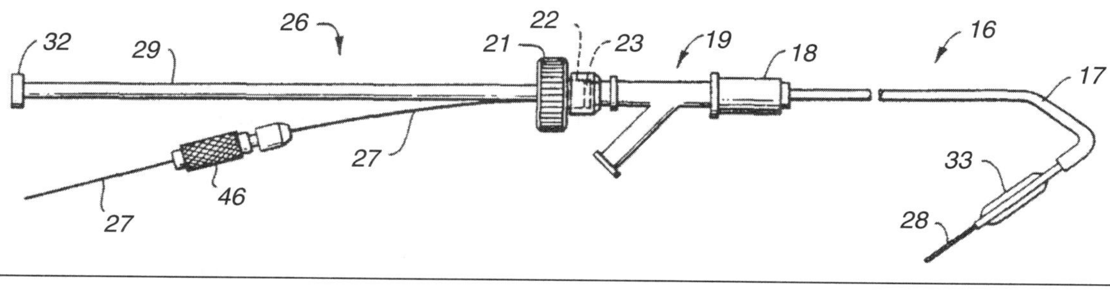

FIGURE 4.1.8
The face page of a patent includes a summary of important information (from US utility patent 5,061,273; reprinted with permission from Paul G. Yock; the complete patent is shown in Appendix 4.1.2).

Stage 4 Concept Selection

Preparing a utility patent

Utility patents are the "workhorse" form of intellectual property in the medical technology industry. As such, it is important for the inventor to understand them well. For practical purposes, drafting of the patent should almost always be performed by an expert attorney or patent agent. However, the inventor can save considerable time and expense by providing the attorney or agent the detailed background information required for the specification, drafts of the drawings, and the results of the prior art search with analysis and comments.

The inventor will also want to perform a detailed review and editing of the application that is drafted by the attorney or agent prior to filing. For most inventors, the most foreign and difficult part of this process is trying to understand the claims. When reviewing claims, an inventor should seek to ensure that they cover all commercially relevant aspects of the invention. This includes an invention's physical form, materials, and methods of use. Too often, inventors emphasize the structural elements of the device without paying adequate attention to the clinical significance or the important functionality of the invention. For example, it would not be uncommon for an inventor to submit a claim such as, "A device comprising X, Y, Z ... ." However, a much more valuable and strategic approach would be to think about the clinical advantages that the device enables (e.g., it is faster, easier to use, less expensive, etc.) and reflect in the IP the functionality that reveals those unique advantages. For example, a device that enables faster surgical times because it deploys elements simultaneously, as opposed to individually and sequentially, may read, "A device for use in X procedure wherein the device enables simultaneous deployment of elements ..." Similarly, a device that is easier for a surgeon to use because it requires only one hand may read, "A device for use in X procedure wherein the device may be fully operated with a single hand ..." In addition to protecting the invention in a way that is more difficult for competitors to circumvent, this approach to drafting claims also protects the inventor as the device evolves through ongoing product development. If an inventor has patent claims that cover the functionality of its device rather than just the structural elements of "X, Y, and Z," then it will be in a better position to enforce those claims if "Z" changes slightly over time.

When developing claims, it is also important to try to understand the IP activities of the inventor's competitors (other individuals and companies working in the same field). Actively review what information is being published and what patents are being issued in the field and evaluate this information against the inventor's established and desired claims. More information about developing an ongoing patent strategy is presented in 5.1 Intellectual Property Strategy.

Additionally, inventors should think about how enforceable their claims would be if they were infringed. Because each element of a claim must be proved to be infringed and the burden of proof falls to the patent holder, an inventor should be careful not to make claims that require time-consuming, complex, and/or expensive measurements to prove. For example, making claims where infringement would have to be proven through in vivo clinical trials is discouraged (unless those trials have been completed already as part of the FDA regulatory process). Whenever possible, claims should be based on relatively simple ex vivo measures (e.g., bench top or non-clinical data – see Appendix 4.1.2 for a sample utility patent that includes more examples of claims).

Filing and review of utility patents in the United States

A patent application can be filed by mail or electronically. For context, in fiscal 2006, the USPTO managed over 440,000 patent applications, more than double the number of applications filed in fiscal 1996.[14] This large number of filings has resulted in a substantial backlog and delays in reviews (the backlog was estimated to have reached approximately 600,000 applications).[15]

When a utility patent application is submitted, it is opened, sorted, given a serial number, scanned, and checked for completion by the Office of Initial Patent Examination (**OIPE**). A USPTO drawing inspector also reviews the quality of the figures for clarity and conformity with guidelines. Complete applications (that meet all USPTO submission guidelines) are then routed to a "technology center," which is the organizational unit of the USPTO responsible for that type of invention. For medtech inventors, the main technology center is 3700 (Mechanical Engineering, Manufacturing and Products), which includes 37D (Medical and Surgical Instruments, Treatment Devices, Surgery and Surgical Supplier), 37E (Medical Instruments, Diagnostic Equipment, Treatment Devices), and 37F (Body Treatment, Kinestherapy, Exercising). If the application is missing information or otherwise incomplete, the OIPE issues a notice to the filer and sets a time period for submitting the missing information. Failure to submit missing or inaccurate information during the time period will result in abandonment of the application.

Within the technology centers, utility patent applications are assigned to a patent examiner who is considered an expert in the subject area. This person researches previous patents and available technical literature to determine whether a patent should be granted based on the parameters outlined above. Because of the backlog in the office, it can take from 18 months to 3 years for the patent examiner to pick up the patent application for initial review.

Utility applications are made public after 18 months from their earliest priority date, usually the filing date of the provisional patent application on which the non-provisional utility patent application is based. Inventors can file a "Non-publication request" to the USPTO to prevent having their inventions published; however, this process requires an inventor to forego any foreign patent protection for the invention.

Once a thorough examination has been performed, the USPTO issues a first **office action** (OA) to the filer. A patent is almost never approved in the first OA. The letter from the examiner will instead outline his/her objections to the application (e.g., it may cite prior art that, in the examiner's opinion, renders the new concept obvious or not novel; it may list the claims that are rejected; or it could call out any problems in the specification and drawings). The filer is given three months to respond to the office action in writing (an additional three months will be granted if an extension fee is paid). In the response, the filer can either argue against the office action or make the required changes in a document called an amendment, which modifies the claim language in a direction intended to make the claims acceptable in light of the examiner's analysis. The office action/amendment process may be repeated (on balance, this occurs twice in the patent application process). Finally, the examiner rejects or allows the patent application. If a **notice of allowance** is granted, the USPTO issues the patent upon receipt of required issue fee. The process from the initial filing to an allowance usually takes about three years. For applicants who ultimately do not get the approval they seek, there is a rarely used appeals process through the USPTO Board of Patent Appeals and Interferences. Decisions from the Board may be further appealed to the US Court of Appeals for the Federal Circuit.

One confusing feature of an issued utility patent is the priority date, which is effectively the date against which prior art challenges to a patent will be adjudicated. For international filings, any prior art published before the priority date can be used to help invalidate the patent. In the United States, non-patent publications by the inventor dated less than a year before the priority date cannot be considered prior art. For example, if a provisional patent was filed on January 1, 2009, any publication by the inventor subsequent to January 1, 2008 would not be considered as prior art. In practice, this means that an inventor who publishes a paper generally has a year in the United States to obtain patent protection.

If an inventor has filed a provisional patent and then follows with a utility patent, the date of filing the provisional patent is the priority date. If the original filing is made as a utility patent, the utility patent filing date becomes the priority date.

There are several layers of costs associated with utility patents. Attorney costs are reviewed in more detail in 5.1 Intellectual Property Strategy but, in general, will run from $6,000 to $14,000 for the filing of an uncomplicated medtech utility patent. The filing fees paid to the USPTO depend on the number of claims in the patent and some other factors, but are in the range

Stage 4 Concept Selection

of $950. If the patent is granted, there is an additional USPTO issue fee followed by three maintenance fees at 3–3.5 years, 7–7.5 years and 11–11.5 years. The total of the issue and maintenance fees are on the order of $8,500.[16]

International patenting

The protection provided by a patent issued by the USPTO stops at the US borders. To obtain patent protection outside the United States, an inventor must file patent applications in each country in which patent protection is desired. Under a treaty called the Paris Convention, most countries in the world (with the notable exception of Taiwan) give patent applicants one year to file a corresponding patent application in their patent office and use the US priority date as the effective date of the foreign filing. In addition, through the international Patent Cooperation Treaty (**PCT**), an inventor can file a unified patent application to seek IP coverage in each of a large number of contracting countries. As of mid-2008, there were 139 countries participating in the PCT, representing most major nations (excluding Iran, Iraq, Pakistan, Taiwan, and Thailand, among others).[17] The PCT is administered under the auspices of the World Intellectual Property Organization (**WIPO**), a specialized agency of the United Nations, which manages this program and over 20 other international patent treaties. As with other Paris Convention filings, international inventors have one year from filing an international application under PCT to file a utility patent application in the United States. (This timing is parallel to that for the US inventor who files a provisional patent, which means that the inventor will often be filing US utility cases and foreign and/or PCT cases simultaneously on the one-year anniversary of the provisional filing date.)

Some patent applicants file a PCT first and enter the USPTO through the PCT in the national stage (discussed below). If the PCT is the first patent application filed, the international filing date becomes the priority date in the United States.

Under the PCT, a single filing of an international application (called a PCT or international application) is made with a Receiving Office in a single language (the receiving offices are typically the patent offices of the PCT contracting states). An International Searching

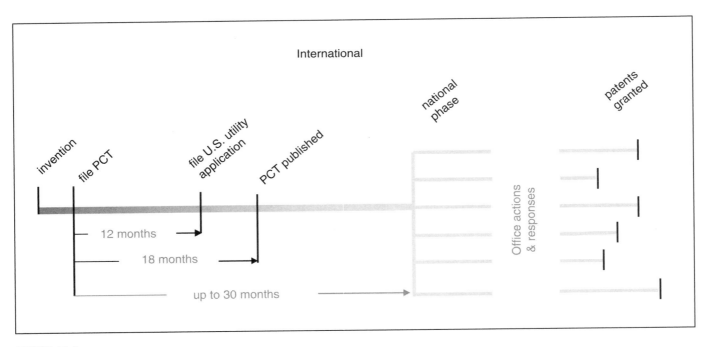

FIGURE 4.1.9
Typical timing associated with international patent filings.

Authority (**ISA**) performs an extensive international patent search and delivers a written opinion regarding the patentability of the invention.[18] Importantly, this is a non-binding opinion. There is no collective "international patent" that results from the PCT process – patents must be reviewed and granted individually by patent agencies in each of the countries in which coverage is sought. What the PCT provides is a way of starting the application process in an efficient and relatively economical way (typically a few thousand dollars) and receiving a preliminary opinion on patentability from an expert international agency. In some cases, the applicant can further request a preliminary examination by an International Preliminary Examining Authority, whose opinion can supersede that of the ISA.

The process of pursing country-by-country patent coverage is called the national phase or stage. Depending on the number of countries selected, this can be an expensive step, consuming hundreds of thousands of dollars. Some strategies for selecting countries for filing are reviewed in 5.1 Intellectual Property Strategy. The PCT allows inventors to delay national phase filing in PCT member countries for up to 30 months from the priority date, which can allow for substantial development and testing prior to making this major financial commitment (see Figure 4.1.9).

The international application is published by the International Bureau of WIPO, 18 months after the filing date. The WIPO patent database can be an invaluable source of search information about patents in process, particularly those being pursued outside the United States.

Although the national phase must generally proceed country by country, the European Patent Office (**EPO**) constitutes an important exception.[19] The EPO provides a uniform application procedure for individual inventors and companies to be granted patent coverage in up to 37 European countries through a single application process that is consistent with PCT guidelines. Once granted, the EPO patent is the equivalent of a patent in each of the member countries.

Regardless of the specific country in which international patent coverage is being sought, innovators should keep in mind a few important considerations. For US inventors who file a PCT application, the USPTO may serve as the searching agency and will render the patentability opinion. (Some inventors choose to use the European Patent Office or the Korean Patent Office as the search authority in order to have someone other than a US patent examiner perform the search in an effort to uncover different prior art.) As noted above, under the PCT, an inventor can wait until one year from the first patent filing in the United States to file in other countries and still use the US filing date as the effective filing date on the non-US patent applications.[20]

Intellectual property and patenting in India and China

Because of their relatively large size and emerging economic importance, China and India are of particular interest in international patenting. Both countries have had a history of relatively weak IP rights and patent enforcement, which has been a barrier to foreign interest in many technology sectors, including medtech. While both countries are improving, they were on the Priority Watch List of the Office of the US Trade Representative in 2008 (along with seven others). The activities of the nations on this list are carefully scrutinized for the adequacy of the IP protection and enforcement rights provided to inventors.[21]

In India, patenting is managed by the Indian Patent Office (IPO). India began upgrading its IP environment in 1995 as a condition to its admission to the World Trade Organization (WTO). The country joined the Paris Convention and signed the Patent Cooperation Treaty in 1998, allowing inventors to secure coverage in India through this process. Despite ongoing amendments to the country's Patent Act, India's patent processes and regulations still do not meet all global standards. Among other complaints, India has been criticized for the weakness of its patent law and its enforcement, the growing backlog in the patent office, a slow judiciary process, and the theft of proprietary information.[22] Patent protection in pharmaceuticals has been particularly contentious, with Indian generic makers retaining significant scope for copycatting

Stage 4 Concept Selection

patented Western drugs without penalty and Western companies seeing relatively few of their patent applications approved.[23]

In China, the State Intellectual Property Office (SIPO) governs patenting, as well as the country's membership in the PCT. The closest thing to a US utility patent in China is an invention patent, which protects "any new technical solution relating to a product, a process or improvement" for a period of 20 years from the date of filing. As with India, China moved to strengthen its IP position following its ascension to the WTO in 2001. The country successfully strengthened its legal framework and amended its IP laws and regulations, but critics charge that China has been remiss in enforcing these new rules. The country has a first-to-file system and inadequate processes for investigating patent applications. In combination, these factors make it relatively easy for non-inventors to secure patents on technologies that are not their own.[24] Slow and antiquated court systems also make it difficult to rectify such situations. Tension over IP rights in China reached a peak when the United States asked the WTO to intervene in August 2007. Among other concerns, the United States objected to the fact that China's threshold for the criminal prosecution of IP-related cases was so high that few were ever prosecuted, even when it came to commercial counterfeiting.[25]

Despite the challenges faced by India and China, both countries remain committed to upgrading their IP environments to world standards. In 2008, the Indian government completed the first phase of a modernization effort, spending more than $35 million to set up integrated IP offices in four major cities and launch electronic filing of patent applications. Another $75 million was earmarked to be spent on a second phase of IP-related improvements.[26] In China, the government announced a National Intellectual Property Rights Strategy in June 2008 to support the government's scientific and technology development goals. As part of an aggressive plan to bring the country's IP environment further into compliance with WTO requirements, it accelerated the drafting of a third amendment to its patent law. Included in the proposed amendment were changes meant to strengthen IP enforcement in China.[27]

Other important IP concepts

In addition to understanding issues around patents per se, there are a number of basic points about intellectual property that the innovator should understand well.

Documentation

To establish invention dates under the *first to invent* system, inventors must begin keeping an innovation notebook early in the innovation process. Julio Palmaz, inventor of the coronary stent, emphasized that his early documentation proved to be critically important in litigation surrounding the stent, advising innovators to, "Get in the habit of putting down in writing what is in your mind."[28] This information can also be used to establish the prior art cut-off date against which the novelty and non-obviousness of an invention are measured. Information within the innovation notebook can also be used to defend the precise time when an invention was conceived by the inventor. Refer to 1.2 Observation and Problem Identification for more information about guidelines that should be followed to ensure that the entries in an innovation notebook are legally defensible.

Inventorship and ownership

Any patent application must include all of the inventors who contributed to the creation of an innovation, as defined by the claims of the patent. To be considered an inventor, an individual must have contributed to the idea and participated in its reduction to practice. If someone adds to any part of an invention as it is being claimed, then s/he is a joint inventor. Whether s/he is instrumental to a core concept or only helped with some detail is irrelevant. However, if someone provided only labor, supervision of routine techniques, or other non-substantive contributions, then s/he is not to be considered an inventor.[29] With medical devices, this issue often surfaces when a prototyping engineer is involved in building the first versions of the device. If the builder of the device is operating entirely under the inventor's instructions, the builder is not an inventor. On the other hand, it is frequently the case that the builder adds a key insight. For example, when one of the editors of this textbook was developing the Smart

Needle™ concept (a needle with a Doppler transducer), the engineer who was building the prototype pointed out that moving the transducer to a different position could improve the signal. This was an important addition to the design, and the engineer became an inventor by virtue of this contribution.

It is important to be aware that in the absence of any employment or consulting agreement obligating an inventor to assign his/her invention to another entity (such as a university or an employer), each inventor has both an ownership interest in the patent and the right to license, commercially develop, or sell the invention *as an individual*, with or without the knowledge or permission of the other inventors. While most inventors prefer to work together to maximize the value of their patent in any licensing deal, the ability of each inventor to act alone is something to consider when involving multiple individuals in the innovation process.

Understanding the implications of ownership for an invention is essential prior to initiating a collaboration. In the case of medical devices, it is common for the inventors of a new technology to seek out a physician expert to validate their early ideas. It can be irresistible for the physician to add some potential improvements to the core idea. These suggestions may be useful and important but, in the process of receiving the feedback, the original team may have added a new inventor – at least for some of the claims of the patent. This is particularly problematic if the new inventor has a special relationship with a company that presents a conflict later on, or if s/he is a faculty member of a university that will automatically assume ownership of the IP (see below).

If it is discovered retroactively that the inventorship on a patent is inaccurate, and the mistake was made in bad faith, the patent can be invalidated.[30] While this happens infrequently, litigators challenging a patent will often start with inventorship as possible grounds for invalidating it.

Student, faculty, and employee inventors

Inventors working in an academic setting must keep in mind that if an invention is conceived and/or developed using significant university resources, the university may assert its rights in taking ownership of the invention. Generally the interpretation of "significant resources" does not include desks, computers, the use of conference rooms, or the use of other commonly available equipment or facilities. University policy regarding IP ownership comes out of the institutions' need to avoid situations in which public resources (such as grants from the NIH) are used to support private enterprise without some oversight regarding the fairness and best deployment of the technology. Among all science and technology areas, the biomedical sciences are, in fact, the area of highest concern for government regulators with respect to university inventors. Unfortunately, in the past decade, there have been some high-profile and tragic examples of conflicts of interest related to ownership of biomedical innovations (e.g., the 1999 Jesse Gelsinger case).[31] As a result, universities and government agencies have become extremely attentive to issues of IP ownership of medical technologies.

Some universities have language in their policies that states, in effect, that inventions in the faculty member's areas of expertise also belong to the university. For student inventors, policies vary across academic institutions. In some universities, students are treated exactly like faculty; in others, there are no IP regulations. A wide range of policies exists between these two extremes. It is important for any university-based inventor to have a clear understanding of the relevant rules. If the university guidelines are not clear, it can be useful to enlist the assistance of an outside lawyer who has experience with that particular university and its inventorship policies (more information about licensing back one's own invention from a university's Office of Technology Licensing can be found in 6.4 Licensing and Alternate Pathways).

The same principles apply to employees or consultants working for private companies. It is common for an employee to be obligated to assign ownership of inventions to the company, provided that the inventions are within the employer's business or if the inventors used the employer's resources in developing the invention. For anyone working in more than one company, or for a

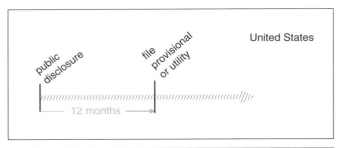

FIGURE 4.1.10
Public disclosure has different implications for patenting in US and international markets.

student or faculty member working for a company part-time as an employee or as a consultant, it is important to understand whose ownership rights apply to any inventions.

Public disclosure

Many inventors underestimate the importance of confidentiality in the patenting process. In the early phases of a new project it is natural to want to share the excitement of the new approach or present the concept to academic colleagues. As noted earlier, in the United States an inventor has a grace period of one year to file a patent application from the date that s/he first publicly discloses the concept, as shown in Figure 4.1.10 (i.e., before the inventor's own disclosures become prior art that can be used against the patentability of his/her invention).

Public disclosure can include publication, presentation, and announcement, as well as commercial use, offerings, or sales. Be aware that casual conversations with colleagues are similarly considered to be a form of disclosure. Medical technology inventors (especially academic engineers or physicians) run a particular risk of submitting an abstract or manuscript that describes their invention assuming that the "confidential" editorial review process protects their patent rights. This situation is at best in the gray zone for public disclosure, and it is wise to file at least a provisional patent application before sending in a manuscript or abstract.

Internationally, there is no grace period. If the innovator publicly discloses the invention prior to filing a patent application, the disclosure is likely to preclude patent protection outside the United States, even if the disclosure is within the one-year grace period provided by US law. As a result, experts in the field strongly suggest filing a patent application before making any public disclosures.[32] For medical devices, issues with public disclosure have led many inventions to have coverage in the United States only. Depending on the device, this has not necessarily been a significant problem, given the relatively large size of the American device market compared to that of other countries. With the expansion of the global markets, however, international coverage is becoming increasingly important, making issues related to public disclosure a serious matter.

Non-disclosure agreements

As Rodney Perkins, founder of multiple companies including ReSound Corporation, Laserscope, Collagen Corporation, Novacept, and Pulmonx, put it:[33]

> Loose lips sink ship. You want to keep your ideas fairly tightly held until you have some protection on them. This actually enhances the value of the idea and improves your ability to finance the development.

To keep an idea or invention confidential, inventors are strongly encouraged to use non-disclosure agreements (**NDAs**), alternatively known as confidential disclosure agreements. These are legally binding documents that enable an entity to record the terms under which confidential information is exchanged (thus preventing the interaction from qualifying as a public disclosure). Such agreements can be useful any time an individual or a company must disclose the details of its technical secrets, including (but not limited to) discussions with potential investors, partners, third-party suppliers, employees, contractors, and consultants. For students, the need for confidentiality can extend to faculty members and fellow students, as well as other mentors, family, and friends. In discussions with an IP attorney, confidentiality is legally protected (with or without an NDA) under attorney–client privilege.

The key issue to address in an NDA is *use* (how the party to whom the confidential information is disclosed can use the information that is shared). It is not enough to ask someone to keep quiet about an invention or other confidential information. A strong NDA must explicitly restrict the ways in which the information can be used.

An example of an NDA is provided in Appendix 4.1.3. In general, an NDA will have several components:

1. It specifies who the legal parties are in the agreement.
2. It defines the content and scope of the confidential information.
3. It provides for some reasonable exclusions from the obligation of confidentiality (for example, when the receiving party can prove they already knew about the information or when it was public knowledge).
4. It details what the nature of the obligation is for the receiving party – that is, how the information can and cannot be used (e.g., restrict access of employees to the information; do not publish or otherwise disclose the information).
5. It specifies a time period that confidentiality is in force (typically some number of years).

Non-disclosure agreements are used to protect disclosures made in multiple scenarios – individual to individual; individual to company; company to individual; and company to company. Typically, innovators and companies are discouraged from using NDAs that require the discloser to document in writing what was disclosed orally in the meeting. Although it is always a good idea to summarize what information was shared, the innovator or company should not be required to do so within a specific time frame and should not be constrained by what may or may not have been included within a summary document.

An NDA does not guarantee that a breach of confidence will not occur. However, it does give the company or innovator a stronger legal position if a problem arises. Some entities will not want to sign an NDA and the inventor must decide how much s/he is willing to disclose without protection. In some situations, disclosure may be still be warranted. For example, venture capital firms routinely abstain from signing NDAs in favor of a generally accepted principle that sensitive information will be kept confidential. In this case, innovators are advised to mark all documents confidential (including each slide of PowerPoint presentations) and disclose only what is absolutely necessary to get the firm interested. In other scenarios, where the level of trust is less certain, it is best not to reveal confidential information without an NDA.

When in doubt, inventors are encouraged to "file first, disclose later,"[34] using NDAs to help control the sharing of any confidential information before the filing date. Whenever innovators disclose information before filing a provisional or utility patent application, the risk exists that the person(s) they have disclosed to might develop a competing invention based on the revealed information. Using a "file first" strategy before disclosing can help minimize this risk.

The following case example on iRhythm Technologies, Inc. demonstrates how the essential information presented in this chapter can be brought together to help an innovator and his/her team understand the IP landscape for a concept under consideration and begin to establish a strong IP position.

FROM THE FIELD

iRHYTHM TECHNOLOGIES, INC.

Assessing the IP landscape and making preliminary filings

When Uday N. Kumar, John White, Kityee Au-Yeung, and Joseph Knight were fellows in Stanford University's Program in Biodesign, they began working on the need for *a better way to detect potential rhythm disturbances in non-hospitalized patients with suspected arrhythmias.* Arrhythmias are abnormal heart rhythms (atrial fibrillation, described at length in 2.1 Disease State Fundamentals, is one common type). Although some arrhythmias have few negative health consequences, others lead to serious heart disease, stroke, or sudden cardiac death.[35]

According to Kumar, a cardiologist specializing in cardiac arrhythmias, "The vast majority of people who have symptoms that could be due to arrhythmia first go to their primary care physician or an emergency room doctor. But physicians in those settings are not enabled fully to diagnose the problem since current cardiac rhythm monitoring technologies are complicated and have many drawbacks that limit their use by these doctors. So, many people never get assessed until they have a very severe presentation, such as passing out or experiencing a cardiac arrest. Unless something relatively serious happens, patients are often told, 'You had palpitations two days ago. But you look fine now. Let's wait and see if it happens again.'"[36] In contrast, if these patients were referred to a cardiologist, advanced monitoring technologies could be used to diagnose the problem and facilitate more proactive treatment, in many cases, before more serious symptoms occurred. "We figured out that enabling better diagnosis through more accessible and simple approaches was the key to getting more people into treatment earlier in the process," said Kumar.

Once the team understood the parameters of the need and the key need criteria (as articulated in the need specification), its members held a series of brainstorming sessions. Together, they generated dozens of potential solutions and screened them against the need. "After we had a solution that we thought would best meet the need spec," Kumar recalled, "we had to figure out if people had done this before so we conducted IP searches. Our focus was on patents that had already been issued as well as pending patent applications," he added. Importantly, the team did not spend too much time looking at medical journals and other general sources since, as a field, cardiac rhythm monitoring had not experienced a great deal of innovation in recent years. "There just weren't many recent publications out there talking about cardiac rhythm monitoring," Kumar recalled.

The initial patent searches were meant, in part, to establish the utility, novelty, and non-obviousness of the team's idea. Demonstrating the utility of a device designed to perform outpatient cardiac rhythm monitoring was not a primary concern. However, the team was intent on clearly identifying that its solution was novel and non-obvious. Kumar and his teammates performed many searches using a wide variety of search terms related to the description of the device concept, its functionality, appearance, interaction with the body, and links to the disease. Patents on existing cardiac rhythm monitoring therapies were also searched for relevant information. The team used different combinations of search terms to help ensure that the scope of the research was not too narrow and that nothing was inadvertently overlooked. "You have to be broad in your searches," said Kumar. "The information you uncover really depends on how you search. I used a thesaurus to figure out different ways to describe the same thing. People use different approaches, and you can miss a whole patent if you don't expand your thinking about search terms." The team also realized that being too broad in its search strategy led to many irrelevant results. Understanding the limitations of being too narrow or too broad allowed the team to develop a search strategy that gave them a good sense of what was out there.

As the team reviewed patent documents it became increasingly clear that their solution had several important elements that made it unique. To address the non-obviousness of the device, they spent significant time and energy figuring out where the cardiac rhythm monitoring field was trending. "If you understand where the field is heading, you can determine how you fit in," Kumar commented. "If we're going in one direction and everyone else is going another, it shows that we're on a different path, which makes it difficult for the patent office to say that what we're doing is really obvious."

At that point, in parallel with other activities in the biodesign innovation process, the team was eager to seek input and opinions about its solution from other people in the field, "so we filed a provisional patent application," said Kumar (see Figure 4.1.11). "It wasn't just a paper napkin with a sketch. It had a detailed background section and drawings." It also included a single claim the team believed captured the device's utility. Filing a provisional application secured a priority date in the patent office and protected the team from third-party disclosure (see Figure 4.1.11).

As biodesign fellows during this preliminary IP assessment, the team did not have the financial means to engage an IP attorney. However, upon completing the program, "I made the decision to start a company," said Kumar. "That's when I knew it was time to get a patent attorney involved." To prepare for his interactions with an IP lawyer, Kumar revisited the team's early IP assessment and initiated a more detailed review of the prior art. "I spent a lot of time going back to the original patents we had examined. I also did more searching, brushed up on what else was out there, and dived much more deeply into the analysis of specific claims," he remembered. Kumar and team had started a claims analysis but realized they needed outside expertise to complete it. For this reason, they decided to file the provisional application and come back to claims analysis later, if any of them decided to pursue the project beyond graduation. "At the end of the day, the claims are what matters," Kumar note. "In the first instance, it can be discouraging to read descriptions and see figures in another patent that look similar. But 'similar' doesn't mean 'the same.' You have to closely examine and interpret the claims, especially the independent claims, to determine whether a potential problem exists." Kumar looked at both device and method claims, since his product and business model were directed towards a monitoring device and an approach for using the monitoring technology.

Based on his preliminary claims analysis, Kumar narrowed the list of patents to those that were closest and most relevant to the proposed solution. "Then, I went back to look at the text to understand the arguments that were made to support the claim sets. This also helped me think about and refine my own arguments for why the solution

FIGURE 4.1.11
The team with its first provisional patent filing (courtesy of Uday N. Kumar).

FROM THE FIELD: iRHYTHM TECHNOLOGIES, INC.

Continued

was novel and non-obvious compared to what was already out there."

With his own analysis complete, Kumar was ready to engage an attorney. As a fellow, he had been introduced to Ben Glenn, who regularly volunteered time to be an "IP coach" to teams in the Program in Biodesign. "I recalled meeting Ben and the positive experience our team had in talking with him. He had deep knowledge and experience in medtech, so I sought his help," said Kumar. He shared his assessment with Glenn as the first step in establishing a close and highly collaborative relationship." Over time, I spent countless hours with Ben, really explaining exactly what we were doing and bringing him up to speed. This way he could help direct how I could be most helpful as we parsed up the work. Even with a great IP lawyer, the innovator really has to stay involved and understand what's happening because you know better than anyone else what the technology is intended to do." In addition to validating the team's assessment of the IP landscape, Glenn orchestrated a conversion of the provisional application that emphasized the differences and advantages of the device and the method of use. The result was an IP foundation with three utility patent applications (covering the device and associated methods), as well as a PCT application. "In the end, it's hard to overstate the importance of having an experienced IP lawyer such as Ben on my side. He definitely spent a great deal of time trying to understand the space so that he could thoughtfully put together claims and send them to me. He'd ask, 'Is this really what you mean?' and we'd go back and forth, trying to define different elements or combinations of elements to distinguish over the prior art. Being pushed and questioned by Ben was the key to developing claims that really got to the essence of what was patentable about the solution," Kumar explained.

One important tool that Kumar recommended in developing a preliminary patent position was the USPTO Patent Application Information Retrieval system (or PAIR database), which allows users to access the status of current patent applications. "It gives you all of the history on what's gone on between a patent applicant and the patent examiner," he said. "By reading these interactions, you can see where applications have been rejected by an examiner, and for what reason. It shows you what claims were rejected, the prior art used in the rejection, and the rationale behind it. With Ben's help, I was also able to appreciate which of the many documents in the file for a given patent application really were significant. This information helped me to think more specifically about where Ben and I might expect push-back in the examination of our own applications, and it allowed us to realistically address these potential issues."

The company that Kumar founded around the IP that he worked on and licensed from Stanford became the foundation for iRhythm Technologies, Inc., based in San Francisco, California. In addition to developing the device and establishing the methods to use it, the company has continued to monitor the IP landscape to stay aware of salient new developments in the field and has also started to file additional patent applications.

4.1 Intellectual Property Basics

The process of background searching and then preparing a patent filing can often seem monumental to the first-time inventor. Fortunately, the "upstream" parts of the biodesign process will prove to be extremely useful in getting started. Allocate plenty of time for the search and be prepared to learn a huge amount about inventions in the field.

See **ebiodesign.org** for active web links to the resources listed below, additional references, content updates, video FAQs, and other relevant information.

1. Compile background information

1.1 What to cover – Begin by assembling complete background information on the concept/invention and the area in which it will be practiced. This will generally include the disease state and a review of existing treatment technologies/approaches. Consider writing draft claims that cover the basic invention. These can be in plain language, not "legalese." Brainstorm about keywords for the search. Try to identify as many of these as possible (see example in text). A thesaurus and medical dictionary can be helpful.

1.2 Where to look – Refer back to information collected as part of 2.1 Disease State Fundamentals and 2.2 Treatment Options. Augment this data with additional research as needed (using previously listed sources).

2. Search the prior art

2.1 What to cover – Start with patent database searches as the most efficient way to see a broad range of prior art (before searching non-patent sources). For patentability searches, scan all parts of any patents that are of interest; for freedom to operate searches, concentrate on the claims. Consider hiring a patent agent or attorney to assist with the prior art search. Many inventors benefit greatly from a more formal "prefiling search" or patentability assessment. An attorney can help identify the areas where patent protection is most likely and can help focus the content of the patent application covering the invention.

2.2 Where to look
- **The US Patent and Trademark Office Web Patent Databases** – US patents issued from January 1976 through present are full-text search enabled. Patents from 1790 through 1975 are searchable only by patent numbers and current US classifications. Full-page images are available for most patents from 1970 to present. A tutorial on how to use the database is available on the site. Note that issued patents and patent applications are in separate search fields.
- **US Patent and Trademark Office Official Gazette** – Weekly publication to announce those patents being issued and those trademarks being registered or published for opposition. Users can search or browse from 1964 to present to stay informed regarding new patents and changes in the field.
- **Google Patent Search** – Provides searchable access to the full text of the US patent corpus to find patents of interest.
- **Delphion Intellectual Property Network** – A subscription database that allows individuals to search the bibliographical and claims portions of US patents from 1974 to present. This service recently added European Patent Office and World Intellectual Property Office (WIPO) text and images. All patent references are hyperlinked for citation searching. Classification codes are also hyperlinked to assist in finding similar patents granted.
- **Derwent Innovations Index** – A database of international patent information. On the site, "equivalent" patents are available in multiple

Stage 4 Concept Selection

languages. This database is provided by the Institute for Scientific Information.
- **Free Patents Online** – Allows quick searching through US patents, US applications, and European patents. The most useful aspect of the site is being able to download PDFs of actual patents.
- **International Patents Databases**
  - WIPO Patent Scope
  - Canadian Patent Database
  - The European Patent Register Online
  - The Franklin Pierce Law Center Intellectual Property Mall
  - Italian Utility Models
  - Patent Abstracts of Japan
  - World Patent Search
- **Other databases** – Be sure to search PubMed and other medical sites for articles and abstracts, as well as perform general Google searches for more obscure references to relevant information.

3. Identify relevant prior art for patentability

3.1 What to cover – The first step is to determine which of the patents and other materials that have been located via the prior art search are most relevant to the patentability of the invention. The draft claims from section 1.1 above are useful here. The more elements of the claims that are taught by the prior art, the more relevant this patent is to the search. Aim for the most relevant 10 to 20 patents in the first cycle. The second step is to broaden the search again, using either citation search or classification index from the patents of interest identified in the first cycle. This should result in some new patents, and exclude some others from the first step. Repeat this cycle of broadening and filtering. Once the list of patents of interest stabilizes at some relatively small number (probably 5 to 15), these are key patents. Remember that new art is published daily, however, so searching is an ongoing process

3.2 Where to look – Use the information gathered from the prior art search to complete this analysis. When necessary, go back to the sources listed above to perform additional research.

4. Prepare a patent application

4.1 What to cover – For provisional patent applications, the application can be prepared by the inventor, but it is wise to have it reviewed by a patent attorney or patent agent. Although any material can be filed as a provisional patent application (including a drawing on a napkin or a PowerPoint presentation), it is essential to have a thorough description of the invention, including the main sections of a provisional application described in the chapter. For a utility patent application, a patent attorney or agent needs to be involved in writing the application, but the inventor can save considerable time and expense by drafting the background, providing rough drawings, and drafting key claims in plain language (see 5.1 Intellectual Property strategy for more information about selecting an IP attorney and the costs involved). If draft claims have already been written (as in step 1.1 above), modify these based on the prior art to demonstrate advantages provided by the new invention. Identify possible variations. Make sure all possible ways of putting together and carrying out the invention have been covered. Consider drafting both method and device claims. Prepare all necessary figures – the USPTO requires that drawings show each and every element claimed in the patent application. Next, draft the specification. With the new claims and the drawings as a guide, write a description of the invention. Be sure to include all of the elements of the invention, and all variations that have been brainstormed. *Now* there can be an effective meeting with a patent attorney or agent.

4.2 **Where to look** – Even if an inventor is planning to work with a patent attorney to file, the following resources can be helpful in understanding and preparing for the patent application process:
- **USPTO Guide to Filing a Non-Provisional Utility Patent Application** – Step-by-step instructions made available by the USPTO.
- **USPTO Provisional Application for Patent** – General information about completing a provisional patent application.
- **USPTO PAIR Database** – The Patent Application Information Retrieval system allows users to access the status of current patent applications.
- *Patent It Yourself* **by David Pressman** (Nolo Press) – A comprehensive reference for understanding the end-to-end patent process (as well as broader, important issues in the IP arena).

5. File a patent application

5.1 **What to cover** – Both provisional and utility patent applications can be filed online using the USPTO's system, EFS-Web. Documentation of receipt will be provided by the site. To file the application by mail, download the coversheet and transmittal form from the USPTO website. Make sure to fill out a return receipt postcard (this will be the *only* documentation of the priority date provided if the filing is by mail). It is wise to use a stable return address (students may want to use the address of their university's office of technology licensing). It is best to mail by United State Postal Service (USPS) Express Mail rather than FedEx or some other express carrier. If the document gets lost, only USPS handling guarantees that the mailing date will be the filing date.

5.2 **Where to look** – Refer to the USPTO website and the EFS-Web system for more information.

Credits

This chapter was based on lectures and mentoring in Stanford University's Program in Biodesign by Jim Shay and Ben Glenn of Shay Glenn LLP, and Mika Mayer and Tom Ciotti of Morrison & Foerster LLP. The editors particularly appreciate the careful editing of this material performed by Jim Shay and Jessica Connor of Shay Glenn LLP and Jeffrey Schox of Schox PLC. Many thanks also go to Darin Buxbaum and Asha Nayak for their help in developing this chapter.

Appendix 4.1.1

Sample provisional patent application[37]

PROVISIONAL PATENT APPLICATION

Inventors:
Benedict James Costello
Mark Zdeblick
For Proteus Biomedical, Inc.
Filed 10/5/04

Amplified Compliant Force Pressure Sensors

BACKGROUND OF THE INVENTION

1. Field of the Invention.

The present invention is related to microsensing devices, and, more particularly, to sensing devices with increased sensitivity.

2. Prior Art.

Advances in micromachined sensor technology has developed whole new applications for medical devices, particularly implantable devices, such as for cardiac disease treatment. Engineering on this highly miniaturized scale, with its many unique applications, has also brought with it new challenges.

A number of research groups have reported on different approaches for improving the performance of micromachined sensors. For micromachine pressure sensors, this effort has focused on modifying the dimensions of the various components of the pressure sensors. This work has resulted in improved designs by providing modifications to the sensing membrane components size, shape, and thickness.

Other research approaches to optimization effort have resulted in insights and improvements in the placement of the pressure and the strain-measuring elements on the membrane. The strain-measuring elements on current pressure sensing membrane elements are typically piezo resistors. However, the strain-measuring elements can also be vibrating beam strain sensors.

When designing the geometry of pressure sensor membrane, engineers have several parameters which can be modified in order to achieve the greatest improvement in sensor performance. By example, design considerations can includes modifications in the proportions of the diameter of the sensing element, providing a round, oval, or other shape to this sensing membrane. Researchers will also attempt optimization of sensor function by modifying the membrane thickness. The thickness can be modified in an omnibus manner, or made to vary in different regions of the membrane.

Unfortunately, there are outside physical limitations on how far the pressure sensor designer can change sensor membrane physical parameters to increase the sensitivity. By example, to increase the sensitivity of a pressure sensor, the engineer has the option of providing further thinness to the diaphragm or increasing the diameter of the sensor. These factors can render the membrane unacceptably fragile, and give unwanted larger dimensions to a device which is preferably of a very small size. Also, when the membrane is made larger at the same time membrane thinnest is increased, the burst pressure will be reduced dramatically. All these factors in the pursuit of increased sensitivity can potentially result in a substantially less robust device.

In other areas of micromachining, researches have sought various approaches to improving sensitivity. By example, Pedersen et al have reported the use of amplifier mechanisms for resonant accelerometers (Pedersen et al, J. Micromech. Microeng. 14 (2004) 1281-1293.

It would be a very useful advancement in the art of microsensor technology if the sensitivity of pressure sensing devices could be improved without resort to modifying the dimensions of the sensor membrane component to the extent other strengths of the device are compromised.

SUMMARY OF THE INVENTION

The present invention provides, for the first time, an innovative design which substantially increases the sensitivity of pressure sensors through the inventive used of a beam element in the pressure sensor design. This innovative design approach represents a dramatic advancement over prior art sensitivity methods which have relied on larger, thinner sensor membranes, or bosses. Using the present innovative approach of optimizing compliant force through the use of beam elements in the pressure sensor design provides, for the first time, pressure sensor devices of unprecedented small dimensions and robust character while achieving uniquely fine sensitivity levels.

The inventive devices have special applications in microsensing use, particularly in medical devices. A particularly useful application of the innovative devices is for use in implantable medical devices, most especially cardiac devices. There are many other medical applications for the present inventive pressure sensors, such as detecting pressure in the eye and in spinal fluid.

The invention provides an unprecedented increase in signal output for pressure sensors for a given amount of pressure. In this way, the present invention provides sensing devices which, while constrained in size, are able to provide highly accurate pressure readings at very small changes in pressure. The force amplification of the present invention means and methods increases the capacity for sensitivity of micromachined pressure sensors by about 1-1,000 times preferably about 50-500 times, and most preferably about 150-250 times (see Fig. 5). When combined with other, standard sensitivity design modifications, these sensitivities can reach even higher levels.

The present inventive devices and design methods provide the sensor design engineer a tool by which a which the apparent strain on the sensor membrane can be magnified or amplified. This allows a given membrane deflection due to a pressure difference to be dramatically amplified. With the inventive approach of employing a beam element, the strain-measuring elements will experience a larger strain without distortion. As a result, the electrical sensor signal generated by the sensor will be correspondingly increased.

BRIEF SUMMARY OF THE DRAWINGS

Fig. 1 provides a cross sectional view of a prior art pressure sensing device.

Fig. 2 provides a cross sectional view of the present invention.

Fig. 3 provides a cross sectional view of an alternate embodiment of the present invention.

Fig. 4 is a flow diagram of a simplified inventive fabrication sequence.

Fig. 5 provides a view of an inventive in-plane and mechanical amplification.

Fig. 6 provides a diagrammatic view of one embodiment of the present invention.

Fig. 7 provides a flow diagram of one embodiment of the present inventive fabrication method.

DETAILED DESCRIPTION OF THE INVENTION

The inventive use of beam elements to amplify pressure signals in micromachined sensor devices is a unique and important advancement in the field of micromachining. Following the present inventive teaching, for the first time engineers will be able to substantially increase the sensitivity of pressure sensors through the used of a beam element.

The innovative design approach enjoys a number of advantages over prior art efforts to increase sensitivity of micromachined pressure sensors. Prior art methods relying on larger, thinner sensor membranes, or bosses. By optimizing compliant force through the use of beam elements in the pressure sensor design, the present invention provides for pressure sensor devices of unprecedented small dimensions while providing robust sensitivity.

When designing the geometry of pressure sensor membrane, engineers have previously several parameters to modify in order to achieve the greatest improvement in sensor performance. These can be employed with the advancement of the present invention to further increase the sensitivity of the inventive pressure sensor devices, often with impressive synergistic improvement in sensitivity. By example, design considerations can includes modifications in the proportions of the diameter, providing a round, oval, or other shape to the membrane, and modifying the membrane thickness, which can be made to vary in different regions of the membrane.

However, there are outside limitations on how far the designer can change these physical sensor membrane parameters to increase the sensitivity. By example, to increase the sensitivity of a pressure sensor, the engineer has the option of providing further thinness to the diaphragm. Another option to increase the sensitivity of a pressure sensor would be to increase the diameter of the sensor. Unfortunately, there are physical limitations to both of these optimization approaches. The present invention allows the use of these methods without resorting to extreme measures, with their concomitant risks and disadvantages.

The present invention allows the use of prior art methods without the prior risks incurred in taking them beyond their appropriate physical limits. By example, if the strain for a given amount of pressure exceeds a certain value, typically on the order of 0.01% or 10^{-4}, that the membrane deflection and membrane strain is no longer a linear function of the applied pressure. This nonlinearity that can be accounted for and calculated but complicates the measurement circuit considerably. As a result, previous effective pressure sensor designs required that the device operate within the linear strain versus pressure regime. Prior to the present invention, devices were thus limited by how large and how thin a pressure sensor membrane could be effectively constructed.

Another serious limitation to prior micro pressure sensor designs is the burst pressure that the pressure sensor is required to withstand. In prior art devices, there was typically a safety factor over which the devices operating pressure which will be rated as its burst pressure. As the membrane is designed on a larger scale or where the membrane is made thinner to improve sensitivity, the burst pressure which it can tolerate is reduced dramatically.

The burst pressure would have to be provide that would accommodate the largest possible pressure that the device is likely to experience in the environment in which it would function over a specific required lifecycle. Prior to the present invention, the burst pressure represented a severely limiting factor for the pressure sensor designer.

With the advent of the present invention, these two limitations of linearity and burst pressure no longer represent limits on how large a strain can be generated for a given pressure sensor specification. Because of the much improved sensitivity provided by the inventive beam elements, these standard methods need no longer be taken to inappropriate extremes.

The use of bosses on the pressure sensing membrane represent another prior art method used to maximize the optimizations now achievable by the present inventive beam element. Bosses on the sensor membrane have the effect of concentrating the strain due to pressure applied to the membrane. These currently employed elements have the effect of concentrating the pressure on a few locations on the membrane. The prior art approach of bosses allows the strain-measuring elements to be placed at those locations to optimize sensitivity. The strain for a given applied pressure, accordingly, will be greater if the bosses are placed in the correct location.

The previous use of bosses alone, however, has engineering limitations on how far sensitivity can be optimized. From an electrical measurement standpoint, it is always preferable to have the largest possible strain for a given amount of pressure at the sensor elements. Thus, use of the inventive beam element associated with a boss provides a signal to noise ratio that is as large as possible. Therefore, use of the present invention allows the detect of smaller and smaller differences in pressure. The present invention allows the detection of pressures in the range of about 0.01 to 100,000 mmHg, preferably about 0.1 to 10,000 mmHg, and most preferably 1 to 1000 mmHg.

For a given plate bending, it is possible to calculate the position of the where the center of the curvature. It is also possible to calculate the radius of the curvature of the plate bending. From mechanical texts and from standard engineering analysis, the practioner will be able to locate the strain at any given location within the membrane. This strain is typically equal to the distance of that point from the neutral plane of membrane divided by the radius of curvature.

The beam dimensions in the present invention can range from about 1-1,000μm, preferred about 5-500μm, most preferred about 10-100μm. Additional, in the present invention, multiple inventive beams can be used on a sensor membrane, for instance about 1-100 beams, preferred about 3-50 beams, and most preferred about 4-5 beams.

The designer of the inventive sensors will readily be able to optimize the structure to achieve as small an arc as is practically possible in order to achieve optimal results. By applying bosses to the sensor membrane and changing the membrane dimensions to reduce to radius of curvature, one must consider that a larger strain will result

Fig. 1 provides a cross-sectional view of a segment of a membrane or plate undergoing a deflection. This diagrammatical representation is of a section of pressure-sensing membrane that is experiencing a pressure difference, causing it to bow.

From the discussion above, the formula which will be employed by the practioner in practice of the present invention will effect the prior art device of **Fig. 1** in the following manner. The largest strains will be when z is the largest. However, since the strain element has to be connected to the plate, the greatest possible z occurs at one or the other surface of the plate.

The present invention is particularly suitable for use in the multiplex systems previously developed by some of the present inventors. This system is described in part in currently pending patent applications US Patent Application No.10/764429 entitled "Method and Apparatus for Enhancing Cardiac Pacing", US Patent Application No.10/764127 entitled "Methods and Systems for Measuring Cardiac Parameters", and US Patent Application No. 10/764125 entitled "Method and System for Remote Hemodynamic Monitoring" all filed 01/23/2004, and US Patent Application No.10/734490 entitled "Method and System for Monitoring and Treating Hemodynamic Parameters" filed 12/11/2003. These applications are herein incorporated into the present application by reference in their entirety.

In **Fig. 1** it can be observed from the top surface of the membrane in the example shown is equal to the thickness divided by 2. On the bottom surface, z is equal to the negative of the thickness divided by 2. This puts a limitation on the maximum strain that the sensor element can experience for a given radius of bending.

Fig. 2 show the effect of the inventive design which serves to displace the strain-measuring elements from the membrane as shown. The section **201** of the pressure-sensing diaphragm is shown in this view bending about the center of radius **202**. Offset elements **203** are provided which serve to displace strain-measuring element **204** from the surface of the membrane.

From this depiction, one can observe that z-prime, the distance of strain-measuring element **204** from the neutral axis **205**, is larger than the thickness divided by 2. In fact, as practiced in the present invention, z-prime can be any arbitrary value. As will be understood by the practioner, z-prime may in some cases be limited by some practical considerations such as fabrication techniques.

Fig. 3 provides an example of an alternate embodiment of the present invention. This figures shows offset elements **303** placed on either side of membrane **301**. In this case, because the offset is below the membrane, the z-prime has a negative value. However, this effect does not affect the engineering principle shown in this case.

In a specific embodiment of the present invention, if one were to take a pressure-sensing membrane with typical dimensions of a thickness of 1.5μm and in the prior art, the maximum z would be half of that, or 0.75μm. If these standoff elements were manufactured using an additional 1.5μm, the z-prime would now be 1.5 + 0.75, or 2.25μm. This engineering modification can be accomplished simply and with ease using known fabrication techniques.

Using the above inventive engineering advances, the inventive devices has effectively increase the sensitivity of this prior art pressure sensor design by 3-fold. This provides a simple exemplification of the present invention. However, using the present inventive techniques, amplification values up to 10 or more times can be easily achieved. The, the present invention increases sensitivity by about 1-100 times, preferably about 10-80 times, and most preferably about 20-40 times.

Fig. 4 provides a flow diagram of a simplified fabrication sequence for making one on the present inventive devices. **Fig. 4a** shows a starting with etch-stop layer **401**, and membrane layer **403**. The etch-stop layer **402** is optional. In a typical device, wafer **401** then will be a silicon. Etched-up layer **402** would typically be silicon dioxide, and membrane layer **403** would also typically be silicon.

In **Fig.4b,** offset layer **404** is deposited on top of wafer **401**. In **Fig.4c** offset layer **404** is patterned to make openings or features **405** in offset layer **404**. In **Fig.4b** a strain-sensing material **406** is deposited on top of the offset layer **404**. Strain-sensing material **406** can be a piezo resistive metal such as platinum. Alternatively, strain-sensing material **406** can be an diffused resister into a silicon layer. In **Fig.4e,** a hole is etched through the back of chip **407** to define the sensing membrane. **Figs. 4f, 4g, and 4h** provide planer views of the constructs illustrated in **Figs. 4a, 4b, and 4c,** respectively.

Practical considerations limit the amplification factor using this simpler embodiment of the inventive technique to amplifications of about 10. However, as shown in **Fig. 5**, by extending the inventive concept further, in a more advanced, sophisticated embodiment using in-plane amplification, much larger amplification ratios of the strain are possible. In this case, 100 or several hundred fold increase is available using the present inventive approaches.

Fig. 5 provides a planer view of pressure sensor chip **501**, with a pressure sensor membrane **502**. Amplifying structures **503** and **504** are deposited on the pressure sensor chip surface. **Figs. 5b and 5c** provide cross-sections through this device at different locations marked by the A and A-prime and B and B-prime. As shown in **Figs. 5a, 5b and 5c, the** force-amplifying structures contact the surface of the chip in some locations but do not contact it in others, that is are freestanding above the surface in those locations.

An example of an inventive force amplification structure is provided in additional detail in **Fig 6**. Pad **601** is a location that is attached to one part of the pressure sensor membrane and pad **602** is attached to a second part of the pressure sensor membrane. Using the method of the present invention, these locations will be chosen such that there are locations that experience a large displacement when membrane deflects due to an applied pressure.

Using the example of beam **603**, if location **601** were to move away from location **602** when a positive pressure was applied, beam **603** would get pulled toward pad **601**. This movement would cause a rotation of beam **604** whose one end is anchored to pad **602**. However, a mid-point is attached to beam **603**. That rotation would cause a tension on beam **605** which is then attached to a fixed pad **606**. Fixed pad **606** is attached to some portion of the chip that would not move. This stationary portion of the chip can be, by example, in the periphery of the membrane.

Beam **604** is provide with a segment **607**. Comparing the length of segment **607** to the length of the segment **608**, if these lengths are unequal, it will result in either magnification or a reduction in the amplitude of the relative motion of pad **601** or pad **602**. For instance, if segment **607** were 10μm long and segment **608** were 100μm long, then the end of beam **604** would move 10 times as much as the displacement between pad **601** and pad **602**. This inventive design provides a 10-fold multiplication in the amplitude of the motion. This improvement translates to a 10-fold increase in the strain in beam **605** and a 10-fold increase in the electrical output of the sensor for a given amount of pressure.

As an example, this particular structure is provided with a mirrored structure. As such, pad **606** has its mirror image in pad **609**. This inventive design is a convenient approach to fabrication. It also meets standards of good mechanical practice by providing symmetry. This inventive embodiment has the additional advantage that if, for instance, a strain measuring element **605**, was a piezo resister, the resistance between pad **606** and pad **609** can be measured. By observing the change in the resistance, a measure of the strain is provide in those elements, and hence a measure of the pressure.

The above provides one example of using the inventive lever principle to amplify the force. It will be appreciated by the ordinary skilled artisan that there are many variations on a lever. Equally, how to make levers has been provided in at a previously unavailable level of sophistication by computer methods for determining the optimum shape of levers for micromachine structures. In the prior art, such approaches have been applied to applications like accelerometers and to sort of the micromachine equivalent of a Pantograph. In the latter example, the motivation is to apply a large displacement and cause a very precise motion.

Otherwise, the device is employing a force generator that has only a very small displacement which must be amplified.

Fig. 7 is a flow diagram depiction of one embodiment of the present inventive fabrication method to make the inventive in-plane lever structure. In **Fig.7a**, the fabrication begins with wafer **701**. Wafer **701** is preferably a silicon wafer. Deposited on wafer **701** is etch-stop **702**. Etch-stop **702** can be silicon dioxide. Etch-stop **702** is surfaced with membrane layer **703**. Toping these layers is sacrificial layer **704**. In **Fig. 7b** sacrificial layer **704** is patterned to form a series of features **705**.

The features **705** in the sacrificial layer **704** represent areas where mechanical structure of the inventive device will not touch the underlying membrane. The holes **706** in the sacrificial layer **704** are positioned in places where the lever layer **707** will be attached to the membrane. In **Fig. 7c**, lever layer **707** is deposited. Lever layer **707** preferably is constructed of polycrystalline silicon.

In **Fig. 7d** the intermediate chip is patterned into structures **708**. Structures **708** represent the various lever arms and anchor pads described in the previous figure. In **Fig. 7e**, the sacrificial layer **704** is etched away. If, by example, silicon dioxide is used as the sacrificial layer **704**, it can be etched away with hydrofluoric acid. In whatever manner sacrificial layer **704** is etched, freestanding lever structures **708** are produced.

Fig. 7f describes the last step in this embodiment of the present inventive fabrication method. In the backside of the chip **hole 709** is etched to define the membrane area.

Amplified Compliant Force Pressure Sensors

CLAIMS

Claim 1. An amplified compliant force pressure sensor comprising a pressure sensor membrane with one or more beam features.

Claim 2. The amplified compliant force pressure sensor of Claim 1 wherein said beam features have dimensions from about 1-1,000μm.

Claim 3. The amplified compliant force pressure sensor of Claim 2 wherein said beam features have dimensions from about 5-500μm.

Claim 4. The amplified compliant force pressure sensor of Claim 3 wherein said beam features have dimensions from about 10-100μm.

Claim 5. An amplified compliant force pressure sensor comprising a pressure sensor membrane with multiple inventive beam features.

Claim 6. The amplified compliant force pressure sensor of Claim 5, wherein said pressure sensor membrane is provided with about 1-100 beams.

Claim 7. The amplified compliant force pressure sensor of Claim 6, wherein said pressure sensor membrane is provided with about 3-50 beams.

Claim 8. The amplified compliant force pressure sensor of Claim 7, wherein said pressure sensor membrane is provided with and most preferred about 4-5 beams.

Claim 9. A micromachined pressure sensor which can detect pressures in the range of about 0.01 to 100,000 mmHg.

Claim 10. The micromachined pressure sensor of Claim 10 which can detect pressures in the range of about 0.1 to 10,000 mmHg.

Claim 11. The micromachined pressure sensor of Claim 10 which can detect pressures in the range of about 1 to 1000 mmHg.

Claim 12. A amplified compliant force pressure sensor with an increase in sensitivity over planer micromachined pressure sensors in of about 1-100 times.

Claim 13. The amplified compliant force pressure sensor of Claim 12, where the increase in sensitivity is about 10-80 times.

Claim 14. The amplified compliant force pressure sensor of Claim 13, where the increase in sensitivity about 20-40 times.

1

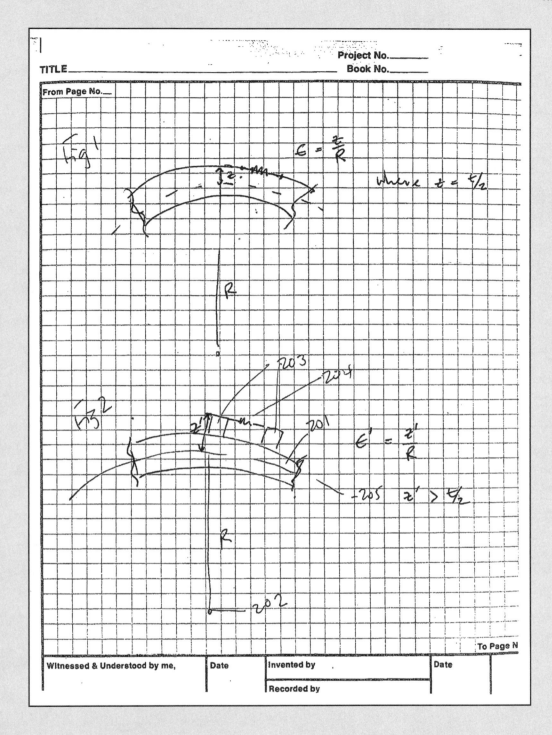

Stage 4 Concept Selection

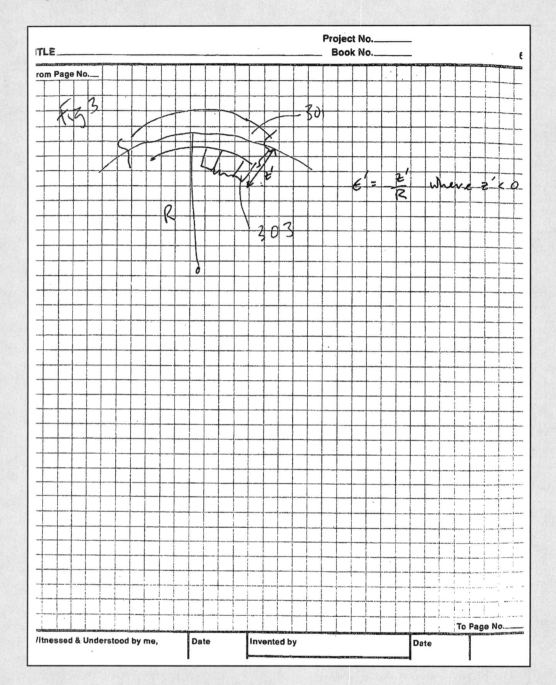

$\epsilon' = \dfrac{z'}{R}$ where $z' < 0$

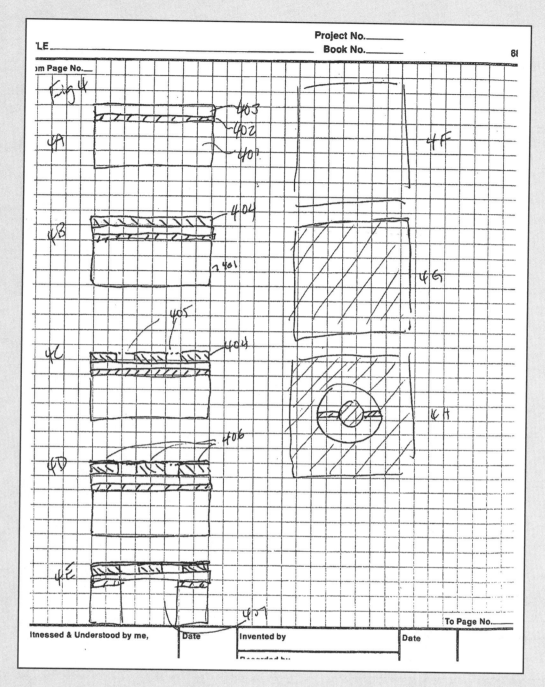

Stage 4 Concept Selection

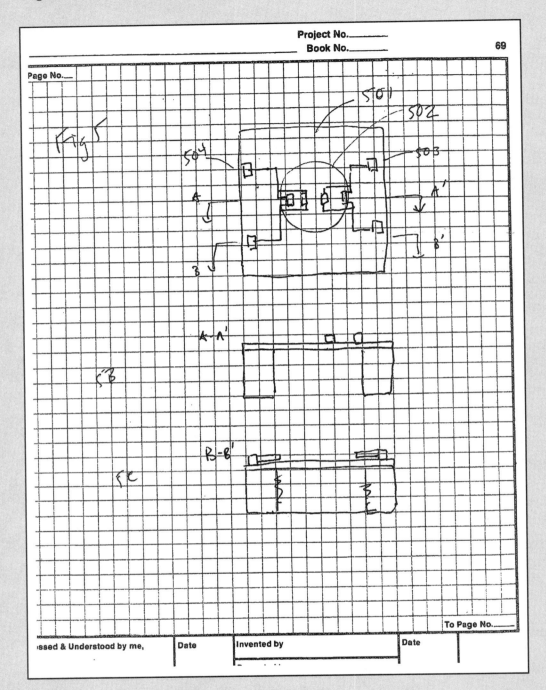

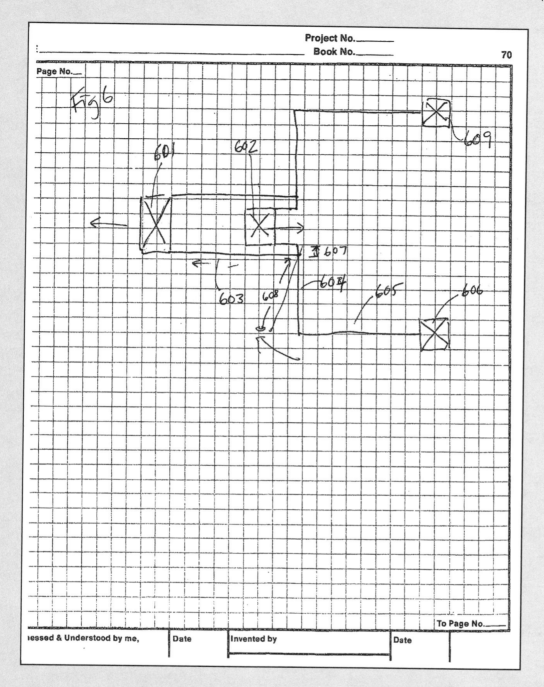

Stage 4 Concept Selection

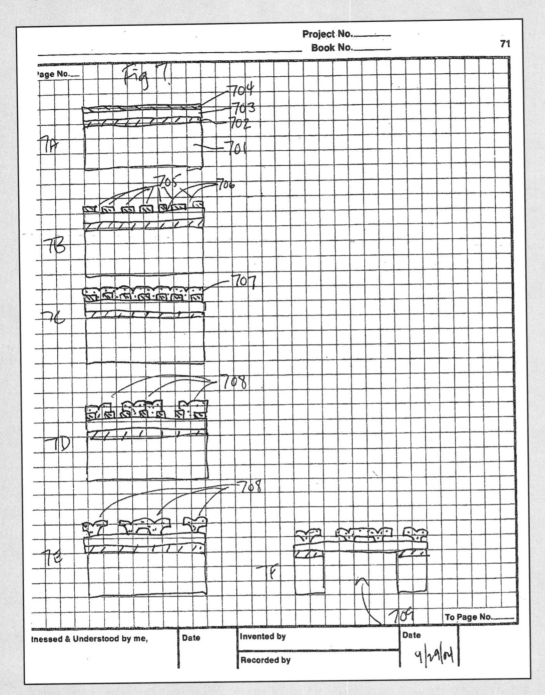

256

Appendix 4.1.2

Sample utility patent[38]

United States Patent [19]
Yock

[11] Patent Number: 5,061,273
[45] Date of Patent: Oct. 29, 1991

[54] ANGIOPLASTY APPARATUS FACILITATING RAPID EXCHANGES

[76] Inventor: Paul G. Yock, 1216 San Mateo Dr., Menlo Park, Calif. 94025

[21] Appl. No.: 548,200

[22] Filed: Jul. 5, 1990

Related U.S. Application Data

[63] Continuation of Ser. No. 361,676, Jun. 1, 1989, abandoned, which is a continuation of Ser. No. 117,357, Oct. 27, 1987, abandoned, which is a continuation of Ser. No. 852,197, Apr. 15, 1986, abandoned.

[51] Int. Cl.⁵ A61M 25/00
[52] U.S. Cl. 606/194; 604/96
[58] Field of Search 604/96, 101, 102; 606/192, 193, 194

[56] **References Cited**
U.S. PATENT DOCUMENTS

2,043,083	6/1936	Wappler	128/303.11
2,687,131	8/1954	Raiche	128/349
2,883,986	4/1959	de Luca et al.	128/351
2,936,760	5/1960	Gants	128/349
3,731,962	5/1973	Goodyear	128/351
3,769,981	11/1973	McWhorter	128/348
3,882,852	5/1975	Sinnreich	128/4
4,195,637	4/1980	Gruntzig et al.	128/348.1 X
4,198,981	4/1980	Sinnreich	128/344
4,289,128	9/1981	Rusch	128/207
4,299,226	11/1981	Banka	128/344
4,367,747	1/1983	Witzel	128/344
4,468,224	8/1984	Enzmann et al.	604/247
4,545,390	10/1985	Leary	604/95
4,569,347	2/1986	Frisbie	128/344
4,610,662	9/1986	Weikl et al.	128/348.1 X
4,616,653	10/1986	Samson et al.	128/344
4,619,263	10/1986	Frisbie et al.	128/344
4,652,258	3/1987	Drach	604/53

FOREIGN PATENT DOCUMENTS

591963 4/1925 France.
627828 10/1978 U.S.S.R..

OTHER PUBLICATIONS

A New PTCA System with Improved Steerability, Contrast Medium Application and Exchangeable Intracoronary Catheters, Tassilo Bonzel, PTCA Proc. Abstract, Course 3, Center for Cardiology, University Hospital, Geneva, Switzerland, Mar. 24–26, 1986.
Nordenstrom, ACTA Radiology, vol. 57, Nov. 1962, pp. 411–416.
Nordenstrom, Radiology, vol. 85, pp. 256–259 (1965).
Diseases of the Nose and Throat, at pp. 776–794–797 (S. Thomson) (6th Ed., 1955).
Achalasia of the Esophagus, at pp. 122–147 (F. Ellis, Jr. et al.) (1969) (vol. IX in the Series Major Problems in Clinical Surgery, J. Dunphy, M.D., Ed.)

Primary Examiner—Michael H. Thaler
Attorney, Agent, or Firm—Fulwider, Patton, Lee & Utecht

[57] **ABSTRACT**

Apparatus for introduction into the vessel of a patient comprising a guiding catheter adapted to be inserted into the vessel of the patient and a device adapted to be inserted into the guiding catheter. The device includes a flexible elongate member and a sleeve carried by the flexible elongate member near the distal extremity thereof and extending from a region near the distal extremity to a region spaced from the distal extremity of the flexible elongate element. The device also includes a guide wire adapted to extend through the sleeve so that the guide wire extends rearwardly of the sleeve extending alongside of and exteriorly of the flexible elongate element into a region near the proximal extremity of the flexible elongate element.

6 Claims, 3 Drawing Sheets

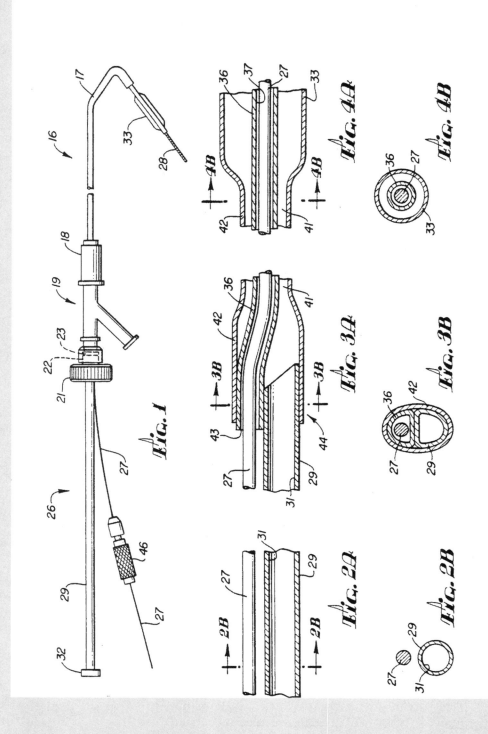

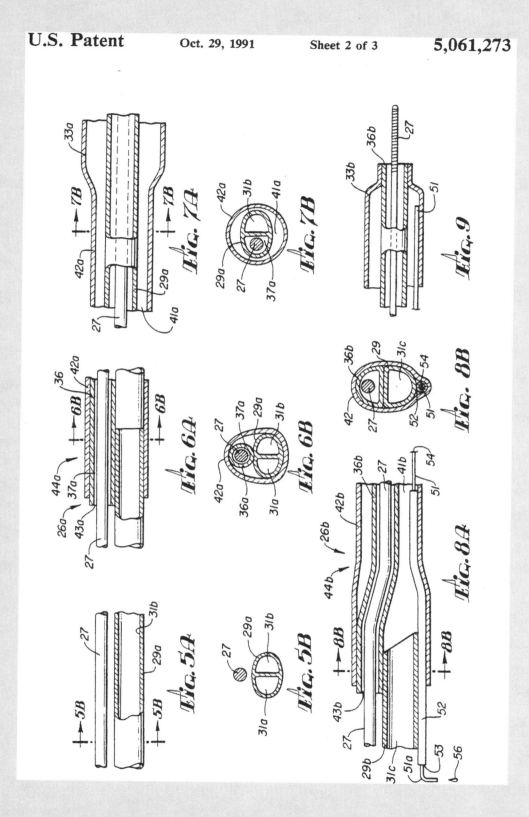

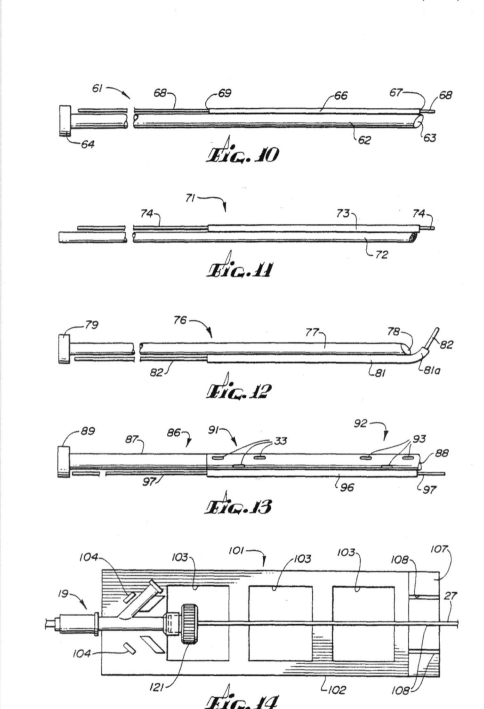

5,061,273

ANGIOPLASTY APPARATUS FACILITATING RAPID EXCHANGES

This application is a continuation of application Ser. No. 361,676, filed June 1, 1989, now abandoned, which was a continuation of Ser. No. 117,357, filed Oct. 27, 1987, now abandoned, which was a continuation of Ser. No. 852,197 filed Apr. 15, 1986, now abandoned.

This invention relates to angioplasty apparatus facilitating rapid exchanges and a method for making rapid exchanges of angioplasty devices.

At the present time in practicing angioplasty, it is often necessary to exchange one dilatation catheter for another. In doing so, it has been necessary to utilize long exchange wires having a length of approximately 300 centimeters which typically requires two operators to perform the procedure. During this procedure, it is necessary that the operators communicate with each other which makes the procedure time consuming. In addition, since the exchange wire is so long it often is awkward to handle and for that reason may come in contact with the floor or become contaminated which necessitates removing the entire apparatus being utilized for the angioplasty procedure. There is therefore a need for a new and improved angioplasty apparatus which overcomes such difficulties.

In general, it is an object of the present invention to provide an angioplasty apparatus and a method which facilitates rapid exchanges of various types of devices.

Another object of the invention is to provide an angioplasty apparatus and method of the above character which greatly facilitates exchanges of dilatation catheters.

Another object of the invention is to provide an angioplasty apparatus and method of the above character which can be utilized for the positioning of flexible elongate members.

Another object of the invention is to provide an angioplasty apparatus and method of the above character which can be utilized with various types of devices utilizing flexible elongate members.

Another object of the invention is to provide an angioplasty apparatus and method in which dye injection and pressure measurements can be made.

Additional objects and features of the invention will appear from the following description in which the preferred embodiments are set forth in conjunction with the accompanying drawings.

FIG. 1 is a side elevational view of an angioplasty apparatus incorporating the present invention.

FIGS. 2A, 3A and 4A are partial cross sectional views of the shaft, transition and balloon regions of the balloon dilatation catheter utilized in the embodidment of the invention shown in FIG. 1.

FIGS. 2B, 3B and 4B are cross sectional views taken along the lines 2B—2B, 3B—3B and 4B—4B of FIGS. 2A, 3A and 4A respectively.

FIGS. 5A, 6A and 7A are cross sectional views corresponding to FIGS. 2A, 3A and 4A of another embodiment of a balloon dilatation catheter incorporating the present invention.

FIGS. 5B, 6B and 7B are cross sectional views taken along the lines 5B—5B, 6B—6B and 7B—7B of FIGS. 5A, 6A and 7A respectively.

FIGS. 8A and 9 are cross sectional views of the transition and balloon regions of another balloon dilatation catheter incorporating the present invention.

FIG. 8B is a cross sectional view taken along the line 8B—8B of FIG. 8A.

FIG. 10 is a side elevational view of a dedicated dye injection/pressure measurement catheter incorporating the present invention.

FIG. 11 is a side elevational view of a fiber optic cable incorporating the present invention.

FIG. 12 is a side elevational view of a dedicated dye injection/pressure measurement catheter incorporating the present invention and having specific guiding means for facilitating entering acute bends in arterial vessels.

FIG. 13 is a side elevational view of a bail out catheter incorporating the present invention.

FIG. 14 is a plan view of a holder utilized in connection with the present invention.

In general, the angioplasty apparatus of the present invention is designed for introduction into the vessel of a patient. It consists of a guiding catheter which is adapted to be inserted into the vessel of the patient. It also consists of a device which is adapted to be inserted into the guiding catheter. The device includes a flexible elongate member, a sleeve is secured to the flexible elongate member near the distal extremity thereof and extends from the distal extremity into a region spaced from the distal extremity of the flexible elongate member. The device also includes a guide wire which is adapted to extend through the sleeve from the distal extremity of the flexible elongate element, through the sleeve and rearwardly of the sleeve alongside of and exteriorly of the flexible elongate element.

More particularly as shown in FIGS. 1-4, the angioplasty apparatus 16 for facilitating rapid exchanges of dilatation catheters consists of a conventional guiding catheter 17 which is provided with a rotatable hemostatic adapter 18 mounted on a proximal end and a y or two-arm connector or adapter 19 which is mounted on the rotatable adapter 18. The y-connector 19 is provided with a knurled knob 21 which carries a threaded valve member 22 that carries an O-ring 23 which is adapted to be urged into sealing engagement with a balloon dilatation catheter 26 and a guide wire 27 extending through the y-adapter 19 and through the guiding catheter 17 as shown in FIG. 1.

The balloon dilatation catheter 26 is of a single lumen type and is provided with a flexible elongate tubular member 29 which has a lumen 31 extending therethrough. The flexible tubular member 29 can be formed of a suitable material such as plastic. A Luer-type fitting 32 is mounted on the proximal extremity of the flexible tubular member 29 and is adapted to be connected to a syringe or other type of instrument for introducing a radiographic contrast liquid into the flexible tubular member 29. A balloon 33 is mounted on the distal extremity of another flexible tubular member 36 which also is formed of a suitable material such as plastic. The distal extremity of the balloon 33 is bonded to the distal extremity of the flexible tubular member 36 to form an air-tight and liquid-tight seal with respect to the same. The balloon 33 is coaxial with the tubular member 36 or sleeve as shown in FIG. 4B. The flexible tubular member 36 is provided with a guide wire lumen 37 through which the guide wire 27 carrying its flexible tip 28 can extend.

Means is provided for forming a balloon inflation lumen 41 substantially concentric with the flexible tubular member 36 and extends toward the distal extremity of the flexible tubular member 36. As can be seen from FIGS. 3B and 4B, the balloon inflation lumen 41 is formed by a flexible tubular member 42 which can be formed integral with the balloon 33. The flexible tubular member 42 extends into a transition region 44 which overlies the distal extremity of the flexible tubular member 29 so that the lumen 31 therein is in communication with the balloon inflation lumen 41. As can be seen particularly from FIG. 3A, the flexible tubular member 36 makes a transition and extends out of the tubular member 42 and provides an opening 43. The proximal extremity of the tubular member 36 overlies the flexible tubular member 31. The guide wire 27 exits through the opening 43 and extends alongside and exteriorly of the flexible tubular member 29 from the proximal extremity of the flexible tubular member 36 to the proximal extremity of the flexible tubular member 29.

The transition region 44 should be positioned at least approximately 10-15 centimeters from the distal extremity of the balloon dilatation catheter 26. This is important for two reasons. One is that the transition region be kept at a point where when the balloon dilatation catheter 26 is utilized in a procedure, the transition region remains in the guiding catheter 17 and out of the coronary artery. The spacing from the distal extremity of the dilatation catheter for the transition region is also advantageous in that it permits the person performing the procedure to pull the balloon dilatation catheter 26 out of the guiding catheter 17 until the transition region 44 clears the y-connector 19 so that all of the portion of the guide wire 27 which is exterior of the balloon dilatation catheter 26 is proximal of the y-connector. While this is being done, the operator can then utilize the knurled nut 21 to again close the o-ring to form a hemostatic seal between the y-connector and the balloon dilatation catheter to minimize the loss of blood from the patient.

The flexible tubular member 42 can be formed of a suitable material such as a heat shrinkable plastic so that it can be shrunk onto the distal extremity of the flexible tubular member 29 and onto the proximal extremity of the flexible tubular member 36 to form liquid-tight and air-tight seals with respect to the same. From the construction shown it can be seen that the guide wire 27 exits from the balloon dilatation catheter 26 in a region which is relatively close to the distal extremity of the balloon dilatation catheter 26 and extends exteriorly of the balloon dilatation catheter to the proximal extremity of the same. As shown in FIG. 1, the guide wire 27 and the balloon dilatation catheter 26 extend outwardly from the y-connector 19.

A torquer 46 of a conventional construction is secured to the guide wire 27 for rotating the guide wire as hereinafter described.

Operation and use of the angioplasty apparatus shown in FIG. 1 may now be briefly described as follows. The guiding catheter 17 is inserted into the coronary artery in a conventional manner. The balloon dilatation catheter is prepared for insertion into the guiding catheter 17 in a conventional manner. The balloon 33 can be inflated outside the body by the use of a balloon flushing tube of the type described in U.S. Pat. No. 4,323,071 and inflated by introducing a radiopaque liquid through the fitting 32 into the lumen 31 and through the lumen 41 into the balloon 33 to flush all of the air in the balloon 33 through the balloon flushing tube to fully inflate the balloon. After the balloon 33 has been inflated, the balloon can be deflated by removing the radiopaque liquid from the balloon.

The guide wire 27 is then introduced into the balloon dilatation catheter 26 by a back loading technique. Without the torquer 46 on the guide wire, the proximal extremity of the guide wire 27 is inserted backwardly through the tip of the balloon dilatation catheter through the guide wire lumen 37. The guide wire is advanced rearwardly by holding the distal extremity of the balloon dilatation catheter in one hand and advancing the guide wire 27 rearwardly with the other hand until the guide wire 27 exits through the opening 43 at the transition region 44 of the dilatation catheter. As soon as the guide wire has cleared the opening 43, the guide wire can be grasped by the hand and pulled rearwardly paralleling the balloon dilatation catheter 26 until its proximal extremity is near the proximal extremity of the dilatation catheter and so that the distal extremity of the guide wire 27 with its flexible or floppy tip 28 protrudes at least partially from the distal extremity of the balloon dilatation catheter.

At this point in time, the O-ring 23 in the y-connector 19 is opened by operation of the knurled knob 21. The distal extremity of the balloon dilatation catheter 26 having the flexible tip protruding therefrom is then introduced to the y-connector past the opened o-ring 23 and slid down the guiding catheter 17. The balloon dilatation catheter 26 and the guide wire 27 are grasped between the fingers of a hand and are advanced parallel into the guiding catheter 17. This procedure is continued until a substantial portion of the balloon dilatation catheter is disposed in the guiding catheter 17.

The torquer 46 now can be attached to the guide wire 27 near the proximal extremity of the same. The guide wire 27 is then advanced ahead of the balloon dilatation catheter until it enters the arterial vessel of the patient. The balloon dilatation catheter 26 is held stable by the fingers of the hand while the guide wire 27 is being advanced. The positioning of the guide wire 27 in the desired arterial vessel can be observed under a fluoroscope by using x-ray techniques well known to those skilled in the art. As is well known to those skilled in the art, the torquer 46 can be utilized for rotating the guide wire 27 to facilitate positioning of the flexible tip 28 in the desired arterial vessel so that the distal extremity of the guide wire can be advanced into the stenosis which it is desired to open or enlarge.

As soon as the guide wire 27 is in the desired location, it can be held stationary by two fingers of the hand and at this point in time, the balloon dilatation catheter 26 is advanced over the guide wire until the deflated balloon 33 is across the desired lesion or stenosis. If any difficulty is encountered by the person conducting the procedure in introducing the balloon dilatation catheter so that the balloon 33 resists crossing the lesion or stenosis, the guide wire 27 can be retracted slightly. The person then can observe under the fluoroscope to see that the tip 28 of the guide wire is wiggling in the blood stream indicating that it is free to move in the blood stream. Then the person can grasp both the guide wire and the dilatation catheter in one hand and advance them as a unit so that they can cross the stenosis as a unit. It has been found by utilizing such a procedure, greater pushability can be obtained in advancing the balloon dilatation catheter across the stenosis. In other words, more force can be applied to the balloon to cause it to cross the stenosis or lesion in case the opening therein is very small.

After the balloon 33 has crossed the stenosis or lesion, the balloon 33 can be inflated in a conventional manner by introducing a radiopaque contrast liquid through the lumen 31. After the inflation has occurred and the desired operation has been performed by enlarging the opening in the stenosis, the balloon dilatation catheter 26 can be removed very rapidly by the person performing the procedure by grasping the guide wire 27 by two fingers immediately proximal of the y-connector 19 after the torquer 46 has been removed. The balloon dilatation catheter 26 can be removed in several seconds in comparison with the much longer time required for removing the balloon dilatation catheter utilizing prior art exchange wire procedures. As soon as the balloon dilatation catheter 26 has been removed from the guiding catheter 17, another injection of radiographic contrast liquid can be introduced through the guiding catheter 17 to observe whether or not the balloon dilatation procedure which has been performed on the lesion or stenosis has in fact opened the lesion or stenosis to the satisfaction of the person performing the procedure.

If it is ascertained by the person performing the procedure that additional dilation of the stenosis is desired and that a larger balloon should be inserted into the stenosis, this can be accomplished very rapidly by selecting the desired size of balloon dilatation catheter.

As the balloon dilatation catheter 26 is being retracted out of the guiding catheter 17 and as soon as the transition region 44 has cleared the y-adapter 19, the o-ring 23 can be tightened down to form a seal over the balloon dilatation catheter to minimize the loss of blood of the patient. Thereafter, if desired, the remainder of the balloon dilatation catheter 26 can be removed from the guiding catheter 17 until the proximal extremity of the guide wire passes through the opening 43 and passes through the end of the balloon dilatation catheter 26. As soon as this has been accomplished, a new balloon dilatation catheter can be loaded onto the guide wire in a rearward direction by introducing the proximal extremity of the guide wire 27 into the tip of the balloon dilatation catheter. As this is being done, the index finger of the hand performing the procedure can be utilized for opening the o-ring by adjusting the knurled knob 21. The guide wire 27 is grasped by the fingers of the hand and the balloon dilatation catheter 26 can be advanced rapidly over the guide wire into the guiding catheter 17 and advanced across the lesion in a manner hereinbefore described with respect to the smaller balloon dilatation catheter which had been utilized. The balloon of the new dilatation catheter can be inflated in the same manner as hereinbefore described. If necessary even another exchange procedure can be readily accomplished in the same manner as hereinbefore described utilizing a still larger balloon dilatation catheter if that turns out to be necessary.

It has been found that an exchange utilizing the present angioplasty apparatus can be performed in less than 10 to 15 seconds whereas in the past utilizing a prior art guide wire exchange procedure required an average of approximately two minutes.

After the desired amount of dilation of the stenosis or lesion has been accomplished, the balloon dilatation catheter 26 can be removed and thereafter, the guiding catheter 17 can be removed.

Another embodiment of an angioplasty apparatus incorporating the present invention is shown in FIGS. 5A and 5B, 6A and 6B and 7A and 7B in which an additional dye/pressure lumen has been incorporated into the apparatus in order to enable an injection of a distal dye through the balloon dilatation catheter and also to enable the measurement of pressures at the tip of the balloon dilatation catheter. The construction which is utilized is very similar to that shown in the balloon dilatation catheter 26 shown in FIG. 1. The corresponding parts of the balloon dilatation catheter 26a shown in FIGS. 5-7 is very similar to that hereinbefore described and for that reason the corresponding parts have been given the same corresponding numbers with letters being added to the numerals where changes are present in the parts or components. Thus the tubular member 29a, rather than having a single lumen 31 is provided with dual lumens 31a and 31b disposed side by side in the shaft region of the balloon dilatation catheter as shown in FIGS. 5A and 5B. In the transition region 44a, the two lumens 31a and 31b are still disposed side by side with the lumen 37a for the guide wire being disposed above the lumens 31a and 31b. In the balloon region, the lumen 31a has been terminated and extends into the balloon lumen 41a. At the transition region 44a, the guide wire lumen 37a inclines downwardly and sidewise and adjoins the lumen 31b through the distal extremity of the balloon dilatation catheter 26a. The lumen 31b extends to the distal extremity of the balloon dilatation catheter.

The balloon dilatation catheter which is shown in FIGS. 5-7 can be utilized in the same manner as the balloon dilatation catheter shown in FIG. 1. It can be seen that the guide wire 27 extends out of the opening 43a in the transition region 44c and parallels the balloon catheter to its proximal extremity. A balloon dilatation catheter of the type shown in FIGS. 5-7 can be utilized initially in an angioplasty procedure. However, it should be appreciated that if a very small opening is present in the stenosis or lesion, it may be desirable to utilize a balloon dilatation catheter of the type shown in FIG. 1 first because it can be constructed with a smaller diameter than a balloon dilatation catheter of the type shown in FIGS. 5-7 because of the additional lumen which is provided for dye injection and pressure measurements. After a smaller balloon dilatation catheter has been utilized, a balloon dilatation catheter of the type shown in FIGS. 5-7 can be used utilizing the exchange procedure hereinbefore described to make dye injection and/or pressure measurements through the use of the additional lumen 31b. It is particularly desirable to make such a pressure measurement before conclusion of the angioplasty procedure to be sure that the proper dilation of the lesion or stenosis has occurred and that there is adequate blood flow through the lesion or stenosis.

Still another embodiment of the angioplasty apparatus incorporating the present invention is shown in FIGS. 8A and 8B and shows the transition region of a balloon dilatation catheter 26b which incorporates a vent tube 51 which is utilized for venting air from the balloon during inflation of the balloon and before insertion into the patient with radiopaque liquid to ensure that all the air is exhausted from the balloon. As shown in the transition region 44b in FIG. 8A, the guide wire 27 extends through an opening 43b provided in the transition region and extends through a flexible tubular member 36b out the end of the balloon dilatation catheter as shown in FIG. 9. A balloon filling lumen 31c is provided by the flexible tubular member 29b and terminates in the transition region 44b where it opens into the balloon filling lumen 41b that opens into the interior of the balloon 33b. A relatively short sleeve 52 formed of a suitable material such as plastic is also provided in the transition region 44b and as shown in FIG. 8A underlies the flexible tubular member 29b and extends from a region forward of the flexible tubular member 42b and terminates distally within the balloon inflation lumen 41 as shown in FIG. 8A.

The sleeve 52 is provided with a lumen 53 through which the vent tube 51 extends. The vent tube 51 can be formed of a suitable material such as metal and is also provided with a lumen 54 of a size so that gas can escape therethrough. The proximal extremity of the vent tube 51 is provided with a portion 51a which is bent at right angles to the main portion of the vent tube 51 to ensure that the vent tube will be removed from the balloon dilatation catheter 26b prior to insertion into the guiding catheter 17. As shown in FIG. 9, the vent tube 51 extends into the balloon 33 into a region near the distal extremity of the same.

Operation of the balloon dilatation catheter 26b shown in FIGS. 8A, 8B and 9 may now be briefly described as follows. With the vent tube 51 in place in the balloon dilatation catheter, radiopaque contrast liquid is introduced through the balloon inflation lumen 31 and through the balloon inflation lumen 41b to introduce the liquid into the balloon. As the liquid is introduced into the balloon, any air in the balloon is discharged through the vent tube 51. Pressure is maintained on the radiopaque contrast liquid introduced into the balloon until droplets 56 of the liquid exit from the proximal extremity of the vent tube 51 which serves to indicate that the balloon has been completely filled with the radiopaque contrast liquid and that all of the air therein has been exhausted therefrom. As soon as this occurs, the vent tube 51 can be withdrawn completely from the balloon dilatation catheter. The sleeve 52 which carries the vent tube collapses upon withdrawal of the vent tube and will remain collapsed to provide a valve to prevent the escape of any additional radiopaque contrast liquid from the balloon 33b. The sleeve 52 remains collapsed because when a high pressure is being introduced through the balloon inflation lumen 31c, the flexible tubular member 29b will force collapsing of the sleeve 52. Alternatively, when a negative pressure is being applied to the balloon 33b as, for example, when the balloon is being deflated, the positive atmospheric pressure on the exterior of the flexible tubular member 42b will again cause collapsing of the sleeve 52. Thus in effect there is provided a double valve system in which positive pressures on the interior will collapse the sleeve and when there is negative internal pressure the positive exterior atmospheric pressure will collapse the sleeve.

In all other respects, the balloon dilatation catheter 26b can be utilized in the same manner as the balloon dilatation catheters hereinbefore described in connection with exchanges on the guide wire 27.

Still another embodiment of an angioplasty apparatus incorporating the present invention is shown in FIG. 10 in which there is disclosed a dedicated pressure/dye catheter 61. The pressure/dye catheter 61 consists of an elongate flexible tubular member 62 formed of a suitable material such as plastic which is provided with a pressure dye lumen 63 extending therethrough. The proximal extremity of the tubular member 62 is provided with a Luer-type fitting 64 to which devices having Luer-type fittings can be attached. A sleeve 66 formed of a suitable material such as plastic is secured to the exterior of the flexible tubular member 62 by suitable means such as an adhesive. It is provided with a guide wire lumen 67 extending therethrough. It should be appreciated that the sleeve 66 can be formed integral with the flexible tubular member 62 if desired. The sleeve 66 extends for a distance of at least 10 to 15 centimeters from the distal extremity of the catheter 61 so that the transition region where it terminates at its proximal extremity is be within the guiding catheter 17 so that the transition region does not enter into the arterial vessel of the patient. A guide wire 68 is provided which extends through the guide wire lumen 67. The guide wire 68 can be of the same type as the guide wire 27. It is inserted into the sleeve 66 by taking the proximal extremity of the guide wire which is relatively stiff and inserting it into the distal extremity of the sleeve and then pushing it backwardly or rearwardly through the sleeve until it clears the opening 69 at the proximal extremity of the sleeve 66. The guide wire 68 is then pulled so that it extends in a direction parallel to the flexible tubular member 62 into a region near the proximal extremity of the tubular member 62.

It can be readily seen from the foregoing description that the pressure/dye catheter 61 can be readily introduced into a guiding catheter 17 and that the distal extremity of the pressure/dye catheter can be positioned in a desired location in the arterial vessel by utilizing the guide wire 68 to position the same. It also should be appreciated that a torquer of the type hereinbefore described such as the torquer 46 can be utilized on the proximal extremity of the guide wire 68 to cause rotational movement of the guide wire to facilitate positioning of the guide wire in the desired arterial vessel and to thereafter have the tubular member 62 follow the same. The desired picture and/or dye measurements can then be made by utilizing the lumen 63 provided in the tubular member 62. As can be seen from FIG. 10 the distal extremity of the tubular member 62 can be slanted and rounded as shown to facilitate entry into the stenosis in the arterial vessel. This is desirable because of the eccentricity created by the addition of the sleeve 66.

Another embodiment of an angioplasty apparatus is shown in FIG. 11 and takes the form of a fiber optic device 71. An encased fiber optic bundle 72 which is generally circular in cross section is provided. A sleeve 73 of the type hereinbefore described formed of a suitable material such as plastic is secured to the distal extremity of the fiber optic bundle 72 which is adapted to receive a guide wire 74. As in the previous embodiments, the sleeve 73 extends from the distal extremity for a distance of approximately 10 to 15 centimeters after which the guide wire exits from the sleeve and extends alongside and exteriorally of the fiber optic bundle 72 for substantially the entire length of the fiber optic bundle. As with the previous devices, the guide wire 74 is threaded into the sleeve by taking the proximal extremity or stiff end of the guide wire and inserting it at the distal extremity of the sleeve 73 and pushing it from the rear towards the forward extremity of the sleeve. The fiber optic device 71 can then be inserted into a guiding catheter 17 and advanced to the desired location through the use of the guide wire. The fiber optic bundle then can be utilized for angioscopy for looking directly at the blood vessel or alternatively, for delivering energy to plaque in the blood vessel to perform laser angioplasty. It should be appreciated that steerable systems can be utilized for directing the distal extremity of the fiber optic bundle if that is desired.

It should be appreciated that the concept of using a relatively short sleeve extending from the distal extremity of the device to a region approximately 10 to 15 centimeters to the rear and then having the guide wire extend externally of the device is applicable for a number of medical devices as well as other applications. For example, ultrasonic catheters for imaging ultrasound and for measurement of Doppler velocity can be utilized to provide various types of dedicated devices having the guide sleeve with the guide wire therein for facilitating positioning of the same in arterial vessels. The apparatus of the present invention is particularly useful in devices where multiple re-entries are required in order to complete the procedure.

In FIG. 12, there is disclosed another embodiment of an angioplasty apparatus to provide a pressure dye catheter **76** having additional steering capabilities. It consists of a flexible tubular member **77** formed of a suitable material such as plastic which is provided with a lumen **78** extending through a slanted and curved end. A Luer-type fitting **79** is provided on the proximal extremity. A sleeve **81** formed of a suitable material such as plastic is secured to the distal extremity of the flexible elongate member **77**. The sleeve is provided with a curved portion **81a** which extends slightly beyond the distal extremity of the flexible elongate member **77** and curves over the end of the flexible elongate member **77**. The guide wire **82** extends through the sleeve **81** as shown. The catheter shown in FIG. 12 can be utilized in situations where there is an acute bend in the arterial vessel. By using the catheter shown in FIG. 11, the guide wire can be directed into the acute bend by rotation of the catheter **76** to help direct the guide wire into the acute bend. After the acute bend has been negotiated by the guide wire, the distal extremity of the catheter can follow the guide wire and negotiate the acute bend. The desired pressure and/or dye measurements can then be made. If by chance a guide wire should enter the wrong vessel, the guide wire can be retracted into the sleeve and then the catheter itself can be reoriented to have the distal extremity of the sleeve **81** directed into the proper region so that the guide wire will enter the proper arterial vessel. The catheter **76** shown in FIG. 12 can be introduced into the guiding catheter **17** in the same manner as the other catheters hereinbefore described.

Still another embodiment of the angioplasty apparatus of the present invention is shown in FIG. 13 in the form of a bailout catheter **86**. The bailout catheter **86** consists of a flexible tubular member **87** formed of a suitable material such as plastic which is provided with a lumen **88** extending therethrough. A Luer-type fitting **89** is secured to the proximal extremity of the tubular member **87**. The distal portion of the tubular member **87** is provided with two sets **91** and **92** of holes **93** which are spaced circumferentially and apart longitudinally of the tubular member. A sleeve **96** formed of a suitable material such as plastic is secured to the distal extremity of the tubular member **87** and extends from the distal extremity of the tubular member **87** into a region 10 to 15 centimeters from the distal extremity and is adapted to receive a guide wire **97** which extends through the same. The guide wire **97** is inserted into the sleeve by taking the proximal extremity of the guide wire and inserting it into the distal extremity of the sleeve and pushing it rearwardly into the sleeve until it exits from the sleeve. The guide wire **97** is then pulled in a direction generally parallel to the flexible tubular member **87** until it is adjacent the fitting **89**.

The bailout catheter **86** is utilized in situations where an obstruction has occurred in a blood vessel and stops the flow of blood. In order to reestablish the flow of blood, the bailout device is inserted into the guiding catheter **17**. If a guide wire is already in place, the bailout device can be placed on the guide wire by introducing the proximal extremity of the guide wire into the sleeve and then pushing the bailout catheter on the guide wire into the guiding catheter **17** until it passes through the obstruction in the arterial vessel. The distal extremity of the bailout device is so positioned so that the obstruction is disposed between the two sets of holes **91** and **92**. When the bailout catheter is positioned, blood can still flow through the holes **93** past the obstruction which is held out of the way by the bailout catheter.

Thus it can be seen that the same principle utilizing a guide tube and an external guide wire passing through the guide tube can be utilized for positioning the bailout device. As pointed out previously, the bailout device can be utilized for positioning other types of devices in arterial vessel, as for example, atherectomy devices particularly where multiple re-entries or reintroductions of the devices are required.

In FIG. 14 there is disclosed additional angioplasty apparatus in which a holder **101** is provided which serves as a support structure for a y-type connector **19** of the type hereinbefore described in conjunction with the angioplasty apparatus shown in FIG. 1. The holder **101** consists of a rectangular member **102** which is generally planar. The member **102** can be formed of a suitable material such as plastic and is provided with a plurality of rectangular openings **103** extending longitudinally of the same to lighten the same. Posts **104** are provided on the forward extremity of the member **102** and are adapted to receive the y-type connector **19** and to hold it in place on the member **102**. When positioned in the posts **104**, the knurled knob **21** extends into one of the openings **103** so that it can be readily operated. A block **107** is carried by the other end of the member **102** and, if desired, can be formed intergal therewith. The block is provided with a plurality of spaced apart slots **108** which are adapted to frictionally engage and receive the guide wire **27**. The friction block **107** should be positioned a suitable distance as, for example, 15 to 20 centimeters from the o-ring carried by the y-connector **19**.

Use of the holder **101** shown in FIG. 14 may now be briefly described as follows. The holder **101** can be placed on the operating table near the region where the guiding catheter **17** has been inserted into the patient, as for example, in a femoral artery in the leg of a patient. After the guide wire has been inserted into the guiding catheter, the proximal end or, in other words, the stiff end of the guide wire can be placed in the friction clamp **108**. When it is desired to utilize a dilatation catheter, the end of the guide wire which has been positioned in the clamp can be lifted out of the slot **108** and inserted into the sleeve carried by the distal extremity of the dilatation catheter by taking the proximal end and advancing it from the tip rearwardly through the sleeve. As soon as the guide wire has been introduced through the sleeve, the proximal extremity of the guide wire can be repositioned in the slot **108**. Thereafter, the dilatation catheter can be advanced independently without the operator needing to pay any attention to the guide wire which is held in the desired position by the holder **101**. Similarly, the holder can be utilized to keep the guide wire in place while the dilatation cathether is being briskly withdrawn.

More than one of the slots **108** has been provided in the holder **101** in order to make it possible to accommodate two wire or two balloon dilatation catheters in which one of the other slots **108** can be utilized for accommodating the additional guide wire. This prevents the guide wires from becoming entangled with each other.

It is apparent from the foregoing that there has been provided an angioplasty apparatus which greatly facilitates the exchange of devices which utilize flexible elongate elements as a part thereof. Rapid exchanges are possible with only one person being necessary to make the exchanges. The need for long exchange wires has been eliminated. One device can be readily substituted for another utilizing the same guide wire which has already been positioned. It can be seen from the foregoing that a relatively simple and expedient solution has been provided with eliminates the need for long exchange wires and the danger of those exchange wires becoming contaminated.

Although the present invention has been described principally in conjunction with catheters having coaxial lumens, it should be appreciated that the invention is as applicable, if not more applicable, to catheters having side-by-side lumens.

What is claimed is:

1. An elongated dilatation catheter which is adapted to be inserted over a guidewire through a patient's blood vessel to perform angioplasty procedures therein, comprising:
   a) an elongated catheter body having an inner inflation lumen extending therethrough, a proximal portion which is adapted to extend out of the patient during angioplasty procedures and which has means to direct inflation fluid into the inner inflation lumen and a distal portion which is adapted to be disposed entirely within the patient's vascular system during angioplasty procedures and which has a transition region with a proximal port therein opening to the exterior of the catheter and located at least 10 cm proximally from a distal port in the distal tip of the catheter body and a substantial distance distally from the proximal portion of the catheter body which extends out of the patient during an angioplasty procedure;
   b) an inflatable balloon on the distal portion of the catheter body which is distal to the proximal port and which has an interior in fluid communication with the inner inflation lumen therein; and
   c) an open-ended tubular member fixed with respect to the catheter body and having an inner lumen which is substantially shorter than the catheter body, which is disposed essentially in its entirety within the distal portion of the catheter body and which extends proximally within the catheter body from the distal port at the distal tip of the catheter body, through the interior of the balloon, to the proximal port in the transition region, said inner lumen of the tubular member adapted to slidably receive a guidewire therein so that a distal portion of the guidewire extends distally out of the distal port and a proximal portion thereof extends proximally out of the proximal port to the exterior of the catheter body and generally parallel and exteriorly to the catheter body over most of the length of the guidewire and out of the patient.

2. The catheter of claim **1** including an additional sleeve disposed within the tubular member which extends from the interior of the balloon to the proximal extremity of the tubular member and which has a vent tube removably disposed therein.

3. In an elongated dilatation catheter assembly for performing angioplasty procedures within a patient's arterial system, which assembly includes a guiding catheter adapted to be positioned within a patient's arterial system with the proximal end thereof extending out of the patient during the angioplasty procedure, a balloon dilatation catheter adapted to be positioned within an inner lumen of the guiding catheter with the proximal end thereof extending out of the proximal end of the guiding catheter during the angioplasty procedure and a guidewire adapted to be positioned within an inner lumen of the dilatation catheter during the angioplasty procedure, an improvement in the dilatation catheter of the assembly, comprising:
   a) an elongated catheter body having an inflation lumen which extends therein, a proximal portion which is adapted to extend out of the proximal end of the guiding catheter during angioplasty procedures and which has means to direct inflation fluid into the inflation lumen of the catheter body, and a distal portion which is adapted to extend partially out of the distal end of the guiding catheter into the patient's artery during angioplasty procedures and which has a distal port in the distal end thereof, a transition region in the distal portion which is adapted to remain within the inner lumen of the guiding catheter during angioplasty procedures and which has a proximal port therein disposed a substantial distance distally from the proximal portion which extends out the proximal end of the guiding catheter, and at least 10 cm proximally from the distal port;
   b) a relatively short tubular member fixed with respect to the distal portion of the catheter body having an inner lumen which is adapted to slidably receive the guidewire therein and which extends through the interior of the distal portion of the catheter body between the proximal port and the distal port, with a distal portion of the guidewire extending out the distal port and a proximal portion of the guidewire extending out of the proximal port and generally parallel and exteriorly to the catheter body within the inner lumen of the guiding catheter; and
   c) an inflatable balloon located on the distal portion of the catheter body distal to the proximal port having an interior which is in fluid communication with the inflation lumen in the catheter body.

4. A dilatation catheter assembly for performing an angioplasty procedure in a human patient, wherein the assembly is adapted to be disposed during the procedure within an inner lumen of an elongated guiding catheter which has a distal end inserted into an ostium of the patient's coronary artery and a proximal end extending out of the patient, said assembly comprising:
   a) a dilatation catheter which is adapted to extend out of the distal end and the proximal end of the guiding catheter during the angioplasty procedure, said dilatation catheter comprising an elongated catheter body and an inflatable balloon on the distal portion thereof, the catheter body having a first, relatively long inner lumen for directing inflation fluid from the proximal end of the catheter body to the interior of the balloon and a second inner lumen, much shorter than the first inner lumen,

which is defined at least in part by a tubular member fixed with respect to the catheter body and extends through the interior of the balloon between a distal port at the distal end of the catheter body and a proximal port disposed proximally from the balloon, at least 10 cm proximally from the distal port and a substantial distance distally from the portion of the dilatation catheter which extends out of the proximal end of the guiding catheter so that the proximal port is adapted to remain within the guiding catheter during the angioplasty procedure; and

b) an elongated guidewire which is disposed within the second inner lumen of the catheter body and which has a distal portion adapted to extend out of the distal port and a proximal portion adapted to extend out of the proximal port, generally parallel and exteriorly to the portion of the catheter body disposed within the guiding catheter and out the proximal end of the guiding catheter.

5. A dilatation catheter for performing coronary angioplasty procedures which facilitates the rapid exchange thereof with another catheter without the need for exchange wires or guidewire extensions, the catheter comprising:

a) an elongated catheter shaft having proximal and distal ends and a first inner lumen extending therein;

b) an inflatable balloon disposed at the distal end of the catheter shaft having an interior in fluid communication with the first inner lumen extending within the catheter shaft; and

c) a second inner lumen which is defined at least in part by a tubular member extending through the interior of the balloon, which is adapted to slidably receive a guidewire therein, which has a first port in the distal end thereof and which has a second port spaced at least about 10 cm proximally from the distal port and spaced distally a substantial distance from the proximal end of the catheter shaft.

6. A dilatation catheter assembly for performing an angioplasty procedure within a patient's coronary artery, comprising:

a) a guiding catheter having proximal and distal ends which is adapted to be disposed within a patient's cardiovascular system with the distal end seated within an ostium of the patient's coronary artery and the proximal end extending out of the patient;

b) a guidewire with proximal and distal ends which is adapted to be disposed within the guiding catheter with the distal end of the guidewire extending out the distal end of the guiding catheter into the coronary artery of the patient and the proximal end of guidewire extending out the proximal end of the guiding catheter which extends out of the patient;

c) a dilatation having an elongated catheter body with a proximal end adapted to extend out of the patient, a distal end and adapted to be advanced through a stenosis, an inflatable balloon on a distal portion of the catheter body, a first elongated inner lumen within the catheter body which is adapted to direct inflation fluid from the proximal end of the dilatation catheter to the interior of the inflatable balloon and a second much shorter inner lumen which is defined at least in part by a tubular member extending through the interior of the balloon and which is adapted to slidably receive the guidewire, the second inner lumen having a distal port located in the distal end of the catheter body and a proximal port located at least 10 cm from the distal end of the catheter body and a substantial distance from the proximal end of the catheter body.

* * * * *

Stage 4 Concept Selection

UNITED STATES PATENT AND TRADEMARK OFFICE
CERTIFICATE OF CORRECTION

PATENT NO. : 5,061,273
DATED : October 29, 1991
INVENTOR(S) : Paul G. Yock

It is certified that error appears in the above-identified patent and that said Letters Patent is hereby corrected as shown below:

Column 14, line 20, after "dilatation" insert --catheter--.

Signed and Sealed this

Eleventh Day of July, 1995

Attest:

BRUCE LEHMAN

Attesting Officer   *Commissioner of Patents and Trademarks*

Appendix 4.1.3

Sample non-disclosure agreement[39]

CONFIDENTIALITY AND NONDISCLOSURE AGREEMENT

THIS CONFIDENTIALITY AND NONDISCLOSURE AGREEMENT ("Agreement") is entered into as of _____ ("Effective Date") by and between _____ ("Disclosing Party"), as an individual and _____ ("Recipient").

1. <u>Purpose</u>. This Agreement is intended to prevent Recipient from disclosing or using certain confidential information that Disclosing Party may disclose to Recipient in connection with exploring a business opportunity of mutual interest relating to _____, for any purpose other than engaging in discussion related to, and evaluating such business opportunity.

2. <u>"Confidential Information"</u> means any tangible or intangible information and/or materials disclosed to Recipient by Disclosing Party, or by others under the direction of Disclosing Party, either directly or indirectly, in writing, orally, visually, electronically, by inspection of tangible objects, or in any other manner. Confidential Information includes, without limitation, Disclosing Party's non-published patent applications, the terms and conditions of this Agreement (and the existence and nature of any discussions related thereto), as well as information, data, know-how, ideas, concepts, trade secrets, technical knowledge, procedures, techniques, methods, designs, and any other information that Disclosing Party treats as confidential. Confidential Information may also include information disclosed to Disclosing Party by third parties. Tangible Confidential Information shall be marked "confidential" or the like at the time of disclosure. Confidential Information shall not include any information which Recipient can establish (i) was publicly known and made generally available in the public domain prior to the time of disclosure to Recipient by Disclosing Party or those acting under its direction; (ii) becomes publicly known and made generally available after disclosure to Recipient by Disclosing Party or those acting under its direction through no action, inaction, or disclosure by Recipient; or (iii) was in the possession of Recipient without confidentiality restrictions at a time immediately prior to the disclosure by Disclosing Party or those acting under its direction, as evidenced by Recipient's written files and written records at a time prior to the time of disclosure; (iv) is received by Recipient at any time from a third party without breach of a non-disclosure or confidentiality obligation to Disclosing Party; (v) is developed independently by Recipient as shown by contemporaneous written documentation; (vi) is required by law to disclose, provided that Recipient gives Disclosing Party reasonable advance notice of its intent to disclose such information so that Disclosing Party may contest the disclosure and/or seek a protective order.

3. <u>Non-use and Non-disclosure</u>. Recipient agrees not to use any Confidential Information for any purpose except to evaluate and engage in discussions concerning a potential business opportunity between Recipient and Disclosing Party. Recipient further agrees not to disclose any Confidential Information to third parties or to employees of Recipient, except to those employees who are required to have the information in order to evaluate or engage in discussions concerning the contemplated business opportunity and who are subject to confidentiality agreements and have been apprised of the confidential nature of the Confidential Information.

4. <u>Maintenance of Confidentiality</u>. Recipient agrees that it shall take all reasonable measures to protect the secrecy of and avoid disclosure and unauthorized use of Confidential Information. Without limiting the

foregoing, Recipient shall take at least those measures that Recipient takes to protect its own most highly confidential information and shall have its employees, if any, who have access to Confidential Information sign a non-use and non-disclosure agreement in content substantially similar to the provisions hereof, prior to any disclosure of Confidential Information to such employees. Recipient shall reproduce Disclosing Party proprietary rights notices on any copies, in the same manner in which such notices were set forth in or on the original. Recipient shall immediately notify Disclosing Party in writing the event of any unauthorized use or disclosure of the Confidential Information.

5. <u>No Obligation</u>. Nothing herein shall obligate Disclosing Party or Recipient to proceed with any transaction between them, and each party reserves the right, in its sole discretion, to terminate the discussions contemplated by this Agreement concerning the business opportunity.

6. <u>No Warranty</u>. ALL CONFIDENTIAL INFORMATION IS PROVIDED "AS IS". Disclosing Party MAKES NO WARRANTIES, EXPRESS, IMPLIED OR OTHERWISE, REGARDING ITS ACCURACY, COMPLETENESS, PERFORMANCE, OR FITNESS FOR A PARTICULAR PURPOSE. RECIPIENT ACKNOWLEDGES THAT Disclosing Party SHALL NOT BE LIABLE FOR ERRORS, OMISSIONS, OR INACCURACIES OF ANY KIND IN THE CONFIDENTIAL INFORMATION AND RECIPIENT SHALL BE RESPONSIBLE FOR VERIFYING THE ACCURACY AND CORRECTNESS OF THE CONFIDENTIAL INFORMATION. RECIPIENT ACKNOWLEDGES THAT NO WARRANTY OF ANY KIND IS GIVEN REGARDING THE CONFIDENTIAL INFORMATION. THE FOREGOING IN NO WAY MODIFIES THE RETENTION BY Disclosing Party OF ALL RIGHT, TITLE AND INTEREST IN THE CONFIDENTIAL INFORMATION.

7. <u>Return of Materials</u>. All documents and other tangible objects containing or representing Confidential Information and all originals and copies thereof in the possession of Recipient shall be and remain the property of Disclosing Party and shall be promptly returned to Disclosing Party upon written request.

8. <u>No Other Obligations</u>. This Agreement imposes no obligation on either party to disclose Confidential Information or to purchase, sell, license, transfer or otherwise make use of any technology, service, or products, or to enter into any other agreements. Nothing in this Agreement is intended to grant any intellectual property rights to Recipient, nor shall this Agreement grant Recipient any rights in or to Confidential Information except as expressly set forth herein.

9. <u>Continuing Nondisclosure and Confidentiality</u>. Whether or not Recipient Disclosing Party enter into or continue a business relationship, the covenants pertaining to confidentiality, nondisclosure and non-use in this Agreement shall nevertheless remain in full force for a period of five (5) years following the date of disclosure.

10. <u>Remedies</u>. Recipient agrees that its obligations hereunder are necessary and reasonable in order to protect Disclosing Party, and expressly agrees that monetary damages would be inadequate to compensate Disclosing Party for any breach of any covenant or agreement set forth herein. Accordingly, Recipient agrees and acknowledges that any such violation or threatened violation will cause irreparable injury to Disclosing Party and that, in addition to any other remedies that may be available, in law, in equity or otherwise, Disclosing Party shall be entitled to obtain injunctive relief against the threatened breach of this Agreement or the continuation of any such breach, without the necessity of proving actual damages or posting any bond. Should Disclosing Party prevail in an action to enforce the provisions of this Agreement by obtaining substantially the relief sought, Disclosing Party shall be entitled to attorney's fees and court costs.

11. <u>Recipient Information</u>. Disclosing Party assumes no obligation, either express or implied, with respect to any information disclosed by Recipient.

12. <u>Miscellaneous</u>. This Agreement shall bind and inure to the benefit of the parties hereto and their successors and assigns. This Agreement shall be governed by the laws of the State of California, without reference to conflict of laws principles. This document contains the entire agreement between the parties with respect

to the subject matter hereof and supersedes all prior and or contemporaneous communications, understandings and agreements. Any failure to enforce any provision of this Agreement shall not constitute a waiver thereof or of any other provision hereof. This Agreement may not be amended, nor any obligation waived, other than in a writing signed by both parties hereto.

13. <u>Severability</u>. In the event any term of this Agreement is found by any court to be void or otherwise unenforceable, the remainder of this agreement shall remain valid and enforceable as though such term were absent upon the date of its execution.

14. <u>Term</u>. This Agreement shall expire one (1) year from the Effective Date.

DISCLOSING PARY

By: _____

Title: _____

Date: _____

RECIPIENT

By: _____

Title: _____

Date: _____

Notes

1. David Pressman, *Patent It Yourself* (Berkeley: Nolo Press, 1991), p. 1/9.
2. Ibid.
3. Ibid., pp. 1/2–1/19. The following information is drawn from Chapter 1 of *Patent It Yourself*.
4. William H. Eilberg, "Frequently Asked Questions About Trademarks," http://www.eilberg.com/trademarkfaq.html (October 22, 2008).
5. Pressman, p. 5/3.
6. David Pressman, *Patent It Yourself* (Berkeley: Nolo Press, 2008), p. 160.
7. "2138.05 "Reduction to Practice" [R-5] – 2100 Patentability," USPTO, http://www.uspto.gov/web/offices/pac/mpep/documents/2100_2138_05.htm (December 4, 2008).
8. Patent Search Tutorial," op. cit., http://www.stanford.edu/group/biodesign/patentsearch/goals.html (November 16, 2006).
9. Ibid., http://www.stanford.edu/group/biodesign/patentsearch/benefits.html (November 16, 2006).
10. Based on the approach outlined by David Hunt, Long Nguyen, Matthew Rodgers in *Patent Searching: Tools & Techniques* (Wiley, 2007).
11. Inventors can also produce these drawings themselves using resources such as Jack Lo and David Pressman's *How to Make Patent Drawings Yourself* (Nolo Press, 1999).
12. "Claim (Patent)," www.wikipedia.org, http://en.wikipedia.org/wiki/Claim_(patent) (March 13, 2007).
13. Ibid.
14. David Dawsey, "Medical Device Industry Patent Litigation Likely to Rise?" Ezine@articles, http://ezinearticles.

com/?Medical-Device-Industry-Patent-Litigation-Likely-to-Rise?&id=542615 (September 7, 2007).

15. Andrew Brandt, "Patent Overload Hampers Tech Innovation," *PCWorld*, February 27, 2006, http://www.pcworld.com/article/124826/patent_overload_hampers_tech_innovation.html (October 2, 2005).

16. "Revision of Patent Fees for Fiscal 2009," USPTO, August 14, 2008, http://www.uspto.gov/web/offices/com/sol/notices/73fr47534.pdf (October 2, 2008).

17. "PCT Contracting States," World Intellectual Property Organization, July 3, 2008, http://www.wipo.int/pct/guide/en/gdvol1/annexes/annexa/ax_a.pdf (October 2, 2008).

18. Applicants are sometimes given the choice of having a search done on the invention at a patent office other than at the Receiving Office. Furthermore, not all Receiving Offices are authorized to act as International Searching Authorities.

19. See The European Patent Office's website at www.epo.org.

20. The full text of the Paris Convention for the Protection of Industrial Property can be found at http://www.wipo.int/treaties/en/ip/paris/trtdocs_wo020.html. However, the treaty is difficult to understand. Consult a patent attorney for practical advice about how to obtain international patent protection.

21. "USTR Issues 2008 Special 301 Report," Office of the U.S. Trade Representative, April 25, 2008, http://www.ustr.gov/Document_Library/Press_Releases/2008/April/USTR_Issues_2008_Special_301_Report.html (October 2, 2008).

22. Eric S. Langer, "Understanding India's New Patent Laws," BioPharma International, April 1, 2008, http://biopharminternational.findpharma.com/biopharm/India+Today/Understanding-Indias-New-Patent-Laws/ArticleStandard/Article/detail/507465 (October 2, 2008).

23. Ibid.

24. Thomas S. Babel, "Patents in China – Is There Any Real Protection?" LocalTechWire.com, April 23, 2008, http://localtechwire.com/business/local_tech_wire/opinion/story/2776264/ (October 2, 2008).

25. Nate Anderson, "US Hauls China Before WTO Again: Less Counterfeiting, Piracy Please," Ars Technica, August 14, 2007, http://arstechnica.com/news.ars/post/20070814-us-hauls-china-before-the-wto-wants-less-counterfeiting.html (October 2, 2008).

26. K.C. Krishnadas, "Patents Granted in India Double from Previous Year," *EE Times*, April 22, 2008, http://www.eetimes.com/news/latest/showArticle.jhtml?articleID=207001275 (October 2, 2008).

27. "China's New Draft Patent Law," Grain, September 25, 2008, http://www.grain.org/bio-ipr/?id=555 (October 2, 2008).

28. From remarks made by Julio Palmaz as part of the "From the Innovator's Workbench" speaker series hosted by Stanford's Program in Biodesign, February 10, 2003, http://biodesign.stanford.edu/bdn/networking/pastinnovators.jsp. Reprinted with permission.

29. "Patent Search Tutorial," Stanford Biodesign Program, http://www.stanford.edu/group/biodesign/patentsearch/inventor.html (November 16, 2006).

30. Pressman, 1991, op. cit., p. 16/3.

31. Gelsinger was the first person publicly identified as having died in a clinical trial. An FDA investigation concluded that the scientists leading the trial took shortcuts that may have contributed to this outcome. In this case both the scientists and their university had equity in the company conducting the trial. See Kristen Philipkoski, "Perils of Gene Experimentation," *Wired*, February 21, 2003, http://www.wired.com/techbiz/media/news/2003/02/57752 (December 4, 2008).

32. Pressman, 1991, op. cit., p. 5/2.

33. From remarks made by Rodney Perkins as part of the "From the Innovator's Workbench" speaker series hosted by Stanford's Program in Biodesign, April 14, 2003, http://biodesign.stanford.edu/bdn/networking/pastinnovators.jsp (September 17, 2008). Reprinted with permission.

34. Greg Gardella, "Keeping Your IP Options Open," IEEE Spectrum Online, http://www.spectrum.ieee.org/careers/careerstemplate.jsp?ArticleId=i020303 (March 13, 2007).

35. "Arrhythmia," American Heart Association, http://www.americanheart.org/presenter.jhtml?identifier=10845 (November 6, 2008).

36. All quotations are from interviews conducted by the authors, unless otherwise cited. Reprinted with permission.

37. From Benedict James Costello and Mark Zdeblick. Reprinted with permission.

38. From U.S. utility patent 5,061,273 by inventor Paul G. Yock. Reprinted with permission.

39. From Tom Ciotti and Mika Mayer of Morrison & Foerster LLP. Reprinted with permission.

40. Based on an approach outlined by David Hunt, Long Nguyen, and Matthew Rodgers in *Patent Searching: Tools & Techniques* (Wiley, 2007).

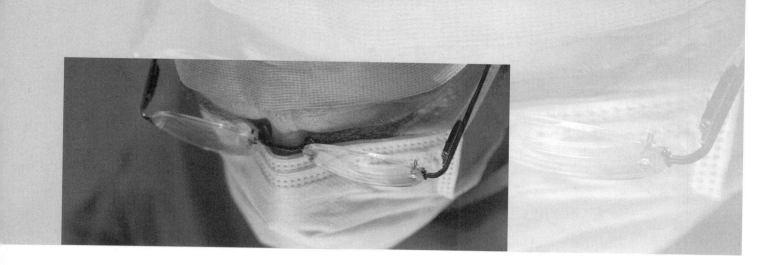

4.2 Regulatory Basics

Introduction

Without regulatory clearance by the FDA (or the equivalent agency abroad) even the most innovative and important breakthrough in medical technology will never reach patients. The issues involved in determining the safety and effectiveness of a new technology are often complex – and the data on which decisions are made are never perfect. Innovators can lose patience with a process that seems vague, arbitrary, and interminable, while FDA reviewers can lose sleep over the prospect of approving a device that may someday do unexpected harm to patients.

Because of the critically important role that regulatory issues play in the ultimate success of a new technology, understanding the regulatory pathway is imperative, even in the early stages of developing a device. In practice, innovators almost always employ an expert to actually write and manage their regulatory submissions. However, regulatory issues are so deeply embedded into product design and development that innovators need to understand regulatory processes and nomenclature in order to provide effective leadership in the biodesign innovation process.

This chapter provides an overview of regulatory terminology and a primer on basic regulatory processes. A second chapter, 5.4 Regulatory Strategy, describes the more nuanced implications of regulatory requirements and how they can become a source of competitive advantage (or disadvantage) to an innovator or company.

OBJECTIVES

- Understand the basic goals of the FDA and how the agency is organized.

- Learn about the FDA medical device classification system and how it relates to the two main regulatory pathways for medical devices: 510(k) and PMA.

- Develop a basic understanding of requirements for regulatory approval outside the United States.

Regulatory fundamentals

The US Food and Drug Administration (FDA) is a regulatory, scientific, and public health agency with a vast jurisdiction, overseeing products that account for roughly 25 percent of all consumer spending in the United States.[1] Products under FDA jurisdiction include most foods (other than meat and poultry), human and animal drugs, therapeutic agents of biological origin, radiation-emitting products, cosmetics, animal feed, and, of course, medical devices.[2] The FDA is the leading regulatory agency in the world, although other important device markets either have or are developing robust regulatory systems (regulatory approaches in countries outside the United States are addressed later in this chapter).

Food and Drug Administration background

The modern era of the FDA began in 1906 with the passage of the Federal Food and Drugs Act, which created the regulatory authority for the agency. This law was replaced in 1938 by the Food, Drug, and Cosmetic Act, which focused primarily on drug safety.[3] Remarkably, devices were essentially not regulated until the Medical Device Amendments Act (MDAA) of 1976. The device amendments were stimulated, in part, by a therapeutic disaster in which thousands of women were injured by the Dalkon Shield intrauterine device.[4] The new law provided for three classes of medical devices based on risk, each requiring a different level of regulatory scrutiny (see below for details). The Safe Medical Devices Act was added in 1990 to increase the scope of FDA responsibility to include regulations involving **design validation** and recall authority.

At the most basic level, the FDA's goal is to protect the public health by assuring the *safety* and *effectiveness* of the products under its supervision.[5] A strong focus on safety is at the core of the FDA's mission and culture. In most cases, the laws that have shaped the regulatory authority of the agency have come in response to high-profile incidents involving unintended, harmful effects of drugs or devices. Even though the FDA's mission also refers to the goal of "advancing the public health by helping to speed innovation," the mandate to protect the public health takes precedence. It is important for the innovator to understand this priority and how it influences the reviewers who make key decisions about new technologies. These reviewers are government employees who share a deeply-held motivation to serve the public by spotting problems before they get to patients. It follows that there is less incentive to approve submissions quickly than there is to be as certain as possible about the safety of the device.

To understand what the FDA means by effectiveness it is important to realize that the end result of a successful FDA submission is that the agency clears or approves the marketing and sale of a device for certain, specific clinical indications. For example, a pacemaker might be approved to treat symptomatic bradyarrhythmias (slow heart rates causing dizziness and other symptoms). The exact language the FDA approves for the use of the device is reviewed in great detail by the agency and results in a statement of "indications for use" (**IFU**) that is included in the product packaging and advertising. To judge effectiveness, the FDA must decide that the device functions as specified by the IFU. Exactly what kind of evidence is required by the FDA to prove effectiveness depends on the risk associated with the device. While a device with minimal risk will be exempt from any type of FDA premarket clearance (no evidence is required), a device that treats a life-threatening condition will typically require a large-scale, controlled clinical trial for approval. It is important to understand that the criteria for clearing or approving a device are not fixed, but evolve with time in response to a number of factors, including new clinical science and accumulating experience in the marketplace with medical technologies. As mentioned, the general tenor of regulation can be strongly influenced by major safety failures that attract media and public attention. For instance, recent issues with Vioxx® on the drug side of the FDA's purview, and with pacemaker leads on the device side, have created a climate of particular caution and scrutiny at the agency.

Because device regulation was implemented relatively recently (compared to other areas within the FDA's scope), the agency's practices and policies continue to evolve as it gains more experience in medtech. More change is anticipated (and, in many cases, desired) by industry insiders in the coming years. For example, as John Abele, cofounder of Boston Scientific, pointed out:[6]

> I think a lot of incremental enhancements are absolutely necessary ... That's the way the device world works. This has presented a major problem for the FDA because assessing the value of incremental enhancements is hard. The FDA still has a pharmaceutical mindset, which says, 'Is it good or bad?' as opposed to, 'Is it better than what we had before?'

There are two broad and important areas in which the FDA does not exert regulatory control. First, *cost effectiveness has no part in the FDA's assessment*. Data about cost are not part of the submission and there is no mandate for the FDA to be involved in the determination of prices or reimbursement. In the federal government, reimbursement levels for medical technologies are determined by the Center for Medicare and Medicaid Services (CMS) – see 4.3 Reimbursement Basics. The CMS generally awaits FDA approval before making a positive reimbursement decision, but this is not an absolute requirement. Second, *the FDA does not regulate or otherwise monitor the practice of individual physicians*. Once a device is cleared for sale in the United States, a physician can use it as s/he sees fit.[7] If something goes wrong, the physician may be sued for malpractice if the device has been used in a manner that is not in accordance with the clinical standard of care in the medical community. However, the FDA has no jurisdiction in the matter.

The FDA is headquartered in Silver Spring, Maryland, in the new White Oak Federal Center. This campus hosts approximately 6,000 agency employees, with another 4,000 deployed across the country for regional inspections and audits. In 1980, the FDA became part of the Department of Health and Human Services. The agency is periodically reauthorized by Congress and is subject to Congressional oversight, which in practice is distributed across a large number of committees. The proposed FDA operating budget for fiscal year 2009 was $2.4 billion.[8] The FDA commissioner is nominated by the president and confirmed by the senate. For the most part, however, the FDA has relatively few political appointees compared to other government agencies, so it is less subject to internal staff changes with turnover in political administrations.

The FDA's Center for Devices and Radiological Health

The FDA is organized into several centers according to types of products they regulate. The main unit for medical devices is the Center for Devices and Radiological Health (**CDRH**), which also regulates radiation-emitting products (including X-ray and ultrasound instrumentation). The Center for Drug Evaluation and Research (**CDER**) regulates pharmaceuticals, while the Center for Biologics Evaluation and Research (**CBER**) oversees biologics (e.g., vaccines, blood products, and biotechnology derived products). Within the CDRH, the Office of Device Evaluation (ODE) is the entity responsible for review and approval of most devices, while diagnostic technologies fall under the Office of In Vitro Diagnostic Device Evaluation and Safety. Quite recently, in 2002, the Office of Combination Products was established to deal with drug–device, drug–biologic, and device–biologic therapies.

The FDA defines a medical device as "an instrument, apparatus, implement, machine, contrivance, implant, in vitro reagent, or other similar or related article, including a component part, or accessory which is ... intended for use in the diagnosis of disease or other conditions, or in the cure, mitigation, treatment, or prevention of disease, in man or other animals, or intended to affect the structure or any function of the body of man or other animals, and which does not achieve any of its primary intended purposes through chemical action within or on the body of man or other animals and which is not dependent upon being metabolized for the achievement of any of its primary intended purposes."[9] Chemical action and metabolic change are hallmarks of drugs and biologics. This definition says, in effect, that if the product is not a drug or biologic, it is a device. Of course, there are therapies that fall in the gray zone of this definition (e.g., biodegradable implants and drug eluting stents). If there is ambiguity about whether or not a new product is a device, the FDA will ultimately make the determination. As discussed further in 5.4 Regulatory Strategy, it is generally advantageous for a company to get a

Table 4.2.1 *Device classification has direct implications for the number and complexity of requirements imposed by the FDA.*

Class	Examples	Description	FDA requirements
I	Bandages, tongue depressors, bedpans, examination gloves, handheld surgical instruments	**Class I** devices present minimal potential harm to the person they are being used on and are typically simple in design.	With class I devices, most are exempt from premarket clearance. There is no need for clinical trials or proof of safety and/or efficacy since adequate predicate experience exists with similar devices. However, they must meet the following "general controls": • Establishment registration with FDA. • Medical device listing. • General FDA labeling requirements. • Compliance with quality system regulation (**QSR**).
II	X-ray machines, powered wheelchairs, surgical needles, infusion pumps, suture materials	**Class II** devices are often non-invasive, but tend to be more complicated in design than class I devices and, therefore, must demonstrate that they will perform as expected and will not cause injury or harm to their users.	Class II devices must generally be cleared via the 510(k) process, unless exempt by regulation. They must also meet all class I requirements, in addition to the "special controls," which may include: • Special labeling requirements. • Mandatory performance standards. • Post-market surveillance.
III	Replacement heart valves, silicone breast implants, implanted cerebellar stimulators, implantable pacemakers	**Class III** devices are typically implantable, therapeutic, or life-sustaining devices, or devices for which a predicate does not exist.	Class III devices must generally be approved by the PMA regulatory pathway, although a small number are still eligible for 510(k) clearance. They must also meet all class I and II requirements, in addition to stringent regulatory approval requirements (see below), before they can be used in humans.

product characterized as a device rather than a drug or biologic. The Office of Combination Products is now making the assignments in most of these ambiguous situations; however, it usually involves consultation with drug or biologic reviewers.

Although the primary responsibility of the CDRH is to review requests to market and sell medical devices, it serves several other functions. For example, the group collects, analyzes, and acts on information about injuries caused by medical devices (or radiation-emitting products). It also sets and enforces standards for good manufacturing practices, monitors compliance and surveillance programs, and provides technical assistance and advice to small manufacturers of medical products.[10]

Device classification

Once it is clear that a new product is properly characterized as a medical device, the next major consideration is to determine its risk profile based on the three-tier safety classification put into force with the 1976 Medical Devices Amendment Act. Class I addresses devices with the lowest risk, while class III includes those with the greatest risk.[11] This categorization serves as the basis for determining the regulatory pathway that the device must take before being cleared for human use, as described in more detail in Table 4.2.1.

Innovators developing devices are initially required to make a "best guess" selection of the appropriate classification of their device in consultation with their expert regulatory consultants. This classification will be reviewed by the branch of the CDRH that evaluates the technology. At any given time, roughly half of all medical devices fall within class I, 40 to 45 percent in class II, and 5 to 10 percent in class III.[12] If the technology is so novel that the innovator cannot judge the likely classification, then informal or formal discussions can be pursued with the FDA to clarify the classification.

Table 4.2.2 *The ODE is organized into five major divisions with branches (22 in total) that are based on medical specialties (compiled from the US Food & Drug Administration's "CDRH Management Directory by Organization").*

Office of Device Evaluation (ODE)	
Division	**Branch**
Division of General, Restorative and Neurological Devices	• Plastic and reconstructive surgery • General surgery • Orthopedic joint devices • Orthopedic spine devices • Restorative devices
Division of Cardiovascular Devices	• Pacing, defibrillator and leads • Cardiac electrophysiology and monitoring • Interventional cardiology • Circulatory support and prosthetic devices • Peripheral vascular devices
Division of Ophthalmic and ENT Devices	• Vitreoretinal and extraocular devices • Intraocular and corneal implants • Diagnostic and surgical devices • Ear, nose, and throat devices
Division of Reproductive, Abdominal and Radiological Devices	• Obstetrics/gynecology devices • Urology and lithotripsy devices • Gastroenterology and renal devices • Radiological devices
Division of Anesthesiology, General Hospital, Infection Control and Dental Devices	• Anesthesiology and respiratory devices • General hospital devices • Infection control devices • Dental devices

Through the CDRH, the FDA regulates more than 100,000 medical devices ranging from simple thermometers, tongue depressors, and heating pads to pacemakers, intrauterine devices, and kidney dialysis machines.[13] These devices are organized into 1,700 different categories of technology, which are managed within ODE by 5 "divisions" with a total of 22 "branches" that are based on medical specialties (see Table 4.2.2).[14]

Each of these branches has separate teams of reviewers who are experts in that area. The primary reviewer will typically have at least an undergraduate degree in engineering or one of the biomedical sciences. Reviewers work on a dozen or more submissions at a time. As mentioned, diagnostic devices are evaluated under a separate Office of In Vitro Diagnostic Device Evaluation and Safety, which has three divisions (Chemistry and Toxicology Devices, Immunology and Hematology Devices, and Microbiology Devices).

The innovator can choose which branch to target for his/her device submission (although the FDA will ultimately determine which branch reviews the device). The process for making the initial selection of a branch is described in more detail in the Getting Started section below, but basically involves either: (1) searching the FDA's classification database, or (2) browsing the device precedents to determine where other, similar devices have been assigned. Using these tools, innovators are able to determine an appropriate classification by finding the description that best matches their own device. For instance, if an innovator created a new type of steerable colonoscope, s/he could go to the classification database and do a search on "colonoscope." Alternatively, s/he could access the device

Table 4.2.3 *The three regulatory pathways for medical devices vary in their requirements based on the level of risk associated with the device (compiled by authors from CDRH's Device Advice).*

Pathway	Description
Device exemption	These are devices for which the risk is so low that they are exempt from regulatory clearance. Most class I devices take this pathway.
510(k)	This is the largest category of medical device applications, in which clearance is based on a device being similar to existing devices in clinical use. Some class I devices and most class II devices take this pathway.
Premarket approval (PMA)	This is the most stringent pathway, used for devices that are significantly different from existing technologies and/or represent the highest risk to patients. The vast majority of class III devices take the PMA pathway, although a few remain eligible for 510(k) clearance.

classification panels, choose Gastroenterology/Urology Devices, select Diagnostics Devices, and then review the description for Endoscope and Accessories.

Regulatory pathways

There are three major pathways for medical device regulation by the CDRH; **exemption**, **510(k)**, and **PMA** (see Table 4.2.3). These pathways are based on the three-level risk classification, although there is not a one-to-one correspondence between the classification and the pathway. Which pathway the device takes is extremely important to the innovator and any company developing the technology because the effort, time, and cost associated with these different alternatives vary significantly.

Relatively speaking, few medical devices are required to receive premarket approval. For example, in fiscal year 2007, the ODE received 3,193 510(k) submissions, but only 31 original PMA applications.[15]

Exempt devices

Roughly three-quarters of class I devices are exempt, meaning that they do not require FDA clearance to be marketed.[16] Examples of exempt class I devices include elastic bandages, tongue depressors, bedpans, and surgical gloves. A much smaller number of class II devices are exempt (less than ten percent), based on the agency's determination that they represent a minor safety risk.[17] No class III devices qualify for exemption. In addition to devices that pose little or no risk to patients, there are some other special circumstances under which an exempt classification is given, such as finished devices that are not sold in the United States or custom devices ("one-off" devices made for a specific patient or application).[18]

Even if a device is determined to be exempt, it still must comply with a minimum set of FDA requirements called "general controls" that also apply to the other two regulatory pathways. These requirements oblige companies to register their facilities with the FDA, fill out a form listing the device and its classification, comply with general FDA labeling and packaging requirements, and adhere to the FDA's Quality Systems Regulation (QSR) – a set of guidelines for safe design and manufacturing (see 5.5 Quality and Process Management). A limited number of class I exempt devices are also exempt from QSR requirements.

The 510(k) process

The 510(k) review process applies to devices of moderate risk where there is some similarity to an existing technology already in use. This is the pathway required for most class II devices. A device that passes FDA scrutiny by 510(k) is said to be *cleared* and to have achieved *premarket notification* (in contrast, a class III device that follows the PMA pathway is said to be *approved*). The company making a submission for either a 510(k) or a PMA is referred to as the *sponsor* of that submission.

The "510(k)" nomenclature refers to the section of the federal regulations that describes this pathway, as put into effect by the 1976 Medical Device Amendment Act. In creating a new system for medical device regulation, the MDAA took into account the fact that there were a number of existing, moderate-risk devices that were already widely and safely in use. These devices were essentially grandfathered into approval by the act and are now described as *pre-amendment devices*. The MDAA also enacted a mechanism for approving new devices based on similarities to these existing pre-amendment devices that had been "road tested" prior to 1976. A further provision allows for new devices to be approved based on comparison with other devices that have received 510(k) clearance subsequent to 1976 or a device that is exempt from 510(k) clearance. In any of these cases, the pre-existing device to which the new device is compared is called the *predicate device* (see Figure 4.2.1).

To support the 510(k) pathway, the MDAA put into place a set of standards, known as *substantial equivalence*, to allow the FDA to compare a new device to its chosen predicates. In essence, in order for a new device to be substantially equivalent to a predicate device, it must: (1) have the same indication for use; (2) have technological characteristics that are similar to the existing device, and (3) not raise any concerns about safety and effectiveness in those areas where there are differences with the predicate device.[19] Specifically, this means that the device must be comparable to the predicate device in terms of its intended use, design, energy used or delivered, materials, chemical composition, manufacturing process, performance, safety, effectiveness, labeling, **biocompatibility**, standards, and other characteristics, as applicable.[20] Substantial equivalence does not mean that the new and existing devices are identical. However, it does require that the new device provides a relevant comparison (a concept that is ultimately decided by the FDA) and is at least as safe and effective as the predicate device.

Another important aspect of the 510(k) pathway is that a sponsor may choose *more than one predicate device* to make the argument of substantial equivalence. This provides a considerable degree of freedom in developing a proof of equivalence. For example, the 510(k) clearance of the mechanical intravascular ultrasound (IVUS) catheter by Cardiovascular Imaging Systems was based on two predicate catheters: a diagnostic Doppler ultrasound catheter and an atherectomy catheter, which had a rotating cutter at its tip. The new IVUS catheter used ultrasound energy (like the Doppler catheter) and had a spinning element at its tip (like the atherectomy catheter). The combination did not raise any new safety concerns for the FDA and the new device was shown to be effective in its intended use of generating images of the inside of a vessel.

Substantial equivalence for 510(k) approval usually can be demonstrated on the basis of bench and animal testing. However, approximately ten percent of 510(k) submissions include clinical data.[21] The FDA will advise innovators and companies whether data collection is necessary. If required, the clinical studies to collect such data are typically much smaller, faster, and less expensive than the trials required for the PMA pathway (see 5.3 Clinical Strategy).

Recently, the FDA's application of the 510(k) process has been under scrutiny from the media and Congress.[22] A forthcoming Government Accounting Office (GAO) report is expected to criticize the FDA's application of the 510(k) process as overly liberal, allowing too many products to be cleared via this pathway with minimal clinical testing as opposed to the more rigorous PMA

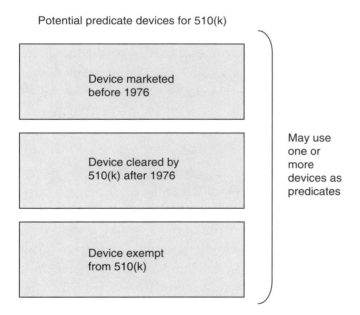

FIGURE 4.2.1
510(k) clearance can be obtained based on either a pre-amendment predicate device, a post-amendment device that has already been cleared via the 510(k) pathway, or a 510(k) exempt device.

pathway. Thus, in the future, there may be increasing challenges in convincing the FDA that a novel device may be cleared via the 510(k) pathway, particularly where a combination of predicates are needed to support the 510(k) argument. There may also be increasing data requirements for 510(k)s, which in some cases may approach the type of data more typically required for a PMA.

De novo 510(k) clearance

In 1997, a new category of 510(k) clearance was introduced called the "de novo" pathway. This category is intended for devices that do not have the major risks of a class III device, but for which no predicates exist. The de novo 510(k) generally requires a higher level of proof of efficacy than a standard 510(k), but much less evidence than for a PMA. One example of a de novo 510(k) clearance is the Given Imaging Pillcam™ endoscopic capsule. This capsule contains a tiny video camera that transmits images from inside the intestines as the capsule works its way through the gastrointestinal tract. The company provided convincing data to the FDA that the capsule provided acceptable images to aid in diagnostic evaluation as an adjunct to standard endoscopy procedures. The FDA did not regard the device as sufficiently similar to any existing predicate to support 510(k) clearance, but was willing to grant the de novo 510(k) because of the favorable safety profile of the capsule and the demonstration of effectiveness in imaging.

Two other categories of 510(k) clearance are worth a brief mention. A company can pursue a "special" 510(k) when it has modified its own device and is seeking clearance for this modification. An "abbreviated" 510(k) can be used when the company is responding to a special "guidance" or "control" issued by the FDA with explicit requirements about the characteristics of the device. In this case, the company submits a declaration that the device is in conformance with these standards and does not need to submit the detailed test reports required for a traditional 510(k).

Mechanics of 510(k) submissions

In almost all circumstances an innovator will use an expert regulatory consultant to prepare a 510(k) submission. The necessary documentation can be many hundreds of pages long (or more), and the approach to choosing predicates and arguing substantial equivalence requires experience (see 5.4 Regulatory Strategy for more strategic considerations for engaging an expert).

While there is no standard 510(k) application, the requirements for a 510(k) submission are relatively well defined.[23] The heart of the submission is a section in which the new device is compared to the predicate(s). This requires a detailed and scientific comparison that includes device performance characteristics, data from **bench testing**, and, in some cases, the results of animal and clinical tests. A second important section is the indications for use – a list of the clinical indications for which clearance is sought. The submission also includes a copy of all printed material and labeling to be distributed with the device or provided to patients. The sponsor may also submit sample advertising and educational materials. If biocompatibility or shelf-life/stability data are required, these results are provided. Finally, the submission includes a 510(k) summary, which is a public statement that will be posted on the FDA website once the device is cleared.

Numerous device-specific guidance documents are available via the FDA's guidance document database. These guidance documents contain detailed information about what should be included in a 510(k) submission to enable the FDA to determine substantial equivalence for the new device,[24] as well as the format to be used[25] and how to be sure that nothing is overlooked.[26]

510(k) review process and timeline

As noted, each 510(k) submission received by the FDA is assigned to one of the agency's primary divisions for review. Not all divisions have the same approach to doing business or working through the review/approval process and there may be benefit in trying to direct the submission to a particular group (see 5.4 Regulatory Strategy).

The FDA is required to review a traditional 510(k) submission within 90 days of its receipt. This does not necessarily mean that the FDA must issue a decision within 90 days, but it is obligated to provide feedback within that period. If no decision is made, the feedback typically comes in the form of a request for additional information. There may be multiple such requests and, each time the FDA requests information, the 90-day review period begins anew. Ultimately, the FDA issues a substantially equivalent (**SE**) determination or a not substantially equivalent determination (**NSE**). The notification of substantial equivalence comes in the form of a letter from the FDA stating that the device can be marketed in the United States. Not substantially equivalent decisions put a company "back to the drawing board" and are a major liability not only for that device but for the company in general since they potentially jeopardize the reputation of the company for future submissions.

There is no standard timetable for 510(k) clearance. For the most part, a straightforward clearance can be obtained in several months. Some 510(k)s can drag on for more than a year, based on the complexity of the analysis and number of requests for additional information. The process can also be slowed by substandard preparation on the part of the sponsor or poor communication between the sponsor and the reviewer.

The PMA process

The PMA pathway is required for devices that represent the highest risk to patients and/or are significantly different from existing technologies in use within a field. There are also certain strategic reasons for an innovator or company to pursue a PMA versus a 510(k); these are reviewed in 5.4 Regulatory Strategy. A PMA is based on a determination by the FDA that sufficient, valid scientific evidence exists to assure that a device is safe and effective for its intended use(s) before it is made commercially available. Approval of a PMA device is made based on the merits of that device alone, regardless of any predicate devices that may exist. The large majority of PMA applications are submitted for class III devices. There are certain class III devices that can still be cleared by 510(k) pathways (intra-aortic balloon pumps are an example), but these are unusual.

Pursuing a PMA is considerably more complicated than the 510(k) path, primarily because the sponsor needs to provide data from a pivotal study. Such studies are typically large, multicenter, randomized clinical trials (see 5.3 Clinical Strategy) and often represent the single largest expense – and the biggest risk – in the entire biodesign innovation process. As a rough rule of thumb, these studies involve hundreds of patients and cost millions of dollars. In turn, PMA submissions often reach thousands of pages in length (see Figure 4.2.2).

If a device is life sustaining, it will require PMA submission whether or not there are similar devices that are already approved. For example, coronary stents are a type of class III device. Even though predicate coronary stents exist in the market, all new stents must follow the PMA pathway. A device that received premarket approval *cannot* be used as a predicate device for 510(k) regulatory clearance.

FIGURE 4.2.2
Informally, the length of a PMA submission is often measured in feet rather than pages (by Joan Lyons, courtesy of Emphasys Medical).

Mechanics of PMA submissions

The PMA submission begins with a summary of safety and effectiveness data. The core of the submission is the clinical study report, which includes the study design and protocol, patient enrollment and exclusion data, primary and **secondary endpoints** of the study, data from all patients entered into the trial, and detailed statistical analysis of the results. There is also a major section on technical data, which includes bench-top testing results and data on biocompatibility, stress and fatigue, shelf life, and other relevant nonclinical tests. The proposed labels and instructions for use are included in the submission along with a physician training plan. There is a section on compliance to quality system regulations and a report on a preapproval facility inspection by the FDA. There is also a section that outlines a negotiated agreement of additional studies that will be required following approval of the device (called post-market surveillance or post-approval studies). As with the 510(k), there is no preprinted PMA application form. However, guidelines regarding the contents and format for a submission are clearly outlined.[27]

Premarket approval applications are reviewed by a special panel, a group of 5 to 15 physicians, statisticians, and other experts (all non-FDA employees) who serve a three-year term. In addition to the core experts, the panel can add topic experts on a case-by-case basis and also has non-voting industry and consumer members. After the PMA is submitted, the panel convenes to hear presentations from the company sponsor, its expert consultants, and from the FDA team. The panel votes to recommend whether the technology should be approved, approved with conditions, or disapproved. The recommendation is non-binding, but generally carries great weight in the approval process. The final decision is based on the analysis of the CDRH branch team, subject to the approval of the director of the Office of Device Evaluation.

There are several different types of PMA routes.[28] The **traditional PMA** is an all-in-one submission that, at this point in time, is used primarily when a device has already been approved by a regulatory agency in another country and the clinical testing has already been completed. The **modular PMA** is increasingly being used by sponsors in the United States. With this approach, the complete contents of a PMA are broken down into well-delineated components (or modules) and each component is submitted to the FDA as soon as the sponsor has completed it, compiling a complete PMA over time. The FDA reviews each module separately, as soon as it is received, allowing companies to receive timely feedback during the review process. This approach may lead to a quicker approval, though this is not guaranteed.

The product development protocol (PDP) method is an alternative to the PMA process, and is essentially a contract between the sponsor and the FDA that describes the agreed-upon details of design and development activities, the outputs of these activities, and acceptance criteria for these outputs. Ideal candidates for the PDP process are those devices for which the technology is well established in the industry. Note that although the PDP mechanism has existed for years, it has not been widely used, and the FDA is currently exploring ways to make the process more efficient and more attractive to sponsors.[29]

PMA review process and timeline

The formal FDA review period for PMA submissions is 180 days. As with 510(k) applications, the FDA will approve, deny, or request additional information from the company upon review of its application. There are usually at least two cycles of requests and responses before a decision is made. If the applicant submits new information on his/her own initiative or at the request of the FDA, the agency can extend the review period up to 180 days. Final approval comes in the form of a letter from the FDA and represents, in effect, a private license granted to the applicant for marketing a particular medical device (see Appendix 4.2.1).[30] There is no standard approval time for PMA submissions, but it is by no means unusual for it to take a year or longer.

Investigational device exemptions

In order to begin human testing in advance of regulatory clearance or approval, official permission must be granted to the innovator, either by the FDA, the supervising institutions where studies will be conducted, or both. Any study involving a new medical technology that does not constitute a **significant risk** to patients can be approved by the Institutional Review Boards (**IRBs**) in the hospitals in which the study will be conducted (see 5.3 Clinical strategy for more information about IRBs). The determination that a device is of non-significant risk is made by each IRB, without required review by the FDA or any other agency. If even one of the IRBs does not approve the study, however, the FDA must become involved.

For significant-risk devices that are headed for a 510(k) or PMA submission, clearance to begin clinical testing must be granted by the FDA via an Investigational Device Exemption (**IDE**). No US patients can be enrolled in one of these studies before an IDE is granted by the FDA. Unlike the submissions required for 510(k)s and PMAs, the IDE application process is manageable enough that the innovators who are interested in testing the device may make a submission without using an expert regulatory consultant (although a review by an expert is advisable).[31] Importantly, an IDE does not allow a company to market or sell the device, but only to test it under well-defined and carefully controlled circumstances.

To obtain IDE approval by the FDA, sufficient data must be presented to demonstrate that the product is safe for human clinical use; this may require mechanical, electrical, animal, biocompatibility, or other supportive testing. In addition, the patient consent form (also called the **informed consent** form) to be used in the study must be approved by the FDA. Sponsors have the opportunity to meet with the FDA in an informal pre-IDE session, during which the **clinical protocol** and any pre-clinical studies can be reviewed. Strategies for approaching these and other FDA meetings are described in 5.4 Regulatory Strategy. Following submission of an IDE application, the FDA has 30 days to respond. The application is approved, disapproved, or conditionally approved. If the response is a conditional approval, the sponsor has 45 days to respond to the FDA with the revised device and/or clinical trial proposal.

Humanitarian device exemptions

The FDA recognizes that certain devices have a limited application in terms of numbers of patients affected, but still are important medically. For these devices, termed humanitarian use devices (HUDs), there is a special approval pathway, the Humanitarian Device Exemption (HDE). The agency defines an HUD as a device that would be used in 4,000 or fewer patients in the United States per year. The first step in pursuing an HDE is for the sponsor to apply for an HUD designation for its device from the Office of Orphan Products Development. The sponsor then must have IRB approval before submitting an application for an HDE. Because clinical data for these devices is so difficult to obtain, the HDE pathway does not require the same type or size of trials as for a PMA (in general safety must be assured, but effectiveness requires a lower standard of proof than for a typical PMA). The sponsor must also make a convincing argument that it cannot develop the product except by using the HDE pathway and that no existing device can be used as effectively for the same clinical purpose. The FDA expects that the company will not make a profit from the device except in certain narrow circumstances, and requires the sponsor to document that the cost of the device does not exceed the costs of research and development, fabrication, and distribution. Under a recent change in the law, a profit can be made under certain circumstances for HDEs intended for use in pediatric patients.[32] Congress made this change in the law to stimulate the development of pediatric devices.

The review period for an HDE is 75 days. An example of a device receiving HDE approval is the Amplatzer® PFO occluder for the treatment of patent foramen ovale (PFO). This condition results from the incomplete closure of the septal wall between the right and left atria (upper chambers) of the heart, a problem that allows blood to partially bypass the natural filtration process provided by the lungs. The new technology was originally developed to target the small group of patients who suffer from strokes related to PFO. However, new exploratory data suggested that PFO closure could potentially help relieve migraine headaches. If confirmed, this would, of course, affect a significantly larger patient population. In the face of this type of new opportunity, the Amplatzer device would not be eligible to address the broader market unless the company obtains approval for a PMA with supportive data for the broader use. Off-label promotion of an HDE-approved product for broader use can lead to withdrawal of the HDE.[33]

Costs of FDA submissions

The Medical Device User Fee and Modernization Act (**MDUFMA**), enacted in 2002, established user fees for the first time in the medical device industry. Pharmaceuticals and biologic applicants had been paying user fees for some time, but medical devices were historically reviewed "for free." The Medical Device User Fee and Modernization Act was created to generate resources to help the FDA address increasing review times and to facilitate quicker access to market for medical device applicants. In 2007, the MDUFMA was reauthorized from 2008 to 2012. It was also amended to significantly reduce the user fees that were originally enacted.

In general, the fees for a 510(k) submission are modest when the time and expertise required by the FDA for review of these applications is taken into account. For fiscal year 2008, the standard application fee was $3,404 and the small business fee (for companies with less than $100 million in sales) was $1,702. Premarket approval applications are considerably more expensive ($185,000 for a standard application; $46,250 for small businesses).[34] Even these costs are small, however, compared to the costs of actually conducting the trials, collecting the information, and preparing the application.

One upside of the new fee schedule is that the act provides for increased measurement and accountability of the FDA in terms of its performance and review times. Effectively, this act creates a more "businesslike" model in which there is an agreement between the applicant and the FDA to complete a regulatory review. Importantly, however, the MDUFMA does not hold the FDA to specific timelines associated with applications. The FDA is not obligated to clear a 510(k) in 90 days or PMA in 180 days. Yet, Congress measures the FDA's performance to these goals and the agency's continued funding depends on its performance, thereby creating an incentive for the FDA to complete timely reviews and approvals. To date, FDA performance seems to have improved as a result of these changes.[35]

Deciding on a regulatory pathway

In practice, dealing with the FDA to confirm an appropriate regulatory pathway for a device can be a complex and iterative exercise. While making a determination can be relatively straightforward for some devices, others can encounter unexpected twists and turns along the way. The Perclose story below demonstrates the importance of carefully understanding regulatory pathway requirements and being able to respond quickly and effectively under unforeseen circumstances.

FROM THE FIELD | **PERCLOSE**

Adapting to FDA regulatory requirements in a competitive market

Hank Plain joined Perclose, a company founded in Sunnyvale, California, as its president and CEO in February 1993. At the time, Perclose was developing a technology designed to percutaneously close the arterial puncture site created when the femoral artery was accessed during catheter-based therapeutic and diagnostic cardiology procedures. The first device for which it would seek FDA approval in the United States was the Prostar® – a sealing device that used flexible needles and standard sutures to surgically close the arterial access site, without the need for additional exposure of the artery to place the sutures. Unlike the Prostar, which was designed to be an active method to achieve secure closure, the standard method consisted of applying pressure at the femoral artery access site for 15 to 30 minutes to form a clot to seal the artery; usually at least 4 to 6 hours of bed-rest was needed after this to ensure that the clot would not fail. A secure, suture-based closure would allow patients to ambulate earlier and, thus, be discharged more quickly (see Figure 4.2.3).

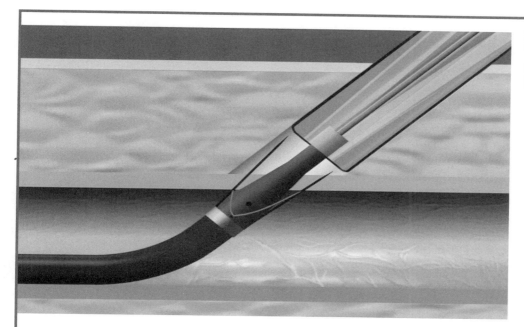

FIGURE 4.2.3
The Prostar device in an artery (courtesy of Hank Plain).

Based on the fundamental needle and suture technology of the company's device, Perclose anticipated that it would be allowed by the FDA to follow a 510(k) regulatory pathway, which potentially would have led to market clearance approximately 90 to 180 days after submission.[36] "We believed we had a straightforward 510(k) because our device was basically needles and sutures," explained Plain.[37] "The FDA looks at intended use and predicate devices when determining whether something is substantially equivalent to other 510(k) devices. We thought we had a very good argument because needles and sutures had been the definitive procedure used for 50 years to close puncture sites in the femoral artery – we just eliminated the need to have a surgeon make a four-inch cut to expose the artery so the arterial access site could be sutured closed."

Representatives from Perclose met with the FDA in May 1994 in an effort to validate the 510(k) regulatory pathway. The FDA's Interventional Cardiology Devices Branch and the General Surgery Devices Branch were both invited to attend the meeting. The Perclose team indicated that it believed the needle and suture predicates dictated that it should be working with the general surgery division and that it intended to submit clinical data from a 100-patient registry to support its 510(k) submission.[38] In response, "The cardiology branch told us that if they were the responsible group, they would require a 500-patient randomized trial," said Plain.

In December 1994, Perclose submitted an IDE for the 100-patient registry trial. However, the FDA disapproved the application, informing the company that the cardiovascular branch would be managing its regulatory process and that it expected to see results from a 500-patient randomized trial (randomized to the device or standard compression – the same study requirements imposed on two competitor companies, Datascope and Kesney Nash, that developed closure methods using collagen plug sealants). The FDA acknowledged that the suture technology was, in fact, similar to existing devices, but maintained that Prostar's percutaneous delivery mechanism was novel, thereby necessitating the study. The FDA also indicated that it would consider a 510(k) with clinical data once the results of the 500-patient randomized study were reviewed.

Perclose performed the trial as requested, meeting its desired endpoints and then sharing the results with the FDA in May 1996. "At this time," recalled Plain, "the cardiovascular branch reaffirmed that it was open to reviewing our 510(k) submission." In July 1996, the company made its submission. Yet, in October 1996, "Our regulatory counsel, Howard Holstein, received a call from the FDA to inform us that Perclose would be receiving a 'not substantially equivalent' or NSE determination for our 510(k) submission, meaning our device would have to follow the more rigorous PMA pathway for product approval," Plain said. "The FDA also said that since the Prostar system was indicated for a patient population that was not served by the approved collagen devices[39] and there were compelling clinical data, the agency would expedite its review of the application." The Perclose team quickly pulled together a PMA submission and presented it to the cardiology branch

FROM THE FIELD: PERCLOSE

Continued

just one month later, in November 1996. Approval of the device was granted six months later, in May 1997. During this period, Plain said, "We were an early stage public company [Perclose went public in November 1995]. You can imagine how this regulatory rollercoaster impacted our investors and the company stock."

According to Plain, "We never fully understood why we were required to do the PMA. But it was our suspicion, although we could never verify it, that it was because of pressure from our collagen competitors arguing that it should be a level playing field." With competition in the vascular closure market intensifying rapidly, Datascope and/or Kesney Nash may have argued that the regulatory pathway for this device should follow the precedents set by competitive products VasoSeal and AngioSeal (which were both subject to a PMA pathway). VasoSeal, from Datascope, was a collagen-based plug used to seal arterial punctures. This product was first to market in the United States when FDA approval was granted in late 1995.[40] AngioSeal, by Kesney Nash, received FDA approval in August 1996. This device used collagen in combination with an absorbable anchor that allowed the collagen to be pulled down to the puncture site by a suture.[41] "The difference," explained Plain, "was that collagen had never been used in the history of medicine to close femoral artery puncture sites and, therefore, had no basis to claim substantial equivalence to other 510(k) devices." Other possible reasons for the FDA's determination were that Prostar introduced new intended use to existing technology and/or new questions of safety or effectiveness compared to predicate devices.

Interestingly, Datascope challenged the FDA's expedited approval of Perclose's PMA application. Datascope first took up the matter with the agency, "but the FDA rejected the petition and stood behind our approval," said Plain. Perclose next learned that Datascope had elevated the matter to the Energy and Commerce Oversight and Investigation Subcommittee, led by Congressman Joe Barton (R-TX), on the grounds of "inequitable treatment." This committee was conducting extensive hearings on food and drug laws at the time that later led to the FDA Medical Device Modernization Act.[42] "In the summer of 1997, we had a meeting with the chief counsel to the House subcommittee," said Plain. "In this session, we tried to educate him about Perclose, the rigor of our clinical trial data, our FDA path, and why Datascope's claims were unfounded." Dr. Richard Kuntz, representing the independent data analysis center that assessed the Prostar trial data, reviewed the strength of the clinical data and the company outlined the events in Datascope's campaign of "competitive harassment." By this time, Perclose had already sold 75,000 devices worldwide and the Prostar was outselling the collagen devices in international markets. Fortunately for the company, the investigation never gained traction. "Eventually, we heard through our attorneys that the House subcommittee had decided not to pursue the matter," noted Plain, "and the issue just went away."

In hindsight, Plain emphasized the importance of having the clinical trial data to back up the company's FDA submission. Without it, he acknowledged, Perclose never would have been able to respond so quickly to the FDA's change in regulatory requirements. Furthermore, the company would not have had such a strong case when Datascope tried to have Prostar's PMA approval overturned. Reflecting on the experience, Plain commented: "We tried everything we could to persuade the FDA to allow us to do the 100-patient registry and provide a 510(k) clearance. And we did that because it's the fastest and least expensive way to get the trial done and get the product to market. But, in some ways, the FDA really did us a big favor when you look at how things played out. If we didn't have the more extensive randomized clinical trial data analyzed by an independent data management group, which proved the safety and efficacy of the device, we would have ended up playing into our competitor's hands. Even if the registry might have been sufficient for approval, it could have allowed Datascope to gain traction with the Congressional subcommittee based on a level playing field argument. I think there is a high probability that they would have been successful in getting Congress to put enough pressure on the FDA that the FDA might have felt that they needed to withdraw our approval or, at minimum, force us to do a randomized study to justify staying on the market. We were certainly nervous at the time and didn't know how it would play out with the FDA or Congress. But both organizations got it right in our view. And we were ultimately vindicated when the Perclose devices became the top-selling closure device systems and the company was acquired by Abbott Laboratories in 1999."

Regulatory approval outside the United States

Every innovator and company seeking regulatory approval outside the United States (**OUS**) should conduct in-depth research to understand the unique regulatory requirements they will face. Working with a seasoned regulatory expert with experience in each target country is also advisable.

Traditionally, the most significant OUS medical device markets have been in Europe, Canada, Australia, and Japan (based on their size and revenue potential). Regulatory requirements in these countries are generally well understood and have begun converging through the efforts of a Global Harmonization Task Force (GHTF) for medical devices. However, many developing countries are becoming increasingly important medtech markets, particularly Brazil, Russia, India, and China. While the regulatory policies of these nations are not yet as well defined, significant strides are being made to upgrade them to global standards.

Europe

Medical devices are regulated in the European Union (EU) by three EC Directives.[43] The main directive, which covers the vast majority of medical devices from surgical gloves to life-sustaining implantable devices such as heart valves, is the Medical Devices Directive (**MDD**). The MDD is supplemented by an older directive on active implantable medical devices (the **AIMDD**) and a newer directive on in vitro medical devices (the **IVMDD**). These three EC directives have been enacted ("transposed") into the national laws of each EU member state, resulting in a legislative framework comprising literally dozens of medical device laws. Regulatory approval in the EU is signified by a **"CE" mark** of conformity. "CE" is not an official acronym, but may have originated in the terms *Communauté Européenne* or *Conformité Européenne*, meaning European community or conformity.[44]

The medical device directives are known as "new approach" directives. Based on these directives, medical devices bearing the CE mark can circulate freely and be sold and marketed according to approved indications throughout the 27 countries in the EU without any barriers. The underlying principle of the new approach is that each of the medical device directives contains a legislative list of "essential requirements" that must be met by any product falling within its scope. In the MDD, the list of essential requirements can be broken down into two groups: (1) a set of general requirements for safety and performance that apply to all devices, (2) a list of specific technical requirements with regard to design and manufacturing that may or may not apply, depending on the nature of the device. For example, the technical requirement for electrical safety will not apply to a urinary catheter. Compliance with the technical requirements is generally demonstrated by using the relevant European harmonized technical standard. The need for clinical data in the CE marking process arises from the general requirements for demonstrating safety and performance.[45]

Under the MDD, there are four classes of devices that generally correspond to the US device categories, as shown in Table 4.2.4.[46] A key aspect of medical device regulation in the EU is that the responsibility for ensuring that devices meet the essential requirements lies with the manufacturer. For low-risk devices (class I), such as tongue depressors or colostomy bags, the manufacturer is allowed to self-declare conformity with the essential requirements. For medium- to high-risk devices (classes IIa, IIb, III), the manufacturer must call on a third party to assess conformity. To some degree, the manufacturer may choose among methods for "conformity assessment" of the device and/or manufacturing system. The end result is a certificate of conformity that enables the manufacturer to apply the CE mark to the product.

Another major aspect of the CE marking process is that, contrary to the United States, the "conformity assessment" for medical devices in Europe is not conducted by a central regulatory authority. The CE marking system relies heavily on third parties known as "notified bodies" to implement regulatory control over medical devices. Notified bodies are independent commercial organizations that are designated, monitored, and audited by the relevant member states via the national "competent authorities." Currently, there are more than 70 active notified bodies within Europe. A company is free to choose any notified body designated

Table 4.2.4 *The four device classes in the EU generally correspond to the three classes defined by the FDA in the United States.*

EU class	Description	US equivalent
Class I	Devices that present a relatively low risk to the patient and, except for sterile products or measuring devices, can be self-certified by the manufacturer. Typically, they do not enter the human body.	Class I
Class IIa	Devices that present a medium risk to patients and may be subject to quality system assessment. Generally, they are invasive to the human body, but only via natural body orifices. This category may also include therapeutic diagnostics and devices for wound management.	Class II
Class IIb	Devices that present a medium risk to patients and may be subject to quality system assessment, as well as third-party product and system certification. They are usually either partially or totally implantable and may modify the biological or chemical composition of body fluids.	Class II
Class III	Devices that present a high risk to patients and require design/clinical trial reviews, product certification, and quality system assessment conducted by a European notified body (see below). In most cases, they affect the functioning of vital organs and/or life-supporting systems.	Class III

to cover the particular class of device under review. After approval, post-market surveillance functions are the responsibility of the member states via their competent authority.

Clinical trials may be carried out in the EU prior to CE marking; however, such studies must comply with "Clinical Investigations of Medical Devices in Human Subjects, Parts 1 and 2," and the national laws in the member states where the trials will be conducted. This involves obtaining ethics committee approval and making necessary notifications to the relevant competent authorities.

The medical devices directives (MDD and AIMD) were amended by a new Directive in 2007, which comes into full effect in 2010. This amendment is significant as it is anticipated to increase the requirements for clinical data in the EU in the pre-approval and post-market phases. For more information, see the European Commission's guidelines relating to the Medical Device Directives.[47]

Canada

In Canada, medical devices are regulated by Health Canada's Therapeutic Products Directorate and are subject to the Medical Devices Regulations under the country's Food and Drugs Act. Before they can be sold in Canada, all medical devices must be classified and most must be licensed. Health Canada has enacted four device classes (I, II, III, or IV), which vary based on the level of risk associated with their use. Factors evaluated to determine the level of risk include the degree of invasiveness, duration of contact with the patient, energy transmission hazard, and consequences of device malfunction or failure. As in the United States and EU, class I devices present the lowest potential risk and do not require a license. Class II devices involve moderate risk and require the sponsor to make a declaration of device safety and effectiveness. Class III and IV devices involve substantial risk and, for this reason, are subject to in-depth regulatory scrutiny before licensing.[48] Guidelines regarding the type of submission needed to secure a device license are provided on the Health Canada website.[49]

Australia

Formal processes for regulating medical devices in Australia were enacted in 1989, under the Therapeutic Goods Act (TGA).[50] The Australian Register of

Therapeutic Goods (ARTG) is the central point of control for the legal supply of devices and other therapeutic goods in the country. Most devices have to be entered in the register before they can be made commercially available (although 2002 legislation now allows a small number of devices to be excluded from entry into the ARTG while others may be given specific approval for a special purpose without being entered).[51]

Australia maintains five device classes – I, IIa, IIb, III, and active implantable medical devices (AIMDs). The classification of a medical device determines the conformity assessment procedures that it must undergo before it can be sold in the country. Higher class devices undergo a more stringent form of conformity assessment than lower class devices.[52] Because the Australian regulatory system is modeled after the system used in the EU, some devices can be certified by the TGA or by an overseas notified body authorized to award a CE mark. More about the Australian regulatory system can be found on the TGA website.[53]

Japan

Japan's regulatory authority is the Ministry of Health, Labor, and Welfare (MHLW), which, as of 2004, works in cooperation with the independent Pharmaceutical and Medical Device Agency (PMDA) to award certification (*ninsho*) or approval (*shonin*) for medical devices. Medical devices are regulated under the Pharmaceutical Affairs Law (PAL), enacted in the 1940s and revised most recently in 2002 (with an effective date in 2005). Revisions to the law were made to allow the MHLW to operate more effectively while increasing the country's regulatory conformity relative to world standards and addressing the lengthy and complicated regulatory processes that have plagued Japan for years.[54] Among other changes, the revised PAL implemented a recategorization of medical devices from four to three classes based on the risks inherent in their use.

The lack of forms available in English and the country's complex registration process can make Japan a more challenging and time-consuming market for device manufacturers to enter than Europe, Canada, and Australia. For this reason, the advice and guidance of a regulatory consultant is particularly important for companies seeking to do business in the country.

The BRIC countries

As noted, the large and developing markets of Brazil, Russia, India, and China are becoming increasingly important medical device markets.

Brazil

The Brazilian National Health Surveillance Agency (ANVISA) is an independent agency that works in cooperation with the country's Ministry of Health (MOH) under a management contract. ANVISA has responsibility for the regulation of all medical products and pharmaceuticals. A 2001 resolution pertaining to medical devices outlines the specific documents necessary in order to register devices and equipment with ANVISA before making them commercially available.[55] Although the process for registration of medical products has been harmonized across the MERCOSUR countries (Argentina, Brazil, Paraguay, and Uruguay) in the past few years, it can still be lengthy, taking from six months to two years between filing for registration and final approval by the government.

In addition to registration, companies bringing medical devices to Brazil must comply with the Code of Consumer Protection and Defense. This code ensures consumers that equipment is safe and will be used correctly by requiring companies to provide sufficient documentation to demonstrate the safety of their products. As in other countries, medical devices covered by this code are classified into one of four risk-based categories (I–IV),[56] each with different safety requirements. Additionally, companies are required to have an import license for each shipment of medical devices into Brazil. These licenses are used to control imports into the country and are separate from the manufacturer's MOH registration and safety hurdles.[57]

Russia

Two major government entities oversee the regulation of medical devices in Russia. The first is the Ministry of Health, which works to ensure the clinical safety and effectiveness of products through the process of issuing registration certificates. The second is the State Committee for Standardization, Metrology and Certification (Gosstandart), which is focused on confirming product conformity with established technical and safety standards through the issuance of certificates of conformity.[58]

Before pursuing a certificate of conformity, a company must register its product with the MOH and have it added to the national register. The Council for New Medical Equipment within the MOH scrutinizes product documentation, as well as the results of technical, safety, toxicology, hygienic, and clinical tests to make its registration recommendations. If the council determines that foreign products have analogs in Russia, it may deny registration of the foreign product. Once registered, the company can seek a conformity assessment by submitting a declaration-application to the authorized certification organization of its choice. The focus of a conformity assessment is on end-product testing.[59]

India

India has not had an agency for the regulation of medical devices, but is poised to create one in the near future. Up until this point, medical devices have been imported into India essentially without restriction, based on the purchaser's evaluation of quality, with FDA- and CE-approved products receiving preferential treatment. Any form of medical device regulation has consisted of classifying certain devices as "drugs" and then applying drug laws to them under the purview of the country's Drug and Cosmetic Act (DCA).[60]

In 2006, ten medical devices were classified as drugs, including cardiac stents, drug eluting stents, catheters, intraocular lenses, heart valves, and orthopedic implants. Since then, the government has begun developing specialized regulations for these devices through the Central Drug Standards Control Organization (CDSCO) within the Ministry of Health. As India's chief drug regulator, this organization was given the responsibility of establishing standards for manufacturing, technical staff qualification, and testing processes for each device. India's Department of Science and Technology (DST) also drafted legislation on a regulatory framework for medical devices, which would establish a separate system for medical device validation and registration.[61]

China

China passed its first set of laws for the regulation of medical devices in 2000. Regulation is managed by the State Food and Drug Administration (SFDA). To sell a medical device in China, a company must register its product with the SFDA. Devices are organized into three product classes (I–III). The SFDA is primarily focused on regulating class III products, delegating oversight of class I and II devices to provincial government agencies.[62] SFDA registration requires the submission of 12 forms of documentation that may include clinical trial data and specified test reports. Importantly, prior to registration in China, a device must receive marketing approval in its country of origin.[63]

In some cases, devices may also require approval by either the Ministry of Health or the General Administration of Quality Supervision, Inspection and Quarantine (AQSIQ). AQSIQ conducts mandatory safety registration, certification, and inspection for certain devices. Once certified, devices are awarded a "China Compulsory Certification" (CCC) mark, which serves as evidence that the product can be imported, marketed, and used in China.[64]

OUS clinical trials and the FDA

Many US companies perform their initial clinical studies and/or seek preliminary regulatory clearance OUS, driven by the less cumbersome regulatory processes found in some other countries (e.g., those in South America). In some cases, regulatory bodies outside the United States accept compilations of key literature and a written analysis of those papers to make a case that a device is expected to be safe and efficacious in lieu of expansive clinical data. As a result, approval can be much quicker, easier, and less expensive in other countries than it is in the United States. The FDA does not prohibit this practice, but expects to see all patient data obtained from such studies. Many times, companies use OUS trials to support an IDE application in the United States. In some cases, OUS data may be sufficient to support a 510(k) submission although, depending on the disease state, the agency may have concern about potential differences in the response of American patients to the device. However, almost all PMA applications and the majority of 510(k) submissions that require patient data will require at least some studies to be performed in the United States.

As noted throughout this chapter, regulatory issues require expert assistance fairly early in the biodesign innovation process. On the other hand, it is extremely useful for the innovator to develop a first-hand sense of which regulatory pathway is likely to be required for a particular device concept. This initial regulatory perspective can be of considerable help in the concept selection process, since the approval pathway has major implications for time and cost.

See **ebiodesign.org** for active web links to the resources listed below, additional references, content updates, video FAQs, and other relevant information.

1. Confirm the appropriate regulatory branch

1.1 What to cover – Determine which branch of the FDA is most likely to regulate the device under development (for most medical devices this will be CDRH).

1.2 Where to look
- **Is the product a medical device?** – Information on the FDA website that defines requirements for medical device categorization.
- **Federal Food, Drug, and Cosmetic Act** – Chapter II of the Food, Drug, and Cosmetic Act specifies the precise definition of what the FDA considers a drug or device.

2. Classify the device

2.1 What to cover – Search the FDA's device classification database and the device classification panels to ensure that the best possible classification decision can be made. Be aware that comparable products may have several different classifications. Finding as many comparable products as possible will help narrow the range of classification options and submission types for the product under development. Collect the following information for each one: device class name, seven-digit regulation number, class number, submission type, physician review panel. Even if a device is quite novel, it may be possible to classify components of the product or utilize similar products as a guide. If the above methods do not result in a classification, use the CDRH Super Search. When performing a search, use the manufacturer, brand name, or common name for one or more similar products.

2.2 Where to look
- **FDA Classification Database** – An FDA search page that enables a user to classify a device based on the device name, review panel, regulation number, product code, submission type, third-party eligibility, and device class.
- **Device Classification Panels** – This FDA website lists the medical specialties used to review medical devices. Selecting the most likely specialist that would use the product under development enables the user to see the seven-digit device classifications categorized within the specialty.
- **CDRH Super Search** – Allows individuals to search through multiple FDA databases, including the device listing database, 510(k) and PMA databases, Products Classification database, the Code of Federal Regulation (CFR) Title 21 database, and the Clinical Laboratory Improvement Amendments (CLIA) database. For device classification purposes, the first four databases listed above will prove most useful.

3. Determine the regulatory pathway

3.1 What to cover – Depending on the device classification, determine if it is necessary and appropriate to pursue a 510(k) or a PMA path (or if the device may be exempt). The key question is: can a reasonable argument be made that the new product combines features of products that were on the market prior

Stage 4 Concept Selection

GETTING STARTED

to 1976, products that themselves have been cleared through the 510(k) pathway, or exempt products? If so, the 510(k) pathway is probably appropriate. In practice, it can be difficult to be certain which pathway to take and help from a regulatory expert will ultimately be necessary.

3.2 Where to look
- **Device advice** – CDRH's self-service site for obtaining information concerning medical devices and the application/submission processes.
- **CDRH guidance documents** – Documents prepared for CDRH staff, regulated industry, and the public that relate to the processing, content, and evaluation of regulatory submissions, design, production, manufacturing, and testing of regulated products, as well as CDRH inspection and enforcement procedures.
- **CDER and CBER guidance documents** – Guidance documents (as defined above) for investigational new drugs.

4. Secure a regulatory consultant

4.1 What to cover – Look for regulatory advice from an expert early – at least by the time the team enters into the development strategy and planning stage of the biodesign innovation process.

4.2 Where to look
- **Regulatory Affairs Professionals Society** – Ask for a contact in the local chapter nearest to the location of the company.
- **Personal networks** – Network with professionals in the field and ask for assistance in meeting and choosing a regulatory consultant. Board members and venture capitalists may have contacts in this area.

Credits

The editors would like to acknowledge Darin Buxbaum, Trena Depel, and Asha Nayak for their help in developing this chapter. Many thanks also go to Janice Hogan and Howard Holstein, of Hogan & Hartson, and Michelle Paganini, of Michelle Paganini Associates, for lecturing on Regulatory Basics in Stanford University's Program in Biodesign. This chapter is based, in part, on their presentations. Jan Benjamin Pietzsch and Janice Hogan deserve special recognition for their extensive editing of this material, as does Hank Plain, of Morganthaler Ventures, for his assistance with the case example. Thank you also to Sarah Sorrel, of MedPass International, for contributing to the information on international regulatory requirements.

Appendix 4.2.1

Sample FDA PMA approval letter[65]

DEPARTMENT OF HEALTH & HUMAN SERVICES Public Health Service

Food and Drug Administration
9200 Corporate Boulevard
Rockville MD 20856

Ms. Jane E. Beggs   APR 30 1997
Regulatory Affairs Manager
Perclose, Incorporated
199 Jefferson Drive
Menlo Park, California 94025

Re: P960043
    Prostar® Percutaneous Vascular Surgical (PVS) System
    Filed: November 26, 1996
    Amended: December 5, 6, and 23, 1996; January 14, 27 and 31;
        February 7, 10, 10, 11, 24 and 26; March 3 and 7,
        and April 18, 1997

Dear Ms. Beggs:

The Center for Devices and Radiological Health (CDRH) of the Food and Drug Administration (FDA) has completed its review of your premarket approval application (PMA) for the Prostar® Percutaneous Vascular Surgical (PVS) System. The Prostar® PVS System consists of the Prostar® PVS Device (9 and 11 French sizes) and the following accessories: a Prostar® Pre-Dilator (9 and 11 French sizes), a Perclose® Knot Pusher, a Prostar® Transition Guidewire, and a Perclose® Arterial Tamper. The Prostar® PVS System is indicated for the percutaneous delivery of sutures for closing the common femoral artery access site and reducing the time to hemostasis and ambulation (time-to-standing) of patients who have undergone interventional procedures using 8 to 11 French sheaths. We are pleased to inform you that the PMA is approved subject to the conditions described below and in the "Conditions of Approval" (enclosed). You may begin commercial distribution of the device upon receipt of this letter.

The sale, distribution, and use of this device are restricted to prescription use in accordance with 21 CFR 801.109 within the meaning of section 520(e) of the Federal Food, Drug, and Cosmetic Act (the act) under the authority of section 515(d)(1)(B)(ii) of the act. FDA has also determined that to ensure the safe and effective use of the device that the device is further restricted within the meaning of section 520(e) under the authority of section 515(d)(1)(B)(ii), (1) insofar as the labeling specify the requirements that apply to the training of practitioners who may use the device as approved in this order and (2) insofar as the sale, distribution, and use must not violate sections 502(q) and (r) of the act.

Page 2 - Ms. Jane E. Beggs

Expiration dating has been established and approved at six months for the Prostar® PVS System and at 18 months for the Perclose® Knot Pusher, Prostar® Transition Guidewire and the Perclose® Arterial Tamper accessory devices, which are packaged separately. This is to advise you that the protocol you used to establish this expiration dating is considered an approved protocol for the purpose of extending the expiration dating as provided by 21 CFR 814.39(a)(8).

CDRH will publish a notice of its decision to approve your PMA in the FEDERAL REGISTER. The notice will state that a summary of the safety and effectiveness data upon which the approval is based is available to the public upon request. Within 30 days of publication of the notice of approval in the FEDERAL REGISTER, any interested person may seek review of this decision by requesting an opportunity for administrative review, either through a hearing or review by an independent advisory committee, under section 515(g) of the Federal Food, Drug, and Cosmetic Act (the act).

Failure to comply with the conditions of approval invalidates this approval order. Commercial distribution of a device that is not in compliance with these conditions is a violation of the act.

You are reminded that, as soon as possible and before commercial distribution of your device, you must submit an amendment to this PMA submission with copies of all approved labeling in final printed form.

All required documents should be submitted in triplicate, unless otherwise specified, to the address below and should reference the above PMA number to facilitate processing.

> PMA Document Mail Center (HFZ-401)
> Center for Devices and Radiological Health
> Food and Drug Administration
> 9200 Corporate Boulevard
> Rockville, Maryland 20850

Page 2 - Ms. Jane E. Beggs

Expiration dating has been established and approved at six months for the Prostar® PVS System and at 18 months for the Perclose® Knot Pusher, Prostar® Transition Guidewire and the Perclose® Arterial Tamper accessory devices, which are packaged separately. This is to advise you that the protocol you used to establish this expiration dating is considered an approved protocol for the purpose of extending the expiration dating as provided by 21 CFR 814.39(a)(8).

CDRH will publish a notice of its decision to approve your PMA in the FEDERAL REGISTER. The notice will state that a summary of the safety and effectiveness data upon which the approval is based is available to the public upon request. Within 30 days of publication of the notice of approval in the FEDERAL REGISTER, any interested person may seek review of this decision by requesting an opportunity for administrative review, either through a hearing or review by an independent advisory committee, under section 515(g) of the Federal Food, Drug, and Cosmetic Act (the act).

Failure to comply with the conditions of approval invalidates this approval order. Commercial distribution of a device that is not in compliance with these conditions is a violation of the act.

You are reminded that, as soon as possible and before commercial distribution of your device, you must submit an amendment to this PMA submission with copies of all approved labeling in final printed form.

All required documents should be submitted in triplicate, unless otherwise specified, to the address below and should reference the above PMA number to facilitate processing.

    PMA Document Mail Center (HFZ-401)
    Center for Devices and Radiological Health
    Food and Drug Administration
    9200 Corporate Boulevard
    Rockville, Maryland 20850

Notes

1. "The Food and Drug Administration Celebrates 100 Years of Service to the Nation, U.S. Food & Drug Administration, January 4, 2006, http://www.fda.gov/bbs/topics/NEWS/2006/NEW01292.html (October 21, 2008).
2. "FDA History," U.S. Food and Drug Administration, http://www.fda.gov/oc/history/default.htm (December 17, 2007).
3. Ibid.
4. Ibid.
5. "The FDA's Mission Statement," U.S. Food and Drug Administration, http://www.fda.gov/opacom/morechoices/mission.html (December 17, 2007).
6. From remarks made by John Abele as part of the "From the Innovator's Workbench" speaker series hosted by Stanford's Program in Biodesign, February 2, 2004, http://biodesign.stanford.edu/bdn/networking/pastinnovators.jsp. Reprinted with permission.
7. If a device is used in a way that deviates from the indication on the FDA-approved label, it is called "off-label" usage. While physicians may elect to engage in off-label use, the manufacturer of the device is not allowed to promote any methods of use other than those cleared by the FDA.
8. "President's FY 2009 Budget Advances Food and Medical Product Safety, and the Safety of FDA-Regulated Imports," U.S. Food & Drug Administration, February 4, 2008, http://www.fda.gov/bbs/topics/NEWS/2008/NEW01789.html (October 21, 2008).
9. "Is the Product a Medical Device," CDRH Device Advice, http://www.fda.gov/cdrh/devadvice/312.html (October 21, 2008).
10. "Overview of What We Do," U.S. Food and Drug Administration, http://www.fda.gov/cdrh/overview2.html (December 17, 2007).
11. "Classify Your Medical Device," CDRH Device Advice, http://www.fda.gov/CDRH/DEVADVICE/313.html (September 25, 2007).
12. Estimates made by authors based on publicly available information.
13. Carlos Rados, "FDA Works to Reduce Preventable Medical Device Injuries," *FDA Consumer Magazine*, July/August 2003, http://www.fda.gov/fdac/features/2003/403_devices.html (October 21, 2008).
14. "CDRH Management Directory by Organization," U.S. Food & Drug Administration, http://www.fda.gov/cdrh/organiz.html#ODE (November 7, 2008).
15. "Office of Device Evaluation Annual Report," CDRH Device Advice, Fiscal 2007, http://www.fda.gov/cdrh/annual/fy2007/ode/report.pdf (November 7, 2008).
16. "Classify Your Medical Device," CDRH Device Advice, http://www.fda.gov/cdrh/devadvice/313.html#determine (March 22, 2007).
17. Class II devices are subject to the limitations on exemptions as described at www.fda.gov/cdrh/modact/frclass2.html (October 21, 2008).
18. An innovator can check the possibility of achieving exempt status (as well as any limitations that may apply) through parts 862–892 of the code of federal regulations; see http://www.accessdata.fda.gov/scripts/cdrh/cfdocs/cfcfr/CFRSearch.cfm?CFRPartFrom=862&CFRPartTo=892 (October 21, 2008).
19. See FDA Program Memorandum K86-3.
20. "510(k) Premarket Notification," CDRH Device Advice, http://www.fda.gov/cdrh/devadvice/314.html (December 12, 2006).
21. Aaron V. Kaplan, Donald S. Baim, John J. Smith, David A. Feigal, Michael Simons, David Jefferys, Thomas J. Fogarty, Richard E. Kuntz, Martin B. Leon, "Medical Device Development: From Prototype to Regulatory Approval," *Circulation*, 2004, http://circ.ahajournals.org/cgi/content/full/109/25/3068 (December 4, 2006).
22. See, for example, "Quickly Vetted, Treatment is Offered to Patients," *The New York Times*, October 26, 2008.
23. For more specific guidelines on making a 510(k) submission, see "Screening Checklist for All Premarket Notification [510(k)] Submissions," U.S. Food & Drug Administration, http://www.fda.gov/cdrh/ode/checklist-f102.html (November 7, 2008).
24. "How to Prepare a Traditional 510(k)," U.S. Food & Drug Administration, http://www.fda.gov/cdrh/devadvice/3143.html (October 21, 2008).
25. See "Guidance for Industry and FDA Staff Format for Traditional and Abbreviated 510(k)s," U.S. Food & Drug Administration, http://www.fda.gov/cdrh/ode/guidance/1567.html (October 21, 2008).
26. See "Screening Checklist for all Premarket Notification [510(k)] Submissions," U.S. Food & Drug Administration, http://www.fda.gov/cdrh/ode/checklist-f102.html (October 21, 2008).
27. See "Application Contents," CDRH Device Advice, http://www.fda.gov/CDRH/devadvice/pma/app_contents.html#required_element (October 21, 2008).
28. "Application Methods," CDRH Device Advice, http://www.fda.gov/cdrh/devadvice/pma/app_methods.html (February 25, 2007).
29. Ibid. See also "Product Development Protocol," U.S. Food & Drug Administration, http://www.fda.gov/cdrh/pdp/pdp.html (November 7, 2008).

30. "Information on Premarket Approval Applications," FDA, http://www.fda.gov/cdrh/pmapage.html (December 4, 2006).
31. See "Application," CDRH Device Advice, http://www.fda.gov/cdrh/devadvice/ide/application.shtml (October 21, 2008) for guidelines regarding IDE application submissions.
32. See "Draft Guidance for HDE Holders, Institutional Review Boards (IRBs), Clinical Investigators, and FDA Staff – Humanitarian Device Exemption (HDE) Regulation," U.S. Food & Drug Administration, http://www.fda.gov/cdrh/ode/guidance/1668.html (November 7, 2008).
33. See, for example, "NMT Medical Voluntarily Withdraws CardioSEAL(R) PFO HDE and Receives FDA Approval for New STARFlex(R) PFO IDE Study," *Medical News Today*, August 17, 2006, http://www.medicalnewstoday.com/articles/49797.php (November 7, 2008).
34. "Medical Device User Fees Have Been Reauthorized for Fiscal Years 2008–2012," U.S. Food & Drug Administration, September 28, 2007, http://www.fda.gov/cdrh/mdufma/092807-reauthorized.html (October 21, 2008).
35. See "Medical Device User Fee and Modernization Act (MDUFMA) of 2002, U.S. Food & Drug Administration, http://www.fda.gov/cdrh/mdufma/ (October 22, 2008).
36. By law, the FDA has 90 days to review a 510(k) submission and respond. Most products are cleared for marketing by the agency after one or two 90-day review cycles.
37. All quotations are from interviews conducted by the authors, unless otherwise cited. Reprinted with permission.
38. A patient registry is a clinical trial in which treatment is not randomized and there is no control arm.
39. The Prostar device was used to closed arterial access sites that measured 9/11Fr versus the collagen plug devices, which were studied in access sites measuring 8Fr and smaller.
40. "Datascope Corporation," Funding Universe, http://www.fundinguniverse.com/company-histories/Datascope-Corporation-Company-History.html (November 12, 2007).
41. Ibid.
42. "Rep. Joe Barton," *The Almanac of American Politics*, June 22, 2005, http://nationaljournal.com/pubs/almanac/2006/people/tx/rep_tx06.htm (November 30, 2007).
43. See "Medical Device Sector–Legislation," European Comission, http://ec.europa.eu/enterprise/medical_devices/legislation_en.htm (October 21, 2008).
44. "CE Mark," www.wikipedia.org, http://en.wikipedia.org/wiki/CE_mark (December 16, 2007).
45. Ibid.
46. Les Schnoll, "The CE Mark: Understanding the Medical Device Directive," September 1997, http://www.qualitydigest.com/sept97/html/ce-mdd.html (October 5, 2007).
47. See "Guidelines Relating to Medical Devices Directives," http://ec.europa.eu/enterprise/medical_devices/meddev/index.htm (October 21, 2008).
48. Ibid.
49. "Guidance Documents," Health Canada, http://www.hc-sc.gc.ca/dhp-mps/md-im/applic-demande/guide-ld/index-eng.php (October 21, 2008).
50. "Medical Device Regulatory Requirements for Australia," Export.gov, September 25, 2002, http://www.ita.doc.gov/td/health/Australiaregs.html (October 21, 2008).
51. "An Overview of the New Medical Devices Regulatory System," Therapeutic Goods Administration, 2003, http://www.tga.gov.au/docs/pdf/devguid1.pdf (October 21, 2008).
52. "Medical Device Registration Requirements," U.S. Department of Commerce, June 8, 2005, http://www.advamed.org/NR/rdonlyres/EC646067-88E4-4FF8-BB41-1660314FAC06/0/medical_device_registration_requirements.pdf (October 21, 2008).
53. See "Regulation of Medical Devices," Therapeutic Goods Administration, http://www.tga.gov.au/devices/devices.htm (October 21, 2008).
54. "Recent Revisions in Japan's Pharmaceutical Affairs Law for Medical Devices," Pacific Bridge Medical, September 2003, http://www.pacificbridgemedical.com/publications/html/JapanSeptember03.htm (October 21, 2008).
55. "Medical Device Regulatory Requirements for Brazil," Export.gov, September 23, 2008, http://www.ita.doc.gov/td/health/Brazil%202008%20Medical%20Device%20Profile.pdf (October 22, 2008).
56. "Brazilian Medical Device Import Regulations," DRW Research & Information Services, March 2008, http://www.drw-research.com/newsletter/Mar%2008.htm (October 22, 2008).
57. "Latin American Medical Device Regulations," *Medical Device Link*, July 2000, http://www.devicelink.com/mddi/archive/00/07/005.html (October 22, 2008).
58. "Medical Device Regulatory Requirements for Russia," Export.gov, October 24, 2008, http://www.trade.gov/td/health/russiaregs.html (October 22, 2008).
59. Ibid.

60. Ames Gross and Momoko Hirose, "Regulatory Update: Asia's Largest Medical Device Markets," Pacific Bridge Medical, January 2008, http://www.pacificbridgemedical.com/publications/html/AsiaRAFocusDevJan08.htm (October 22, 2008).
61. Ibid.
62. Ames Gross and Rachel Weintraub, "China's Regulatory Environment for Medical Devices," Pacific Bridge Medical, 2005, http://www.advamed.org/MemberPortal/Issues/International/Asia/china_reg2005.htm (October 22, 2008).
63. "China Medical Device Registration," Export.gov, August 2004, http://www.ita.doc.gov/td/health/regulations.html (October 22, 2008).
64. Ibid.
65. From "PMA Final Decisions Rendered for April 1997," U.S. Food and Drug Administration, http://www.fda.gov/cdrh/pdf/p960043.pdf (January 6, 2008).

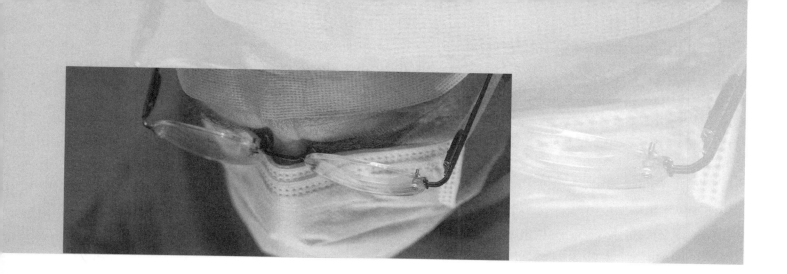

4.3 Reimbursement Basics

Introduction

Patients interested in a new technology or treatment; doctors committed to delivering it: in most industries, this would be a formula for success. But in the healthcare field, one critical factor is missing: the role of payers, or those third-party private or public insurance companies that make the decisions whether or not to pay for (or reimburse) a new medical device. With healthcare costs escalating and few patients able to afford their own medical expenses without insurance coverage, payers (through their reimbursement decisions) are exercising unprecedented levels of influence and control over the adoption of new technologies and, in turn, the direction of patient care.

The purpose of reimbursement analysis is to determine whether or not the existing healthcare payment infrastructure will accommodate a solution to a need, if such a solution becomes available. It addresses issues such as whether there will be adequate payment for the physicians using the solution, as well as for the facilities where patients would be treated. It also explores whether or not the coverage would be applicable to a large enough segment of the target market to make the development of the solution financially viable. If not, in today's healthcare environment, the innovator must either anticipate slow (or no) adoption of the device or consider pursuing a change to the established payment infrastructure to accommodate the new solution.

This is the first of two chapters on reimbursement. While this chapter focuses on understanding the existing reimbursement landscape for a solution and assessing how a new solution might fit in, 5.6 Reimbursement Strategy explores the process required to expand the existing payment infrastructure to accommodate a new device if the established payment and coverage levels are inadequate.

OBJECTIVES

- Obtain a high-level understanding of the reimbursement system for medical devices in the United States.

- Learn how to identify appropriate codes supporting the reimbursement of existing medical devices relevant to a need.

- Understand the status of reimbursement for existing medical devices that address the medical need under consideration, including reimbursement amounts, restrictions on types of patients covered, the location where the device will be used, and the frequency at which service is provided.

- Evaluate differences between US-based private and public payers, as well as international public payers.

Reimbursement fundamentals

Thomas Fogarty, the renowned innovator of the embolectomy balloon catheter and dozens of other medtech devices, stated emphatically that securing reimbursement, in combination with regulatory approval, is one of the most critical obstacles for new medical devices to overcome:[1]

> If you look at the things that are on the critical path now that are making it difficult, it's regulatory and reimbursement. If you look at the regulatory end from the device side, it's not unusual for a given device to be outdated within a seven- to ten-year period. Well, it may take you five to seven years to get through the regulatory path. It may take you another two years to get reimbursement. So the rapid advances of technology have outstripped the ability of certain agencies to respond to it.

In the United States, reimbursement for medical devices is handled by both public and private insurance programs. Public healthcare programs, such as Medicare, the US public insurance program for the elderly and disabled, which is managed by the Centers for Medicare and Medicaid Services (CMS), make up roughly 46 percent of all US healthcare spending and provide a template for private sector reimbursement (with private healthcare spending accounting for the balance).[2] This chapter focuses primarily on Medicare's reimbursement policies, while providing some details on how to obtain additional information on private payer policies and payers outside the United States. However, because Medicare's policies are complex and somewhat arcane, partly reflecting the political compromises that made the Medicare program possible (see Figure 4.3.1), the innovator should not expect to become an expert on the reimbursement policies relevant to the need s/he is examining just by reading it. Rather, the main goal of the chapter is for the innovator to become sufficiently familiar with the complicated reimbursement system so that s/he can better benefit from the assistance of reimbursement experts.

To understand the importance of reimbursement analysis as the innovator moves forward in the biodesign innovation process, consider the example of EndoNav, a company that developed a novel disposable device that could be inserted over a standard colonoscope to normalize complex colonoscopies and minimize the adverse effects traditionally associated with the procedure. Diagnostic colonoscopy is covered by Medicare, as well as most private payers, and has two primary

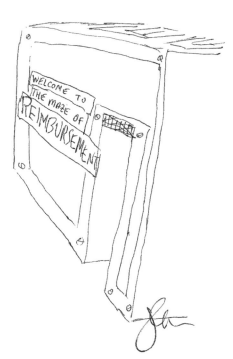

FIGURE 4.3.1
After regulatory clearance, the innovator still must navigate a complex maze of challenging reimbursement policies.[35]

reimbursement components: one for the physician performing the procedure and one for the facility where the procedure is performed. Depending on the details of the procedure, a physician chooses one of five common procedural terminology **CPT) codes** (explained in more detail later in this chapter) and is reimbursed from $407 to $512 if the procedure is performed in-office, or between $214 and $294 if it is performed at a facility, such as a hospital or ambulatory surgical center. Hospital outpatient facilities are reimbursed at $490 per procedure, whereas ambulatory surgical centers are reimbursed at $333 or $446, depending on which code is used.[3]

In this situation, EndoNav, like every other medical device start-up, needed to decide whether to operate under existing reimbursement codes or if it would seek to obtain new codes. With existing codes, the cost of its disposable device would have to be absorbed within the existing reimbursement rate paid to facilities (or to the physician in cases where the procedure is performed in a doctor's office). The company would need to develop a compelling value proposition to facilities and physicians regarding the benefits of adopting the device without additional reimbursement. It would also have to ask itself whether the revenue that could be generated using existing reimbursement codes is adequate to support the company's own existence. If not, EndoNav would have to justify the need for a new/separate CPT code and face the uncertainty associated with this risky and time-consuming route (see 5.6 Reimbursement Strategy for more information about the process for obtaining new codes). Seeking new codes can be particularly challenging for medical device manufacturers because codes are issued for new procedures, not new technologies. This distinction can make it difficult for device makers to justify new codes under certain circumstances.

The remainder of this chapter is focused on providing information about how to identify the relevant reimbursement codes and policies that are most likely to affect a new device, and to assess the reimbursement landscape as an input into final concept selection.

Reimbursement in the United States

Both Medicare and private payers in the United States follow the same general process for reimbursing medical services, which is described below. However, this discussion must be preceded by a cautionary note regarding the rather byzantine nature of US reimbursement for medical devices. The information that follows introduces new language and processes that may seem somewhat convoluted and without a compelling underlying logic. Although a much better way of managing reimbursement is clearly possible, this approach has been developed in the United States over multiple decades and is the result of decisions made when no one could anticipate the complexity of the modern medical system. As a result, innovators and healthcare professionals are left to struggle with a difficult and sometimes confusing reimbursement system.

Within the US reimbursement system, there can be multiple payments for what may appear to be a single medical encounter. A medical encounter typically involves a physician performing a service either in his/her office, in an outpatient clinic setting (e.g., a catheterization laboratory or in a hospital without overnight stay), or in an inpatient setting (e.g., requiring an overnight stay). Payments in the United States include a component for the physician and a component for the facility if the treatment is not provided in the physician's office. There can be multiple payments to the physician and the facility if the encounter involves multiple services, except for payments to the facility in the case of an inpatient stay.

Once services are performed, providers match them to an appropriate *code*. The payers, in turn, must make a positive *coverage* decision to result in a *payment*.

Coding

In order for physicians and facilities to be paid efficiently they submit bills to payers using standardized codes to document the diagnoses and procedures performed (see Table 4.3.1). Different codes are chosen depending on the setting. **ICD-9-CM** codes are used to describe both diagnoses and procedures for facility bills in the inpatient setting in order to justify the bill submitted to a payer. For facility bills in the outpatient setting, so-called HCPCS codes (an acronym for the Healthcare Common Procedure Coding System, which is pronounced "hicks-picks") are used. These include two levels of

Stage 4 Concept Selection

Table 4.3.1 *Different codes are used by different parties for the reimbursement of different procedures, services, and supplies.*

	Hospital or other facility	Physician
Inpatient setting	ICD-9-CM codes to indicate diagnoses and procedures	HCPCS level I codes (CPT) for medical professional services
Outpatient setting	HCPCS level I (CPT) codes for medical professional services HCPCS level II codes for products used outside the physician office	HCPCS level I codes (CPT) for medical professional services

codes. Level I codes are the CPT codes maintained by the American Medical Association (AMA). CPT codes are used to denote procedures and services provided by medical professionals (e.g., physician bills). **HCPCS Level II** codes are for products, supplies, and services not included in the CPT codes that are used/provided outside a physician's office. For example, level II codes would be used to submit bills for ambulance services and durable medical equipment, prosthetics, orthotics, and supplies. Level II codes are maintained by CMS. ICD-9-CM codes are used to justify the HCPCS codes used. Both for inpatient and outpatient facility payments, the ICD-9-CM codes and HCPCS codes are eventually mapped into another set of codes for payment purposes. For hospital inpatient payments, ICD-9-CM codes are mapped into so-called MS-DRG (described in the Payment section of this chapter). For outpatient facility payments, HCPCS codes are mapped into **APC** codes (also described in the Payment section).

Coverage

A properly coded bill submitted to an insurance company does not automatically translate into a payment. To grant reimbursement, the payer needs to have policies in place that state that the procedures, services, and supplies described in the bill are covered. These policies specify conditions under which a procedure is covered and provide details about the codes that should be used to submit and justify the bills. These policies are not uniform and can vary dramatically across payers. In Medicare's case, there are both national policies (called national coverage determinations, or **NCDs**) that cover the whole of the United States, and local policies that cover specific regions (Medicare has more than 17 regions in the United States, each empowered to make local coverage determinations, **LCDs**). National coverage determinations typically apply to payments for inpatient services while local decisions tend to be for outpatient services. Private payers have their own coverage policies, which they list on the Internet.

Payment

With appropriate codes and coverage decisions in place, the final ingredients needed for reimbursement is a payment. With Medicare, payment levels are linked to another set of codes that introduce even more acronyms to the reimbursement process. Inpatient facility payments are based on the "Medicare severity-diagnosis related group" (MS-DRG) number for the bill. The ICD-9-CM codes included in the bill are mapped into 745 MS-DRG codes based on the procedures performed and the disease severity.[4] Then, the payment to the hospital is based on the MS-DRG and adjusted to reflect local labor costs and conditions.

For outpatient facility payments, including payments to ambulatory surgical centers (ASCs) that do not involve an overnight stay, the CPT codes are mapped to an ambulatory payment classification (APC) code, which is then translated into a payment. For newly covered devices, there may also be a technology add-on (or pass-through payment) in effect for two years until the cost of the new devices is rolled into the standard payment levels set for the appropriate APC code.

For physician payments, the Medicare physician fee schedule is used to translate CPT codes into actual payments. For the same CPT code, the payment to the physician may vary depending on whether the service was performed in the physician's office (called a non-facility setting by Medicare), or in a facility setting. Payments may also vary geographically. Information about how to ascertain exact payment rates for specific codes is given in the Getting Started section of this chapter. Appendix 4.3.1 provides a summary of all the codes introduced above. The working examples provided below are also meant to clarify how, when, and why different codes are used to secure reimbursement payment.

More about physician reimbursement

Physicians use CPT codes, also called HCPCS level 1 codes, to describe and bill for medical, surgical, and diagnostic procedures, regardless of the setting in which they are performed. In parallel, they use ICD-9-CM codes to describe patient medical conditions and diagnoses to support their insurance billings and the chosen CPT code.[5] Medicare and private payers, in turn, use the ICD-9-CM codes to audit physician claims and validate the appropriateness of the billing codes used, based on the diagnosis and treatment performed.

> **Working example: Atrial fibrillation ICD-9-CM and CPT codes**
>
> When a patient with atrial fibrillation seeks treatment, the physician would use ICD-9-CM code 427.31 to indicate that the patient has this particular heart rhythm disorder. This code would serve as a means of justifying the charges made to a payer for any procedures performed, tests ordered, or services provided related to the management of atrial fibrillation. If the patient has multiple other conditions, s/he may have many ICD-9-CM codes noted by the healthcare provider. Again, this is because ICD-9-CM codes are descriptors of some or all of a patient's medical conditions relevant for a particular healthcare encounter.
>
> If the physician performs a cardiac ablation procedure for atrial fibrillation (using catheters maneuvered through the blood vessels to the heart), s/he would use CPT code 93651 in billing the payer for this procedure. If additional procedures are performed beyond cardiac ablation, the physicians would use multiple CPT codes to receive payment for all the procedures performed. See Figure 4.3.2 (steps 1 and 2) for an illustration of the various codes used by the physician.

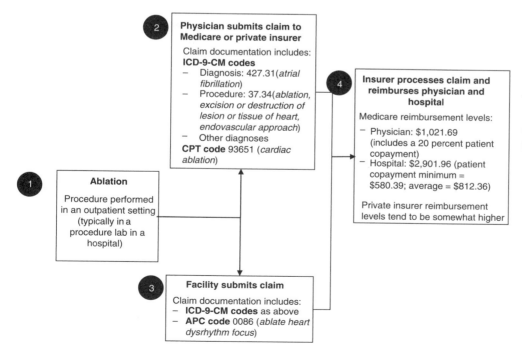

FIGURE 4.3.2
This example for the reimbursement of ablation for atrial fibrillation demonstrates how to trace a procedure through the reimbursement process. The payment values reflect Medicare averages across all geographic locations.

Physician reimbursement payments under Medicare are determined using the Resource Based Relative Value System (RBRVS) defined by the Health Care Professionals Advisory Committee Review Board (HCPAC). Each CPT code is assigned a number of relative value units, or **RVUs**. Relative value units provide a standardized measure of the resources needed to provide a particular procedure and include three components: (1) a work component, which reflects the time the physician spends on the procedure; (2) a practice expense, which includes nursing time, overhead, and supplies used in the procedure; and (3) a malpractice component that covers liability insurance for the procedure. The number of RVUs is then multiplied by a monetary conversion factor that is determined annually by the Health Care Financing Administration. The component RVUs may be further multiplied by factors to account for the cost of practicing in different geographic locations. The result is the amount Medicare will pay for the procedure or service performed in a particular geographic area. Medicare has used the RBRVS for physician reimbursement since 1992. The main rationale for using RVUs is that they standardize the measurement of resources needed for a particular physician service across specialties. Depending on where the procedure or service is performed, there may be two RVUs: a non-facility one for services performed in the physician's office and a facility one for all others. Changes to CPT codes and to the RBRVS are effective January 1st of each year.[6]

> **Working example:** Atrial fibrillation
> **Reimbursement payment**
>
> As shown in Figure 4.3.2, step 4, the reimbursed amount for performing a cardiac ablation (CPT 93651) in a facility setting, such as an inpatient hospital lab in San Francisco (locality 3114005), would be based on total RVUs of 26.24, a conversion factor of 37.8975, and various geographic modifiers resulting in a reimbursement of $1,021.69. This reimbursement payment includes a 20 percent copayment by the patient as mandated by Medicare coverage policies (payment and copayment amounts from other payers will vary).

> Note that for cardiac ablation performed in an outpatient setting, a single APC code is rarely submitted by itself. In most claims, the codes are "stacked" to reflect all of the services (and costs) associated with a procedure. The result is that the one APC code shown in the figure contributes roughly $2,901.96 to an overall charge that may typically be greater than $4,000 to $5,000.

More about hospital inpatient coding and reimbursement

Hospitals apply ICD-9-CM codes to cover both diagnoses and procedures for inpatient admissions. For Medicare, inpatient hospital reimbursement is addressed under the MS-DRG system. More than 10,000 ICD-9-CM diagnosis and procedure codes are grouped into 745 MS-DRGs. Patients within each MS-DRG category are considered clinically similar in terms of their resource use, and the hospital receives the same payment even if different procedures are performed and different diagnoses are made. For each admission, only one MS-DRG is assigned, regardless of the number of potentially relevant codes or the duration of the patient's stay.[7] Each MS-DRG has a unique relative weight, which is then converted into the payment amount. Medicare has used DRG codes for hospital inpatient reimbursement since 1983. DRGs were introduced to help control escalating hospital costs. Prior to the introduction of this system, hospitals were paid separately for each day of a patient's hospital stay and for all of the various procedures that were performed. By some accounts, this created an incentive to overuse procedures and keep the patient in the hospital for too long. Changes to DRG codes are effective October 1st of each year, along with updates to the ICD-9-CM procedure and diagnosis codes. In October 2007, Medicare introduced the MS-DRG system to expand the earlier DRG system by adding more than 200 new codes that reflect the presence or absence of major complications or comorbidities (MCCs). Under the new system, there are multiple MS-DRG codes for the same procedure to indicate the presence or absence of MCCs.

4.3 Reimbursement Basics

> **Working example: Atrial fibrillation MS-DRG code**
>
> If the patient in the atrial fibrillation example is hospitalized for a complex cardiac surgical procedure to treat his/her disease (e.g., a Maze procedure that requires many days of post-operative recovery in the hospital), the hospital would use MS-DRG 227 if the patient had major complications or comorbidities, or MS-DRG 228 if the patient was without major complications and comorbidities. These codes encompass all of the relevant patient care issues (and, thus, relevant ICD-9-CM codes) that would come up during the procedure and subsequent inpatient hospitalization. Alternatively, if the patient has a cardiac ablation procedure requiring a brief inpatient stay, the hospital might use the ICD-9-CM code 37.34 (ablation of cardiac tissue) as the description of the inpatient service. If Medicare is the payer, this code can get assigned to either MS-DRG code 250 (percutaneous cardiovascular procedure without coronary stent, or acute myocardial infarction but with major complication or comorbidity) or 251 (percutaneous cardiovascular procedure without coronary stent, or acute myocardial infarction and without major complication or comorbidity). (Note that MS-DRG codes are broad, with individual codes encompassing a wide array of procedures.)

More about hospital outpatient coding and reimbursement

Hospitals and other outpatient facilities use CPT or HCPCS II codes for claiming outpatient services. Under Medicare, each CPT and HCPCS II code is assigned to an Ambulatory Payment Classification (APC) group with a unique relative weight, which is then converted into a payment amount. Unlike the MS-DRG system, multiple APCs can be assigned and paid for as part of a single outpatient encounter, depending on the procedure performed. Medicare has used APCs for hospital outpatient reimbursement since 2000, with changes to APC procedure codes effective each January 1st.[8]

> **Working example: Atrial fibrillation APC code**
>
> If the patient undergoes a cardiac ablation procedure not requiring an inpatient admission, the hospital could use APC code 0086, which covers ablation, as shown in Figure 4.3.2, step 3. If electrical mapping of the heart was also performed, APC code 0087 (mapping) could be submitted, too (although, when multiple APCs are submitted, some are reimbursed at a 50 percent discount).

New technology add-ons

When a new technology or medical service is used in the outpatient setting, Medicare can temporarily establish a special additional payment. A new technology qualifies for such additional payment (referred to as an "add-on" or sometimes a "pass-through" to the APC or MS-DRG code)[9] if it represents a substantial clinical improvement, and if data reflecting the cost of new technology is not yet available. When data becomes available, it is used by Medicare to recalibrate the MS-DRG and/or APC payment, after which the add-on is discontinued.

Add-on payments for devices are supported by HCPCS c-codes. Typically, add-ons are only allowed for two years before they are discontinued (the cost of the new technology is eventually bundled into the APC and/or DRG payment). When the add-on payment is discontinued, then the c-codes are still used for documentation purposes. As a result, depending on when a new c-code was implemented, it may still be associated with an additional payment (if implemented in the last two years) or the additional device cost might already have been embedded in the APC and/or DRG payment. In the latter case, the c-code is used only for informational purposes.

> **Working example: Atrial fibrillation New technology add-ons**
>
> On August 1, 2000, an add-on payment was established for electrophysiology catheters used in conjunction with a cardiac ablation procedure not requiring an inpatient admission (billed under APC code 0086). Facilities were instructed to use HCPCS c-code

> **Working example:** *Continued*
>
> C1730 to receive the add-on payment. This additional payment was discontinued in December 2002, by which time the APC payment had been updated to reflect the new cost. The C code was nevertheless retained and is still used for documentation purposes.

Reimbursement by private payers

As noted, private or commercial health insurance companies have traditionally used the same **coding** system and approach to reimbursement as Medicare. In years past, there was an adage in the US healthcare industry that "what Medicare does, the rest will follow." Today, while private payers certainly look to Medicare for direction, they are increasingly likely to make independent decisions based on the goals and objectives of their individual plans. This means that reimbursement decisions can differ not only between Medicare and private payers, but also among the thousands of private payers in the United States. This has led to a reimbursement environment in which medical device companies must often seek to establish reimbursement coverage (and appropriate reimbursement rates) on a payer-by-payer basis – a rather daunting undertaking, particularly for young start-ups with limited resources.

Unlike the government, most private insurance companies are in business to make a profit. Even non-profit insurers, such as Blue Cross/Blue Shield (**BCBS**), have goals to maintain their financial health through mechanisms such as increasing their cash reserves (the equivalent of profit to a non-profit company).[10] While patients would like to think that health insurance companies have their best interests at heart, the decisions of these companies may actually be driven by a series of interrelated and complex factors. Patient well-being is certainly among the considerations, but other issues such as profitability, efficiency, and risk management also come into play. As one professional association stated on its website:[11]

> Insurers are not necessarily in business to assure that everyone receives access to care. Nor are they in business to guarantee that all qualified healthcare providers are fairly and adequately compensated for their services. Healthcare providers often try to assign 'moral obligations' to insurance companies, but they are not obligated to accept them. Although the hope is that insurance companies have a basic concern about the health needs of the general public and fair payments to practitioners, it should not be expected that this is their primary consideration.

Generally, as a guiding philosophy, commercial insurance plans state that they cover services that are deemed "medically necessary" by medical doctors. However, the concept of medical necessity is highly subjective and open to interpretation across payers and types of plans.

The two most common types of health insurance plans within the United States are health maintenance organizations (HMOs) and preferred provider organizations (PPOs). Health maintenance organizations bring together healthcare providers (e.g., doctors and hospitals) that have contracted with an insurance company to offer their services at a fixed price. Typically, HMOs are one of the most affordable insurance options available to healthcare consumers. Premiums are relatively low and copayments are inexpensive (or free). However, in order to offer services at a low, fixed price, HMOs depend on a high volume of patients and are notorious for being restrictive. Health maintenance organizations have stringent rules regarding which physician a patient can see and where service can be delivered. From a reimbursement perspective, they tend to place more limitations on what services will be covered, at what rate, and for which patient groups.[12] Preferred provider organizations, on the other hand, tend not to be as restrictive as HMOs.[13] While PPOs also contract with medical providers, they try to manage medical expenditures through financial incentives (e.g., charging different copayments for preventive versus corrective treatments, reimbursing certain medical procedures at rates higher than others). In addition, patients have greater choice in choosing a physician, selecting a facility, and managing their medical care in exchange for higher premiums.

Yet, even within the HMO/PPO paradigm, the policies of all insurance providers are not created equal.

Kaiser Permanente is one major HMO that uses stringent evidence-based criteria for adopting new technologies.[14] As a result, the company often approves and covers new technologies only *after* they have been thoroughly studied in the post-approval environment and when there is strong evidence (ideally from controlled clinical trials) that the new device improves clinical outcomes. Because the process of conducting studies and publishing the results in reputable medical journals can take years to accomplish, Kaiser physicians may be reluctant to use the latest generation of a technology if the evidence behind it is insufficient.[15] Consider, for example, implantable cardiac defibrillators (ICD) used to treat patients with life-threatening cardiac rhythm disorders. While it is undisputed that ICDs help prevent sudden cardiac deaths, new generations of the device include features for which the clinical benefit is not fully supported by the results from extensive clinical trials. An article published in the *Permanente Journal* pointed out that while new features, such as dual-chamber and rate responsive pacing, had driven up the cost of ICDs, most patients do not necessarily benefit from this specialized functionality. For this reason, it advised physicians to consider whether the additional cost associated with the latest technology was justified relative to the potential benefit that each individual patient would receive.[16]

This statement suggests that payers who utilize strict evidence-based guidelines may be more resistant to covering new medical device innovations if the value of the innovation is not fully supported by clinical trials.

In general, innovators and young companies are advised to consider their market in determining how much time to invest in understanding the specific reimbursement policies of private payers. If there is a significant Medicare market for the device, understanding Medicare reimbursement practices will be sufficient in most cases. However, if most reimbursement will come from commercial payers, then it will be necessary to invest more time in investigating private payer policies, reimbursement behaviors, and the precedents that have been set by other devices. Some commercial insurers have established groups that specifically evaluate new medical technologies before reimbursement decisions are made. The Blue Cross/Blue Shield Technology Evaluation Center (**TEC**) is one well-known example.[17] Many individual plans also post their medical policies regarding coverage and reimbursement on the Internet. Individual payment rates for private insurers, however, can be difficult to obtain from public sources.

The self-pay reimbursement model

Even though more than 87 percent of total health expenditures in the United States are covered by public or private payers, slightly more than 12 percent are paid for directly by patients. These out-of-pocket expenses can include copayments that are a common part of regular insurance coverage, but also self-pay expenditures for elective procedures, such as laser eye surgery, that are typically not covered by insurers.

An innovator should examine carefully whether or not the need on which s/he is working is appropriate for a self-pay model. This approach can be advantageous since it removes the obstacle of reimbursement. However, it also has certain risks and challenges. For instance, the amount of money consumers may be willing to spend out-of-pocket is likely to be much smaller than the amount third-party insurers typically pay for procedures. Nevertheless, the option of self-pay should be examined, especially for needs related to cosmetic improvements and other procedures designed to enhance quality of life rather than to address a dire medical condition.

Preparing a reimbursement analysis

Before selecting a final concept, an innovator should have a detailed understanding of the reimbursement environment associated with each solution idea. Such an analysis summarizes the reimbursement landscape for similar or related innovations, including all relevant codes, coverage decisions, and payment levels that will potentially serve as precedents for a new technology in the field. The purpose of performing a reimbursement analysis is best illustrated through an example. The story of Metrika, Inc. exemplifies many of the issues discussed in this chapter, as well as illustrating how basic reimbursement analysis serves as a bridge to the more complex exercise of developing a reimbursement strategy (see 5.6 Reimbursement Strategy).

FROM THE FIELD: POINT-OF-CARE DIAGNOSTICS IN DIABETES

Evaluating the adequacy of reimbursement under established codes

In the care of diabetes, it is recommended that all patients have a hemoglobin A1C (HbA1C) test twice a year. This test shows the average amount of sugar in the patient's blood for the three months preceding the test. The test results are then used by the patient's physician to adjust treatment and modify nutritional guidelines. Traditionally, the HbA1C test involved drawing a patient's blood in the lab. As a result, some patients skip testing because of the inconvenience. Michael Allen, a California BayArea innovator and entrepreneur, came up with a vision for a disposable, convenient, hand-held point of care test for HbA1C (see Figure 4.3.3), which would enable patients to have their HbA1C tested when they visit their physicians rather than having to take the test in a laboratory setting. Being able to perform the test in the doctor's office had several potential advantages: it was more convenient for patients, it would likely increase patient compliance since there was no second step to the testing process (i.e., the visit to the lab), and it would potentially lead to better management of diabetes, because the results would be available immediately so that the physician and patient could discuss them face-to-face in the office.

Allen founded Metrika, Inc. in 1994 to commercialize his idea (Bayer later acquired Metrika in 2006). When considering the reimbursement environment for a new POC test, the manufacturers presumably performed a reimbursement analysis.[18] The first step in this process would likely have been to understand whether HbA1C tests were currently being reimbursed, at what level, and whether the reimbursement criteria would be met by the new POC test. In 1994, Medicare was reimbursing HbA1C lab tests using the CPT code 83036 (Hemoglobin; glycosylated). That is, labs ("facilities" in reimbursement terms) performing HbA1C tests were using this code to submit their claims to Medicare and were reimbursed at the rate of approximately $13 per test.

The next question facing any manufacturer in the space was whether or not it could make use of this code to cover its test. The answer (with perfect hindsight) appears to have been relatively straightforward: if the manufacturer could demonstrate that its POC test was equivalent to the existing HbA1C tests performed in the lab, then it could probably take advantage of the existing code (there is no reason why Medicare would

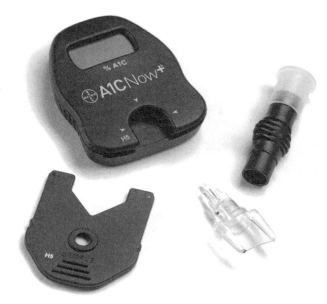

FIGURE 4.3.3
The A1CNow+® point of care HbA1C monitor (courtesy of Bayer HealthCare).

not be willing to pay the same amount for an equivalent test performed at a different setting, although careful consideration of any stakeholder issues arising from a potential threat to a lab's business should be evaluated).

The third question to be addressed was whether or not the reimbursement level associated with the existing code was adequate. To determine this, the manufacturers had to calculate a rough estimate for their **manufacturing cost** per test plus a reasonable mark-up (i.e., margin on that cost). Without knowing the exact numbers for any manufacturer of an HbA1C POC test (since this information is proprietary), one can perform a quick "back of the envelope" calculation using what is known as the "50 percent rule" of approximation. To be viable, the $13 reimbursement rate needed to be sufficient to cover the end user's (physician's) cost of acquiring and performing the test, plus a potential mark-up for the manufacturer. Applying the 50 percent rule, one can infer that the cost of supplies for the physician should be no more than approximately 50 percent of the total reimbursement rate. This means that the price a physician would pay to purchase the test at wholesale would be approximately $6.50 (including shipping and handling). Applying the 50 percent rule again,

> one can infer that a wholesaler's mark-up is typically 50 percent, which means that the price a manufacturer could charge the wholesaler for distributing the device would be about $3.25. Assuming that the manufacturer wants and/or needs to make a 50 percent margin on its costs to justify development of the product, the company's production cost would need to be no more than $1.62. This means that the manufacturer should be able to produce the test at a cost of no more than $1.62. If such a cost level is not technically feasible, then any manufacturer must consider seriously whether it should seek a new CPT code and a different reimbursement rate. Alternatively, even if this cost is technically feasible, the manufacturer may want to evaluate whether a value-based argument could be created to support a higher reimbursement and a different code to reflect both the innovation inherent to the POC test and the benefit arising from it.
>
> In the case of the HbA1C POC tests, what actually happened seems to suggest that the reimbursement level associated with the existing CPT code was inadequate to cover the manufacturing costs, manufacturer's margin, and distributor's profits (or that the team felt the innovation in the POC test could justify higher reimbursement). In 2005, the retail price for Metrika's test was $24 (including shipping and handling).[19] Assuming a rough 50 percent profit margin at any part of the value chain (as described above), this means the wholesale price was about $12 per test, the price paid by the wholesaler to Metrika was approximately $6, and production costs per test were about $3. In 2006, the AMA issued a new CPT code (83037). As of 2007, Medicare was providing reimbursement for the Metrika device and the products of its main competitors at $21.06 in most states.[20] Importantly, however, this achievement took seven years from the first FDA approval (September 2000) to the time that the new CPT code was approved by the AMA, accepted by CMS, and associated with a standard reimbursement amount.

A comprehensive reimbursement analysis includes information on the topics described in Table 4.3.2. The innovator can then use this information to compare the concepts in terms of how difficult or simple it may be to obtain reimbursement for them. S/he can also use the information gathered through this research to refine and/or validate key market assumptions (see 2.4 Market Analysis), since it results in new data on the number of procedures performed and total reimbursement granted per procedure.

The steps for completing a reimbursement analysis are outlined in the Getting Started section of this chapter. Table 4.3.3 shows a high-level reimbursement summary for pulmonary vein ablation (cardiac ablation used to isolate the pulmonary vein).

A note on global reimbursement

Medical reimbursement outside the United States (OUS) has a reputation for being even stricter than inside the country. Innovators intending to market products OUS have an obligation to carefully understand the reimbursement systems and processes in the countries they are targeting. This section provides a general framework that innovators can use to approach global reimbursement. Some of the key issues and challenges in major developed and emerging markets are also presented.

To understand the reimbursement process for a medical device in any country, the innovator first needs to investigate the basis for financing and delivering healthcare. In most developed nations, there is public financing of healthcare from taxpayers so the government is ultimately responsible for deciding how healthcare funds are distributed to providers. Delivery of healthcare could be public (as in the United Kingdom) or private (as in France). Providers may be allocated a fixed budget from which they are to cover the expenses of all medical services, or they may be paid from the government for each service they provide (using payment processes analogous to the ones in the United States). Another model is a hybrid system in which capital expenses (e.g., purchases of expensive equipment or capacity expansions) are covered by a budget allocation, with additional payments granted for each service provided (e.g., variable costs such as physician salaries and supplies). Across the globe, the hybrid system appears to be the model that is most widely utilized.

It is important for the innovator to recognize that the method used to allocate funds to providers is critical, since it is directly related to the adoption of a new

Table 4.3.2 *Reimbursement analysis should provide the innovator with an overview of the current reimbursement landscape for technologies related to the need being studied, in the following topic areas.*

Topic	Description
Location of procedure	Describe the setting in which the new device would be utilized and/or procedure would be performed.
Coverage decisions and technology assessment by Medicare and/or private payers	Summarize the coverage decisions for comparable devices and/or procedures, including how long it took to achieve reimbursement and any constraints or exceptions that may affect coverage.
CPT code and payment amount	Define the appropriate CPT code and patient copayment under Medicare or determine if a new code is needed.
DRG, APC, and/or other codes and payment amounts	Define the appropriate DRG, APC, or other code (e.g., HCPCS Level II) and patient copayment under Medicare or determine if a new code is needed.
Facility costs	Summarize the anticipated facility-related costs for the device and/or procedure.
Number of procedures	Summarize the number of procedures performed and reimbursed, reimbursement per procedure, and payer mix.

Table 4.3.3 *A sample reimbursement summary for pulmonary vein ablation.*

Topic	Description
Location of procedure	Electrophysiology lab
Coverage decisions and technology assessment by Medicare and/or private payers	According to a 2006 TEC Assessment, the available evidence was insufficient to permit conclusions on the effect of cardiac ablation of atrial fibrillation using a pulmonary vein isolation (PVI) procedure on outcomes of atrial fibrillation.[36] This meant that the procedure would not routinely be covered by many payers.
CPT code and payment amount	CPT code 93651 Total payment (including copayment) $1,021.69
DRG, APC, and/or other codes and payment amounts	APC code 0086 Medicare payment per procedure $2,901.96
Facility costs	Median facility cost $2,829.20
Number of procedures	9,622 (Medicare) Number includes all ablation procedures covered by Medicare.

technology. As one moves from a fixed budget allocation to a per-service system, the incentives of the providers to use a new technology go up, as long as the technology is adequately reimbursed. Another key issue is to understand whether the purchasing decision and the price paid for a device will be determined through direct negotiations with each provider, or through some global purchasing agreement that involves multiple providers and possibly the government.

High-level reimbursement questions that the innovator should ask, for each global market s/he is considering for device sales, include the following:

- Is healthcare financing public or private?
- Is delivery public or private?
- How are providers paid for health services and for capital expansions?
- Is the volume of services delivered by each provider regulated?
- Are the prices for devices regulated?
- Who negotiates the price of a device and what is the process used (e.g., direct negotiations between providers and manufacturers, government contracting, or contracting with an alliance of providers)?

- Do new devices have to undergo technology assessment before they can be used and reimbursed?
- Is there a list of approved devices that are reimbursed?

With a solid understanding of the answers to these general questions, the innovator should next consider the unique aspects of the reimbursement systems in the countries being evaluated. A sample of these issues is provided below for three European[21] and two Asian nations. Innovators must carefully understand the local reimbursement policies and practices of the countries in which they seek to do business, particularly if they anticipate the need to negotiate with government agencies. One way to quickly and effectively learn about national practices is to partner with a local distributor. A strong relationship with an established local player can be a valuable asset in securing reimbursement, as well as setting up an effective sales and distribution infrastructure overseas (see 5.8 Sales and Distribution Strategy).

France

France is the second largest medical device market in Europe. Healthcare is publicly financed and delivered by both public and private providers. All employed individuals in France, as well as their children and spouses, are covered by the national health insurance plan called *securité sociale*. Individuals who are not entitled to participate in this program (e.g., affluent individuals who are not employed) must purchase special coverage, known as *assurance personelle*. Many people covered by the state-run program also choose to purchase additional insurance to supplement their basic coverage.[22]

Outpatient procedures are reimbursed if they are listed on the Liste des Produits et Prestations Remboursables (LPPR) found on the securité sociale website.[23] Inpatient procedures are reimbursed using a DRG-system referred to as the Groupes Homogenes de Sejours (GHS), which is available on the same website. Payments may vary according to whether the procedure is performed in a private or public setting, and the list of approved outpatient procedures may vary according to the setting. Expensive devices may receive an add-on payment. Public hospitals have a fixed budget to cover all their capital expenses, so these do not have to be funded by revenue from fees.

When a new device that is not included in the existing lists for the outpatient or inpatient system enters the market, it undergoes a health technology assessment by the Commission d'Evaluation des Produits et Prestations (CEPP, France's products and services assessment commission). The process may take as long as three to four years, and there is no provision for interim payments. In addition, to manage the total costs of expensive devices, the number of procedures performed in each region of the country is regulated.

Purchasing is both centralized and decentralized, with larger hospitals using a bidding process and smaller hospitals relying on a central purchasing organization, the Union des Groupements d'Achat Publique (UGAP). The acquisition of expensive medical equipment requires approval from the health ministry.

Germany

In Germany, the country's *statutory health insurance system* (SHI) provides health insurance coverage to approximately 90 percent of the population, with the remainder covered through a private health insurance system.[24] Healthcare expenditures are covered by these insurers with the exception of capital acquisitions and expensive equipment (greater than €150,000), which are covered by state infrastructure funds.

In January, 2006, the German reimbursement system was reformed and now follows reimbursement policies similar to the French system. There is an approved list of outpatient procedures, which can be found in the Einheitlicher Bewertungs-maßstab (EBM, or uniform value scale).[25] Payments for these procedures vary by region. Inpatient procedures are grouped according to a DRG system.[26] Diagnosis related group reimbursement covers the cost of a procedure and any device used to perform it, but not capital expenses and equipment.

New devices to be used in the outpatient setting need a positive health technology assessment before they can be reimbursed. In contrast, new devices used in the inpatient setting can be reimbursed as long as there is not a negative health technology assessment. Technology assessment is performed by the Institute for Quality and Efficiency in Healthcare (IQWiG).[27]

Private hospitals and physicians purchase medical devices directly from the suppliers. Municipal and teaching hospitals use an open bidding process. Expensive equipment is also purchased through public tender.

United Kingdom

In the UK, the majority of healthcare is provided by the National Health Service (NHS), a publicly funded healthcare system established in 1948.[28] The NHS is organized in Primary Care Trusts and Hospital Trusts with responsibility in their geographic areas. The Department of Health allocates funds to these trusts, which are used to provide the necessary services and to invest in infrastructure. Trusts reimburse providers of services using a pay-for-results system introduced in 2002. General practitioners (GPs) play a critical role in the UK health system because they serve as gatekeepers for specialized care. Medical devices reimbursable in the outpatient setting are listed in the drug tariff list. Inpatient procedures are reimbursed using a DRG system referred to as HRG (Healthcare Resource Groups).[29]

All new devices must undergo a rigorous technology assessment by the National Institute of Health and Clinical Excellent (NICE). The process may take 12 to 24 months. Devices that receive a positive assessment are then added to the drug tariff lists and an appropriate HRG. Before that, the device manufacturer may be able to negotiate reimbursement on a pass-through basis directly with the trusts.

China

China's healthcare sector is growing rapidly – one estimate predicted that China's healthcare expenses would exceed 10 percent of its GDP by 2015.[30] Consequently, the Chinese market provides exciting opportunities for biodesign innovators.

In recent years, there has been a gradual shift in the healthcare system with more autonomy being granted by the government to local hospitals and healthcare providers. While the vast majority of hospitals are administered by China's Ministry of Health (MOH), they are now expected to generate revenue to cover as much as 90 percent of their operating expenses.[31] This creates incentives for hospitals to overuse services they can charge for, such as dispensing prescription drugs.

However, most individuals in China lack health insurance and pay for their medical expenses out of pocket (53 percent of health expenditures are paid for out of pocket, while 38 percent are covered by public funds, and the remainder is funded by private insurance).[32] This dynamic continues to keep modern healthcare beyond the reach of most of China's citizens, although Chinese officials have acknowledged the problems associated with high out-of-pocket expenses and are working on plans to improve the availability of health insurance for both urban and rural residents.[33]

Overall, the Chinese system remains in flux. However, the medtech industry is characterized by repeated attempts by the MOH to introduce policies that place downward pressure on medical device prices. Expensive capital equipment (greater than the equivalent of $250,000) is purchased through a centralized, government-run bidding process. A similar process was proposed in 2003 for expensive implantable devices. Other cost containment programs have been the subject of experimentation at the local and regional level.

India

India is the fourth-largest medical device market in Asia. The delivery of healthcare is provided by entities in both the public and the private sectors. Private healthcare developed rapidly over the last decade and has established a reputation for delivering care of superior quality to the public sector. As a result, private providers are responsible for delivering 65 percent of primary care and for managing more than 40 percent of India's hospitals. In terms of expenses, the public sector covers 17.3 percent of total healthcare expenditures, while out-of-pocket funding accounts for a full 77 percent of healthcare spending.[32]

As of 2008, the Indian government was considering including price negotiations for medical devices as part of its process for regulatory/marketing approval. Under such a system, the National Pharmaceutical Pricing Authority, which maintains a list of approved drugs and their agreed price ceilings in India, would gather information on the prices of medical devices in India and in other markets. This information would then be used as part of the negotiations with device makers to establish a cap on the prices of their new products.[34]

As noted, understanding reimbursement in the United States and around the world can be challenging, particularly for those innovators new to the medical field. However, the following process can be used to help the innovator effectively come up to speed and prepare a reimbursement analysis for the concepts under consideration.

See **ebiodesign.org** for active web links to the resources listed below, additional references, content updates, video FAQs, and other relevant information.

1. Confirm location of procedure

1.1 What to cover – Determine the setting in which the procedure will be performed.

1.2 Where to look – Based on what is known about the innovation, determine the most appropriate setting for the procedure. Use the following sources to benchmark the settings used for similar or related procedures and to justify the determination:

- **PubMed** – A database of the US National Library of Medicine that includes more than 16 million citations from MEDLINE and other life science journals back to the 1950s.
- **UpToDate** – A database of evidence-based clinical information.

2. Research coverage decisions and ICD-9-CM/CPT/APC/MS-DRG codes

2.1 What to cover – For the most similar and relevant existing procedures, research the coverage decisions made by Medicare and private insurance, as well as any technology assessment decisions. Determine when, how, and why they received reimbursement and any technology assessment recommendations. Identify the relevant ICD-9-CM codes for the procedure and associated diagnoses, the assigned CPT codes for each procedure, and any related MS-DRG, APC codes, and technology add-ons. Typically, it is easiest to identify the CPT code and then map that to MS-DRG/APC codes and technology add-ons using the information obtained in this step, or in steps 3 and 4 below. (Note: keep in mind that there may be multiple relevant CPT codes. Conversely, some devices may not be assigned a CPT code(s) and/or may not have been awarded reimbursement coverage. If an appropriate CPT code cannot be identified in this step, proceed to the next step.)

2.2 Where to look

- **Private insurance medical policy websites** – Such as Regence, Wellmark, or Aetna.
- **Medicare site** – Be sure to search for both local and national coverage decisions.
- **Technology assessment sites** – Be sure to identify all technology assessment reports generated by the Blue Cross/Blue Shield Technology Evaluation Center, NICE in the UK, and other related technology assessment groups.
- *Ingenix CPT Expert* (Thomson Delmar Learning, 2005) – This book is the recognized standard for hospital coding. It contains all CPT codes categorized by organ systems. Since many codes could potentially be used for one procedure, it is helpful to review the filtered list of codes with someone familiar with the codes being assessed, such as a clinician or office practice manager.
- **Treatment research** – 2.2. Treatment Options may include technology assessment data for the different treatment options.

3. Investigate reimbursement information for non-covered devices

3.1 What to cover – If no coverage decision information or CPT codes are available, look for reimbursement-related information on manufacturer websites. Often, manufacturers will provide status updates on their progress toward receiving reimbursement to educate potential customers and keep them interested in their devices. They also might include information detailing the CPT codes they

Stage 4 Concept Selection

are pursuing for reimbursement, the time frame within which they expect to receive reimbursement, the process for appealing coverage decisions, and/or the number of appeals that have been won if reimbursement has been granted on an exception basis. (Note: if candidate CPT codes are identified in step 3 but not in step 2, go back to step 2 and research those codes and their relevant coverage decisions.)

3.2 Where to look
- **Device manufacturer websites** – Companies such as Medtronic offer general reimbursement assistance on their website for certain practice areas (e.g., for cardiac rhythm management). Similar information can be found by searching other sites for major manufacturers. Guidelines are often provided by product or treatment area to help physicians choose the most appropriate code(s) for maximizing their reimbursement.
- **Federal Register**
- **HCPCS Physician Fee Schedule** – Download the fee schedule from Medicare's website.

4. Identify reimbursement rates

4.1 What to cover – Find the Medicare reimbursement rates for the relevant CPT code that has been identified. Be sure to account for physician and facility reimbursement using the resources listed below.

4.2 Where to look
- **HCPCS Physician Fee Schedule Look Up** – Multiply the RVU for the appropriate CPT code by the conversion factor to get the Medicare payment to the physician. If the procedure is performed in a facility setting, use the RVU listed under the fully implemented facility total. Otherwise, use the fully implemented non-facility total or use the resource below.
- **Medicare Hospital Outpatient Prospective Payment System** (OPPS) – Available online on the Medicare website. Search for the HCPCS code in the document, find its corresponding APC code, and then obtain the facility payment.
- **Cost of procedures covered by APC and number of procedures performed** – This Excel file is available online. It displays median costs, by APC group, for services payable under the OPPS in calendar year 2007. The data are based on claims for hospital outpatient services provided January 1, 2005 through December 31, 2005.
- **Ambulatory surgical center file** – If the payment cannot be located in one of the outpatient files, then try the ambulatory surgical center file and search by HCPCS code.
- **Hospital DRG file** – If reimbursement rate information still cannot be located, try the hospital DRG files on the Medicare website. Be sure to download the *final version* of the list of all DRGs. Using this file, identify the DRG and its "relative weight." Next, this will need to be multiplied by a base payment rate that consists of a labor and non-labor component. Access these files from the Medicare website under Acute Inpatient – Files for Download.

5. Identify number of procedures

5.1 What to cover – Identify the number of procedures performed per year, reimbursement per procedure, and payer mix.

5.2 Where to look
- **HCUPnet** – A free, online query system based on data from the Healthcare Cost and Utilization Project (HCUP). It provides access to health statistics and information on hospital stays (inpatient encounters) at the national, regional, and state levels.

- **Medicare Part B Physician/Supplier Extract Summary File** – This file summarizes the number of procedures, total submitted charges, and total payments by HCPCS code. The file can be ordered from the Center of Medicare and Medicaid Services.
- **Other databases** – Verispan and National Patient Profile are subscription services that provide procedural data before HCUP releases it.
  - **National Patient Profile** – A desktop-enabled database (i.e., on a CD) provided by Verispan. It is searchable by diagnosis or procedure codes (ICD-9-cm only) and includes information similar to that on HCUPnet, except that it is often more up to date and has an interesting feature to show the frequency of different ICD-9-cm diagnoses for each procedure.
  - **Verispan** – A subscription service that provides access to data on the frequency of procedures carried out for patients who are diagnosed with specific conditions. The data is provided in such a way that in-depth customization and analysis can be performed.

Credits

The editors would like to acknowledge Darin Buxbaum, Trena Depel, and Asha Nayak for their assistance in developing this chapter. Many thanks also go to Mitchell Sugarman, of Medtronic, for lecturing on Reimbursement basics in Stanford's Program in Biodesign. This chapter is based, in part, on his presentations. We are also grateful to Michael Allen for bringing point-of-care testing in diabetes to our attention.

Appendix 4.3.1

Common reimbursement codes used to support inpatient/outpatient and facility/non-facility charges

Code	What it stands for	What it is	How it is used for reimbursement	Who manages the system
ICD-9 *Example* 427.3 for atrial fibrillation	International Classification of Diseases, 9th Revision	Codes for **classifying patient diagnoses** and mortality data from death certificates	ICD-9 codes are the international standard from which the US clinical modifications are derived and used in the reimbursement process (see below)	Managed by the CMS and the National Center for Health Statistics (**NCHS**) in collaboration with the World Health Organization
ICD-9-CM *Example* 427.31 for atrial fibrillation	International Classification of Diseases, 9th Revision, Clinical Modification	Codes for classifying morbidity data and **describing patient diagnoses and procedures** associated with utilization of health services in the United States	Used by physicians to justify the CPT and/or DRG codes they use to bill for a particular patient encounter (see below for more information on CPTs and DRGs); also used by public and private payers for auditing the use of healthcare codes by providers	Managed by the CMS and the National Center for Health Statistics in collaboration with the World Health Organization
CPT *Example* 93651 for cardiac ablation	Current Procedural Terminology; also known as HCPCS Level I codes (under the Health Care Financing Administration's Common Procedure Coding System)	Codes for **describing medical, surgical, and diagnostic procedures** (but not diagnoses themselves), regardless of whether they are performed in a clinic, inpatient hospital, or outpatient hospital setting	Used by Medicare for making physician reimbursement decisions, but not for covering facility-related reimbursement; does not include codes needed to report and cover medical items or services billed by suppliers other than physicians	Maintained by the American Medical Association (AMA) on behalf of the CMS
MS-DRG *Example* 227 for a Maze procedure if the patient had major complications or comorbidities; 228 without them	Medicare Severity – Diagnosis Related Group	Codes for **describing the procedures and other services performed in a hospital** setting	Used by Medicare for making non-physician related reimbursement decisions for procedures performed in an inpatient hospital setting	Maintained by the CMS

Continued

Code	What it stands for	What it is	How it is used for reimbursement	Who manages the system
APC *Example* 0086 for heart ablation	Ambulatory Payment Classification	Codes for **describing hospital outpatient services**, including diagnostic procedures, cancer therapies, ambulatory surgery, clinic and ER visits, partial psychiatric hospitalization, and surgical pathology	Used by Medicare for setting a specific, uniform amount that all hospitals will be paid for the same outpatient services	Maintained by the CMS
HCPCS Level II *Example* C1730 for electrophysiology catheter for diagnostic and/or ablation other than 3D or vector mapping (19 electrodes or less) Note: Often not billed separately beyond base facility fee unless an extra device is required during the procedure	Health Care Financing Administration's Common Procedure Coding System – Level II	Codes used **for supplies and services obtained outside the physician office that are not covered by a CPT code or APC code**, such as ambulance services and durable medical equipment, disposable supplies, injectable drugs, prosthetics, and orthotics	Used by hospitals, as well as other non-hospital and non-physician suppliers, to submit claims; also used by Medicare to make reimbursement decisions for procedures, services, equipment, and supplies utilized in an outpatient setting but not typically billed by physicians	Maintained by the CMS Note: HCPCS II codes may apply to supplies used as part of a procedure, but may not be separately reimbursable beyond the APC code

Notes

1. From remarks made by Thomas Fogarty as part of the "From the Innovator's Workbench" speaker series hosted by Stanford's Program in Biodesign, January 27, 2003, http://biodesign.stanford.edu/bdn/networking/pastinnovators.jsp. Reprinted with permission.
2. "Report to Congress: Medicare Payment Policy," MedPac, March 2006, p. 9.
3. "Polypectomy/Foreign Body Removal: Coding and Payment Quick Reference," Boston Scientific, 2005, http://www.bostonscientific.com/templatedata/imports/collateral/Endoscopy/rmbgde_2005%20Polypectomy%20billing%20guide_us.pdf (August 8, 2007).
4. Federal Register, Department of Health & Human Services, August 22, 2007, http://www.bkd.com/docs/industry/082207finalrule.pdf (October 22, 2008).
5. "Classification of Diseases and Functioning and Disability," National Center for Health Statistics, http://www.cdc.gov/nchs/icd9.htm (October 30, 2006).
6. "2006 Physician Coding: Reference Guide for Cardiac Rhythm Management Procedures," Medtronic, http://www.medtronic.com/crdmreimbursement/downloads/2006Physician_Ref2069aENp6.pdf (October 27, 2006).
7. "2006 Hospital Coding: Reference Guide for Cardiac Rhythm Management Procedures," Medtronic, http://www.medtronic.com/crdmreimbursement/downloads/2006_Hospital_Code4aENp4.pdf (October 27, 2006).
8. Ibid.
9. "List of Device Category Codes for Present or Previous Pass-Through Payment and Related Definitions," Centers for Medicare and Medicaid Services, http://www.cms.hhs.gov/HospitalOutpatientPPS/Downloads/DeviceCats_OPPSUpdate.pdf (January 9, 2008).
10. "Private Insurance Companies," American Academy of Physician Assistants, http://www.aapa.org/gandp/priv-insur.html (August 8, 2007).
11. Ibid. Reprinted with permission.

12. Eric Wagner, "Types of Managed Care Organizations," *The Managed Health Care Handbook* (Aspen Publishers: Maryland, 2001).
13. Ibid.
14. Ken Krizner, "Health Plans Apply Scientific Evidence to New Technologies," Managed Healthcare Executive, July 1, 2006, http://www.managedhealthcareexecutive.com/mhe/article/articleDetail.jsp?id=357680&sk=&date=&pageID=2 (October 31, 2007).
15. Mitchell Sugarman, "Permanente Physicians Determine Use of New Technology: Kaiser Permanente's Interregional New Technologies Committee," *Permanente Journal*, Winter 2001, http://xnet.kp.org/permanentejournal/winter01/2001winter.pdf (October 31, 2007), p. 46.
16. Michael R. Lauer, "Clinical Management for Survivors of Sudden Cardiac Death," *Permanente Journal*, Winter 2001, http://xnet.kp.org/permanentejournal/winter01/2001winter.pdf (October 31, 2007), p. 24.
17. See "Technology Evaluation Center," Blue Cross/Blue Shield Association, http://www.bcbs.com/betterknowledge/tec/ (October 20, 2008).
18. This case example was developed from information available through public sources.
19. "Cost and Contact Information for Available Point-of-Care Glycosylated Hemoglobin Monitors," *The Annals of Pharmacotherapy*, http://www.theannals.com/cgi/content/full/39/6/1024/TBL1 (September 13, 2007).
20. "2007 Clinical Diagnostic Laboratory (Medicare) Fee Schedule," American Medical Association, http://www.metrika.com/media/files/2007%20Clinical%20Diagnostic%20Laboratory%20Fee%20Schedule%2004%2D24%2D07.pdf (September 13, 2007).
21. Information drawn from "Reimbursement of Medical Devices in Western European Markets," Frost & Sullivan, June 30, 2008, unless otherwise cited.
22. "Healthcare in France," National Coalition on Healthcare, http://www.nchc.org/facts/France.pdf (August 8, 2007).
23. See www.ameli.fr (October 21, 2008).
24. J. M. Schmitt, "Reimbursement and Pricing of Medical Devices in Germany," *HEPAC* (2000): 146–148.
25. See www.kbv.de (October 21, 2008).
26. See www.g-drg.de (October 21, 2008).
27. See www.iqwig.de (October 21, 2008).
28. "About the NHS: How the NHS Works," NHS Choices, http://www.nhs.uk/aboutnhs/howtheNHSworks/Pages/HowtheNHSworks.aspx (July 2, 2007).
29. See www.dh.gov.uk and www.ic.nhs.uk for more information on both programs.
30. "China Healthcare Sector, Credit Suisse, February 2007.
31. "Medical Device Reimbursement in China," International Trade Administration, http://www.ita.doc.gov/td/health/medical%20reimbursement%20in%20china%202007.pdf (October 21, 2008).
32. "World Health Statistics," World Health Organization Statistical Information System, 2007, http://www.who.int/whosis/en/ (October 21, 2008).
33. Ibid.
34. "India Considers Price Caps on Medical Devices," *Medical Devices Today*, August 26, 2008, http://www.medicaldevicestoday.com/2008/08/india-considers.html (October 21, 2008).
35. Cartoons created by Josh Makower, unless otherwise cited.
36. "Pulmonary Vein Isolation and Ablation as a Treatment of Atrial Fibrillation," Regence Group, http://www.regence.com/trgmedpol/surgery/sur138.html (October 22, 2008).

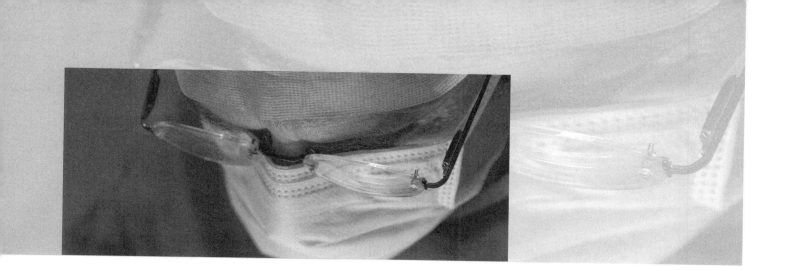

4.4 Business Models

Introduction

"It seemed like such a good idea, but why did it fail?" asks the frustrated engineer. "Hospitals are capable of purchasing massive million-dollar imaging systems in this specialty, so why were they so resistant to buying disposables?" Even though the total cost per year of this innovator's idea was comparable to (or even less expensive than) the established device solution, there was a fundamental problem that he failed to overcome. The flaw was in the business model. Sometimes an idea can be medically compelling, capable of clearing all regulatory hurdles, and manufacturable at a profit, but if the business model cannot work, the idea will fail. Innovators must consider the business model as one of the key factors that affects an innovation's success, and an issue that requires as much vetting as the feasibility of the device.

OBJECTIVES

- Understand the different types of business models that are typically utilized in the medical device field, including their relative advantages and disadvantages.

- Determine how to choose an appropriate business model based on the unique characteristics of the innovation and its customers.

A business model broadly refers to how an offering (e.g., a product or service) is defined and the way it will generate revenue and deliver value to customers. In the medtech field, common business models include disposables, reusables, implantables, and capital equipment, although many others exist. With each of these models, offerings make money in different ways and, in the process, pose different challenges to the innovator in terms of how the company organizes its operations, resources, and processes. The business model also dictates, to some extent, how interactions with customers and other external stakeholders should be managed to achieve mutually beneficial results. Just as a company needs to assess the intellectual property (IP), regulatory, reimbursement, and technical feasibility of its ideas, it should evaluate the appropriate business model before selecting a final concept.

Business model fundamentals

The business model is often the forgotten axis of innovation, arrived at by default after all other aspects of the product or service are determined. If an innovator recognizes that considering alternative business models is yet another design variable to be utilized in achieving the customer's needs, the innovation will have its greatest chance of success.

The defining characteristics of any business model are: (1) the innovation, which could be a product, service, or blend, (2) the customer, and (3) the primary interface between the two, or the way they interact. The primary factors that affect interactions between the customer and the innovation include those shown in Table 4.4.1.

When an innovator chooses a business model, s/he must take these factors into account. The idea is to design an offering that plays to the strengths and capabilities of the company while providing the customer with a desirable interaction.

Choosing a business model

During the concept selection stage of the biodesign innovation process, selecting a business model starts with an evaluation of the match between the unique characteristics of the innovation and the defining aspects of the different business models. The revenue stream and manner in which the innovation will get into the customers' hands are two other primary considerations. The information necessary to develop the business model will be extracted from the other steps in the process as the innovation continues to progress.

An appropriate business model must allow the company to extract value for its innovation in a way that makes sense to the customer. For example, if a new company tries to market a magnetic resonance imaging (MRI) machine by charging for the equipment and then requiring hospitals to buy a disposable platform for each patient tested, it might have difficulty generating interest.[1] Even if the company offers to service the machine for free (creating an ongoing revenue stream through the sale of the disposables rather than through a service contract, as is typical in this field), it is likely that the buyer would object. Most customers understand the value of paying for a service contract to keep their expensive, high-utilization equipment in good repair. However, it is much harder to convince them to pay for something, such as a disposable in this case, for which they do not understand the need or appreciate the value – especially if the MRI machines from competitors do not have the same associated charge.

Acclarent, a company that manufactures and markets endoscopic, catheter-based tools to perform what is known as *Balloon Sinuplasty*™ a procedure in which a balloon is used to dilate various areas of the sinuses, provides another, real-world example. In the ear, nose, and throat (ENT) specialty, companies have traditionally pursued models that are dependent on selling high volumes of inexpensive disposable materials and smaller numbers of moderately priced reusable products. Acclarent sought to change the field's dominant business model by selling higher value disposables that cost more per patient but led to improved results. Such a change is difficult, but not impossible if the innovation's performance is substantially better against one or more metrics that are important to the customer. As one board member described:[2]

> Whenever you try to extract more value from an established market, you have to be sure you can deliver more value through your product or service. The reason Acclarent has been successful in moving the market is that the products provide improved clinical performance and save money in the overall treatment of the disease.

Of course, any time a company seeks to change a dominant business model, it automatically takes on a market development challenge. Incremental investment (and often a significant commitment of time) is required to adjust customer expectations, modify their perceptions, and/or alter their behavior. However, the possibility of "changing the game" in an industry or field can also be the deciding factor that makes an opportunity interesting. Hypothetically, selling Acclarent's technology as a low-margin disposable might not have made a compelling business proposition. However, through the lens of a different business model (higher margins, lower volume), the opportunity became viable.

Importantly, many diverse business models are relevant within the medical device field. Investors have historically favored models that generate an ongoing

Table 4.4.1 *Each of the above factors can vary dramatically based on the business model chosen for an innovation. Understanding the impact of the business model choice and its influence on these factors can allow an innovator to determine which business model is the most favorable and compatible with success.*

Factor	Explanation
Revenue stream	How revenue is generated and its frequency.
Price	How much the business can charge for its products or services.
Margin structure	The profit to the company from sales (and its adequacy to support the inherent characteristics of the chosen business model).
Sales investment	The required mechanism for getting the innovation into customers' hands.
Customer training requirements	The extent to which specialized training is required to utilize the innovation.
Competitive differentiation	The degree to which the innovation is unique.
Intellectual property	The importance of IP protection to the success of the business model.
Other barriers to entry	Factors that could serve as barriers to adoption (e.g., high switching cost, brand or customer loyalty, access to distribution channels, etc.).
Clinical hurdles	The complexity and duration of clinical requirements (e.g., trials) before commercialization can begin.
Reimbursement	The way physicians, surgical centers, and hospitals are paid.
Financial requirements	The level of investment necessary to develop and commercialize the innovation.
Culture/geography	The extent to which customer needs related to the same clinical area differ across geographic boundaries or different cultural environments.

revenue stream or an annuity (e.g., disposables, implantables, or capital equipment with an associated service contract), as opposed to pure capital equipment businesses. However, numerous medtech innovators have proven that different approaches can be successful. In some cases, these alternate business model innovations have been the factor that permitted an innovation to succeed. For example, a company might provide access to a technology at little or no cost but then charge on a per-use basis to help get customers interested. Or, a device may be offered in a "resposable" format that allows a certain number of reuse cycles before it is thrown away. Because some individuals and companies have shied away from novel business models in the past (seeking to attract investor interest more easily), many innovative approaches to medical device businesses have inevitably been overlooked. For this reason, as technologies continue to advance, innovators and investors alike should anticipate the emergence of an increasing number of non-traditional medtech business models in the coming years.

Types of business models

As noted, there are multiple types of business models for medical device innovators to consider. When evaluating each one, the innovator should be looking for a good fit between the innovation and the business model such that a meaningful, growing, and profitable business can be envisioned. The nine business models included in Figure 4.4.1 and described in this chapter in more detail are among the most commonly employed medtech models.

Disposable products

Disposable products are those goods that are used and then discarded without being reused. Low-cost disposables include items such as paper examination gowns and stopcocks (used with intravenous tubing), both of which might cost pennies per unit. A surgical stapler is an example of a more expensive disposable, which might cost $100 or more for a single per-patient use. Disposables can also be attached to major medical equipment, such as ablation catheters used in

Stage 4 Concept Selection

Business model	Margin structure	Sales investment	Importance of IP	Barriers to entry	Customer training	Clinical hurdles	Financial requirements
Disposable							
Low cost	Low	Low	Low	Low	Low	Low	Neutral
High cost	High	High	High	High	High	Neutral	Neutral
Reusable							
Pure	Low	Low	Low	Low	Low	Neutral	Low
Implantable							
Mid cost / high-cost	High	High	High	High	High	High	High
Capital equipment							
Pure or combined	High	Neutral	High	High	Low	Neutral	High
Service							
Pure or attached to product	Low	Low	Low	Low	Neutral	Low	Neutral
Fee per use							
Pure or combined	Neutral	High	High	Neutral	High	High	Neutral
Over the counter (OTC)							
Pure or combined	Neutral	High	Neutral	Neutral	Low	Neutral	Neutral
Prescription							
Pure or combined	High	High	High	High	High	High	High
Physician-sell							
Pure or combined	Neutral	High	Neutral	Neutral	Neutral	Neutral	Neutral

Note:
Neutral = Importance of issue could go either way.

FIGURE 4.4.1
Each common medtech business model listed above carries with it an expected set of opportunities and challenges that dramatically impact the plan for the business.

combination with generators that produce the energy for ablation. Additionally, they can be coupled with reusable devices (used several times before requiring replacement), such as disposable razor blades for reusable surgical shavers.

Whether they are associated with the corresponding use of capital equipment or reusables, low-cost disposables:

- Require high sales volumes (to compensate for low margins).
- Must be easy to use.
- Should be marketable through low-cost distribution channels (e.g., medical equipment catalogs).

Higher cost disposables typically require specialized training to use and, as a result, require significantly higher margins in order to support a technical sales force and ensure the maintenance of a reasonable level of IP coverage. To justify a higher cost disposable, the innovator must achieve competitive differentiation (e.g., enabling superior clinical results or establishing key barriers to entry to the competition). Gross margins are usually favorable for most high-cost disposables and can be in the 70 to 80 percent range (or better). If pricing for a disposable is pegged in such a way that it is unrealistic to achieve margins close to this range, it may be an indicator that this is the wrong business model for a given technology.

In terms of their advantages, disposable products generate a regular revenue stream since customers must acquire them on an ongoing basis. In addition, as the volume of procedures increases (a goal of most healthcare providers), so too does the volume of the disposables used. Often, their value is directly correlated with a specific event, so it is easy for customers to understand – for example, every time a provider draws blood, a disposable syringe is required. Finally, based on their relatively low cost and rapid turnover, there is little risk to the buyer in trying a disposable, which can make them easier to sell.

On the downside, there are ethical issues to consider that are associated with the environmental consequences of disposable medical devices. Innovators must also be aware that, in some cases, disposable products can easily be displaced by reusable products that meet the same need, if their value proposition is compelling.

The Concentric Medical example below highlights some of the issues relevant to a disposables business model.

FROM THE FIELD — CONCENTRIC MEDICAL

Pioneering devices in the stroke market using a disposable model

According to Gary Curtis, CEO of Concentric Medical from 2002 to April 2008, stroke affects more than 700,000 people in the United States each year. Approximately 85 percent of those patients experience an ischemic stroke (caused by blockage in an artery supplying blood to the brain). Yet, despite the large number of people suffering from ischemic stroke each year, "The standard care for 95 percent of these patients is aspirin and a dark room," said Curtis in a 2007 interview. Intravenous recombinant tissue plasminogen activator (t-PA), a medication used to dissolve the clots that cause blockages, can be used in some cases of ischemic stroke, but only if it can be administered within three hours from the onset of symptoms. For everyone else, Curtis continued, "they have to wait passively to see how the stroke resolves so that then, a day or two later, a neurologist can consult with the patient's family and tell them their Aunt Martha is lucky enough to go home. For Uncle Harry, on the other hand, his stroke didn't resolve itself and he's going to a nursing home. Dad is kind of in between. He'll need some rehab and then maybe he'll get some of his normal function back."

Concentric Medical, which was founded in 1999, is seeking to change that model. "We are trying to change passive to active," explained Curtis. Using the company's Merci Retrieval System™ (see Figure 4.4.2) to restore blood flow in ischemic stroke patients by removing blood clots in the neurovasculature, "We can now intervene within a reasonable time period. Our trials showed that if we intervene and reopen an artery within zero to eight hours, less damage is done. We can change the course of events. It's the first time that's been done," he continued.

When Curtis became Concentric's CEO, the company was working on multiple projects but was having difficulty getting funding. Recognizing the stroke market as the

FIGURE 4.4.2
The Merci Retrieval System (courtesy of Concentric Medical).

largest unmet need with the most compelling technology in development, he quickly cut all other products in the development pipeline, gained funding for the Merci Retrieval System by promoting the company's improved sense of focus, and led Concentric, in 2004, to the first FDA approval of a device to remove thrombus in ischemic stroke patients.

> **FROM THE FIELD — CONCENTRIC MEDICAL**
>
> *Continued*
>
> In terms of a business model that would best support Concentric's product, the company's choices were somewhat limited due to the sterilization issues associated with products that come into contact with the bloodstream. "It has to be guaranteed sterile when you use it," said Curtis, which led Concentric to a single-use (disposable) model. "We never even contemplated another model. The device also has to have mechanical performance expectations that are exactly the same every time. Device performance can be compromised as you're cleaning, sterilizing, or repackaging it. If we offered a reusable, all those activities would lessen the ability of the physician to predict how the product is going to perform."
>
> According to Curtis, a business model built around a single-use, disposable device offers many advantages. "Clearly, the recurring revenue is a benefit. Once you get surgeons trained in how to perform the procedure safely, they're going to order the device again and again. We had spent roughly $35 million by the time we completed our first clinical trial to get the product approved before we got our first dollar of revenue. We're now trying to recoup that cost. Knowing that there would be this predictable revenue stream is one way we got the venture capital community to invest."
>
> On the downside, he noted that single-use devices are costly to the healthcare system: "A hospital will pay $5,000 to us for every patient treated with our device. If they had a reusable product, they wouldn't have to pay that much." However, he commented, the costs associated with the disposable device are more than offset by the extreme long-term expense of caring for stroke patients "who survive, but survive poorly." That said, competitors are working on reusable technology to address the same need that is targeted by Concentric's device. For example, in 2007 one company was trying to pioneer a device to break up the clots non-surgically by using focused ultrasound. "People are trying to develop reusable solutions to make treatment less costly," said Curtis, "but no one has proven that you can."
>
> When asked if he worried about the environmental impact of disposable products, Curtis responded quickly: "I don't think two seconds about it. Not when I'm saving a life."
>
> Importantly, Curtis pointed out that there is a difference between the business model used to support high-cost, single-use devices versus low-cost, high-volume disposables. With low-cost disposables, there is a constant, never-ending pressure from investors and customers regarding ways to reduce the cost of the product. "It's all about how you change your manufacturing costs from 5 cents to 4 cents, so you can make that extra margin," said Curtis. "That's not our business model." Instead, Concentric focuses on what he calls a value creation model. "All the pressure I have is how to make it more effective. How do I change from 50 percent to 60 percent to 70 percent to 80 percent success rate in restoring flow in these patients? No one here is spending two seconds worrying about how much it costs us to make the product. It's all about making it better." He went on: "That's the niche that I've always played in. And, if I can, that's where I'll continue to play."

Reusable products

Reusables are multi-use products with a moderate lifespan, but their cost is orders of magnitude smaller than a capital equipment item. Scalpel holders, laparoscopic graspers, and endoscopes are examples of products that can be reused for a period of time before they eventually wear out and need to be replaced. Reusables can be attached to a disposable (e.g., the surgical shaver, mentioned earlier, that is used with disposable razor blades) or a service (e.g., servicing of flexible endoscopes).

Business models built solely around a reusable product tend to have no sources of recurring revenue other than replacement. As a result, they generally cannot support a specialized sales force and are commonly sold through medical catalogs and/or through major distribution companies. Although the margins for reusables can actually be quite good – also in the 70 to 80 percent range – the lack of a sustained flow of cash, as compared to disposables, usually makes the size of the associated business opportunity much smaller. Reusable products also carry with them a higher level of risk since customers use them for an extended period. The product is not "factory fresh" each time it is used (as is the case with disposables), so the user faces a greater chance of failure. For example, if surgical scissors become dull or are damaged during a procedure, they may not perform effectively the next time they are used. Many providers of reusables attempt to address this concern by providing service for their products. However, because the products have a finite lifespan (which is much shorter than the lifespan for capital equipment), customers are typically unwilling to pay a great deal for maintenance. This contributes to the factors that make reusables a difficult business model to profitably sustain and grow. Some innovators have attempted to force a disposable business model onto a technology that is clearly reusable (and there is no good case for disposability). However, this can be a risky move. If the reusability of an item is discovered, as it usually is, the business model can quickly change from disposable to reusable, outside the innovator's control. This, in turn, can have a negative impact on the business and significantly affect perceptions and expectation of the company.

On the plus side, customers intuitively respond to reusable products and often favor them over disposables when the business case makes sense. For example, one would not consider anything other than a reusable weight scale or reusable stethoscope in a doctor's office – the cost/benefit ratio of these devices is essentially unbeatable. Once reference devices such as these are introduced into a marketplace, a new technology that seeks to challenge the business model, even at a modest cost increase, is often highly scrutinized and has a greater chance of failure. This should not deter one from investigating improvements that deliver dramatically better outcomes under these types of circumstances, but the innovator needs to realize that the bar will already be set relatively high.

Implantable products

Implantable products are typically mid- ($1,000 to $5,000) to high-cost (>$5,000) items. An example of a mid-cost implantable device is a coronary artery stent. Examples of high-cost implantable devices are pacemakers and artificial joints. Implantable products can be pure, or they can be associated with a service, such as pacemaker follow-up service. Implantable products tend to have high margins – in the 80 percent range or better – and these prices are supported due to the high barriers to entry associated with the technology, such as the regulatory pathway and IP requirements.

Implantable products require the highest level of clinical validation and, as a result, can present significant clinical hurdles to their developers. Because a major investment will be required to support comprehensive, long-term clinical trials, the market for implantable devices must be significant so that investors can recoup relatively sizable returns over a long-term payback cycle. Being able to ensure ongoing IP coverage is another essential aspect to the business model for implantable products.

One benefit of a business model focused on implantable devices is that there is a direct pairing of the value proposition and the procedure – every patient that receives the procedure gets one or more implants. As a result, implantables have an ongoing revenue stream whose growth is linked directly to the increase in the number of procedures performed each year. Additionally, since certain devices eventually wear out due to continuous use (e.g., heart valves) or battery consumption (e.g., pacemakers), patients may need replacement devices and, thus, provide a source of recurring revenue (although this may take many years to capture).

From a risk perspective, however, implantables represent a recurring liability to the company that manufactures and markets them. Many devices have a limited life, even as an implantable. The challenge to the company is to design an implant that can be replaced or otherwise taken out of service before it malfunctions. Many implants serve a primary function, which is useful for a period of time and then yields to other physiologic processes that persist even after the implant is no longer functional (i.e., drug eluting stents or resorbable drug delivery systems). However, even in these cases, the residual impact of the presence of the implant may present some risks that need to be managed and accounted for by the company. In addition, implantable products often require a direct sales force at the point of care to stay in touch, answer questions, and provide follow-up – a requirement that can be costly for a company to maintain. Although there may be some limited protection offered by recent legislation in the United States protecting companies that have gone through the post-market approval (PMA) process from certain types of liability, there is still likely to be continued liability risk for companies offering implantable products after this dust has settled. The St. Francis story highlights key issues associated with an implantable business model.

FROM THE FIELD — **ST. FRANCIS MEDICAL TECHNOLOGIES**

The challenges of being a pathfinder with an implantable device

St. Francis Medical Technologies, founded in 1997 in Alameda, California, focused on the discovery, development, and commercialization of novel treatments for degenerative spinal disorders until the time of its acquisition by Kyphon in January, 2007. The company's first product, the X-STOP® interspinous process decompression system, was developed to alleviate the symptoms of lumbar spinal stenosis (LSS). Lumbar spinal stenosis is a common spinal problem that primarily affects middle-aged and elderly adults, causing significant pain in the back and legs.

Traditionally, the most common surgical solution to LSS was laminectomy. In this procedure, the surgeon trims and removes part of the bone of the vertebrae to reduce the pressure on the spinal nerve root (which causes the pain and other debilitating effects associated with stenosis). According to Kevin Sidow, president and CEO of St. Francis from 2004 to 2007, the idea for the solution that eventually became the X-STOP emerged when two orthopedic spine surgeons, Jim Zucherman and Ken Hsu, were pursuing *the need for a less invasive way to treat the symptoms of spinal stenosis*. "They had a couple of older patients who had experienced short episodes of dementia as a result of the general anesthesia," explained Sidow. Based on this undesirable side effect of the procedure, "They were hoping to find a way to treat patients under a local anesthetic," he added.

All of the solutions that Zucherman and Hsu conceptualized to address the need involved the use of an implant (see Figure 4.4.3). As a result, when the experienced spinal device executives who cofounded St. Francis began thinking about a business model based on Zucherman and Hsu's leading concepts, they knew that they would be dealing with an implant business. Highlighting the benefits of this model, Sidow said, "The upside is that implants have a very straightforward revenue recognition process. There are also strong distribution networks that you can easily tap into in order to market the product, that have great credibility with spine surgeons."

FIGURE 4.4.3
The X-STOP implant (courtesy of Medtronic Spinal & Biologics).

On the other hand, he acknowledged, there are some sizable risks associated with implants that need to be considered early in the biodesign innovation process. Among those he mentioned, "The first is the regulatory hurdle, especially with a brand-new therapeutic option. This is really hard because everyone is incented to say 'no' to new things or extend the pivotal trial timelines," Sidow noted. When he joined St. Francis in 2004 from Johnson & Johnson (J&J), the company expected to receive FDA approval shortly thereafter.

> In fact, just two months later, St. Francis was notified by the FDA that its PMA application for the X-STOP implant had been turned down. Under Sidow's leadership, the company completely regrouped its US business to address the issues raised by the FDA. Reflecting on the situation, Sidow explained, "It really was a function of a lack of understanding. People at the FDA have an incentive to turn things down if they don't understand them perfectly. Nobody at the FDA gets rewarded for getting products to market quicker to help patients, but people get punished for Vioxx®-like results." When he elevated the matter to a higher level within the agency, "it was a much more objective, straightforward process." The X-STOP was approved by the FDA in November, 2005.
>
> Sidow emphasized another risk: that new implantable technologies have to anticipate issues associated with getting physicians in the target population to adopt the product. "Laminectomy was a big operation that was well reimbursed," he said. "The surgeons were very skeptical of this little company coming out with this new device. To make matters worse, all the big players in the business – J&J, Medtronic, Stryker, etc. – have very close relationships and tremendous credibility with these doctors. And they each had hundreds of sales reps telling the physicians that the device was a gimmick and that we were a nobody." Extensive efforts were required of St. Francis to overcome this skepticism and resistance in the field. When a new implant hits the market, "the large, incumbent companies have a strong incentive to battle it," Sidow reiterated.
>
> In addition to anticipating regulatory and adoption hurdles, Sidow advised companies with new implants to think carefully about reimbursement before getting too far in the biodesign innovation process. "It is becoming more and more critical to be proactive regarding reimbursement, all the way back to the point that you're designing your clinical studies," he said.
>
> Ultimately, the device that St. Francis brought to market could be surgically implanted via a less invasive procedure that could be performed in under an hour. According to Sidow, in the company's first year following FDA approval, it had $58 million in worldwide sales. At the time of the Kyphon acquisition, the X-STOP device had been implanted in more than 10,000 patients. St. Francis sold the business to Kyphon for $725 million.[3]

Capital equipment products

Capital equipment products are, in essence, another form of reusable products. They require customers to make a capital expenditure in order to obtain a technology that they will use repeatedly. Capital expenditures come from funds used by companies or entities, such as hospitals, to acquire or upgrade physical assets to maintain or increase the scope or competitiveness of their operation.[4] In the medical field, capital expenditures are often made to obtain equipment such as magnetic resonance imaging (MRI) or computer tomography (CT) scanners,[5] blood analytics equipment, or ultrasound machines. Capital expenditures can also be made for important software programs, as well as facility-related expenses. Some products requiring a capital expenditure operate externally to the patient and, therefore, require shorter term clinical testing and can be more regularly updated without the need for significant product approvals.

A pure capital equipment business model depends on the sale of the equipment, with little or no ongoing interaction between the company and the customer until it is time to purchase new equipment. However, as mentioned earlier, these products can also be associated with a service (e.g., technical support and maintenance contracts) or with disposables. In these scenarios, it is not uncommon for the equipment to be sold at or near cost, with the expectation that greater, recurring revenue will come from the sale of the service and/or disposables.

Capital equipment purchasing decisions usually occur at the administrative level in the healthcare system. For example, while physicians (as the end users of the products) are the targets for the sale of many disposable, reusable, and implantable technologies, hospital purchasing committees often become involved in buying decisions for capital equipment. These transactions still require the buy-in and support from doctors in one or more specialties. However, the decision to invest the funds will be made by committee, based on the broader priorities of the facility. As a result, the sales cycle can be long (as much as 18 months) and may require the vendor to support the sale with a careful business case for the purchasing decision (a plan for how the purchaser will recoup the investment in the capital equipment).

One benefit of capital equipment businesses is that sale of the technology usually represents a long-term commitment by the purchaser. The **switching costs** of moving from one MRI provider to another are extremely high. As a result, unless there are significant problems with the equipment, customers tend to be loyal to the company once the buying decision has been made.

The example of NeoGuide Systems below demonstrates some of the dynamics associated with capital equipment models, as well as how they can be made more attractive to prospective investors.

FROM THE FIELD — NEOGUIDE SYSTEMS

Mitigating the downsides of a capital equipment model through a blended approach

During his residency at the Tel Aviv Medical Center in Israel, Amir Belson discovered that he was "a lousy endoscopist." In one of his first colonoscopy procedures, the scope got stuck only 20 cm into the patient's five-foot colon. "I was pushing harder and the patient was screaming louder. It just wasn't right," he recalled. Through this experience he realized that, especially in colonoscopy, "forcing a flexible tube through a more flexible tube" was an unnecessarily complicated and painful approach to an important procedure. Colorectal cancer (CRC) is the second leading cause of cancer deaths in the United States, yet colonoscopy allows for the early detection and prevention of the disease.[6] To perform this procedure, the endoscopist advances a scope by controlling a bendable tip through the many curves within the colon (except for the tip, the remainder of the scope is not controlled). When the endoscope gets to a curve it often pushes against the colon wall, stretching the colon and its surrounding tissue rather than advancing. This phenomenon, known as "looping," is responsible for the majority of patient pain in colonoscopy, which can be so severe that sedation is required.[7] Looping occurs in more than 90 percent of colonoscopy procedures.[8]

Belson came to the United States with the idea for a computer-controlled flexible endoscope. In late 2000, he cofounded NeoGuide Systems, with Matt Ohline and Craig Milroy of Stanford University's mechanical engineering department. He initially assumed the role of the company's CMO and later became NeoGuide's president. Based on Belson's concept (and his broad IP coverage in the space), NeoGuide designed a colonoscope where the path of the steerable tip is followed precisely by the entire insertion shaft, preventing

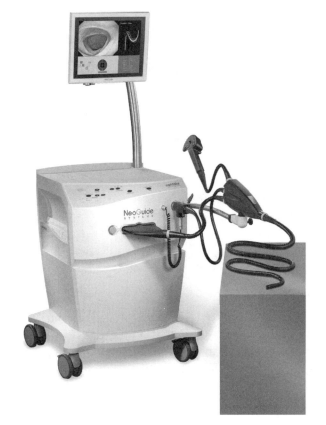

FIGURE 4.4.4
The NeoGuide colonoscope (courtesy of Amir Belson).

looping and allowing a less complex procedure with greatly reduced need for patient sedation. A sensor placed at the anus measures the depth of the scope as the surgeon navigates the colon, combining this information with tip angle to create a real-time, 3-D map of the colon and to guide the individual shaft segments.

According to Belson, doctors no longer need to "feel around in the dark" when guiding the scope through the patient's lower gastrointestinal (GI) tract, but can actively guide the scope, avoiding unnecessary contact and pressure (see Figure 4.4.4).

As with many colonoscopy products in the current market, NeoGuide's system would be considered capital equipment, requiring an up-front investment from buyers of more than $200,000. However, as Belson put it, "Working in the world of capital equipment can be a big disadvantage because it requires big decisions by people in the administration hierarchy – people who are not the users – to make the purchasing decision." He also pointed out that it can be difficult to get investors interested in capital equipment without having other sources of recurring revenue built into the business model. NeoGuide's control unit and four scopes would have a lifespan of approximately three to four years, which could cause major peaks and valleys in the company's sales.

"In our case," he said, "the investors liked the product *despite* the fact that it was capital equipment." The strength of the fundamental technology was one way that NeoGuide was able to spark investor interest. The company also broadened its business model by incorporating service and disposable elements into its system. While a service organization can be challenging to manage, service contracts are usually lucrative and also provide an important source of recurring revenue. "If you look at Olympus," said Belson, referring to a potential competitor to NeoGuide, "they sell contracts on their systems for about $5,000 per year and realize a 70 percent gross margin on the service." Additional revenue would be realized through sales of the disposable depth measurement instrument that was required for every procedure and would be priced between $20 and $30 per unit. "Investors always want to see something in the garbage can at the end of a procedure," noted Belson. The company had plans to further expand its revenue-generation opportunities by offering software enhancements and upgrades between equipment sales cycles. "Anything we can do to get a little more out of the capital equipment model makes our position a little stronger," he said.

Reflecting on one more benefit of a blended capital equipment model, Belson commented: "By combining equipment with service or disposables, it enables you to take the initial price of the system down and get the product into more accounts where more users and patients can benefit from it." However, as a caution to innovators pursuing equipment-based solutions (blended or not), he noted that the higher the price of the system, the higher customers' expectations in terms of its reliability and performance. "The customer demands placed on companies selling million-dollar equipment are exponentially higher than those placed on companies selling bag fillers," he said.

Service

A service is work performed by one person or group for the benefit of another. A long-term care facility is an example of a *pure* service model. Dialysis service centers commonly provide another example of a pure service model. However, in some cases, the companies that provide these services also play a role in providing capital equipment and disposables. As a result, organizations in this market may have a blended product/service business model. In such circumstances, the services provided are actually attached to certain consumable or reusable products. A service plan to maintain and support an MRI machine after it has been installed represents another example of a blended product/service model.

A notable characteristic of blended product/service models is the fact that companies may frequently sell a product at or near cost (with very low margins) in order to generate service revenue. It is not uncommon for companies such as General Electric, Philips, and Siemens to maintain relatively slim margins on their capital equipment in order to stimulate the sale of more service contracts, which typically have high margins and provide a recurring revenue stream. Once in place, these contracts tend to be extremely stable given the high switching costs mentioned above. In some ways, service contracts on equipment can be thought of as insurance policies. Customers buy them to ensure that their equipment will be properly maintained and quickly repaired in the event of a problem. However, both the customer and the company benefit if little or no service (beyond regular maintenance) is actually required because this means that the equipment experienced no unintended downtime, which can be costly and a source of risk for the customer.

In contrast to their blended counterparts, pure service models can be challenging. They are highly dependent on having the right management capabilities, organization tools, and resources on staff to make the model a success. Furthermore, customers tend to be sensitive to changes in management and company leadership, often valuing their personal relationships with individuals within the company higher than the service the company provides. Additionally, there are few economies of scale that can be realized in a service business. Unlike the production of a physical device (which allows a company to simply "turn up" manufacturing as it adds new customers), service businesses must add staff to keep pace with the acquisition of new customers. As a result, they face increasing risk in managing their costs during periods of volatility. Although service contracts for capital equipment can be lucrative (with high profit margins), healthcare services have low margins that are often squeezed by third-party public and private payers. Moreover, services that focus on or utilize technologies or procedures that the company itself does not manufacture or control are further at risk if there is a major change in the external environment. For example, if a company sets up a business servicing some piece of capital equipment (but does not produce the machine itself), its business would evaporate if the technology were suddenly replaced by a new innovation in the market. While this model is relatively uncommon in the device field, it highlights the need for companies using service models to stay attuned to changes in the external business environment and be prepared to adapt their business models.

The Bariatric Partners story provides an example of a service business model.

FROM THE FIELD — BARIATRIC PARTNERS

Building a service model in the weight loss field

Edmund Bujalski was an entrepreneur in residence at New Enterprise Associates (NEA) when he learned about an opportunity to get involved in the development of ambulatory surgical centers (ASCs) designed specifically to serve the weight loss and healthcare needs of the obese population on an outpatient basis. Two colleagues, Steve Puckett and Todd Johnson, were spearheading the idea in collaboration with NEA. While heading Hospital Partners of America, they were approached to enter into a transfer agreement with an ASC engaged in weight loss surgery. (All freestanding ASCs are required to have transfer agreements with an inpatient hospital in case a patient must be admitted.) After entering into this agreement, "they observed the success of that ASC over time," said Bujalski. "As they watched a little line item called 'bariatric surgeries' increase each month, they determined that there was a significant opportunity to roll this out on a national basis. In other words, the basic concept was already out there, but they made the decision to replicate and improve upon it."

Bujalski performed his own due diligence before formally deciding to join the venture that would become known as Bariatric Partners. "I first looked at whether this was medically appropriate, clinically sound, and did something good for patients," he explained. The second issue he investigated was whether such a business could demonstrate a return on investment to health plans and self-insured employers. "I knew that, long term, this business would be driven by insurance and third-party payers. Unless we could show a benefit to them, it didn't matter what we could do for the patient." Bujalski continued: "But then I did something that, frankly, no one else here had done. I talked to patients who had been through weight loss surgery because, at the end of the day, these are the people who would be our customers." He interviewed a number of morbidly obese individuals – most of whom had gone through gastric bypass surgery and a few that had received laparoscopic banding – to learn about their experiences. As a result of these discussions, he became passionate about the opportunity. "Not one person had a regret about it," he recalled. "The improvements in health were remarkable." In terms of the associated change in lifestyle, patients describe the procedures as "transforming" or "life-changing."

Furthermore, through these interviews, he became convinced that a change in the treatment paradigm was required. As people talked about their experience in receiving treatment, "Most of them had embarrassing stories of humiliation. One gentleman had to be weighed on the loading dock because the scale couldn't handle his weight. Others didn't fit into wheelchairs that were too small, or chairs in the waiting room that couldn't accommodate them because of their size," recounted Bujalski. "When I asked, 'If you could go to an outpatient center that was dedicated to your needs, understood you, and was designed around your pre-operative and post-operative requirements, would you have an interest in that?' people were like, 'My God, is there such a place? Why didn't I know about such a place?' After that, the decision was easy."

Bariatric Partners invests jointly with physicians in outpatient surgery centers. Each ASC is set up as a limited liability company (**LLC**) with a common ownership model and management structure. Bariatric Partners owns at least 51 percent of the LLC so that it can ensure high quality and consistent standards across centers. It also secures a contract to manage and operate each LLC – the function at the heart of the Bariatric Partners service model. "Our physicians provide all of the clinical input and program design, and they obviously perform the procedures," explained Bujalski. "We handle all of the business functions, including the marketing, which is a critical piece of the business." See Figure 4.4.5.

In reflecting on the differences between offering a product and a service, Bujalski highlighted flexibility and adaptability as the primary benefits to a service business. Bariatric Partners can adopt new weight loss treatments and methods as technology in the field evolves, so it does not run the risk of technological obsolescence. It can also reduce risk and increase revenue by adding other specialties to the mix of surgeries and procedures that it offers. For example, in some parts of the country, the company found that consumer awareness of laparoscopic banding (and "readiness" for the procedure) was relatively low. As a result, a significant amount of consumer education would be required before Bariatric Partners would be able to build a sizable enough procedure volume to profitably operate an ASC offering this service. By diversifying into other specialties and treatments, all under the same service/management model, the center could more quickly become cash-flow positive.

According to Bujalski, "Another big advantage in healthcare is that the need for medical services is constantly growing. Well-run service organizations that deliver real value to

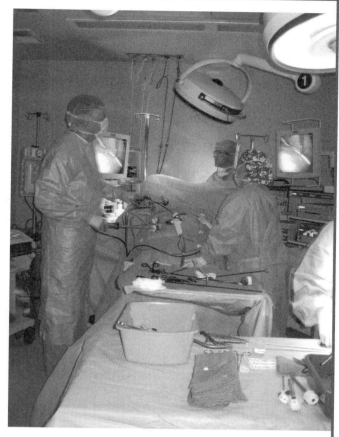

FIGURE 4.4.5
A laparoscopic adjustable gastric banding procedure being performed at a Bariatric Partners, Inc. affiliated ambulatory surgery center (courtesy of Bariatric Partners, Inc.).

patients are likely to attract loyal followings. There is some pricing advantage to being able to deliver high patient satisfaction and good outcomes. And there is definitely a 'word of mouth' marketing advantage," he noted.

On the other hand, service models have some unique challenges. "Differentiation from one's competitors is a challenge," said Bujalski. "It takes a long time to build a good reputation. Many of the competitive advantages of a service business are 'soft' and harder to quantify. A product can more readily be differentiated and its attributes delineated. At the fundamental level, a product works or it doesn't work, it fills a need or it doesn't. Service businesses can exist for a long time even if they are delivering mediocre quality. To some extent, it takes consumer education and experience before they can differentiate quality among service organizations."

Bujalski's goal is to create a brand by bringing in excellent surgeons who will, in turn, attract more excellent

> **FROM THE FIELD — BARIATRIC PARTNERS**
>
> *Continued*
>
> surgeons. However, he acknowledged, "Getting the right people can be a challenge." If demand for a product peaks, device manufacturers can often rapidly respond by ramping up production. "Increased production is more difficult to achieve with a service if the right people are not readily available," Bujalski said. The business is also vulnerable to good people leaving. "In a service business, the team is essentially the 'product.' If one loses the team, there may be nothing to sell," he said. The departure of skilled staff negatively affects patient satisfaction and also creates the risk of increased competition. Because there are relatively few barriers to entry and virtually no IP protection, former employees/partners looking to branch out on their own can replicate the service model. "It's not quite that easy to try to replicate a patented device or technology," Bujalski commented.
>
> Finally, Bujalski noted that it can be difficult to maintain high margins in a service business. Without significant IP protection or regulatory requirements (which help companies justify and sustain their margins in the device and pharmaceutical arenas), many service businesses face downward pressure on margins. "One way for someone to compete is to offer a price reduction to overcome a lack of experience," he said. Consumers often contribute to this pressure, as well, since many do not understand the differences between services offered at different centers. To justify its pricing, Bariatric Partners has to "sell" a complicated value proposition – that the quality of its services is higher and customers get more service for their money through follow-on and post-operative support that is bundled into the surgery charge. As pricing and margins decline, few economies of scale can be realized in the service business to help compensate. As Bujalski put it, "People can only provide so much of a service until they're maxed out."
>
> Despite these challenges, Bariatric Partners rapidly expanded after its founding in 2005. "In some ways," said Bujalski, "I'd much rather come up with a great device than try to build a service business. But this happens to have been my life for 30 years, so I'll stick with it."

Fee-per-use

A fee-per-use business model can be appropriate for innovations that sit squarely at the intersection of products and services. For example, laser eye surgery requires a capital expenditure to cover the cost of the equipment. However, practitioners are also charged a fee every time they perform a surgery using the machine. The event that triggers a payment to the company is nothing other than use of the machine.

Another form of the fee-per-use business model in the medical device field is referred to as a capitated model. In a capitated model, a medical provider is given a set fee per patient, regardless of treatment or equipment required. For example, a company that offers all the disposables necessary to perform a laparoscopic gallbladder removal may charge a fixed price per patient, regardless of which disposables are actually used. Similarly, a company with cardiology products may charge a fixed fee per patient for all the stents, balloons, and catheters required to perform certain predefined procedures. This approach allows companies with broad product lines to achieve an advantage over those that offer a smaller number of related products – the bundling of products makes it more difficult for less diverse competitors to penetrate accounts. This model can be appealing to customers seeking increased certainty in their costs since the per-patient payments are fixed regardless of the complexity of the individual cases they perform (e.g., hospitals have traditionally lost money on procedures requiring the use of more than two or three drug eluting stents in a single patient).

On the downside, customers sometimes resist paying the fixed cost for cases that require few devices (e.g., one stent), which can require extra time and effort of the company to defend its business model. Also, unless a business generally has a high degree of IP protection or other barriers to entry, this type of business model is often challenged by other businesses trying to compete in more traditional ways.

The VISX example below is illustrative of the challenges and benefits associated with a fee-per-use business model.

> **FROM THE FIELD**
>
> **VISX**
>
> Developing the fee-per-use model in the vision correction field
>
> In the late 1980s, Charles Munnerlyn, founder and CEO of VISX, and Allen McMillan, the company's COO, were considering how to commercialize the first system for photorefractive keratectomy (PRK), the original method used for laser vision correction. They knew there must be a better approach than the traditional capital equipment model. The cyclical nature of capital equipment sales would leave little money available for research and development since they would sell one model to as many customers as possible and then have to wait for years until the same customers were ready to invest in the next generation. They were also concerned that the high purchase price of capital equipment could make market penetration quite slow for a new company entering the field.
>
> Seeking a way to realize a more consistent revenue stream and quickly build a commercial presence, Munnerlyn and McMillan conceptualized a business model that would allow the company to sell equipment and service contracts to its customers, but also charge a fee each time the system was used. "The laser vision correction industry was really the first in the medical device field, and certainly in the United States, to use a true per-procedure model," explained Liz Davila, CEO of VISX from 2001 to 2005.
>
> In the late 1980s and early 1990s, VISX, and its main competitor, Summit Technologies, were each developing excimer lasers for eye surgery and had patents on different aspects of the technology. In order to avoid patent litigation, they combined their IP in 1992 and created a third entity, Pillar Point Partners, which licensed the combined IP. Pillar Point Partners would then sub-license the patents back to VISX, Summit, and any company that wanted to license the IP. All licensees paid Pillar Point a per-procedure royalty. Pillar Point profits were then distributed to VISX and Summit. Since the licensees were paying on a per-procedure basis, it was logical that they would charge their customers by procedures. The result was that the US laser vision correction industry adopted the fee-per-use business model.
>
> VISX's shareholders immediately appreciated the value of the new model, but there was a period of time when the surgeons were resentful and antagonistic toward the company. VISX originally charged upward of $500,000 for the equipment, plus $250 per eye treated. The $250 charge was positioned as a licensing fee. As Mark Logan, the company's CEO (following Munnerlyn but before Davila), explained in an interview for the *Tribune Business News*, "We've spent $52 million bringing this to market, and there's no way we could sell these lasers for $500,000 and ever recoup what we've put into it. If it weren't for the $250 fee, we'd have to sell these systems for $3 million each and have a very small market, which isn't good for anyone."[9] VISX also asserted that these fees enabled the company to sustain its research and development efforts so that it could continue improving its technology to treat new indications.
>
> The surgeons accepted the need to pay for the disposables required by the procedure because they had a utilitarian function, but they did not understand the value of the VISX keycard (which tracked equipment usage and calculated the associated licensing fees). According to Davila, "What eventually got them to quiet down was that they fairly quickly realized how much money they were making." Previously, as corneal specialists, most of the VISX surgeons performed cataract surgeries, which were primarily covered by Medicare and reimbursed at a total of about $1,600 per procedure. In contrast, Davila continued, "By expanding their practices to include vision correction surgery, the surgeons could now charge $1,500 to $2,500 *per eye* and were making money as they had never made money before." Eventually, they accepted the $250 per-use fee as a necessary expense associated with the more lucrative laser surgery market.
>
> Since VISX and Summit were the first two companies to offer this technology in the United States, they exercised a high level of market control (approximately 70 percent and 30 percent market share respectively). Eventually, however, other competitors began to enter the market. Several took licenses to the patents. However, a company called Nidek refused to license the IP from Pillar Point Partners. Despite the fact that the company offered inferior technology, it was able to gain market share because it also did not charge a per-procedure licensing fee. To protect itself against substantial market erosion, VISX reduced its per-procedure fee from $250 to $110. Surgeons responded positively to the change, and many doctors who had switched to Nidek eventually came back to

> **FROM THE FIELD** — **VISX**
>
> *Continued*
>
> VISX because of the company's superior technology. Over time, Nidek's business declined as the company lagged technologically further and further behind. As VISX continued to improve its technology, further outdistancing its competitors, the company was able to increase its per-procedure fee back to $250.
>
> The Nidek story demonstrates that a fee-per-use model has the greatest likelihood of success in an industry or market where everyone is playing by the same rules. Alternatively, the company employing the model must sustain such a strong competitive advantage that customers view the license as being worth the additional cost. "When people ask me if the VISX model would work for them," said Davila, "I ask them if their technology is truly revolutionary. Because if it's not, it will be very difficult to get medical professionals to pay a per-procedure fee."
>
> Davila also underscores the importance of having IP that can be protected. Among the reasons that VISX reduced its usage fee in the face of competition from Nidek was the slowness of the patent enforcement process. VISX did sue Nidek for patent infringement, but the court process required several years and VISX decided that it was too costly to wait.
>
> Finally, she points out that eye surgery is an elective procedure that is not reimbursed through Medicare and infrequently covered by private insurance providers. The fact that VISX did not have to take third-party payers into account simplified its decision to pursue an alternative business model. The company only had to convince providers and patients that the value provided by its offering would outweigh any non-traditional costs – a somewhat easier hurdle to clear than the cost effectiveness requirements increasingly imposed by payers.
>
> VISX (which is now owned by Advanced Medical Optics) is the worldwide market leader in the design, manufacture, and sale of laser vision correction systems. Using the VISX system, ophthalmologists have performed over 6 million laser vision correction procedures.

Over-the-counter products

An over-the-counter (**OTC**) business model depends on patients' ability to choose a treatment path and then acquire the product(s) for themselves. Over-the-counter products are more typically seen with drugs (e.g., Motrin®, Benadryl®), but can also include devices (e.g., steam machine for treating sinus congestion, home blood-pressure cuff devices, or glucose monitors). Over-the-counter products can, at times, be combined with services. For example, companies selling at-home blood pressure monitoring devices may offer data analysis and feedback services to users who upload their blood pressure readings via a computer.

Because physicians often are not involved in recommending OTC treatments, they must be relatively simple and easy to use. Advertising is usually targeted directly to consumers, and generic retail outlets (e.g., Walgreen's) can be used to support sales. On the upside, OTC products rarely require an expensive direct sales force. On the other hand, they require huge marketing budgets to promote brand recognition, which can be a sizable barrier to overcome. In this model, the goal is to make the consumer aware of the product and how it works through conventional advertising channels (e.g., television, magazine, radio). Within this construct, companies at times find it challenging to differentiate their products since consumers spend less time than physicians understanding the clinical benefits of one product over another.

Models that require doctors to specify an OTC product to their patients can be even more challenging. In general, the smaller and more specific a physician population is, the easier it is for a company to reach. However, physician-specified OTC products tend to rely on the recommendations of general practitioners – a vast market that is nearly impossible for any single company to reach effectively with a reasonable investment of time, money, and resources.

Prescription products

Prescription drugs provide a classic example of prescription-based medical products. However, prescriptions can also be used for devices and combined with services (e.g., physical therapy), disposables (e.g., blood glucose monitoring testing strips, drug cartridges to a delivery system), and hardware (e.g., nerve stimulator pain control units, certain inhalers). With this type of business model, the physician selects the treatment and directs the patient toward it, but the patient is still required to act on the physician's instructions.

Because the physician is directly involved in selecting the treatment, prescription products are often more complicated and require more specialized training to understand than OTC alternatives. However, a more clinically oriented sales approach can be used to differentiate products, usually through a direct sales force. The size of the sales force is typically proportional to the number of physicians the company is trying to reach. For example, a sales force focused on primary care physicians would be the largest, whereas a smaller, but still significant, sales force would be required to support specialty products, such as anti-nausea medications used for patients receiving chemotherapy. To support these requirements, the products must command high enough margins, as is the case with most prescription medications and devices. Direct-to-consumer advertising is sometimes used in parallel with the physician-focused sales effort, because physicians are expected by many patients to respond to their desires and requests.

Because the large pharmaceuticals companies have so much advanced expertise in marketing, prescription markets can be difficult to penetrate for small start-ups. Companies are often advised to enter the prescription business only if their product is clearly differentiated from the competition and/or if they can enter into a codevelopment partnership with a large pharmaceutical company that has an established sales force. "Me too" products have little chance of success without the marketing "muscle" and deep pockets of a major partner.

Physician-sell products

Physician-sell products are those treatments that are sold directly through physicians. With this business model, the physician essentially becomes a retailer for the product and usually receives some direct incentive for helping to promote and provide the treatment. Common examples include BOTOX injections, teeth whitening products, and hearing aids. Physicians can also sell disposables, such as contact lenses or solutions.

Typically, physician-sell products are offered on an outpatient basis and are often paid for by the patients (rather than by insurance). Once again, the margins for products sold through this channel must be high enough to cover not only the compensation to the company, but also to the physician for being a distributor. While some physician-sell products can be quite profitable, the primary downside to this model is the potential for ethical conflict. Any time a physician receives a direct incentive to steer a patient toward a particular treatment, questions may arise about whether the physician is truly keeping the patient's best interests in mind. For this reason, physician-sell products tend to be limited to non-essential, elective treatments that patients may desire but are not medically indicated.

Validating a business model

When presented with an array of business model choices, innovators will almost always choose the most profitable and sustainable path. However, unless they have carefully evaluated all of the available alternatives, they cannot be sure which model will have the highest likelihood of success. Once the innovator identifies a business model that seems to be the best fit with the innovation, s/he is well served to validate the approach by seeking input from other innovators, entrepreneurs, and business advisors. A few important questions to ask include the following:

- Can the technology or therapy be delivered via a different business model?
- If so, and if this was to be done by a competitor, would it be a threat to the business model or is the model still sound?
- If the other business model could be a threat, are there significant barriers to allowing customers or other businesses from executing that business model against the one the innovator has chosen?

If the business model chosen by the innovator is the most profitable and sustainable one available, it should rise to the top regardless of the competitive scenarios outlined above. Because it is so important not to choose a business model in a vacuum, it should be tested with customers (to ensure that the model works in their reimbursement and coding environments) before a final decision is made. Additionally, the innovator should consider the primary risks associated with the model and develop specific plans for managing them. A framework such as Michael Porter's five forces can be helpful when evaluating business models within the medtech industry (see 2.4 Market Analysis).

Operationalizing a business model

Fundamentally, the steps in the biodesign innovation process that come after a business model has been chosen are focused on helping innovators operationalize the business model they selected. As soon as a business model is chosen, it is important to begin thinking about what kind of expertise will be required to implement it. Ideally, a young company will hire someone who has previously built the kind of business it is pursuing. For example, if an innovator has chosen a service model, such as Bariatric Partners, s/he might look for someone with deep experience managing hospitals or other major health systems. If a disposable model has been chosen, it might be time to augment the engineering team with someone who has successfully pioneered disposable products in the past.

There are many different ways to align a company's expertise with its business model. One approach, of course, is to directly hire individuals with the right experience. Depending on the stage of the company's development, however, consultants can also be leveraged. Another approach is to seek this expertise in the form of advisors who might sit on the company's board of directors. When considering what type of expertise is required, it is important to consider both technical competencies and business competencies, both of which will play a critical role in successfully operationalizing a company's business model.

Sometimes, as a company grows, a new product is considered that differs from the model used by the other products in the portfolio. The first and most important step is to recognize this fact before proceeding with development. Typically, for instance, companies that are organized around large capital equipment devices with service contracts have a hard time transforming themselves to act as a disposable company, and vice versa. While there are some notable exceptions, such as Intuitive Surgical, which have executed well on both models, the list of companies able to master dual-business models is short. Unless the managers of a business are uniquely prepared for the challenge of executing multiple models, the best choice, at least initially, is to "keep it simple" and focus on one model at a time based on an evaluation of the company's total overall opportunity.

GETTING STARTED

The choice of the business model is as important as the choice of the technology that the innovator decides to take forward. Use the following process to guide the selection of an appropriate business model.

See **ebiodesign.org** for active web links to the resources listed below, additional references, content updates, video FAQs, and other relevant information.

1. **Understand characteristics of different business models**

   1.1 **What to cover** – Develop an in-depth understanding of each type of business model. Determine which characteristics are the most important (i.e., those that truly define each model). These are the characteristics that will be most important to consider when selecting a business model in the next step.

   1.2 **Where to look** – Unlike many other steps in the biodesign innovation process, there are not many comprehensive references that can be used to assist in understanding various business models. Internet searches and business texts can be a good source of information regarding the general concept of business models. When seeking medtech-specific information, seek the counsel of other innovators, entrepreneurs, and advisors with experience in the field. Venture capitalists can also be a good source of information based on their experience in evaluating the business models of different device companies.

2. Choose a business model

2.1 What to cover – Look at the most important characteristics or criteria for each business model and compare these requirements to the innovation under development, as well as the underlying need. Immediately eliminate those business models that are an obvious mismatch with the given innovation and need (e.g., do not pursue a business model that relies on strong IP protection if such coverage is impossible to achieve). Then, perform more detailed analysis and research among the remaining business models until the best choice can be made.

2.2 Where to look – See the guidelines in Appendix 4.4.1 for a list of critical business model elements to help with this important decision and general rules-of-thumb to validate each approach. Again, seek the counsel of other entrepreneurs, advisors, and venture capitalists before deciding on which business model to adopt.

3. Validate the preferred business model and identify risks

3.1 What to cover – Perform a secondary analysis of the chosen business model to identify any "fatal flaws" that may be associated with the approach. Seek input from experts to validate that the chosen model represents the best possible approach. Identify any "soft spots" in the model and develop plans for minimizing, eliminating, and/or managing these risks.

3.2 Where to look – Leverage personal networks to help validate the intended approach. Utilize standard risk management constructs to begin thinking about and minimizing risks.

4. Determine what new expertise will be required to operationalize the model

4.1 What to cover – Assess the current skills sets and expertise that the innovator and any other members of the team currently possess. Evaluate these competencies against the critical competencies necessary to implement and execute the chosen business model. Initiate a search to identify individuals with the skills and experience that make them suitable for hire, to consult, or advise the company as a member of the board.

4.2 Where to look – Involve all members of the team in assessing its skills and competencies and identifying needs. Refer back to 1.1 Strategic Focus for information about strengths and weaknesses. Again, leverage personal networks to make contact with candidates with the right skills and experiences to assist.

Credits

The editors would like to acknowledge Darin Buxbaum, Jared Goor, and Joy Goor for their help in developing this chapter. Many thanks also go to the individuals who contributed to the case examples: Amir Belson, Edmund Bujalski, Gary Curtis, Maria Sainz, Liz Davila, and Kevin Sidow.

Appendix 4.4.1

Decision Tree and Rules-of-Thumb for Choosing a Business Model

Step 1 – Scale down the possible universe of business models
Determine the unit (or units, if taking a blended approach) which defines the revenue flow to the company.

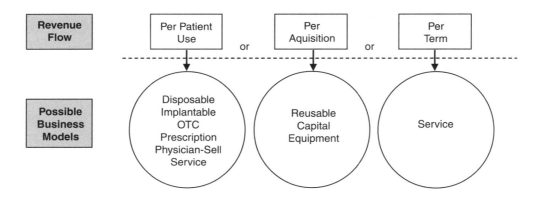

Step 2 – Set unit pricing
All models typically have an associated price/unit. Some models may also include after-sale revenue for each term in use.

Step 3 – Calculate the potential market size (revenue/time)
Multiply expected units per time by the pricing/unit. For example, there may be 500,000 patient candidates for a particular implant per year (units/time) x the implant pricing (set in Step 2). Another example is 2,000 hospitals need this equipment installed over the first 10 years it is available (units/time) x the pricing (set in Step 2). See 2.4 Market Analysis for a more detailed discussion of how to calculate market size.

Step 4 – Determine the cost per unit

Step 5 – Calculate the gross margin = (price/unit – cost/unit) divided by price/unit
The following rule-of-thumb can be used for validation.
Financing types available:
1. If revenue/time is greater than $500 million per year, it is typically backable by a venture capital firm and may lead to an **IPO**.
2. If revenue/time is greater than $100 million but less than $500 million, it may be backable by a venture capital firm and may not have IPO as an option.
3. If revenue/time is less than $100 million, it is probably not likely to be backed by a venture capital firm, but may be interesting to individual (or angel) investors.

Operating expense coverage (the ability of the business to support high sales costs like a direct sales organization, big advertising budgets, or to repay the cost of extensive clinical trials):

1. If gross margin is greater than 70 percent, the business can support higher operating expense costs.
2. If gross margin is less than 50 percent, the business most likely will not be able to support higher operating expense costs and may need a partner to support key activities such as clinical trials or marketing.

When additional brainstorming might be needed:

If the business requires an extensive clinical trial to reach the market, and therefore a significant amount of capital, but the revenue/time is less than $100 million per year, then there is a problem. If it is greater than >$500 million, then it is probably acceptable.

If a product requires a direct sales force to sell it, and it is a low-margin product, then there is a problem. If the product cannot be reimbursed at a higher level or the cost cannot be reduced, it may not have a future.

If a product requires high margins (e.g., due to a direct sales force, customer training, or large research and development or clinical investments) and has little IP and low barriers to entry, then there is a problem.

If one wants to sell a product OTC, or if insurance providers try to push a product OTC, and the product requires technical training, then there may be a problem. If simplifying the current product is not a possibility, then the product may not be feasible.

Notes

1. An MRI machine, which stands for magnetic resonance imaging, provides a non-invasive method for visualizing various structures and disease functions within the body.
2. All quotations are from interviews conducted by the authors, unless otherwise cited. Reprinted with permission.
3. "Kyphon Stock up 30 Percent on $725 Million St. Francis Medical Tech Acquisition," *Silicon Valley/San Jose Business Journal*, December 4, 2006, http://sanjose.bizjournals.com/sanjose/stories/2006/12/04/daily2.html (October 9, 1008).
4. "Capital Expenditure," Investopedia.com, http://www.investopedia.com/terms/c/capitalexpenditure.asp (December 11, 2006).
5. CT scanners, which perform computed tomography, provide another type of diagnostic imaging.
6. "Evaluation of a Breakthrough Computer-Assisted Colonoscopy System Presented at Digestive Disease Week 2005," *Medical News Today*, May 18, 2005, http://www.medicalnewstoday.com/articles/24642.php (December 14, 2007).
7. Ibid.
8. S. Shah, *et al.*, "Patient Pain During Colonoscopy: An Analysis Using Real-Time Magnetic Endoscopic Imaging," *Endoscopy* (2002): 435–440.
9. "VISX, Incorporated," Answers.com, http://www.answers.com/topic/visx-incorporated?cat=biz-fin (November 26, 2007).

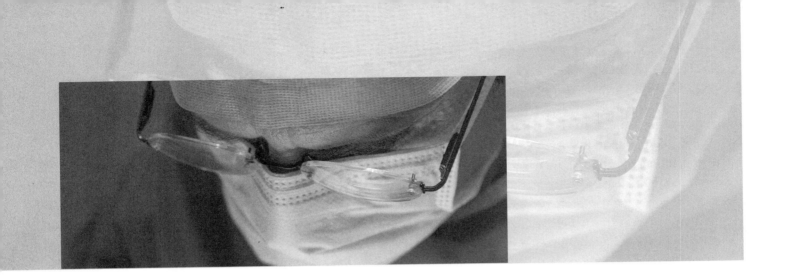

4.5 Prototyping

Introduction

If "a picture is worth a thousand words," then in the medtech field, "a prototype is worth a thousand pictures." Fundamentally, there is no substitute for taking all of the conceptual, abstract thinking that has been performed to date and giving it a physical form. Simple concepts can be fraught with problems or result in elegant, effective solutions. Complex ideas can lead to revolutionary results or be impossible to achieve. The only way to find out is to start prototyping.

The goal of prototyping is to translate a promising concept from an idea into a rudimentary design, and then into a working form. Prototyping is an essential step through which the innovator learns about functionality, explores features, gathers preliminary feedback from target users, and answers questions that can only be resolved through the manifestation of the design. Prototyping plays a role at multiple stages of the biodesign innovation process. Early on, prototyping can be performed quickly and inexpensively to help the innovator evaluate multiple solution ideas against more specific design criteria before deciding on a final concept. As the innovator moves forward, prototype requirements, designs, and models become more advanced. These more robust prototypes are used to test functionality and features, often in conjunction with tissue and animal testing. Near-final prototypes that meet design requirements are used in gathering data for quality documentation and preparation for manufacturing. Across these different stages, the underlying principles of effective prototyping remain the same.

OBJECTIVES

- Understand how to approach prototyping and the types of questions it can help address.
- Learn how to think about and translate concepts into functional blocks for more efficient and directed prototyping.
- Become familiar with prototyping tools and techniques and how to apply them in different engineering disciplines.
- Understand how to use prototyping to create design requirements and generate high-level specifications related to technical feasibility.
- Know how to use the biodesign testing continuum to transform a concept into increasingly advanced prototypes.

Prototyping fundamentals

Concept screening helps innovators narrow their focus from hundreds of concepts to a few (see 3.2 Concept Screening). In some ways, prototyping causes them to broaden their focus again by exploring dozens of different ways to technically realize the few concepts under consideration. However, both activities have the same goal: to enable the innovator to choose a final concept and a single way of addressing an important need.

As shown in Figure 4.5.1, some prototypes are quick, easy, and inexpensive to develop, while others are more complex, costly, and time consuming. This is determined primarily by the stage in the biodesign innovation process and the objectives the innovator is seeking to address.

At its most fundamental, prototyping is iterative, with each successive prototype built to answer questions that arise from the performance successes and deficiencies of previous versions. For prototyping to be effective, the innovator should be clear regarding what information is already known about a concept, its function, and its interaction with human biology, as well as where gaps exist in his/her understanding. As the innovator's knowledge evolves, so should the specific objectives for each prototype.

Guidelines for effective prototyping

One way to effectively prototype in an iterative and focused manner is to break a concept down into smaller blocks that correspond to its different functions. Rather than prototyping the whole concept at once, the innovator focuses on proving the feasibility of smaller, essential components before testing them as a system. While this approach makes it easier for the innovator to define highly specific objectives for each preliminary prototype, it also help him/her determine which aspects of a concept are truly novel and, therefore, may represent the highest risk in terms of needing to be "proven." Additionally, for large-scale concepts (e.g., a new type of X-ray scanner), smaller blocks may be all that can be prototyped at an early stage.

Through the creation of an iterative series of prototypes, the innovator refines his/her understanding of how the concept will work and gains an appreciation of issues related to the technical feasibility of the concept.

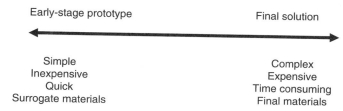

FIGURE 4.5.1
Different types of prototypes can be used to accomplish different objectives at varying stages of the biodesign innovation process.

As the smaller functional blocks are successfully prototyped, the innovator can then start to put them together to understand how the solutions to the various blocks work together. As this process unfolds, the innovator identifies some of the key design requirements that are most likely to satisfy the need in the user's mind. In response, s/he can continue to iterate on the prototypes until the underlying need is ultimately solved.

In taking this approach, the innovator has a clear path:

1. Identify the questions or issues to be addressed through prototyping.
2. Recognize that different categories of prototypes can be used to address different questions.
3. Isolate the functional blocks of a solution.
4. Understand what is known about each functional block and what must be learned or proven.
5. Use what is learned through prototyping to define more detailed design requirements and technical specifications.

To help illustrate this approach, an example of an innovator working on a solution to the need for ***a way to prevent strokes in patients with atrial fibrillation caused by left atrial appendage (LAA) thrombus*** is followed throughout the chapter. As described in 2.2. Treatment Options, the LAA is a small, pouch-like structure attached to the left atrium of the heart in which blood can coagulate to form clots, or thrombi, in patients with an abnormal heart rhythm (such as atrial fibrillation). These clots can dislodge and travel to the brain where they have the potential to create a blockage in an artery, leading to a stroke. In this example, the innovator focuses on the concept of preventing LAA thrombus by filling the LAA with a material that can be delivered as a liquid and then increase in viscosity to a solid-like consistency to eliminate the space where thrombi can form.

1. Identify the questions to be addressed

When an innovator prepares to develop a prototype, s/he must define a specific question or issue to address. Clearly articulating this for each prototype helps to identify what critical elements of the concept should be included in the model. Including "extra" or extraneous features in a model complicates the end result, distracts from testing issues on the critical path, and can unnecessarily increase the time and cost required to build a model. The point is to construct the simplest model possible that will adequately address the key question or issue.

One of the most critical issues to address through the prototyping process is whether or not a concept is technically feasible. Most early prototyping work is focused exclusively on proving the technical feasibility of an idea. For instance, it is important to determine whether or not the invention will work with living anatomy. To answer this question, a crude mock-up in conjunction with animal tissue could be used to demonstrate that the basic approach is feasible. In the LAA example, a cow's heart could be accessed by means of an existing device, such as a catheter. With only minor modifications, the device could be used to inject dye into the LAA to demonstrate that the general approach of filling the LAA with liquid is feasible. While the functionality of this preliminary model is severely limited, the innovator learns through the experiment that it is possible to fill the LAA with liquid. In doings, s/he was not distracted by other features that were not necessary to answer this question.

As the biodesign innovation progresses, other questions, beyond technical feasibility, become increasingly important. Prototyping can be equally effective in resolving these issues. A sample of the many different types of questions prototyping can be used to answer is shown in Table 4.5.1. Continuing with the LAA example, the innovator might want to answer questions about what components are required to make the concept feasible. The crude catheter and cow's heart could be used to try various materials that start as a liquid and then solidify, and to find one that might work. Next, the innovator could seek to demonstrate to key stakeholders that this can be done in a live animal. S/he may further develop or modify a catheter, then deliver the previously identified material in an animal study, and later study its pathology post-mortem. Finally, the innovator may want to translate this design and everything that has been learned about the technical specifications to a human scenario, which would include additional design requirements. The innovator may do additional research and make more modifications to ensure that the catheter and material are inert and will not have side effects in the human bloodstream.

As shown in this example, the question or issues that prototyping can address evolve as the design progresses. Each prototype is built on the learnings of the previous models. In this fashion, prototyping and its associated questions and issues are highly iterative, similar to much of the biodesign innovation process.

2. Recognize the best category of prototype

Based on the question or issue that is identified as being important at a given stage of the prototyping process, the following types of models can be used:

- **Works-like model** – Demonstrates how the device works. It may not look or feel quite right, but it can be used to demonstrate basic technical feasibility and/or to assess whether the customer would be interested in a device that functions in this new way. The bulk of prototyping often is performed using works-like models.
- **Feels-like model** – Made of the final material or a surrogate material to demonstrate ergonomics, grip, weight, size, etc. (e.g., surgical tools) and to serve as important input into human factors considerations.
- **Is-like model** – Performs the desired function and works as intended. An is-like model may not resemble the final form, but it could be used clinically. Depending on the nature of the device and the likely regulatory requirements, these models might be used in animal or possibly early human testing. They are often used to transition

Table 4.5.1 *There are many different and important questions that can be answered by prototyping.*

Issue	Process for resolving issue	Related chapters
Will the concept work? Is it technically feasible and will the product function as designed?	Understand the underlying fundamentals of the clinical problem and then build the prototype and test it.	2.1 Disease State Fundamentals 5.2 R&D Strategy 5.3 Clinical Strategy
Is the innovation novel and unobvious? Can it be patented?	Having sketches and/or prototypes of the idea will help to define the claims of the invention.	4.1 Intellectual Property Basics 5.1 Intellectual Property Strategy
Will customers adopt and use the product?	Take the prototype to thought leaders and target users in the market. Let them touch and feel it to provide feedback and confirm their interest.	5.3 Clinical Strategy 5.7 Marketing and Stakeholder Strategy
Can it be manufactured?	The development of sketches and/or CAD drawings will help determine manufacturability. It can also be helpful to find precedents and reverse-engineer them. Discuss the prototype with materials and manufacturing vendors to understand their input, ideas, and concerns.	5.2 R&D Strategy
When can it be made available?	Use vendor estimates/quotes (volume, price, time frame) to develop a project plan to outline when it is realistic to expect a finished product. Consider hiring a consultant to assist with this process.	5.2 R&D Strategy 6.1 Operating Plan and Financial Model

from the design phase to human testing and the manufacturing phase.

- **Looks-like model** – Demonstrates what the device will look like in terms of its shape, color, size, and/or packaging. Though this is important nearer the end of the development stage when user feedback and marketing decisions are made, a crude looks-like model may be useful early in the development stage to help communicate to others what one is trying to achieve.
- **Looks-like/is-like model** – Provides a combination of the previous two models. This type of model would both function as and look like the final device. This step may be skipped or undertaken by manufacturing (e.g., if the is-like model simply incorporates the looks-like elements as part of the technology transfer process from design to manufacturing).

Again, it is generally most effective to start with simple works-like prototypes that *only* convey basic information about the concept and are narrowly focused on answering a single, specific question. In the LAA scenario, the innovator's first step was simply to determine if the LAA could be accessed and filled with liquid. The exact shape, size, weight, look and feel, and complete functionality of the device were irrelevant to answering this basic question. As a result, a simple works-like model was not only appropriate, but relatively easy and inexpensive to construct. It also did not introduce unnecessary complexity into the process of answering the relatively simple question the prototype was designed to address.

The more questions a single prototype seeks to answer, the more risk an innovator faces in understanding the results of the model (i.e., the specific cause of a particular technical problem). This is not to say that advanced prototypes that represent more finalized designs should be avoided. On the contrary, they are essential to proving the total concept. However, they should be built in such a way that they represent the summation of all the work done to date (and all the questions that have been answered by previous

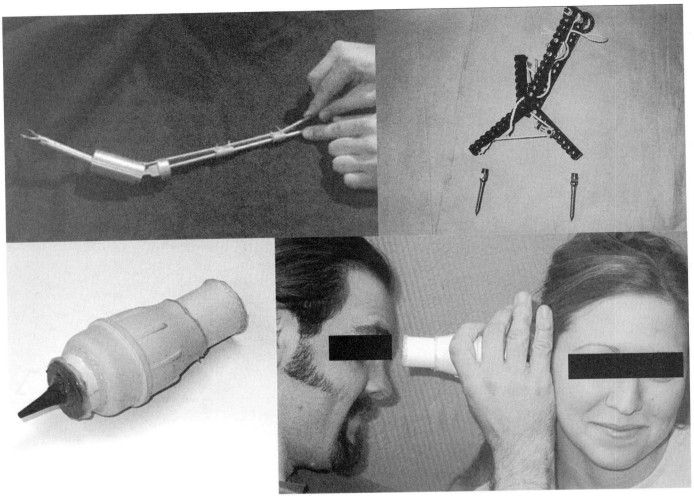

FIGURE 4.5.2
Everyday household materials, such as Legos, string, turkey basters, and wire, can be used make effective preliminary prototypes (courtesy of Stanford University's Product Realization Laboratory).

prototypes). In this way, advanced prototypes can still be built to focus on resolving one question or issue, although it may be a higher order one, such as, will the model work in a living system?

Just as developing a prototype that is unnecessarily complex can be distracting to the innovator, it can also be distracting to users when asked to give their feedback. For example, if an early works-like prototype looks too much like a finished product, users may concentrate on how the device looks and feels instead of focusing on critical issues related to its fundamental functionality. The level of the prototype must match the question or issue being considered and only incorporate as much complexity as is needed to find this answer. Crude prototypes, using common materials, can provide a highly time- and cost-effective mechanism for testing the basic functionality of a device. Some examples of rudimentary, but effective, prototypes are shown in Figure 4.5.2.

3. Isolate the functional blocks of a solution

Often the initial questions or issues an innovator wants to answer with a prototype are too broad to be easily addressed with one prototyping step. One technique that innovators often use as they prepare to prototype a concept is to first divide the concept into "functional blocks." The original need criteria, along with more specific functional design requirements that may have emerged through the analysis performed in Chapters 4.1 through 4.4 can be used to establish the boundaries for each block. Each block should represent one aspect of the concept that could be prototyped. Each block can usually be tied to a distinct engineering discipline based on its characteristics. Some common examples of the engineering disciplines addressed by different functional blocks include mechanical engineering, materials science, and electrical engineering.

Once the relevant functional blocks for a concept are defined, an innovator can prototype each one independently, based on which questions s/he seeks to answer, before bringing the blocks together to prototype the concept as a whole. Importantly, the key elements that could demonstrate the viability of an idea may be represented by only a few blocks (or sometimes a single one). The primary functional blocks for a prototype for the LAA example might be:

1. A material that is liquid when injected into the LAA via a delivery system.
2. A material that can fill up the LAA completely to prevent blood flow from subsequently getting into the LAA.
3. A delivery system that can deliver this material to the LAA and be removed without causing perturbation of the material-filled LAA.
4. A material that "hardens" or polymerizes in the LAA without leakage into the bloodstream.

As shown by these functional blocks, the original concept has a mechanical engineering component (#3) and materials science components (#1, 2, and 4). While some concepts reflect a "pure" solution within a single engineering discipline, it is not unusual for innovative concepts to combine different types of engineering science. Instead of breaking the blocks related to a materials science into several blocks, another approach might be to classify all of the blocks related to materials into a single block with multiple properties. While this may be appealing if the blocks are closely related, by breaking the concept into a more granular level of detail, with multiple blocks for a given engineering discipline, the innovator can more easily determine which of these smaller blocks should be prototyped first, based on what might be novel and what represents the highest degree of risk that needs to be addressed. As the innovator learns about what blocks can be addressed with readily applicable existing solutions, there is no harm in combining them together if it makes sense.

Importantly, the emergence of combination products, such as the drug eluting stent or biologic-coated stent for treatment of coronary artery disease, now demands even more sophisticated prototyping. These new devices often have even more complex functional blocks that interact with each other in complicated ways. Biologic and pharmacologic blocks add other mechanisms of action to consider that may be mechanical, electrical, or chemical in nature. In this case, expertise in pharmacology and molecular biology would be essential to the prototyping team. In addition, lab resources for both macro- and microscopic evaluation of mechanisms would be required. This may not only involve cell and tissue culture facilities, but also new technologies to measure these mechanisms, such as small and large animal imaging.

4. Understand what is known

Once the functional blocks of a concept have been defined, an innovator can evaluate what is already known about each one. This is essential since it helps an innovator determine which blocks to focus on first, in order to address the ones representing the most risk or embodying the most novelty. In some cases, many elements of a solution concept may already be well understood, while just one or two aspects still need to be proven through prototype development. For instance, in the LAA example, the innovator might choose to focus on a catheter-based delivery system that provides a well-understood mechanism for reaching the heart and is acceptable to certain stakeholders, such as cardiologists. In this scenario, much greater uncertainty exists surrounding the material and its interaction with the heart. As a result, the innovator might decide to focus the earliest prototyping efforts on the materials science functional blocks #1, #2, and/or #4 before moving on to #3.

Examining the functional blocks even further, the innovator might also determine that there are many examples of materials that can fill an anatomical space (such as the LAA) and be delivered in a liquid form via a catheter. In contrast, s/he may be uncertain about the existence of materials that can polymerize while in contact with the bloodstream. In this case, the innovator could choose to focus on this latter functional block first to ensure that the unknowns can be resolved before investing significant time in other areas. Note that under these circumstances, it makes sense for the innovator to divide the materials science aspects of the solution into different functional blocks, rather than considering them all together since many more unknowns exist in one area versus the others.

5. Define more detailed design requirements and technical specifications

Just as breaking a solution concept into functional blocks can simplify how to approach the prototyping process, it can also aid with defining more precise design requirements and technical specifications. Once the original need criteria (which are usually fairly high level) have been considered, additional criteria, relevant to each of the functional blocks, can be defined based on what has been learned to date. For instance, in the LAA example, a more precise design requirement for block #3 is that a catheter-based delivery system should be less than 8 French[1] in diameter and steerable to deliver the chosen material to the LAA. This is based on the determination that cardiologists, not cardiac surgeons, would probably perform the procedure and would prefer methods and tools similar to what they already use. Such specifics can be used at this point to guide prototype development along a pathway that takes into account relevant outside factors. At the same time, detailed design requirements help further define the questions that should be answered through each successive prototyping exercise.

Once an innovation or a key element of an innovation has been given a working form, it is easier for an innovator to solicit specific input that impacts the ability to make improvements affecting the usefulness and marketability of the idea. Real users – members of the target audience for a device – play an important role in guiding prototype development by identifying detailed design requirements that may be unknown to the innovator. Sometimes the term "design requirements" is interchanged with "user requirements," which emphasizes the importance of the user in creating these requirements. Similarly, they can raise issues and risks to which the innovator has become "blind" through his/her proximity to the concept.

Having a crude prototype that meets the need criteria and some basic design requirements of key functional blocks (as understood by the innovator) is the easiest way to start this process. Again, going back to the LAA example, the innovator can gain important user input by showing them a model based on a standard catheter used by cardiologists, a substance such as glue, and a cow's heart. In this scenario, the innovator could demonstrate how the glue can be introduced via the catheter to fill the LAA, after which it polymerizes and hardens within one hour. The cardiologist might then point out, for instance, that the glue would need to polymerize in less than 20 minutes, as one hour could be too long to keep a catheter in the left atrium. The next prototype in the iterative cycle would then have a goal to address this more refined design requirement. Consulting with multiple users can help the innovator avoid the biases of any single individual.

When interacting with users regarding prototypes, an innovator should follow some general guidelines:

- **Ask users open-ended questions** – What do they think of this? What could be better? Why don't they like it? Is the reason they do not like it functional, design-related, or practical in nature?
- **Observe users as they handle and use the prototype** – What features give them the greatest difficulty as they use it on the bench or in animals? What seems to attract or appeal to them?
- **Ask users to imagine using the device in a clinical setting** – If it is not possible to observe users trying the device on patients (i.e., due to regulatory constraints), ask them about the other factors that must be considered in designing the device to integrate it into an operating room, physician's office, catheterization laboratory, etc.? Are there other equipment/environment considerations that affect its use?

While it is unlikely that all of the feedback gathered from multiple users will be consistent, the opportunity to consider a wide range of opinions can reveal the full spectrum of strengths and deficiencies associated with a prototype. The innovator will then have to determine, based on his/her own observations, what user feedback to act on, as well as which elements of the prototype to further refine or modify, and which ones to accept.

Through the process of developing increasingly detailed prototypes and refining design requirements, an innovator will naturally develop an appreciation of key technical specifications. Technical specifications represent the important engineering parameters that a solution must satisfy in order to meet the need and important design requirements, provided that these requirements are within the bounds of technical feasibility. Stated another way, technical specifications capture and explain how a solution must function. These specifications typically focus on technical features (e.g., what loads the device must tolerate, what material a device must be made of, how durable it must be), but may also address "softer" attributes desired by the user that are related to device functionality (e.g., compatibility with other devices, how it is operated, ergonomic features). All such specifications warrant careful consideration during device design as they will likely impact product development and engineering activities later in the biodesign innovation process.

Technical specifications can also emerge from the successes and failures that the innovator experiences in developing different prototypes. The LAA example can once again be used to illustrate this point. Prototyping in a live animal may show the innovator that a catheter has to have a certain rigidity so that it does not collapse as it is inserted into a blood vessel. Through trial and error, the innovator would gradually come to understand the appropriate load factor that it has to withstand. Once this determination is made, the technical specification would be considered essential for the solution to be technically feasible.

While little emphasis is placed on rationalizing design requirements and technical specifications early in the biodesign innovation process, an innovator must begin to acknowledge necessary trade-offs between these parameters as design work progresses. For example, users might want a device that has many technically complex features, but also is miniature in size. When a design is initiated, the innovator may be faced with limits on the degree of miniaturization that is possible while meeting the requirements for technically complex features. If the innovator does not have a clear sense of which requirements are most critical, more user input may be needed to help balance these requirements against relevant technical specifications. Alternatively, multiple designs can be created to emphasize different combinations of key requirements. Target users can then be asked to respond to the specific designs instead of helping to prioritize requirements in the abstract – an exercise that many individuals find challenging. Regardless of which approach is chosen, the innovator must carefully monitor the interplay between design requirements and technical specifications to ensure that a solution can be feasibly engineered.

Timing of prototyping

While there is not necessarily a "right" time to start prototyping, many experts encourage innovators to begin as early as possible so that they learn from the process and apply that learning to iterative design and development. If there is a question that can be answered through prototyping, that exploration can be done through engaging in the prototyping process. As Tom Kelley, general manager of design firm IDEO, explained:[2]

> Quick prototyping is about acting before you've got the answers, about taking chances, stumbling a little, but then making it right. Living, moving prototypes can help shape your ideas. When you're creating something new to the world, you can't look over your shoulder to see what your competitors are doing; you have to find another source of inspiration. Once you start drawing or making things, you open up new possibilities of discovery [by] doodling, drawing, [and] modeling. Sketch ideas and make things, and you're likely to encourage accidental discoveries. At the most fundamental level, what we're talking about is play, about exploring borders.

The downside of prototyping too early in the biodesign innovation process is that an innovator can get swept up in creating a working model before s/he is

Table 4.5.2 *Regardless of when an innovator initiates prototyping, s/he should remain aware of the associated advantages and disadvantages of early prototyping to help avoid common pitfalls.*

Advantages of early working models	Disadvantages of early working models
• Flexible; can be relatively easily changed. • Relatively inexpensive (compared to later prototypes). • Can help identify critical design requirements from users early on and incrementally throughout the development cycle. • May provide the proof of concept necessary to attract funding. • Can demonstrate key technical challenges and issues regarding feasibility. • Can help give users an idea of what the final product will look like and how it will operate. • Can encourage active engagement in the development process by users. • Can motivate the innovator(s) and drive an increase in product development speed.	• User needs and design requirements may not yet be adequately understood, rationalized, and reflected in the designs. • Early designs may not adequately address user expectations and may, therefore, disappoint the target audience. • The innovator may become prematurely attached to certain aspects of an early prototype that are counterproductive to upstream (user requirements) or downstream (technical feasibility and manufacturing) considerations. • The innovator might waste time and money prematurely prototyping an incomplete or flawed design.

certain that the solution meets key need criteria and/or design requirements. Another pitfall is becoming committed to a concept once it takes a working form, despite fundamental flaws that may not be technically feasible in a final product. Issues such as these may lead innovators to invest money and to spend days, weeks, or months defining and refining a concept that may ultimately be inadequate to satisfy the underlying need. Innovators must make a careful judgment about the ideal time for initiating (and sustaining) prototype development, since there are many pros and cons (see Table 4.5.2).

Prototyping as part of the biodesign testing continuum

Beyond concept selection, prototyping plays an integral and significant role in the biodesign testing continuum for medical devices (as shown in Figure 4.5.3). The use of props to demonstrate basic concepts while brainstorming represents an extremely early form of prototyping, before any real design work has been done (see 3.1 Ideation and Brainstorming). The more formal prototypes that begin to take shape after designs have been developed go through increasingly rigorous levels of testing from bench testing to human clinical trials. Early prototypes (as addressed in this chapter) are used primarily to perform bench testing, simulated use testing, and tissue testing to drive design evolution and gather early user feedback. These steps are usually taken multiple times throughout the iterative design process. Live animal testing, cadaver testing, and human testing are performed during the later stages of the biodesign innovation process when works-like and is-like models are available. Throughout the testing continuum, the innovator continues to collect information about key design requirements and understand increasingly important technical specifications. Key considerations for bench, simulated use, and tissue testing are outlined below (and in 5.3 Clinical Strategy).

Bench testing and simulated use testing

Bench testing refers to the testing of materials, methods, or functionality in a small-scale, controlled environment, such as on a laboratory workbench. These early tests typically serve to help identify key elements of a design in terms of materials or functionality. This allows the innovator to make optimal choices about how to build more robust prototypes for later stages of testing. In bench tests, individual components or subassemblies are put into a test loop where all of the associated variables can be independently controlled,

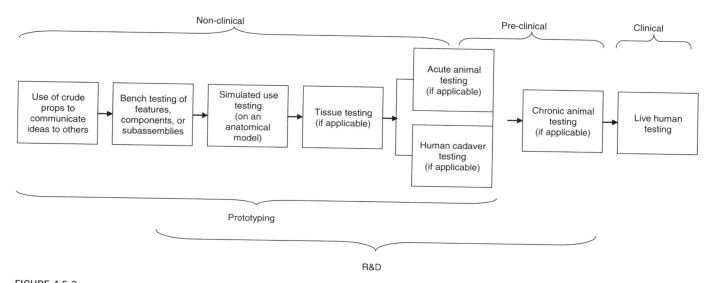

FIGURE 4.5.3
The biodesign testing continuum spans non-clinical, pre-clinical, and clinical testing, with prototyping playing a key role in non-clinical studies.

measured, and recorded.[3] Bench tests should be designed as simply as possible. For example, in a catheter device with an articulating arm and handle mechanism, the innovator should test the arm mechanism independently of catheter tracking or lubricity to ensure that each of the individual features works independently. This makes it easier to understand the test results when combinations of features are eventually tested together. In simulated use testing, various features or functions are tested as a system on an anatomical model or proxy.

When developing bench tests or simulated use experiments, an innovator should use simple methods and generic materials or machines (tensile testers, weights, levers, calipers, etc.). S/he should test in real tissue later in the testing continuum, as many less costly surrogates are readily available to refine and develop prototypes to a point where the design has a better chance of working in real tissue. During preliminary tests, materials that can provide a substitute for tissue include silicone, chamois, nylon, sponge, foam, or ceramic pieces (for bone).

As a rule, an innovator should seek to use tests consistent with the stage of prototyping that is being undertaken. As the prototypes and models grow more sophisticated (through design evolution) and risks are retired, tests can become more complex such that they begin to more closely resemble actual use. Eventually (and often rapidly), the design gets to a point where tissue testing is required to generate the next set of design requirements.

Tissue testing

Tissue testing is an essential component of medical device prototyping that should be undertaken after a concept has been proven in simpler tests. This is not to say that tissue testing cannot be undertaken earlier, but by answering many of the straightforward questions through bench and simulated use testing, results are less likely to be confounded by non-critical data. An innovator can quickly evaluate a device's mechanism of action using animal tissue available from the local grocery store – for example, substituting the skin on chicken feet to emulate human skin when testing the ability of a new suturing device to penetrate tissue. Electrical mechanisms may be evaluated in much the same way. For instance, a skinless chicken breast can be used to assess the ability of a new ablation catheter energy source to create tissue injury. With chemical mechanisms, a similar approach can be used. As shown in the LAA example, developing a glue that would adhere to the endocardial (inside) surface of the heart is critical. For a preliminary tissue test, an inexpensive cut of beef could be used to evaluate the interaction between the glue and the tissue.

Although mechanisms of action can often be evaluated using simple, widely available animal tissue (like beef tongue, steak, gelatin, or poultry), device feasibility must almost always be confirmed using animal or cadaver tissue that is more representative of the anatomical or physiologic properties that the device intends to address. Swine and sheep organs can be obtained from many commercial slaughter houses. These tissues can be shipped fresh or be frozen for later use. When using animal organs or body parts, the innovator's goal is usually to assess the basic functionality of the prototype. While animal parts are good substitutes for many (but not all) prototyping projects, human tissue testing will eventually become essential. Human tissue can be drastically different from animal anatomy for certain organ systems, so more advanced tests are necessary to evaluate a device. In an effort to better understand the size, shape, and contour of the space in which a device needs to operate before a human tissue test is designed, an innovator may find that a trip to an anatomy lab is invaluable. Additionally, with proper licensing and disclosure of intent, human cadaver tissue can be obtained from commercial sources (note that when working with human tissue, proper lab facilities and processes for its safe use and disposal are essential).

Keep in mind that dead tissue has obvious limitations compared to living tissue. However, for purposes of initial anatomical and physiologic properties, it is a good place to start. Once tissue fed by oxygenated blood, at body temperature, and with certain intact anatomical relationships is required, then tissue testing may not be adequate. Nonetheless, even with dead animal tissue, there are certain guidelines that need to be followed with respect to proper handling and disposal.

In-house versus outsourced prototype development

When an innovator or company is ready to begin prototyping, it is necessary to decide whether prototypes will be developed in-house or by a third party that specializes in prototype development. This decision depends in large part on the nature of the concept, the cost-efficiency of each approach, and the skills and resources of the team. While third parties can provide access to vast experience and specialized equipment and materials, a company may give up an important learning opportunity if it decides to outsource the development of its prototypes. However, this may be essential for more complicated and sophisticated projects. Innovators, especially relatively inexperienced ones, are encouraged to entrust at least some portion of their prototyping efforts to the company's in-house engineering team to capitalize on the invaluable learning that can take place. The advantages and disadvantages of these two different approaches are summarized in Table 4.5.3.

Whenever possible, innovators should make an effort to establish a basic prototyping lab of their own to perform early prototyping activities. Then, as the complexity of the models increases and more specialized equipment and materials are required, specific steps or models can be outsourced on an as-needed basis. This hybrid approach is often the most sensible and affordable way for a lone innovator or young start-up company to proceed. However, there are also times when one might choose to establish a more comprehensive prototyping facility. For example, Perclose, a pioneer in the femoral vascular closure space, took this approach and gained vast efficiencies in its design cycle, thereby realizing a significant return on its investment. In this scenario, the cost of setting up the prototyping lab was built into the initial business strategy and financial model (see 6.1 Operating Plan and Financial Model and 6.2 Business Plan Development).

The Evalve story below highlights the advantages that can be realized through in-house prototyping and also demonstrates how one company followed the general approach outlined in this chapter.

Table 4.5.3 *Innovators should consider these pros and cons before deciding to do in-house or outsourced prototyping.*

	In-house prototyping	Outsourced prototyping
Advantages	• Hands-on prototyping experience contributes to the expertise of in-house engineers. • Company engineers directly learn what works and what does not and can apply those learnings directly to the design to test changes or add/delete features. • Direct prototyping experience allows the iteration process to occur faster. • In-house prototyping can save a great deal of time if the design proves to be not feasible. • Overall, in-house prototyping is usually cheaper than outsourcing (except for complex processes and custom parts). • Captures intellectual property and proprietary information about what may or may not work. This information may be used to block competitors.	• Outside specialists possess high levels of expertise on specific processes and equipment. • Outside shops have seen thousands of prototypes for other companies and can offer invaluable advice and design assistance, especially in the prototyping of non-critical, proven components. • In some cases, outsourcing can be less expensive than purchasing the required equipment (and undergoing training) in-house.
Disadvantages	• Some equipment may be too costly for a small start-up to purchase. • Some processes may require expertise that is not available in-house. • The team typically cannot be built up quickly or scaled down easily based on prototyping needs, which may have significant time and money implications.	• Outsourcing can often take more time since the shop has to meet the needs of multiple customers. • An outside shop may not possess unique or vast knowledge about the innovation. • Issues related to intellectual property must be managed carefully in every outsourcing deal. • Design iterations and risk retirement can take longer. • Can be costlier than in-house prototyping, especially for simple designs.

FROM THE FIELD — EVALVE, INC.

Understanding prototyping as part of the biodesign innovation process

Evalve, Inc. was founded in 1999 to develop percutaneously delivered devices and tools for repairing the mitral valve, one of the heart's four valves.[4] Mitral valve regurgitation (MR) occurs when the valve's leaflets or flaps do not close completely, resulting in the backflow of blood from the left ventricle into the left atrium and into the lungs, dilation of the left atrium, and the eventual enlargement of the left ventricle. Approximately 2 percent of the US population suffers from this condition.[5] Despite the significant sequelae of untreated mitral valve regurgitation, which can include atrial fibrillation, heart muscle dysfunction, congestive heart failure, and an increased risk of sudden death, just 50,000 individuals undergo surgery each year to correct the problem.[6] Three to four times that number of patients experience symptoms and complications serious enough to warrant the traditional therapy used to treat MR – open heart surgery – but do not receive the procedure, either because they are too sick for heart surgery or are in a position of "watchful waiting." Evalve's founder, Fred St. Goar, a Stanford-trained interventional cardiologist, was familiar with the morbidity associated with open heart surgery and saw a market opportunity for a less invasive solution. After brainstorming several different approaches focused on suturing the valve leaflets, St. Goar and other Evalve team members, including vice-president of R&D Troy Thornton, hit on the idea of a clip that would stabilize the leaflets and also hold part of the valve closed, allowing the leaflets to function properly.

FROM THE FIELD: EVALVE, INC.

Continued

Evalve's small team of engineers and technicians began building prototypes early in the development process. First, they made rudimentary sketches by hand. Then, technicians and engineers developed crude prototypes in ones and twos in Evalve's lab, using pig hearts from a local butcher to experiment with what worked and what did not. According to Thornton, "Eliminating options is an important part of prototyping,"[7] and simple experiments with early prototypes were used to rule out certain concepts and ideas that might have consumed significant time and money down the road. Additionally, even though they started working with pig hearts, the team recognized the need to understand the disease in humans as quickly in the process as possible. Many development paths that may seem promising early on can be derailed by the unique intricacies of the human anatomy. "We sent a few of the engineers out to the pathology lab to look at human hearts that had mitral valve disease," said Thornton. "That was enlightening and important. Surgeons can tell you what it looks like and you can see pictures in a book, but seeing the problem in person is quite different." With a strong first-hand understanding of the disease in humans, the team was able to be more efficient in its approach to prototype development, relatively quickly eliminating paths that proved not to be fruitful.

As Evalve pushed ahead, it discovered that there was too much prototyping and testing for a single engineering team to handle. In addition to the clip, Evalve needed to develop a guiding system and a clip delivery device, so it developed three engineering groups to help break down the problem into smaller, more manageable pieces. "From a conceptual standpoint, we realized that the project was too big to tackle all at once," recalled Thornton. "Breaking the product into manageable chunks became critical for a complex project like this." One team, which was focused exclusively on the guiding system, spent many months developing different designs and conducting animal studies just to isolate and understand the functionality required to address this variable. Another team worked on the clip delivery catheter that would be able to actuate the clip once inside the heart. That left one additional engineering group to focus solely on designing the best clip they could to hold the mitral valve leaflets closed and restore normal heart function. While breaking into separate teams was effective in terms of advancing Evalve's understanding of the device's core components, it required a certain amount of additional coordination. "Everything has to come together," said Thornton. "All three engineering teams have to talk to each other constantly. They have to communicate well because if somebody changes something on part A, it could affect system B or C." Due to the interdependencies of the different components, each team also needed to maintain and document specific design requirements for their components to keep the effort synchronized. Adjustments to resources had to be made at various times during the project to assure that all three teams could move forward at the same fast pace. Despite these challenges, however, the narrower focus of each of the engineering teams allowed Evalve to innovate much faster and more efficiently than when they were a single working group.

Another important input to Evalve's prototyping efforts was the collection of physician feedback as early and often as possible. St. Goar said, "The downfall of more than a few companies has been getting an idea and then having a group of engineers go off to create something very elegant, very sophisticated, and very complex that the physicians just can't imagine using in clinical practice." For Evalve, the downside of physicians potentially being turned off by flaws in early, crude designs was far outweighed by the upside of having their input to take into account in the design process. Having physicians who were able to be involved all the way through the process, looking at many versions of prototypes and speaking to the positives and negatives of different design features was especially critical. In addition, although the new, less invasive procedure would ultimately be performed by interventional cardiologists ouside of the operating room, St. Goar and Thornton made a point of involving surgeons in their prototyping work. St. Goar explained, "We were trying to match what the surgeons did. If we simply addressed the need from an interventional cardiologist perspective, we might miss the therapeutic mandate. The surgeons knew what needed to be accomplished, and we certainly didn't want to lower their clinical standards. Having surgical input early on was invaluable."

Eventually, after its engineering workstreams started coming together, Evalve had to make the decision about when to begin testing in humans. "There is a little bit of a leap of faith that's required," St. Goar said. "You've got to bite the bullet and say, 'Okay, this is it, we're going to go.' But that's not always easy to do." For Evalve, having positive, longer term results from animal tests was one factor that helped the team make this decision. It was also important that the design had been shown to meet its critical design requirements. Additionally, executive management played a key role. "Our president and CEO, Ferolyn Powell, was very helpful. An engineer by training, Ferolyn had been intimately involved in the development of the technology, which allowed her to fully comprehend the risks," said St. Goar. In this case, the board of directors also weighed in. Thornton remembers one director saying, "You can do all the animal studies you want, but you're not going to learn a damn thing until you get into people. You could spend another six months refining this thing, but it's time to move forward."

This decision was closely related to a development challenge that many inventors and companies struggle with: **design creep**, or the tendency for engineers to spend too much time changing minor aspects of a design or prototype that are unlikely to affect the overall performance of the device. Clearly defining what the "must-haves" versus the "nice-to-haves" are for a device is one strategy that inventors often use to keep the iteration process under control and avoid design creep. Fully satisfying the must-have requirements should be a signal to the team to take the next step forward. According to St. Goar and Thornton, the FDA has particularly strict rules regarding design changes that occur after human trails are initiated, especially for a permanent implant, so design creep can be highly problematic for firms at this stage. "There are always ways to make a device better than it is today, but you really have to be committed, once the decision is made to move into human studies, to stick with your design," said St. Goar.

Although Evalve developed its prototypes in-house, the company also worked with an outside consultant with experience in medical device design, development, and manufacturing. "I think there's sometimes a little hesitancy on the part of the engineers to bring in an outside consultant," said Thornton, "but you've just got to get over it and say, 'Hey, maybe we've got something to learn from somebody else.'" He continued, "We brought in the consultant to work with our top three engineers who were working on the clip design to help them refine it. We used him extensively for brainstorming and developing ideas. He had years and years of experience and knew these specialized vendors who could do

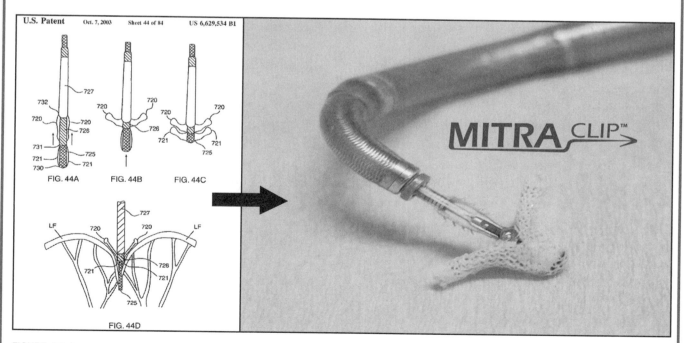

FIGURE 4.5.4
From the early design captured in patent drawings to the finished product, the MitraClip evolved significantly over time (reprinted with permission of Evalve, Inc.).

> **FROM THE FIELD**
>
> **EVALVE, INC.**
>
> *Continued*
>
> micro-stamping processes, which none of us had known about before. Bringing in outside expertise at the right time in the project can be a pretty critical step."
>
> When offering advice to other inventors and companies, both Thornton and St. Goar stress perseverance as being essential to prototyping, as well as the larger biodesign innovation process. In the case of Evalve, as Thornton described, "There were a couple years of trial and error before it really took off." St. Goar added that patience also comes in handy. "We spent a period of time trying a number of different approaches, none of which looked anything at all like where we are today," he said (see Figure 4.5.4).
>
> Furthermore, despite being founded in 1999, and treating its first patients in 2003, Evalve was still enrolling participants in a Phase II (pivotal) clinical trial for its MitraClip™ device as of 2008. "To give you some perspective as to how long medical devices take to develop compared to other types of technology, this company started the same year as Google," St. Goar remarked. Despite these challenges, St. Goar acknowledged that the benefits to working in the device field were sizable. "We already have a lot of people out there who have dramatically benefited from our device. This has to be the ultimate goal and measure of satisfaction."

General tips for effective prototyping

As innovators gain experience with prototyping, they build confidence and become increasingly effective in designing and developing models. Those without vast experience, however, will be well served to consider the following advice as they get started:

- **Consider multiple factors** – While designing and prototyping, be sure to think about the function, form, material, manufacturing, cost, and feasibility of a prospective device.
- **Play with scale** – When prototyping small devices, consider scaling up five times to ten times to make the initial exploration process easier during preliminary experiments. Look at things (especially moving parts) under a microscope to be certain to achieve a detailed understanding of how they function. Just remember that it will be necessary to find solutions that can be replicated at the size appropriate for their actual use.
- **Iterate** – Anticipate that the prototyping process will require multiple iterations and plenty of trial and error. Keep ample resources/materials on hand for practicing and correcting mistakes.
- **Measure twice, cut once** – For mechanical prototypes, never cut material until verifying that the measurement is correct. It is far more difficult to add materials than to remove material after cutting. Always use a sharp blade to minimize errors during cutting. When using adhesives, let everything dry properly before handling.
- **Consider the effects of reuse** – Construct reusable prototypes and test them for degradation with time and repeated use. Consider computer modeling and/or the use of specialized testing machinery to subject prototype devices to repeated physiologic loads. This process (and the associated test data) is often required for regulatory submissions. For example, devices implanted into the vasculature must endure tens of thousands of simulated heartbeats without failure or breakdown. Some computer models are commercially available to test this. Others are being developed for specific disease or device applications.
- **Take pictures** – Photograph everything! A great deal can be learned from understanding and keeping a record of how each prototype has progressed based on lessons learned from earlier models. This also serves as compelling history for the company and customers as the product is introduced into the market and the company's story evolves.
- **Save all work** – Keep all prototypes as a physical record of development.

- **Maintain an innovation notebook** – Keep detailed notes (measurements, techniques, materials, etc.) in an innovation notebook regarding exactly how the first prototype was constructed so that the process can be replicated. Use a disciplined version control system for all drawings (as described in 1.2 Observation and Problem Identification) so that the design can be replicated later, as needed.
- **Be cautious regarding confidentiality** – Because there may be a significant amount of invention that occurs during prototyping, be sure to contract with a product designer who agrees to relinquish all rights to any intellectual property generated from the design process. Protect all technology that is developed as a result of the prototyping process. Non-disclosure agreements should be signed with any contract suppliers/shops used (see 4.1 Intellectual Property Basics).
- **Be safe** – Follow all standard safety precautions when working with tools and/or machinery.

Appendices 4.5.1, 4.5.2, and 4.5.3 provide more information and tips for prototyping specifically in the disciplines of mechanical, materials science, and electrical engineering.

GETTING STARTED

In conjunction with the steps below, the guidelines described in the chapter can be used to help initiate an effective prototyping effort. Just remember that the iterative nature of process may require the innovator to circle back through these steps many times over.

See **ebiodesign.org** for active web links to the resources listed below, additional references, content updates, video FAQs, and other relevant information.

1. Identify the questions or issues to be addressed through prototyping

1.1 What to cover – Before translating a design for a solution concept into a working model, clarify the purpose of the prototyping activity to be certain that the goal can be achieved. Define *specific* questions that must be answered through the creation of the prototype (e.g., will this specific functionality work?) to guide test development. Keep in mind the environment in which the device will be used (e.g., hospital, home use, portable, fixed). This will help define the questions. The answers to these questions should prove a specific element of the design and, in so doing, reduce risk along the development cycle. Resist the temptation to raise questions that are not on the critical path. Remember that these questions can (and should) evolve over time as the invention progresses from idea to product. Each time a prototype is developed, the questions it is designed to answer must be revisited and revised to maximize the benefit of the effort. The answers to these questions should be translated into increasingly specific design requirements.

1.2 Where to look – Go to trade shows in the therapeutic area and look at competitive products. Generate ideas for questions by looking at competitive devices as well as those entirely outside the therapeutic space, and use this as input to help define prototyping goals. Also, study the company's own designs carefully to isolate the most critical questions to resolve at each stage of prototype development. Interact extensively with the engineering team to define prototyping goals and allow this team to drive the process.

2. Design the minimal model needed to answer those questions

2.1 What to cover – Determine the best category of prototype to develop in order to answer the first question that has been defined. Strip out anything and everything from the model that does not explicitly need to be tested. Focus only on the critical path in defining the model. Look at what other companies have done and how they have designed their prototypes. For example, well-known companies have developed highly effective prototypes using nothing more than foam core and paper clips or two syringes and a wood-working clamp. A model does not have to be expensive, complex, or made from specialized materials to be effective. Custom prototyping is costly and time-consuming when certain concepts can be proven with ready-made, easily available, and inexpensive materials.

2.2 Where to look – Interact extensively with the engineering team (if one exists) and/or seek input from others who have experience in designing and developing prototypes.

3. Identify and prioritize functional blocks

3.1 What to cover – Identify the functional blocks associated with the solution being studied. Each block should represent one aspect of the concept and will likely be tied to a distinct engineering discipline based on its characteristics (e.g., mechanical engineering, materials science, and electrical engineering). Prioritize which functional blocks to start with, based on what is known about each one and where the greatest risk/uncertainty exists.

3.2 Where to look – Use the original need criteria, along with more specific functional design requirements that may have emerged through previous analysis (see Chapters 4.1 through 4.4) to establish the boundaries for each block and to prioritize them.

4. Build the model

4.1 What to cover – Create the prototype focusing only on those elements necessary to address the key question that has been defined (e.g., functionality, interaction of parts, whether the clinical problem can be solved with the design). Prove the concept, retire the most significant risks, and iterate the design along the way. When considering materials, start with basic, inexpensive materials and then gradually progress to more complex, expensive alternatives (e.g., for a mechanical prototype, an innovator might advance from paper and wood to plastics and metals). Similarly, use purchased parts and other off-the-shelf materials whenever possible, reserving the need for specialty or manufactured parts until the later stages of prototype development. As the complexity of models progresses, consider what can realistically be accomplished in-house and what services might require specialized third-party assistance. Given the fact that multiple engineering disciplines may be involved in creating a single prototype, coordination of the efforts across multiple shops may be required.

4.2 Where to look
- **Medical Device Register** – Database that can be used to identify manufacturers of medical devices and see what companies are working in specific areas of interest.
- **Medical Device Link** – Online information source that includes a database of suppliers.
- **Medical design and manufacturing shows** – The best place to talk to suppliers and review latest prototyping technology. Shows are held annually in Southern California, the East Coast, and Minneapolis.
- **Mechanical engineering resources**
  - **GlobalSpec** – A search engine and information source designed especially to serve the engineering, manufacturing, and related scientific and technical market segments.
  - **McMaster-Carr** – A supply company with more than 450,000 products.
  - **Local machine shops**
- **Materials science resources**
  - **ASM International website** – A searchable database of modules on materials commonly used for medical device development. Some content may require a license. Has clinically approved devices, including FDA information and literature reviews.

- **MatWeb** – Database of materials. Does not tell which devices use what materials, but it can be used to check the grade of materials.
- **Society for Biomaterials**
- **Materials Research Society**
- **American Society for Testing and Materials (ASTM) International**
- **International Organization for Standards (ISO)**
- Buddy D. Ratner, Allan S. Hoffman, Frederick J. Schoen, and Jack E. Lemons, *Biomaterials Science: An Introduction to Materials in Medicine* (Academic Press, 2004).
- M. N. Helmus, *Biomaterials in the Design and Reliability of Medical Devices* (Kluwer Academic/Plenum Publishers and Landes Bioscience, 2003).
- **Electrical engineering resources**
  - John Peatman, *How to Program a PIC* (Prentice Hall, 1997).
- **Programmable Interface Controller (PIC) Microcontroller website**
- **LabView website**
- **The MathWorks website** – For information about MatLab and Simulink.

5. Test/refine prototype to develop design requirements and technical specifications

5.1 What to cover – Use an appropriate bench, simulated use, or tissue test to prove the concept being studied. Define a test requirement (this is often a guess, but grounded in good sense) and develop baseline results. This can determine whether redesign is necessary or redundancy is required. Document the test method and procedures and record results, including technical specification related to fatigue, tensile, force, electrical, and other important issues. Remember that virtually no bench or animal model replicates the human model, so be thoughtful and vigorous in trying to establish a model that is tougher than any probable clinical use. Based on test results, modify the prototype to address clinical, mechanical, and electrical needs. Build a new prototype. Test again. As this process progresses, identify and document increasingly refined technical specifications and design requirements based upon design, user input, and other standards for the device. For more advanced prototypes, use experienced practitioners (doctors, nurses, technicians) to handle the device in the clinical environment and diligently collect their feedback.

5.2 Where to look
- **Tool shops and outside suppliers** – To design simple test fixtures and models.
- **Vasodyn** – Glass anatomical models.
- **Limbs and Things, USA** – Medical simulation models.
- **FDA human factors considerations** – Increasing emphasis on human factors, the study of how people use technology, should be considered throughout the prototyping process, but not at the expense of rapid development if human factors can be built into later designs. These issues are particularly important in developing feels-like/looks-like/is-like models toward the end of product development.

Credits

The editors would like to acknowledge Trena Depel, Steve Fair, Craig Milroy, and Asha Nayak for their help in developing this chapter. Many thanks also go to Kityee Au-Yeung, Gary Binyamin, Scott Delp, Mike Helmus, and Ben Pless for sharing their expertise, as well as Fred St. Goar and Troy Thornton for sharing the Evalve story.

Appendix 4.5.1

Special considerations for the development of mechanical device prototypes

Mechanical engineering prototypes are three-dimensional objects made out of metal, plastics, or other physical materials using common manufacturing schemes. Typically, mechanical engineers (and/or mechanical engineering principles) play an important role in the development of these prototypes, which are often made in machine shops or product realization laboratories.

Mechanical engineering prototyping begins with design drawing. When thinking about mechanical engineering solutions, innovators should not commit to any single feature or combination of features until they begin to put their ideas down on paper. In this way, abstract thinking represents a creative stage of the prototyping process. It is also the least expensive, so innovators should be encouraged to spend plenty of time considering a breadth of different alternatives and then narrowing down the list of potential approaches before they begin to sketch their ideas.

Two-dimensional design drawing begins the process of formalizing an idea. A design allows the innovator to represent a concept and communicate it to others in a more concrete (rather than abstract) manner. In this way, drawing serves as a bridge from concept to prototype. Too often, innovators hesitate to draw their ideas because they worry about their drawing skills. However, especially early in the design process, it is not important for the drawings to be exact. Rough sketches, no matter how imprecise, can be a powerful tool for communicating an idea, identifying potential issues, and stimulating new possibilities.

The most common types of design drawings used in mechanical engineering are isometric drawings, cross-sectional drawings, layout drawings, orthographic projection, and computer-aided design (CAD). At a high level, drawings are often created in a sequence. In this example, the drawings show different views of a single-valve, one-way syringe.

Isometric drawing

Isometric drawing, also called isometrics projection, is a method of visually representing three-dimensional objects in two dimensions. Isometric means "equal measurement" so the true dimension of the object is used to construct the drawing. The height of the object is drawn along a vertical line. The width and depth of the object are drawn at a 30-degree angle to the horizontal plane such that the three coordinate axes appear equally foreshortened and the angles between any two of them are 120-degree (see Figure 4.5.5).[8]

Layout drawing/cross-sectional drawing

Layout drawings can be done in isometric form or cross-sectional form. These drawings show the number of parts, how they are sized with respect to one another, and how they fit together into a complete assembly. A layout drawing provides a rendition or blueprint of the concept that can later be used as the basis of more detailed drawings.[9] By cutting away a portion of the object, as in a cross-sectional drawing (see Figure 4.5.6), the hidden components in a device can be shown. For example, imagine a plane cut vertically through the center of the object shown below to expose its inner workings. Diagonal lines (cross-hatches) can be used

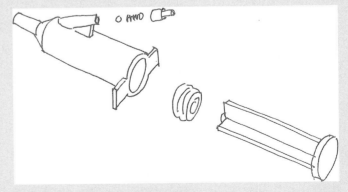

FIGURE 4.5.5
An isometric view of the syringe concept (courtesy of Craig Milroy).

4.5 Prototyping

to show regions where materials have been cut by the cutting plane.[10]

Orthographic projection

Orthographic projection is a method that provides multiple views of the same object. To create an orthographic projection, as shown in Figure 4.5.7, imagine that an object is suspended inside a glass box. Draw the object on each of three faces of the box, as seen from the different directions. Then, unfold the box for a multi-view perspective of the object. Up to six such views can be created for any object – choose the views that reveal unique details about the object (all six views may not be necessary to describe the object fully).[11]

Computer-aided design

Computer-aided design drawings can be made by using a wide range of computer-based tools (e.g., SolidWorks, Solid Edge). CAD drawings are especially useful in early design work and prototyping – both as a way to communicate ideas about design and intended functions, but also to communicate with the machinists, shops, and vendors that will contribute to the prototyping effort. Computer-aided design is mainly used for detailed engineering of three-dimensional models and/or two-dimensional drawings of physical components, but it is also used throughout the engineering process from conceptual design and layout of products, through strength and dynamic analysis of assemblies, to definition of manufacturing methods of components. Computer-aided design offers many benefits, including lower product development costs and greatly shortened design cycles (because designers can lay out and develop their work on screen, print it out, and save it for future editing rather than having to create drawings from scratch).[12] See Figure 4.5.8.

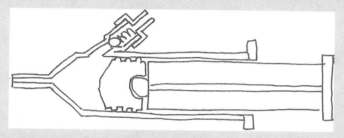

FIGURE 4.5.6
A cross-sectional view of the syringe concept (courtesy of Craig Milroy).

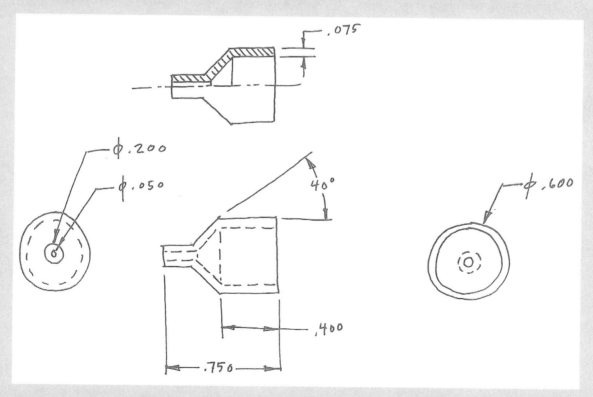

FIGURE 4.5.7
An orthographic projection of the syringe concept (courtesy of Craig Milroy).

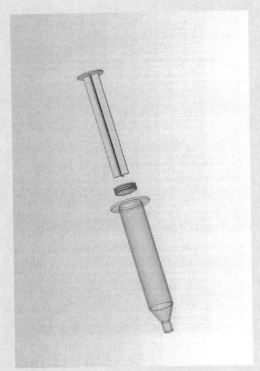 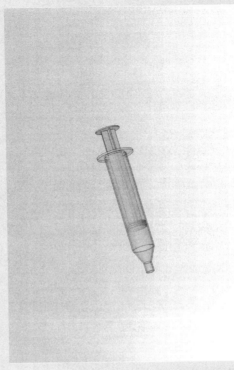

FIGURE 4.5.8 A CAD drawing for the syringe concept (courtesy of Craig Milroy).

Note that mechanical prototyping and design drawing are not linear, mutually exclusive activities. Mechanical prototypes can be constructed to discover new information that should then be taken into account within the design. Similarly, as the prototypes become more advanced, more detailed drawings can be created that allow for certain models to be replicated, as required by the innovator.

Setting up a lab for basic mechanical engineering prototyping is likely to require an initial investment of at least $50,000. Innovators are encouraged to allow the engineering team to determine how this money should be spent (with appropriate oversight), since each engineering leader will have his/her own experiences, preferences, and ideas for setting up a lab. These individuals will also have contacts and resources for procuring equipment (e.g., in the used equipment market). More information about setting up an engineering lab as part of the research and development process is covered in 5.2 Research and Development Strategy.

When developing mechanical prototypes, innovators are encouraged to start with materials that are easy to work with before investing in the use of specialized or more complicated materials. Wood, for example, is an extremely versatile, inexpensive material that can be used in early models. Even if the finished product would never be made of wood (e.g., a scalpel), the innovator can explore issues related to the size, shape, and other dimensions of the device using nothing more than hand tools and a wood block to create the models. Once these issues have been worked out, the innovator will have amassed some experience in creating the prototypes (in addition to resolving certain design considerations) and will be able to apply these learnings more effectively to models that use different, potentially more expensive materials.

Appendix 4.5.2

Special considerations for the development of materials science device prototypes

Materials science, or materials engineering, investigates the relationship between the structure of materials and their properties. It is a multidisciplinary area of study that includes elements of applied physics and chemistry, as well as chemical, mechanical, civil, and electrical engineering.[13] If a solution will involve materials engineering, the innovator should define the materials science pieces of the concept and the functional blocks to which they relate. Expanding the design requirements based on certain chemical, mechanical, biological, electrical, etc. property requirements (e.g., needs to be biocompatible with the blood) makes it easier to understand the exact functions of the materials science component and the acceptable ranges of interaction with patient biology, which may be captured as technical specifications. This will help shed light on surface interaction biology and biocompatibility (e.g., what if a material disintegrates over time, breaks down, or is converted by the body into another substance?), as well as general properties such as strength, fatigue, and other such factors.

Once the innovator understands what a material needs to do, s/he should seek to identify materials with the performance characteristics to potentially meet these requirements. This can be multifactorial, especially when the requirements are complex, such as finding a material that is tough, but highly elastic. Various databases are available to assist innovators in their searches, including the American Society of Metals (ASM) International database and MatWeb (see the Getting Started section for more information on these sites). Innovators can also look at the medical literature, patents, and regulatory guidelines for ideas on materials that may meet their objectives. From patents, one can identify trends in materials used, as well as materials that may be "blocked" from use by others based on patent protection. Established materials are typically commodities that are often used in innovative medtech concepts, and may be approved for use by the FDA in other applications. However, novel materials, which may be synthesized, can contribute another dimension to an intellectual property portfolio, which may serve as a barrier to entry by adding further burden to regulatory and testing processes (often creating additional risk to the company as well). Information about materials from the technical literature sometimes sheds light on failure modes associated with various substances. Regulatory guidelines and established standards will provide information on the testing and characterization needed if certain materials are used within a device (an important downstream consideration).

After looking at these various resources, an innovator will be in a position to generate a short-list of materials intended to meet the design requirements of the functional block under study. However, obtaining the desired materials for testing may be difficult or expensive. In these cases, the innovator can try to find a "proxy" material. **Proxy materials** have similar characteristics to the property that needs to be tested for proof of concept. However, they may not have *all* of the same characteristics. For instance, in the example focused on filling the LAA with a material such as glue, key design requirements include finding a glue that can anatomically fill the LAA, be delivered through a catheter, and polymerize in less than 20 minutes. If such a glue cannot readily be identified, a substitute material delivered through a catheter could still demonstrate the process and answer some key questions. Though not the ideal solution, it could prove the feasibility of this approach by serving as a proxy for at least one desired trait of the final material. Once some preliminary "book-work" and proof-of-principle testing with proxy materials is completed, it may be necessary to involve someone with specific expertise to perform more complex tests with more advanced materials.

In terms of the specific tests that innovators use to assess different materials, there are many from which to choose. Among other tests, innovators can use finite element analysis to theoretically model how materials may act/interact in response to mechanical perturbation. There are also certain tests for materials, such as testing purity (note that having a "clean" interaction surface is critical with this form of testing). Analytical techniques, such as microscopy and chemical analysis/characterization can be particularly useful as well as in vitro and in vivo biocompatibility (ISO 10993–1). Furthermore, the mechanical properties, chemical reactivity/stability, bulk material properties, and surface interactions (chemical/biological) of a material should all be tested.

Appendix 4.5.3

Special considerations for the development of electrical device prototypes

An increasing number of medical devices involve components that require electronics and signal processing, which generally is related to the discipline of electrical engineering. Examples of devices that utilize electronics include pacemakers, oxygen saturation sensors, fiber-optic endoscopes, and blood analysis machines. While much of the discussion that follows may be difficult for anyone not trained as an electrical engineer to appreciate, the goal is to highlight that concepts requiring expertise in electrical engineering can be approached in much the same way as any other concept. Whether or not innovators personally have electrical engineering experience, they should understand what needs to be done in the prototyping phase and whom to recruit to assist with this work.

If a concept being considered by an innovator has one or more functional blocks that involve electronics design, it is important to expand on the design requirements to specify what the electronics components need to do within the context of the solution. For example, as part of a device to treat slow heart rhythms, there may be a functional block focused on sensing the heart rate from the surface electrocardiogram (ECG) signal. With this functional block in mind, the innovator can create an electronics block diagram that outlines the high-level input/output relationships and flow of information within this functional block. (Using the term "block diagram" – a standard convention in electrical engineering – to describe the components that make up a "functional block of the concept" may be confusing, but it emphasizes how the overall approach is similar at various levels of the prototyping and development process.) For the ECG example, a block diagram may look like the one shown in Figure 4.5.9.

By creating this block diagram, the innovator can understand the various electrical functions and use them to create more specific design requirements. For instance, the innovator may define requirements such as: (1) the need to filter or amplify the sensed signal; (2) a need to convert the signal to a digital form; (3) the need to include a microprocessor to perform a logic or decision step about the incoming signal; or (4) the type of output that is required. Additionally, how the sequence of blocks is laid out can be important to creating a good design. While there are generally accepted conventions for the elements and sequence of the components of a block diagram, engineers may produce diagrams that vary in terms of the sequence and level of detail of the listed blocks.

Without knowledge of electrical engineering or a background in signal processing, it may be difficult to come up with a simple block diagram. However, through some basic research, the innovator should be able to break the high-level functional blocks of the concept (e.g., sensor to determine heart rate) into the high-level electrical components of an electronics block diagram (e.g., amplifier, analog-to-digital converter). Electronics design textbooks and published patents (particularly expired ones) can be of some help in constructing a new block diagram or modifying an existing one. These resources contain many useful and commonly used examples of electronics designs that

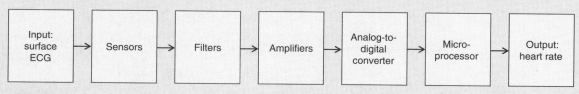

FIGURE 4.5.9
An example of an electrical engineering block diagram (courtesy of Kityee Au-Yeung).

can be leveraged across various solution concepts with minor fine-tuning.

Once a block diagram has been created or modified for the solution concept, an innovator should determine how to prototype each component of the diagram. For the blocks that do not perform logic steps (e.g., making decisions), the function can be simulated using software programs or by building a hardware version. For the hardware version, a drawing of a circuit diagram showing the required hardware should be made first. At a basic level, a circuit is the design implementation of a block diagram. For example, in the block diagram shown in Figure 4.5.9, for the amplifier block, a potential circuit diagram showing components such as resistors (R) could be drawn as shown in Figure 4.5.10.

Again, without knowledge of electrical engineering, it would be challenging for an innovator to create this kind of output alone. However, many of these blocks are commonly used, so a survey of an electronics design textbook could make it easier to find or develop a diagram such as this. The help of an electrical engineer will also be invaluable. Using these resources, the innovator could then obtain the various components outlined in the circuit diagram (e.g., purchase various resistors, capacitors, etc. and a **breadboard** – a reusable solderless device usually used to build a temporary prototype of an electronic circuit and/or experiment with circuit designs)[14] and build this circuit. A general example of a breadboard and its components would look like that shown in Figure 4.5.11.

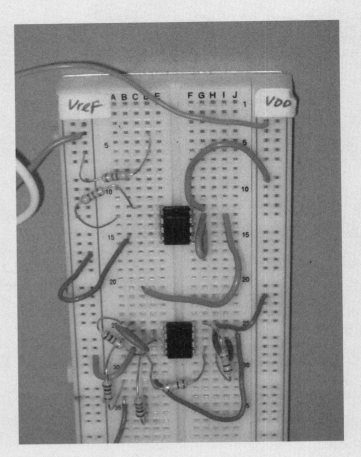

FIGURE 4.5.11
An example of an electrical engineering breadboard (reprinted with permission of iRhythm Technologies, Inc.).

Another, potentially simpler approach to simulating the function of the block is to use a program such as Simulink. Simulink is a MatLab (engineering software suite) add-on that can be used in prototyping to simulate signal processing algorithms during the design phase. Simulink is useful for determining and fine-tuning design parameters. For example, an innovator can simulate the performance of a filter in Simulink, and by changing the variables in the equation representing the filter, s/he can effectively determine what resistor/capacitor value should be used. Simulink should be considered because it allows for rapid prototyping – and without the need for hardware components, it is possible to exactly determine what the output of the design block will be (which serves as the input for the next block). Once the simulation is finalized, it can then be built using the identified hardware, but this can be delayed until all the steps requiring simulation have been completed.

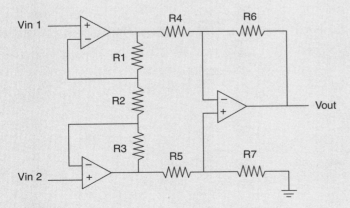

FIGURE 4.5.10
An example of an electrical engineering circuit diagram (courtesy of Kityee Au-Yeung).

While not all designs involve a logic operation, if there is a step requiring this type of operation, then a microprocessor is usually needed. In the example shown above, a microprocessor is used to take the digitized data from the analog-to-digital converter and determine what the heart rate is. To develop the hardware that can perform this type of operation, an innovator first needs to pick a development environment – typically a software environment in which code can be written to perform the desired logic and the associated hardware chip onto which this software code is programmed. This results in a chip that can then be incorporated onto a breadboard with the other components to create the entire circuit represented by the block diagram. Examples of development environments include digital signal processing (**DSP**), application specific integrated circuits (**ASIC**), programmable interface controller (**PIC**), or field-programmable gate array (**FPGA**). An innovator would choose a particular environment based on his/her familiarity with it, as well its specific pros and cons. For example, some are more costly, others are good for more complex tasks, and still others have built-in functions that do not need to be programmed (i.e., they are already programmed). In general, the choice of a development environment is made based on the desired functionality.

Once all the pieces representing the components outlined in the block diagram have been simulated or created, they can be assembled on a breadboard. This breadboard may contain some of the blocks already prototyped as described above, or may be populated with all the simulated circuits and programmed chips at once. While it could be theoretically possible to try to test everything together using a combination of simulations and software code, it is generally more effective to put everything together in a hardware setting, since issues that cannot be simulated may arise when everything has to work together in a hardware environment. Further, it is often more efficient to start the design process in a simulated environment to test, refine, and prove all the functional blocks, and to create the hardware equivalent later. The innovator can also set up certain hardware elements to create a specific simple output (e.g., turning on an LED) if needed, but in many cases, a more robust output environment is required.

Finally, with a breadboard in hand, the innovator can gather the required input and display the desired output. For the input, there are usually straightforward ways to sample the signal that can be directly fed into the breadboard. In the ECG example, electrodes could be attached directly to a person and the signal could be fed into the breadboard itself. For other concepts in which the sensor or input signal itself is novel, this may be its own functional block, but the idea here is that it is usually feasible to directly input a signal into the breadboard. By connecting a computer to the breadboard via an adapter, a program such as LabView can be used to create an interface for displaying the output from the circuit. In the ECG example, it would be used to display the heart rate determined by the microprocessor. The system might look something like that shown in Figure 4.5.12.

In general, it is important to create a working breadboard early in the process so that the innovator can observe the solution's interface with the biological signal. As a rule of thumb, when combining predictable systems, such as circuits and sensors, with dynamic systems, such as human physiology, problems may

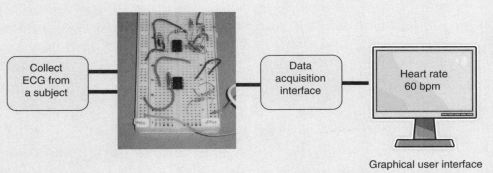

FIGURE 4.5.12
An example of how these steps come together (courtesy of Kityee Au-Yeung).

arise that can only be identified by observing their interaction in a working prototype.

When first getting started with prototyping the electronics components of a concept, it makes sense to use as many existing, off-the-shelf components and solutions as possible in order to minimize the development time required to get to the stage of testing with animals or humans. As with all prototyping, the importance of testing an electronics solution in a biological setting early in the process cannot be emphasized enough in order to ensure that the final electronics design meets all the needs criteria and critical design requirements. The effort to create a custom solution or an integrated circuit can be expensive and time-consuming, so it is crucial to verify at an early stage of development that the effort will be worthwhile.

Notes

1. The French catheter scale is used to measure the outside diameter of a cylindrical medical instrument. 1 French (Fr) is equivalent to approximately 0.33 mm.
2. From Vadim Kotelnikov, "Prototyping," 1000ventures.com, http://www.1000ventures.com/business_guide/new_product_devt_prototyping.html (February 11, 2007). Reprinted with permission from Tom Kelley.
3. Bill Adams, "Bench vs. System Testing: The Pros and Cons," Overclockers.com, October 16, 2002, http://www.overclockers.com/articles638/ (February 26, 2008).
4. The heart's mitral valve lies between the left atrium and the left ventricle.
5. Brian Griffin and Emil Hayek, "Mitral Valve Regurgitation," Cleveland Clinic, http://www.clevelandclinicmeded.com/medicalpubs/diseasemanagement/cardiology/mitralvalve/mitralvalve.htm (February 28, 2008).
6. "About Evalve, Inc." http://www.evalveinc.com/pages/about_overview.html (February 25, 2008).
7. All quotations are from interviews conducted by the authors, unless otherwise cited. Reprinted with permission.
8. From "Isometric Projection," Wikipedia.org, http://en.wikipedia.org/wiki/Isometric_projection (January 29, 2008) and "Isometric Illustrations," TekIt, http://www.foothillsgraphics.com/iso.htm (January 29, 2008).
9. "Layout Drawing," Answers.com, http://www.answers.com/topic/layout-drawing?cat=technology (January 29, 2008).
10. Ernesto E. Blanco, David Gordon Wilson, Sherondalyn Johnson, and LaTaunynia Flemings, "Cross-Sectional Views," *Design Handbook: Engineering Drawing and Sketching,* http://pergatory.mit.edu/2.007/Resources/drawings/index.html#xsection (January 29, 2008).
11. Ibid.
12. "Computer-Aided Design," Wikipedia.org, http://en.wikipedia.org/wiki/Computer-aided_design (February 22, 2007).
13. "Materials Science," Wikipedia.org, http://en.wikipedia.org/wiki/Materials_engineering (February 14, 2008).
14. "Breadboard," Wikipedia.org, http://en.wikipedia.org/wiki/Breadboard (March 10, 2008).

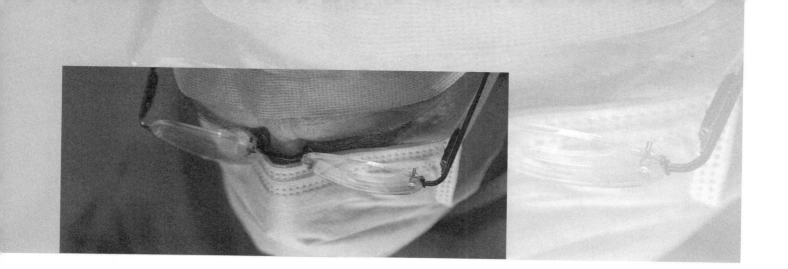

4.6 Final Concept Selection

Introduction

The team started with duct tape, tubing from the hardware store, and the mesh from a flour sifter. Now it has three working prototypes that prove the solution is feasible. But there is only enough time, money, and manpower to take one forward into detailed development. In a move that can feel a lot like betting everything on a single hand of poker, innovators must decide which solution is most likely to come to fruition, challenge the "gold standard" in the current treatment landscape, capture the interest of key stakeholders, and support the development of a sustainable business.

The purpose of final concept selection is to use everything that has been learned about all concepts under consideration in order to choose the one that will be brought forward into development. While a final concept will continue to be refined, tested, iterated, and improved during development and implementation, the innovator will focus on pursuing that single concept rather than dividing his/her attention across multiple potential solutions. While some experienced innovators are able to make this decision based on gut instinct, most benefit from a more formalized process, such as the Pugh concept selection method. While a structured approach may not initially appeal to everyone, it helps to use a process to be sure that key user and design requirements are kept at the forefront of decision making.

OBJECTIVES

- Understand how to use the data gathered to this point in the biodesign innovation process to effectively evaluate solution concepts.

- Recognize how to apply an approach, such as the Pugh method, and develop a concept selection matrix to decide on a final concept.

Concept screening fundamentals

Selecting a final concept is considered by some innovators to be one of the most crucial steps in the product development process.[1] Innovators should take their time and exercise care in deciding on a final concept to avoid the mistake of anchoring on a single design concept too early, without fully exploring the available options. An innovator chooses a final concept from the short-list of one to five solutions generated in 3.2 Concept Screening. This decision is made by considering the intellectual property (IP), reimbursement, regulatory, and business model implications associated with each one (see Chapters 4.1 through 4.4), building on what has already been learned about issues such as relevant markets and stakeholders. Additionally, the innovator should consider the data gathered from the prototyping exercise since these results are important to assessing the technical feasibility of the ideas and the likelihood that they can reasonably be brought to fruition to address the defined need (see Chapter 4.5). Importantly, innovators must remember that any concept that is chosen, regardless of how well it satisfies user and design requirements, must still address all the original need criteria. Otherwise, they may find themselves with a fundamentally flawed concept under development, a product that does not effectively address user needs, or other problems that can threaten to send them back to the drawing board.

At this stage in the biodesign innovation process, innovators often wonder why (and how) certain individuals and companies in the medtech field are able to make consistently good decisions leading to solutions that have a high likelihood of success, while others seem to choose the wrong concepts and/or develop solutions with fundamental flaws that come back to haunt them. According to different sources in the medical device field, one of the most important factors dictating an innovator's ability to choose the right solution concept is the use of a structured, objective approach to concept selection. A structured approach helps the innovator and his/her team to look beyond the most obvious or readily understandable concepts, in order to select a concept with a better chance of success. Some seasoned innovators may argue that they do not need a formalized approach. However, when probed about their reasons for making certain choices, it is often apparent that they have their own internal process for picking a winning concept. While possibly more informal than some of the well-defined methods used in the medtech field, their favored process has likely been honed through years of experience but is nonetheless systematic enough for evaluating important factors – such as market size, regulatory pathway, likelihood of reimbursement, match between product and user type, etc. – before deciding on a final concept to pursue.

For less seasoned innovators, starting with a structured approach is the best substitute for experience in eliminating risks.

The Pugh method of concept selection

While numerous structured approaches exist, the Pugh concept selection method is the most widely referenced and easily understandable in the medtech field (with others tending to be more highly mathematical). It was developed in 1981 by Stuart Pugh, a professor at the University of Strathclyde in Glasgow, Scotland. Fine details of the method can be found in Pugh's book *Total Design*.[2]

The Pugh method applies particularly well to final concept selection in the context of the biodesign innovation process because it explicitly links concept selection to user and design requirements, as well as the need criteria defined in the needs finding and needs screening stages. It is also a relatively simple, yet effective approach that can be quickly and easily applied by experienced practitioners and novices alike. Other benefits of the Pugh method include the following:[3]

- It compels the design team to review the user and design requirements in detail and to understand how the requirements apply to the concept.
- It requires the team to look beyond the obvious first concept and fully explore a wider range of concepts.
- It provides an objective way to evaluate concepts.
- It results in a concise, auditable document for the product's design history file that is easily understood and defensible (see 5.5 Quality and Process Management).

At its most basic, the Pugh method has four steps as shown in Figure 4.6.1. Through this process, the

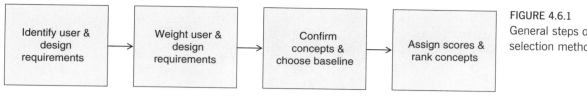

FIGURE 4.6.1
General steps of the Pugh concept selection method.

innovator creates a two-dimensional decision matrix to facilitate the quantitative evaluation of the leading solution concepts against a baseline concept. The idea is that by comparing each potential solution to a baseline (rather than assessing each one against every other alternative), the analysis is significantly simplified.

Another factor to keep in mind is that, even though this method is useful in selecting a final concept, the process of iteration is by no means complete once a final concept is selected. In fact, as the concept undergoes development, design modifications will be generated. The Pugh method can repeatedly be used to select from among different variations of the concept, using increasingly specific requirements. Certain early concept-specific technical specifications that have been learned through prototyping can also be incorporated into these later rounds of selection. This point underscores how iteration and progressive refinement are hallmarks of the biodesign innovation process.

To illustrate how the Pugh method works, the example introduced in 4.5 Prototyping will continue to be followed. This example involves an innovator working on the need for **a way to prevent strokes in patients with atrial fibrillation caused by left atrial appendage (LAA) thrombus**. As noted, the LAA is a small, pouch-like structure attached to the left atrium of the heart in which blood can coagulate to form clots, or thrombi, in patients with an abnormal heart rhythm (such as atrial fibrillation). These clots can then dislodge and travel to the brain, where they have the potential to create a blockage in an artery that supplies blood to the brain, leading to a stroke.

Step 1: Identify user and design requirements

An innovator should begin final concept selection by referring back to the need criteria outlined in the need specification. These criteria continue to represent the most important, overarching requirements that the final concept must meet in order to satisfy its target audience. Recall that need criteria are defined *before* the innovator begins considering specific solutions to address a need. This is why they play such a central role in brainstorming and then screening and refining the list of ideas that is generated (see 3.2 Concept screening). Ideally, all of the potential solutions taken forward into concept selection should satisfy the need criteria.

Once the innovator feels comfortable that the need criteria have been verified, s/he should compile the most important user and design requirements for each solution. In contrast to need criteria, user and design requirements emerge as individual solutions that are investigated through iterative brainstorming, research, and prototyping. For this reason, they are typically much more detailed than need criteria and may become specific to a particular concept. For instance, in the LAA example, the idea to fill the LAA with glue would eventually have design requirements that are quite distinct from those related to the concept of placing a mesh in the aorta, even though both concepts support an approach to obstructing LAA thrombi. At a fundamental level, user and design requirements begin to resemble product specifications.

When choosing which requirements to use for final concept selection, a careful balance is required. Using too many requirements that resemble precise, product-like specifications at this stage (e.g., the product must be able to deliver a liquid that turns to a solid once released, the product must be able to be visualized only using X-rays without the need for contrast dye) can skew the comparison of the concepts toward one particular idea. On the other hand, though uncommon, including only general requirements that closely resemble the need criteria will not allow the innovator to adequately differentiate between concepts. While there is no clear way to know exactly how detailed and specific the design

Table 4.6.1 *Sample user and design requirements and assigned weighting.*

Requirement	Weight
Device must be delivered percutaneously and not through a surgical incision	5
Device must be able to be used in a cardiac catheterization environment	3
Device must not lead to iatrogenic thrombi development	5
Device must be less than 12 French[7] in diameter	1
Device must allow procedure to be performed in less than half-an-hour	3
Device or solution must be reversible at time of procedure	4

requirements should be for this exercise, the goal is to define a well-rounded set of requirements that can be met by more than one of the solutions being considered, yet still illuminate important distinctions between the concepts being evaluated. No more than three to seven design requirements should be chosen to keep the assessment manageable and sufficiently encompassing.

Although many user and design requirements are discovered through prototyping, the innovator should also take into consideration the new information that has been uncovered through the activities described in Chapters 4.1 through 4.5, both to understand the limitations of certain concepts as well as to help shape how some of the requirements are worded and weighted. This information, while not usually directly incorporated into the Pugh method, can be invaluable to quickly eliminating or including certain concepts in the analysis. For example, if an innovator learns that a great deal of IP has already been filed that is relevant to one concept under consideration, whereas the other concepts are relatively clear from an IP perspective, this may suggest that this particular concept has key limitations. In another example, if an innovator learns that pursuing a premarket approval trial for regulatory clearance is something that certain stakeholders would not support (e.g., patients might not want to participate if existing treatments are readily available), the concepts that could utilize a 510(k) pathway might be favored. Intelligence such as this should be used to help shape the way user and design requirements are worded.

As innovators investigate the Pugh method, it is worth noting that in an alternate version of the approach, user and design requirements are determined first, before any concepts have been generated. Concepts focused on satisfying these requirements are then developed through ideation and brainstorming. Although this is somewhat different from the biodesign innovation process presented here, the two approaches are complementary. Requirements that are created prior to concept generation are similar to the initial need criteria referenced in 2.5 Needs Filtering (in that they identify the most important, high-level specifications that must be addressed by any concept). Concept generation activities aimed at meeting these requirements can then be considered a focused version of brainstorming. Furthermore, when innovators use the Pugh approach at later stages of the biodesign innovation process to choose between variations of the final concept that have been built to satisfy increasingly specific requirements, they are also using this alternate version of the Pugh method.

Step 2: Weight user and design requirements

Once a meaningful set of requirements has been defined, the design team must then assign weightings to reflect their relative importance. At a high level, this process is similar to that performed in 2.5 Needs Filtering, although in this case weightings are applied to user and design requirements instead of the important factors used to assess needs. For example, using a scale of 1 to 5, an inventor might assign a "5" to requirements that are "essential" and a "1" to those considered "nice to have." Each user or design requirement (along with its weighting) should be captured in rows down the left side of the selection matrix. For the LAA example, the requirements might be chosen as shown in Table 4.6.1.

Assigning weightings is a relatively subjective process. The innovator should do so by considering all that

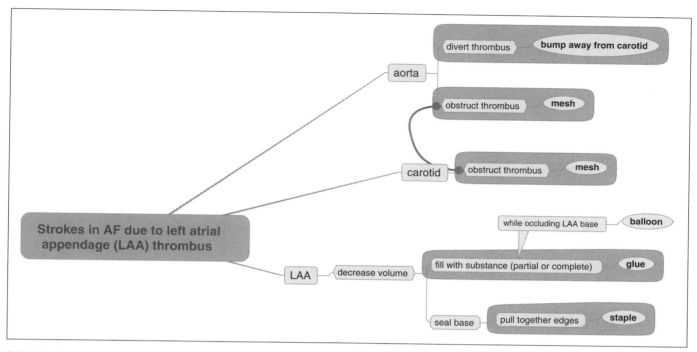

FIGURE 4.6.2
Key concepts from a sample mindmap for preventing strokes in patients with atrial fibrillation caused by LAA thrombus (provided by Uday N. Kumar).

s/he has learned from the prototypes that have been built and from the information gathered about IP, reimbursement, regulatory issues, and business models. In the sample weightings shown in Table 4.6.1, the first requirement that a "device must be delivered percutaneously and not through a surgical incision" may have been weighted so heavily because it is directly linked to a key need criterion and is also linked to research showing that a percutaneous solution would be more likely to utilize a 510(k) regulatory pathway, which is preferred by key stakeholders.

Step 3: Confirm the concepts and choose a baseline

The next step is to confirm which concepts will be evaluated. This should include the most promising solutions from the concept screening process that have been undergoing research and prototyping by the innovator and his/her team. As noted, no more than three to five concepts should ideally be under consideration at this point.

For the LAA example, Figure 4.6.2 shows a list of the potential solutions that were chosen via concept screening for further investigation. It is up to the innovator to decide whether all of these concepts should be included in the Pugh analysis, or if one or more should potentially be eliminated prior to final concept selection based on what has already been learned.

At this stage of the Pugh method, it is also essential to choose a baseline concept against which all solutions can be compared. This could be the solution that the team believes is most likely to achieve success, the "gold standard" among existing treatment options, or a competitor's solution.[4] The important thing is to gain enough information about the baseline to effectively evaluate each new solution concept against it. For the LAA example, a baseline could be provided by the WATCHMAN®, depicted in Figure 4.6.3. This device (discussed in 2.2 Treatment Options), while not yet FDA approved, has been used in human clinical trials.

Step 4: Assign scores and rank concepts using the selection matrix

Each concept should next be rated against the user requirements based on how likely it is to exceed the performance of the baseline concept (+1), lag behind it (–1), or perform at a comparable level (0). Because it

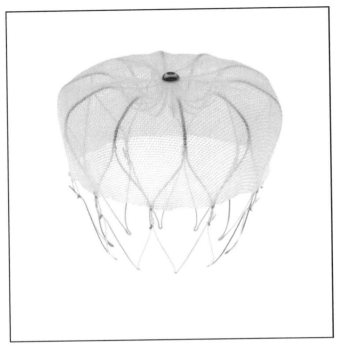

FIGURE 4.6.3
The WATCHMAN LAA device incorporates a permeable fabric over a nitinol frame. It is implanted using a catheter to deliver it into the LAA (courtesy of Atritech; reprinted with permission).

serves as the point of comparison, the ratings for the baseline solution should be set at (0) for each requirement. Even if the baseline does not perform particularly well against a certain requirement, it should still be given a (0) rating since this level of performance represents the current standard.

The total score for each concept can be obtained by multiplying the score for each user criterion with its weight, and then totaling the results. The concept with the highest total score is the leader. Graphically, a matrix can be constructed with each solution concept listed in its own column, with the baseline solution furthest to the left. For the LAA example, a completed selection matrix could appear as shown in Table 4.6.2.

In this example, the LAA glue concept emerges as the clear leader. Of course, the innovator would complete such an assessment using detailed information on each concept that is not provided within this chapter.

In some cases, numerous concepts may score poorly against the baseline. Assuming that the requirements are generally sound, this may suggest that the baseline concept meets both the need and the currently defined user and design requirements relatively well (indicating that the innovator may face a difficult market) or that the concepts have not been developed to a degree of detail such that they could possibly satisfy the chosen requirements. On the other hand, sometimes more than one concept scores well or no clearly superior concept emerges from the exercise. In these cases, the innovator should reexamine both the user and design requirements, as well as the concepts chosen for evaluation.

The reason for revisiting user requirements is to be sure that the most relevant ones have been identified and weighted appropriately. One warning signal is if many of the requirements have been given low weightings. If the requirements are mostly "nice-to-haves," this could cause all concepts to yield a similar score, such that no clear leader emerges from the Pugh assessment. The innovator should also check to see if the requirements are too general, making all the concepts satisfy every requirement equally. In this scenario, the innovator and his/her team may have started the concept selection phase too early without sufficient user input. This can be remedied by going back to the process outlined in Chapters 4.1 through 4.5 to collect more information.

In terms of reexamining the concepts chosen for evaluation, the innovator and his/her team can perform additional brainstorming to be certain that the most promising concepts have, indeed, been identified. Also, successful elements of one concept could be applied to other concepts to make them stronger and boost their likelihood of success. Additional prototyping, in an iterative loop, will provide valuable insights about these questions.

After taking these steps, the innovator can make necessary adjustments to the requirements, weightings, and concepts in the selection matrix. Then, s/he can again try assigning a score to each concept, repeating this process until the leading concept is clear. Ultimately, the innovator and his/her team will choose the final concept, taking into account the outcome of the formal assessment as well as their insights from research regarding IP, reimbursement, regulatory issues, business models, and prototypes. Their collective experience, professional judgment, and strategic focus (as defined in Chapter 1.1) should also come into play in making a final decision. Over time, as more experience is gained with such decisions, an innovator may place

Table 4.6.2 *Sample selection matrix showing requirements, weightings, ratings, and total rank scores.*

Requirement	Weight	WATCHMAN (baseline)	Aortic bumper	Aortic mesh	Carotid mesh	LAA glue	LAA stapler
Device must be delivered percutaneously and not through a surgical incision	5	0	0	0	1	1	−1
Device must be able to be used in a cardiac catheterization environment	3	0	0	0	0	0	0
Device must not lead to iatrogenic thrombi development	5	0	0	−1	−1	0	1
Device must be less than 12 French in diameter	1	0	−1	−1	1	1	0
Device must allow procedure to be performed in less than half-an-hour	3	0	1	0	0	1	0
Device or solution must be reversible at time of procedure	4	0	−1	0	0	−1	−1
Rank score			−2	−6	1	5	−4

less emphasis on the formal methodology and more on his/her intuition. However, the Pugh method can provide a solid framework for making such decisions, regardless of the experience level of the innovator.

The ACS story below provides an example of how one set of experienced innovators approached final concept selection. Note that although they did not formally use the Pugh method, there are strong similarities between Pugh's approach and the one they employed – well-defined criteria were used to evaluate the concepts against a baseline solution before making a decision.

FROM THE FIELD: ADVANCED CARDIOVASCULAR SYSTEMS

Driving concept selection through intuition and an intimate understanding of user requirements

Advanced Cardiovascular Systems (ACS) was the first cardiovascular medical device start-up in California's Silicon Valley. Founded in 1978 by cardiologist John Simpson and entrepreneur Ray Williams, it pioneered the development of percutaneous balloon angioplasty catheters and over-the-wire catheter systems. Carl Simpson, who was working as the lead technician at Stanford's catheterization laboratory when he met John Simpson (no relation), later became the company's first full-time employee. Together, with the help of cardiologist Ned Robert and a team of engineers, they revolutionized the treatment of coronary heart disease and helped to create the field of interventional cardiology.

John Simpson, who was training in cardiology at Stanford at the time, became intrigued with angioplasty when he heard a presentation by Andreas Gruentzig, the creator of this innovative new technique. Gruentzig's approach used a fixed-stylet catheter system and balloon to widen a coronary artery that had narrowed or become obstructed through the build-up of atherosclerotic plaque. Although initially skeptical, Simpson went to Switzerland to learn about the procedure and became convinced of its benefits. Back in the United States, he ran into several problems. First, given that the procedure was considered quite radical in its approach, he did not receive much support from colleagues to perform it. Second, as a young

Stage 4 Concept Selection

FROM THE FIELD — ADVANCED CARDIOVASCULAR SYSTEMS

Continued

cardiologist assumed to be lacking in experience and training, he had difficulty obtaining the parts and tools he needed to perform the angioplasty procedure when he ordered a catheter system from Gruentzig. Nevertheless, with a strong commitment to the therapeutic benefits of coronary angioplasty and faced with no other alternative, John Simpson and his colleague, Ned Robert, began experimenting with materials to build a balloon angioplasty system of their own. In the process, they came to believe that Gruentzig's device was somewhat difficult to use, especially due to the fixed-stylet system. They also recognized that his technique required such a high skill level that it would be challenging for an average cardiologist to replicate.

As John Simpson began developing prototypes and testing concepts, Carl Simpson was managing Stanford's cardiac catheterization laboratory. He helped the young innovator by connecting him with various engineers within the hospital, such as people in the machine shop. John Simpson continued to "tinker" and eventually met Ray Williams, to whom he showed some of his prototypes. Williams saw the market potential of Simpson's ideas and put together a network of angel investors to fund the creation of ACS. At its core, ACS was focused on developing the tools and techniques that would provide *a better way to perform coronary angioplasty*. After ACS was created, John Simpson contacted Carl Simpson, who had left Stanford for Hewlett-Packard. He joined ACS to accelerate the process of designing, prototyping, and testing device concepts. The team would ultimately have to develop three primary components for its system – a balloon catheter, a guidewire for catheter navigation, and a guiding catheter. Several points related to the development of the balloon catheter and guidewire provide a good example of how understanding the clinical need and key requirements was essential to deciding on a final concept.

One of the crucial things the ACS team did early in the biodesign innovation process was to define clear user requirements that its balloon catheter would have to meet in order to provide a significant improvement over the available technology. According to Carl Simpson, these requirements included creating a balloon that had a relatively low profile (thin enough to penetrate severe lesions), was relatively non-compliant (did not keep expanding beyond the maximum desired diameter), and could sustain a certain pressure without rupturing. "That was our focus. We knew what we had to do," he said. "You might have ten ideas for how to make it better, but you can't try to do everything. You have to make a choice on what's most important and stick with it. Because if you don't, that's the kiss of death."[5]

Likewise, in understanding how a guidewire should perform, they realized that it first had to be very fine in order to navigate the small branches of the heart's coronary arteries. Given the complex and tortuous anatomy of the coronary arteries, the guidewire would also have to be flexible, but in a way that could be controlled. Through many unsuccessful attempts to navigate the coronary anatomy with shaped guidewires, they ultimately determined that one of the guidewire's most important design requirements was to be "torqueable," with the tip responsive to guided control from the other end. This key design requirement (which was integral to the clinical problem) eventually helped make ACS guidewires the standard in the industry.

While ACS initially had an entire "tree" of potential user requirements, Carl Simpson advised innovators to be wary of becoming distracted by "nice-to-haves," especially in the early stages of product development. "You're never big enough to do more than three or four things right at any given time," he noted. ACS identified its top priority user and design requirements, including the need for a torqueable guidewire, by using its prototypes to facilitate detailed and extensive interactions with cardiologists. ACS routinely sent its engineers to the catheter lab to observe catheter-based procedures and study what competitive products, like the Gruentzig system, could and could not do well. "You need to understand better than the physicians what the real needs are," he said. "The company has to have a really high IQ when it comes to understanding what drives user requirements."

Once these requirements were identified, the ACS team was able to weight them to assist the company in making trade-offs during design and prototyping. "We weighted them based on what we learned from being so involved on a daily basis with what the customer needs next," said Carl Simpson. For example, in the development of the balloon catheter, the requirement for a thin balloon profile was given highest priority. "Profile was always number one," Simpson explained, "because if you can't get across the lesion, you can't dilate it." If necessary, he recalled, the team was willing to make sacrifices around other requirements such as inflation pressure, to maximize the desirability of the balloon's profile. Carl Simpson reiterated, "Trade-offs must be driven by the clinical need. That's why the engineers must understand the medical need so well that they can make those decisions intelligently."

Fundamentally, because there was only one competing product in the market, Gruentzig's device was the baseline against which all of ACS's test results were compared. The company worked through multiple concepts before finally deciding on one to take into development. When asked how the team made that decision, Carl Simpson replied, "We just knew it." When the prototype of the balloon catheter concept could perform better than the baseline device on all three of the fundamental user requirements the company had chosen, ACS knew it was ready to move forward. At that stage of the process, Carl Simpson said, "If it's not intuitively obvious, you need to get some help. Talk with other successful entrepreneurs that have a track record in this business." He also advised inventors to go back to physicians and get more input until they felt certain that the concept would satisfy their most important requirements. See Figure 4.6.4.

Carl Simpson acknowledged that developing this kind of intuition was not easy. "You learn by doing it," he said. However, even for new innovators, "after a certain accumulation of knowledge, the right solution should start to present itself." He also issued a reminder that success is often built on a series of failures. "There is no right formula, other than knowing what the patient and the customer really needs." Simpson personally dedicated himself to helping other inventors develop this sense of

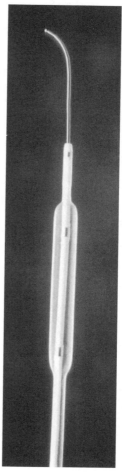

FIGURE 4.6.4
An early Simpson-Robert version of ACS's first catheter (courtesy of Carl Simpson).

intuition. During his tenure at ACS, the company became informally known as the University of Medical Devices for Silicon Valley, giving birth to over 100 medical device companies.[6] ACS was sold to Eli Lilly in 1984 and spun off to Guidant in 1994.

Stage 4 Concept Selection

GETTING STARTED

The final decision about which concept to pursue can be a source of anxiety for some innovators. Using a well-structured, rigorous approach, like the one outlined below, can help boost innovator confidence and increase the likelihood of a successful result.

See **ebiodesign.org** for active web links to the resources listed below, additional references, content updates, video FAQs, and other relevant information.

1. Identify user and design requirements

1.1 What to cover – Get started by validating the concepts chosen through concept screening (see Chapter 3.2) against the defined need criteria. Apply the new information gathered through the processes outlined in Chapter 4.1 through 4.5 (as well as other steps in the biodesign innovation process) to compile more detailed user and design requirements. Pay particular attention to information gathered through prototyping, as this may affect the overall feasibility of a concept. Find an appropriate balance between being so general that the requirements are meaningless and so specific that they apply only to a single concept. Choose the top three to seven requirements.

1.2 Where to look – Refer to Pugh's book *Total Design* (Addison-Wesley, 1991), or the follow-up *Creating Innovative Products Using Total Design* (Addison Wesley Longman, 1996), for more information about the methodology. In addition, revisit the output from the following chapters:
- 2.5 Needs Filtering
- 3.2 Concept Screening
- 4.1 Intellectual Property Basics
- 4.2 Regulatory Basics
- 4.3 Reimbursement Basics
- 4.4 Business Models
- 4.5 Prototyping

2. Weight user and design requirements

2.1 What to cover – Define a weighting scale (e.g., a scale of 1 to 5, where 5 is the most important). Then prioritize the concepts based on how critical they are to the potential success of the final solution.

2.2 Where to look – Refer back to the chapters listed above for input in assigning appropriate weightings.

3. Confirm the concepts and choose a baseline

3.1 What to cover – Choose the most promising concepts emerging from the analysis in 4.1 through 4.4 and the prototyping exercises described in 4.5, eliminating those that have proven to be infeasible due to technical, market, regulatory, IP, or other important constraints. As noted, no more than three to five concepts should be included as part of the selection matrix. Select a baseline concept that reflects the established (or most promising) treatment alternative in the medical field. This can be the current standard of care, an emerging competitive product, or the most promising solution concept being considered by the team. Because all other concepts will be compared against this baseline, be sure that enough information is known about the baseline that the other concepts can be effectively evaluated against its performance relative to the requirements.

3.2 Where to look – Refer back to 2.2 Treatment Options in choosing a baseline concept. Network with experts in the field to validate that the appropriate baseline has been chosen.

4. Assign scores and rank the concepts using the selection matrix

4.1 What to cover – Set the score for the baseline concept to (0) for each requirement. Then, carefully evaluate the individual concepts against the defined criteria in the matrix, assigning a (1) to concepts that outperform the baseline, a (–1) for concepts that underperform, and a (0) for those that achieve parity. Calculate the total score for each concept by multiplying the score for each requirement against its defined weight. The concept with the highest score is the leader. If no concepts score particularly well, or multiple concepts score the same, reconsider the requirements and the way they have been weighted. Additionally, evaluate the concepts to determine if one or more elements from disparate solutions might be combined to create a stronger overall solution concept. Additional prototyping can also be an invaluable source of information for making adjustments to the selection matrix. Perform the assessment again, in an iterative manner, until a clear leader emerges. Exercise judgment to make the final decision on which concept to pursue.

4.2 Where to look – If it becomes necessary to revisit the requirements and/or the way they are weighted, network with experts and users in the field for additional input, as needed. These individuals can also serve as effective sounding boards for solution concepts that may combine different elements of previously prototyped concepts. Note that additional prototyping of these hybrid solutions may be required before a final decision can be made.

Credits

The editors would like to acknowledge Steve Fair for his help in developing this chapter. Many thanks also go to Carl Simpson, of Coronis Medical Ventures, for sharing the ACS story.

Notes

1. David Warburton, "Getting Better Results in Design Concept Selection," *Medical Device Link*, http://www.devicelink.com/mddi/archive/04/01/006.html (February 12, 2008).
2. Stuart Pugh, *Total Design: Integrated Methods for Successful Product Engineering* (Addison-Wesley, 1991).
3. Ibid.
4. Ibid.
5. All quotations are from interviews conducted by the authors, unless otherwise cited. Reprinted with permission.
6. "Board Members," Uptake Medical, http://www.uptakemedical.com/board.php (February 29, 2008).
7. The French catheter scale is used to measure the outside diameter of a cylindrical medical instrument. 1 French (Fr) is equivalent to approximately 0.33 mm.

Acclarent Case Study

Stage 4 Concept selection

To determine whether the idea of using balloons to address chronic sinusitis was feasible, Makower and Chang had to take the idea through steps described in the concept selection stage of the biodesign innovation process. Even though they were not directly comparing the balloon solution against another idea, they had to be sure it would clear all appropriate hurdles before making a decision as to whether or not this would be their final concept. "We realized that there were a million questions that needed to be answered once we had cleared the simple hurdle of widening a sinus passageway. But none of these questions were worth asking if we did not know if we could find a technology that worked," Makower said.

4.1 Intellectual property basics

To assess the IP landscape in ENT, Makower and Chang spent considerable time researching patents in the ENT field. They were particularly focused on understanding existing patents related to the nose. "We looked at this area in detail," Chang summarized, "and it seemed pretty open." He elaborated:

> There was some IP around epistaxis [nosebleeds]. There was a fair amount of IP around turbinate reduction, cutting things, reducing things, and a lot of energy delivery. But the field was generally pretty wide open in terms of interventional devices, as well as methods and different procedures. We found some patents related to drug delivery, but not really specific to ENT. This was important because we recognized from early on that drug delivery could eventually become part of our solution. As we started to hone in on balloons and catheters, we knew that these devices had been around for a long time in cardiology and other spaces, so we focused on aspects of these technologies that were unique and novel to this application and started to create a massive set of disclosures.

As they performed more research and consulted with an IP attorney, Makower and Chang became convinced relatively quickly that their solution would be patentable. They decided that the first step in this process was to focus on building a picket fence around their unique method of using balloons and other tools designed specifically for the nose. According to Chang, "We recognized from the get-go that the method was where we would really shine." Then, as they learned more about the devices and how they would be used, they began to build increasingly stronger IP protections around what made them uniquely different from existing balloons and other tools for cardiovascular interventions.

"So we filed a series of patent applications," said Chang. "Our first ones were not provisionals. We worked extremely hard and fast with our IP attorney and decided to put in a tremendous effort up-front into our patents." He described using a comprehensive approach to these early applications. "While we clearly described our preferred technology and preferred way of using it, we also put in many other technologies and methods to lay the groundwork for future products. There are multiple approaches to achieve a desired outcome and if you don't cover them, someone else will after they see what you are doing." See Figure C4.1.

One unexpected outcome from these IP efforts was the decision to operate in what the team called "stealth mode." As they learned more and more about the opportunity, Makower became increasingly concerned about another company learning of the idea and trying to beat them to market with a similar solution. They intentionally decided to disclose their ideas to as few people as possible, and agreed to use carefully constructed non-disclosure agreements (NDAs) when a disclosure was absolutely necessary. Chang explained: "With the balloon idea, we were basically working with well-understood technology. There are a lot of companies that have experience with balloons, guides, and wires. There are smart people out there who could potentially do this and they might get a hint about what we're doing and invent in front of us. The longer we could keep the potential competition in the dark, the better."

4.2 Regulatory basics

When thinking about the regulatory pathway associated with the leading solution, Makower referred back to the mantra "no science experiments." This meant that the solution would need to be a candidate for the 510(k) pathway. "We wanted something with a relatively simple regulatory path," said Chang, along with limited requirements for extensive clinical data. "So, we set up a meeting with a regulatory consultant and started talking about how we could get clearance with this kind of a product." They found a regulatory consultant who had some experience in the ENT field. "His impression was that most ENT devices fall within Class I or Class II," Chang recalled. "From there, we pulled out the regulations to make sure we really understood them. After poring through this information with the consultant, they felt confident that the decision to pursue a 510(k) pathway was directly in keeping with the guidelines of the Food and Drug Administration (FDA). The key would be to identify a highly effective set of predicate devices upon which to base their submissions.

4.3 Reimbursement basics

When it came to reimbursement, the team took great care in understanding the codes, coverage, and payment decisions associated with FESS. William Facteau, who joined the team in November 2004 as the CEO of the soon-to-be company, described how the ENT specialists performing FESS procedures were heavily invested in the way the procedure was reimbursed:

> Functional endoscopic sinus surgery is the gold standard for the surgical treatment of chronic sinusitis and it is reimbursed by Medicare, as well as all private insurance plans. In its early days, though, there were challenges in getting the procedure reimbursed – this conflict is often referred to as the 'FESS mess.' The leading physicians in the field used a significant amount of political capital in order to get these codes established. And they are some of the most sacred codes throughout all of otolaryngology – they pay very well, they pay individually by sinus, and physicians also receive additional reimbursement for the follow-up treatments, called debridements [the surgical removal of scar tissue from a wound], which is pretty unique. Due to this favorable reimbursement, sinus surgery is one of the most profitable outpatient procedures for both the physician and the hospital. For this reason, leaders in the field, as well as the societies, are very protective.

Because the physicians had fought long and hard to establish codes, coverage, and payment levels, the team knew not to proceed if it seemed to be a questionable fit. On the other hand, if this technology were considered just another device to be used in sinus surgery, it could fit comfortably within the existing reimbursement guidelines and the team would be well positioned

Stage 4 Concept Selection

US007462175B2

(12) **United States Patent**
Chang et al.

(10) Patent No.: **US 7,462,175 B2**
(45) Date of Patent: **Dec. 9, 2008**

(54) DEVICES, SYSTEMS AND METHODS FOR TREATING DISORDERS OF THE EAR, NOSE AND THROAT

(75) Inventors: **John Y. Chang**, Mountain View, CA (US); **Joshua Makower**, Los Altos, CA (US); **Julia D. Vrany**, Sunnyvale, CA (US); **Theodore C. Lamson**, Pleasanton, CA (US); **Amrish Jayprakash Walke**, Milpitas, CA (US)

(73) Assignee: **Acclarent, Inc.**, Menlo Park, CA (US)

(*) Notice: Subject to any disclaimer, the term of this patent is extended or adjusted under 35 U.S.C. 154(b) by 524 days.

(21) Appl. No.: **11/037,548**

(22) Filed: **Jan. 18, 2005**

(65) **Prior Publication Data**
US 2006/0095066 A1   May 4, 2006

Related U.S. Application Data

(63) Continuation-in-part of application No. 10/829,917, filed on Apr. 21, 2004, and a continuation-in-part of application No. 10/912,578, filed on Aug. 4, 2004, now Pat. No. 7,361,168, and a continuation-in-part of application No. 10/944,270, filed on Sep. 17, 2004.

(51) Int. Cl.
A61M 31/00   (2006.01)
(52) U.S. Cl. **604/510**
(58) Field of Classification Search 604/509–510, 604/94.01, 103.1, 103.04, 103.05, 96.01, 604/164.01, 164.09, 164.1, 164.11, 164.13, 604/164.08, 171, 173, 101.02
See application file for complete search history.

(56) **References Cited**

U.S. PATENT DOCUMENTS

705,346 A   7/1902 Hamilton
2,525,183 A   10/1950 Robison

(Continued)

OTHER PUBLICATIONS

Strohm et al. Die Behandlung von Stenosen der oberen Luftwege mittels rontgenologisch gesteuerter Balloondilation Sep. 25, 1999.*

(Continued)

Primary Examiner—Kevin C Sirmons
Assistant Examiner—Deanna K Hall
(74) *Attorney, Agent, or Firm*—Robert D. Buyan; Stout, Uxa, Buyan & Mullins, LLP

(57) **ABSTRACT**

Sinusitis, mucocysts, tumors, infections, hearing disorders, choanal atresia, fractures and other disorders of the paranasal sinuses, Eustachian tubes, Lachrymal ducts and other ear, nose, throat and mouth structures are diagnosed and/or treated using minimally invasive approaches and, in many cases, flexible catheters as opposed to instruments having rigid shafts. Various diagnostic procedures and devices are used to perform imaging studies, mucus flow studies, air/gas flow studies, anatomic dimension studies and endoscopic studies. Access and occluding devices may be used to facilitate insertion of working devices such as endoscopes, wires, probes, needles, catheters, balloon catheters, dilation catheters, dilators, balloons, tissue cutting or remodeling devices, suction or irrigation devices, imaging devices, sizing devices, biopsy devices, image-guided devices containing sensors or transmitters, electrosurgical devices, energy emitting devices, devices for injecting diagnostic or therapeutic agents, devices for implanting devices such as stents, substance eluting or delivering devices and implants, etc.

21 Claims, 44 Drawing Sheets

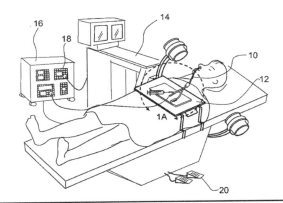

Figure C4.1
The first core Balloon Sinuplasty patent created by the team was finally issued to Acclarent in December, 2008.

CPT code	Description	Hospital Outpatient		Ambulatory Surgical Ctr		Physician Payment	
		APC	APC paymt	ASC grp	ASC paymt	Total RVUs	Ave paymt
31254	Nasal/sinus endoscopy, surgical, with ethmoidectomy, partial	0075	$1,112	3	$510	7.92	$296
31256	Nasal/sinus endoscopy, surgical, with maxillary antrostomy	0075	$1,112	3	$510	5.71	$213
31276	Nasal/sinus endoscopy, surgical, with frontal sinus exploration	0075	$1,112	3	$510	14.76	$551
31287	Nasal/sinus endoscopy, surgical, with sphenoidotomy	0075	$1,112	3	$510	6.73	$251
61795	Stereotactic computer assisted volumetric (navigational) procedure	0302	$345.20	N/A	$0	7.05	$263.23

Figure C4.2
Sinus surgery was reimbursed by Medicare and private payers under a series of existing codes (provided by Acclarent).

from a reimbursement perspective. The codes being considered included those shown in Figure C4.2.

Lam Wang performed a detailed analysis of the reimbursement landscape. While the target population of chronic sinusitis patients was largely covered by private insurance, she used Medicare data as a baseline for payment levels and rates since it was readily available, and also because most private payers followed CMS's lead. For physicians, CMS (Centers for Medicare and Medicaid Services) paid 100 percent for the highest weighted code, and then 50 percent for all additional codes. Use of an image-guidance system was addressed separately as an add-on payment at 100 percent. Facilities were also reimbursed for the first sinus at 100 percent, and at 50 percent for all others. However, while image-guidance received an add-on payment at 100 percent in a hospital outpatient setting, no additional payment was granted in an ambulatory surgical center (ASC) setting. CMS data at the time indicated that approximately 70 to 90 percent of all sinus surgeries were performed in a hospital outpatient facility. Importantly, existing FESS codes could not be used in an office setting. If the team wanted to use the devices in an office-based procedure, it would need to apply for new codes (as well as prove that the treatment was safe to administer in an office setting).

To validate what she had learned and secure a formal opinion, Lam Wang contracted with a professional reimbursement consulting firm – Clarity Coding. In its report, Clarity indicated that the team should qualify to use existing FESS codes, since the objective of the procedure when using balloons was essentially equivalent to that of FESS using other instruments and fell directly within the language of the code. "In other words," said Facteau, "the codes for functional endoscopic sinus surgery are broad enough that they do not specify how you create an opening in the sinus or what tools you use. It leaves it up to the physicians to decide whatever tool, device, or instrument is appropriate to make the opening." The company's technology would be just one more option available to surgeons when performing this procedure.

Encouraged by this outcome, but still concerned about the sacred nature of the FESS codes, Facteau sought the opinion of another reimbursement consultant when he joined the team. He explained:

Reimbursement was a big concern of mine. I worked closely with a legal/reimbursement expert at Reed Smith at a previous job, so one of the first things I did was call her and ask her for her opinion. She did some research and came back with the same opinion as Clarity Coding. In fact, she said, 'Bill, I really enjoyed working with you at Perclose. Unfortunately, we're not going to have to do much work on this one. It's pretty straightforward. The existing codes apply and you guys are all set. So, good luck.'

4.4 Business models

Another issue on Facteau's mind when he joined forces with Makower, Chang, and Lam Wang had to do with the business model. "One of the biggest business risks that I saw was just the fact that it was in ENT. I struggled with whether you could build a sustainable company solely dedicated to ENT. When I researched the space, it was not very impressive. The field just hadn't seen a lot of innovation over the years. It was primarily a reusable and capital equipment environment," he said. Following the path established in interventional cardiology, the devices being designed by Makower and Chang were intended to be disposable. Given the nature of the procedure and the devices themselves, this model made practical sense. More importantly, the team saw this approach as one way to help inject the field with greater innovation. According to Lam Wang, "When you shift from reusable instruments to a disposables platform, you open a field up to much more innovation. That's because the doctors are not stuck with the same tools they used the last year." Facteau elaborated: "Disposables drive innovation. We have seen it in general surgery, with laparoscopy and surgical staplers, orthopedics, spine, and cardiology – they were all driven by implants and innovative disposable technology. With a disposables model, we can innovate very, very quickly. The question was, were ENT physicians willing to adopt new disposable technology? To some extent, the business model was predicated on our ability to get doctors addicted to innovation and be able to provide them with new, value-added technology every six to nine months."

One obstacle to introducing this kind of a paradigm shift would be in getting the target population of physicians to give up the reusable instrument sets to which they were so attached. The team also anticipated some resistance related to price – many ENT specialists were known for being somewhat cost-conscious. "As a general rule of thumb," Lam Wang noted, "disposables command higher margins." However, convinced that the benefits would outweigh the costs, and encouraged by the relatively straightforward pathways identified in the areas of IP, regulatory, and reimbursement, the team felt comfortable taking on the challenge of building a business in ENT based on a disposable model. Basically, said Chang, "This didn't stop us. We just said, 'Okay, let's remember that. That's going to be a challenge.'"

4.5 Prototyping

In parallel with these other efforts, Makower and Chang had begun developing basic prototypes in earnest. Their goal was to prove the feasibility of the concept. Chang recalled how they got started:

> We asked ourselves, 'Why can't they treat these sinuses better?' One thing was that they operated via endoscopy – in other words, required direct visualization. So that led us to consider alternative visualization methods since our devices would at some point be placed around and behind structures, out of view of most common endoscopes. We looked at fluoroscopy as the most reasonable way to navigate the anatomy, similar to the way interventional cardiologists guide catheters in the heart. So don't remove the anatomy; look around or through it with fluoroscopy. And then there was the fact that there were no flexible instruments for FESS. When you examined what they were doing, they were just chopping, cutting, or ripping the anatomy to gain access and make the opening bigger. We asked ourselves if there were less traumatic ways to accomplish this, ways that wouldn't lead to so much bleeding and the scarring cascade that often brought patients back for a revision surgery. We looked at using wires and flexible instruments. And we also thought, well, maybe you could dilate the ostium. Would fracturing the bone via balloon dilation be less traumatic? Was it going to cause the mucosa to necrose and cause more problems? We didn't know. We wanted answers to these questions. And so that led us to prototyping.

"We'd never read or heard of anyone doing any dilating of bone," Chang continued, "so there were a lot of things that we just didn't know." Using readily available materials, they built a preliminary working model and then went back to the cadaver lab to try it (importantly, they decided that there were no animal models that would serve as an effective proxy for human sinuses). According to Chang, "Not knowing what would work, we were like MacGyver, with duct tape and boxes of materials and supplies for cutting, melting, bonding, and shaping existing things. We were ready to make changes right there in the cadaver lab."

Makower described their initial experience using a prototype:

> John and I went in with very crude guides, wires, and balloons. We didn't even have a scope for our first study. We went in and poked around. And, to our amazement, we were able to wire all the major sinuses in a cadaver head, and pass balloon catheters up over those wires, and deploy stents, and deliver balloons. It was unbelievable. The first time we tried, we got it to work. So, in terms of the feasibility, we said, 'Okay, this is doable!' Now, what we didn't have was any real way of assessing what we had done. And we didn't know what pressures to inflate the balloons to. We saw that at some point the balloons would crack open, and we saw that we were dilating bone. We just didn't know exactly whether we were doing something bad or good. (See Figure C4.3)

Encouraged by these early results, Makower and Chang eagerly pressed forward. "We brought Dave Kim in, under a confidentiality agreement, to help with those early exploratory studies. And he brought a scope with him. For the first time, we got to see what it looked like, and it looked pretty good. So, that's when we realized it was time to really talk to some rhinologists and figure out whether this thing had any merit."

They put together a brief presentation on their concept that included the need specification, an overview of the prototyped technology, and the approach, results, and X-rays from the preliminary feasibilities studies. Tapping into a member of their network, they arranged a meeting with Dr. Mike Sillers, who was president of the American Rhinological Society (ARS) at the time. According to a friend who had pitched various ideas to Sillers in the past, he had a reputation for being a tough critic, so they were prepared for the worst. Makower recalled the meeting:

> When we sat down with Mike [who agreed to sign an NDA], he was silent for most of the presentation. When we got to the end, he pushed himself away from the table and looked at us and said, 'If this works, it's going to change everything we do.' So then I asked the question, 'What would you need to see before you'd be willing to try it on a patient?' And he said, 'Well, I'd want to do it on a cadaver head.' So, we said, 'Okay. We'll get you out to the cadaver lab. What next?' He said, 'Well, try it on some patients.' And then I said, 'Well, how many patients would you need to see, and what kind of study would you need to see before you think we could commercialize it?' And his first response was, 'You know, this is just a tool. If I see that with this tool, in a handful of my own patients, I can make a wider opening, I don't need any clinical studies beyond that. I know that in the right patients, if I made a wider opening, it's going to be good for the patient.'

Figure C4.3
An early prototype of the stabilized guide access concept (provided by Acclarent).

4.6 Final concept selection

With this feedback from Sillers, Makower and team had reached a decision point. He explained: "After that meeting with Mike, we knew we had hit upon something. If an important potential critic couldn't come up with anything to kill the concept, then we knew we wanted to move forward with balloons. We looked at each other and said, 'I think we found the one. Let's go for it.'"

Because all of the key issues to consider in the concept selection process had panned out relatively favorably (IP, regulatory, reimbursement, business model, and prototyping), they felt comfortable making this decision informally, in lieu of a structured concept selection exercise. However, just to be sure, they took the idea to one more leading specialist in the field – Dr. Bill Bolger, a famous surgeon and anatomical expert in ENT. When Bolger (who also signed an NDA) was equally impressed with their presentation and willing to play an ongoing role in the development of the device, the team committed itself to proceed, initiating the development of the devices that became known as *Balloon Sinuplasty*™ technology.

IMPLEMENT Development Strategy and Planning

IDENTIFY

STAGE 1 NEEDS FINDING
1.1 Strategic Focus
1.2 Observation and Problem Identification
1.3 Need Statement Development
Case Study: Stage 1

STAGE 2 NEEDS SCREENING
2.1 Disease State Fundamentals
2.2 Treatment Options
2.3 Stakeholder Analysis
2.4 Market Analysis
2.5 Needs Filtering
Case Study: Stage 2

INVENT

STAGE 3 CONCEPT GENERATION
3.1 Ideation and Brainstorming
3.2 Concept Screening
Case Study: Stage 3

STAGE 4 CONCEPT SELECTION
4.1 Intellectual Property Basics
4.2 Regulatory Basics
4.3 Reimbursement Basics
4.4 Business Models
4.5 Prototyping
4.6 Final Concept Selection
Case Study: Stage 4

IMPLEMENT

STAGE 5 DEVELOPMENT STRATEGY AND PLANNING
5.1 Intellectual Property Strategy
5.2 Research and Development Strategy
5.3 Clinical Strategy
5.4 Regulatory Strategy
5.5 Quality and Process Management
5.6 Reimbursement Strategy
5.7 Marketing and Stakeholder Strategy
5.8 Sales and Distribution Strategy
5.9 Competitive Advantage and Business Strategy
Case Study: Stage 5

STAGE 6 INTEGRATION
6.1 Operating Plan and Financial Model
6.2 Business Plan Development
6.3 Funding Sources
6.4 Licensing and Alternate Pathways
Case Study: Stage 6

Failing to plan is planning to fail.

Alan Lakein[1]

Always in motion is the future.

Yoda[2]

By this point, you've taken steps to *Identify* and *Invent*. Now it's time to *Implement*. As Tom Fogarty, surgeon inventor extraordinaire, says: "A good idea unimplemented is no more worthwhile than a bad idea. If it doesn't improve the life of a patient, it doesn't count."[3]

This is by far the longest and most complex stage, with good reason. Regardless of the validity of the need, the ingeniousness of the concept and the size and scope of the market, at the end of the day, sound business underpinnings are essential if a product is to be delivered to the bedside, in a box, with a 1–800 number on the side for sales and service.

Getting to this reality requires a balanced consideration of the rules of the road from Stage 4 (intellectual property, reimbursement, regulatory, and business models) with the addition of a series of overlying and overlapping strategies. These strategies focus more deeply on the following key areas: (1) intellectual property integrated with ongoing research and development and clinical plans, (2) regulatory strategy including process management and quality, (3) reimbursement strategy, (4) basic business blocking and tackling – marketing, sales, and distribution, and (5) combining all assets and strategies together to develop a sustainable competitive business advantage.

Heretofore, a small team may have the wherewithal to work independently; now is the time to involve "varsity players." Use them as consultants or mentors, retain them on a part-time basis, or hire them. Without deep technical knowledge, failure is likely.

What to do first? Gordon Bethune's book chronicling the turnaround of Continental Airlines was entitled *Everything, All At Once*. The lessons are that few things are simply sequential, that nothing can be considered in a vacuum, and that time waits for no one. Given the fast pace of change in the medtech space, a six-month delay may leave you behind your competitors.

Notes

1. Unsourced quotation widely attributed to Alan Lakein, author of *How to Get Control of Your Time and Your Life* (New American Library, 1973).
2. From the movie *Star Wars: Episode V – The Empire Strikes Back*.
3. From remarks made by Thomas Fogarty as part of the "From the Innovator's Workbench" speaker series hosted by Stanford's Program in Biodesign, January 27, 2003, http://biodesign.stanford.edu/bdn/networking/pastinnovators.jsp. Reprinted with permission.

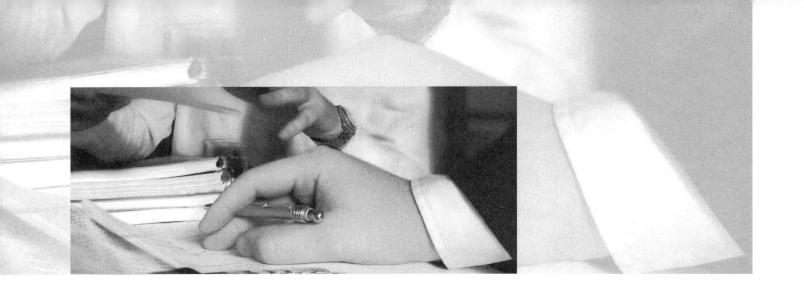

5.1 Intellectual Property Strategy

Introduction

On the third day of depositions in a patent infringement suit, the opposing counsel returns to the room with a fax in his hand and a look of satisfaction on his face. "Doctor," he says, "were you aware that what you have called your invention was published in the Annals of Physiology in 1947?"

Intellectual property (IP) is a particularly important strategic business asset in the medical device field. Not only can a strong IP position help mitigate costly legal settlements, but it can serve as a barrier to entry for competitors, an early source of potential revenue through licensing agreements, and an important overall form of collateral for an entrepreneur or business. Further, an innovator's IP position is a major factor considered by investors in deciding whether or not to fund a technology or its associated company. The best way to develop an effective patent strategy is to establish a solid working knowledge of the patent process, employ experienced IP counsel, work diligently in pursuing patents, and – importantly – understand that patenting is an ongoing activity that requires continual monitoring of the IP landscape and adjustments to one's patent portfolio.

While 4.1 Intellectual Property Basics provides an overview of fundamental IP concepts, the purpose of this chapter is to help readers understand the value of taking a strategic approach to IP and developing an effective patent portfolio.

OBJECTIVES

- Understand strategies involved in filing provisional and utility patent applications.
- Appreciate how to conduct and utilize a freedom to operate (FTO) analysis.
- Understand when and how to hire a patent attorney.
- Review basic strategies of international patent filing.
- Appreciate the importance of an ongoing patent management strategy based on continuous internal and external monitoring.
- Consider different defensive and offensive IP portfolio strategies.
- Understand the basic features of patent litigation in the medtech field.

Intellectual property strategy fundamentals

With more than 2 million patents in effect in the United States alone,[1] effectively navigating the IP landscape is undoubtedly complicated. The number of IP lawsuits increased by nearly 60 percent between 1995 and 2005 (from 1,700 to 2,700 respectively).[2] The medtech industry sector is especially litigious, with large teams and significant budgets directed toward pursuing and protecting patents.

Against this backdrop, there are two main categories of IP strategy to consider – first, issues that the innovator should keep in mind while pursing patentability and freedom to operate and, second, strategies to establish and maintain an integrated and ongoing IP management process.

Strategies in pursuing patent coverage

There are many strategic issues to consider regarding patentability, but a handful of them are common concerns, particularly for first-time innovators. One frequent question is whether to file a provisional patent or proceed directly to a full utility application. In general, a carefully prepared provisional patent has the advantage of gaining the inventor an effective extra year of protection, comparable to the situation of an inventor filing outside the United States. The utility application has a 20-year life and does not need to be filed until a year after the provisional. So the provisional can be used to firmly establish the priority date; then, once the device is in sales (typically a few years later), the protection offered by the utility patent will last a year longer than if the utility was filed immediately. There can be a cost advantage to a provisional patent compared to a utility application if an attorney is not used to draft the provisional application. This can be a false economy, however, if the resulting provisional patent application is not thorough and of high quality. As described more fully in 4.1 Intellectual Property Basics, a careless provisional application can create enormous downstream trouble and expense for an innovator and his/her company.

A second question that comes up frequently is whether it is possible to "improve" a provisional patent application once it is filed. The short answer is that it is not possible to add to or modify a provisional application. If there is a substantial new discovery related to the initial disclosure, the innovator has the option to file a new provisional application or to initiate a utility filing that includes both the original and the new material. However, the priority date of a new provisional patent application will become the date of the second filing.

Innovators often wonder whether to bundle together all aspects of their invention into a single patent, or try to divide it up into multiple patents. For a provisional patent, this issue is usually not critically important. At the time of filing a utility patent application, an IP attorney can generally advise the innovator on whether or not to file more than one application. At this point, there are several considerations that may come into play. On the downside of filing multiple patent applications, there is the issue of the cost (which is magnified greatly if there will be international filings). However, multiple filings may be attractive if some piece of the original concept is more clearly novel and unobvious than the rest. Creating a separate filing for this piece of the original provisional patent could help the innovator gain approval more quickly for a particular subset of claims, which could be desirable, for instance, in advance of an upcoming product release. Similarly, if the innovators have a possibility of licensing some part of their IP to an existing company for **royalties**, it may make sense to isolate that aspect of the technology in its own patent filing to make the licensing agreement more straightforward.

Separating a patent application into parts can be accomplished by a continuation application, described in more detail later in this chapter. Another way that carving up a patent application occurs is in response to an office action (OA) by the patent office in which there is a restriction requirement. This comes about when the patent examiner deems that the original filing covers more than one invention (patent law allows only one invention per patent) such that it needs to be divided into several applications. The filing mechanism for dividing a patent into parts in this setting is called a **divisional**, which is a new patent application that claims a distinct or independent invention based upon pertinent parts carved out of the specification in the original patent.

A more sophisticated patent filing strategy issue – one that is more often dealt with by companies than individual inventors – is the question of whether to pursue trade secret protection rather than a patent. As described in Chapter 4.1, a trade secret is information, processes, techniques, or other knowledge that is not made public but provides the innovator with a competitive advantage. In the medtech field, trade secrets often have to do with how a particular device is made (materials, manufacturing process, etc.). There is no application for a trade secret and no official granting of this right by the government. Innovators or companies assert trade secret protection by suing another entity that has wrongfully obtained this information. In order to prevail, the company with the trade secret rights must demonstrate that it has taken reasonable precautions to protect the information (for example, having its employees sign an agreement not to disclose, creating a special, high-security room where the trade secret is protected, and so on).

In some circumstances trade secret protection is much more effective than a patent, something which first-time innovators tend not to realize. For a start, if an innovation is kept as a trade secret, no competitor should become aware of it. In contrast, a utility patent publishes 18 months after filing. Perhaps more important, a properly kept trade secret can potentially last forever, as long as the company is careful to keep the information secret (the most famous example of this is the formula for Coca-Cola®). Although trade secret protection does not involve any filing fees up front, the costs and organizational effort involved in keeping a process secret can be considerable (as the Coca-Cola example demonstrates).

Freedom to operate strategies

As described in 4.1 Intellectual Property Basics, it is a common misconception among first-time inventors that gaining a patent on a device means that the device is free and clear to commercialize. The device may indeed have a patentable feature, but it may also have other features that infringe on claims of existing patents that are in force. With medical devices, inventors often discover potential freedom to operate (FTO) problems once a careful analysis is performed for a new invention. In certain heavily trafficked areas of medical devices – vascular stents, for example – it is extremely difficult to come up with an idea that has complete FTO.

An FTO analysis begins with a comprehensive search of patents that are currently active (generally in the past 20 years). The inventor is looking for prior art that is in the same area as the new invention. Importantly, a thorough search will almost always turn up prior art that is potentially of concern. It is easy to get discouraged by a superficial look at the abstract or drawings of an existing patent. However, inventors should not automatically assume that there is no FTO if a device pictured in a patent or described in the abstract looks or sounds similar to the device they invented. *The key is in the claims.*

The way to understand whether the new invention is distinct from the prior art is to carefully assess whether one or more claims of the potentially infringed patent apply to the new invention. In order for an existing patent to prevent an inventor from going forward with an invention, every part of one of the independent (stand-alone) claims has to fit the invention. For example, suppose an inventor has developed an angioplasty catheter (a catheter with a balloon used to dilate a narrowing in a blood vessel) where the key feature is that the balloon is wider in the middle than at the ends. In a patent search, the inventor might find an existing patent where the claim is: "A tubular member with a balloon at the distal end in which said balloon has a larger dimension in the middle region than at both ends of the balloon and said middle region has a generally convex outward shape." The good news is that if the new invention does not have a convex shape in the middle, it is not covered by this particular claim. In fact, this claim may suggest that these inventors thought that anything besides a convex shape would not work, leaving room for a different shape to be patented. The bad news is that the existing claim covers the general idea and, thus, would potentially make the new invention obvious. Furthermore, the inclusion of specific language about the convex shape raises the possibility that somewhere in the patent world is a more general claim covering the broader case (otherwise, it might not have been necessary to make this independent claim so specific). In this situation, it would be helpful to look at the prior art cited in the patent to see where this claim might lie. In practice, it can be difficult to determine whether or not there is an FTO issue. A patent attorney with medical device domain expertise can be extremely helpful in clarifying the situation. Ultimately, the final decision about FTO may be made in court through the litigation process.

During an FTO analysis, the innovator should keep in mind that, under US patent law, inventors may "act as their own **lexicographer**" in a patent application. This means they can give common words or phrases meanings that are specific and different from their normal definition.[3] However, all such terms must be described or defined in the patent specification. If a term is defined in the specification, it must be construed in the remainder of the patent and claims in accordance with that definition; otherwise, the plain meaning of the word (as determined by a how a regular person in the field would understand the term) will be used. In medical device patents, it is common to list examples to help clarify the use of a term without limiting it. For example, in the classic Palmaz stent patent, the key claims referred to an "expandable intraluminal vascular graft." What this term means for this patent is clarified in the second paragraph by examples: "Structures which have previously been used as intraluminal vascular grafts have included coiled stainless steel springs; helically wound coil springs ... and expanding stainless steel stents ..."[4] Sometimes patent attorneys will define a key term by saying it "includes *but is not limited to* X, Y, and Z" to keep the definition as open as possible.

There is a range of possible outcomes from an FTO analysis, which vary from low to high risk as shown in Figure 5.1.1. If a thorough analysis suggests that there may be an FTO problem, there are a number of conditions to check that may help the inventor avoid the issue:[5] (1) the patent may not have been applied for, or granted, in the country where the company is seeking to operate, (2) the patent may have lapsed (e.g., if the patent-holder has not kept current with the required maintenance fee payments), (3) the patent may have expired (be sure to check expiration dates), (4) the patent may be invalid (sorting this out will take legal expertise).

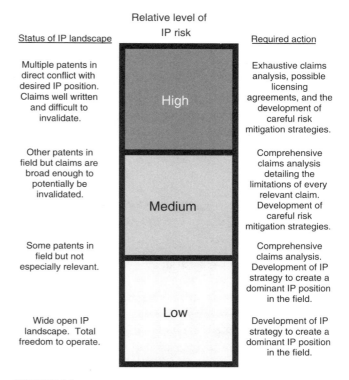

FIGURE 5.1.1
The level of risk associated with protecting a technology has profound strategic implications on the inventor and his/her company (developed in collaboration with Mika Mayer and Tom Ciotti of Morrison & Foerster LLP).

If there is a clear, high-risk FTO problem, the innovator still has potentially fruitful options to pursue. First, s/he can try to modify the invention so that the patentability is maintained but the features that infringe current patents are removed or altered. This is a routine part of the biodesign innovation process, which most inventors go through with most inventions. It is a perfect opportunity to convene brainstorming sessions and in particular to bring in fresh advisors to see if there are other approaches that avoid infringement. In many cases, this process of re-inventing leads to important and unanticipated improvement in the fundamental invention.

If the feature revealed in the FTO analysis is absolutely essential to the new invention, the second option is to pursue licensing of the problematic patent from the entity that owns it. Medtech patents owned by universities may be relatively readily available for an innovator to pick up, since these can be complicated and time-consuming for the university to license to large companies. Licensing of a key patent from a company is generally more challenging. It is sometimes possible to get a license to a limited *field of use* (usually a clinical area) that the company is not interested in pursuing. A company may also be willing to trade IP, such that the innovator gives rights to the new invention (perhaps in a limited field of use) in exchange for rights to the critical patent from the FTO analysis. See 6.4 Licensing and Alternate Pathways for more information on licensing strategies.

The Spiracur example below illustrates how one emerging company managed a complex patent landscape analysis as part of developing its IP strategy.

FROM THE FIELD — SPIRACUR

Building an effective IP strategy

Moshe Pinto, Dean Hu, and Kenton Fong (see Figure 5.1.2) began working together while students in Stanford University's Program in Biodesign to address the need *to promote the healing of chronic wounds*. This team, which eventually became Spiracur, initially developed an IP strategy through extensive prior art searching and the filing of a provisional patent application. However, these steps were just the beginning of Spiracur's activities.

The team also went to work right away on an in-depth patent analysis. Spiracur recognized the importance of working with strong IP counsel to guide its IP activities and the development of a holistic IP strategy. However, the team also felt strongly about performing much of the early work themselves, so they could save money on fees by providing the IP attorneys with a concrete, comprehensive analysis from which to start. This work would also help them become more knowledgeable in the field and increase their credibility with investors and other stakeholders. For these reasons, the three entrepreneurs initiated the analysis on their own, performing extensive searches using resources such as the USPTO database and Free Patents Online.

In terms of performing patent searches, Pinto commented, "You have to be savvy about the search strings you use."[6] While many start-up companies use a funnel approach – starting broad and then getting more specific – Pinto offered another suggestion, which he called "inverting the funnel." The idea is to start with highly specific search strings that yield a relatively small number of relevant patents. After examining these patents, the team can then use the reference charts within the patents to identify other relevant patents. It can also mine this information for clues for developing new, slightly broader search strings that yield a greater number of results. "Through this approach, the inventors quickly gain familiarity with the IP landscape in their space and can more effectively separate the wheat from the chaff as they continue their analysis with the broader search terms," Pinto explained.

To support the patent analysis, the team carefully analyzed each search result. Without any patent law training, the team was at somewhat of a disadvantage in analyzing them. "Dean was an engineering PhD and Kenton was a resident of plastic surgery," explained Pinto. The team, however, worked out a process. Hu and Fong would utilize their scientific training to screen the initial search results for subject matter that might have some threshold interest to them. These interim results were then passed to Pinto, who had more of a business background, and later to their attorneys.

Pinto admitted that the patent analysis could, at times, seem daunting. "If, for example, you run a search on the words 'pressure sores,' you'll get 60,000 patents. There is an initial evaluation of the patents that almost anyone can do just by looking at abstracts. But when it's time to derive conclusions based on what you find, it's important to have someone who is legally savvy to perform this work."

In total, Pinto estimated that the team reviewed the abstracts and drawings of more than 1,000 patents and identified roughly 250 to 300 patents for their IP attorney to evaluate and use to help them develop a patent strategy for their technology. When Spiracur initiated its Series A funding round, the team was prepared for the due diligence performed by the investors' IP attorneys and was able to successfully close the funding round.

FIGURE 5.1.2
Hu, Pinto, and Fong of Spiracur (courtesy of Spiracur).

In terms of offering advice to others who are building an IP strategy, Pinto emphasized that IP needs to be a primary concern for entrepreneurs. "Inventors need to realize that their exit from a medical device venture is pretty much constrained or determined by the strength of their IP position," he said. "If there is one thing I learned, in retrospect, it is that we should have dedicated more time and effort to the entire IP endeavor than we did." When asked to estimate the amount of time a start-up company should invest, he commented: "The investors have it calculated. If we look at it from a budget perspective, they want to see the ongoing time of an IP lawyer at roughly $20,000 dollars a month. That translates to about 25 to 30 hours of an IP expert per month."

Pinto also stressed that one's focus on IP had to be continuous and progressive. "What I suggest is that even after you submit a provisional with claims with the help of a legal expert, revisit it after a month. Don't wait for the end of the 12-month period to say, 'Now, let's see how we can convert this into a utility patent application.' Do the iterations sooner, as more information is gathered." Finally, he said, "Remember that quantity does matter." A sound IP strategy must seek to consistently improve the coverage and protection surrounding the company's technology. "When later-stage investors asked what we had done in IP, I told them that we had eight provisionals and three pending utility patent applications," said Pinto. "This kind of ongoing activity assures them that you are paying sufficient attention to IP."

Hiring legal help

While medtech innovators sometimes file provisional patent applications on their own, it can be highly beneficial to have an expert attorney prepare these or at least review them before filing. To file a utility patent application, it is almost always essential to hire a professional. The cost of an IP attorney varies significantly. Usually an initial consultation can be obtained without any charge. Hourly rates thereafter range from $300 to $800. This means that the costs of preparing a utility application on a relatively simple medical device will be in the $6,000 to $14,000 range, including a relatively narrow patentability search. A utility patent application on a complicated technology may run as much as $30,000. A search with a patentability opinion might be $3,000 to $5,000. A complete, formal FTO opinion may cost $25,000 to $100,000 (this expense not only reflects extensive search work, but the fact that the attorneys issue a detailed, written opinion on FTO). Some medtech patent attorneys offer an intermediate approach, consisting of a careful search for prior art and a discussion with the inventors (and potential investors) about the broad scenario for FTO. The cost of this less formal analysis may be in the $5,000 to $10,000 range, depending on the complexity of the patent landscape.[7]

Law firms recognize that inventors are cash challenged in the early stages of the biodesign innovation process and some will offer alternate payment strategies, including deferring payments or taking equity (stock). Because it is fairly easy to run up a large patent "tab" quickly, many established firms are not willing to defer payments or take equity alone. Another option, which can often be less expensive than working with a patent attorney, is to engage a patent agent. Patent agents are registered to practice *only* before the USPTO and, as such, cannot conduct patent litigation in the courts or perform services that the local jurisdiction considers as practicing law (for example, providing opinions on patentability, ownership, or validity, or performing FTO analyses).[8] Usually, patent agents are engineers who have successfully passed the USPTO patent office exam, or have served for four years or more as a US patent examiner before entering private patent practice.[9] Patent agents can offer a cost-effective approach when the innovator's IP needs are relatively straightforward, there is broad FTO in the field, and/or the innovator does not anticipate litigation. Note that many law firms employ patent agents and patent attorneys in their IP practices, so that a client company can benefit from a full range of services, for example, by using agents to draft applications and associates and partners to help with IP strategy and litigation.

There are several guidelines to consider in finding a patent attorney or agent.[10] It is important to seek someone with experience filing similar medtech patents (in the same technology and clinical area, if possible). The better medtech patent firms will have previously performed detailed patent landscape analyses in many areas, and these provide attorneys in these firms with substantial background. In some situations, an attorney may not be able to work on a filing because his/her firm has another case that represents a potential conflict of interest. The law firm will make this determination before agreeing to begin the contract. As with engaging any consultant, it is helpful to get recommendations from experienced inventors and entrepreneurs in the field. Note that faculty or student inventors may be steered to a particular attorney by the university's office of technology licensing (**OTL**). However, it is worthwhile doing an independent validation of the expertise of the attorney before proceeding.

Working with an attorney or patent agent who is geographically located near to the innovator is of course convenient. Yet, given the ease of electronic communication, it is generally more important to find someone with expertise in the technical and clinical domain of the invention than someone who is co-located. Make sure there is an explicit understanding of how (and how much) the attorney will charge for his/her time to avoid unpleasant surprises. Ask for a face-to-face consultation to gauge the attorney's approach and assess the "chemistry" between the attorney and the individual(s) with whom s/he will work most closely (open, clear communication is essential).

In meeting with the attorney, do not ask general questions that can be addressed by reading a book or performing online searches. Invest time in advance to become educated on IP issues so that the attorney's time can be used more effectively to address unique and complex issues (rather than basic questions). Consider creating an initial draft of the application and/or the claims before getting too deeply into the patent drafting process. While the attorney will surely modify the draft, it will likely save time and money by ensuring that everyone has the same understanding of what the patent application will look like.

International filing strategies

An international IP strategy should be developed with the costs and benefits associated with the filings outside the United States (OUS) clearly in mind. Every OUS patent application will require a significant financial commitment (e.g., $250,000 to $500,000 over the life of the patent, not including the cost of any potential litigation required to enforce it). This high cost is driven primarily by the need for specialized legal and translation services for each country (the European Patent Organization is an exception, see 4.1 Intellectual Property Basics). To justify such a sizable financial commitment, an innovator or company must be certain that the benefits of the OUS filings will be measurable and central to the overall business strategy. Particularly during the early stages of a company's development, OUS filings should be made only when there is a clear value proposition (e.g., the product is anticipated to generate a sizable percentage of its revenues internationally). The selection of countries varies from case to case. For many medical devices, it is common to include Europe, Japan, Australia, Canada, and Israel in international filings of US patents. Increasingly the BRIC countries (Brazil, Russia, India, and China) are of interest because of their strong emerging economies. Innovators should also take into account the fact that they may ultimately want to *manufacture* in a different country. To the extent they can anticipate this, it is critically important to file in the candidate locations (major medical device manufacturing exists in Belgium, Costa Rica, Ireland, Taiwan, and several other countries). If an innovator is targeting a particular large company as a potential acquirer of his/her technology, it is also worth considering where that company has its deepest roots. For example, a German medical technology company such as Maquet might have some hesitation in acquiring a technology for which no patent rights have been pursued in Germany.

Strategies in managing a patent portfolio

Many early-stage innovators put much more emphasis on the filing and obtaining of an initial patent than on developing a comprehensive and ongoing patent strategy. This kind of approach to IP carries with it sizable risks that can negatively affect an innovator's ability to raise funding, not to mention threatening the viability of the company once it gets started. In contrast, innovators and companies that strategically manage their IP portfolios stand to minimize their IP-related risks, differentiate themselves from their peers, and increase the value of their businesses. By taking an active approach to managing IP, the value a company receives from its IP investment can evolve from merely protecting its inventions to controlling (and even pre-empting) competition, building markets, delivering revenue, and driving business strategy.[11] It is useful to think of these benefits in the form

FIGURE 5.1.3
As innovators and companies take a more strategic approach to managing IP, the value they receive from their IP investments increase dramatically (based on Ron Epstein, "Building a Business Relevant IP Strategy," IPotential, LLC, 2006; reprinted with permission).

of a hierarchy, in which protecting inventions is the base upon which increasingly important benefits are built (see Figure 5.1.3).

There are two basic features of an effective and valuable patent portfolio. First, the patents must actually cover the products that are developed. This sounds trivially obvious but, in practice, there is a real risk that the development and refinement of a medtech device will outrun the patent coverage. This happens due to a lack of continued attention to IP during the development process (i.e., in response to prototyping and testing, the team adds key features to the design that may or may not be patentable). In the hurry and pressure to get a product to market, the team may simply forget that new IP has been created (or that the new features may infringe existing patents).

The second characteristic of an effective patent portfolio is that it creates real barriers to entry. It is not sufficient for the innovator to patent the features of his/her new device. It is essential to anticipate who might be able to develop a similar device or a next-generation product and then develop patents to block that from happening. This requires an in-depth understanding of the competitive landscape – not only the patent portfolios and filings of the main competitors, but their business strategies as well.

The mechanism for developing and maintaining a valuable patent portfolio is *continuous internal and external monitoring* of IP. Internally, IP must be integrated into the ongoing research, development, and business planning strategies – that is, it must be an active part of the awareness and work of all team members. Time and attention need to be explicitly allocated to correlate the progress being made in prototyping and testing with the approach to patenting. As new team members are added, it is essential that they be educated in the importance of keeping careful innovation notebooks and in the dangers of public disclosure. External monitoring means that at least some member(s) of the team must focus on surveying and understanding what competitors and potential competitors in the field are doing on a regular basis. Effective ways to gather this information include talking with physicians and attending symposia, courses, and trade shows. This will also help the team stay up to date on the other relevant developments in the field (including the clinical science).

The ongoing monitoring of internal and external IP activity leads to two basic types of proactive patent strategies: *defensive* and *offensive*.

Defensive strategies

Defensive strategies represent the more familiar of the types of approaches innovators can pursue. They refer to developing patents and claims that will exclude others from making, using, or selling the technology that the innovator has invented. The key issue is to understand where the innovation is going as it moves toward clinical practice. This means that in thinking about an invention it is essential to consider not only the device in isolation, but how it will be used. Is it part of a system (and if so, are there other parts of that system that also need to be protected)? How will it be packaged? Will it come in a kit? Are there certain methods of use that should be patented along with the device claims? Are there other indications (other places in the body or alternate procedures) in which the device could be used? If so, are there other components or aspects that need to be added? Using these considerations to create a series of secondary patents and claims that serve as

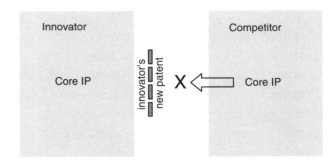

FIGURE 5.1.4
A picket fence strategy gets its name because it uses adjunctive patents to surround the core intellectual property, keeping a competitor outside of the innovator's core patent territory.

a barrier to competitors is sometimes called a picket fence strategy (see Figure 5.1.4).

To illustrate this approach with an example, consider a new vascular stent that has the property of being bioresorbable (it dissolves within the body over time). The inventors are clearly interested in protecting the stent itself – the special, resorbable material and the strut configuration that gives the stent enough radial force to prop open the artery. This is the core IP. But there are other, potentially patentable aspects of using the stent that could also be covered. How is the stent expanded (e.g., with a balloon or some other deployment mechanism)? Does the catheter delivery system have special features, perhaps related to the deployment method chosen? Does the packaging need to be specialized, for instance, to preserve the shelf life of the new stent material? Is there a need for this kind of stent in some other application, e.g., to open the esophagus in children with atresia (narrowing)? Pursuing patents in some (or all) of these different areas would provide the innovators with a defensive shield around the core IP.

One extremely important tool in building a defensive patent strategy is through continuation patent applications, which allow inventors to refine the claims structure during the patent prosecution process. Basically, a continuation is an application that is filed to pursue additional claims to an invention using the identical specification from the original patent. A continuation must list at least one of the same inventors as the original patent. Claims filed as part of a continuation will receive the same filing date priority as the original application.

The power of the continuation process for defensive strategy is obvious. As the team gains experience with the original concept, including further prototyping and early-stage testing, important aspects of the invention become clearer. With time, the team is also likely to gain a more informed idea of what potential competitors in the field are doing and what will be necessary to protect the core IP. The continuation process then allows the team to submit new claims that provide a sharper focus on the invention, based on continued experience, thinking, and surveillance of the marketplace. Although obtaining new claims for an invention can seem like "getting something for nothing," keep in mind that claims are meant to define clearly what constitutes the invention. The invention – as described in the specification – does not change with continuations. The new claims provide a way to communicate more exactly what the invention is. Moreover, the continuation process does not extend the life of the patent – the priority date goes back to the filing date of the original patent specification and the expiration date is based upon the priority date.

In developing a defensive strategy for patent protection, it is essential to promote an active dialog between the patent attorney and the development team. Early teams tend to be worried about the expense of frequent meetings with an attorney. This can be mitigated by having a regular but brief check-in, perhaps by phone, where the team has prepared a succinct update of new test results, development milestones, and the concepts that have come from these. In addition, as the team grows, every member needs to be educated about the basics of intellectual property, either by the attorney or by savvy members of the team. This kind of communication and education is the best way to prevent the loss of rights of a patent either in the United States or internationally.

One mechanism to employ the creative talents of the team in building the defensive patent portfolio is to organize invent-around brainstorming sessions. Have the team pretend they are working for a competitor and challenge them to design approaches that will bypass the existing patent claims. These sessions are of course worthwhile for the patent attorney to attend. The results can be used to fill in the gaps in the defensive position, either through new patent filings or new claims through the continuation process.

With respect to claims, an important approach in developing a defensive strategy is to create a range of broad and narrow claims, sometimes called "layered" protection. Broad claims are of course desirable, but keep in mind that they are also at higher risk of being declared invalid in litigation. Narrow claims do not offer as much protection (i.e., they are easier to work around), but they can be granted more quickly. A layered combination of broad and narrow claims provides the best chance to maintain an effective defensive barrier after the patent prosecution process. The timing of product entry into the market can be an important factor in suggesting an appropriate mix of broad and narrow claims. If FDA approval is anticipated to be relatively quick (a 510(k) clearance, for example) and the product can be brought to market easily, it may be strategically important to have some narrow claims already issued on the device to protect its introduction to the marketplace.

To benefit from the defensive protection of a patent, the owner must threaten or take legal action (e.g., filing a lawsuit) against anyone who infringes on his/her IP. In the realm of medical devices, the company that has been assigned the IP or licensed the patents (from a university, for example) is generally the one that takes on the responsibility of pursuing potential infringers. Occasionally, universities will also take on this role, although in medtech there are a relatively small number of "blockbuster" patents that rise to this level of attention on the part of the university.

There is a sophisticated variation on the picket fence defensive patenting strategy that some innovators and companies employ – namely, to intentionally publish information on incremental improvements and new uses related to a core IP asset in order to create a buffer of publicly disclosed prior art to keep competitors away. This has the same effect as additional patent coverage in terms of keeping competitors from owning these improvements, but has the advantage of being fast and free.

Offensive strategies

Offensive patent strategies refer to approaches that look outward to the competitive landscape to try to take advantage of the patent position of other companies. These strategies tend to be less intuitive and more difficult than defensive strategies for first-time innovators (who are typically focused on developing and protecting their inventions). However, understanding the direction of competitors in the marketplace and moving in an intelligent and proactive way in reaction to (or anticipation of) their IP portfolios is an absolutely essential means of ensuring that the innovator's product will successfully make it into clinical care.

The first offensive strategy is to anticipate the direction that a competitor may be headed and create IP that blocks that approach (see Figure 5.1.5). Sometimes an innovator can see the competitive pathway clearly enough to create new IP just for the purpose of blocking a particular competitor. In other cases, the block arises from both groups independently seeing the same general technical direction, and one group inventing earlier than the other. In the case example below on Intuitive Surgical, Dr. Fred Moll recounts a classic situation in which Intuitive Surgical, a robotics company, became concerned about patents owned by a competitor, Computer Motion. The patents, which described fundamental software-controlled mechanisms of translating hand movement to instrument tip movement, were directly in the path that Intuitive wanted to pursue in developing its da Vinci® robot. After a great deal of wrangling, this offensive block

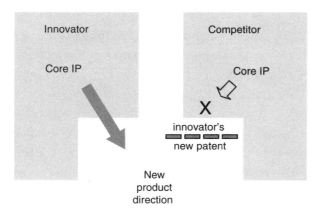

FIGURE 5.1.5
One offensive IP strategy is to pre-empt a competitor's ability to move toward a new product design by means of a blocking patent.

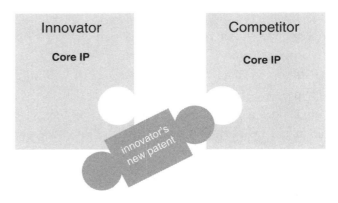

FIGURE 5.1.6
An offensive bridging strategy that complements and/or fills a gap in a competitor's IP portfolio can be effective for an innovator seeking to license or sell a technology to a larger company.

ultimately led to the acquisition of Computer Motion by Intuitive.

A second offensive strategy is also based on an intimate knowledge of a competitor's IP portfolio and business strategy, but in this case involves explicitly linking into the competitor's IP by building a bridge to it with a new patent, as shown in Figure 5.1.6. This strategy should be used specifically when there is a candidate company that the innovator thinks would be a good fit to acquire the new technology. It requires the innovator to have a good understanding of both the basic IP position of the acquiring company and at least some sense of the company's priorities with respect to new technology directions. In practice, this latter information may only become apparent as the inventor has discussions with the company about potential licensing of the core technology.

All of these offensive and defensive strategies require a high level of diligence in pursuing information about competitive companies. With respect to IP, it is important to perform regular and ongoing searches of US and international patent databases using the appropriate keywords, competitor companies, and known inventors in the field. In addition to patent searching, the team must keep up with the clinical literature in the area, stay current on new product releases, and periodically perform Google searches around the technology itself, key companies, inventors, and also clinicians. In an active technology area, it is not uncommon to find

50 or more "hits" a week of potentially interesting new information.

Strategies related to timing of patents

Real patent coverage begins when a utility patent is issued. Between the date of publication and the date of issuance of a utility patent, an innovator cannot prevent another company from operating in the space of the invention. Having additional companies operate in the same space, even if just for a short while, can erode prices, thus, impacting long-term revenues even once the other companies are restrained. Ideally, an innovator will have patent coverage (under an issued utility patent) at the time the product goes to market. As part of an effective IP strategy, the innovator should seek to coordinate the timing of its patent issuance with the product's entry into the marketplace. Innovators often seek to delay filing to extend the patent term as long as possible and in order to keep competitors from learning of their devices and methods, since a patent application is not public knowledge until it is published. However, the delay should not be so long that a competitor is able to negatively affect the innovator's desired IP position. There is also a risk that the company may inadvertently disclose the invention prior to filing.

A slightly technical but important timing issue is called "provisional protection" and refers to the fact that if claims are issued in substantially the same form in which they were published, a company can seek back damages starting from the date of publication. That is, if a company B infringes the claims of company A, company A may collect damages dating back to the time those claims were initially published (generally 18 months after filing) rather than the date the claims were granted (which may be several years after filing). This makes it important to write at least some claims in the initial filing that have a strong chance of being granted (as opposed to trying only to get the broadest conceivable claims in the first round).

In some cases, it is possible to extend patent coverage beyond the conventional 20 years when an innovator or company can make the case that the introduction of the product in the marketplace has been delayed for legitimate reasons. This strategy has been relatively widely used in the pharmaceutical industry where development times are so long that the 20-year life of a patent is perceived as a major limitation (by the time a drug or biologic is approved by the FDA, there may be only ten years or less remaining of patent coverage). Under these circumstances, a company can apply to the patent office for an extension of the term of the patent, which is calculated based on the length of time the development was held up in clinical trials and regulatory approval. In the past, medical devices – which tend to have a much quicker approval process than drugs or biologics – generally have not had patents extended through this mechanism. Recently, however, with the evolving complexity of medical devices (particularly with drug–device combinations) this approach to extending patent covered is being pursued with increasing frequency in the medtech industry. For instance, Abbott Vascular applied for an extension of the Rapid Exchange™ delivery for the Xience drug eluting stent based on the relatively long FDA approval process for this device

A related IP timing strategy employed by some innovators and companies is colloquially called "**evergreening**" – a process of introducing modifications to existing inventions and then applying for new patents to protect the invention beyond its original 20-year patent term. Again, this strategy has been much more widely applied in the drug industry compared to medtech. For example, a company reaching the end of the patent life of a popular drug may choose to develop a time-release formulation that is patentable. The company then phases out the original formulation and replaces it with the time-release capsule, which has new patent protection. It is reasonable to expect similar strategies to appear in drug–device combinations, although there are no high-profile examples at present.

Patent litigation

Strategies for patent litigation are beyond the scope of this text since they generally come into play at a later stage of product development within a company and involve company management and legal experts in the decision making. However, a few basic issues are important for the medtech innovator to understand at the outset of developing an IP strategy.

There are several potential outcomes to be aware of in medical device patent litigation. In rare cases, the courts may perceive a case to be clear enough that they will issue an **injunction** against the party that is accused of infringement (violating the patent rights of the plaintiff). Consequently, the company accused of infringing the patents must stop the manufacturing and sales of the device. Much more commonly, the case goes to trial or is settled without an injunction. The outcome of a trial may be decided by either a judge or a jury, depending on the nature of the lawsuit. If the patent holder wins (that is, proves that there was infringement), that entity awarded damages that typically reflect a reasonable royalty based on the sales of the infringing company (e.g., something on the order of 5 percent of net sales). In recent years, there have been a number of multimillion-dollar medical device patent infringement settlements paid by large companies to individual inventors who hold key patents, particularly in the cardiovascular and orthopedic fields. If the courts make a determination of "willful infringement" – that is, the courts believe that the infringing company acted in bad faith, with full knowledge that it was infringing and no mitigating circumstances – then the damages can be tripled (treble damages).

Attorneys who perform patent litigation in court are generally different from the attorneys who write and "prosecute" (pursue the approval of) patents, although both types of attorneys may work for the same firm. It is not unusual for venture capitalists or companies that are doing serious due diligence on an innovator's IP to pay a litigator to evaluate the robustness of the patent protection, particularly in a space that is known to be litigious.

Patent litigation is a time-consuming, expensive, and emotionally draining experience for the participants. The big companies that dominate the medtech industry have large "war chests" for IP litigation. At present, it is part of the culture of the industry sector that important products are, more often than not, embroiled in patent litigation. Practically speaking, if an innovator develops an important medical device s/he can essentially rely on being involved in significant litigation at some stage later on. The burden of litigation, in terms of productivity alone, can be considerable. The inventors and other key contributors to product development are required to participate in depositions in advance of a trial, which are formal opportunities for counsel – from both sides – to discover essential facts about the case. These can be highly adversarial and tricky encounters, which require a great deal of careful preparation. The discovery process may go on for many months before a trial or a settlement occurs, with the process consuming hundreds or even thousands of hours of potentially productive time from inventors and other personnel.

The Intuitive Surgical story below provides an example of the importance of IP strategy in effectively managing competitive threats. Although, in general, innovators do not face litigation until some years into development of their product (typically after sales begin), the seeds for successful litigation can be planted from the beginning of the biodesign innovation process. Focusing on the basics, as presented in Chapter 4.1, provides a good chance of avoiding the biggest problems in litigation. Careful and honest record keeping, engaging a competent attorney early in the process, and diligence in uncovering prior art and publications are the key success factors.

FROM THE FIELD — INTUITIVE SURGICAL

Using IP strategy to manage risk and opportunity

Intuitive Surgical was founded in 1995 by physicians Fred Moll and John Freund, as well as engineer Robert Younge, to pioneer the field of minimally invasive, robotic-assisted surgery. The company's primary product, the *da Vinci*® Surgical System (see Figure 5.1.7), translates the surgeon's natural hand and wrist movements on instrument controls on a console into the corresponding micro-movements of specialized surgical instruments positioned inside the patient through small incisions.[12] By combining the technical skill of the surgeon with

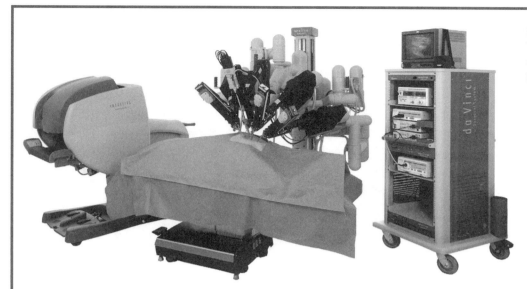

FIGURE 5.1.7
The da Vinci system (courtesy of Intuitive Surgical).

computer-enhanced robotic technology, *da Vinci* enables a minimally invasive approach to procedures such as prostatectomy, hysterectomy, myomectomy, gastric bypass, and mitral valve repair.[13]

Moll, who became the company's first CEO, learned of the technology through a non-profit research organization called the Stanford Research Institute (now SRI International), which had developed an initial prototype under contract to the US Army. Seeing the potential of such a system to accelerate the viability of minimally invasive surgery for a broad range of procedures, Moll and team licensed SRI's core IP surrounding surgical robotics[14] and founded Intuitive Surgical. According to Moll, "We were lucky early on in that we licensed a significant patent portfolio that had early dates on inventions that were important to the system and the core capabilities that we wanted to develop in surgical robotics." By adding some of its own proprietary technology and collaborating with IBM, MIT, and other institutions and thought leaders in the field, they developed the *da Vinci* system,[15] which received its preliminary FDA clearance in 2000.

Intuitive Surgical placed great importance on its IP strategy early in the company's existence. Regarding the licensing agreement with SRI, Moll explained the steps the company took to ensure it would maintain adequate control over the IP: "When you license intellectual property from the inventors, you can do it in various forms. But the important points are to get an exclusive worldwide license, which we were able to get. We also secured the ability to manage the patents ourselves. By that, I mean that we took over the prosecution of the IP." This allowed Intuitive Surgical to add to the patent portfolio and protect it in the way that best covered the specific product it was developing. "If we didn't have this ability, it could have been a real disadvantage," said Moll.

In terms of expanding the company's IP position, Moll noted, "We were able to do follow-on continuations to the patent portfolio in certain areas that were important to continuing to protect the Intuitive product. By that I mean we used the dates of these early patents to expand the breadth of the filings to cover features and capabilities that were anticipated but not specifically covered in the first filings by SRI." The company also took a systematic and proactive approach to managing its IP position. "You need to have a disciplined way to continuously look for prior art that would somehow modify your view of what patent positions are possible or the strength of your patents, as well as continuing to have a strong system to collect invention disclosures and turn them into a continuous stream of new filings that protect the newer aspects of the technology that you're developing on a weekly or monthly basis," Moll explained. To accomplish this, Intuitive added an in-house IP attorney to its team in the company's early days. "It makes a lot of sense, even for start-up companies, to hire in-house IP counsel sooner rather than later," Moll said. "In-house counsel is always going to understand better than an outside law firm what's going on within the company and what's most important to the product because they're just a lot closer to it." Relative to the value that medtech companies extract from their patent portfolios five or more years into their existence, "most people tend to

> **FROM THE FIELD** | **INTUITIVE SURGICAL**
>
> *Continued*
>
> under-invest in IP, which can be a big mistake," Moll added.
>
> Although Intuitive Surgical established a leadership position in the surgical robotic market, the field quickly became competitive with the entry of companies such as Computer Motion. While there were important differences between *da Vinci* and Computer Motion's Zeus® system, they competed head to head. Before long, the companies became entangled in a series of patent disputes. Moll described the issue at the core of the conflict: "Computer Motion had some early patent filings with a number of broad claims that had to do with fundamental ways of controlling rigid tools and how you translate hand movement to instrument tip movement if the connection is electromechanical and software related rather than directly mechanical ... And, arguably, a couple of these patents had the potential to inhibit our ability to control instruments the way we wanted to control them to make sense to the user."
>
> The Intuitive Surgical team believed that it had a strong IP position and a good chance of invalidating the Computer Motion patents. However, "The risk, not only the cost, associated with going through the legal process to find out whether their patents were valid was very significant. If we went through a litigation process and were unsuccessful, meaning that we were judged to be infringing on their very broad claims about how you control an instrument, it would be a difficult situation from a business standpoint and put the company at risk of having to redo a lot of what we had done," Moll said.
>
> Despite this risk, "there was no settlement in sight," recalled Moll. Intuitive Surgical briefly considered cross-licensing, but determined that this was not an attractive solution. According to Moll, "It progressed to the point where it was clear that it would probably end up in a jury trial. As a lot of litigators will tell you, a company's chances of winning a patent dispute that makes it all the way to a jury trial is about 50 percent. Most juries have a difficult time understanding a highly technical dispute. So you don't want to get all the way to a jury because your chance of winning isn't entirely related to the strength of your case."
>
> The team continued to believe it could win the suit, but the risk of losing was so great that it decided to pursue one more option. "It became clear that there was an opportunity to buy the company," Moll recalled. "The upside of an acquisition was not that we would be getting a new product line, because we didn't consider the Computer Motion product line additive to ours. It was competitive and, we believed, not as clinically useful. But what we would be getting was a very strong position with the combined IP portfolios in the area of surgical robotics controlling rigid instruments." Ultimately, he explained, "It came down to a judgment call on how much it was going to cost us to litigate the dispute versus how much it would cost to buy the company, in addition to the certainty of outcomes on each path." Fortunately for Intuitive Surgical, the company's balance sheet was strong enough that acquiring Computer Motion was a viable alternative. The two companies joined forces in March 2003.
>
> The breadth of the combined IP position gave Intuitive strength in the surgical robotics market that Moll believes has deterred some competition. "We always worried that the Japanese would enter the market for surgical robotics. To date, they haven't in a significant way. I think some of that is due to our success in technical development and the know-how associated with building the Intuitive system. And, I think some of it is also due to the perception of a very strong IP barrier, or picket fence, around the system that we developed," he said. Since the sale of its first *da Vinci* system, Intuitive Surgical has expanded its installed base to more than 700 hospitals and has sustained its position as the global leader in the field (see Figure 5.1.8).[16]
>
> Reflecting on his experiences, Moll underscored the importance of a strong IP strategy: "You can't just start a company, file a patent, and hope for the best. You need to file, as early as possible, continuations of the invention that surround the initial technology effort. In other words, you should have a rigorous process of invention disclosure by employees, new filings, and refinements of existing filings that's continuous, throughout the life of the company. The idea of continuously monitoring prior art and other inventions going on within the field is also very important. At almost every company I've been involved with, there are problems with intellectual property that the company doesn't own. If this blindsides you late in the game, it can be devastating."

5.1 Intellectual Property Strategy

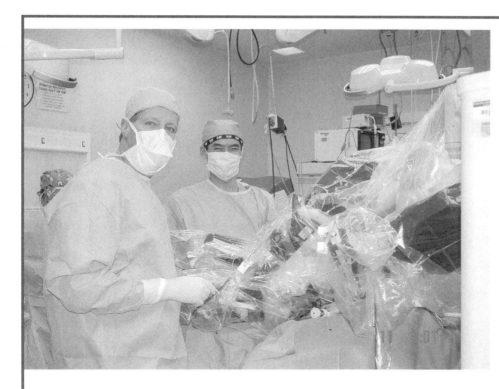

FIGURE 5.1.8
Chief of surgery Thomas Krummel and surgical resident David Le use the da Vinci system at the Lucille Packard Children's Hospital (courtesy of Thomas Krummel).

Finally, he said, "IP gets more and more important every year in the medical device business for principally one reason: the cost associated with clarification of who owns what has gotten so enormous. If you start with and maintain a strong IP strategy, it helps ensure that you don't get into situations where you're spending as much on IP litigation as you are on product development."

GETTING STARTED

Developing and maintaining an effective IP strategy is a challenging but essential step in preparing to take an innovation to market. References and resources concerning strategies for patentability can be found in the Getting Started section in Chapter 4.1. The following process can be used to develop an overall IP strategy.

See **ebiodesign.org** for active web links to the resources listed below, additional references, content updates, video FAQs, and other relevant information.

1. Understand the IP landscape

1.1 What to cover – A key part of assessing how one should define an IP strategy is knowing what patent references exist in the area of invention. Creating an assessment of IP landscape begins with patent searches to identify key patents and patent applications. Patent searches should be performed early in the biodesign innovation process to help an innovator understand the IP landscape and assess freedom to operate in the field. The output from these searches can help shape ways of thinking about the claims for the invention, as well as developing strategies to work around prior art, challenging existing patents, obtaining licenses, or avoiding wasted investment. An explanation of how to perform these searches is included in 4.1 Intellectual Property Basics. Because new IP is published

GETTING STARTED

every week, and other prior art is published daily, keeping an IP landscape up to date is a constant challenge, but it is extremely important in understanding competitive IP.

1.2 Where to look – Refer back to the resources and references listed in Chapter 4.1 for more information.

2. Validate freedom to operate

2.1 What to cover – The goal of an FTO analysis is to find any patent claims from another inventor that are currently in force and that explicitly cover the new invention, a component of it, or methods of making or using it. For any claim that appears to read on a feature of the new device, assess each word and clause of the claim carefully to make certain that it applies completely. Interpret words in the claim in the context of their definitions (and diagrams) in that patent. If a claim does appear to block the new device, make sure to consult an IP attorney to confirm that there is in fact an FTO issue. If FTO is indeed in jeopardy, conduct brainstorms to see if there are ways that the troublesome feature can be circumvented. If not, consider options for licensing the patent at issue.

2.2 Where to look – Use the USPTO website, Google Patents, or another search engine. For keywords refer back to information collected through prior art searches and the assessment of the competitive landscape.

3. Hire IP counsel

3.1 What to cover – Invest early in hiring an IP attorney or patent agent. This person (or team of people) must have deep, first-hand experience with patent filings – ideally in the domain of the invention – as well as with IP strategy and protection.

3.2 Where to look – The best way to identify a qualified IP attorney is through a referral from a trusted source. Network with professionals in the field and ask for assistance in meeting and choosing an IP attorney. Other sources to consider include:

- **IP professional societies** – Associations such as the American Intellectual Property Law Association can be a resource for referrals.
- **American Bar Association (ABA) Lawyer Locator** – Another resource for finding IP attorneys in different areas of the United States.
- **USPTO Patent Attorneys/Agents Search** – A searchable database of contact information for attorneys and agents with licenses to practice before the USPTO.
- **University office of technology licensing** – If a company or entrepreneur is affiliated with a university, the on-campus office of technology licensing (OTL) should be able to provide information as well as referrals related to IP issues.

4. Devise defensive and offensive IP strategies

4.1 What to cover – The key to developing defensive and offensive strategies is to maintain an up-to-date understanding of how the technology is evolving within the team (internal monitoring) while continuing to evaluate the technology and IP of competitors (external monitoring).

4.2 Where to look

- **Internal** – For internal monitoring, it is essential to calendar regular meetings between the technology group and IP counsel to assess what new technical directions are underway (and how these may be creating new IP and/or moving away from the original coverage).
- **External** – External monitoring is time consuming and difficult, but critically important. Set up Google alerts for any news items or published IP relevant to the

GETTING STARTED

5. Develop a comprehensive IP strategy and implementation plan

5.1 What to cover – Begin by specifically outlining what would constitute an ideal patent portfolio for the company. Formalize this vision into a comprehensive IP strategy that clearly lays out what the team needs to do in order to develop this portfolio. Work closely with an IP attorney in these first steps. Create a work plan that outlines the critical milestones for operationalizing the strategy, linking them to other important milestones (e.g., R&D accomplishments, funding needs). Use this strategy and work plan as a guide for continually strengthening the IP portfolio. Revisit the strategy and plan to keep it current. Make adjustments frequently, based on changes in the internal and external environments.

technology. Talk with physicians in the field. If possible, visit hospitals where competitive technologies are being used. Attend medical meetings, making sure to spend time in the exhibit halls, visiting competitor's booths, attending evening sessions, and being alert to gleaning information in the hallways.

5.2 Where to look – Work together as a team in collaboration with an IP attorney who truly understands the company's business strategy and plans for the future, as well as the timing of its operating milestones. Validate the company's plan with the board of directors and/or other trusted advisors and experts in the field.

Credits

This chapter was based in part on lectures by Tom Ciotti and Mika Mayer of Morrison & Foerster LLP (who also provided extensive editing consultation). The editors would also like to acknowledge Jeffrey Schox, of Schox PLC, for his review. Additional thanks go to Fred Moll, Gary Guthart, and Moshe Pinto for their assistance with the case examples.

Notes

1. Ron Epstein, "Building a Business Relevant IP Strategy," IPotential, November 10, 2006, www.ipedinc.net/isroot/ipedinc/powerpoints/Building_a_Business_Relevant_IP_Strategy.ppt (October 16, 2007).
2. David Dawsey, "Medical Device Industry Patent Litigation Likely to Rise?" Ezine@articles, http://ezinearticles.com/?Medical-Device-Industry-Patent-Litigation-Likely-to-Rise?&id=542615 (October 16, 2007).
3. "Claim (Patent)," Wikipedia.org, http://en.wikipedia.org/wiki/Claim_construction (October 25, 2008).
4. "Expandable Intraluminal Graft, and Method and Apparatus for," http://www.google.com/patents?id=iioBAAAAEBAJ&dq=4733665 (April 7, 2008).
5. "What Does Freedom to Operate Mean?" *Patent Lens*, http://www.patentlens.net/daisy/patentlens/g4/tutorials/2768.html (March 14, 2007).
6. All quotations are from interviews conducted by the authors, unless otherwise cited.
7. Based on estimates provided by Tom Ciotti and Mika Mayer of Morrison & Foerster LLP, Fall 2008.
8. "Attorneys and Agents," USPTO, http://www.uspto.gov/web/offices/pac/doc/general/attorney.htm (April 9, 2007).
9. "Patent Agents and Patent Attorneys," Wilson Enterprises, http://www.wilsonenterprises.org/AgntAtty.htm (April 9, 2007).
10. Sujith Pillai, "Working with a Patent Attorney," ezine@articles, http://ezinearticles.com/?Working-With-A-Patent-Attorney&id=476457 (March 13, 2007).
11. Epstein, op. cit.
12. "Intuitive Surgical's da Vinci Surgical System Receives First FDA Cardiac Clearance for Mitral Valve Repair Surgery," *Business Wire*, November 13, 2002 http://www.

allbusiness.com/medicine-health/medical-treatments-procedures/5958479–1.html (March 20, 2008).
13. Ibid.
14. "About Intuitive," Intuitive Surgical, http://www.intuitivesurgical.com/corporate/companyprofile/index.aspx (March 21, 2008).
15. Gary Singh, "The Robot Will See You Now," *Metroactive*, October 2004, http://www.metroactive.com/papers/metro/10.06.04/scalpelbots-0441.html (March 21, 2008).
16. "About Intuitive," op. cit.

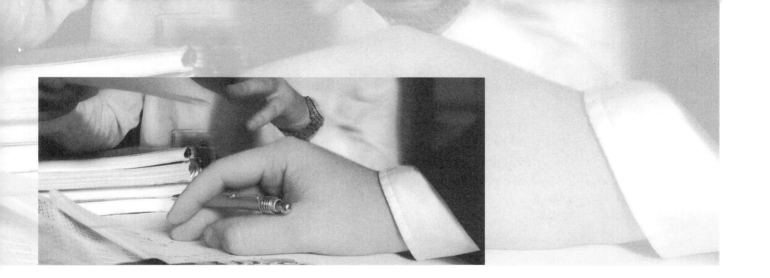

5.2 Research and Development Strategy

Introduction

Initial prototyping of the chosen concept has shown that it is technically feasible and can address what users want. But now the solution has to be made real. Successive iterations must lead to a device that is safe, performs effectively in humans, and can be efficiently manufactured. Thousands of hours will be spent on engineering and testing, as expensive equipment and scores of raw materials are used to produce a final product. This is R&D.

In the medtech field, research and development (R&D) typically refers to the scientific and engineering work required to take a concept from an early-stage prototype to a user-ready final device.[1] Whereas early prototyping focuses on proving the general feasibility of an idea, the goal of R&D is to develop a series of progressively advanced working models until all critical user requirements and core technical specifications have been met. Along the way, bench, tissue, and animal tests are employed to confirm that a product can be safely and effectively used in humans. The entire process can be lengthy and complicated, and calls upon many different engineering disciplines and skills. As such, it is essential to have a cohesive strategy for approaching R&D. An R&D strategy defines key milestones that need to be achieved to demonstrate development progress, identifies and prioritizes the technical challenges that must be addressed to achieve each milestone, and calls out the engineering activities, resources, and testing necessary to validate the solutions to these challenges. At its core, an effective R&D strategy, and the tactical R&D plan that supports it, seeks to resolve the greatest risks associated with developing an innovation as early as possible, with the most efficient commitment of capital, time, and effort. Ultimately, the articulation of a sound R&D strategy with clear milestones (and an understanding of how to tactically achieve them) is essential to the successful development of an innovation.

OBJECTIVES

- Understand the importance of defining strategic R&D milestones.
- Recognize how to identify and prioritize the key technical challenges associated with each milestone.
- Learn to outline the engineering activities, resources, and tests required to address these challenges as part of a high-level R&D plan.
- Appreciate the importance of retiring risk early in the creation of an effective R&D strategy.

Research and development strategy fundamentals

Research and development is sometimes referred to as "engineering" in the medtech field, since the design and development of medical devices is heavily dependent on activities related to one or more engineering disciplines (see Figure 5.2.1). In the context of the biodesign innovation process, R&D typically includes all engineering and testing activities beyond early-stage prototyping and final concept selection to the point when a product is ready to be released into production (for more information about the activities that precede R&D, see 4.5 Prototyping and 4.6 Final Concept Selection). Overall, a company's ability to transform an initial "proof-of-concept" prototype into a final product is central to its viability. However, there are many other tactical and strategic reasons why R&D is important. From a practical, near-term perspective, R&D:

- Plays a role in defining how the original need is ultimately addressed.
- Provides the engineering framework for developing a company's technology with the least amount of risk.
- Helps manage a primary driver of cost early in the company's life in terms of how personnel and other resources are used and managed.
- Lays the foundation (i.e., processes and culture) that helps the company continue innovating and developing future product iterations.
- Can lead to important insights related to the firm's intellectual property (IP) position.

In the longer term, a strategic approach to R&D can also help a company:

- Continually increase product differentiation and thereby mitigate market risk.
- Drive growth through new product innovation.
- Create a product development pipeline, which can make the company more attractive to investors and/or prospective acquirers.

As shown by these benefits, R&D strategy is inextricably linked to other important aspects of the business and, thus, is best developed by keeping these aspects in mind. Additionally, the R&D strategy must be built with a clear focus on the clinical need and how it can best be solved from the perspective of the target audience, not just from a technical point of view. As Guy Lebeau, a physician and businessman who became company group chairman of Cordis Corporation, explained:[2]

> One big issue is the lack of connection that can arise between the clinical need and the pure engineering desire. And that means you have to tell your engineers, 'Be sure you're in contact with people who are using the tool every day.'

The linkage between R&D strategy and planning

An R&D strategy can be thought of as a company's overall approach to addressing the key engineering challenges anticipated in the development of the final product. Usually, the innovator already has a general idea of what the final concept should do and how it will solve key user requirements and satisfy important technical specifications based on the work performed as part of prototyping and final concept selection (see Chapters 4.5 and 4.6). An R&D strategy picks up from there by articulating how the team or company will actually develop the product, in particular by considering the obstacles that it will need to overcome and how long that is expected to take. Accordingly, an R&D strategy can be developed by focusing on three key

FIGURE 5.2.1
Testing of a prototype with electronic components as part of the R&D process (courtesy of iRhythm Technologies, Inc.).

Table 5.2.1 *The output associated with key R&D milestones demonstrates to a company's stakeholders that important risks have been retired.*

Important R&D milestones
Proof-of-concept that addresses the scientific and technical feasibility of the concept
First working prototype that performs effectively in a bench model
First prototype that performs effectively in tissue testing
First prototype that performs effectively in live animals
First prototype that results in long-term safety in live animals
First prototype that performs safely and effectively in humans
Pre-production device that demonstrates manufacturing feasibility
Production device that supports scalable manufacturing

activities: (1) defining high-level R&D milestones, (2) using these milestones to identify and prioritize anticipated technical challenges, and (3) creating an early R&D plan that identifies the engineering work, testing methods, resources, and time required to solve these challenges.

The creation of an initial R&D plan can be considered part of R&D strategy since understanding high-level issues related to engineering personnel, resources, and timelines is important to determining when certain milestones can be achieved. More detailed information about R&D planning can be found in 6.1 Operating Plan and Financial Model. It is addressed in that chapter since the specifics of an R&D plan are usually articulated as part of an integrated plan that accounts for all of the personnel and resources required to take a final solution into commercialization.

Quality (and the implementation of a quality system) is another facet of development that is closely linked to R&D strategy and planning. Quality systems play a key role in informing the engineering methods that address the development and manufacturing requirements necessary to receive regulatory approval for a new medtech innovation. Specifically, quality systems lay out processes for properly capturing user requirements and technical specifications. They also detail key testing and validation methods, which are important considerations when thinking about the timeline for engineering R&D work. However, because R&D strategies vary significantly from innovation to innovation and are usually relatively high level, the requirements governing the implementation of a quality system, which tend to be more consistent across companies and encompass many more aspects than those just related to R&D, are discussed separately in 5.5 Quality and Process Management.

Defining high-level R&D milestones

As noted, the first step in creating an R&D strategy is to define essential R&D milestones. Such milestones usually correspond to significant events in the R&D process that correspond to the retirement of risks. Accomplishing key R&D milestones is one way that a company can demonstrate value in the eyes of its investors and potential end users. Certain R&D milestones also represent well-accepted **valuation** points to investors. For example, a firm is viewed as more credible among clinicians and more attractive to investors after it demonstrates the feasibility of a device on the bench, and then again once it develops a pre-production model that is effective in animal testing. By thinking about milestones early, innovators can also use R&D to their strategic advantage by sequencing and timing these types of milestones to correspond with other important events in the company's evolution (e.g., financings, the formation of clinical advisory boards).

The specific R&D milestones chosen by a company as part of its R&D strategy can vary, but they typically include those shown in Table 5.2.1. In the company's overall operating plan, these R&D milestones may be

Table 5.2.2 *Key technical challenges associated with the LAA example.*

#	Technical challenge
1	Creating a prototype that can deliver the material to the LAA in less than one hour (timeframe based on a user requirement by physicians who would use the device). (Note: the solution to this particular challenge does not necessarily have to be percutaneous, as this requirement is associated with a subsequent milestone.)
2	Ensuring that the substance from which the delivery prototype is made does not react with the material being delivered.
3	Finding a material that can remain in a liquid form for delivery into a living animal.
4	Finding a material that, once released, can reliably and predictably become solid when required within the conditions of a living body (e.g., blood flow, body temperature, clotting factors).
5	Finding a material that will not migrate from the place where it is released or finding a method to prevent migration.

"rolled up" into a smaller number (see Chapter 6.1). Regardless, innovators should recognize their serial, progressive nature – no single milestone can be achieved until those before it have been successfully addressed. Furthermore, although milestones such as proof-of-concept and working prototype may already have been achieved as part of the prototyping process (before much thought may have been given to an R&D strategy), they are still important R&D milestones to note as having been accomplished. Given the iterative nature of R&D, it is important to understand that the process of achieving these milestones is likely to take multiple, successive attempts on a development pathway that is not necessarily linear.

To help illustrate the steps of this process, the example introduced in 4.5 Prototyping will be revisited. This example involved injecting the left atrial appendage (LAA) with a material that can be delivered in liquid form and then increase in viscosity to a solid-like consistency to eliminate the space where thrombi can form via a percutaneous approach as a way to solve the need for ***a way to prevent strokes in patients with atrial fibrillation caused by LAA thrombus.***

Assuming that an innovator pursuing this need has already used some crude works-like prototypes and bench tests to show that the concept is feasible, the next most important strategic milestones to achieve might be to: (1) show effectiveness of the material in a live animal, (2) show long-term safety of the material in an animal, (3) show effectiveness using a percutaneous method of delivery in an animal, and then (4) show safety and effectiveness in a human.

Importantly, the milestones chosen and the way they are sequenced reveal information about the project and the innovator leading it. For instance, this particular sequence of milestones would seem to indicate the innovator's belief that demonstrating effectiveness and long-term safety in live tissue is of higher importance from a risk standpoint than developing a percutaneous solution. This is why it is addressed earlier in the development path. As it relates to R&D planning, this sequence further indicates that resources and personnel aligned with developing the percutaneous component of the solution will not be needed until later stages of the R&D effort.

Identifying and prioritizing technical challenges

After selecting relevant R&D milestones, an innovator can next consider the key technical challenges that need to be addressed to achieve them. The innovator must think about all the important engineering issues that s/he will likely encounter and then prioritize them. Typically, those challenges that involve the greatest uncertainty should be addressed sooner rather than later, so that a minimal amount of time is wasted if a particular challenge cannot be overcome.

Technical challenges obviously vary greatly by project. Each may have a different focus, depending on the milestone they are associated with in the overall development path. However, identifying engineering challenges is similar to thinking about the functional blocks of a concept during prototyping (see Chapter 4.5). Often, the challenges that need to be solved break down along the lines of different engineering disciplines (e.g., mechanical engineering, materials science, and electrical engineering). In this way, technical challenges represent the critical questions that the functional blocks of a prototype seek to address.

Using the LAA example, and focusing on the milestone of showing effectiveness in an animal, the innovator might come up with a short-list of key technical challenges, as shown in Table 5.2.2. With these technical challenges identified, the innovator can think about how to prioritize them. From the table, it is clear that there are three important challenges related to the material intended to fill the LAA, one related to a delivery mechanism, and one related to the interaction of the two. To prioritize them, the innovator can think about which challenges have been solved before and which have not. For example, many innovators working on other projects have accessed the LAA of an animal. In contrast, it is unclear whether anyone has ever found a material that meets the challenges outlined in items 3–5. For this reason, the innovator might sequence the work as follows: 4, 3, 5, 2, 1. This order makes finding a material that reacts in a specific manner with living tissue – which is likely the greatest challenge with the most inherent uncertainty – the innovator's top priority.

This example demonstrates that the innovator will need to speculate on which technical challenges will be the most difficult to address. In doing so, s/he must not forget to draw on any foundational knowledge in the various engineering disciplines that s/he may have and, more importantly, leverage the expertise of other engineers. While some of this knowledge would likely have come to bear in the early prototyping process related to the feasibility of certain elements of a final concept, it is nonetheless important to tap into this knowledge when considering all of the technical challenges together. Depending on the innovator's background and level of experience, s/he may have difficulty making the determination about which technical challenges actually will be the most important (or difficult) to address. For instance, with the LAA concept, an innovator with a strong mechanical engineering background might have a clear understanding of the challenges associated with developing a percutaneous delivery system, but may not be able to predict with great accuracy the challenges involved in finding a material to deliver to the LAA. The best way to address gaps in one's knowledge base and avoid the effect of personal biases is to involve individuals from all of the engineering disciplines that play a part in developing the total solution. These individuals can help the innovator understand the degree of engineering complexity and uncertainty associated with each challenge and support the prioritization process. However, even with input from a comprehensive and capable team, unanticipated technical challenges will almost always arise in any R&D effort. By considering known and anticipated challenges at the outset, having plans in place to address them, and staying on the look-out for unexpected problems, the innovator will be in the best possible position to overcome whatever challenges arise.

As innovators and companies start developing an R&D strategy, they must carefully consider how to individualize the milestones and testing strategies they choose for their specific project. The importance of thinking critically about these assumptions and testing the most important ones is illustrated by the ArthroCare story.

FROM THE FIELD: ARTHROCARE CORPORATION

Tailoring an R&D strategy

Hira Thapliyal, one of the founders of ArthroCare Corporation, is an electrical engineer by training. He started at Corning Inc. and subsequently worked at various cardiovascular start-ups, including Devices for Vascular Interventions (DVI) and Cardiovascular Imaging Systems (CVIS). During his time at DVI, a company focused on the field of cutting plaque out of arteries, he noticed that many of the arteries they acquired from the morgue for performing device tests were severely occluded. With this observation in mind, he teamed up with one of his former colleagues, Phil Eggers, to work on the need for *a better way to treat chronic total occlusions of the arteries*. One of Thapliyal and Eggers' ideas was to use an electrical current, directed to electrodes in contact with an arterial plaque, to "melt" away these occlusions. Not knowing where this idea would ultimately take them, they broadly filed several patent applications to use this technology in various disease states. Early work with arterial samples seemed promising, so Thapliyal decided to seek funding. However, the early reaction of investors was lukewarm. He recalled, "I talked to some VC friends and they were very polite, but they didn't see a big market in this. This was 1992, before the field of coronary interventions had exploded with the introduction of stents."

Thapliyal's inability to raise funding called into question his assumption that a novel solution for treating arterial occlusions would be well received. To help him understand if there were other needs that would benefit from his solution, Thapliyal and Eggers began seeking advice from experienced colleagues in the medtech field. Among others, they talked with Bob Garvey, "a marketing guy who had spent part of his career in arthroscopy." When Garvey heard about their experience with treating occlusions, he suggested to Thapliyal and Eggers that they should look at various orthopedic applications since there were still important unmet needs related to tissue removal. At the time, the tools used to perform arthroscopy, a minimally invasive orthopedic procedure involving a small endoscope and tools inserted via small incisions in a joint to examine and treat various joint conditions, were still relatively crude. Taking this advice, they spent some time understanding the need for *a better way to remove joint tissue* and tried their device in a chicken meniscus, which was similar to "the gasket between the bones of a joint." "It worked very well," he said. "It eliminated the tissues, which just melted away."

Based on the results of these early tests using a prototype of the device, Thapliyal, Eggers, Garvey, and another colleague from Thapliyal's past, Tony Manlove, were able to raise some funding to develop the arthroscopic application. As part of this effort, they realized that they would have to lay out an R&D strategy that would make sense to investors and give the project the best chance to achieve regulatory clearance in the shortest possible time frame. Thapliyal and his colleagues made a series of new assumptions regarding the sequence of events that needed to occur. "The recipe is the same for all medical devices. Bench-top, some live animals, cadaver types, and then maybe on to humans. But it has to be looked at from the context of the regulatory framework. If you have a PMA device, you structure it differently than a 510(k). If it's a 510(k), then you need to understand more about the 510(k) requirements. Some are straightforward substantial equivalence. Many require live animal data. But you don't want to be second-guessed there. By the time you have filed your regulatory submission, you have put in a lot of resources, time, money, and sweat. You don't want to get caught without the right data. Review all your early assumptions."

The team sought advice from various sources and used that information to decide on the best course of action. They first determined that they would likely be able to file a 510(k) based on substantial equivalence to the current electrosurgical cautery tool used in arthroscopy. Their regulatory advisor told them to "keep it a ho-hum 510(k). Don't make a big deal of it. Electrosurgery tools are already in the market. You want to be in the market. So just show, with some live animal data, that your device is no worse than what's already available." With this regulatory approach in mind, they next sought to understand which R&D milestones would be most important and how to prioritize them based on the technical challenges they anticipated.

They had already achieved an early technical feasibility milestone through their work in chicken menisci. For their next milestone, they had two options. They could either seek to show efficacy in live tissue through live animal testing or perform cadaver studies to show feasibility with human anatomy. For many devices, especially ones using combinations of electrical and mechanical components, animal studies are done earlier in the R&D path in order to optimize the electrical engineering functionality in live tissue. Once this is achieved, and the risk associated with using the device in live tissue is eliminated, the technology can then be refined for use in human anatomy. However, in Thapliyal's case, he and his team felt confident about the electrical components of their solution and felt strongly that effective arthroscopy was anatomy dependent. Therefore, they decided to focus on the mechanical components of their device and optimize them for human anatomy using cadavers before performing animal tests.

To accomplish this, they teamed up with a physician they knew from San Diego, Dr. James Tasto, who helped them to work on cadaver samples. "We linked up with him and he said, 'Sure, you can come to where I'm working in the evenings, after hours, and test some menisci with your tool.'" Several months of testing and design iterations on cadavers allowed them to validate the feasibility of using the technology in human anatomy and to optimize the design. "We used the cadaver data to make sure that what we did was correctly sized and designed for human use," Thapliyal said. For example, it was essential to understand the correct angles to build into their probe to make it compatible with human joints.

The next challenge was to find an appropriate animal model. The model needed to provide similarities to human anatomy and needed to be a widely available resource. Thapliyal recalled, "We settled on a goat model. Goat models were well published by that time and they had important similarities to humans. The size of the joint was a little smaller, but the anatomy and presentation of the meniscus was similar. Plus, goats were available and we could do a lot of them."

Over the course of a few months, the team performed experiments on about a dozen goats, with the objective of generating in vivo data to support regulatory clearance. "We didn't have to do a chronic study; we just had to acutely show that we can remove tissue in a way that was equivalent to what was currently on the market. Our tests were all statistically designed to show non-inferiority. We were not targeting that our tissue damage would be lower, but we were amazed that we saw virtually no tissue damage."

During most of the goat studies, the team used contract manufacturers to develop and produce its prototypes. They had decided to outsource this activity early in the development of their R&D strategy. According to Thapliyal, "Seek out known entities with expertise in a specific area. In the early stages, you can get a lot of work done very quickly when you outsource. You just need to have a smart group of people inside the organization who can communicate and convey to your vendors what you need." However, as part of its plan, the team knew it would eventually have to bring development in-house. By the time of their FDA filing, Thapliyal said, "We had a director of R&D, a couple of engineers, and some quality assurance people. We had started bringing disposable device development in-house because we knew we had to have control of it since this would allow us to iterate more quickly."

Based on the results of the goat tests, regulatory clearance was granted for the company's arthroscopy device. Once in the market, the technology caught on relatively quickly because it operated at a lower temperature and was more precise than traditional surgical tools so that damage to healthy tissue surrounding the target area could be minimized.

In terms of its ongoing R&D strategy, the team committed 80 percent of its time, resources, and engineering staff to development opportunities in arthroscopy and 20 percent on exploring new areas. Over time, ArthroCare expanded into the fields of spine and ENT. According to Thapliyal, "What we found is that we have a platform. It's not just arthroscopy. We believed it would be useful in other areas and we wanted to explore the limits of our technology." However, he offered a word of caution to innovators as they manage their R&D strategies: "Guard against projects that can take away from the focus of your company. If they become too strong, do something about it to protect the core business."

Reflecting on his experiences, Thapliyal suggested that innovators should "take a lot of advice." Moreover, he said, "Don't be seduced by the elegance of your assumptions. You have to go back and review them and test them from time to time." This philosophy allowed Thapliyal to effectively recast his focus when he ran into difficulties, and also to keep a complex technical development effort on track through the achievement of important company milestones.

Stage 5 Development Strategy and Planning

Developing an initial R&D plan

With specific R&D milestones and related technical challenges identified and prioritized, the innovator can next develop an initial R&D plan. As part of an R&D strategy, this initial plan should be kept at a relatively high level and should consider: (1) R&D personnel, (2) engineering resources, such as equipment and facilities, (3) testing methods, such as bench and animal work and target outcomes from these testing methods, and (4) overall timelines. More specific details will be reserved for inclusion in the company's integrated operating and staffing plan (see Chapter 6.1).

High-level discussions about these issues are necessary as part of the R&D strategy process to allow the innovator to make more accurate forecasts of when R&D milestones can realistically be met. As noted, once the timing of key milestones is determined, the innovator can strategically position R&D as a contributor to other important development milestones, such as funding and regulatory clearance.

R&D personnel

Carefully considering the types and skill sets of necessary R&D personnel is critical because the number of engineers on the team has a direct impact on when R&D milestones can be achieved. Furthermore, R&D staffing costs can be the highest R&D budget expense and, thus, can affect financing events.

To determine what types of engineering and technical resources a company will need, an innovator should assess each technical challenge that must be addressed to achieve a milestone. Each challenge should be considered individually, in terms of the relevant engineering skills required to address it. Then, the challenges should be evaluated as a whole, paying particular attention to the interactions between key aspects of the solution and any unique engineering skills needed to address them.

Assessing the difficulties involved in a particular technical issue can be a challenge in itself, particularly for innovators working outside their area of expertise. However, developing reasonable estimates is important since it can affect the number (and skill levels) of engineers required to get an activity completed in a reasonable amount of time. The best approach to overcoming this hurdle is to query others with varied backgrounds to get a rough idea of the time and effort needed to address a challenge, assuming that knowledgeable resources are found. As a rule of thumb, the length of time spent on engineering depends heavily on the complexity of the innovation being developed. Even with a clear plan, achieving certain milestones will simply take longer for a complicated solution.

During prototyping, the innovator might have been able to develop working models that demonstrated a concept's basic feasibility with little more than hard work, book research, and some guidance and/or coaching from those with relevant expertise. However, the transition from early-stage prototyping to R&D usually necessitates much deeper knowledge and expertise in each of the relevant functional blocks or engineering disciplines. Unless the company intends to outsource the development of its innovation, it must hire a team of appropriately skilled engineers and map out how that team will grow over time as it works to achieve successive milestones.

To address the technical challenges shown in Table 5.2.2 for the LAA example, the innovator would recognize the need for both mechanical engineering and materials science expertise. Based on how these challenges were prioritized, s/he would also surmise that getting a materials science expert on board first would be advantageous to retire the risk of finding a material with the desired properties before thinking about the mechanical aspects of the solution. Later, a mechanical engineering resource could be brought in to address the development of a delivery system. A third resource, perhaps an expert in how the material and delivery system interact, could be added next, at the appropriate point in time. In a resource-constrained environment, it would not be practical to hire the latter two resources until enough is known about the core material used in the solution, since this will affect the design of the delivery mechanism and the substance from which it is made. Understanding these types of sequential interactions allows the innovator to manage R&D staffing more effectively within prevalent resource and time constraints.

Other projects may lend themselves more readily to parallel development efforts. For example, recall the ACS case example in 4.6 Final Concept Selection, in which parallel development efforts focused on a balloon catheter, a guidewire for catheter navigation, and a guiding catheter are described. In this scenario, the company made a strategic decision to run multiple engineering work streams rather than working in sequence. In some types of project, and where time and money allow, this alternative is available to the innovator.

The point is that, especially in a start-up company with scarce resources, it may be possible for a company to stagger its approach to hiring based on the R&D strategy and product development pathway it defines. However, this decision should still be predicated on ensuring that the greatest technical challenges and those with the longest development time are addressed early in the process so that adjustments in the development pathway can be made as soon as possible.

Another reason to think about staffing as part of the R&D strategy process is that hiring itself poses an element of risk. For example, if a highly skilled electrical engineer with many years of experience in a given field will be needed in three months to help solve a key technical challenge, actually finding and hiring the right resource may be unrealistic in that time frame, especially if the demand for such engineers is high and availability is low. Understanding this may help an innovator sequence key technical challenges more realistically. This, in turn, may affect the timing of certain strategic milestones, causing further impact on other factors, such as company funding. Furthermore, while more people on the R&D team does not necessarily lead to faster resolution of critical issues, planning for adequate staff eliminates at least one constraint that can interfere with a company's R&D progress.

Importantly, when hiring R&D personnel, companies face a trade-off. They can spend more money to attract and hire engineers with deep, often specialized experience. Or, they can pay less to add bright but relatively inexperienced staff members to the team. Inexperienced engineers might be easier to find and hire, but they typically introduce more risk from a development standpoint (in terms of needing more time to solve a technical challenge). However, while experienced engineers may represent less risk from the standpoint of knowledge, they are more expensive and harder to find. That said, it is usually necessary to have at least one experienced manager with product development expertise to oversee the R&D process and ensure that product development is launched in the right direction (e.g., in the role of vice-president or director of R&D). Beyond that, a company should consider the type of work it must accomplish in order to achieve its goals, as well as the likelihood of hiring experienced engineers in a reasonable time frame. With complex technologies in treatment areas that are relatively unexplored, it usually makes sense to hire experienced engineers, even if that means the company can bring on fewer people and needs more time to hire. In contrast, when working on incremental technologies in areas where speed to market may be an important driver, a somewhat greater number of less experienced individuals can more effectively swarm the development challenge. As a general rule, however, start-up companies almost always benefit from hiring the best people they can – particularly in assembling a core of experienced engineers to ensure that product development is launched in the right direction.

Another approach that can help a company carefully manage its investment in R&D personnel is to use consultants to assist with certain tasks. Hiring engineers on a contract basis can aid the company in several ways. First, it can be an effective mechanism for gaining access to expensive, specialized expertise required to address a specific engineering challenge, but not required on an ongoing basis. For example, in the LAA scenario described above, an expert materials consultant could help guide the materials scientist(s) working at the company to find an appropriate material, but then may not be required once this effort has been completed. In some cases, seeking such expertise on a contract basis is more feasible than trying to hire a full-time employee because many engineers with critical, yet highly specific skills often gravitate to consulting simply due to the nature of their knowledgebase (i.e., their value-added skills are not required for long-term projects). Second, hiring consultants can allow the company to respond to temporary or short-term peaks in the engineering workload without having to hire (and then lay off) dedicated resources. For instance, as the LAA delivery system moves toward regulatory submission, there may be multiple, short-term tests that can be accomplished simultaneously. Hiring contract labor may be beneficial to execute the tests without distracting in-house engineers from their other priorities. Finally, consultants can help bring a fresh perspective to the resolution of challenging problems.

When using consultants, however, be aware that they may work differently from in-house staff and require different incentives. Occasionally, consultants may not exercise the same sense of urgency as dedicated, full-time employees in performing their tasks. They also may not be knowledgeable about all the intricacies of the development process that would allow them to move at a fast pace. Moreover, issues related to confidentiality and IP must be adequately addressed before any contract work is initiated. Finally, consider the fact that when consultants finish their assignments, much of the working knowledge that has been accumulated through their involvement leaves the company with them. At a minimum, the company should orchestrate a proactive knowledge transfer and ensure the completion of relevant documentation prior to their departure.

Engineering resources

In addition to R&D personnel, a company will be required to make an investment in facilities, equipment, and other resources to support R&D (see Figure 5.2.2). While there is less risk associated with the acquisition and utilization of these resources from a timeline perspective (i.e., most of these other resources are readily available), some thought must be given to this from a strategic development standpoint so that the resources obtained match the availability of R&D personnel and both are fully and efficiently utilized. This is particularly important given that decisions related to engineering resources can have significant financial implications.

In the LAA example, if the company decides to develop a delivery system via an in-house engineering effort, it will need to purchase enough equipment to support this activity. However, if finding a mechanical engineer with skills in developing percutaneous systems takes too much time, the development work could instead be contracted to an outside firm. Such a

FIGURE 5.2.2
Engineering work space and equipment can be costly, requiring a significant investment by the company (courtesy of IDEO).

move would obviate the need to purchase the equipment required to develop this portion of its device. In another scenario, the company might be able to hire an engineer, but decide to not purchase any equipment and have the engineer work with an outside vendor that possesses the machines and equipment to develop the product. All of these choices have an impact on the timeline to develop the delivery system. They also directly affect issues related to resources, which must be taken into account when developing the overall operating plan. Finally, the ultimate decision can affect the level of funding required (as different investments are needed to equip and maintain a machine shop versus contracting with an outside firm).

When the need for specialized equipment, parts, materials, or processes is anticipated in the development of a product, lead time issues often have to be considered. If a company requires hard-to-get items, this may affect the hiring of personnel, the sequence of addressing challenges, and the time required for milestones to be achieved. For example, creating a certain electronic circuit may require many circuit board revisions to reach a final board design. Outsourcing the fabrication of each of its prototype board designs with each revision may necessitate a substantial lead time of weeks to months for each of the boards to be fabricated. By understanding and anticipating such needs, a company can determine whether a different development pathway is more efficient (e.g., outsourcing design as well as fabrication to the same board house). Such a decision could, again, have significant strategic ramifications.

Other resource issues to take into account include laboratory space, as well as the specific lab equipment that will be required. These questions are largely driven by the type of engineering work being performed, along with what functions might be outsourced by the company. For an example of the basic equipment required to set up a mechanical engineering R&D laboratory, see Appendix 5.2.1.

Research and development testing methods

Once an innovator understands the scope of the engineering work and the personnel and resources required to accomplish it, s/he must consider the tests to perform on the resulting prototypes for a medtech project and the purpose of those tests. At a high level, testing is first required to iteratively assess and learn about various designs in order to eventually create a solution that performs as desired and is safe. Second, testing is required as part of quality system processes to validate and verify that the final solution actually meets key user requirements and technical specifications (see Chapter 5.5). This is important since each type of testing and the number of expected rounds of testing can have an impact on the product design and timeline to achieve a milestone. Furthermore, the testing and development methods that a company adopts directly influence its R&D culture and, therefore, have an ongoing effect on the nature of its R&D strategy.

The scope of testing used in the R&D process is illustrated in the biodesign testing continuum (first introduced in 4.5 Prototyping). As shown in Figure 5.2.3, the prototyping, R&D, and clinical work streams overlap and support one another as the development and testing of a device progress. Many of the stages of the testing continuum are intimately related to key strategic R&D milestones, such as achieving use in an animal or human. As such, understanding how successes and failures at each testing stage can affect progress to the next stage directly impacts how an innovator thinks about key R&D milestones. Early prototyping and bench testing can be thought of as the concept phase of the R&D plan (see 4.5 Prototyping for

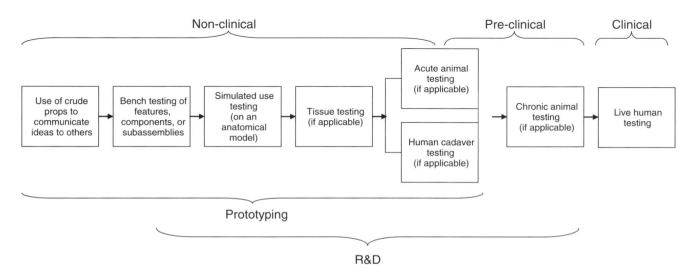

FIGURE 5.2.3
The biodesign testing continuum spans non-clinical, pre-clinical, and clinical testing, with R&D playing a key role from bench testing through animal studies.

Stage 5 Development Strategy and Planning

additional information about these activities). Animal, cadaver, and human tests have dual R&D and clinical objectives and, in this way, are directly linked (see 5.3 Clinical Strategy). A few important notes on the R&D-related aspects of these types of tests are included below.

Testing activities begin once a prototype is available to determine if a technical challenge has been met. Each sequential stage of testing in the R&D process typically represents increasing complexity and time requirements. Bench testing and simulated use testing are often the easiest and quickest to accomplish, since they usually only require models of representative anatomy and no special equipment beyond what is required for the prototype to function. Given this, many iterative rounds of prototyping and subsequent testing to address certain technical challenges can be accomplished rather quickly, especially if R&D personnel with expertise in the engineering discipline required to develop the prototype are directly involved in the tests. With each round of R&D testing, an innovator also gains an appreciation of how to meet important design parameters with each successive prototype.

In the LAA example, an important early technical challenge may be the identification of material that can occlude a simulated LAA using a prototype in a bench-top test. To perform this test, an LAA-like pouch could be created simply by suspending a plastic bag in a chamber filled with moving and heated blood. By attempting to deliver the material into the plastic pouch, the innovator could then determine if it solidified to achieve the occlusion. If it did not solidify, a new plastic bag could be swapped with the old one and testing could be performed again, once a new prototype (e.g., using different material) that addresses a limitation of the failed prototype is created. The failure may have been due to the material, the delivery mechanism, or the artificial set-up.

If it did solidify, in addition to meeting the technical challenge, an innovator might gain an appreciation of the timeline to solidification. If it took many hours, for example, this might be problematic if an important user requirement is that the material solidifies in one hour or less. Hence, in the next round, the innovator could make modifications until this important design requirement is met. Once a prototype that works adequately is created, additional experiments would be needed to further optimize the solution, especially with respect to integrating it with other components of the solution. Bench and simulated use testing allows various technical challenges to be tested in parallel or with one variable altered each time. Needless to say, the benefit of being able to prototype quickly is tempered by the limitations of a simulated model. Nonetheless, the iterative process can continue until a solution is found in which all relevant components work reliably.

Taking the example one step further, if achieving effectiveness in a tissue model is an important subsequent milestone, a similar set-up could be created using a cow's heart instead of a plastic bag. While this might be somewhat more complicated, based on the availability and cost of hearts, this type of testing can be performed iteratively and relatively quickly. The basic point is that to achieve milestones that do not involve live animal testing, the testing methods do not have to be time consuming or expensive.

In contrast, live animal testing, which is typically another important R&D milestone, takes more time to prepare and plan for, is far more costly, and can often only be performed in approved facilities. When working with live animals, it is important to keep in mind that the need for multiple rounds of prototype iteration can have a significant impact on the R&D timeline because of ethical and other requirements associated with these tests (see 5.3 Clinical Strategy). If chronic studies are needed that involve many animals, important controls and specific follow-up protocols to investigate long-term device safety and effectiveness are required. As a result, the time and expense to conduct them will be even greater. Chronic live animal tests should be used only when necessary and should be performed as few times as possible to achieve the desired results.

To progress to milestones related to live human testing, a final prototype should have not only demonstrated its safety and effectiveness in controlled animal studies (if animal testing is required), but likely will also have required transfer of a design to manufacturing. This is because materials that are safe for human use are likely to be the ones used in manufacturing the final device. Hence, this factor should be taken into account when thinking about moving from simulated use and/or animal testing to human studies.

Research and development timelines

As described, estimating R&D timelines is an important aspect of developing an R&D strategy. The R&D plan created as part of the R&D strategy process requires a high-level view of how long the achievement of each milestone will take. This is critical because time is a significant driver of both value and risk given its obvious connection to the consumption of capital and resources.

The key inputs to determining a realistic timeline are the type of R&D personnel required, the number of engineering resources to be employed, and the testing required to prove the effectiveness and/or safety of a design. Proper documentation of user requirements, technical designs, experimental reports, and verification/validation testing (which are all aspects of the quality system – see Chapter 5.5), will also affect R&D timelines. Understanding the trade-offs between different scenarios can help an innovator articulate a realistic timeline that still takes into account the effort required to address key technical challenges and retire risks. This information also allows the innovator to answer the important strategic question: "How long will R&D take?"

While R&D timelines vary dramatically across projects, some engineers believe that it generally takes less time to achieve key R&D milestones for devices based on mechanical engineering concepts and more time for devices that depend on materials science and/or electrical engineering fundamentals. For instance, it might take an innovator as little as a year to develop a mechanical device that is ready for controlled animal testing. On the other end of the scale, five or more years may be needed to reach that milestone for a high-complexity or combination product, such as a drug eluting stent. These variations are primarily driven by the greater number of unknowns that are introduced when chemistry, biotechnology, and other scientific disciplines become part of the medical device R&D process (see Figure 5.2.4).

When creating an R&D strategy, innovators and companies must take an approach that integrates R&D into the context of the broader organization. The connection between R&D strategy and clinical strategy is particularly important throughout the development process, as the NeoTract case demonstrates.

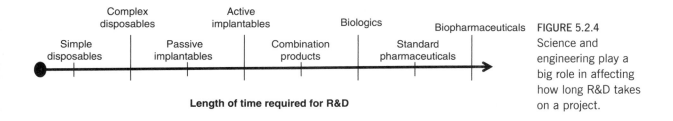

FIGURE 5.2.4 Science and engineering play a big role in affecting how long R&D takes on a project.

FROM THE FIELD

NEOTRACT, INC.

Moving from prototyping into development

According to Ted Lamson, CEO of NeoTract, Inc., a start-up based in Pleasanton, California, there are 4.2 million men in the United States on medication for benign prostate hyperplasia (BPH) and its accompanying urological symptoms. However, each year, 1.3 million of these patients stop taking their medication, with only about 250,000 of them seeking surgical intervention to address the disease. The rest, though still suffering, remain untreated. "We sought to find out through a needs assessment why there is such a huge discrepancy," said

FROM THE FIELD: NEOTRACT, INC.

Continued

Lamson, describing NeoTract's focus when the company was founded in 2005.[3] "We spent our time trying to tease out what it is about the current interventional option that is scary or doesn't fit the risk/benefit profile that these patients are looking for." Over time, the team conceptualized a less invasive, office-based procedure that could be performed relatively quickly and easily, and required only a local anesthetic. To realize this solution would require the development of an implant and the instruments to deliver it.

However, when NeoTract began developing an approach to R&D, it set aside some of its more elaborate designs and set a first strategic milestone focused solely on proving the fundamental technical concept in the clinic. "To develop the complete, more elegant solution would have taken quite a bit of time and money," recalled Lamson. "But we recognized before making this investment that we really hadn't answered the one question that would make it all worthwhile, which is will patients respond well to our approach? So rather than spending time building the perfect device that could potentially do the wrong thing, we asked ourselves what's the quickest way we can answer this question. It turned out that this was to figure out a more invasive, bare-bones way to see if the concept would achieve the result we were after and then bring that forward into an operating room setting. We were going for a clinical proof of concept that would deliver results that were better than the surgery that was currently available, but were not necessarily all the way to our end goal."

The NeoTract team developed a somewhat crude set of surgical instruments to deliver the implant and a procedure performed under general anesthesia that more closely resembled the current surgical treatment. "Of course," said Lamson, "It was still a lot of work. We had to build everything within the guidelines of a quality system. But the intense design work and engineering to build the more elegant solution could be put on hold." The company also had to act ethically, developing a reversible procedure so that the patients receiving it could still undergo conventional therapy if there were any adverse effects. "We did ten patients with that device – only then did we confirm that they responded really positively to our idea," Lamson explained. "At that point, we knew we

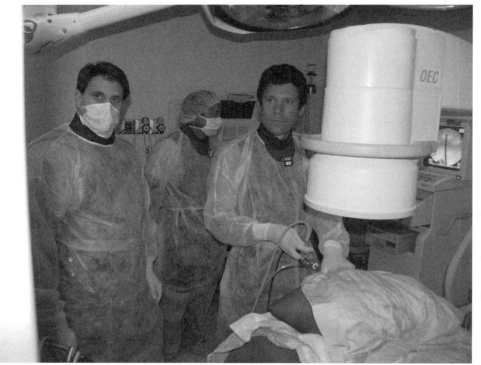

FIGURE 5.2.5
Lamson testing a device (courtesy of NeoTract).

had the right approach, and we were ready to start the program to develop the real device." See Figure 5.2.5.

"We probably spent about six months on this first milestone, which ultimately delayed the final product," he continued. "But this early clinical experience gave us the confidence that this was worth doing. More times than not, I've seen projects develop a very elegant and well-engineered device and procedure only to find that there's a critical flaw once they get into the clinic. When that happens late in the engineering process, it's harder to abandon the project. Or, even worse, you've got so much momentum that you just keep trying to tweak the solution when really there's more of a critical flaw in it and you should really just change course. If I had one piece of advice that served us well, it is that a team needs to identify the single biggest risk in the program and emphasize taking that on first."

Another benefit of the rapid R&D work and early clinical experience was linked to the fact that the device was focused on improving quality of life for sufferers of BPH. "Your ability to eliminate pain and discomfort is not something that you can get from an animal experiment," he said. "These cases allowed us to figure out if what we want to do is feasible and helps patients, but also if it's going to be something patients want or if they're going to say, 'This doesn't feel good. I don't want it,'" Lamson noted.

Additionally, "By getting into the clinic early, we learned a lot about aspects of the therapy that we might have designed incorrectly. For example, at one point, we had the surgeon performing the procedure holding a sterilized force gauge to measure forces on the instruments. By doing that, we found out what force works and could make that a key requirement," he said. The team spent considerable time analyzing the data it had collected in the clinic and then translating the information into important design requirements and technical specifications to guide the R&D effort.

NeoTract's next strategic R&D milestone was to develop a device and a procedure that more completely addressed all of the team's goals and then get those into the clinic. To accomplish this, Lamson focused on identifying the key technical challenges that needed to be overcome along the way and then set up checkpoints or sub-milestones to make sure the project was tracking to them. "When you have a broad project where people are doing different subassemblies and it's supposed to all come together at the end, you realize that the end is often too far away to make sure everything works. A lot of times we break up the development work into functional elements. But before we get too far down the road on developing a certain configuration, we try to identify the key things we can test in some simple mode to know that we have the right basic design. This approach can keep you from going too far down the wrong path." When an engineer reached a key checkpoint, the entire R&D team would get together to review the work, ask questions, and help identify any fatal flaws. "Often," said Lamson, "the people who aren't working on that element have most valuable insights about what may present a problem."

Another approach that Lamson advocated was for his team to become "amateur surgeons." "To the extent that you can actually do your procedure and operate your device, it helps you understand what a clinician worries about and doesn't worry about." Lamson and team performed approximately 80 to 90 percent of their own pre-clinical procedures, enabling them to capture key learnings. That said, "I always bring in at least one clinician early on to do early experiments, teach us about his or her concerns, and help mark out the procedure," noted Lamson. Then, experts are involved periodically throughout the development process to "make sure we're not being biased in what we're seeing and to help give us a fresh look at our work," he added.

In terms of planning for R&D costs, Lamson advised innovators to allocate at least $1–2 million to get from concept into humans. "You can do it for $1 million, plus mistakes. There are always mistakes," he said. "That's where NeoTract came in but, of course, there are many projects that require much more because the pre-clinical work is so much more extensive." One of the primary drivers of costs is people. "Try to be honest about how many people you need. There is such a thing as critical mass," Lamson said. "You can't just say 'It's you and me and we'll get there when we get there.' You can take too long on these things." To go from concept to the first human studies took NeoTract six people working nine months, whereas to develop the minimally invasive device from concept to human studies took NeoTract approximately 16 people working over the course of 1.5 years.

Lamson also advised start-ups to be aware that "prototypes are always more expensive than you think." He issued a similar warning regarding the cost of pre-clinical studies. "It's actually quite expensive to do animal studies. Animals can run anywhere from $10,000 to $15,000 each, just to get one data point. If you need to do five or six in two or three different experiments, it

Stage 5 Development Strategy and Planning

FROM THE FIELD: NEOTRACT, INC.

Continued

starts to add up. And you also have to figure that you're going to have some re-dos in there," he said. Finally, Lamson noted that the consultants can also be costly. "My goal is to never have to lay off anyone for any reason. In order to do that, I try to hire people that can do multiple things. I also use quite a few consultants because there are aspects of development that are project based and then they go away. Counting on $100,000 or $200,000 per year toward consulting is not outrageous," he said. For example, NeoTract used consultants to help with its regulatory filings, some market and patient analysis, and some aspects of its prototyping and design work.

When asked about common pitfalls in the R&D process, Lamson reiterated the importance of "being disciplined in assessing your key unknowns and in being sure you are addressing them." Get into the clinic as soon as possible, he said, and "be clever about answering your key clinical issues." He also underscored the need to "build a bed of testing and quality that applies to everything as it moves forward." He continued, "It's always a delicate balance to assess how well your previous testing applies to the next generation device – what can you take from it, and what you have to re-do. It requires a fairly constant and continuous conversation because enough improvement is going on that we're always weighing whether it creates the need for additional testing, or different testing, or that sort of thing."

Finally, Lamson offered, "I truly believe one of the key strategies to pursue when doing a start-up is to constantly seek expert advice. Ask yourself what you critically need to know and who you can ask. People will generally share information and you usually don't have to disclose much to get your answer. You can say, 'I'm building something that looks like this, this is what I think the team looks like, and here's our proposed development approach.' If you have the rapport with the person, they'll generally challenge you on a few things and give you important information to think about," which can be used to refine and improve the R&D strategy.

GETTING STARTED

While the following steps can be used to help an innovator or company develop an R&D strategy and begin thinking about issues related to the development of an R&D plan, remember that all R&D activities must remain closely linked to prototyping and clinical efforts and should not be developed in isolation.

See **ebiodesign.org** for active web links to the resources listed below, additional references, content updates, video FAQs, and other relevant information.

1. Determine strategic R&D milestones

1.1 What to cover – Identify critical R&D milestones necessary to achieve the company's R&D goals, as well as its broader goals for the development of a commercial-ready product. These milestones should correspond to outcomes that allow the company to retire important risks. Such milestones might include proof-of-concept, first prototype that performs effectively in a bench model, first prototype that performs effectively in tissue testing, first prototype that performs safely and effectively in live animals, first prototype that performs safely and effectively in humans, pre-production device that demonstrates manufacturing feasibility, and/or production device that supports scalable manufacturing. Consider different sets of critical milestones, alternative ways of sequencing them, and the strategic implications of changing the sequence and order of the milestones.

1.2 Where to look – Refer to 6.1 Operating Plan and Financial Model for more information about setting milestones. Network with other experienced innovators and professional advisors for assistance in defining milestones and estimating a reasonable time frame for achieving them.

2. Identify and prioritize key technical challenges

2.1 What to cover – Define the significant technical, clinical, and manufacturing design challenges that must be solved to achieve each milestone. Once identified, these technical challenges should be prioritized based on the level of uncertainty and risk they present to the project. Seek to resolve those challenges with the greatest risk as early in the engineering process as possible, working within whatever constraints may exist within the company.

2.2 Where to look
- Refer to 4.5 Prototyping for ideas and information relevant to identifying engineering and potential manufacturing challenges and/or consult references, such as Paul H. King and Richard C. Fries, *Design of Biomedical Devices and Systems* (CRC, 2008).
- Refer to 5.3 Clinical Strategy for help identifying clinical and testing challenges.

3. Develop an initial R&D plan

3.1 What to cover – At a high level, assess the R&D and testing activities, staff, other resources, and timeline needed to support the company's strategic R&D milestones. Understand the strategic implications associated with these tactical decisions. This information should be used to guide the R&D strategy and iterate/refine the R&D milestones. They should be outlined in increasing detail as the R&D process progresses to support the creation of the detailed R&D plan, as well as the development of an integrated operating plan and financial model.

3.2 Where to look – In addition to referring to 6.1 Operating Plan and Financial Model, it may be helpful to review other books on R&D, especially when considering requirements related to engineering personnel and resources:
- Steven C. Wheelwright and Kim B. Clark, *Managing New Product and Process Development* (Free Press) 1992 – see Chapters 6, 8, and 9.
- Karl Ulrich and Steven E. Eppinger, *Product Design and Development* (McGraw-Hill/Irwin, 2003), Chapter 13.
- Theodore R. Kucklick, *The Medical Device Handbook* (Taylor & Francis Group: Boca Raton, 2006).
- Paul H. King and Richard C. Fries, *Design of Biomedical Devices and Systems* (CRC, 2008).
- Gerard Voland, *Engineering by Design* (Pearson Prentice Hall: Upper Saddle River, 2004).

Credits

The editors would like to acknowledge Jared Goor for his help in developing this chapter. Many thanks also go to Scott Delp and Ron Jabba for editing the material, as well as Hira Thapliyal and Ted Lamson for contributing the case examples.

Appendix 5.2.1

Building an R&D laboratory

Required laboratory space, as well as specific lab equipment, is dependent on the type of engineering work being performed. For example, a company with a mechanical engineering-based device that it plans to develop in-house might need a basic machine shop that includes the following equipment:

- **Mill** – A metalworking machine that operates by moving a cutting bit.
- **Lathe** – A machine tool that spins a block of material to perform various operations such as cutting, sanding, knurling, drilling, or deformation. Effective for creating objects that require symmetry around an axis of rotation.
- **Drill press** – A standing drill designed for making holes with a high level of repeatable accuracy and control.
- **Grinder** – A versatile machine used to perform a variety of grinding operations or for the grinding of complex shapes.
- **Ultraviolet (UV) curing box** – Device used in the curing of UV reactive adhesives and compounds.
- **Hand tools** – Common tools without a motor (e.g., hammer, saw, wrench, etc.).
- **Assorted supplies** – A wide variety of hardware and other supplies, including nuts, bolts, springs, tubing, etc.

More specialized equipment and test fixtures may also be required based on the specific type of device being developed. For instance, if the engineering team intends to use computer-aided design (CAD) programs (e.g., SolidWorks, SolidEdge), appropriate computer equipment will also be needed.

Laboratory space must be calculated for all required equipment, as well as for shared (e.g., bench-top surfaces) and personal workspace for the engineers hired. Another factor to take into account is the cost of raw materials. Some materials may be acquired relatively inexpensively (such as wood and plastic), although materials for more precise and/or finished products can be costly.

Materials science engineers often require significantly different equipment from that of mechanical engineers. For example, a wet lab for extensive tissue testing will likely be needed (e.g., sinks, burners, precise scales and measuring equipment, ovens, fume hoods and exhaust, refrigeration), as well as space for analyzing, mixing, and testing chemical reagents, polymers, and other compounds. Raw materials and tissue samples can be moderately expensive, especially for certain polymers.

For electrical engineering-based projects, more advanced computer equipment and different software programs (MatLab, LabView) are needed, but fewer machine tools must be purchased. Additionally, no raw materials are needed. In general, for electrical engineering, the total cost of a lab set-up depends on how this functional block (see 4.5 Prototyping) ties into the main engineering effort, as this is usually an important adjunctive piece for many medtech devices.

Notes

1. Research is a planned activity aimed at the discovery of new knowledge with the intent of developing new or improved products and services. Development is the translation of research findings into a plan or design for new or improved products and services. See "Research and Development," StartHereGoPlaces.com, http://www.startheregoplaces.com/glossary/#R (November 18, 2008).
2. From remarks made by Guy Lebeau as part of the "From the Innovator's Workbench" speaker series hosted by Stanford's Program in Biodesign, April 5, 2005, http://biodesign.stanford.edu/bdn/networking/pastinnovators.jsp. Reprinted with permission.
3. All quotations are from interviews conducted by the authors, unless otherwise cited. Reprinted with permission.

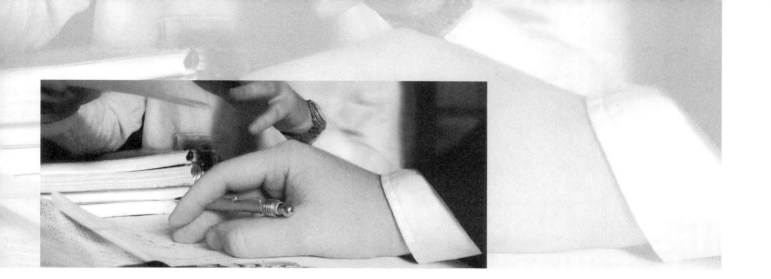

5.3 Clinical Strategy

Introduction

Human device testing is at the pinnacle of the medtech innovation process. The stakes are high. Two to three years to perform patient recruitment, enrollment, investigation, and follow-up. Costs as great as $100,000 per patient studied. Regulatory approval and reimbursement hanging in the balance. The credibility and commercial viability of the company on the line. And, most importantly, patient lives in the innovator's hands. These are the types of issues that make human device testing a risky and challenging (yet essential) element of the biodesign innovation process. For these and many other reasons, a clinical strategy is essential not only for planning human testing, but for the entire clinical development pathway.

Clinical trials are broadly defined as human studies performed to determine specific outcomes based on the use of a new medical treatment. Traditionally, the objective of clinical trials has been to demonstrate that a new device offers measurable, clinically important benefits to patients in terms of its effectiveness and safety.[1] Such evidence can be essential to supporting the regulatory approval of the device. However, clinical studies are increasingly being performed for other reasons, including reimbursement and market adoption. Because it can be difficult to design a single study that is capable of meeting all desired regulatory, reimbursement, and marketing endpoints, innovators should think about study design in the context of a larger clinical strategy that includes a progressive series of trials to achieve the company's overarching goals. A clinical strategy, thus, helps define the overall pathway for the clinical development of an innovation, and also may help a company define and prioritize each study's endpoints.

OBJECTIVES

- Recognize the primary goals of a clinical strategy.
- Understand the importance of establishing an overarching clinical strategy for early non-clinical, pre-clinical, and human clinical studies.
- Appreciate the different types of clinical studies and how their designs relate to the overall goals of clinical testing (including regulatory, reimbursement, and marketing considerations).
- Learn the process involved in planning and initiating a human clinical trial, including the different participants and their functions.
- Understand tactical considerations for developing safe, cost-effective, and statistically robust clinical trials.

The creation of a medtech clinical strategy begins in parallel with research and development (R&D) since the clinical development pathway must be tightly integrated with device engineering (see 5.2 Research and Development Strategy). In addition to maintaining a tight linkage to R&D, an effective clinical strategy must be synchronized with the other strategic work streams in the development stage – see 5.4 Regulatory Strategy, 5.6 Reimbursement Strategy, 5.7 Marketing and Stakeholder Strategy, and 5.8 Sales and Distribution Strategy.

Clinical strategy fundamentals

Every innovator strives to develop a medical innovation that ultimately addresses the defined clinical need. Clinical studies, performed as part of a comprehensive clinical strategy, provide the mechanism for ensuring that a new medical technology offers a safe, effective solution to the need.

As described in 5.2 Research and Development Strategy, research and engineering are focused on mitigating technical risk in device development to build value for the organization. While closely linked to R&D, clinical strategy has a different focus – it defines a prospective approach through which the organization can anticipate and manage the clinical risk associated with the project. A well-developed clinical strategy also helps ensure that clinical activities are tightly coupled with other important efforts underway within the company as part of the biodesign innovation process. Just as a company's clinical efforts must complement its R&D work, it must also be coordinated with regulatory, reimbursement, marketing, and sales to achieve important milestones, build value, and use scarce time and resources in the most efficient manner possible.

Innovators and companies often feel both excited and apprehensive as they prepare to execute clinical trials. In recent years, the requirements for clinical data to address regulatory, reimbursement, and adoption concerns have become increasingly stringent, and they were expected to become even more complex. According to Frank Litvack, former chairman and CEO of Conor Medsystems:[2]

> The complexity, size, and expense of clinical trials for important medical devices are going to do nothing but get bigger before they get smaller. And I think that has implications for everybody who's in the start-up business. It means it's going to take longer and you're going to need more money. And there's going to be more risk. That's just the world in which we live.

A clinical strategy can help innovators successfully tackle this growing challenge. However, a flexible strategy is required to allow for adjustments as new information is discovered and changes are revealed in the dynamic regulatory, reimbursement, and commercial environments.

Before undertaking a detailed overview of clinical strategy development and tactical trial planning, which follows in this chapter, it is important to emphasize the ethical responsibility that innovators assume when initiating clinical studies. Individual innovators and companies alike must commit themselves to the ethical treatment of all living animal and human **subjects** involved in their studies and use this commitment to guide the development and implementation of a clinical strategy. Care should also be taken to avoid conflicts of interest and other ethical dilemmas that have the potential to negatively affect actual (and perceived) study results.

Clinical study goals

The development of a clinical strategy begins with the definition of the objectives (desired outcomes) of the strategy and of the study(ies) that the company intends to undertake. The innovator should ask him/herself questions such as: "What results are needed to support regulatory approval?" "Are economic outcomes important to support reimbursement decisions?" "Will data be necessary to help market the device to physicians and/or patients?" Trial design and execution will be based on the answers to these types of regulatory, reimbursement, and marketing inquiries.

While it can be tempting to try designing a single trial that supports all of these objectives, it is important to remember that the complexity of a study increases its cost, makes management of the trial more difficult, and can make the outcomes less clear. As with any large project, the majority of time should be spent *up front* considering the most critical objective of the trial in order to design the most cost-efficient study that can meet the company's desired end results. Companies should also keep in mind that their objectives can be achieved progressively – it is not uncommon for a company to stage its studies based on its strategy. For example, trials with specific objectives (regulatory, reimbursement, or marketing) are synchronized with the milestones in a company's operating plan such that the data generated by the trials becomes available as they are most needed (see 6.1 Operating Plan and Financial model). Implantable cardiac defibrillators (ICDs) provide a good example of a staged approach to clinical testing. Initially, ICDs were studied and approved for use as a form of secondary prevention in patients who had experienced cardiac arrest due to ventricular fibrillation. However, after this approval, the MADIT trial demonstrated that patients with a history of coronary artery disease and heart pump dysfunction could benefit from ICDs as a form of primary prevention (i.e., even if they had not yet experienced a cardiac event caused by ventricular fibrillation).

Regulatory considerations

For the large majority of medical device clinical studies, regulatory considerations are the primary drivers of trial design. (The basic requirements for the two primary regulatory pathways in the United States – 510(k) or PMA – are described in 4.2 Regulatory Basics. More strategic regulatory considerations are addressed in 5.4 Regulatory Strategy.) Usually, a final trial design will result from a negotiation between the company and the US Food and Drug Administration (FDA) or the regulatory agency in another country. In certain areas, the interests of the company and the agency are aligned with regard to the details of the trial design; that is, both parties want to achieve valid data that will be publishable in high-quality medical journals, while providing maximal safety to the research subjects who participate in the trial. However, the company has the additional imperative to achieve an endpoint that is favorable to the product and deliver the trial results as quickly as possible and with the minimal possible expense.

Reimbursement considerations

Increasingly, over the past ten years, medical device trials are being designed to provide cost-effectiveness outcomes and economic evidence for attaining reimbursement (particularly from the Centers for Medicare and Medicaid Services (CMS) in the United States and technology assessment authorities outside the United States – see 5.6 Reimbursement Strategy). This may include the incorporation of a formal cost analysis into the trial, along with measures of the financial impact of the outcomes and stricter endpoints than those the FDA would normally require. On the upside, this can lead to more rigorous trials that result in more robust data and more formidable barriers to entry for competitors. However, it also makes trial design more complex since the desired outcome is not only to demonstrate the safety and efficacy of the device (as required by the FDA), but also to show equivalence or even superiority to an existing technology or procedure on the market as this latter outcome could be deemed necessary to justify reimbursement.

Marketing considerations

A third objective of many clinical trials is to generate data that help give the new technology an optimal launch in the marketplace. Such marketing considerations can have a major impact on trial design. It may be that the physician group targeted by the new technology will be most convinced by a certain type of trial – for example, a **randomized trial** that compares the new technology to a current standard of practice. Or, if there is a choice to test the technology in different patient groups, it may be advantageous from a marketing standpoint to target a certain population (e.g., younger patients who are willing to pay directly for the technology and, thus, provide an early revenue stream for the company). The choice of which investigators to include in the trial can also be important. Companies frequently pursue investigators from among "**marquee**" physicians and key opinion leaders (KOLs) – high-profile practitioners who give talks frequently and are influential among their colleagues (see 5.7 Marketing and Stakeholder Strategy).

Non-clinical, pre-clinical, and clinical phases of evaluation

Prototyping, device R&D, and pre-clinical studies often overlap, with the boundaries for where one begins and

Stage 5 Development Strategy and Planning

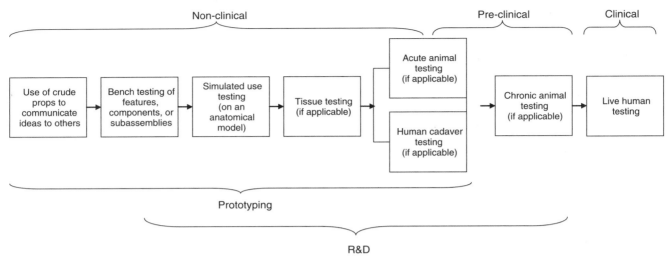

FIGURE 5.3.1
The biodesign testing continuum corresponds to the spectrum of non-clinical, pre-clinical, and clinical evaluation.

another ends varying significantly from project to project. To help make sense of how these efforts interrelate, one can think about the steps in the medical device testing continuum (introduced in 4.5 Prototyping) as generally corresponding to three different types of effort: non-clinical, pre-clinical, and clinical studies. Non-clinical studies begin with basic prototyping and continue into device R&D studies (see Figure 5.3.1). The focus of non-clinical trials is primarily device feasibility. Pre-clinical testing usually refers to **hypothesis**-driven animal studies that seek first and foremost to evaluate the safety and/or efficacy of a device in an animal model of the disease or condition. Clinical studies involve human device testing for the evaluation of specific outcomes, often specifically focusing on safety and effectiveness.

Non-clinical studies

The studies performed during prototyping, which typically include bench, simulated use, and tissue testing, are commonly referred to as *non-clinical* studies. When a company seeks to answer critical feasibility questions that extend beyond the prototyping of basic device functionality, it typically extends its ex vivo non-clinical efforts into human cadaver or live animal testing. Usually, non-clinical live animal tests are performed as acute animal studies. In vivo animal tests enable researchers to perform a preliminary evaluation of whether or not the device will function as desired in a living system. For example, an ablation catheter might successfully destroy tissue in simple tissue tests (i.e., using steak as a proxy for heart tissue and/or animal hearts from a butcher). However, until the device is placed in a living system, the inventor cannot be certain regarding fundamental device feasibility (e.g., Is the device able to reach the correct anatomic area? Does its interaction with the animal's blood affect the device's ability to effectively ablate tissue?).

An acute animal study is often an ideal method for evaluating device feasibility. These studies are performed in order to accomplish a specific, important question and then the animal is euthanized immediately following the test. Acute animal studies allow researchers to assess the device in a small, finite number of living systems prior to accepting the risks and expense of initiating a more extensive, systematic pre-clinical study. As with any animal study, researchers must be certain to apply for approval in advance of performing the tests, and should adhere to defined protocols for the ethical treatment of animals (see section on **Institutional Animal Care and Use Committees** for more information).

The first step in initiating any in vivo animal testing is to establish if an animal model of the human disease or disorder exists. If multiple animal models have the potential to provide a viable test, then consider secondary factors such as which animal models have been used by similar existing technologies in the past. For instance, there is a long history of using a swine model for evaluating coronary stents. However, not all models are this straightforward. For example, no animal model may be available for an obstetric device due to the unique characteristics of the human pelvis and the high hormone levels experienced during pregnancy.

Pre-clinical studies

Once researchers have successfully addressed critical issues related to device feasibility through non-clinical trials, they can transition into more systematic pre-clinical studies. In contrast to the acute animal studies, which are primarily focused on the viability of the technology, pre-clinical studies typically involve larger numbers of specimens and seek to prove or disprove a specific predefined hypothesis in small or large animal models (e.g., mice or pigs, respectively). The safety and effectiveness of the device also become important objectives since the overarching purpose of these experiments is to gather evidence to justify research in human subjects.

Given the emphasis on safety and effectiveness, animal survival is almost always included among the important endpoints in a chronic study. Careful attention to analgesia and post-operative comfort and mobility is essential. Methods for monitoring the animals must be established and procedures defined for ethical euthanasia for any animals that are suffering unduly, regardless of how this affects study outcomes.

One challenge associated with chronic animal studies is that animals involved in the tests are usually young and healthy. As a result, it may be difficult to get an accurate assessment of safety when compared to the risk of the device in a patient with a disease, several comorbidities, and/or advanced age.

Some pre-clinical studies look beyond safety and general measures of efficacy (e.g., does the device perform as intended?) by seeking to statistically establish the efficacy of a particular endpoint. **Efficacy endpoints** are selected and then placebo or "sham" procedures (designed to simulate the risk of the device procedure without providing active therapy) are used to analyze the effectiveness of a device treatment. These tests require careful attention to design since the data are only of value if the animal model reflects human disease characteristics and if the endpoint chosen is meaningful for a human study. For example, if the goal of a new device is to improve heart function through the transplantation of stem cells into an injured heart, efficacy could be evaluated by examining the change in pumping function following cell transplant. In this example, an injection of saline could be used to assess the effect of the injection alone, while the difference between the effect of the injection of saline on heart function and the injection of stem cells on heart function could be used to evaluate stem cell efficacy. Having a clearly defined clinical strategy in place prior to embarking on device testing helps ensure that researchers choose the most appropriate endpoint for each pre-clinical study. This can help them to maximize the value of the data when it is time to transition into human trials.

Results from chronic pre-clinical animal studies are usually helpful to the company in obtaining approval for first-in-man (**FIM**) testing. They can also help establish the potential clinical value of the device so that physicians are willing to enroll patients in the study. Additionally, they expand the researchers' base of knowledge so that the probability of causing harm to patients (or of not meeting desired endpoints) is reduced.

Institutional Animal Care and Use Committees

According to US federal law, any company or institution that uses laboratory animals for research or instructional purposes must establish an Institutional Animal Care and Use Committee (**IACUC**) prior to beginning research. The purpose of the IACUC is to oversee and evaluate all aspects of the company's animal care and use program. Such a committee is assembled by the institutions performing animal testing and often includes administrators, veterinary experts, and members of the public. All research and teaching activities involving live or dead vertebrate animals must be reviewed and approved by the IACUC *before a study is launched*. In most cases, approval is obtained by submitting a protocol, which establishes the reason for the study, justification for why an animal study is necessary to evaluate the medical problem, and the processes put in place to ensure the ethical treatment of the animals in the study.

Importantly, over the last decade, the intensity and quality of IACUC oversight has evolved substantially. Members of the committee are experts dedicated to both the safety and comfort of the animals, as well as the quality of the science resulting from the trials. The committees themselves are under careful, routine scrutiny by government agencies.

It is worth noting that many members of the public have grave ethical concerns about animal experimentation. Keep in mind that this may include members of the development team. Importantly, all other options for testing a device must be fully explored before any animal is caged, anesthetized, operated on, or potentially sacrificed. Furthermore, it is essential to have a discussion about these issues prior to conducting animal tests to allow any team members with ethical objections to voice their concerns and opt out of the studies.

Good Laboratory Practices

When conducting non-clinical and pre-clinical studies, researchers must also determine whether or not to follow Good Laboratory Practices (**GLP**), as defined by the FDA. In essence, GLP is a set of guidelines that describes in detail how studies are performed and data are collected. Data generated using GLP are almost always required when a company seeks approval for an FIM study, 510(k), or premarket approval (PMA) from the FDA. The guidelines specify minimum standards for safety protocols, facilities, personnel, equipment, test and control activities, quality assurance, record keeping, and reports used in conducting the trial. Good Laboratory Practices requires the laboratory to have an extensive written set of operating guidelines (called standard operating procedures or SOPs – see 5.5 Quality and Process Management) for conducting the study. The decision to perform a GLP study has a profound effect on how a company conducts its development and engineering processes, since these must also be rigorously documented.

Importantly, not every non-clinical or pre-clinical study needs to be conducted in accordance with GLP guidelines. In general, a company should be sure to follow GLP when it anticipates submitting the results of the study to the FDA, otherwise it can exercise its own discretion. While GLP studies force a company to follow through on every trial detail, they tend to be more time consuming and expensive due to the extra paperwork and additional rigor that is required to support the experiments. For example, data from clinical lab work, which could be collected on an as-needed basis in a non-GLP study, would have to be collected at multiple points in time, cataloged, and trended under GLP requirements. The additional structure imposed by GLP guidelines may not be appropriate for many non-clinical studies and even some pre-clinical work. Synchronization of the clinical strategy with the company's developmental timeline will help establish the need for GLP studies to support key milestones.

Human clinical studies

Progressing from pre-clinical animal studies to human clinical studies represents a major milestone for most device companies. The opportunity to test and evaluate a new medical device in a human is both an opportunity and risk (see Figure 5.3.2). Medical devices that fail in a human (or, more importantly, harm a patient) are unlikely ever to be used again. Several different types of studies can be performed for human device testing, depending on the nature of the device, the clinical problem being addressed, and the stage of testing. Each study type may provide an opportunity to further advance the technology.

First-in-man studies

Before a company can initiate large-scale clinical studies, it first must complete FIM studies. The most important outcome of these small-scale, preliminary human studies is safety, although investigators are also looking to see whether the device performs as intended (even though a specific efficacy endpoint may not be defined). Because FIM studies are not designed to establish a clinical benefit, efficacy is anecdotal. Careful thought must be given to the appropriate time and place to perform FIM studies, given the significance of using a device in a human. Ensuring that the device design has been optimized for human patients

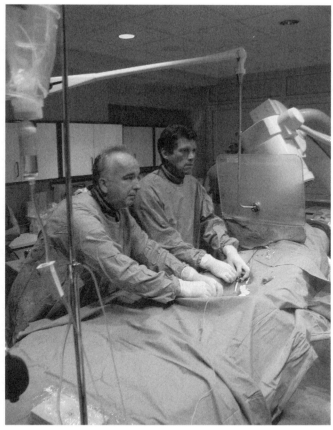

FIGURE 5.3.2
Human clinical trials can take many different forms, but all require careful planning and the highest of ethical and safety standards (courtesy of ExploraMed).

used in the pilot phase of testing a new device – typically, the first 10 to 100 cases following FIM where the company, investigators, or the FDA may be attempting to learn enough about a device's performance to design a definitive trial. A registry of device implants performed overseas, for instance, could be used in support of a 510(k) application in which the burden of clinical proof is modest (i.e., the device is known to be comparatively safe).

An increasingly important use of registries is to monitor the outcomes of a device after it is approved by the FDA and as it is launched into more widespread clinical use. This approach has resulted from recent, high-profile examples of devices that showed unanticipated complications following FDA approval (e.g., implantable cardioverter defibrillators, drug eluting stents). Registries are generally much less complex and expensive than randomized trials, but have a lower power to demonstrate important discernible differences.

Case control studies

A case control study statistically compares outcomes of a group of patients treated with a new device or procedure to a matched group receiving no treatment (or a standard treatment). Patients in both groups must be well matched on characteristics that are known to influence outcomes associated with treatment for the disease state, such as age or comorbid conditions. Case control studies are often accomplished by searching a large database of patients treated with the standard (or no) device or procedure to find a group of patients that matches the characteristics of those treated with the new device or procedure. A retrospective comparison is then performed to identify which group experienced better outcomes. For example, a new device to treat lumbar disc herniation could be compared to fusion surgery (an accepted technique) by finding patients in a database who have undergone fusion and are matched to patients receiving the new treatment according to age, gender, weight, duration and severity of back discomfort, location of disc herniation, etc. Once a comparable population is identified, researchers could then analyze the medical outcomes of both groups to determine whether the new device or fusion surgery led to more favorable results.

(not animals) will improve the researchers' chance of success and minimize the risk of causing harm. There are many reasons that may explain device failure in early human studies, but it is often due to too much reliance on prior animal studies in which the anatomy was overly forgiving and free of disease. This risk can be minimized through the careful evaluation of human anatomy and human cadaver studies as part of the clinical strategy.

Registries and observational studies

A **registry** is a collection of cases performed in a real-world setting (rather than patients treated in a stringently designed study), which may be accumulated either prospectively or retrospectively after a number of cases have been performed. Prospective registries are also called **observational studies** and are often

The primary advantages of case control studies are that they are less expensive than randomized studies, can be carried out by smaller teams of researchers, and take less time to complete than prospective studies. The main disadvantage is that the results are not as definitive as prospective, randomized, **controlled** studies and almost always necessitate further research. Although data from case control studies used to be acceptable for supporting a PMA regulatory pathway, the FDA now requires a randomized trial in most cases.

Prospective, randomized, controlled (blinded) trials

Prospective, randomized, controlled trials are the "gold standard" for medical device testing. They are increasingly required in the United States for both approval by the FDA and for reimbursement by CMS (see 5.6 Reimbursement Strategy). The term *prospective* refers to the fact that the trial is designed before any devices are tested. The patients are divided into treatment and comparison, or control, groups by a statistically *random* assignment. Interpretation of the results is based on the outcome of the group treated with the new device or procedure, relative to a *control* group that may receive no treatment or treatment with a more established approach. The advantage of prospective, randomized, controlled trials is that they have the greatest statistical **power** to discriminate whether or not the outcome and safety profile of the new technology is, indeed, superior to the control group. The main disadvantage of these kinds of trials is the considerable time and expense required to complete the study compared to the simpler trial structures described above.

In some cases, additional rigor can be added to a randomized, controlled trial if it is possible to **blind** the study participants (patients, physicians, nurses, data analysts) to the device being used. The term "double blind" is used when both the patient and the physician are blind to the treatment. For example, a double blind trial can be performed to assess a bare metal versus a drug eluting stent if it is not possible for the physician or the patient to tell the difference between the two stents based on appearances or deployment. For the trial, both stents would be provided in non-descript packages with code numbers that would eventually be used to determine which type of stent was placed in which patient. Sometimes in device trials, it is not possible to blind the physician to the treatment (e.g., in the comparison of two implants) and only a single blinded study can be performed.

Other trial nomenclature

The term **pilot** is used as a general term to describe early clinical trials, usually conducted as a registry. The definitive clinical trial conducted for FDA approval (and perhaps reimbursement) is known as a **pivotal** trial. **Postmarketing** studies refer to trials performed after the commercial approval of the device. These studies are often required as a condition of approval for a PMA, but also can be used for other purposes as part of a staged clinical strategy. (See Appendix 5.3.1 for more information about pilot, pivotal, and post-marketing studies and a comparison to trials in the pharmaceutical industry.)

Keep in mind that a company may require a clinical strategy that includes a combination of more than one type of study listed above to support its different objectives. In this case, the company must remember to carefully define and sequence the trials that it intends to undertake. In most instances, no single study design can achieve all ends without becoming so complex that trial outcomes may be compromised.

Before deciding on what study(ies) to conduct, researchers are advised to complete a *thorough* review of available literature for the condition being studied and understand *in detail* what has been studied to date. They should also consider outcomes that are similar to those of previously published studies so that the results can be compared. This is particularly important for observational studies, but is also useful in randomized, controlled studies. Being able to demonstrate how an outcome is significantly better in a trial testing a new device compared to the same outcome in a previous trial of an older device can have a significant impact. While this is harder to do for an entirely novel treatment, consultation with the FDA, users, and payers (although payer requirements are often different from those of the FDA) can help generate the best study design possible, with modifications made as input is gathered.

Clinical trial costs

Based on the purpose and complexity of the study, the costs for medical device clinical trials range from hundreds of thousands to tens of millions of dollars. The cost of conducting clinical trials can be one of the largest expenditures required of a company to get its product to market. Study costs are driven by:

- The cost of the device(s) being used in the trial.
- The cost of performing the procedure, including physician costs and hospitalizations, if needed.
- The costs of follow-up clinical visits and/or tests to evaluate the safety and efficacy of the medical device.
- The cost of paying investigators and institution study coordinators to perform the clinical studies.
- The cost of conducting the trial, including training, monitoring, and data management.
- Patient recruitment costs, including advertising and potential payment to patients.
- In-house management and personnel costs.
- The cost of trial support and other resources provided by contract research organizations (**CROs**).
- Institutional review board (IRB) costs.
- Consulting expenses for data safety monitoring boards, physician advisory boards, and **core laboratories** to independently evaluate trial results.

At the most extreme, clinical trial costs may range from as little as $2,000 per patient (e.g., for a non-implantable device with a short follow-up period) to as much as $100,000 on a per patient basis (e.g., for an implantable or therapeutic device with a lengthy follow-up period).

Within start-up medical device companies that expect significant clinical trial program costs driven by a PMA regulatory pathway, financing rounds are often coordinated with clinical trial milestones. It is not unusual for series B or C (early- to mid-stage) financing to correspond to milestones for first-in-man or pilot studies. Series C to E (mid- to late-stage) financing tends to correspond to milestones related to pivotal trials (see 6.1 Operating Plan and Financial Model and 6.3 Funding Sources). The close interplay between financing and clinical trial milestones underscores the importance of staging clinical studies and gathering results to build sequential value to a company, which can, in turn, justify subsequent rounds of funding. It is also indicative of the significant financial burden that is associated with a clinical program. To efficiently manage this sizable expense, as well as the other important issues related to clinical studies, careful strategic planning is essential.

Location of clinical studies

Researchers also must consider whether or not to conduct clinical trials within the United States or outside the United States (OUS), based on their objectives, particularly with regard to where device approval is ultimately desired. The most common reasons that OUS studies are favored, particularly for FIM studies, are for increased patient access, quicker enrollment, and reduced cost. Therefore, a sizable percentage of clinical device testing has shifted OUS. For example, as much as 75 percent of first clinical use testing for cardiovascular devices is now performed OUS.[3] Importantly, though, these studies are almost exclusively FIM and pilot studies. Clinical trials performed for FDA PMA regulatory approval are predominantly performed in the United States. Companies may include OUS sites in large pivotal studies to increase enrollment and gain footholds into important markets where they intend to market the device. However, where the United States is the ultimate target market, the majority of clinical sites are based there. Importantly, the FDA often looks favorably upon OUS data as a precursor to an IDE to validate initial safety testing of the device. As a result, OUS studies can be a useful strategy in this regard for companies preparing to conduct pivotal trials in the United States.

Another factor contributing to this shift toward OUS trials is the fact that FDA regulatory requirements in the United States have traditionally been more extensive and demanding. Initiating a US trial is often more time consuming than launching a trial in another country. One estimate indicated that seeking FDA approval to start a clinical study through the IDE process added three to six months to the process of device development. Subsequent review by the IRB at the clinical site(s) was shown to add another three to six months to this timeline, putting US investigators at a 6- to 12-month disadvantage compared to overseas researchers.[4]

Additionally, the regulatory and study monitoring protocols for OUS trials were once less burdensome than those in the United States, which often meant that there were fewer challenges in getting a trial underway. However, this is becoming less so, with most countries adopting international guidelines on the performance of clinical trials. Nonetheless, depending on the disease state/condition being studied, patients can be enrolled more quickly in OUS locations where the patient volume is simply higher.

Clinical trials managed in the United States usually are more expensive than those conducted OUS, but this depends on the travel required of company personnel and the payments required to the investigators.

When considering a clinical strategy that involves OUS trials, it is important to note that the FDA may not consider the patient population treated overseas to be equivalent to a population in the United States. In order for OUS clinical data to be relevant for the US approval process, the patient population must be comparable (ethnically, as well as in terms of treatment regimens) to that of the US population. As a result, careful guidance should be pursued prior to the initiation of a trial to ensure that OUS data will be accepted by the FDA.

Basic issues in designing a trial

At this point, the focus of this chapter shifts to the design of specific clinical trials.

Hypothesis

The design of a clinical trial starts with a hypothesis, which the study will prove or disprove. Generally, this is in the form of a comparison of the outcomes (endpoints) achieved on patients treated by a new device or procedure relative to the control group. For example, a hypothesis for an ophthalmology trial might be that a new intraocular lens will provide superior visual acuity and an equivalent safety profile to an existing lens.

For the purposes of FDA approval in the United States, an innovator or company might define a hypothesis that the device being studied is equivalent or "non-inferior" to an existing standard: that is, the outcomes are statistically indistinguishable. From a statistical standpoint, equivalence can usually be demonstrated in a much smaller trial than would be required to show superiority. A company might undertake an equivalence trial even if it thinks its device is better than the competition's because it could save a substantial amount of money and still sell effectively against the competition based on the performance of its device.

Endpoints

The efficacy of a new device or procedure is measured in the form of *endpoints*. Endpoints are the prospectively identified and quantifiable parameters that a study is designed to meet. For well-designed studies, there are one or more **primary endpoints** that the company, investigators, and the FDA agree will be the main criteria on which the device or procedure is evaluated and the one(s) that the study will focus on from a power standpoint (see below). There also may be *secondary endpoints*, which are of interest scientifically and clinically, but will generally not lead to approval or clearance by the FDA (e.g., endpoints to determine cost effectiveness for reimbursement). Selection of the proper endpoints for any study is exceedingly important and, for FDA trials, is a major point of negotiation between the company and the agency.

To be useful, an endpoint must be measurable and as unambiguous as possible. For instance, even in situations in which the outcome is improved quality of life (e.g., a new treatment for tennis elbow), some method of measuring that outcome should be utilized (e.g., a quantitative questionnaire designed to quantify the degree of improvement). Another important consideration is that the trial endpoints must be achievable in a reasonable span of time. A new cardiac device may have the potential to prolong life, but the company will not be interested in a trial that takes 40 years to prove this point statistically. In this situation, a **surrogate endpoint** must be selected and then accepted by the FDA. For instance, in the cardiac example, the surrogate endpoint might be a measure of the pumping function of the heart and a nuclear medicine measurement of its blood supply – variables that have been shown in previous studies to correlate with survival.

Importantly, the statistical power of an endpoint is maintained only when it is identified prospectively. Post-trial data analysis ("data dredging") may suggest that some unanticipated subgroup of patients seems to gain special benefit from the device. However, to prove this rigorously, it will be necessary to test this endpoint prospectively in a new trial.

Before deciding on the endpoints for a study, the innovator or company that will sponsor the trial should complete a thorough review of available literature for the condition being studied and understand in detail what has been studied to date. Additionally, it will be helpful to select outcomes that are similar to those of previously published studies so that the results can be compared. This is particularly important for observational studies, but is also useful in randomized, controlled studies. Being able to demonstrate how an outcome was significantly better in a trial testing a new device compared to the same outcome in a previous trial of an older device can have a significant impact. While this is harder to do for an entirely novel treatment, consultation with the FDA, users, and payers (although payer requirements are often different from those of the FDA) regarding the appropriate endpoints will help generate the best study design possible.

Statistical power and trial size

One of the most important issues in trial design is the number of patients to include in the study. A properly designed trial uses a "**power calculation**" to determine how many patients are required in the treatment and control groups to adequately test the hypothesis. This calculation is based on a best guess of the impact of the new therapy on the primary endpoint. For example, suppose a new technology for obesity has been tested in a pilot observational study and has been shown to reduce body mass index (BMI) in the study group by twenty percent. The company and investigators might propose a primary endpoint of ten percent BMI reduction. This means that if the device truly reduces BMI by ten percent, the results from the trials will prove that. The clinical trial will then collect data to perform a hypothesis test. The null hypothesis is that there is no difference and the alternative is that there is a ten percent improvement. The null hypothesis will be rejected if the difference between the BMI in the control group and treatment group has a **statistical significance** of five percent. Because of the randomness in the data, it is possible, especially if the number of patients enrolled is small, that the null hypothesis will not be rejected even if the device has an effect. The statisticians will help determine the sample size such that the chance of erroneously failing to reject the null hypothesis will be small (typically ten percent). The endpoints and power calculations will be reviewed by the FDA and approved or modified as its experts see fit.

It may be possible to achieve the goals of the power analysis using other forms of **randomization** besides a 1:1 model. For instance, when the control group will be treated with a relatively well-understood procedure or technology, it may be possible to perform a 2:1 randomization (that is, the group receiving the new therapy will be twice the size of the control group). This approach reduces the size and expense of the trial without negatively affecting the significance of the results.

See Appendix 5.3.2 for a basic primer on statistical design.

Choosing investigators and centers

An innovator or company, as the sponsor of a clinical trial, will select one or more *principal investigator(s)* (**PIs**) for a trial based on a number of considerations, including experience in leading trials, track record of publishing trial results, stature in the field, quality of support personnel and – an increasingly important criterion – freedom from conflict of interest. The PIs will typically help the company select additional investigators to conduct the trial. These investigators will be chosen based on their technical skill with similar devices or procedures, their ability to enroll patients quickly (generally meaning bigger practices), their prior experience with clinical trials, their reputation among their colleagues, the effectiveness of their research support staff, and their geographic location.

Finding highly productive investigators can be facilitated by searching the literature (looking for contributors who appear on the prior studies that are similar to the one being designed). Polling companies and contract research organizations that have worked in the specific therapeutic area can also be helpful. However, newcomers (physicians who are new to a particular field) should not be overlooked as they often can make significant contributions to the trial. Many up-and-coming physicians are hungry to make a mark. Being associated with marquee physicians through a clinical trial can be highly motivating to these new physicians and, as a result, can yield significant enrollment from lesser-known sites.

Typically, all investigators involved in a trial will sign a contract with the trial sponsor that specifies their obligations, including accurate recording of data and timely reporting of complications. These agreements also address indemnification and the assignment of ownership rights of new discoveries (IP) made in the course of the study. The investigators and/or their centers are reimbursed for their costs in conducting the study, which includes their time (and staff time), extra equipment, tests, hospital time, and other expenses. If the investigators are faculty members in a university, the contract is made with the university and an "indirect" charge (typically an additional 25 to 50 percent) is added to support general university infrastructure. Innovators and companies should expect contracts to take anywhere from two to twelve months to finalize, depending on the type of institution with which they are working.

The number of investigators and centers required depends on the size of the trial and the expected rates of enrollment. In practice, enrollment often turns out to be much slower than the investigators believe will be the case. A rough rule of thumb to help manage this risk is for the sponsor to budget for the pace of enrollment to be approximately half as fast as expected and total enrollment from each center to be half as large as projected.

Investigational device exemption

As noted in 4.2 Regulatory Basics, many clinical studies supporting a regulatory submission in the United States require an IDE, which allows a premarket device to be tested in humans such that the necessary safety and effectiveness data can be collected to support the FDA submission. For a PMA, clinical studies supported by an IDE are always required. In contrast, an IDE is only needed to support the small percentage of 510(k) submissions that require clinical data. An IDE can also be used to cover the clinical evaluation of certain modifications and/or new intended uses of legally marketed devices[5] (e.g., off-label usage).

The purpose of the IDE process is to ensure that researchers "demonstrate that there is reason to believe that the risks to human subjects from the proposed investigation are outweighed by the anticipated benefits to subjects and the importance of the knowledge to be gained, that the investigation is scientifically sound, and that there is reason to believe that the device as proposed for use will be effective."[6] Once approved, an IDE clears a device to be lawfully used in conducting clinical trials without the need to comply with other FDA requirements for devices in commercial distribution. All clinical trials that include investigational devices must have an approved IDE *before* the study is initiated, unless the device is determined to be exempt from IDE requirements.[7]

To obtain an IDE, researchers must complete an IDE submission. As discussed in 5.4 Regulatory Strategy, early communication with the FDA and a pre-IDE meeting can help facilitate more efficient approval through the IDE process, as well as assisting with the development of a study design that supports the desired/necessary endpoints for regulatory approval. The FDA is mandated to respond to every IDE application within 30 days. Though an IDE is rarely approved in the first 30-day period, the FDA must provide the applicant with feedback and/or request additional information within this time frame. Often a series of back-and-forth communications with the agency is required before an application is approved. If, after 30 days, the FDA has not responded to an IDE application, a company is authorized by default to proceed with the study.

During the study, the researchers must maintain compliance with specific IDE requirements, which include (but are not limited to):

- Obtaining advance approval from the institutional review boards (IRBs) where the study will be conducted and working with the IRB through the execution of the trial.
- Obtaining informed consent from all patients involved in the study.
- Labeling the device for investigational use only.
- Carefully monitoring the study.
- Completing all required records and reports.

In rare circumstances, the FDA allows for investigational devices to be used in patients that are not part of an IDE-approved clinical trial. Such usage is allowed under clearly defined conditions of emergency use, compassionate use, treatment use, or continued access[8] (see Appendix 5.3.3 for more information).

Institutional Review Board approval

Once an IDE is obtained, researchers must next seek IRB approval before the study is initiated (note that if the study is IDE exempt, IRB submission is the first step following the design of the trial protocol). It is a federal mandate that each site where a clinical trial will be conducted has an IRB that is responsible for protecting the rights, safety, and welfare of research subjects. The IRB can be developed and managed in-house or contracted from a third-party provider. In either scenario, IRBs are regulated by the FDA and their policies and practices are subject to periodic review and certification. An IRB is generally made up of clinicians, nurses, and one or more hospital administrators. Optimally, the IRB will also include a statistical expert, an expert in medical ethics, and one or more "lay" representatives (such as community advocates, clergy members, or a working professional). The IRB is responsible for monitoring complications of the study and, in many cases, also serves as the screening point for issues of conflict of interest on the part of the investigators or their institutions.

The lead investigator is responsible for preparing the application to the IRB at each institution where the study will be conducted. This application describes the device, outlines the proposed clinical study, including trial endpoints, and includes a sample patient consent form. The IRB formally reviews this application, often requesting changes as needed, before approving the study. As a rough average, this process can take up to three months to complete. It can require more or less time, depending on how often the members of the IRB meet and how they work together.

It is worth noting that local IRBs often have additional policies and restrictions beyond the general requirements specified by the FDA. Institutional Review Boards have come under increased scrutiny recently, in part as a result of the 1999 death of a patient enrolled in a gene therapy trial at the University of Pennsylvania. In this case, involving 18-year-old Jesse Gelsinger, it was alleged that the IRB did not adequately review the safety of the study and the protections put into place. Furthermore, investigators did not adequately counsel participants on the extent of the risks involved in the study[9] and were not following all of the federal rules requiring them to report unexpected adverse events associated with the gene therapy trials.[10] The Gelsinger case, in conjunction with other events, served as a catalyst for IRB reform and the improved protection of human research subjects. In June, 2000, the Office for Protection from Research Risks (**OPRR**) was officially renamed the Office for Human Research Protections (**OHRP**) and moved from the National Institutes of Health (NIH) to the Department of Health and Human Services (which also oversees the FDA). The move was intended to increase both the visibility and accountability of human research protections, as monitored by the federal system.[11] To carry out this mission, the OHRP has established formal agreements ("assurances") with nearly 10,000 universities, hospitals, and other research institutions in the United States and abroad to comply with the regulations pertaining to human subject protections.[12]

Because each IRB maintains its own governance, trial sponsors may have an inconsistent experience from institution to institution. Sponsors and investigators should be prepared to modify their clinical protocols (and especially the informed consent document) to meet the requirements of each IRB. It is the sponsor's responsibility to ensure proper document control for each location, a practice that can present complex control/management issues for trials that utilize multiple sites.

It is also important to note that many IRBs charge for their services (i.e., initial review and approval of the study, as well as ongoing reviews). An initial review can cost anywhere from $2,000 to $4,000, while ongoing reviews may range from $1,000 to $2,000 every 6 to 12 months, with these fees paid for by the sponsor. The Ventritex story below highlights the increasing number of budgetary and contracting issues that must be considered in working with IRBs.

FROM THE FIELD — VENTRITEX, INC.

The increasing complexity of clearing pre-trial clinical hurdles

Ventritex was founded in 1985 by a seasoned medical device team: Bill Starling, Ray Williams, Geoffrey Hartzler, Roger Winkle, Ben Pless, and Michael Sweeney. Recognizing the promise of implantable cardioverter defibrillators (ICDs), as well as the shortcomings of the only ICD approved for sales in the United States at the time, this team sought to develop improved ICD technology. Implantable cardioverter defibrillators are small battery-powered devices that monitor patient heart rhythms and deliver electrical shocks when they detect ventricular tachycardia or ventricular fibrillation, dangerous heart rhythms that are the leading causes of sudden cardiac death (SCD). Approximately 300,000 people experience SCD each year in the United States, with 75 to 80 percent of these incidents linked to ventricular fibrillation.[13]

Frank Fischer joined Ventritex in July 1987 as its president and CEO. "They were about half-way through the development of the product at the time," he recalled.[14] By 1989, the Ventritex device was ready for clinical testing. Although the clinical study completed by Cardiac Pacemakers, Inc. (CPI) – the first mover in the market – could be followed as a precedent for trial design, Fischer indicated that "our clinical study was most informed by our scientific advisory board." The trial would involve hundreds of patients across multiple sites. In terms of endpoints, "We looked at SCD mortality as compared to a historic control," he said, noting that it would be virtually impossible to get the same study approved today (since most trials for class III devices must be randomized and blinded, with a built-in control group). Once the study was designed, the next step was to seek IDE and IRB approval, which the company gained relatively quickly.

However, as Fischer emphasized, innovators and companies today can face much greater challenges in clearing these hurdles prior to launching a trial. "Now, when you're preparing to run a study, you not only have to get IDE approval, but work through complex budgeting and contracting issues with the site. When we did our clinical study on our first defibrillator, there was no issue with the hospitals from a budget and contractual standpoint. They just wanted to do the study. We sold our devices to the participating hospitals and they paid for all the expenses incurred for those procedures. Basically, there was so much excitement around this state-of-the-art product that the hospitals and physicians were vying for the opportunity to participate." He continued, "But today, with hospitals having been squeezed from a financial perspective, they have to try to recover costs or generate revenues in any fashion that they can. So, they look for significant payment to participate in studies – that's the budgeting issue. The contractual issue is related to responsibility. In addition, many sites have looked enviously at intellectual property that has been generated and the ability to get royalties, so now they're sensitized to those issues. However, it has gotten to a ridiculous point where some institutions believe that IP should be theirs regardless of whether or not they were involved in the generation of a new idea through the study. And then there are issues related to who owns the study data and who gets publication rights. Twenty years ago these issues were handled on a more collegial and gentlemanly basis. Now they're things that have to be stipulated from a contractual vantage point."

For innovators and companies facing these pre-trial challenges, Fischer offered this advice: "Recognize that

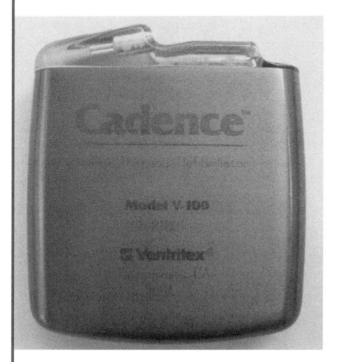

FIGURE 5.3.3
The Ventritex Cadence® V-100 ICD (courtesy of Frank Fischer).

there are no shortcuts. Recognize that it's going to take time and money to get through clinical trials and you'd better have both of them planned into your projections, otherwise you're not going to be able to succeed." He also advised start-ups to factor consultants into their budgets. "Make sure that you get good advice. There are clinical consultants and regulatory experts that have lots of good experience that can be brought to bear. Make an investment in getting good advice as early as you can in the process. The worst thing is to find out after the fact that your study approach or outcomes data is not acceptable." Finally, Fischer encouraged innovators and companies to interact early and often with the FDA: "Nowadays, a smart thing to do is to take advantage of what the FDA has to offer. Schedule a pre-IDE meeting, present what you intend to do in your clinical studies, and get feedback. It can save a lot of time bouncing back and forth in putting a submission together and then possibly having it rejected."

Ventritex ultimately made its submission to the FDA in 1991 and received a unanimous recommendation for approval from an FDA advisory panel in 1992.[15] The company's Cadence V-100 ICD (pictured in Figure 5.3.3) quickly became the "envy of the industry" when it was cleared for sale in the United States.[16] Ventritex was acquired by St. Jude Medical in 1997.

Patient enrollment

Patients are screened for studies by the investigator and his/her staff. In the modern era of careful trial design, IRB review, and patient advocacy, the ability of researchers to enroll subjects has become the rate-limiting resource in clinical research.[17] Despite this challenge, innovators and companies must carefully adhere to all guidelines and requirements for enrolling their trials (see Figure 5.3.4).

There is no doubt that that patients are increasingly cautious about agreeing to participate in trials of new devices and drugs. A recent article summed up the patient's perspective about participating in a cardiac defibrillator trial this way: "Would you sign up for an experimental heart device trial, where there's a 50 percent chance your device will be inactive ('control group'), and there's an FDA-approved device already on the market?"[18] For those involved in sponsoring and conducting clinical trials, it is important to keep in mind a key fact: volunteering for a clinical study is an act of generosity and public service on the part of patients and their families. As a result, study participants should receive all the respect and gratitude they deserve.

The process of enrolling patients into a study must be meticulous and is often highly time consuming. Subjects must be evaluated carefully and thoroughly to determine if their conditions and health profiles match the targeted audience for the device. This is accomplished through detailed screening using:[19]

- **Inclusion criteria** – Characteristics or indications that subjects must have in order to participate in the clinical trial. For example, if a new spinal disc is being tested, participants might be required to demonstrate a specific site of disc degeneration in the back to be included in the study.
- **Exclusion criteria** – Characteristics or contraindications that eliminate subjects from

FIGURE 5.3.4
In today's environment, innovators are advised not to take chances when it comes to performing human research.[41]

participating in a clinical study. For example, individuals who have had fusion surgery or advanced spinal arthritis might be excluded because their spines already show significant differences from those that a device is intended to treat.

Meticulous patient screening involves multiple participants in the care continuum. First, the patient must be correctly diagnosed. Then, his/her physician must be familiar with the clinical studies available for which the patient might qualify. Researchers often impose other requirements and testing procedures on the patient to ensure that s/he fits the trial criteria. At the same time, the patient may seek counseling from another healthcare professional, particularly if s/he is dealing with stressful or life-changing news regarding the illness.

It is worth keeping in mind that patients can benefit from trial participation in several ways. The new technology or procedure may indeed represent a substantial improvement in care. Even if allocated to the control group, the patient may benefit from the increased level of medical attention provided to participants in the trial, including additional tests and provider visits. Patients also appreciate the chance to learn more about their condition by participating in a study. Some studies will reimburse patients for their participation in the trial, particularly if there is a requirement for the patients to return for further evaluation or testing. Finally, and perhaps most importantly, many patients are genuinely motivated by a desire to serve others. Having suffered themselves, they are inspired to do what they can to reduce or eliminate the medical problem for subsequent patients.

Informed consent and patient protection

Before entering a study, patients must provide written *informed consent*. In practice, this means that someone (generally not the physician performing the procedure) reviews the IRB-approved consent form with the patient, explains the risks and benefits of the study, answers any questions, and obtains the patient or guardian's signature on form. The consent form provides the opportunity for the patient to withdraw at any point in the study and identifies a third party who can be contacted if there are perceived irregularities or other issues with the trial.

Standards governing the protection of human research subjects, including mandatory informed consent, arose from several shameful experiments conducted in the early part of the twentieth century (most notably in Nazi Germany and in Tuskegee, Alabama). The standard of informed consent emerged in the 1940s as the field of clinical research became codified and well established. Today, requirements for informed consent in research involving human subjects exist in every country, as defined by international standards.

Because new investigational devices have not, by definition, been tested previously in humans, a certain level of risk exists for the subjects involved in any clinical trial. As the level of device invasiveness and procedure complexity increases, so does the level of risk. According to the FDA, "Although efforts are made to control risks to clinical trial participants, some risk may be unavoidable because of the uncertainty inherent in clinical research involving new medical products. It's important, therefore, that people make their decision to participate in a clinical trial only after they have a full understanding of the entire process and the risks that may be involved."[20] This philosophy is at the heart of the FDA's policy for informed consent, which states:[21]

> No investigator may involve a human being as a subject in research covered by [FDA] regulations unless the investigator has obtained the legally effective informed consent of the subject or the subject's legally authorized representative. An investigator shall seek such consent only under circumstances that provide the prospective subject or the representative sufficient opportunity to consider whether or not to participate and that minimize the possibility of coercion or undue influence. The information that is given to the subject or the representative shall be in language understandable to the subject or the representative. No informed consent, whether oral or written, may include any exculpatory language through which the subject or the representative is made to waive or appear to waive any of the subject's legal rights, or releases or appears to release the investigator,

the sponsor, the institution, or its agents from liability for negligence.

Data entry and monitoring

In many medical device trials, the patient undergoes a procedure in which the device is used or implanted. The time at which the device is used represents a critical point in the study where intensive data collection is typically performed. Generally, this is accomplished by a study nurse or other trained expert who collects all of the pertinent information about the encounter. The primary data are kept carefully in files or notebooks, which are subject to auditing by study monitors from the sponsoring company as well as the FDA (see Figure 5.3.5).

Core laboratories

Certain endpoints of a clinical study may best be analyzed in a special *core laboratory*, staffed by clinical scientists with specialized equipment and expertise to make the quantitative measurements required. Core laboratories may be in universities or private clinical research organizations. For example, clinical trials of coronary stents routinely use core angiographic laboratories, where radiographic images of the arteries (angiograms) before and after stent placement are carefully measured. Core laboratories typically charge on a per-patient basis and are subject to audit by the FDA.

Data safety and monitoring board

Significant trials will generally have an independent data safety and monitoring board (**DSMB**) to review the preliminary results at prespecified time points during the trial in order to ensure that patients are not being inadvertently harmed by the study. The DSMB may decide that the initial trial results are within the range of expected outcomes and allow the trial to continue; it may terminate the trial based on unexpectedly bad outcomes in the treatment group; or it may terminate early based on unexpectedly good results (when it is no longer ethical to continue the control treatment).

Clinical trial management

Once the trial has been launched, a number of ongoing management considerations must be addressed to ensure the timely and effective completion of the trial.

Research site staff

The resources required at each center to perform the high-quality research necessary for a randomized, controlled trial are formidable. In addition to the resources needed to initiate and manage a clinical trial program

FIGURE 5.3.5
Clinical research nurses and coordinators must keep detailed records on every enrolled patient (courtesy of Todd Brinton).

at the site (including physicians, facilities, and equipment), the study center must devote research nurses to each study. Research nurses play an instrumental role at all phases of the trial, including general study management, IRB process management, and accurate completion of case report forms. It is the sponsor's responsibility to make sure each clinical site has the necessary resources in place to fulfill the demands set forward in the protocol.

Sponsor personnel

Most sponsors will have at least one or two in-house employees managing a clinical trial, if not an entire team (depending on the level of control and support the company determines to be appropriate). Clinical trial management is not something that should be entirely outsourced. Different models commonly adopted by trial sponsors include the following:

- Companies may engage a contract research organization (CRO) to manage the on-site elements of the study, as well as data management. A CRO typically provides the infrastructure required to recruit, qualify, and audit sites. A CRO is particularly beneficial when the scope of the trial is large and the CRO has experience in the type of study being conducted. Also, because of the transient nature of clinical trials, it may not make sense for a company to hire an entire staff of full-time researchers who will not have roles when the trial ends.
- Alternatively, some companies prefer to maintain tight control over a clinical study by managing it in-house. In these cases, certain elements of a trial still may be outsourced (some CROs offer "a la carte" services, such as monitoring only, data management, or regulatory [IRB, IDE, etc.] responsibility), but the overall management of the trial is still performed by the internal team.

The sponsor, or its CRO, has responsibility for monitoring the progress of the trial, compliance with the protocol, and various other requirements. When using a CRO, the external organization should be carefully managed. All contracts also should be explicit with respect to expectations and deliverables. Outsourcing can be risky since the contractor does not have as much at stake as the sponsor. On the other hand, CROs have vast experience and expertise that can be valuable to young, start-up companies. Close, proactive management of any external partners can help minimize this element of risk while capitalizing on its benefits.

Data management

Data management is perhaps one of the most important elements of a clinical trial. However, this factor often is overlooked until the data already have been captured. There are a multitude of systems to consider that can assist with trial data management – from the costly but sophisticated Oracle Clinical or SAS systems (appropriate for large, pivotal trials) to the customized, in-house solutions that rely on a database platform such as Microsoft Access or dBase. Increasingly, web-based data capture is being used, which minimizes the back and forth of case report forms through fax or e-mail. Another advantage of these web-based systems is that they can be designed to accept only data that fall within expected limits (i.e., data that make sense), prompting the person entering the data to recheck the values that violate defined expectations.

Many companies (especially young start-ups) make the mistake of deciding on a data management system long after the data has begun accumulating. This can be problematic because inputting the data as the trial progresses can reveal problems in the data collection process that will be time-consuming and difficult to correct retrospectively. In addition, any interim analysis will require an accurate data set – and it is best not to leave this until a "fire drill" arises, if/when safety issues are raised within the study.

Good clinical practices for conducting clinical trials

When conducting a clinical trial in the United States under an IDE, the investigators, sponsors, IRBs, and the devices themselves are all subject to Good Clinical Practices (**GCP**) regulations. These guidelines are synonymous with the FDA's GLP guidelines for the execution of pre-clinical studies. They outline specific standards for the design, conduct, performance, monitoring, auditing, recording, analyses, and reporting of clinical trials to provide assurance that the data and reported results are credible and accurate, and that the rights, integrity, and confidentiality of trial subjects are protected. Resources for learning more about GCP regulations can be found in the Getting Started section.

Importantly, IDE-exempt trials are subject to their own regulations (outlined in the FDA's exemption regulation), which are not quite as robust as GCP requirements. However, it is worth noting that even exempt trials should comply, at a minimum, with all GCP requirements regarding the protection of human subjects if their credibility is to be maintained.

A note on ethics: conflicts of interest in clinical trials

Conflicts of interest in the context of clinical trials refer to a situation where an individual or group has potentially competing interests in the outcome of the trial – for example, wanting results that benefit the patients in the trial but also seeking outcomes that benefit the company that has worked hard to develop the new technology. There are different kinds of competing incentives for different stakeholders in the clinical trial process. For instance, the lead physicians may be strongly motivated to have positive study results that can be published in an important medical journal. However, without question, the conflicts that receive the most scrutiny are those related to financial ties to the success of the product.

Financial conflicts of interest are essentially an unavoidable part of the testing of new medical technologies. Clinical trials of new technologies are so expensive that governmental health agencies like the NIH or the FDA can only sponsor a tiny fraction of the trials that need to be conducted. In a market-driven economy (e.g., the United States), companies are expected to take on the financial burden of the trials. In doing so, they automatically assume a conflict of interest in the process. As a result, sufficient regulatory checks and balances must be maintained to ensure that these conflicts do not significantly distort trial data.

It is important to understand that conflicts of interest occur at all organizational levels. Any individuals who participate in the trial could have a potential conflict if they stand to benefit financially (or by reputation, publications, etc.) from the outcomes. Their institutions also may have a conflict of interest. For example, a university may receive royalties on a new technology that one of its faculty members has invented. In the current climate, many universities would opt not to participate in the clinical testing of the device because of this financial conflict of interest. The national press has recently highlighted a number of examples of physicians and medical centers that have received questionable payments from pharmaceutical and device companies. For instance, a *New York Times* article highlighted a potential conflict of interest involving numerous researchers studying the Prodisc device (an artificial spinal disc developed as an alternative to fusion surgery). Doctors at roughly half of the 17 medical centers participating in the pivotal trial were revealed to have financial ties to the success of Prodisc, and the article questioned whether adequate financial disclosure was made to the FDA prior to the agency's approval of the device.[22] Anticipating the public backlash associated with these kinds of conflicts, there is a strong movement among leading medical centers to distance themselves from questionable circumstances, even to avoid the *perception* of any wrongdoing.

A special situation arises when physicians are involved in the invention and early development of a device and, as a result, play an integral role in the early animal studies. Through this involvement, the clinician/inventor obtains in-depth, first-hand knowledge of the device's performance and its failure modes. Given this experience, it may be most ethical from the standpoint of patient safety for these physicians to perform the first clinical studies. However, because these clinician/inventors often have leadership and/or equity positions in the company developing the device, careful steps must be taken to mitigate and manage conflicts of interest. This situation is recognized in the Association of American Medical Colleges (AAMC) guidelines on conflicts of interest, which allow the conflicted surgeon/clinician with substantial pre-clinical experience to perform the first-in-man studies, but recommend that these individuals not serve as principal investigators in any definitive, multicenter trials.[23] In addition to processes put in place by the various IRBs at the clinical facilities where the studies are conducted, the FDA's IDE process requires disclosure by any investigators with a significant equity or consulting stake in a company. Investigators are not barred from participating in studies involving their devices, but the nature of the relationship with the company must be disclosed to FDA, the IRB, and the subjects in the study.

Tips for designing and managing successful clinical trials

A few final words of wisdom can be helpful for innovators and companies as they prepare to engage in clinical trials:[24]

- Collect data early on enrollment patterns from the different centers. This exercise provides a realistic view of the speed of the overall trial and can help to ensure that each clinical research site is screening all patients and is giving the study top priority.
- Statistics only summarize the data. Examine the real data directly to look for early indicators and issues.
- Beware of drawing inferences from small samples over large populations. Design the pivotal trial keeping in mind that results from the pilot might be overly optimistic. Review all data that have been published to date as a "sanity check" of what has been achieved in the field so far.
- Select only one or two primary endpoints and be sure that those endpoints are the most significant (by comparing to similar trials). Be careful to avoid arbitrary or meaningless endpoints.
- Resist the temptation to develop an overly ambitious study design. Review the company strategy and staging possibilities to ensure a successful study for the desired endpoint.

The DVI example below highlights interactions between the company, scientific advisory board, and principal investigators in designing and executing a clinical trial.

FROM THE FIELD | DEVICES FOR VASCULAR INTERVENTION

The overall importance of clinical strategy and trial design on business success

In the mid-1980s, John Simpson pioneered the concept of directional coronary atherectomy (DCA). The DCA procedure used a device called an AtheroCath® to cut, capture, and remove plaque from the coronary arteries. The need for the device was created by the ongoing desire in the field for a less-invasive, more cost-effective solution to removing obstructive atherosclerotic lesions than cardiac artery bypass grafting (CABG), as well as the perceived failure of percutaneous transluminal coronary angioplasty (PTCA) to yield a durable result for all patients.[25] Simpson believed that the clean removal of plaque through atherectomy would render a larger and less traumatized lumen and, therefore, less restenosis.[26] He founded Devices for Vascular Intervention (DVI) as a mechanism for developing and commercializing this new approach. See Figure 5.3.6.

Allan Will joined DVI in 1986 as vice-president of marketing and sales and became the company's CEO in 1987. "At that point in time, most of our work was focused on what we would need in order to translate our success in peripheral vessels to the coronaries and get the coronary product approved by the FDA," Will recalled. "We knew that eventually we would be involved in a large-scale clinical study." Between 1988 and the time of the company's advisory panel meeting in 1990, clinical testing of the AtheroCath in 873 patients was performed

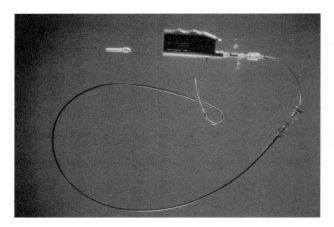

FIGURE 5.3.6
The DVI AtheroCath (courtesy of Allan Will).

at 12 US hospitals. The value of DCA was assessed by means of a large patient registry that cataloged acute and late clinical outcomes and then compared them to a historical control. "Ultimately, our final PMA registry data included roughly 2,000 patients," Will said. "These data were compared against generally accepted angioplasty results. During this initial clinical study, we used the commonly accepted threshold for a significant stenosis – greater than 50 percent residual stenosis – as our success endpoint." Data from the study indicated that DCA achieved comparable or better results than those achieved by balloon angioplasty and pointed to a lower incidence of serious coronary dissection and abrupt occlusion after DCA.[27] Devices for Vascular Intervention received FDA approval of its device in September, 1990, based on the outcome of this study. AtheroCath use grew quickly after its US commercial launch. "Sales grew from approximately $10 million in the year of FDA approval, to $41 million, then $60 million, and finally $84 million in our third full year," Will noted.

In the meantime, however, DVI was acquired by Eli Lilly in 1989. "As part of that acquisition," explained Will, "Lilly wanted to include an **earn-out**, so that the value they ultimately paid would be based upon actual versus a speculated value. And they originally proposed that a very significant amount of the earn-out – I believe it was around $25 million – should be based on the results of a prospective, randomized clinical study demonstrating at least a 50 percent reduction in restenosis compared to PTCA. At the time, we didn't even know the full results of our approval study, let alone how the device would perform in a hypothetical, prospective, randomized clinical study." Moreover, these kinds of studies, while common in pharmaceuticals, were relatively new in the medical device industry. According to Will, "I believe this was the first seriously conducted prospective randomized clinical study of a medical device in the field of interventional cardiology."

Will and team "pushed back" on the proposal from Lilly's management. "There was just so much we needed to learn about how to best use this device before we launched a definitive trial," he said. Devices for Vascular Intervention was ultimately able to negotiate the amount of the earn-out that was linked to the clinical study down to $5 million, but Lilly management insisted that the company conduct it.

In an effort to better understand the procedure and thereby produce the most positive results possible, the DVI team intentionally sought to delay the trial until late in the earn-out period. "We hoped to avoid conducting the study until we learned enough to know how best to use the device," said Will. "We needed answers to questions like how much plaque should you remove, do certain lesions respond differently to atherectomy than other lesions, and do certain vessels respond better than other vessels?" For this reason, DVI waited until 1991 to launch what would become known as the CAVEAT trial.

Will and his clinical staff also worked carefully to propose a study design. "We had become convinced that in order to get an optimal atherectomy result, you needed to remove enough plaque to achieve less than 20 percent residual stenosis. However, when we proposed that as a technical endpoint for a successful atherectomy and as a guideline for conducting the atherectomies in the study, a couple of very vocal investigators argued back that requiring clinicians to reach less than 20 percent residual stenosis would bias the study toward atherectomy. They told us that if we set that endpoint, it would be the equivalent of saying we're only going to measure the successful atherectomies against all angioplasties. Our argument was that the endpoint for a successful angioplasty was quite well known, because angioplasty procedures had been well proven and accepted for quite some time. All that we were doing was establishing a similar standard for a technique that wasn't yet as widely understood in terms of what the clinical endpoint was."

The need for a clearly defined endpoint was driven, in part, by the fact that DCA was a directional procedure, whereas angioplasty was concentric. When an angioplasty balloon was inflated, "it acted circumferentially on the vessel," explained Will. "With DCA, it was relatively easy to obtain a great angiographic result without performing a complete circumferential atherectomy." To adequately assess a DCA endpoint, the vessels had to be evaluated from multiple angiographic views to ensure completion of the procedure. "Otherwise, the physician might think he or she got a great result even though, when looking axially down the vessel, it had only been atherectomized at 12 o'clock and 6 o'clock, and the result had been assessed by looking at the 'wrong' cross-sectional view."

Devices for Vascular Intervention argued its case aggressively, "But unfortunately, we had selected a principal investigator for the trial who was, if anything, going overboard to ensure that he wasn't favoring atherectomy or the corporation. So any hint in his mind of bias needed to be removed. In the end, that was one of the components to our undoing. We lost that argument, because we had effectively given up control over trial

FROM THE FIELD: DEVICES FOR VASCULAR INTERVENTION

Continued

design to the investigator group," Will said. Instead of being able to define an endpoint of less than 20 percent residual stenosis for atherectomy, DVI had to adopt the clinically accepted endpoint at the time for significant stenosis, which was greater than 50 percent. "If you achieved less than 50 percent stenosis, you'd completed a successful procedure as measured by CAVEAT," Will recalled. "We were allowed to 'encourage' clinicians to achieve less than 20 percent, but we were not allowed to require that as a technical endpoint for success."

Devices for Vascular Intervention's troubles continued when its principal investigator initially refused to share the results of the trial with the company in advance of the 1993 American Heart Association meeting where the data was scheduled to be released. "The chairperson of our clinical trial was determined to maintain control over the clinical study until he released the data. We told him, 'We have an obligation to the patients and the clinicians who use our product to understand what the data is and answer questions. When a clinician comes up to us at our booth, what do we say?'" Ultimately, the DVI team secured the data and was able to quickly prepare for its release. But, said Will, "It wasn't pretty because we had not been able to fully analyze the data prior to its release by our PI."

To make matters worse, the results of the CAVEAT trial failed to show any significant improvement in early or late clinical and angiographic outcomes with DCA.[28] Interestingly, according to an editorial in the *Journal of American College of Cardiology*, "The surprise results of CAVEAT led some investigators to question the conduct of the trial. The CAVEAT was a multicenter study with significant variation in operator experience and skills. The failure of DCA to reduce restenosis could not be reconciled with experience at several of the premier DCA centers. A popular theory arose that DCA fared poorly in CAVEAT because the procedure was not performed optimally in all collaborating centers."[29] Two additional studies (BOAT and OARS) were subsequently completed at hospitals with a strong commitment to DCA as a procedure, as well as substantial experience with the technique. In these trials, aggressive atherectomy, usually combined with adjunct balloon angioplasty, led to significant reductions in restenosis rates when compared to balloon angioplasty alone. However, for many, these results were described as being "too little too late."[30]

Reflecting on the CAVEAT trial results, Will acknowledged that "it threw a damper on the business." However, he continued, "What had a much greater impact on the business was that stents were coming on the scene. The major shortcoming in stenting at that time was recognized to be acute thrombosis. Right around then, Dr. Antonio Colombo had discovered that acute thrombosis post-stenting was caused by an incomplete expansion of the stent. So clinicians began to think, 'Oh, now I can place a stent and not have the acute thrombosis problems that we've been experiencing, and the procedure takes me no longer than an angioplasty would. And maybe I get as good results as an atherectomy, but atherectomy's a lot harder work.' [The average atherectomy took roughly twice as long.] So, it's hard to assess the specific impact that CAVEAT had. Clearly it was negative, and it positioned us to be vulnerable to this finding that came out on stenting. If we had definitively proven in that study that we lowered restenosis … perhaps people would have required stenting to have prospective, randomized results before they abandoned atherectomy and ran to stenting. Instead, stenting began to take off even before the release of their prospective randomized clinical studies."

When asked what advice he would offer to entrepreneurs and companies planning clinical trials, Will had this to share: "First, look at trials as incremental building blocks. Don't consider any single trial as the be-all and end-all. Think carefully and strategically about what you need to show in the first study and what you should prove in subsequent studies. Recognize that you're going to learn how the device should be used or optimized and build that into your clinical trial strategy. These days, unfortunately, companies are required by the FDA to prove acceptability in a prospective, randomized study right out of the blocks, with their first large-scale clinical trial. Do what you need to do for the approval study, but don't overreach. Second, be aware that, in order to get paid for these devices, you need to build reimbursement endpoints into your approval study. Third, always opt for likelihood of success and speed of enrollment in your clinical study design. And, finally, to as great an extent as possible, maintain control over the design and conduct of your clinical trial while not giving up objectivity. You need to be fair, you need to be honest, and you need to do good science, but it's still valuable to maintain control over the design and conduct of the clinical study."

Will also commented that, "A whole lot of ground has been plowed between when we began the CAVEAT study and today. There's now more than 15 years of experience in the device industry conducting prospective, randomized clinical studies. As such, it's a much more accepted practice and there is a raft of principal investigators to choose from – good, bad, and in between. Carefully choosing the right principal investigator is of critical importance ... one who believes in the potential of the technology, who will conduct a quality, expeditious trial, and who has no agenda other than to fairly and effectively evaluate the technology considering its stage of development." Will also encouraged entrepreneurs to work with experienced consultants in the field and the FDA to design studies that are objective and likely to lead to clearly defined, positive results.

GETTING STARTED

Remember that a clinical strategy must be developed in close alignment with the company's R&D, regulatory, reimbursement, marketing, and sales strategies. The following steps can be used to initiate the development of an overarching clinical strategy and define a tactical investigation plan for a specific study.

See **ebiodesign.org** for active web links to the resources listed below, additional references, content updates, video FAQs, and other relevant information.

1. Determine the purpose of the clinical strategy

1.1 What to cover – Define what end result the company hopes to accomplish through its trial strategy (e.g., Regulatory approval? Economic outcomes for reimbursement? Market data for physicians? Additional indications?). Take into account important work streams underway within the company, such as regulatory and reimbursement implications of these efforts on the study's objective.

1.2 Where to look – Refer back to 5.2 Research and Development Strategy, 5.4 Regulatory Strategy, 5.6 Reimbursement Strategy, 5.7 Marketing and Stakeholder Strategy, and 5.8 Sales and Distribution Strategy.

For information about regulatory related trials:
- **21 CFR 812, Investigational Device Exemptions** – Covers the procedures for the conduct of clinical studies with medical devices including application, responsibilities of sponsors and investigators, labeling, records, and reports.

For information about reimbursement related trials:
- **Centers for Medicare and Medicaid Services** – Provides information about reimbursement in the public payer system.

For information about marketing related trials:
- **MedTech Insight** – Provides business information and intelligence on new trends, technologies, and companies in the medical device, diagnostics, and biotech marketplace.
- Customer surveys and market analysis (generated in-house or by analysts in the industry).

2. Determine overall study strategy

2.1 What to cover – Based on the technology being studied, determine the optimal trial progression to achieve the company's goals. For example, does it make sense to conduct the entire spectrum of studies, including a non-GLP animal trial, GLP acute and chronic animal trials, a first-in-man human trial, a pilot trial, and then a pivotal trial? Or, can this progression be modified to shorten the company's time and expense, while still resulting in the necessary clinical outcomes?

Stage 5 Development Strategy and Planning

GETTING STARTED

2.2 Where to look – In making this determination, the evaluation of competitive precedents is invaluable and consultation with a clinical research specialist can be very helpful.
- **Clinical Trials.gov** – Provides regularly updated information about federally and privately supported clinical research in human volunteers. Note that this is not a comprehensive list of trials, as registration is voluntary.
- **Competitor websites and other competitive intelligence** – Competitors often provide information about their clinical trials on their websites or in other public documents (including protocol summaries). Conduct a thorough review of publicly available information.

3. Identify clinical research specialist(s) to work/consult with the internal team

3.1 What to cover – It is common for a company to augment its own internal clinical team when preparing to design or execute a trial. In determining which clinical research specialist(s) to work with, look for experience and in-depth knowledge in the therapeutic area, experience with the type of study to be performed, and individuals/entities that can provide the company with access to a vast network of contacts and relationships in the field of study. Companies also should seek consultants with skills sets that complement those of the in-house clinical teams.

3.2 Where to look – In addition to working with recruiters, reviewing the choices of competitors, and seeking referrals from investors, board members, and advisors, contact the following associations:
- **Association for Clinical Research Professionals** – An international association comprising more than 20,000 individuals dedicated to clinical research and development.
- **Society of Clinical Research Associates** – A community for research professionals focused on providing training and continuing education, as well as establishing and maintaining an international certification program.

4. Choose a trial design/model

4.1 What to cover – Based upon the company's objectives for each study, the type of trial can be determined by considering that pilot studies (case control/retrospective or observational), pivotal studies (randomized, controlled), and post-marketing studies (registries) are used to accomplish different results. Is the strategy an incremental progression of all three types, or does one type of study support the objective? What are the regulatory implications of the trial design?

4.2 Where to look
- **Center for Devices and Radiological Health (CDRH) Device Advice** – Detailed guidance documents and information related to medical devices. Be sure to review the IDE Overview. Also, remember to consult with the FDA directly when an IDE is required.
- **Research literature** – Review sites such as PubMed for information about studies that have been conducted previously in the therapeutic area. Use this information to design a comparable study or to determine how to design for a competitive advantage over predecessors.

5. Determine trial endpoints

5.1 What to cover – Carefully select trial endpoints and a study design to support the trial's objective. One or two primary endpoints should be chosen to support the most important objectives, with secondary endpoints to measure related outcomes.

5.2 Where to look
- **Research literature** – Examine the medical literature for clinical studies performed in the same disease state or with a predicate device or treatment. Examine their primary and

secondary endpoints and carefully evaluate if they are appropriate to support the goals of the new trial.
- **Specialty societies** – Identify the leadership of relevant specialty societies to locate "thought leaders" in the field. Thought leaders in specific disease states are critical in the determination of endpoints. Ultimately, these same leaders will evaluate the significance of results and drive device adoption.
- **Books and references on trial design** – For example, *Intuitive Biostatistics* by Harvey Motulsky (Oxford University Press: New York, 1995) provides an overview of statistical principles in non-technical language and focuses on explaining the proper scientific interpretation of statistical tests rather than on the mathematical logic of the tests themselves. This book provides a non-mathematical introduction to biostatistics for medical and health science students, graduate students in the biological sciences, physicians, and researchers.

6. Write research protocol

6.1 What to cover – The study design is translated into a research protocol, which defines, in detail, all of the tactical elements of the trial, including the patient population, inclusion/exclusion criteria, the statistics (sample size and statistical power, if applicable), prior research, and prior known results.

6.2 Where to look – Reference the FDA website for guidelines about various aspects of writing a research protocol (e.g., see "Guidance for the Use of Bayesian Statistics in Medical Device Clinical Trials"). Many books and other references are also available on the topic, including:
- *A Collaborative Clinical Trial Protocol Writing System* by Chunhua Weng, David W. McDonald, and John H. Gennari of the University of Washington, Seattle.
- *Guidelines for Writing Research Protocols* – Information on the subject developed by the NIH's Office of Human Subjects Research (available on the NIH website).
- *Writing Clinical Research Protocols: Ethical Considerations* by Evan Derenzo and Joel Moss (Elsevier Academic Press, 2005).

7. Determine where to conduct the trial(s)

7.1 What to cover – Based on the company's objectives, determine where it makes sense to conduct the trials (e.g., United States versus OUS, academic institution versus private research facility). Consider issues related to access, cost, regulatory hurdles, applicability of data, contracting process, and IRB history. Determine the relevant regulatory requirements and prepare for IDE or foreign submissions. Based on the chosen geographic markets, identify and negotiate with the appropriate clinical sites and assign responsibilities.

7.2 Where to look – Consult with the FDA on issues related to the applicability data for studies conducted OUS. Consult other companies with studies at the institutions under consideration and gather information from them about their experiences. Refer to competitors/precedents to determine what other companies have done. Review research literature for institutions that are most commonly associated with the therapeutic area. Beware of using "marquee" names as they tend to have low enrollment, although they add credibility through name recognition. Rely on community hospitals for large patient recruitment numbers.
- **CDRH's Device Advice**
- **Medical Device Foreign Regulations at Export.gov** – Provides up-to-date regulations in selected foreign countries.
- **FDA's Guidance for Industry Collection of Race and Ethnicity Data in Clinical Trials** – Guidance on developing a standardized approach for collecting and reporting race and ethnicity information in clinical trials conducted in the United States and abroad for certain FDA-regulated products.

Stage 5 Development Strategy and Planning

GETTING STARTED

8. Determine resources required to implement the protocol

8.1 What to cover – Based on the defined research approach and targeted location, identify what resources are necessary to execute the protocol. Given the scope of the defined trial strategy, determine if all required expertise exists in-house, or if a relationship with a CRO will be required. Estimate the required budget to support the plan. In general, it is best to develop estimates based on working with academic institutions, as these almost always cost more than private research facilities. Expect the study to take about twice as long as planned, and at least double the initially anticipated cost.

8.2 Where to look – Consult other members of the company's network, consultants, board members, and other contacts to evaluate "best practices" from similar organizations. Use these as benchmarks in better defining the approach, timeline, and cost of the strategy. There are innumerable CROs that can also provide information – focus on those in the target therapeutic area.

9. Understand and implement GCP

9.1 What to cover – At the appropriate time, implement GCP guidelines to direct the design, conduct, performance, monitoring, auditing, recording, analyses, and reporting of clinical trials to provide assurance that the data and reported results are credible and accurate, and that the rights, integrity, and confidentiality of trial subjects are protected.

9.2 Where to look – More about relevant GCP requirements can be found from the following resources:
- **Investigational Device Exemptions** 21 CFR 812
- **Protection of Human Subjects** 21 CFR 50 – Provides the requirements and general elements of informed consent.
- **Institutional Review Boards** 21 CFR 56 – Covers the procedures and responsibilities for IRBs that approve clinical investigations protocols.
- **Financial Disclosure by Clinical Investigators** 21 CFR 54 – Covers the disclosure of financial compensation to **clinical investigators**, which is part of the FDA's assessment of the reliability of the clinical data.
- **Design Controls of the Quality System Regulation** 21 CFR 820 Subpart C – Provides the requirement for procedures to control the design of the device in order to ensure that the specified design requirements are met.

Credits

The authors would like to acknowledge Trena Depel for her help in developing this chapter. Many thanks also go to Frank Fischer and Allan Will for their assistance with the case examples.

Appendix 5.3.1

Comparison of pilot, pivotal, and post-marketing studies to trials in the pharmaceutical industry

Category of device study	Overview	Comparison to pharmaceutical trials
Pilot studies	Performed to collect initial information about a device's use in humans before a larger-scale clinical study is launched. Also used to validate trial management processes and logistics or to evaluate device design. Often limited to a single center or a few centers and relatively small number of subjects. Carried out when there is little or no experience using the device in humans. However, pilot studies can be used at any point in the life cycle of a product when research, marketing, or design objectives cannot be met without additional clinical data.	Similar to Phase I pharmaceutical trials in which researchers test an experimental drug or treatment in a small group of people (20 to 80 individuals) for the first time to evaluate its safety, determine a safe dosage/manner of usage, and identify side effects.
Pivotal trials	Typically larger, controlled studies designed to test specific hypotheses that support the submission of a PMA application for a new device, a conformity assessment for CE mark registration in Europe, or a 510(k) premarket notification when *significant* clinical data are necessary to establish substantial equivalence.	Similar to Phase II pharmaceutical trials when the experimental study drug or treatment is given to a larger group of people (100 to 300 individuals) to see if it is effective and to further evaluate its safety. Similar to Phase III trials when larger groups of patients (1,000 to 3,000 individuals) are involved and are used to confirm the effectiveness of a treatment, as well as to monitor for side effects, make comparisons to commonly used treatments, and collect information regarding its safe usage. Note: Phase II and III activities are commonly combined into a single medical device pivotal study, but rarely involve as many subjects as a pharmaceutical trial.
Post-marketing studies	Often required as an approval condition for a PMA, but generally not required for a 510(k). Usually imposed to analyze the ongoing safety of a new device, but can also be used to study other issues. For example, a trial with a narrow scope might be used to gather additional data on a specific safety or performance issue. Another study might compare alternative treatments to support comparative effectiveness claims or failure analysis investigations.	Post-marketing studies are similar to Phase IV pharmaceutical trials used to delineate additional information, including the drug's risks, benefits, and optimal use. Post-marketing device studies and Phase IV clinical trials provide the most directly equivalent comparison between device and drug studies.

Appendix 5.3.2

The null hypothesis, Type I/Type II error, p-values, and sample sizes[31]

In designing a study hypothesis, researchers are comparing two groups – usually defined as the study group versus the standard or control group. The "null hypothesis" in clinical research means that there is no difference between the two populations being compared. The general objective of clinical trials is to prove that the study **arm** is indeed different from the control arm, so researchers aim *to reject the null hypothesis* in favor of the "alternative hypothesis," or the outcome that the two groups are different. For example, in a medical device trial, the null hypothesis might be that the incidence of restenosis (an unwanted outcome) for patients receiving ABC stent is not lower than it is for those receiving XYZ stent. The goal of a study is to determine if the null hypothesis can be disproved in favor of the alternative hypothesis (that restenosis occurs less in patients with ABC stent than with XYZ stent) through the performance of a test evaluating this outcome. The result of the test may be that the null hypothesis is true (that there is no difference in restenosis rates between patients receiving the different types of stents). Or, the null hypothesis may be rejected in favor of the alternative hypothesis, which indicates that there *is* a difference in restenosis rates between the two groups.

Statistical error is commonly described as the difference between an estimated or measured value and the true or theoretically correct value that is caused by random and inherently unpredictable fluctuations in the measurement apparatus, methods, and patient response to treatment.[32] The magnitude of the error depends on the amount of variation in measurement accuracy and treatment response.

Understanding statistical error is important because medical device trials are particularly vulnerable to unpredictability, for a number of different reasons:[33]
- Clinical and commercialization pathways often involve relatively small sample sizes (i.e., compared to pharmaceutical trials).
- Smaller device companies often have scarce resources, limited experience, and suboptimal methods for making critical trial estimates.
- Larger device companies may over-accelerate trial evaluations due to the rapid pace of technology turnover and decreasing length of total product lifecycles.
- Human biology and anatomy are unpredictable and affect patient response to a particular treatment.

In testing for the null hypothesis, there are two possible kinds of errors. Type I error, also known as a "false positive" or alpha error, occurs if the null hypothesis is rejected when it is actually true. Stated another way, this is the error of accepting an alternative hypothesis (the real hypothesis of interest) when the results can be attributed to chance. An easy example to understand is when a test shows that a woman is pregnant when she actually is not carrying a child. A test results in Type II error, also called a "false negative" or beta error, when the null hypothesis is not rejected when the alternative hypothesis is true. In other words, it occurs when the researchers fail to observe a difference when, in reality, a difference exists. Table 5.3.1 summarizes these concepts using the example of a pregnancy.

Hypothesis testing is used to determine whether or not the observed difference between two groups can be explained simply through random chance. Traditionally, an acceptable level of Type I error in medical device trials is set at 0.05. This means that there is a five percent chance that the variation observed as a result of the test is due to chance. This probability is called the "level of significance" and is reported as a study's "p-value." Specifically, if the null hypothesis is true (no difference exists between the two populations being compared), then the p-value is the probability that random sampling would result in a difference as big as, or bigger than, the one observed in the sample size actually evaluated.[34] Another common convention when reporting a

Table 5.3.1 *A simple example helps illustrate the difference between Type I and Type II error.*

		Reality	
		Null hypothesis is true	Alternative hypothesis is true
Research	Null hypothesis is true	Accurate (not pregnant)	False negative Type II or beta error (pregnant but not detected)
	Alternative hypothesis is true	False positive Type I or alpha error (not pregnant but incorrectly detected)	Accurate (pregnant)

p-value is to select a confidence interval; typically, this is set at a value of 95 percent.

Although they are closely related, it is worth noting the difference between the p-value and the confidence interval. A confidence interval gives an estimated range of values that is likely to include an unknown value or parameter. The estimated range is calculated from a given set of sample data. If independent samples are taken repeatedly from the same set of values, and a confidence interval calculated for each sample, then a certain percentage (confidence level) of the intervals will include the unknown population parameter.[35] Confidence intervals are usually calculated so that this percentage is 95 percent. Wider intervals may indicate that more data should be collected before anything definite can be said about the value or parameter. Confidence intervals are more informative than the simple results of hypothesis tests (i.e., deciding whether to reject or accept the null hypothesis), since they provide a range of plausible values for the unknown parameter. By applying mathematical theory, confidence intervals can be inferred from small sample sizes. This method is typically used in clinical studies, as confidence intervals are established before researchers acquire large data sets. Meaningful information is predicated upon a random sample, independent observations, accurate assessment, and the assessment of the event that is truly central to the study (e.g., the probability of a life-threatening complication versus the probability of *all* complications).

Importantly, the lower the probability of Type I error in a study, the higher its likelihood of Type II error (and vice versa). When a significant difference exists in the population being studied but the test fails to find this difference (Type II error), the study is said to lack "power"[36] (power is defined as the probability that the test will reject a false null hypothesis).[37] A significant difference in clinical studies is generally considered to be a p-value of 0.05, meaning that if a p-value is 0.0500 *or less*, the two populations being compared are indeed statistically different from one another and the null hypothesis can be rejected.

Sample size determination

Sample size determination is a complex subject, best performed with the careful assistance of a qualified statistician. However, sample sizes generally can be determined in two ways:

1. ***Use of confidence intervals*** – Through this process, one determines how many subjects are needed so that the 95 percent confidence interval has a desired width. This means that if samples of the same size are drawn repeatedly from a population, and a confidence interval is calculated from each sample, then 95 percent of these intervals should contain the population mean.[38]
2. ***Use of hypothesis testing*** – Determine how many subjects are needed so that the study has enough power to obtain a significant difference, given the specified experimental hypothesis.
   - Usually, alpha (probability of a Type I error) is set to 0.05, but if a more significant value is desired (say 0.01), a larger sample size will be needed.
   - Beta (the probability of a Type II error) is typically set at a power of 80 (0.20) to 90 (0.10)

percent. Conventionally, this is the standard imposed, but a company can choose any threshold it deems appropriate based on what is being studied. However, a threshold more lenient than 80 to 90 percent will be questioned and therefore must be justified. It will also have a higher likelihood of a failed trial.

With either method, it is necessary to make an assumption about the so-called delta (the true difference between the two groups being studied) because that delta will determine the sample size that is needed to provide the desired power. The delta is usually the smallest difference that would be clinically important, which can be hard to define.[39] Consulting prior studies will provide useful guidance. Thoughtful consideration of assumptions leading to a decision on sample size cannot be overemphasized, as these assumptions define the study endpoints from a statistical standpoint.

It is also important to consider whether the study is designed to prove non-inferiority (equivalence) or superiority, as this will impact the statistical considerations and required sample size significantly. Non-inferiority requires a smaller sample size than superiority.

Additionally, always base the study protocol on a sample size that exceeds what is statistically required in order to account for patient withdrawals, patients lost during follow-ups, and other exemptions that inevitably occur during the clinical trial. It is best to study more patients than necessary rather than to come up short during the final analysis and fail to prove the statistical endpoint based upon an insufficient sample size.

Appendix 5.3.3

Circumstances under which investigational devices can be used outside an IDE-approved clinical trial[40]

Type of access	Description	Timing
Emergency use	May occur before an IDE is approved if the disease or condition is deemed life threatening, no alternative exists, and there is no time to obtain FDA approval. In these instances, the IRB and a physician not participating in the investigation must review and approve the investigation. The sponsor of the use must also submit a separate IDE application to the FDA.	Before or after initiation of trial
Compassionate use	Refers to the treatment of seriously ill patients using an unapproved device when no other available treatments have been deemed satisfactory *and* the patient does not qualify for the conventional clinical trials being conducted due to unrelated health problems, age, or other factors. Prior FDA approval is needed before compassionate use occurs, and in no case should a physician treat a patient with the device before then. FDA approval for compassionate use is granted via the submission of an IDE supplement, in which the physician requests approval for a protocol deviation. When making a determination about compassionate use, the FDA considers whether the preliminary evidence of safety and effectiveness justifies such use and if it would interfere with the conduct of a clinical trial to support marketing approval.	During clinical trial
Treatment use	Facilitates the availability of promising new devices to larger groups of seriously ill patients as early as possible by making promising new technologies available before the completion of all clinical trials. The FDA approves such usage under a treatment IDE application. Treatment use may begin 30 days after the FDA receives the treatment IDE submission, unless the FDA responds to deny treatment use or to require modifications to the proposal made in the application.	During clinical trial
Continued access	Refers to the continued enrollment of subjects after the controlled clinical trial under an IDE has been completed. This is typically done to provide patients with access to the investigational device while the marketing application is being prepared by the sponsor and/or reviewed by the FDA. Continued access is generally granted if there is a public health need or preliminary evidence demonstrates that the device is effective and there are no significant safety concerns.	After clinical trial

Notes

1. Carol Rados, "Inside Clinical Trials: Testing Medical Products in People," U.S. Food and Drug Administration, http://www.fda.gov/FDAC/features/2003/503_trial.html (March 27, 2007).
2. From remarks made by Frank Litvack as part of the "From the Innovator's Workbench" speaker series hosted by Stanford's Program in Biodesign, March 5, 2007, http://biodesign.stanford.edu/bdn/networking/pastinnovators.jsp. Reprinted with permission.
3. Aaron V. Kaplan, Donald S. Baim, John J. Smith, David A. Feigal, Michael Simons, David Jefferys, Thomas J. Fogarty, Richard E. Kuntz, Martin B. Leon, "Medical Device Development: From Prototype to Regulatory Approval," *Circulation*, 2004, http://circ.ahajournals.org/cgi/content/full/109/25/3068 (March 28, 2007).
4. Ibid.
5. "IDE Overview," CDRH Device Advice, http://www.fda.gov/cdrh/devadvice/ide/index.shtml (March 27, 2007).
6. "IDE Application," CDRH Device Advice, http://www.fda.gov/cdrh/devadvice/ide/application.shtml (March 28, 2007).
7. One of the most common reasons that a trial is exempt from IDE requirements is if the device being studied has been cleared for commercial use under a 510(k), but the company wishes to conduct trials to support reimbursement, market adoption, or consumer preference testing. A complete list of IDE exemptions can be found at "Definitions," Academic Health Center, University of Minnesota, http://www.ahc.umn.edu/research/indide/definitions/home.html (March 27, 2007).
8. "Expanded/Early Access," CDRH Device Advice, http://www.fda.gov/cdrh/devadvice/ide/early.shtml (May 7, 2007).
9. Nicholas Wade, "Patient Dies During a Trial of Therapy Using Genes," *The New York Times*, September 29, 1999, http://query.nytimes.com/gst/fullpage.html?res=9E06EED8173EF93AA1575AC0A96F958260 (January 28, 2008).
10. Larry Thompson, "Human Gene Therapy: Harsh Lessons, High Hopes," *FDA Consumer Magazine*, September/October 2000, http://www.fda.gov/Fdac/features/2000/500_gene.html (January 28, 2008).
11. "A New Broom? From OPRR to OHRP: Transforming Human Research Protections," *Modern Drug Discovery*, November/December 2000, http://pubs.acs.org/subscribe/journals/mdd/v03/i09/html/Clinical1.html (January 28, 2008).
12. "OHRP Fact Sheet," Office of Human Research Protections, http://www.hhs.gov/ohrp/about/ohrpfactsheet.htm (January 28, 2008).
13. Michael E. Zevitz, "Ventricular Fibrillation," eMedicine, July 18, 2006, http://www.emedicine.com/med/topic2363.htm (March 25, 2008).
14. All quotations are from interviews conducted by the authors, unless otherwise cited. Reprinted with permission.
15. Like other departments at the FDA, the Center for Devices and Radiological Health has established advisory committees (known as "panels") to provide independent, professional expertise and technical assistance to the agency on the development, safety and effectiveness, and regulation of medical devices. Each committee consists of recognized authorities in a specific field. These individuals provide recommendations to the FDA, although all final decisions are made by the agency.
16. "Transvenous Leads Electrify Defibrillator Market," BNet, November 1993, http://findarticles.com/p/articles/mi_m3498/is_/ai_14649516 (April 10, 2008).
17. Nancy Stark, "The Clinical Research Industry: New Options for Medical Device Manufacturers," Medical Device Link, January 1997, http://www.devicelink.com/mddi/archive/97/01/029.html (January 28, 2008).
18. Ibid.
19. "Clinical Trials of Medical Products and Medical Devices: What You Need to Know," Spine Health.com, http://www.spine-health.com/research/what/trials01.html (March 28, 2007).
20. Carol Rados, "Inside Clinical Trials: Testing Medical Products in People," U.S. Food and Drug Administration, http://www.fda.gov/FDAC/features/2003/503_trial.html (March 27, 2007).
21. "Protection of Human Subjects," Food and Drug Administration, http://www.accessdata.fda.gov/scripts/cdrh/cfdocs/cfcfr/CFRSearch.cfm?CFRPart=50&showFR=1 (March 27, 2007).
22. Reed Abelson, "Financial Ties Cited as Issue in Spine Study," *The New York Times*, January 30, 2008, p. 1.
23. "Protecting Subjects, Preserving Trust, Promoting Progress II," Association of American Medical Colleges, Task Force on Financial Conflicts of Interest in Performing Clinical Research, October 2002, http://www.aamc.org/research/coi/2002coireport.pdf (January 28, 2008).
24. Adapted from *Intuitive Biostatistics* by Harvey Motulsky (Oxford University Press: New York, 1995).

25. Stephen N. Oesterle, "Coronary Interventions at a Crossroads: The Bifurcation Stenosis," *Journal of the American College of Cardiology* (December 1998): 1853.
26. Ibid.
27. David O. Williams and Mary C. Fahrenbach, "Directional Coronary Atherectomy: But Wait, There's More," *Circulation* (1998): 309.
28. Charles A. Simonton, *et al.*, "'Optimal' Directional Coronary Atherectomy," *Circulation* (1998): 332.
29. Oesterle, op. cit.
30. Ibid.
31. Based on information provided from "Type I and Type II Errors," Wikipedia.org, http://en.wikipedia.org/wiki/Type_I_error#Type_I_error (April 5, 2007) unless otherwise cited.
32. "Error," Answers.com, http://www.answers.com/topic/error (December 10, 2008).
33. Richard Kuntz, "Data Monitoring Committees and Adaptive Clinical Trial Design, Innovation.org, http://www.innovation.org/documents/File/Adaptive_Designs_Presentations/23_Kuntz_Data_Monitoring_Committees_and_Adaptive_Clinical_Trial_Designs_Medica_Device_Considerations.pdf (April 5, 2007).
34. Motulsky, op. cit.
35. "Confidence Intervals," Exploring Data Website, http://exploringdata.cqu.edu.au/conf_int.htm (April 25, 2007). Motulsky, op. cit.
36. "Research Methods: Inferential Statistics," AllPsych Online, http://allpsych.com/researchmethods/errors.html (April 5, 2007).
37. "Statistical Power," Wikipedia.org, http://en.wikipedia.org/wiki/Statistical_power (May 7, 2007).
38. "Confidence Intervals," op. cit,
39. Motulsky, op. cit.
40. "Expanded/Early Access," CDRH Device Advice, http://www.fda.gov/cdrh/devadvice/ide/early.shtml (May 7, 2007).
41. Cartoons created by Josh Makower, unless otherwise cited.

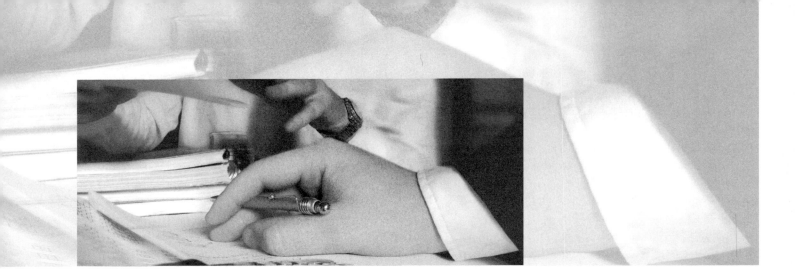

5.4 Regulatory Strategy

Introduction

The team spent 28 months and more than $45 million on the clinical trial. The PMA submission totaled over 5,000 pages and required three detailed responses to questions from the FDA. The panel meeting was a rollercoaster, with several experts voicing significant initial concern about the safety profile of the device. Now the official letter has arrived from the CDRH and the team is savoring one short sentence: "You may begin commercial distribution of the device upon receipt of this letter."

Developing an effective, strategic approach to regulation is of critical importance in the biodesign innovation process because of the time, cost, and effort associated with this work stream. Regulatory strategy is tightly coupled with the competitive positioning of a new technology, as well as the sales and marketing approach, clinical strategy, quality processes, and risk management policies that the innovator puts into place. An effective regulatory strategy establishes the foundation and sets the constraints within which these interrelated issues must be managed.

While 4.2 Regulatory Basics provides fundamental information about the requirements and tactical implications of current regulatory processes, this chapter addresses the strategic aspects of the regulation as part of the broader biodesign innovation process.

OBJECTIVES

- Understand the strategic risks and opportunities associated with the PMA and 510(k) pathways.
- Appreciate how and when to communicate with the FDA.
- Consider how a global regulatory strategy can be integrated with the approach to the FDA.
- Recognize common regulatory mistakes and learn how to avoid them through the creation and implementation of a strong regulatory strategy.

Regulatory strategy fundamentals

When the innovator sets out to develop a regulatory strategy, there are a few key considerations that should guide this effort. First, although it is possible to help inform the thinking of the US Food and Drug Administration (FDA) about the regulatory pathway for a medtech innovation, ultimately the decisions will be made by the agency based on the experience and orientation of the group charged with reviewing the submission. As described in 4.2 Regulatory Basics, the overall culture of the FDA is strongly rooted in its mission of preventing harm to society. How this mission is interpreted when making a decision on a particular submission depends on the context of current events (and can be heavily influenced by recent problems with devices or drugs). In general, regulatory processes for devices are becoming more rigorous and the level of evidence required is moving closer to that for drugs (though the size, complexity, duration, and expense of device trials still tends to be much lower). There is significant risk to innovators and companies working within this constantly changing context (see Figure 5.4.1) – but many important opportunities also surface from the dynamics in the regulatory environment.

The second important "big picture" issue is that FDA clearance or approval by itself is essentially worthless if there is not a pathway to get the new technology reimbursed. As described more fully in 5.6 Reimbursement Strategy, the Centers for Medicare and Medicaid Services (CMS) are responsible for determining whether or not a device is reimbursed by the government which, in turn, influences the practice of thousands of other private insurers. Therefore, developing a reimbursement strategy must proceed in parallel with crafting the regulatory strategy.

Is the device regulated?

One important question to clarify at the start is whether or not the innovation should be regulated as a medical device. In some cases, it is possible to develop claims for the uses of a product that take it outside of an FDA pathway. One example is exercise equipment, where the boundaries regarding whether or not these are medical devices are not completely clear. If a product can be developed and sold directly to consumers (i.e., it does not need Medicare reimbursement), it can be advantageous to take a route that bypasses FDA approval, when possible. However, innovators must be careful about how they promote a device if they do not have regulatory clearance. For instance, an exercise treadmill could be advertised as a way of increasing the pulse rate, but not as a technology that reduces the incidence of heart disease. In some cases, it may be feasible to market a product directly for consumer use while developing a related product that will be regarded by the FDA as a medical device. An innovator can petition the FDA to obtain a formal determination of whether or not a product is a medical device under the FDA definition (this is known as a 513(g) petition).

Which center and division/branch to target?

If it is determined that a product will be regulated by the FDA, the innovator must make an initial determination of which class of product – device, drug, or biologic – is most appropriate, as well as which center – the Center for Devices and Radiological Health (CDRH), Center for Drug Evaluation and Research (CDER), or Center for Biologics Evaluation and Research (CBER) – should evaluate the technology. Most often for medical devices, this is a straightforward decision (see Chapter 4.2 for the FDA's definition of devices) and the CDRH is clearly the appropriate center to perform the evaluation. One practical point is that the CDRH tends to

FIGURE 5.4.1
Approaching the FDA can be daunting, especially for first-time innovators. A well thought out strategy, the right consultant, and effective communication are key elements to success.[16]

be somewhat easier for innovators to deal with than the other centers. If a technology is in the "gray zone" with respect to classification, innovators should push for regulation by the CDRH.

Increasingly, devices are being integrated with drugs or biologics and this has led to the creation of the Office of Combination Products (**OCP**). This group is positioned centrally in the FDA under the commissioner. Its charter is to make decisions about whether or not a new technology constitutes a combination product and, if so, which centers regulate it (and which one takes the lead). The most well-known example of a combined product is the drug eluting stent, in which a basic metal lattice device used to hold open an artery was modified with a polymer coating that releases a drug to inhibit the restenosis (tissue regrowth) that occurred with bare metal stents. In this case, the OCP determined that the mechanical effect of opening the artery was the primary mechanism of action and that the drug effect on restenosis was secondary. As a result, the CDRH took the primary regulatory role, with substantial input from the CDER. All things being equal, the approval pathway for a combination product will be clearer (and the review is more likely to be based at the CDRH) if the drug or biologic that is added to the device is already cleared through its respective center. However, because combination product submissions are so complex, strategies to address them must be developed on a case-by-case basis. Unfortunately, there is limited guidance available at present to help navigate the combination product landscape.

Once a technology is directed to the CDRH, there may be latitude in some cases regarding which division or branch reviews the submission. The various divisions and branches tend to have somewhat different standards of evidence (type and length of trial, number of patients, etc.), based on their traditions and experience. For example, the first company to seek approval for vessel anastomotic devices (devices to attach blood vessels together without requiring a surgeon to suture by hand) was reviewed by the general surgery group. Once the devices reached the marketplace, there were some unanticipated cardiovascular complications. As a result, the market leader – along with a number of other companies that had begun the regulatory process in the interim – were shifted to the cardiovascular division, with new sets of standards that were much more difficult to meet. In practice, it is unusual for the innovator or company to have much latitude in choosing a division, but it is worth understanding the implications of dealing with one division versus another. These dynamics represent the types of issues that regulatory experts track.

Strategy related to IDEs

As described in 4.2 Regulatory Basics and 5.3 Clinical Strategy, an investigational device exemption (IDE) is required before any trials are initiated with US patients. For a device or study with *significant risk*, the FDA must review the protocol prior to IDE approval. For a device or study that the company considers to provide a non-significant risk to patients, the FDA does not need to be involved if the company can convince one or more hospital institutional review boards (IRBs) that there is no significant risk. In this case, the FDA does not review the protocol and the study can proceed (for example, to generate data for a 510(k) submission). It is important to note, however, that the FDA must be notified if any IRB rejects the study, even if several others have approved it.

Pursuing a non-significant risk IDE without discussion with the FDA can lead to a strategically risky situation, because the innovator has no perspective in advance of the 510(k) submission on what the FDA may think of the study once it is presented to the agency. For this reason, it is generally advisable to request a pre-IDE meeting for either a non-significant or a significant risk device. Pre-IDE meetings are also useful for companies intending to launch pilot human studies outside the United States. In these meetings, the company, often accompanied by the lead clinical investigator(s), presents data to the responsible group at the CDRH about the device, the clinical development program, and the intended use after approval. The agency group members review existing bench and animal data, and they make informal, non-binding suggestions regarding the need for additional pre-clinical data, as well as the study design. The FDA will generally have a statistician in the group, so it is advisable to provide a detailed statistical analysis of the study plan

Table 5.4.1 *The advantages and disadvantages of a 510(k) versus a PMA pathway have important strategic implications.*

Pathway	Pros	Cons
510(k) Substantial equivalence to predicate device(s)	• Quicker route to market • Less expensive submission • Easier to modify • Clinical data needed only 10 to 15 percent of the time • Fewer post-surveillance requirements • No facility pre-inspection required	• Competitors can more easily follow company to market • May limit company's ability to market the device as desired (since it must follow the indications of its predicates)
PMA Reasonable assurance of safety and effectiveness established by appropriate clinical trials	• Harder and more expensive for competitors to follow (as they are also subject to PMA requirements) • Can be used to allow a company to market a device for a new or different indication than existing devices • Potentially exempt from product liability cases	• Safety and efficacy must be proven • Longer, more complex application/approval process • Expensive submission and approval process • Requires clinical data (often randomized, controlled) • May require panel review

(and bring a statistician along from the company side to answer questions and defend the statistical methods). Depending on the complexity of the device and study, it may be useful to request two meetings to occur at different stages in the planning process.

Each meeting will generally last an hour and typically requires scheduling at least four to six weeks in advance. It is critically important that the company team making the presentation outlines a clear study plan, with the purpose of having the agency react to its proposal. An open-ended approach to the FDA, asking for guidance in helping to design a study, will likely result in a more complicated and expensive study than the company wants to undertake (and often can harm the company's credibility with the agency). It is important to know that the FDA is not obliged to "approve" any aspects of the study design at this meeting and has the right to change its advice as the study matures. Nonetheless, it is helpful for the company team to try to pin the agency down with questions like: "Does the FDA agree that 100 patients is a sufficient study cohort for this trial?" or "Are nine months of follow-up data sufficient?" Someone from the company team should take detailed notes of the meeting and send a summary to the group leader from the FDA. There is no requirement for the FDA to record notes, although some agency teams will keep minutes.

510(k) versus PMA

The appropriate regulatory pathway within the CDRH – 510(k) versus premarket approval (PMA) – is generally determined by the risk classification of the device. As emphasized in 4.2 Regulatory Basics, it is unusual that an innovator or company has the opportunity to exercise significant influence over the pathway chosen. However, it is worthwhile to be clear about the high-level advantages and disadvantages of a 510(k) versus a PMA, as summarized in Table 5.4.1 and in the discussion of the two pathways below.

Strategies for 510(k) clearance

Clearance under a 510(k) relies upon the concept of substantial equivalence to a predicate device, where a predicate device is defined as a device cleared before 1976, a device already cleared by FDA through the 510(k) process, or a 510(k) exempt device. Substantial equivalence means that the new device is at least as safe and effective as the predicate device or devices. Importantly, according to the FDA, substantial equivalence does not mean the new and predicate devices must be identical. In fact, the predicates used in successful 510(k) applications sometimes appear not to have close similarities to the new device – but, from the standpoint of the FDA, they do provide relevant comparisons. The FDA's criteria for substantial equivalence are reviewed in Chapter 4.2.

Strategically, selecting predicate devices and specifying the intended uses for the new device requires a sophisticated understanding of substantial equivalence – and, in practice, requires input from a regulatory specialist. 510(k) clearance limits the use of a device to a finite set of clearly defined indications (which will ultimately be described in the package insert once the product is sold). For expeditious 510(k) clearance, the indications for the new device must be described in language that is the same as that used to obtain clearance for the chosen predicate device(s). There can be no "creative license" in the language about indications. The only way that new language surrounding indications will be acceptable to the FDA is if data is provided to back up any modifications, while still proving substantial equivalence in terms of the technological characteristics and intended use of the device. For these reasons, companies must think carefully about the predicate devices they choose and the intended clinical uses they intend to promote. Otherwise, they may find themselves in a situation where regulatory clearance of the new device has been achieved, but the new device cannot be marketed to the target clinical population.

It is worth noting that regulatory practice is akin to case law in that the interpretation of substantial equivalence changes according to the accumulated experience of the FDA and the regulatory professionals with whom the agency works. Understandably, the FDA reacts to public and congressional concerns. For example, the association of silicone breast implants with autoimmune conditions in the 1990s brought into suspicion many devices containing silicone elastomer when, in fact, it was liquid silicone that was thought to be the risk factor (note that these concerns were never scientifically proven). Clearance of devices containing silicone – of any type – was stalled for a time while the FDA reacted to the events generated by these concerns.

A company pursuing a 510(k) strategy must consider its desired speed to market in the context of the indications for which it intends to market its device. If speed to market is particularly important, a company may choose the simplest predicate device(s) and indications for use to clear the device quickly, and then pursue additional indications following initial clearance. This may take less time overall than testing the limits of predicate bundling and undergoing multiple rounds of questioning with an FDA reviewer. One example of this was the 510(k) strategy pursued by Intuitive Surgical which developed a robotic technology to provide the precision of a surgeon. As a first step, Intuitive Surgical approached the FDA with the concept of using its device as a surgical assistant to hold tools while the surgeon operated. This required mostly bench data for the 510(k) clearance. In order to obtain clearance for the purpose of actually performing the surgery, the FDA required a randomized, controlled trial. However, the company was able to use the initial 510(k) clearance for the surgical assistant as a predicate for itself in obtaining a second 510(k) clearance for performing the surgery. In the meantime, the initial 510(k) approval gave the company the opportunity to familiarize surgeons with the technology and gain initial revenues from sales of the robot system and equipment for the more limited use.[1]

The amount of time and risk a company is willing to bear in terms of generating all desired indications in one submission (versus pursuing additional indications following initial market release of the product) is an important issue that requires careful thought early in the regulatory process. Rapid product release enables a company to accumulate early experience with the device and understand first-hand which alternative clinical applications are worthy of further study. On the other hand, if the device will be a direct competitor to a product already on the market, the company may benefit from taking more time to gain clearance for additional indications or to gather data that substantiate the superiority of its device.

If a company receives regulatory clearance under a straightforward 510(k) based on substantial equivalence to a competing device, it may be difficult for the company to differentiate its new device from the predicate for marketing purposes. When a device is cleared for use under a 510(k), the company is restricted from

Table 5.4.2 *A company has different alternatives in the way that it approaches the 510(k) regulatory pathway, depending on its strategic priorities.*

510(k) strategy	Pros	Cons
Quickest: use one established predicate device, if it provides the indications for use that are needed to sell to the initial target market	Facilitates the fastest clearance with fewest question rounds (90 days or less)	May not provide the desired indications for use to sell to the target market
Moderate risk: bundle two or more predicate devices to add desired indications	Increases the possible market to which the device can be sold	May be subject to further rounds of questions or additional testing, which will increase review time
Riskiest: push the limits of what is a Class II device by using predicate device(s) that have tenuous substantial equivalence arguments	May enable a company to avoid a PMA	Definitely will be subject to increased FDA questioning, additional data, and possible determination that the device is not substantially equivalent to anything currently on the market

making claims above and beyond those of the substantially equivalent device (see Table 5.4.2). However, it may be possible to differentiate the device through creative marketing techniques. For example, Cardiva Medical has a vascular closure device cleared under a 510(k), yet the company is trying to compete against devices that have been granted PMA approval (such as Abbott/Perclose). Importantly, Cardiva does not make the same claims for its device as the PMA devices, but some of the soft indications granted for the device ("to promote hemostasis at arteriotomy sites as an adjunct to manual compression") allow Cardiva to go after the same market using language that suggests sameness. However, this is a nuanced (and potentially risky) strategy that requires the vigilant involvement of a company's regulatory team in the approval of all marketing materials to be certain that FDA guidelines are not violated. Innovators should also keep in mind that this is an area where strict penalties can be imposed by the FDA.

It is useful to be as broad as possible in describing the features of a device for a 510(k) submission. For instance, if a company makes a type of catheter that comes in a variety of diameters, it should be sure to include a range of sizes in its 510(k) submission that incorporates all diameters of the existing device, as well as those diameters that may be developed in the future. Thus, a guide catheter of the same design can be submitted in 5, 6, 8, and 9 French sizes,[2] even if the original released product is only 6 Fr (though the original submission must bracket all of the sizes for which clearance is sought).

Clinical data in 510(k) submissions

The inclusion of clinical data is only required in approximately 10 to 15 percent of all 510(k) submissions. In some cases the FDA issues specific requirements for clinical data in a guidance document. There are other situations in which submitting clinical data is strategically important. If there are measurable differences between the new device and the predicate device(s), clinical data can be used to demonstrate the safety and effectiveness of the new device. Clinical data can also be submitted to the FDA in order to justify the expansion of the device's intended use (over and beyond what is indicated by the predicate device(s)).

There is a recent trend for the FDA to require clinical data more often for 510(k) clearance and to require higher levels of clinical proof in the study designs. Clinical data can come in a variety of forms. It is not uncommon for companies to submit small human registry-type trials to satisfy the data requirement. In general, in the 510(k) setting clinical data are used to validate bench data and to establish *safety* rather than effectiveness. Trials should be designed to be as small and as simple as possible. Rarely (if ever) are randomized clinical trials required for the 510(k) pathway, and rarely do studies have to be large (20 to 100 subjects is commonly considered adequate). Often, the necessary clinical data can be collected outside the United States (OUS) and the studies, therefore, are exempt from the formal requirements of an IDE. However, they are still subject to international ethical standards. If human studies are performed in the United States, an IDE is likely to be required (see 4.2 Regulatory Basics and 5.3 Clinical Strategy for more information).

Meeting with the FDA

For a 510(k) submission that does not require a clinical trial, it is generally not necessary to have any "premeeting" with the agency to discuss the application. If the technology and predicate device(s) are reasonably straightforward, a request for such a meeting could demonstrate to the FDA that a company does not possess the expertise necessary to follow the path (which could be a liability in terms of how the company is perceived by the FDA). However, a 510(k) pre-meeting might be appropriate under circumstances in which the company knows of problems with the predicate device it plans to use, or is aware of competitors or other companies using the same predicate device(s) and experiencing difficulty with the submission. Similarly, a meeting with the FDA may be appropriate when an innovator or company wants to confirm the appropriate pathway before investing significant time or effort in the development process. As mentioned above, when there is a substantial clinical trial involved, it is wise to meet with the FDA to review the trial design.

Strategies for premarket approval (PMA)

The ratio of PMA to 510(k) submissions ranges roughly from 1:50 to 1:100, reflecting the fact that the time, cost, and risk involved in the PMA pathway is significantly higher than for 510(k) clearance. However, there are some clear advantages to the PMA pathway. An approved PMA is, in effect, a private license granting the owner permission to market the device.[3] This is also sometimes referred to as a kind of "regulatory patent." The PMA provides a barrier to entry such that competitors who desire the same type of device and same indications are required to undergo the longer, more costly PMA process. In early 2008, the US Supreme Court added further weight to the importance of PMAs in its decision in Riegel versus Medtronic, which stated that medical device manufacturers could not be sued for complications if the device used was approved by a PMA or PMA supplement.[4] This ruling was generally viewed as "good news" for the industry from a legal perspective, as it prevents litigants from pursuing legal action on a state-by-state basis in most cases. Companies can still be sued for negligence, however, and the full implications of this decision are subject to further interpretation and clarification.

While it is not at all common for a company to pursue a PMA if a 510(k) can be justified, a company may decide on a PMA pathway in an attempt to get approval for indications that are beyond the scope of those cleared for predicate 510(k) devices. This would usually only be undertaken by large companies with significant resources and an aggressive desire to erect competitive barriers. If successful, the "first mover" can block followers who simply do not have the resources to conduct the necessary clinical trials and overcome the other hurdles arising from the new regulatory path. It is also worth mentioning that Medicare and other insurance plans are increasingly requiring evidence from randomized clinical trials that are similar to the scale and complexity of PMA trials. If the company is obliged to conduct large trials for purposes of reimbursement, the added benefit of PMA approval may come at a relatively small cost. It is also clear that in recent years both the FDA and the Department of Justice are being more aggressive in pursuing medtech companies that promote devices for off-label uses – that is, uses beyond those explicitly approved by the FDA (more information about off-label usage is provided later in this chapter). A PMA approval provides a clear go-ahead for marketing the device for the indications that have been studied.

Premarket approval is required for class III (high risk) devices, even those that are similar to competitors in the field (e.g., stents and pacemakers). Seeking regulatory approval under the PMA pathway as a follow-on device can have both upsides and downsides. For a technology that has a long history of established clinical safety and efficacy (for example, femoral closure devices pioneered by AngioSeal and Perclose), undertaking a PMA for a similar device leaves little guesswork.

The study design and endpoints are well established, and the FDA is familiar with the technology. The PMA process, therefore, is still rigorous, but there are many lessons and shortcuts that can be leveraged as a direct result of the predecessor's experience.

On the other hand, the FDA can impose acquired learning onto the PMA process. This happened with abdominal aortic aneurysm stent grafts (synthetic conduits placed within an aortic bulge to reduce the risk of rupture). The FDA approved two technologies (Medtronic's Aneuryx and Guidant's Ancure) based upon early, promising results. As these two technologies enjoyed wider commercial use, more was learned about performance issues and design problems. Followers in this area, therefore, were hit with significantly increased requirements over and above those faced by the original applicants.

In order to obtain PMA approval, a company is required to validate every indication it seeks with clinical research data. Often, there is temptation to pursue as many indications as possible with the initial submission. However, it is essential to consider how this will complicate clinical trial design. Initial clinical trial design should focus on the simplest, most achievable endpoints possible to maximize the study's chance of success (see 5.3 Clinical Strategy).

A strategy employed by many companies is to target the most important indications for use to capture the initial market, and then add indications following initial approval through a mechanism known as a PMA supplement. A PMA supplement is required by the FDA following PMA approval before a company makes any changes affecting the safety or effectiveness of the device. These changes may include new indications for use of the device, changes in labeling, changes in sterilization procedures or packaging, changes in the performance or design specifications, changes in operation or layout of the device, and use of a different facility to manufacture or process the device.[5] Approval of a PMA supplement can require anywhere from 30 to 180 days of review time.

The strategy of obtaining PMA approval for core indications first (followed by subsequent PMA supplements) allows the company to earn initial revenue to support the development and testing required for the additional indications. Every year, many more PMA supplements are received and approved by the FDA than PMA applications.

Meeting with the FDA

From a strategic perspective, the PMA process should be highly collaborative, with numerous interactions occurring between the FDA and representatives of the company. One of the most important opportunities for collaboration occurs early in the evaluation and submission process.[6] A meeting with the appropriate FDA branch should occur before significant time and expense is put into device development. While these meetings are optional, they are strongly recommended because they help establish a relationship with the FDA and allow the agency to unofficially "buy in" to the selection of safety and effectiveness measures in advance of the submission. Companies may request a formal **determination meeting** before the PMA application process is initiated, in which the FDA is obliged to give an official response to the study that is proposed. However, a determination meeting usually does not turn out to be productive for the company (i.e., the agency does not endorse the study). For this reason, it is generally recommended for companies to work within the context of the more informal meetings.

Premarket approval post-approval requirements

Increasingly, the FDA is interested in collecting data on the impact of new technology *after* PMA approval, as the technology is disseminated into more widespread practice. The recent, high-profile issue of stent thrombosis (clotting) following implantation of drug eluting stents has added impetus to this trend. The FDA approved the first two drug eluting stents based on data that the restenosis rates were lower than with the conventional bare metal stents. Following PMA approval, the stents were deployed in millions of patients, and clinicians began to see a problem with some stents abruptly clotting off completely – a serious event that can occur years after the procedure. This complication was not detected in the PMA trials and led the FDA to require postmarket surveillance for the two companies with approved products, as well as mandating larger and longer trials for the approval of new drug eluting stent products from other companies.

Post-approval requirements, which may include continuing evaluation and periodic reporting on the safety, effectiveness, and reliability of the device for its intended use, are an increasingly common part of the PMA landscape. Evaluation and reporting may be achieved through any of the following measures: a postmarket study or registry to track outcomes; reporting on the continuing risks and benefits associated with the use of the device; maintaining records that will enable the company and the FDA to trace patients if such information is necessary to protect the public health; inclusion of identification codes on the device or its labeling or, in the case of an implant, on cards given to patients in order to protect the public health; submission of published and unpublished reports of data from any clinical investigations or non-clinical laboratory studies involving the device or related devices; and submission of annual post-approval reports.[7]

The Heartport example below demonstrates the important strategic implications associated with PMAs and 510(k)s, as well as the effect of regulatory strategy on other work streams within the biodesign innovation process.

FROM THE FIELD — HEARTPORT, INC.

Managing regulatory strategy as one aspect of a company's go-to-market plan

Heartport was founded in 1991 by Stanford-trained physicians John Stevens and Wesley Sterman. Inspired by the minimally invasive "keyhole" techniques gaining traction in other fields of medicine at the time, Stevens conceived of an idea for performing cardiac surgery using catheters and small incisions (ports) between the ribs instead of the traditional open chest method (see Figure 5.4.2).[8] Together, Stevens, Sterman, and their team developed the Port-Access™ system – specialized surgical instruments and catheter-based devices to perform coronary artery bypass (CABG) and mitral valve repair and replacement procedures. The new system would provide a less painful and invasive alternative to the more than 300,000 patients undergoing CABG operations and 80,000 patients receiving valve surgery each year in the United States in the 1990s.[9]

Initially, the Heartport team believed that it would have to follow a PMA regulatory pathway based on the novelty of the procedure it was proposing. According to Stevens, "This was a pretty new thing. We weren't really sure if there were any predicate devices out there that would be relevant." The company built its operating plan around the PMA approval pathway and also advised its investors to expect a relatively lengthy regulatory process. However, when Bob Chin joined Heartport as its vice-president of regulatory affairs and quality, he proposed a different approach. "Bob had a deep history with the FDA," said Stevens.[10] "He told us, 'Look, there are a lot of predicate devices here that are relevant.'" By evaluating the instruments and devices that made up the Port-Access system on an individual basis (e.g., catheters, cannulae, long-handled versions of common surgical tools), the company identified a series of reasonable predicates against which it could demonstrate substantial equivalence. Now convinced that it had a sound argument for pursuing a 510(k), the company approached the FDA for its input. "The FDA knew exactly what we were doing. It was a very open, constructive dialog," Stevens noted.

The Heartport team made a 510(k) submission, including the results from a small-scale human trial. "I don't recall specifically, but I think we submitted clinical data on ten patients," Stevens said. In October 1996, the FDA granted clearance, giving the company the go-ahead to begin marketing its Port-Access system approximately two years earlier than originally anticipated. Heartport launched the product just three months later, in January, 1997. According to Stevens, the team was highly commercially oriented at this point in time. "The switch to the 510(k) path was a commercial decision more than anything else," he said. "We thought, 'We can get to market faster, we can move quicker, we can get to profitability sooner. The regulatory agency supported this approach. Let's go this way.'"

In 1997, Heartport accounted for nearly half of the $50 million minimally-invasive cardiac surgery market. Its nearest competitor, CardioThoracic Systems, held less than a 20 percent market share.[11] Yet, in 1998, as its competitors continued to grow, Heartport's market share

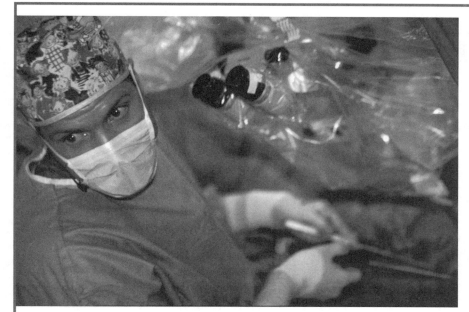

FIGURE 5.4.2
Stevens performing surgery using the Port-Access system (courtesy of Mark Richards, www.markrichards.com).

declined 17 percent and its sales dropped from $23.4 to $18.6 million.[12] While there were multiple issues that contributed to the company's difficulties in commercializing the Port-Access system, Stevens called out its regulatory strategy as one contributing factor. "The change in regulatory path didn't cause the necessary reflection within the company. 'Wow, we're going to be approved here ahead of when we thought. How should that change our strategy?'" Instead, the company rushed to market without thinking holistically about what it needed to do to support the commercialization of the device. "Moving from a PMA to a 510(k), without even the need for clinicals, hurt the overall adoption of the technology," he said, reflecting on Heartport's experience. "It really slowed it down and caused more negative effects than positive. What happened was that suddenly the market said 'You've got a 510(k), you better start selling.' But there really wasn't an adequate body of clinical experience and knowledge on how the product should best be used."

"In retrospect," commented Stevens, "I think the right approach would have been to think about the highest and best use for the product and roll it out slowly in a study context post-approval." The company initiated clinical studies, "but they were all post-marketing," he explained. "The best papers I've seen have come out in 2007 – ten years later. These results much more clearly define where outcomes are best and how best to use the product in a lot of different ways."

When asked what advice he would give to entrepreneurs as they develop a regulatory strategy, Stevens had this to say: "Regulatory strategy needs to be one component of your go-to-market strategy. It can't be the driver and it can't be the response. You have to say, 'OK. How does regulatory strategy fit into a comprehensive, thoughtful, best approach to entering the marketplace?' There are some times when you may have more freedom than you anticipate; you can get a 510(k) and you don't need a lot of clinical data so you can just go out and start selling. That can be to your advantage or it can be to your disadvantage. It's not the end all. It's just one component of a thoughtful approach to the market."

However, despite the challenges experienced by Heartport, he cautioned those with an innovative new technology not to assume that one regulatory pathway is superior to the other. "510(k)s are usually great," Stevens said. "Frankly, pursuing a PMA for the vast majority of 510(k) devices would only be deleterious to patients. The clinicians/inventors should know better than anybody how to thoughtfully, cautiously, and appropriately enter new medical devices into the clinical marketplace. And that should be the driver. It shouldn't be what the regulatory body does or doesn't say, unless you can't get it there. And it shouldn't be the commercial aspects. You always have to make sure you're thinking about the patients first and foremost. Ultimately, that's what's going to make for the best devices, too. If the patients are getting optimal outcomes, then that's the optimal commercial endgame, as well. Then you don't shoot yourself in the foot, like getting a PMA on a device that can get a 510(k) and slowing a device into the marketplace by three years because you decided to be extra cautious."

Strategies regarding off-label device use

One of the more subtle areas of regulatory strategy deals with devices that are used outside of the approved FDA indications (off-label use). Recall that the FDA does not have jurisdiction to regulate the practice of medicine. A physician may use a device in any fashion s/he sees fit, provided this use is in the best interests of the patient and is broadly within the standard of care. In practice, this means that many devices are used outside of the indications for which they are approved by the FDA. A classic example is the biliary stent, which was approved to prop open the bile tracts in the intestine. For years, the large majority of biliary stents were, in fact, used by cardiologists to treat blockages in the coronary arteries, despite the fact that these stents were not approved for this indication (coronary stents were approved in Europe several years earlier than in the United States, convincing US cardiologists that this practice was within a reasonable standard of care).

When companies ultimately sought approval for coronary stents from the FDA, they designed trials strategically with an eye toward proving that the stents were superior to the existing technique of balloon angioplasty. Among other things, this meant targeting vessel sizes that were likely to yield favorable results (smaller coronary arteries, it turns out, have a higher incidence of renarrowing after stenting than vessels of greater than 3 mm in diameter). Once the stents were approved for vessels of the optimal size range, however, cardiologists began using the stents widely in the smaller vessels.

Although companies often consider the off-label potential for their devices, a blatant strategy based on this approach is not advisable. The FDA understands this issue well and is on the look-out for companies trying to exploit this "loophole." Recently, the ten or more companies making biliary stents were effectively censured by the FDA for not being more forthcoming about the dominant use of their products.

Integrating US, European, and other regulatory strategies

Innovators can often achieve important advantages by integrating regulatory strategies across geographic areas. In some cases, regulatory processes may be optimized by leveraging clinical data and regulatory approvals obtained in one market to shorten the time to regulatory approval, reimbursement, or market adoption in another. For example, many US medical device start-ups pursue CE marking of a device subject to a PMA pathway to pave the way for US approval. Conversely, companies may seek the clearance of a 510(k) device in the United States before entering the EU market and then use the US clinical data to gain reimbursement abroad. Another common strategy is to use clinical data from a CE marking trial in lieu of a US pilot study to enable the company to start a US pivotal trial earlier. This data can also be used to obtain regulatory approval in other markets, such as Japan. More information about these integrated strategies is provided below.

Early CE marking of PMA devices

For the majority of devices on a PMA pathway in the United States, CE marking can be obtained more quickly than the PMA. This may be helpful to companies for several reasons: (1) it can provide a valuable revenue stream (although reimbursements and, therefore, product prices tend to be lower in the EU), (2) it provides the company with early user feedback on device performance and adoption, and (3) it provides early clinical data that may be used in subsequent FDA submissions.

The reason why CE marking of class III devices is faster than PMA approval is directly related to the differences in the clinical data requirements between the US and EU regulatory pathways. The FDA requires that class III devices demonstrate reasonable safety and effectiveness, which is typically achieved through prospective, randomized, controlled trials involving hundreds of patients (see 5.3 Clinical Strategy). In contrast, CE marking only requires that devices demonstrate safety and performance. Usually, compliance with the EU requirements, even for class III devices, can be demonstrated with much simpler trials. For instance, the GuardWire® from Percusurge, Inc., which enables debris created during endovascular interventions to be captured to prevent it from embolizing, was awarded the CE marking on the basis of a 22 patient single-arm study[13] demonstrating safety and performance (i.e., that debris was aspirated during the interventional procedure). In the United States, however, the FDA required an 800 patient, multicenter, randomized trial[14] for effectiveness (i.e., that compared the device to the standard care to demonstrate a reduction in complications).

Deferring EU market entry of 510(k) devices

Early European approval of a device that is headed for 510(k) clearance in the United States rarely provides the same value to a company seeking to penetrate the US market as it does for a PMA device, unless the 510(k) device requires clinical data. 510(k) clearances that do not require clinical trials can usually be obtained much more quickly than PMA approvals, allowing a company to begin providing the device to US physicians and building this market. Furthermore, clinical data generated in the United States (before or after 510(k) clearance) may subsequently be used to build markets OUS. This strategy was successfully used, for example, by Kyphon® for its interventional device to perform kyphoplasty for the treatment of vertebral body compression fractures.

Using EU clinical data for US and other regulatory approvals

Pre- and post-CE marking trials may be used for US and/or other regulatory approvals, provided that certain aspects of clinical trial design and conduct conform to the necessary requirements. These include compliance with all relevant local regulations and any applicable Good Clinical Practices (e.g., set forth by the ICH[15]). All such requirement should be designed into the trial from the outset to ensure that the clinical data will be acceptable by the authorities outside of the EU.

It is important to be aware that the FDA looks critically at data obtained from foreign studies in terms of its applicability to the US patient population. If medical therapy or practice differs significantly from that in the United States, the data can be deemed not applicable. For example, in coronary interventions a certain class of blood thinners or anticoagulants (IIb/IIIa inhibitors) is not used as commonly in the EU as in the United States. Therefore, trials involving devices for coronary interventions must be carefully designed to ensure that data acquired overseas will be accepted by the FDA should it be used to support a US approval. Note that the FDA reserves the right to audit foreign clinical sites to confirm whether or not the data is valid for the purposes of obtaining an approval to market a device in the United States. The agency has specialized auditors for this purpose.

Efforts toward the increasing harmonization of regulatory standards have created a dynamic and complex environment that demands a regulatory consultant with strong international experience. The EU is still the largest market outside the United States for most medical devices and is the best understood region in terms of regulatory approvals. Japan is a market that reimburses well, but is notoriously slow to approve new devices, taking much longer than the FDA. Young companies are becoming increasingly familiar with regulatory agencies in South America (especially Argentina and Brazil) since "first in man" studies are often conducted in these countries. As described in Chapter 4.2, regulations in other large markets (e.g., China, India) are evolving. When considering overseas regulatory approval as a means of conducting earlier clinical trials, be cautious about the credibility and applicability of the data to ensure a wise investment.

GETTING STARTED

For the innovator who is considering how to select and begin navigating a regulatory pathway early in the biodesign innovation process, the resources provided in the Getting Started section of Chapter 4.2 will be useful. For regulatory issues that come into play at a more advanced stage of a development program (arising in the context of a start-up or existing company), the following steps are recommended.

See **ebiodesign.org** for active web links to the resources listed below, additional references, content updates, video FAQs, and other relevant information.

1. Validate device classification and regulatory pathway

1.1 What to cover – Before investing in the creation of a detailed regulatory strategy, it is a good idea to validate that the device has been properly classified. If a device classification cannot be determined, classification information can be requested from the FDA (section 510(g) of the Food, Drug and Cosmetic Act describes this process). Due to the growing complexity of the device market, the number of such requests has been steadily increasing over time.

1.2 Where to look – Review the classification determination as outlined in **4.2 Regulatory basics** (if needed). Look at competitors and similar devices used in different therapeutic areas. Attend major specialty meetings and browse the exhibit floor for ideas about device classification and possible predicates. In addition, the following resources can be used to gather relevant information:

- **FDA overview of regulations** – A high-level overview of the FDA's regulatory pathways and policies.
- **Device advice** – The CDRH's self-service site for obtaining information concerning medical devices and the application/submission processes.
- **Determination of device classification** – Information provided by the FDA to help companies classify a device.
- **Device classification database** – Searchable database provided by the FDA to aid device classification.
- **CFR Title 21: Food and Drugs – Parts 1 to 1499**
- **FDA 510(k) Database** for predicate devices.
- **FDA PMA Database** for predicate devices.

2. Develop a regulatory strategy

2.1 What to cover – Involve a regulatory expert and all company functions, especially design, quality, clinical, and marketing, in the development of a regulatory strategy that effectively addresses the chosen path. For either a 510(k) or PMA path, be sure to schedule a pre-meeting with the FDA to launch the strategy on the right foot.

2.2 Where to look
- **Device advice**
- **European Commission medical device sector legislation** – Overview of the European directives that govern medical devices.
- **MHRA medical device directive** – Additional information about regulatory processes in the EU.
- **Regulatory affairs professional society** – An online source for tools, coursework, guidance, and articles on regulatory subjects.
- **Professional articles** – Articles in leading professional journals (e.g., Kaplan et al., "Medical device development: from prototype to regulatory approval," Circulation (2004): 3068–3072) can provide valuable information and timely insights into the dynamic regulatory process.

3. Modify and monitor regulatory strategy

3.1 What to cover – Monitor the regulatory strategy being pursued on an ongoing basis. Be proactive in identifying issues and addressing them before they develop into problems that could affect the company's relationship with the FDA. Consider withdrawing any FDA submission that encounters serious difficulties and could be a candidate for a "not substantially equivalent" (NSE) determination under the 510(k) pathway. Maintain open and ongoing communication with the FDA to avoid surprises and ensure a constructive, mutually beneficial dialog between the agency and the company. Actively work to avoid the common regulatory pitfalls summarized in Table 5.4.3.

3.2 Where to look – Stay abreast of new regulatory-related information made available in the popular press, professional journals, and on the FDA website.

Table 5.4.3. *Regardless of which regulatory pathway is taken, innovators and companies often make mistakes on their regulatory submissions that cause unnecessary delays and/or additional expense.*

Mistake	How to avoid
Documentation errors	Be very clear on requirements around formatting, documents, and procedures. Do not let the submission get hung up based on administrative errors.
Failure to pay attention to FDA guidance documents	The FDA provides substantial information in the form of guidance documents for various technologies. These are available on the web and should be reviewed carefully.
Lack of regulatory involvement in design/development	A company's regulatory professional is required to translate engineering requirements into language for FDA consumption. If the person preparing the submission is not involved in the development process, holes can develop in the "story" that the submission should tell. Regulatory and design/development personnel must work hand in hand to communicate about the submission needs and engineering activities occurring within the organization.
Unnecessarily complex submission	Keep the submission as simple as possible to address the basic requirements. For example, use simple line drawings, not 3-D color images. Use clear language and well-defined arguments for substantial equivalence.
Poor communication with reviewer	FDA reviewers are people who want to help. Company representatives should seek to develop relationships with them as they would with any customer or supplier. Work reasonably and amicably. If the reviewer asks for something the company thinks is unreasonable, try to understand why s/he is asking for that item. Consider if there is opportunity for education. It is far easier to provide the information/data than to argue about it.
Failure to anticipate reviewer questions	A good team anticipates questions in advance and often have data "in their pocket" if/when specific questions arise. The technical team should be "on deck" to conduct testing during the review cycle if this is requested.
Lack of correspondence between marketing claims and approval	Marketing materials must reflect the official summary from the FDA with respect to claims for use. Ensure that the summary is available to and understood by the creators of this material.

Credits

The material in this chapter was based, in part, on lectures by Howard Holstein and Janice Hogan of Hogan and Hartson. The authors would like to acknowledge Trena Depel for her help in developing this chapter. Many thanks also go to John Stevens for his assistance with the case example, as well as Howard Holstein and Jan Benjamin Pietzsch for their substantial contributions to editing the content. We appreciate Sarah Sorrel, of MedPass International, reviewing the information on international regulatory requirements.

Notes

1. From in-class remarks made by Howard Holstein as part of Stanford's Program in Biodesign, Winter 2008. Reprinted with permission.
2. The French catheter scale is used to measure the outside diameter of a cylindrical medical instrument. 1 French (Fr) is equivalent to approximately 0.33 mm.
3. "PMA Overview," CDRH Device Advice, http://www.fda.gov/cdrh/devadvice/pma/index.html#introduction (February 23, 2007).
4. Janet Moore, "Ruling Shields Medtech Firms," StarTribune.com, February 20, 2008, http://www.startribune.com/business/15801117.html (March 3, 2008).
5. "PMA Supplements," U.S. Food and Drug Administration, http://www.accessdata.fda.gov/scripts/cdrh/cfdocs/cfcfr/CFRSearch.cfm?FR=814.39 (March 3, 2008).
6. Additional information on early collaboration meetings can be found at http://www.fda.gov/cdrh/ode/guidance/310.pdf.
7. "Postapproval Requirements," CDRH Device Advice, http://www.fda.gov/cdrh/devadvice/pma/postapproval.html#general (February 25, 2007).
8. "Heartport, Inc.," HBS No. 9-600-020, March 3, 2000, p. 2.
9. Ibid., p. 6.
10. All quotations are from interviews conducted by the authors, unless otherwise cited. Reprinted with permission.
11. "Heartport, Inc.," op. cit., p. 11.
12. Ibid.
13. J. G. Webb, R. G. Carere, R. Virmani, *et al.*, "Retrieval and Analysis of Particulate Debris Following Saphenous Vein Graft Intervention," *Journal of the American College of Cardiology* (1999): 468–475.
14. D. S. Baim, D. Wahr, B. George, *et al.* "Randomized Trial of a Distal Embolic Protection Device During Percutaneous Intervention of Saphenous Vein Aorto-Coronary Bypass Grafts," *Circulation* (2002): 1285–1290.
15. ICH stands for the International Conference on Harmonization of Technical Requirements for Registration of Pharmaceuticals for Human Use. This project brings together experts and regulatory authorities from Europe, Japan, and the United States to discuss scientific and technical aspects of product registration.
16. Cartoons created by Josh Makower, unless otherwise cited.

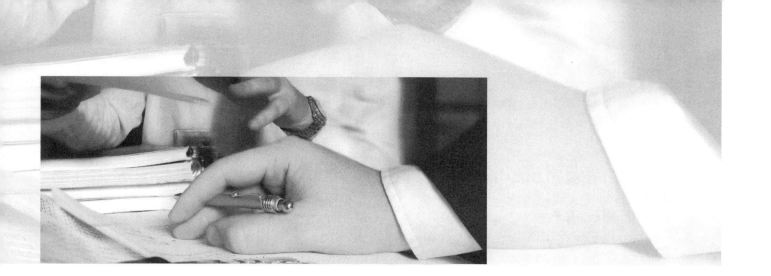

5.5 Quality and Process Management

Introduction

An engineer looks at a monitor over the shoulder of a cardiologist to observe a problem with his device. The nitinol stent is broken and embedded in a patient's artery. Shortly thereafter, several other doctors from hospitals across the country start reporting the same issue, although many others indicate that they have experienced no problems at all. After studying all the films, it is confirmed: despite the fact that all of the cycle testing before commercial launch was completely successful, some of the stents are now breaking in the field. In an effort to understand why some are breaking and others are not, the engineer thoroughly reviews the causes of failure. By checking the detailed records maintained at the company as part of its quality management system, he determines that only stents from certain lots are indicated. Records to support device traceability reveal that the failure is associated only with devices from a particular nitinol. In fact, the failures can be isolated to this one particular material source. Stents from this ingot are then recalled and, after communicating the issue and its resolution to the proper authorities, the company is able to overcome this challenge, eliminating the failures and returning to the market to achieve great success.

Scenarios like this highlight the fact a disciplined approach to development and production is a key enabler of the biodesign innovation process. If the innovator's end goal is to create a new medical device that can be manufactured according to precise specifications and used safely and reliably in medical care, then rigorous process management and quality systems are central to making this happen. Not only are such processes critical for achieving regulatory clearance for a new innovation, but they allow the innovator and team to transition from producing devices one (or a few) at a time to reliably manufacturing batches of products whose performance can be tested and validated. This is essential because patient lives may be put

OBJECTIVES

- Appreciate the critical role that quality systems and effective process management play in allowing an innovator to transition from creating an idea to building a company capable of reliable production.

- Recognize the key components of quality systems and their strategic importance to managing a business.

- Understand the FDA's quality-related regulatory requirements and how they differ from what is required in other countries.

at risk if the quality process guiding development and subsequent production fail to ensure the creation of a product that performs as intended and meets all specified safety requirements.

Quality and process management fundamentals

One of the end goals of the biodesign innovation process is to create a new medical device that can be manufactured according to precise specifications that allow it to be used safely in medical care. In the early stages of the process, there is typically not a "real" company. Instead, an innovator might be working alone or with a small group of engineers to solve problems and produce enough prototypes for preliminary testing. But at some point, if everything goes well, this informal approach to development must incorporate defined operating procedures to ensure that the developed product can achieve regulatory clearance. Further, the team must transition into an entity with precise systems capable of producing devices predictably and reliably, delivering them to the end user in a timely fashion, and monitoring their performance in the clinical environment.

While essential, this transition is fraught with risks. If the systems put into place are inadequate in any way, they can potentially undermine the success of the venture. Rigorous processes are needed to govern all of the activities that must be scaled up as a company prepares for commercialization.

Principles of process management

Process management refers collectively to the disciplined management techniques used to ensure that a company's core activities are performed consistently and deliver outputs that meet precise specifications. At its most basic, process management involves four key steps:

1. **Identify key business processes** – Define all of the actions that need to be taken to deliver an output to a stakeholder. These processes can be focused on production, but may also include broader service and product development activities.
2. **Analyze these business processes**
   - For each process, define the final output, target stakeholders, and their requirements.
   - Clarify the steps in each process, including their inputs and interim outputs, the information needed, management oversight, and key responsibilities. Determine how each step should be performed, specifications for the interim and final outputs, and how the outputs will be verified.
3. **Document the processes** – Record the steps in each process in a way that a third party can replicate and verify the core activities.
4. **Establish mechanisms to manage the processes** – Define mechanisms and metrics to measure and track deviations from the documented processes, identify and implement appropriate corrective actions, and enact process improvements.

A company can use this approach to more efficiently and effectively manage almost any process within the organization.

Process management in product development and manufacturing

Among the most important processes that a company puts in place on its way to scaling up are those related to product development, manufacturing, packaging, labeling, storage, distribution, installation, and service of medical devices. The systems used to manage and monitor these processes, which are regulated by the Food and Drug Administration (FDA) and authorities outside the United States, are collectively referred to as quality management systems (**QMS**). Relevant regulations require that detailed specifications are developed for the devices, that the devices are manufactured according to these specifications, and that the devices perform according to these specifications once distributed and/or installed. Their performance must also be monitored so that problems can be reported to the regulatory bodies as they are identified and corrected. In this context, the term "quality" broadly refers to the activities undertaken by the company to ensure that these regulatory requirements are met and that products delivered to the customer (primarily patients and physicians) are safe and reliable.

Quality systems can be a source of competitive advantage for a company if they lead to rapid product development cycles, superior quality as measured by extremely low defect rates, and low production costs. Moreover, quality systems are required by regulatory bodies, such as the FDA, and they are reviewed as part of the regulatory submission. The FDA demands that "companies establish and follow quality systems to help ensure that their medical devices consistently meet applicable requirements and specifications."[1] The FDA has precise regulations, called Quality System Regulations (QSR), which specify the exact requirements that a company's quality systems will need to meet before a product is cleared or approved for the market.

Traditionally, process management, and especially quality, has carried the stigma of being a "policing" function within many organizations. For instance, quality measures have been perceived as imposing extra work steps and stringent requirements that necessitate a lot of effort for little measurable return. Busy employees sometimes question why the company needs formal quality processes, especially in young start-ups that may not yet have products in the market. However, recent cases demonstrate that the cost of not having an effective quality management system in place can be devastating, no matter what stage of development the company is in.

For example, consider the case of Boston Scientific. In 2000 and 2004, FDA quality inspectors found hundreds of quality control lapses in six of the company's US-based manufacturing facilities. As a result, the FDA issued three warning letters to Boston Scientific. When subsequent inspections in three additional plants revealed quality control and regulatory issues, the FDA issued a broad "corporate warning letter," indicating that the company's corrective actions to address prior violations were inadequate. Such a move by the FDA, which was considered a broad critique of the company's entire quality control systems, is unusual.[2] Disclosure of the letter led to an almost immediate five percent drop in Boston Scientific's share price.[3]

Examples of the quality problems uncovered at the company varied from facility to facility, but all related to the procedures, processes, and timeliness of the company's quality management system. At one plant, employees were unaware that company headquarters had recalled a needle used to treat tumors in cancer patients. In another location, managers missed deadlines for notifying the FDA of reports linking Boston Scientific devices to serious injuries (federal regulations require notification within 30 days).[4] In a third facility, workers were able to override a company computer system to ship devices to a hospital after they had failed a quality inspection.[5] While the corporate warning was in effect, the FDA informed Boston Scientific that it would not approve any new devices that could be affected by the quality problems. At the time the warning was issued, this had potentially damaging consequences for the company since it was preparing to submit its new drug eluting stent for FDA approval.

After receiving the FDA's warning, Boston Scientific launched one of the most systematic and extensive quality system enhancement programs in medical device history. The company dedicated two years, millions of dollars, and hundreds of employees to implementing changes in its manufacturing, distribution, and monitoring systems.[6] For example, one action the company took was to consolidate 23 separate processes for tracking complaints into a single system.[7] In October 2008, based on the company's remediation efforts, the agency lifted a number of the restrictions it imposed in the warning letter, although Boston Scientific was still working actively with the FDA to resolve some remaining issues.[8]

Even in small start-ups, the cost of not having effective quality systems can be high. These costs are known as failure costs, which represent the expenses incurred by a company as a result of having products or services that do not conform to requirements or satisfy customer needs. They are divided into internal and external failure categories by the American Society of Quality (**ASQ**).[9] Internal failure costs occur prior to delivery or shipment of the product (or the furnishing of a service) to the customer. This includes scrap, rework, reinspection, retesting, material review, and downgrading. External failure costs occur after a product or service has been delivered to the customer. These costs include processing customer complaints, customer returns, warranty claims, product **recalls**, and even the risk of being shut down by the FDA. "Soft" costs must also be taken into account, particularly as a company is seeking to establish itself in the marketplace. These include the negative effect of poor quality on a company's reputation, its

FIGURE 5.5.1
Assemblers and quality personnel in a clean room environment (courtesy of Acclarent, Inc.).

ability to attract investors, and its ability to attract and retain valuable employees.

Organizations that proactively address quality early in the design and development of a product have the potential to save significant time, money, and other resources while reducing the likelihood of devastating product safety issues and/or recalls.

What is quality management?

The concept of quality management is best understood by breaking it down into two components. Quality assurance (**QA**) refers to processes that attempt to ensure – in advance – that products will meet desired specifications and will perform according to specifications when delivered to the end user. As defined, QA is a broad concept that covers all company-wide activities, including design, development, production, packaging, labeling, documentation, and service, as well as support activities such as employee training and procedures.[10] Quality control (**QC**) refers to activities performed after these processes have been executed (e.g., once a product has been produced but before it is released to the customer) to confirm that the specifications have, in fact, been met. Unit testing, with the intent of finding defects, is one way that QC is executed (see Figure 5.5.1).[11] Quality control applies to all stages of production from incoming materials, in-process materials/subassembly testing, and finished goods testing.

The primary difference is that QA is *process* oriented (it reflects the four steps of process management as they relate to quality) and QC is *product* oriented. Quality assurance makes sure a company does the right things, the right way, whereas QC makes sure that the results of what the company has done perform as expected.[12] Stated another way, QA is a prospective process and QC is a retrospective process in product development. However, the two are interrelated: QC is a component of a QA system used to identify problems that can then lead to changes in QA practices to prevent the same problems from resurfacing.

Quality management systems

A quality management system (QMS) is the vehicle through which both QA and QC activities are implemented. A QMS includes policies, processes, and procedures for the planning and execution of all quality-related activities within an organization. It also delineates clear responsibilities, starting with the senior

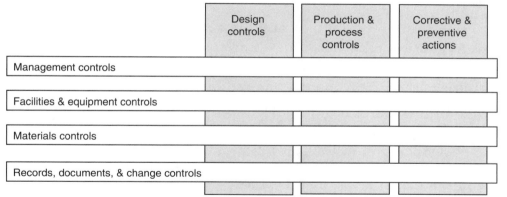

FIGURE 5.5.2
The components of a quality management system and their relationship to one another (based on the FDA's quality management systems guidelines that are part of the agency's QSR).

executive level, and helps drive performance improvement through the measurement and careful management of core business processes.[13] Importantly, quality activities used to be the responsibility of an isolated functional team; however, senior executives are now considered accountable for quality activities by regulatory bodies such as the FDA. Ultimately, all organizational functions have an important role to play in implementing quality management processes.

A typical QMS that would satisfy the basic requirements of regulatory bodies such as the FDA involves as many as seven components (as shown in Figure 5.5.2). In the description that follows, the principles of this type of QMS are outlined at a high level, along with the associated regulatory requirements. More details about the individual components of the system are provided later in the chapter.

At the highest level, **management controls** provide processes and guidelines for administering the complete system. **Design controls** refer to specific processes used to manage design specifications and their modifications. Production and process controls (**P&PC**) ensure that production processes have minimum deviations from their desired performance targets and result in a safe product. Corrective and preventive actions (**CAPA**) refer to the systems used to prevent and correct failures. According to the FDA's inspection techniques, these four subsystems make up the heart of a company's quality system. The three remaining components – **facilities and equipment controls**, **materials controls**, and **records, documents, and change controls** – complement the primary subsystems, but generally refer to a company's approach to managing the equipment, materials, and documentation relevant to the four primary subsystems. Referring back to the QA/QC distinction, QC results are a core part of CAPA, but CAPA also incorporates the process used to act upon the QC results. QA is more of a broad, umbrella concept that reflects the combination of all components of the QMS and how they work together to provide the infrastructure for prospectively delivering a high-quality product.

Importantly, quality practitioners in the field often point out that there is not a single successful quality system model. The best way to evaluate whether a company's quality system is adequate is to verify that it achieves certain core objectives:

- It results in the documentation of all systems and product requirements.
- People are well trained and follow the documented requirements.
- Records are generated to prove that the requirements are consistently followed.
- It establishes proactive systems to deal with the identification and resolution of problems and improvement opportunities.

Innovators should also understand that quality-related regulations by the FDA and other authorities are written as directional guidelines and must be interpreted appropriately for each company's products and business model. How a company interprets and documents its approach then essentially becomes the requirements against which the FDA will audit it for compliance (i.e., the company is audited against its own quality policies and internal documentation).

Management controls

The purpose of the management control subsystem is to ensure that a company's management team provides adequate resources to support effective device design, manufacturing, distribution, installation, and servicing activities. It also establishes mechanisms for ensuring that the quality system is functioning properly, and for allowing management to monitor the quality system and make necessary adjustments on an ongoing basis. According to the FDA, the rationale for this subsystem is that a quality system that has been implemented effectively and is monitored to identify and address problems is more likely to produce devices that function as intended.[14] The key components of the management control subsystem include:[15]

1. Clearly defined, documented, and implemented quality policies and plans.
2. Well-defined quality objectives.
3. An executive in charge of quality management.
4. An organizational structure that includes provisions for resources dedicated to quality management and that enables the organization to fulfill stated quality objectives and requirements.
5. Systems for management reviews to monitor the suitability and effectiveness of the quality system and take corrective action where necessary to bring the system into a state of effectiveness.
6. Audit processes to verify that deficient matters are being addressed.

Design controls

From an engineering perspective, design controls are absolutely essential to any quality system and are likely to be the first aspect of a quality system that an early start-up firm needs to address. The key objectives of design control are that the design itself is reproducible and traceable and is proven to be both safe and effective. The basis for design controls is initiated in 4.5 Prototyping, as detailed user requirements are collected. However, the process for organizing and documenting this information is formalized with the implementation of a design control process.

Design control requirements are implemented to help ensure that every device performs as intended when produced for commercial distribution. To conform with these requirements, device engineers must establish and maintain design plans and procedures that describe design and development activities, define responsibility for implementation, and identify/describe the interfaces with different groups or activities that provide (or result in) input to the design and development process.[16] These plans must be reviewed, updated, and approved as the design and development of a device evolves. These activities are not prescriptive but provide a method for management to exercise appropriate control over early design work, as well as to assign responsibilities that are consistent with the scope of the design effort.

Premarket approval and 510(k) submissions for class III devices typically must include a complete description of the design controls that the company implements. Without this information, the FDA cannot complete its so-called premarket quality inspection (more information about quality inspection is provided later in this chapter). Table 5.5.1 presents guidelines from the FDA that outline the main activities that must be accomplished to achieve an effective design control system.[17]

At times, innovators are confused about the difference between **design verification** and design validation: design verification confirms that the product meets the specifications laid out by the product development team, while design validation confirms that the design specifications meet customer requirements. Any design specification will need to be validated before it is implemented in a production system and verified before the product is delivered to the customer. A simplistic example for explaining the differences involves a company that has decided to produce a cement lifejacket. Following design input and output, the company can verify that its cement lifejacket meets the product specification (i.e., it is made of cement). However, when the company seeks to validate the design against user needs, it encounters problems, since the cement lifejacket clearly will not float (therefore, it will not satisfy user requirements or the lifejacket's intended use).

Production and process controls

The P&PC subsystem is focused on ascertaining that companies develop processes to ensure that

5.5 Quality and Process Management

Table 5.5.1 *Understanding and implementing design controls can be valuable even before the innovator has a complete quality system in place.*

Activity	Description
Identify design requirements for the device	• Establish and maintain procedures to ensure that the design inputs (or requirements) relating to a device are appropriate and address the intended use of the device, including the needs of the user and/or patient. • The procedures should include a mechanism for addressing incomplete, ambiguous, or conflicting requirements. • Design input is the starting point for product design, providing a basis for performing subsequent design tasks and validating the design. Therefore, development of a solid foundation of requirements is the single most important design control activity. If the majority of design time is spent upfront, doing things correctly, later stages of design can be expedited. • Design input requirements must be comprehensive and include functional (what the device does), performance (speed, strength, response time, accuracy, reliability, etc.), and interface (compatibility user, patient, and other external needs) requirements. Almost every device will have requirements of all three types. • Perform risk analysis to identify all possible sources of failures for different components of the device, acceptable failure rates, consequences of these failures, and corrective actions. The more severe the consequences of a failure, the lower the acceptable failure rates and the more robust the corrective actions and back-up systems should be.
Develop the design output or specifications for the device	• Establish and maintain procedures for defining and documenting design output (i.e., the physical manifestation of the design planning and input) in terms that allow an adequate evaluation of conformance to design input requirements. Input requirements generally results in what is called a product specification (a document that details the technical and clinical needs the device should meet to satisfy the design intention). • Design output procedures should contain or make reference to acceptance criteria (in the product specification) to ensure that those design outputs that are essential for the proper functioning of the device are identified.
Verify that the design output meets the design input	• Establish and maintain procedures for verifying that the design output meets the design input requirements (i.e., does the device adhere to the design specification?).
Hold design reviews throughout the design process to identify significant problems with the design or the design process	• Establish and maintain procedures to ensure that formal documented reviews of the design results are planned and conducted at appropriate stages of the device's design development. • Multiple reviews of design verifications are not uncommon for complex devices that undergo successive design iterations, and will occur at each stage of the design process. Reviews must be conducted by cross-functional teams, thoroughly documented, and approved (signed off) by responsible personnel up through the senior management level.
Validate that the design meets defined user needs and intended uses	• Establish and maintain procedures for validating that devices conform to defined user needs and intended uses. This should include the testing of production units under actual or simulated use conditions.[62] • Design validation should also be performed under defined operating conditions on initial production units, lots, or batches, or their equivalents. • Design validation should include software validation and risk analysis, where appropriate.
Transfer the device design to production specifications	• Establish and maintain procedures to ensure that the device design is correctly translated into production specifications.

Table 5.5.1 *Continued*

Activity	Description
Control changes to the design during the design process and changes in the design of products on the market	• Establish and maintain procedures for the identification, documentation, validation or (where appropriate) verification, review, and approval of design changes before their implementation. • This is commonly addressed under a document control or change control system. Especially during the design process, the design teams often manage change and then move the design into the document control systems as part of transferring the design to manufacturing for production.
Document design control activities in the design history file	• The document history file should be set up to contain or reference the records necessary to demonstrate that the design was developed in accordance with the approved design plan and all design control requirements. • The design history file must be made available to the FDA (or other certified inspectors) for review. The FDA will evaluate the adequacy of manufacturers' compliance with design control requirements in pre-approval inspections for class III devices and also during routine quality systems inspections for all classes of devices subject to design control. • A product cannot be legally marketed in the United States without a design history file.

manufactured devices meet their specifications. It also focuses on validating (or fully verifying the results of) those processes and activities to monitor and control them.[18] As a company's R&D efforts move it toward a device that is ready for production, it must consider these types of controls, which are distinct from design controls.

In developing P&PC systems it is important to understand when deviations from device specifications could occur as a result of the manufacturing process or environment and to pay special attention to processes with a high risk of potential deviations. Management should be on the look-out for potential P&PC problems when a process is new or unfamiliar to the company, is used to produce higher-risk devices, or is used for manufacturing multiple devices. Production and process controls problems may also be indicated when a problem with a particular process is indicated through the CAPA subsystem, for example, when a process has a high risk of causing device failures, employs a variety of different technologies and profile classes, or has never been examined or inspected.[19]

Once processes with a higher than average risk for P&PC problems are flagged based on the presence of one of more of these indicators, methods can be developed and implemented for controlling and monitoring them to minimize production deviations.

Corrective and preventive actions

The purpose of the CAPA subsystem is to establish processes for collecting and analyzing information to identify and investigate product and quality problems. The CAPA subsystem is made up of processes such as non-conforming raw materials reporting, production process deviations, and customer complaints. Systems are put into place not only to ensure that deviations, complaints, and other problems are reported and documented, but that appropriate actions are taken to correct and prevent the recurrence of problems and that these actions are verified. Communicating CAPA activities to responsible people, providing relevant information for management review, and documenting these activities is essential to dealing effectively with product and quality problems, preventing their recurrence, and minimizing device failures.[20] As a result, this subsystem is one of the most critical components of the quality system and is viewed as such by the FDA.

Key activities that need to be undertaken in this area include defining, documenting, and implementing robust CAPA processes that ensure the visibility of quality-related problems all the way to the top of the organization. The CAPA subsystem is closely related to medical device reporting (**MDR**). Medical device reporting is the mechanism through which the FDA receives information about significant adverse events from **manufacturers** so they can be corrected quickly. For this reason, CAPA and MDR processes must be tightly integrated (see Appendix 5.5.1 for more information about MDR).

Equipment and facility controls

Equipment and facility controls are meant to ensure that a company's equipment and facilities are qualified (i.e., they are suitable for their intended purposes), and that standard operating procedures have been designed, implemented, and enforced for all equipment and facilities managed by the company. Qualification of equipment involves its installation, ongoing operations, preventive maintenance, and the overall validation of its outputs and related processes. This subsystem applies to equipment and facilities involved in design, production, and post-production activities. These controls allow the company to determine if any quality events are related to equipment at a particular facility so that equipment changes can quickly be made (when needed) to address a quality event.[21]

Material controls

Whether they are related to design, production, or post-production activities, all materials used in a medical device must be carefully controlled. Medical device companies must maintain processes to track materials and their associated suppliers to ensure the quality of those materials and that the final product satisfies the design specifications. All suppliers must be rigorously screened and able to demonstrate that their materials are traceable to qualified and appropriate sources. Detailed records (including traceability of material to suppliers and specific delivery lots) must also be maintained so that any problems that arise can be tracked at the materials level.[22]

Records, documents, and change controls

This subsystem is focused on ensuring that medical device companies maintain a secure, comprehensive, and centralized approach to managing all records and documents related to their quality systems.[23] Clearly defined protocols must be in place to track changes to processes, policies, and the products themselves, manage version control, and make necessary documentation accessible to those who need it during design, production, and post-production activities. Another key principle is that all parties involved agree on how work will be done. When changes to these original work agreements are required, they must be formally evaluated and confirmed.

Implementing a quality system

Implementing a quality system can be time consuming and resource intensive since detailed processes and procedures must be carefully orchestrated and then documented. While regulatory bodies require that quality systems are in place, these requirements focus on the goals of these systems and not on *how* the procedures must be implemented. Therefore, designing a quality system can be somewhat of a creative exercise (and, as mentioned earlier, the company's interpretation becomes the standard against which it is regulated). On the one hand, this allows a company to iterate and develop a system that works with its particular needs and fits the company culture. On the other hand, the degree of interpretation and analysis required to make "appropriate" decisions that are likely to achieve compliance can be challenging. As quality consultants tend to point out, many "gray areas" exist when it comes to achieving compliance.[24]

Another problem that companies struggle with is getting and keeping management's attention when it comes to the quality system. Too often, management has a habit of only caring about quality after a problem has surfaced. This is due, at least in part, to the fact that it is so difficult to quantify the value of a quality system that works, whereas the cost of a quality-related problem once it emerges is all too obvious. Management commitment and attention is also essential to making sure a quality system is maintained and supported by employees throughout the organization. Given that management has to "sign off" on most aspects of the quality system and, thus, share responsibility for its outcomes, it is in management's best interest to ensure that a sound quality system has been put into place.

A third issue that many companies face is characterized by a reactive versus a proactive approach to

Table 5.5.2 *The innovator should remain keenly aware of these common mistakes in order to avoid them.*

Common mistakes when implementing a quality system
Viewing the requirements as a burden, rather than a mechanism for increasing the company's effectiveness.
Entrusting management of the quality system to employees who do not have thorough training and experience in quality systems.
Making quality an isolated, non-essential function rather than integrating it into the business via cross-functional and senior management involvement.
Being too prescriptive early on and not considering the stage of product development or the team.
Not including quality professionals or users in system design.
Not thinking about how the quality system will need to scale and grow to keep pace with growth in the rest of the business.
Not training all personnel on the overall quality system and their essential role(s) in making the system work.
Designing and implementing a quality system, but not maintaining it (which ultimately renders the system useless).
Ineffective and/or untimely action to deal with problems within the system – poor use of CAPA systems.

quality management. Especially with start-up ventures, too many medical device companies postpone or overlook the need to develop a quality system and fail to adequately document their work according to the procedures outlined in the quality plan. As a result, they are forced to scramble when problems arise to retrospectively "fill in the blanks." The quality system can be viewed as onerous, or as a system that supports the ultimate goal of the company – to design and produce an innovative, safe product. Viewing it as the latter yields the best systems in which compliance is part of everyday work life, rather than an additional burden. Moreover, innovators must remember that it is illegal to bring a medical device to market in the United States without having the supporting quality system in place and working properly. Table 5.5.2 summarizes these and other common implementation pitfalls that can jeopardize the effectiveness of a quality system and its return on investment.

Again, senior management sets the tone for quality system design and compliance. If the management team approaches quality as an essential, important, and value-added activity toward the organization's goals, the system is more likely to be valued, followed, maintained, and improved over time to service the company's growth and expansion. When developing and implementing a quality system, companies usually begin by naming an executive to be in charge of quality (often the vice-president of R&D in small companies). This person will sponsor the quality work stream and work with executives and managers in other parts of the business to ensure cross-functional support. A quality engineer, preferably with QSR experience, is also hired to lead the tactical development of the appropriate processes, protocols, and documentation, again with cross-functional involvement. Sessions to educate employees about the quality system, including its importance and specific requirements that affect their work, are another important early step. Then, the company can begin to hold regular review meetings to ensure that it is achieving desired performance levels.

Keep in mind that the US Supreme Court has allowed criminal penalties to be imposed on corporate officers who were in a position to prevent or correct violations, even if they may not have known about or participated in any illegal conduct. As one author described:[25]

> The [Food, Drug & Cosmetic] Act imposes not only a positive duty to seek out and remedy violations when they occur but also, and primarily, a duty to implement measures that will insure that violations will not occur. The requirement of foresight and vigilance imposed upon responsible corporate agents are beyond question demanding and perhaps onerous, but they are no more stringent than the public has a right to expect of those

who voluntarily assume positions of authority in business enterprises whose service and products affect the health and well-being of the public that supports them.

Implications of an increasingly tough enforcement environment

When considering the best approach to implementing a quality system, it is important for innovators to recognize that, in recent years, the FDA and the public have become increasingly conservative and risk averse when it comes to medical devices. One example of how the FDA has been more vigilant about exercising its authority over companies and individuals that fail to meet QSR requirements can be found in the story of the 2007 FDA raid on Shelhigh Inc., a manufacturer of heart valves and other implantable devices for heart surgery. The FDA investigators and US marshals seized all of the company's devices after finding what the FDA deemed to be significant deficiencies in the company's manufacturing processes.[26] According to an FDA statement, the **seizure** followed an FDA inspection of the Shelhigh manufacturing facility, as well as meetings with the company at which the FDA warned Shelhigh that failure to correct its violations could result in an enforcement action. The FDA also alerted the company to its manufacturing deficiencies and other violations in two warning letters.[27]

Another example that illustrates the potential severity of FDA actions to enforce QSR is the story of C.R. Bard Inc. In 1995, the company was found guilty of unlawfully selling and distributing unapproved heart catheters. The company was forced to pay record criminal and civil fines totaling $61 million. It was also required to implement stringent measures to prevent such illegal activities from occurring again.[28] In addition, three former senior Bard executives were convicted of conspiring to defraud the FDA. The three officials were found to be aware of the serious patient complications that resulted and of the company's efforts to change its products without FDA clearance. These men each received 18 months in prison.[29]

Stories such as these demonstrate why the collective attitude about quality is shifting toward increased prevention. These issues are also driving an increased focus on postmarket quality, not just premarket quality requirements.

Quality system regulations in the medical device industry: QSR and ISO

The two most dominant quality systems in the medical devices industry are the FDA's QSR and **ISO 13485**. ISO 13485 was developed by the International Organization for Standardization and is required for devices marketed in the European Union (EU) and other countries recognizing the CE mark. Depending on the target market for their products, companies will follow one or both of these standards. The two systems have elements in common, but are fundamentally separate and regulated differently.

When the FDA began to regulate medical devices in 1976, the agency developed what it called good manufacturing practices (**GMP**) to set forth quality requirements for device manufacturers. In 1997, the FDA revised and expanded the device quality regulations under the QSR rubric (although the term "GMP" still lingers in the device field). The FDA moved to QSR in order to enforce regulation that was more focused on prospectively ensuring quality (QA) as opposed to retrospectively catching quality problems (QC). The QMS outlined earlier in this chapter meets the basic QSR requirements. One of the important changes with the new approach was the introduction of design controls into US quality regulations.

The second important change in the shift from GMPs to QSR was driven by the emergence of **ISO 9001** certification in the late 1980s and early 1990s. The ISO 9001 system was considered a best-in-class model for quality management that was applicable to any industry in any country in the world. In 1994, ISO introduced supplementary guidelines (EN46001) to be used in combination with ISO 9001 to address the unique quality requirements of medical device manufacturers. As medical device companies began adopting ISO certification on a voluntary basis, along with the required GMP standards, the FDA recognized the merits of taking a systems-based approach to quality. This realization dovetailed with the overhaul of the GMP system and stimulated the FDA to make its approach more systems oriented and consistent with the ISO 9001 standard.

Table 5.5.3 *While the QSR and ISO quality systems have a great deal in common, there are important distinctions.*

	QSR	ISO 13485
General requirements/ provisions[63]	"Each manufacturer shall establish and maintain a quality system that is appropriate for the specific medical device(s) designed or manufactured, and that meets the requirements of this part."	"The organization shall establish, document, implement, and maintain a quality management system and continually improve its effectiveness in accordance with the requirements of this International Standard."
Link to regulatory approval	Required in the United States for all class II and III medical devices (as well as some class I where general controls are required).[64]	Voluntary in the United States but required in the EU and Canada as a prerequisite for regulatory approval and increasingly in other countries, often combined with other country-specific requirements.
Auditors	FDA inspectors.	Conformity assessment bodies (CABs) in EU or other third-party ISO-accredited inspectors elsewhere in the world.
Audit frequency	Generally, every two years, but this timetable is rarely maintained by the resource-constrained FDA. Safety-related concerns will merit more frequent visits.	Companies generally follow an annual or semi-annual "maintenance" audit procedure.
Audit scheduling	Audits can be announced or unannounced.	Audits are scheduled in advance.
Cost	There is no cost for an FDA audit beyond the costs the company faces in implementing and maintaining the quality system, dedicating personnel to the audit process, and addressing required corrections.	Costs of initial certification can exceed $30,000 to $40,000, in addition to annual maintenance fees. Cost depends on the size of the organization size, as well as the chosen registrar.
Audit outcomes	The FDA has enforcement power over the organization and audit findings must be acted upon.	Because ISO is voluntary, an "unsuccessful audit" simply results in postponed certification.

In 2001, ISO 13485 superseded the combination of ISO 9001/EN46001 and was launched as a worldwide quality management system designed specifically for medical device manufacturers. ISO 13485 was subsequently updated in 2003. Based on the same basic principles as ISO 9001, ISO 13485 is often seen as a crucial first step in ensuring design and manufacturing processes consistently produce quality products that meet international regulatory requirements. While it is not required in the United States, ISO 13485 certification is a prerequisite for achieving regulatory approval in the EU, as well as marketing in Canada, Australia, and Japan.[30]

Primary differences between QSR and ISO 13485

The FDA participated in the harmonization of QSR with the European ISO standards, but stopped short of adopting the ISO standards outright. The biggest difference between the two systems is that ISO is a voluntary standard in the United States and QSR is not, which makes the compliance process a very different experience for companies. The conventional wisdom is that ISO standards are more stringent and rigid than QSR, but they do not necessarily ensure QSR compliance. A sample of other high-level differences is provided in Table 5.5.3.

An important requirement for US companies is to determine where they intend to market and manufacture the device – in the short term and the long term. They must consider whether to build a quality system that is both QSR and ISO compliant, or just compliant with one or the other. Any distribution of the device in the United States requires QSR compliance and, in general, if a company is to be registered with the FDA it should follow QSR requirements (unless it has an exempt class I device). However, a company's strategy might involve early clinical work performed outside the United States (i.e., in the EU or other foreign countries). If this is the case, ISO compliance may precede QSR compliance. Or, if a company leads with US marketing but intends to later expand to the EU (or beyond), it would be wise to build a system that is both ISO and QSR compliant. Many quality professionals have experience in building such hybrid systems. If an element of one system does not apply at the current time, it is not necessary for all elements to be turned "on" at once. A framework can be established and necessary elements can be brought "online" as needed.

For additional information about ISO 13485 see Appendix 5.5.2. An interpretive summary of the FDA's QSR is provided in Appendix 5.5.3.

Quality systems audits and what to expect

An important component of the regulation of medical devices is the performance of inspection audits of the quality system. To economize resources and maximize the value of its audits, the FDA has developed the Quality Systems Inspection Technique (**QSIT**). This process is based on a "top down" approach to inspections, placing executive management at the core of the quality system. Rather than starting an audit by looking at any specific quality subsystem or potential non-conformance problems ("bottom up"), QSIT begins with an assessment of the company's total system. Then, QSIT sets forth specific guidelines and processes for the evaluation of the four primary subsystems within the company's overall quality system described earlier in this chapter (management controls, design controls, P&PC, and CAPA).[31]

The tone of a quality audit is often set by the company being audited, not the auditor. The FDA inspector has a job to serve the public by ensuring that companies are compliant with QSR. In turn, companies have the obligation to demonstrate to the FDA that they are, in fact, compliant with these regulations. This is largely determined through documented evidence, as well as some facilities inspections. The duration of quality audits depend on the nature and cause of audit, as well as the size of the firm, with brief audits lasting one to two days and lengthy, complex audits requiring one to two months. Auditors typically spend 80 percent of their time during an audit performing records review (looking for proof that the company's quality system requirements are consistently followed). The remaining 20 percent of the time is spent conducting interviews and observations of processes.

When contacting the company for a QSIT inspection, the auditor will generally request a copy of the firm's quality policy, as well as high-level quality system procedures, management review procedures, quality manual, quality plan, or other equivalent documents. (Note that QSIT inspections only apply to preannounced audits, but not all audits are preannounced.) The company is *not* required to supply this documentation. However, the audit will typically progress more quickly if this information is provided in advance and the company assumes a cooperative posture with the FDA. All documentation is returned at the time of the inspection.[32] Within a company that has a well-designed quality system, compiling this documentation should be an efficient, simple, and straightforward process.

Each FDA quality inspection is designed to begin and end with an evaluation of the management control system. Upon the initiation of an audit, the first thing inspectors will likely do is to meet with the executive responsible for quality within the company to get an overview of the overall quality system and to verify that appropriate management controls are in place. Typically, the inspectors will then select a single design project to evaluate through the end-to-end design control process. At times, the inspection assignment will direct the inspectors to a particular design project (i.e., as part of a "for cause" inspection – an audit triggered by some evidence suggesting non-compliance in a particular area). Otherwise, they will select a project that provides the best challenge to the firm's design control system. This project will be used to evaluate the process, methods, and procedures that the firm has established to implement the requirements for design controls outlined in Table 5.5.1.[33]

Based on a discussion with management, inspectors will also choose a manufacturing process to evaluate that seems to be a likely candidate for production deviations. They will then review the specific procedure(s) for the chosen manufacturing process, as well as the methods for controlling and monitoring the process. Their objective is to verify that the process is well controlled and actively monitored.

The agency will also seek to verify that CAPA system procedure(s) have been defined and documented, appropriate sources of product and quality problems have been identified, and data from these sources are analyzed to identify existing product and quality problems that may require corrective action. The inspection also seeks to confirm that the defined CAPA processes are followed when problems arise.[34]

During the audit, employees may be coached to answer only the questions asked by the inspector and not to offer up additional information. It is the auditor's responsibility to follow whatever leads s/he encounters regarding potential quality problems. An audit potentially can be prolonged by an unraveling of issues, and no company wants the FDA on its premises longer than necessary since audits are time consuming and disruptive to the organization. They can also be intimidating to employees. However, a company that has implemented effective, comprehensive quality systems from its inception should have nothing to hide from the FDA. Similarly, a well-designed quality system should be easily and successfully audited. In no case should the company try to hide anything. It is far worse to be caught trying to conceal information (which is illegal) than it is to receive a finding that will help improve the system.

Compliance actions/enforcement

As a result of an FDA audit, the inspector issues what is called a "**483**," which refers to the government form on which the audit report is provided. In this report, the company's results are classified into one of three categories:

- **NAI** – No Action Indicated
- **VAI** – Voluntary Action Indicated
- **OAI** – Official Action Indicated

Rarely is a company designated NAI. It is far more common for the report to include "findings" from the audit, which are actionable items that the company must correct (VAI). Receiving findings on a 483 does not mean that the company has serious quality problems. However, all items cited should be taken seriously and warrant **correction** and/or corrective action. Corrective action is taken to address the cause of the problem, thereby preventing the recurrence of the non-conformance issue.[35]

Generally, corrections and corrective action proposals, along with the documented evidence of those efforts, should be submitted in writing to the FDA. They should contain a detailed description of the action(s) taken and to be taken, in order to bring a given process or product into compliance within a specified time frame. Bear in mind that, just because a company takes voluntary action to correct a problem, this does not preclude the FDA from initiating administrative and/or judicial action against the firm. In determining whether quality systems deviations are sufficient to warrant legal action, the FDA will consider the significance of the device, the company's quality history, and whether the problem is widespread or continuing.[36]

When responding to FDA audit findings, it is important for companies to follow a number of basic guidelines:

- **Accept the findings** – Do not dispute the FDA's audit results unless they are truly egregious.
- **Fix the findings** – A company is not obligated to correct a problem in the way that the auditor suggests. However, it must make an appropriate, sincere effort to address all identified issues and become compliant.
- **Follow up and verify fixes at regular intervals** – Make sure, over time, that the problems have been addressed, as well as their causes. Confirm that all corrections have been sustained over time.
- **Complete all follow-up reports and documentation requested/required by the FDA** – Use this as an opportunity to strengthen the company's relationship with the FDA (by being responsive), not jeopardize it (through sloppy follow-up).
- **Be reasonable, responsible, and friendly to the FDA** – Maintain a collaborative, cooperative tone in all interactions.

More serious findings (those that the FDA believes could impact the health or safety of patients) result in an OAI classification. One of the most common official actions is for the FDA to issue a warning letter, which is a formal

Table 5.5.4 *The FDA can take a series of different actions in response to quality problems.*

Action	Description
Citation	• Formal warning to a company of the FDA's intention to prosecute if violations are not corrected. • A meeting held prior to consideration of criminal proceedings that gives the parties (possible defendants) an opportunity to present their position in the matter.
Injunction	• An order issued by the court requiring a device company to do, or refrain from doing, a specific act (e.g., manufacturing a device). Usually issued if a company has a continuing pattern of significant deviations in spite of past warnings. • If a serious health/safety hazard exists, the FDA may request a temporary restraining order (TRO) to prevent the distribution of devices that have been manufactured under the violative conditions documented by the inspection report.
Administrative detention	• Serves as a temporary "cease and desist" order. • Suspension allows the agency 20 or 30 days to determine what action to take. • Often leads to seizure of a device upon expiration of the administrative detention.
Seizure	• An action taken against a specific device. May include raw materials, labeling, packaging, or the finished device. • Intended for the FDA to take quick control over a product in violation of quality regulation and put it under the possession or custody of the court. • The owner or claimant of the seized merchandise is usually given approximately 30 days to decide on a course of action (i.e., contest the charge or request court permission to bring the product into compliance). If no action is taken, the court recommends disposal of the goods.
Prosecution	• Criminal action directed against the company and/or responsible individuals. • A misdemeanor or felony can result in fines and/or imprisonment.
Civil penalties	• Monetary penalties imposed on a company (or responsible individuals) after an appropriate hearing for violations of the law related to medical devices. • In determining the amount of civil penalty, the FDA takes into account the nature, circumstances, extent, and gravity of the violations, the violator's ability to pay, the effect on the violator's ability to continue to do business, and any history of prior violations.

communication to the company (or individual(s) within the company that control the processes in question) indicating that the FDA considers certain products, practices, processes, or other activities to be in violation of QSR requirements. The warning also states that failure to take appropriate and prompt action to correct the violations may result in regulatory action being initiated without further alerts. Warning letters can result in more severe penalties if the same problems are discovered in a subsequent audit.[37] In 2002, the FDA implemented a new policy to reduce the number of warning letters it issued. The intent of this change was to restore the seriousness of these warnings, eliminate the practice of issuing serial warning letters to the same companies for the same, uncorrected violations, and create a precedent for moving to litigation more quickly.[38]

Other actions that may be taken by the FDA, including **citations**, injunctions, **administrative detentions**, seizures, prosecution, and **civil penalties**, are summarized in Table 5.5.4.

The FDA may also make a legal agreement with a company to force it to make specific changes (or bar it from particular actions). These agreements, which are known as consent decrees, are enforced by the federal courts. They can include fines, government reimbursement for inspection costs, timelines for specific actions, and penalties for non-compliance.[39]

A note about recalls

A product recall is another enforcement action at the FDA's disposal. At some point, nearly every medical device company will experience a recall. Recalls can be triggered from internal discoveries, field failures, or audit actions. They also can be voluntary (initiated by the company) or imposed (mandated by the FDA). If a company voluntarily recalls a product to reduce a health or safety risk or to remedy a violation of an FDA regulation, the company is obligated to report this to the FDA.[40] Recalls that are imposed by the FDA are almost always more serious in nature and are usually the result of negligence on the part of the manufacturer. A company that finds itself in this position may experience damage to its reputation, financial hardship, and strain on its relationship with the FDA.

There are multiple ways that a company can remove a product from market. The appropriate approach depends on the relative severity of the health hazard presented. Different ways for removing a product from the market include:[41]

- **Recall** – Refers to a method of removing or correcting products that the FDA considers to be in violation of the laws it administers and against which the agency would initiate legal action (e.g., seizure). A recall does not include a **market withdrawal** or any form of standard stock recovery.
- **Market withdrawal** – Involves a minor violation that would not be subject to legal action by the FDA or which involves no violation (e.g., normal stock rotation practices, routine equipment adjustments and repairs, etc.).
- **Correction** – Refers to the repair, modification, adjustment, relabeling, destruction, or inspection of a distributed product while it is still under the control of the manufacturer and does not need to be physically removed to some other location.

Recalls are classified by the manufacturer according to the relative severity of the health hazard presented by the product through a health hazard evaluation. The issues considered include whether any disease or injuries have already occurred from the use of the product, degree of seriousness of the health hazard to which individuals would be exposed, likelihood of occurrence, and consequences (immediate or long range) of the health hazard. The three classifications are:[42]

- **Class I** – Reasonable probability that the use of, or exposure to, a violative product will cause serious adverse health consequences or death.
- **Class II** – Use of, or exposure to, a violative product may cause temporary or medically reversible adverse health consequences, although the probability of serious adverse health consequences is remote.
- **Class III** – Use of, or exposure to, a violative product is not likely to cause adverse health consequences.

It is important to note that the classification of recalls is the *inverse* of the FDA's device classification in terms of severity – the higher the recall level, the lower the risk.

The steps for initiating and implementing a recall are fairly prescriptive. Notification to product users is always required. Such notification must include brief reasons for the recall, as well as instructions for returning or disposing of the violative device. A number of notifications and procedures must be followed in collaboration with an FDA district recall coordinator, with the extent of the required effort dependent on the class and scope of the recall.

If a company is faced with a recall, its quality system procedures can play an essential role in helping it effectively and efficiently respond. For example, hazard analysis processes can help in quick and accurate failure analysis of the problem. Similarly, traceability requirements (e.g., material controls) can help a manufacturer identify the lot(s) of product involved in a recall to limit the scope of the recall as much as possible. If appropriate procedures are not in place, a company could be faced with a system-wide recall of product that could leave nothing in the supply pipeline and be devastating to a manufacturer. A sound quality system provides a company with a backbone and an organized approach for handling these kinds of challenges.

Guidant Corporation, which recalled thousands of its implantable cardioverter defibrillators (ICDs) in June, 2005, is one example of a company that faced sizable recall challenges. The Guidant recall was caused by a manufacturing defect that could cause several of its ICDs to short-circuit or malfunction. Thousands of plaintiffs in more than a hundred class action and individual lawsuits claimed that Guidant knew of the problems with the ICDs and failed to publicly disclose the life-threatening defects until the FDA intervened. In total, Guidant recalled or issued warnings for more than 80,000 ICDs, which severely damaged the company financially, competitively, and in terms of its reputation.[43]

On the other hand, many companies experiencing a recall emerge without any major impact to the customer. In some cases, the company may even emerge stronger after the exercise. For example, a small biopsy company initiated two recalls in 2003 (class II) for design-related issues that could have resulted in a threat to patient safety. These issues were discovered before the device was used in a patient, and a recall was initiated. An internal investigation was taken to identify and isolate suspect products and determine whether replacement products could be provided. The suspect products were recalled and examined. In the end, the recall resulted in minimal disruption to the customer, a product improvement that was driven directly by the findings of the failure analysis, and a general improvement in both compliance and the level of attention paid to the quality system, which could help pre-empt such actions in the future.

Note that a quality system cannot entirely eliminate the chance of a recall, but it can minimize the chance that one will occur, while also enabling a company to more quickly and nimbly recover when problems arise to protect business value. The case example below illustrates effective practices for managing an FDA quality inspection to achieve a positive end result.

FROM THE FIELD — DIASONICS AND OEC MEDICAL SYSTEMS

Strengthening quality systems to improve business performance

Allan May joined Diasonics, as its senior vice-president of business development, just as the company was embarking on a major turnaround. According to May, "Diasonics was the company that introduced real-time imaging to the medical device markets. Until then, if you wanted an image, you had to go to the radiography suite or another location where the images could be taken. The idea that you could look at a dynamic image during surgery didn't exist." By introducing innovative new imaging technologies, such as ultrasound and fluoroscopy, the company grew rapidly from its inception in 1978 to its highly successful initial public offering (IPO) in 1983. However, shortly thereafter, it faced a series of performance problems that necessitated the formation of a turnaround team.

When May joined the company, Diasonics was a conglomerate of four or five businesses, each with its own technology. One of the issues facing the turnaround team, he said, "was how to harvest more value out of these various pieces to increase returns to shareholders." May continued, "OEC Medical Systems was the crown jewel of this corporate group. The trouble was that Wall Street wasn't giving Diasonics credit for what we felt was the real value of OEC." OEC specialized in making fluoroscopy C-arms – mobile machines that could be moved into the surgical or special procedures suite for real-time imaging during various laparoscopic or endoscopic procedures (see Figure 5.5.3). "That market was very small in the 1970s," he commented, "but it exploded in the 1980s with the proliferation of minimally-invasive procedures."

The OEC product initially had a reputation for being costly and over-engineered with "needless features." "Fluoroscopy was a Volkswagen market, and OEC was building Mercedes," recalled May. "But as these procedures caught on and the market really started to take off, you had longer procedures, like minimally-invasive cardiac or neurology procedures, that placed new demands on the equipment. Those over-engineered features became critical, and OEC's competitors couldn't duplicate them because they didn't have the more advanced architecture. The new requirements couldn't just be added on as a feature set," he explained. This allowed OEC to capture approximately 70 percent market share against competitors such as GE, Siemens, Philips, and Toshiba. "So here was this little dinky company in Salt Lake City, Utah," said May, "competing globally against these major corporations. Every few years, they would appoint vice-presidents to study and crush this little start-up, but they never broke it. And we held that market share despite their increasing attempts."

FROM THE FIELD

DIASONICS AND OEC MEDICAL SYSTEMS

Continued

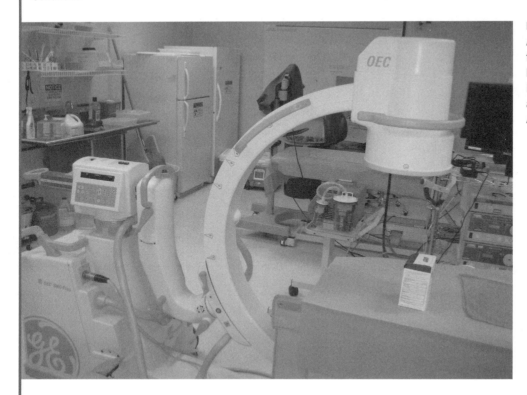

FIGURE 5.5.3
An OEC® C-Arm, similar to the products produced by OEC when it was part of Diasonics. OEC was later acquired by GE (courtesy of Acclarent).

In the midst of OEC's unprecedented growth, the company experienced an unannounced FDA audit. "This was right about the time that a couple of high-profile problems hit the FDA and our sense was that the division offices had been told to step up their inspections," May noted. "The FDA inspector came into the OEC plant in Salt Lake City, inspected it, and said that, in his opinion, it looked like the plant might have to be shut down." During this same period, the US plants of several major competitors had been shut down or their imports restricted.

The FDA's complaints addressed a series of issues, but primarily focused on complaint handling and the company's corrective actions and procedures. "To be fair," May said, "the regulations were really just being put into place when this happened.[44] But the concept of an effective complaint handling system didn't exist at OEC. Basically, someone would complain and one of the engineers would look into the problem and try to fix it. There was no senior-level involvement or review, few defined processes, and no comprehensive complaint handling system tied to corrective action loops. Most problems were considered 'unable to duplicate or verify' and so little was done to redesign the product or change production processes to prevent the problem from recurring. I also don't think we had manufacturing process instructions on the floor. We were making the best product in the world – number one in its class. Everybody knew how to do their specific job. But it didn't occur to people that, gee, if someone got hit by a bus, no one else would know how to do what they were doing."

News of the inspection results were received by employees in the plant with what May described as "outrage" and the perception of "bureaucratic make-work." "Many employees were in just a complete state of denial, or didn't understand what this was all about. So, our first challenge was to get to the core of the problem. We went to the FDA immediately and got involved with the agency. We wanted to make it very clear to them that we were not going to be confrontational. We intended to be completely cooperative in understanding and fixing our problems and would be inclusive of them in our discussions about what to change. Then, we had to go back to the plant and really patiently explain from the top of the organization down through every layer why we needed to do things differently and what those things were," he said. "The key was getting people's attention and focusing them on understanding the real problem. We couldn't just let them slough it off like, 'Okay, we have to fill out a bunch of paperwork, but we're right and the FDA is wrong.'"

As May and team began to investigate the FDA's concerns within the plant, it became increasingly clear that these quality issues were directly linked to the company's business performance. "Our service calls on the equipment had crept up from one or two a year to ten to twelve per year," May recalled. "Our customers were seeing our technicians more than our sales people. We recognized that there were some problems with the equipment but, since it had such a tailwind of being the dominant product in this particular sector, we failed to appreciate the extent to which customers were putting up with it. But they were really starting to grumble. On top of that, our margins were decreasing because it was costing us so much to service the equipment."

Over the next 18 months, OEC launched an expansive root-cause analysis to identify all potential sources of product failures. Based on the outcome, the company completely revamped its quality systems, implementing comprehensive new processes and thorough documentation. They reorganized their entire manufacturing facility and changed many of their process and systems, as well. When the FDA reinspected the plant, it was found to be in full compliance. Moreover, said May, "We ended up getting those service calls back down to one or two a year and driving down our costs. We also dramatically improved the quality of the product, our reputation with customers, and the satisfaction of our customers."

When asked for his advice to innovators implementing quality systems, May commented: "You have to completely forget the idea that this is about paperwork. You can't hire a consultant to draft a bunch of documents that you put in a drawer somewhere. Your quality system is totally about the mentality you use to design and manufacture your product. And I believe a hundred percent that if you do what the quality regulations say, and really understand them, you will design and manufacture a better product with fewer problems that will lead to improved financial performance."

Reflecting further, May added, "Quality should be understood at the top of the organization. When we were going through this, we had a slew of meetings where everyone, right up to the board of directors, had to be involved. The higher you get in the organization, the more people may gripe about this sort of thing. But senior executives can't just delegate this and expect it to be done right. Upper management has got to stay personally involved. The phrase we repeated most often was, 'Quality is a journey, not a destination.'"

Overall, May stressed, the strength of a company's quality systems can either help or hinder the attainment of its strategic goals. In this particular case, Diasonics' senior management team had developed a corporate strategy to spin out OEC (along with two other operating companies) to help "unlock" additional value in the market. However, until OEC's quality issues were addressed and its risk of being shut down by the FDA was eliminated, "We had to keep OEC underneath the protection of the corporate shell," May recalled. This delayed Diasonics' ability to implement its significant corporate restructuring plans. However, some time later, after OEC was found to be in full compliance with the FDA's requirements, Diasonics did, in fact, spin out OEC on to the public markets. Not long after, OEC was able to execute one of the larger corporate exits in the medical device field.

Applying the same rigor and discipline to managing other processes

Because of the involvement of the FDA and other regulatory authorities, the guidelines and requirements for managing quality processes are clearly defined and relatively well understood in the medical device field. However, innovators and companies can benefit by taking a similarly rigorous approach to managing many other processes within their businesses. Proactive, diligent process management of all work streams can help a company increase its efficiency, work more effectively, minimize risks, and improve communications across the organization. Specific areas that can benefit from process management include sales, customer service, and managing reimbursement/payer interactions.

For example, with sales, identifying new potential customers, converting them into paying customers, and then training and supporting them in their use of a new device can be represented as a process. Targets can be specified for the performance of each step in this process and achievement of these targets can be monitored. Knowledge of best practices in customer identification, conversion, and monitoring can be embedded in the process design and documented in sales manuals so that new sales people can learn quickly the best ways to interact with customers and achieve their sales targets. The advantage of representing these activities as a process is that it can standardize an often chaotic system and lead to improved overall performance. A potential pitfall is that it may discourage innovations in sales tactics that can occur when sales people get creative. However, experienced managers can help team members operate within the guidelines of the process while retaining enough flexibility to capitalize on new opportunities and share best practices and innovations with their colleagues (see 5.8 Sales and Distribution Strategy for more information).

Reimbursement provides another example of a work stream that can benefit from a process approach. As described in 5.6 Reimbursement Strategy, one of the key functions to be performed by a new medtech company is to support providers in their effort to get reimbursed for using a new medical device. Again this process can be documented to capture the experience of assisting different providers with different types of payers so that, when an issue arises, the process provides a template for handling it in a timely fashion. Because the goal of the "reimbursement support process" will be to expedite provider reimbursement, performance measures to monitor may include the rate of reimbursement issues, time to resolve them, final resolution outcome, etc. Monitoring these measures can help identify problems in reimbursement and allow a company to manage them more proactively.

GETTING STARTED

It is never too early to start thinking about quality and process management since the choices made early on in the biodesign innovation process can affect a company's speed to commercialization and large-scale manufacturing. Quality consultant experts are essential in this process, but they should not be engaged before the innovator has a general strategy and approach towards quality.

See **ebiodesign.org** for active web links to the resources listed below, additional references, content updates, video FAQs, and other relevant information.

1. Identify quality needs and decide on an approach

1.1 **What to cover** – Begin by understanding in what geographic locations the company intends to distribute its products in the next five to seven years. Then, determine which quality system(s) apply (QSR and/or ISO). If the company is at all uncertain about where it will distribute its products, consider building a quality system that is compliant with both. Some elements can always be "turned off" if the alternative system is not required at the time. Understand the requirements of the chosen standard carefully so that an appropriate approach can be decided upon. There is a wide range of quality publications and software tools designed to help a company understand and implement a quality management system. A range of workshops, seminars, and training courses is also available.

1.2 **Where to look**
- **The FDA's Quality System Regulation Handbook** – Provides an overview of good manufacturing practices for manufacturers of medical devices doing business in the United States.
- **The FDA's Quality System Manual** – Provides the final rule on 21 CFR Parts 808, 812, and 820, effective October 7, 1996.
- **International Organization for Standardization** – Provides access to the ISO 13485 standards, which set forth quality management requirements for manufacturers of medical devices

doing business elsewhere in the world (the standard is for sales on the site). Quality manuals for understanding ISO requirements can also be purchased from many third-party consultants and ISO-accredited certification bodies on the web.
- **American Society for Quality** – Provides general information about quality, including an overview of other quality systems not discussed here (e.g., SixSigma). The ASQ store also sells many helpful tools and references.

2. Hire a quality professional

2.1 What to cover – Because so many potential pitfalls exist in implementing a quality system, and because the consequences of non-compliance can be so devastating to a company, it is often a good idea for a company to hire or contract with an experienced quality professional (or team of professionals, depending on the scope of the need) to lead the implementation and ongoing management of a quality system. Companies often work with external quality consultants because they have deep experience and the most current information about effectively navigating quality standards. All management, however, should be trained in quality systems and motivated to lead quality compliance efforts within their areas. Look for a consultant with expertise in the systems the company desires (ISO, QSR, or both) and familiar with the company's stage of development (start-up, emerging, established). Ask about the consultant's approach to quality system design. For example, does the person have a "recipe" approach (i.e., one size fits all) or does s/he consider the unique circumstances of the company and the team?

2.2 Where to look – In addition to working with recruiters, reviewing the choices of competitors, and seeking referrals from investors, board members, and other advisors, try making connections through the local chapters of quality associations, such as the ASQ.

3. Engage executive and cross-functional management and define quality policies

3.1 What to cover – Responsibility for a quality system begins with executive management. Therefore, it is vital that executive management is involved from the beginning of the process. Meet with executives to ensure their understanding of the management responsibility sections of QSR and/or ISO. Then, initiate the quality system design and implementation process by developing an organizational strategy with top management to engage key players across the organization. Assemble a cross-functional team to define the company's overarching quality policy. The quality policy should set forth the overall intentions and direction of an organization with respect to quality, as established by management with executive responsibility. Validate and reinforce the policy with appropriate employees at all levels within the organization.

3.2 Where to look – Refer to the documentation for the standard being implemented to understand executive management responsibilities. Network with individuals in the field and visit other companies to observe their operations and ask about their approach to quality. Leverage the experience of the company's quality professional(s).

4. Build a "shell" of the quality system and develop elements as the product progresses

4.1 What to cover – Map out the high-level requirements and processes of the quality system from "end to end" (i.e., from design through manufacturing, distribution, and postmarketing surveillance). Then, depending on the stage of the company and its product(s),

Stage 5 Development Strategy and Planning

determine which processes and procedures need to be developed further, at a greater level of detail. Design control is usually one of the first elements to be developed at the tactical level. Records, documents, and change controls often come next, followed by CAPA processes (to guide continuous improvement), and then P&PC. These elements should be built out (following the guidelines provided in the quality policy and working within the construct of the quality system "shell") as they are needed, based on the progress of the company.

4.2 Where to look – Leverage the experience of the company's quality professional(s). Involve the people who will perform the functions where the quality system is being built (e.g., ask R&D engineers to help set up the design control procedures, work with the production manager to design production process and controls). Compliance will be increased if the systems are built by those using them in a way that makes sense in the context of their workflow, and a participative process will yield a better quality system. In addition, a company can leverage a variety of other resources and publications to provide them with relevant information:

- **ASQ Biomedical Division Publications**
- **The FDA and Worldwide Quality System Requirements Guidebook for Medical Devices** – A helpful reference compiled by Kimberly A. Trautman, a GMP/quality systems expert.
- **Textbooks on quality systems**
  - Marie Teixera and Richard Bradley, *Design Controls for the Medical Devices Industry* (CRC Press, 2002).
  - Michael E. Wiklund, *Medical Device and Equipment Design: Usability Engineering and Ergonomics* (CRC Press, 1995).
  - Basem El-Haik and Khalid S. Mekki, *Medical Device Design for Six Sigma: A Road Map for Safety and Effectiveness* (John Wiley & Sons, 2008).
- Richard Fries, *Reliable Design of Medical Devices* (CRC Press, 2006).

5. Assign functional champions to monitor and maintain quality system

5.1 What to cover – While it can be tempting to implement a quality system and then let it run on "auto pilot," a company must proactively monitor and maintain its quality system to ensure ongoing compliance. Quality systems are meant to be dynamic, changing, and improving with the organization as new information, issues, and opportunities are identified. By naming functional champions to oversee the entire quality system, as well as the specific sections most relevant to their individual expertise, the company is more likely to anticipate and prevent challenges and improvements. For this reason, a team approach is typically preferable to assigning a single quality executive to maintain end-to-end oversight of the quality system.

5.2 Where to look – Work with executive managers to identify and empower the appropriate quality champions. The company's quality professional(s) can help structure these roles, responsibilities, and oversight processes.

6. Prepare for audits

6.1 What to cover – Train employees and executives on good audit techniques (discussed in the Quality Enforcement section). This training should be ongoing, not simply performed in advance of a scheduled audit. Provide representatives of the company with the opportunity to practice their audit processes by performing internal audits on a periodic basis.

6.2 Where to look – Many external training courses provided information that can be useful to helping employees and professionals prepare for audits (e.g., offered through ASQ). The company's quality professional(s) can also develop and lead such trainings internally. For more information about the FDA's audit process, refer to the FDA's Quality Systems Manual. ISO training is available through the many ISO-accredited third-party organizations authorized to award ISO certification.

7. Apply a process management approach to other areas of the business

7.1 **What to cover** – Once the company's quality system is under development, look for other work streams where the company will potentially benefit from increased rigor, discipline, and accountability. Apply the same general principles to articulate goals, define processes, develop metrics, and establish monitoring and correction mechanisms.

7.2 **Where to look** – Based on its learnings from implementing a quality system, the team may be able to tackle this step itself (ideally, through cross-functional collaboration). Process consultants can also be engaged to assist.

Credits

The authors would like to acknowledge Trena Depel for her help in developing this chapter. Many thanks also go to Allan May for his assistance with the case example and Michelle Paganini of Michelle Paganini Associates for her lectures in Stanford University's Program in Biodesign, as well as her contributions to the chapter.

Appendix 5.5.1

More information about Medical Device Reporting

Medical Device Reporting (MDR) is the vehicle through which the US Food and Drug Administration receives information about significant medical device adverse events from manufacturers, importers, and user facilities, so they can be detected and corrected quickly.[45] Voluntary reports can also be made by consumers and healthcare professionals through the MEDWATCH program. The FDA has yet to enable online reporting for device manufacturers and importers (although this is expected in the near future), but consumers can submit reports through the FDA website.[46]

Manufacturers and importers of medical devices have been required since 1984 to report all device-related deaths, serious injuries, and certain malfunctions to the FDA. Under the Safe Medical Devices Act of 1990 (**SMDA**), device user facilities also were mandated to report device-related deaths and serious injuries to the FDA and the manufacturer, if known. User facilities also must submit to the FDA a semi-annual summary of all reports made during the time period.[47] Despite these requirements, reports have shown there is widespread underreporting of such device-related incidents. One 1986 report showed that less than one percent of device problems encountered in hospitals were properly reported, with more serious problems increasingly less likely to be reported. A follow-up study in 1989 confirmed that serious shortcomings existed in the MDR system, despite its full implementation.[48]

To manage MDR-related issues and other customer/user concerns, most companies establish a complaint system to help ensure that all complaints are received and processed consistently and appropriately. A company must be vigilant in managing each complaint, especially if it could potentially be considered a report of an adverse event (or an event that *could cause* an adverse outcome, even if it does not). As a company grows, it is important to recognize that the maintenance of complaint and MDR reporting systems will require dedicated resources.

The MDR reporting process is prescriptive and straightforward. Occasionally, a company will receive a request for a follow-up report on an MDR. Or, if a pattern begins to emerge that gets the attention of the FDA, it could trigger a "for cause" audit. However, companies are generally advised to overreport when it comes to actual or potential adverse events. Typically, there are far worse risks and penalties associated with underreporting than overreporting. It is also preferable for a manufacturer to voluntarily report issues, rather than having consumers or healthcare professionals report issues, since this typically results in more swift and severe attention from the FDA. All company employees must understand that reporting any issues into the complaint system is essential in order to avoid underreporting. Furthermore, the complaint system can serve as yet another source for continuous improvement, cost savings, and preventive opportunities if issues and problems are identified early.

Medical Device Reports are reviewed, coded, and tracked on the **MAUDE** system – a searchable database.[49] This system includes voluntary reports since June, 1993, user facility reports since 1991, distributor reports since 1993, and manufacturer reports since August, 1996.[50] Reports generally lag about a month or more behind the date of the event. Because it is publicly accessible, this system can be useful in looking up competitive technologies and understanding patterns of adverse events or device failures.

Appendix 5.5.2

More information about ISO 13485

Background

The International Organization for Standardization (ISO) was founded in 1947 as a means for developing international technical standards. Interestingly, ISO is not an acronym. Rather, it is a name derived from the Greek word "isos" meaning "equal" – the idea being that if two objects meet the same standard, they should be equal. Given the global nature of the organization, this name also eliminates the need for different acronyms as International Organization for Standardization is translated into various languages.[51]

ISO is a voluntary organization whose members are recognized standard-setting authorities from its member countries. In 2006, the organization had 158 members.[52] The American National Standards Institute (**ANSI**) is the US representative to ISO (one representative is allowed from each member country). ISO standards are developed by technical committees made up of experts "on loan" from relevant industrial, technical, and business sectors. These experts may be joined by others with pertinent knowledge, including representatives of governments, testing laboratories, consumer associations, environmentalists, and so on. Every year, as many as 30,000 contributors participate in the development of ISO standards.[53] As of the end of 2006, ISO had developed more than 16,000 standards applicable to a broad spectrum of industries around the world.[54]

Overview of ISO 13485

ISO 13485, published in 2003, is a standard that specifies requirements for quality management systems in the medical device industry. Importantly, all requirements of ISO 13485 are specific to organizations providing medical devices. Within this industry, the standards are applicable to all medical device companies, regardless of the type or size of the organization.

The primary objective of ISO 13485 is to facilitate harmonized medical device regulatory requirements for quality management systems around the world.[55] While ISO 13485 is a stand-alone standard, it is generally consistent with ISO 9001. A few fundamental differences, many of which make the ISO 13485 standard more consistent with the FDA's QS regulation, are listed below:[56]

- While ISO 9001 requires the organization to demonstrate continuous improvement, ISO 13485 only necessitates that it demonstrate that the quality system is implemented and maintained. It was not written to be a business improvement model, but rather as a tool for maintaining effective quality processes.
- ISO 13485 explicitly positions awareness of (and adherence to) regulatory requirements as an executive management responsibility, consistent with US QSR requirements. As a result, it is less focused on customer satisfaction.
- ISO 13485 institutes controls in the work environment to ensure product safety.
- ISO 13485 maintain a focus on risk management and design transfer activities during product development.
- ISO 13485 includes specific requirements for inspection and traceability for implantable devices.
- ISO 13485 includes specific requirements for documentation and validation of processes for sterile medical devices.
- ISO 13485 includes specific requirements for verification of the effectiveness of corrective and preventive actions.

Because of these types of difference, organizations whose quality management systems conform to ISO 13485 cannot claim conformity to ISO 9001, unless they have specifically taken extra steps to ensure compliance with all requirements of ISO 9001.[57]

Why become ISO certified?

In the United States, medical device companies are in no way obligated or required to implement ISO 13485 to manufacture, sell, and distribute their devices domestically. However, if they intend to enter international markets, the ISO quality system becomes significantly more important. As mentioned, compliance with ISO 13485, as granted by a conformity assessment body, is seen as a first step in achieving compliance with European regulatory requirements and being granted a CE mark.[58] ISO 13485:2003 certification is also required for a company to market a medical device in Canada and Japan. Furthermore, ISO certification can be a competitive differentiator, particularly for companies working with (or seeking to become) contract manufacturers or outsourcing partners. In the long run, many companies also believe that implementation of an ISO 13485 quality system saves them time, money, and problems over time. In 2005, 1,310 US firms became ISO 13485:2004 certified, compared to 700 in 2004.[59]

Fortunately for companies seeking ISO 13485:2003 certification, approximately 85 percent of the requirements overlap with the FDA's QSR, according to one consultant's estimate.[60] Technical Report (TR) 14969 is a guidance document for the use and implementation of ISO 13485.

ISO audits

To become ISO certified, a company contracts with a third-party auditor and forms a relationship that is both consultative and compliance-based – first, the auditor helps the company make improvements to its quality systems and prepare for the audit, then certifies the company once it has successfully met all requirements outlined in the standards. Upon conclusion of the audit, the major findings are written up and the company is given time (as well as corrective advice) to fix any problems before a reaudit occurs. In contrast to the FDA, ISO-accredited auditors have no authority to impose fines, initiate product recalls, halt operations, or enforce other corrective penalties. Their objective is to identify problems and help companies fix them. Generally, it is a "friendlier" process than an FDA audit, with the auditor more motivated to help the company become successful. As a result, a company may choose to handle the audit in a more open manner. On the other hand, the company may choose to use the ISO audit as "practice" for an FDA audit and conduct it in the same way it would a visit from the FDA.

Appendix 5.5.3

FDA's 21 Code of Federal Regulations (CFR) Part 820 Quality System Regulations[61]

Interpretive summary – Always refer to the actual regulations for complete requirements and the most recent information. The complete regulations are available on the FDA's website www.fda.gov (use keyword "21CFR Part 820" to search for them).

Subpart A – General provisions: Defines scope of regulation, who must comply, FDA's authority, and the process for exemptions, import requirements. **Definition section** – explains FDA's meaning for keywords and acronyms contained in the regulation.

Subpart B – Quality systems requirements: Management is responsible for providing the structure (policy, organization, resources, defined responsibility and authority) for a working quality system, and must be aware of the effectiveness of the current system. Awareness and corrective action is documented through periodic management review meetings. A quality plan outlining the structure of the system and how it is fulfilled is defined. Internal quality audits are performed and results shared with management. Job descriptions are generated and a sufficient number of qualified employees are trained per a predefined training plan. Persons responsible for quality are given the independence and authority to fulfill their responsibilities.

Subpart C – Design controls: Class II, III and some class I devices are subject to design controls. There is a documented system for developing new products with the primary goals of proving the product to be safe and effective and that the design is both reproducible and traceable. The design is transferred to production in a way that allows production to successfully reproduce the design. The design process requirements are defined in discrete segments: input, output, review, verification (product meets design specifications), validation (that the design specifications meet customer requirements), transfer (to manufacturing), and changes (made to the design). The output of the design history process is a design history file.

Subpart D – Document controls: Document control systems are a mechanism for managing change and defining agreements on how work will be done and/or products will be produced and tested. All processes required by any part of this regulation are defined in writing and documents are reviewed and approved prior to initial release, or subsequent revision. Records of changes are maintained. Documents are revision controlled. Obsolete documents are not available for use.

Subpart E – Purchasing: Documented system for evaluation, approval, control, and monitoring of suppliers. Requires clear purchasing documentation be provided to suppliers. Supplier qualification begins in the design process and applies to both service and material suppliers. Known (and unavoidable) poor quality controls by a supplier translates to tighter receiving inspection criteria and vice versa.

Subpart F – Product identification and traceability: Documented system for identifying materials used in building product, and the product itself, as appropriate, in order to facilitate failure investigation and/or a recall if necessary. The level of detail required is dependent on the type of device. There are special requirements for implantable devices.

Subpart G – Process control: Documented system for conducting, controlling, and monitoring production process to ensure the device conforms to its specifications. Includes instructions for processing activities; controls for equipment (preventative maintenance and installation qualification, operational qualification, process qualification, calibration (ensuring equipment is giving valid results), workmanship standards, manufacturing materials, environmental controls (including personnel), contamination controls, process qualification and monitoring, validation, automated processes (computer controlled). Process validation is required where the results of the process cannot be fully validated by subsequent inspection and test (such as sterile products where testing would be destructive).

Subpart H – Acceptance activities: Documented system for inspection and testing of incoming, in-process, and final testing. Test results are documented; devices are not released for distribution until the release is authorized in writing. Acceptance status is clearly identified throughout all stages of use so that rejected or quarantined materials are not used by mistake.

Subpart I – Non-conforming product: Documented system for control of the non-conforming material or product (quarantine, label rejected, etc.) to prevent inadvertent use, and decision for what will be done with the non-conforming materials or product (rework, scrap, return to supplier, etc.). Rework requires approved and documented processing and reinspection instructions.

Subpart J – Corrective and preventative action: Documented system for collection of information related to system, product, or material problems. Problems are investigated, the cause(s) identified, and solutions to eliminate the problems from happening again are identified and implemented (corrective action). Lastly, solutions are verified as effective. Those responsible for corrective action must be made aware of the issues. Also requires that systems be in place to prevent problems from occurring before they happen (preventative action).

Subpart K – Device labeling and packaging control: Controls for labeling activities, which are intended to prevent inadvertent mislabeling. Device packaging must protect product from alteration or damage during storage, shipping, and normal use conditions.

Subpart L – Handling, storage, distribution, and installation: Documented system for handling and storing raw materials, in-process product, and finished goods so that the materials remain undamaged and the correct materials can be provided as required. Records of distribution are maintained to facilitate possible product recalls. If device is installed, there is a defined system for installations, and documentation of correct installation if performed by the manufacturer.

Subpart M – Records, DMR, DHR, and customer complaints: Documented requirements for what quality records must be maintained and for how long. Records provided to the FDA in an audit can be marked "confidential" as a way of requesting that they are not made available to the public in Freedom of Information requests. Management review reports, internal audit reports, and supplier audit reports are not subject to FDA review during routine FDA audits. A documented general "Quality System Record" of non-device specific documents is required.

A **device master record (DMR)** is generated to include reference to *what is required* to produce a specific device. The DMR is generated from the design control process.

A **device history record (DHR)** is generated *for every lot of product produced*, defining all materials, methods, equipment, personnel, documents, etc. involved in or used to manufacture a specific device batch, lot, or serial number.

The manufacturer has a documented system for receiving, screening, investigating, initiating corrective action, and/or otherwise resolving customer complaints.

Subpart N – Servicing: A documented system for defining and executing service programs. Records of service are maintained. Although the regulation does not explicitly state requirements for technical support programs, controls need to be in place, including links to the complaint handling system.

Subpart O – Statistical techniques: Documented system for using statistical techniques, for example: sampling plans used in incoming inspection, control charting used to monitor production processes, and report trends used in management review meetings.

Notes

1. "Medical Devices; Current Good Manufacturing Practice (CGMP) Final Rule; Quality System Regulation," U.S. Food and Drug Administration, http://www.fda.gov/cdrh/humfac/frqsr.html (May 17, 2007).
2. Ibid.
3. "U.S. FDA Warns Boston Scientific," CBC News, August 23, 2005, http://www.cbc.ca/health/story/2005/08/23/FDA_warns_Boston_Scientific.html (May 17, 2007).
4. Barnaby J. Feder, "Device Maker Moves to Appease the FDA," *The New York Times*, http://www.nytimes.com/2006/02/03/business/03device.html?ex=1296622800&en=7b669a64d72d2e90&ei=5088&partner=rssnyt&emc=rss (May 17, 2007).
5. Heuser, op. cit.
6. John Chesto, "Boston Scientific Reaches a Crucial Milestone," *The MetroWest Daily News*, March 3, 2008, http://www.metrowestdailynews.com/state/x1637135610 (March 3, 2008).
7. Todd Wallack, "A Blockbuster Revisited," *The Boston Globe*, January 13, 2008, http://www.boston.com/business/globe/articles/2008/01/13/a_blockbuster_revisited/ (March 3, 2008).
8. Val Brickates Kennedy, "Boston Sci: FDA Lifts Some of Warning Letter Restrictions," Fox Business, October 22, 2008, http://www.foxbusiness.com/story/markets/industries/health-care/boston-sci-fda-lifts-warning-letter-restrictions/ (October 28, 2008).
9. "Cost of Quality," American Society for Quality, http://www.asq.org/learn-about-quality/cost-of-quality/overview/overview.html (May 18, 2007).
10. "Quality Assurance," www.wikipedia.org, http://en.wikipedia.org/wiki/Quality_assurance (May 18, 2007).
11. "Difference Between Quality Assurance, Quality Control, and Testing?" Test Notes, http://geekswithblogs.net/srkprasad/archive/2004/04/29/4489.aspx (May 18, 2007).
12. "Quality Assurance Is Not Quality Control," http://c2.com/cgi/wiki?QualityAssuranceIsNotQualityControl (May 18, 2007).
13. "Quality Management Systems," www.wikipedia.org, http://en.wikipedia.org/wiki/Quality_management_system (May 18, 2007).
14. "Guide to Inspections of Quality Systems," U.S. Food and Drug Administration, August 1999, http://www.fda.gov/ora/inspect_ref/igs/qsit/qsitguide.htm (April 8, 2008).
15. Ibid.
16. "Quality System: Design Controls," CDRH Device Advice, http://www.fda.gov/cdrh/devadvice/pma/quality_system.html (May 22, 2007).
17. "Design Control Guidance for Medical Device Manufacturers," U.S. Food and Drug Administration, http://www.fda.gov/cdrh/comp/designgd.html (May 22, 2007).
18. Ibid.
19. Ibid.
20. Ibid.
21. "Medical Device Makers Find Solution to FDA Demands," Bnet, December 2007, http://findarticles.com/p/articles/mi_qa5347/is_200712/ai_n21300031 (October 28, 2008).
22. Ibid.
23. Ibid.
24. Jennifer Whitney, "The Century of Quality," Medical Product Outsourcing, May 2007, http://www.mpo-mag.com/articles/2007/05/the-century-of-quality (May 24, 2007).
25. Daniel P. Westman and Nancy M. Modesitt, *Whistleblowing: The Law of Retaliatory Discharge* (BNA Books, 2004), p. 39. See United States versus Park, 421 U.S. 658, 672 (1975) for more information.
26. "FDA Seizes All Medical Products from N.J. Device Manufacturer for Significant Manufacturing Violations," U.S. Food and Drug Administration, April 17, 2007, http://www.fda.gov/bbs/topics/NEWS/2007/NEW01612.html (May 24, 2007).
27. Ibid.
28. Paula Kurtzweil, "Ex-Bard Executives Sentenced to Prison," *FDA Consumer*, December 1996, http://findarticles.com/p/articles/mi_m1370/is_n10_v30/ai_18979009 (May 22, 2007).
29. "Bard Execs Found Guilty," *Health Industry Today*, October 1995, http://findarticles.com/p/articles/mi_m3498/is_n10_v58/ai_17526774 (May 22, 2007).
30. "Management System Certification for Medical Device Manufacturers," www.sriregistrar.com, http://www.sriregistrar.com/A55AEB/sricorporateweb.nsf/0/D1E1DC087A3B2A3D8625729900752FC1/$FILE/ISO+13485+General+Info.pdf (May 21, 2007).
31. "Inspection of Medical Device Manufacturers," U.S. Food and Drug Administration, June 15, 2006, http://www.fda.gov/cdrh/comp/guidance/7382.845.html (May 23, 2007).
32. "Quality Systems Inspections Reengineering," U.S. Food and Drug Administration, http://www.fda.gov/cdrh/gmp/gmp.html (May 22, 2007).
33. "Guide to Inspections of Quality Systems," op. cit.
34. Ibid.
35. N. L. Davis, "Correction Versus Corrective Action," www.isixsigma.com, http://www.isixsigma.com/dictionary/Correction_versus_Corrective_Action-685.htm (May 23, 2007).

36. "Inspection of Medical Device Manufacturers," op. cit.
37. Andrea Lavish and James Wood, "Device Case Processing," http://www.fda.gov/cdrh/oivd/presentations/041905-latish_files/textonly/index.html (May 22, 2007).
38. James A. Dickinson, "FDA Cuts Warning Letters by 70 Percent," Medical Device Link, July 2002, http://www.devicelink.com/mddi/archive/02/07/010.html (May 22, 2007).
39. "Consent Decree," cGMP Enforcement, http://www.cgmp.com/consentDecree.htm (December 9, 2008).
40. "Medical Device Recalls and Corrections and Removals," U.S. Food and Drug Administration, http://www.fda.gov/cdrh/devadvice/51.html (May 23, 2007).
41. "Medical Device Recalls and Corrections and Removals," U.S. Food and Drug Administration, http://www.fda.gov/cdrh/devadvice/51.html (April 8, 2008).
42. Ibid.
43. "Guidant Cardiac Defibrillator Recall and Lawsuit," www.lawyersandsettlements.com, http://www.lawyersandsettlements.com/case/guidant_defective_defibrillator_class_action (May 23, 2007).
44. Recall that the FDA's good manufacturing practices for medical devices were not issued until two years after the Medical Device Amendments of 1976 were enacted. It then took quite some time for these guidelines to have an effect on the processes used to produce products that were already in the market.
45. "Medical Device Reporting – General Information," U.S. Food and Drug Administration, http://www.fda.gov/cdrh/mdr/mdr-general.html (May 30, 2007).
46. See http://www.fda.gov/medwatch/how.htm (October 28, 2008).
47. Ibid.
48. "Medical Device Reporting – General Information," U.S. Food and Drug Administration, http://www.fda.gov/cdrh/mdr/mdr-general.html (May 30, 2007).
49. See http://www.accessdata.fda.gov/scripts/cdrh/cfdocs/cfMAUDE/search.cfm (October 30, 2008).
50. "Search MAUDE Database," U.S. Food and Drug Administration, http://www.accessdata.fda.gov/scripts/cdrh/cfdocs/cfMAUDE/search.cfm (May 31, 2007).
51. Cynthia J. Martincic, "A Brief History of ISO," February 20, 1997, http://www.sis.pitt.edu/~mbsclass/standards/martincic/isohistr.htm (May 24, 2007).
52. "ISO in Figures for the Year 2006," International Organization for Standardization, http://www.iso.org/iso/en/aboutiso/isoinfigures/January2007-p1.html (May 24, 2007).
53. "FAQ," International Organization for Standardization, http://www.iso.ch/iso/en/faqs/faq-standards.html?printable=true (May 24, 2007).
54. "ISO in Figures for the Year 2006," op. cit., http://www.iso.org/iso/en/aboutiso/isoinfigures/January2007-p2.html (May 24, 2007).
55. "ISO 13485:2003," International Organization for Standardization, http://www.iso.org/iso/en/CatalogueDetailPage.CatalogueDetail?CSNUMBER=36786&ICS1=3&ICS2=120&ICS3=10&scopelist (May 24, 2007).
56. "ISO 13485," www.wikipedia.org, http://en.wikipedia.org/wiki/ISO_13485 (May 24, 2007).
57. "ISO 13485:2003," op. cit.
58. "ISO 13485," op. cit.
59. Whitney, op. cit.
60. Ibid.
61. Information provided by Michelle Paganini of Michelle Paganini Associates. Reprinted with permission.
62. See "An Introduction to Human Factors," U.S. Food & Drug Administration, http://www.fda.gov/cdrh/humfac/doit.html (December 9, 2008).
63. "Comparative Matrix for Quality Systems Regulation," The Anson Group, 2004, http://www.ansongroup.com/Docs/ISO%2013485%202003%20vs%20FDA%20QSR.pdf (May 30, 2007).
64. For a list of exempt devices, see http://www.accessdata.fda.gov/scripts/cdrh/cfdocs/cfPCD/315.cfm (October 30, 2008).

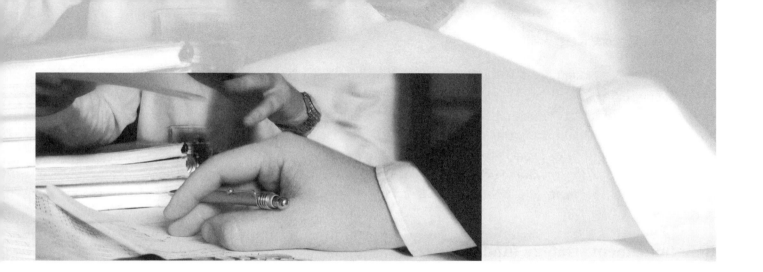

5.6 Reimbursement Strategy

Introduction

A significant proportion of the recent growth in the United States government's total healthcare budget can be linked to the proliferation of just two medical devices: implantable cardioverter defibrillators and drug eluting stents.[1] With medical devices playing an increasingly prominent – and expensive – role in patient care, public and private payers are more rigorously scrutinizing new technologies. The current trend is to provide reimbursement for new devices only if the evidence supporting their use is overwhelmingly strong. In fact, it has reached the point where the most serious hurdle faced by new medical technologies is not gaining approval from relevant regulatory bodies, but obtaining favorable reimbursement decisions by major payers.

With a sophisticated reimbursement strategy, innovators can be proactive. They can anticipate the most pressing questions payers will raise about their new devices and prepare a recommended approach to coding, coverage, and payment, a strong body of supporting evidence, and advocacy from powerful physician groups that can dramatically increase their chances of receiving favorable reimbursement decisions. Given the complexity of the reimbursement landscape in major markets such as the United States, development of a strategy must start early in the biodesign innovation process and its execution may begin as early as 12 to 18 months before product launch. While significant time, effort, and expense are required in the current environment to ensure the successful implementation of a reimbursement strategy, such a strategy is essential to facilitating the rapid adoption of a device upon its commercial release.

OBJECTIVES

- Recognize that regulatory approval does not guarantee reimbursement.

- Identify the key payers that influence reimbursement decisions and understand their criteria and the body of evidence they require for reimbursement.

- Know how to develop a strategy to generate the evidence to support reimbursement.

- Understand the steps involved in requesting and obtaining the necessary codes, an appropriate coverage determination, and adequate payment rates for a new technology when existing codes, coverage determinations, and payment rates are inadequate.

- Learn to develop tactical plans to facilitate successful reimbursement at the time of product launch.

While Chapter 4.3 Reimbursement Basics provides fundamental information about understanding the existing reimbursement landscape and assessing how a new solution might fit in, this chapter focuses on the process required to expand the existing payment infrastructure to accommodate a new device if the established payment and coverage levels are inadequate.

Reimbursement strategy fundamentals

The purpose of reimbursement strategy is to demonstrate that a new device should become the standard of care in order to obtain broad coverage by most payers. This activity is critical because it directly affects the commercialization of a technology. It is clear to industry executives that the absence of clinical data proving that a device ought to be the standard of care can adversely affect market penetration.[2]

Some innovators mistakenly believe that gaining regulatory approval is the final obstacle to overcome before seeking market adoption. However, regulatory agencies, such as the US Food and Drug Administration (FDA), use criteria that are different from those of public and private payers when approving medical products. In fact, the "bar" that regulatory agencies use is frequently set lower than the standards that must be met when seeking reimbursement from payers. For instance, while FDA approval may be achieved by showing that a product has a better effectiveness and safety profile than existing products, payer approval requires that additional criteria be met, such as medical necessity and other favorable outcomes.

To qualify as a medical necessity, a product must be deemed by the payer to be "reasonable and necessary for the diagnosis and treatment of illness or injury or to improve the functioning of a malformed body member."[3] To reach that conclusion, payers require evidence that the treatment should become the new standard of care and that it is effective outside the well-controlled clinical trial environment. Other outcomes used to justify medical necessity must be supported by measurable endpoints related to morbidity and mortality, as well as cost. These outcomes must be meaningful to patients – the Blue Cross/Blue Shield (BCBS) guidelines for reimbursement state that products should "provide value in the eyes of the consumer, not just the seller."[4] Increasingly, they also must bring value to the payers in order to gain reimbursement (see Appendix 5.6.1 for examples of different analyses that can be used to support value to payers). Table 5.6.1 summarizes the different criteria used by regulators and purchasers of a device (payers and facilities) and underscores the fact that FDA approval is only one of many important criteria to address before the commercialization of a device.

The pitfalls of assuming that a favorable regulatory ruling will translate into a positive reimbursement decision are illustrated by the experiences of Genzyme with its Carticel knee replacement implant. When Genzyme launched the Carticel implant as a less invasive alternative to knee replacement, the company assumed that its accelerated FDA approval status would lead to swift reimbursement by payers. Accelerated regulatory approval typically applies only to new technologies that the FDA deems so clinically compelling that they are "fast tracked" in order to more quickly provide patients with access to the improved technology. The cost of Carticel, at $26,000 per procedure, was high, but only about $1,000 higher than total knee replacement, which was the current standard of care.[5] When Genzyme representatives approached payers to confirm coverage, they were surprised to learn that many payers were still considering whether or not to cover Carticel.[6] The BCBS Technology Evaluation Center (TEC), for example, initiated a full assessment of the therapy (see Appendix 5.6.2 for more information about TEC). As a result of its review, TEC concluded that the data submitted to the FDA, while adequate for regulatory approval, was insufficient to justify reimbursement from a cost/benefit perspective. Unfortunately for Genzyme, TEC did not have an orthopedic specialist on its review panel and did not seek input from other orthopedic specialists in the field. According to one report, the review panels instead relied on testimony in a closed meeting given by a "self-avowed critic who failed to review alternatives to total knee surgery in any meaningful manner and resorted to a diatribe in place of a scientific presentation."[7] The final TEC ruling deemed the technology unsuitable for coverage, a move that

Table 5.6.1 *There are notable distinctions between the decision criteria used by regulators, payers, and facilities.*

Regulators For approval for human use of Class II and III devices	Payers For approval for reimbursement	Facilities For approval of use and support within and by facility

• The device must be safe and efficacious
• Manufacturer must present data from one or more well-controlled studies with acceptable clinical endpoints
• Proven quality systems and good manufacturing practices should be in place, along with a plant for post-market surveillance
	• Device must be FDA approved
	• Manufacturer must prove medical necessity, provides cost savings, or improves benefit relative to established alternatives
	• Benefits attainable outside controlled study environment
	• High-quality scientific evidence permitting conclusions regarding costs and benefits
		• FDA approval
		• Quantifiable financial benefits to the facility, including:
		  • Adequate reimbursement
		  • Increased volume of procedures
		  • Reduced costs
		  • Enhanced reputation
		  • Strong physician endorsement/demand

led to unfavorable reimbursement decisions by several other private payers.

In retrospect, Genzyme realized that it had failed to adequately anticipate payer requirements. Data comparing Carticel to established therapies was needed to prove its value, given its higher relative cost. Genzyme had used academic physicians to promote its intervention, while payers preferred to learn about clinical data from healthcare providers in their networks. Most importantly, Genzyme had failed to highlight the benefits of Carticel that were particularly meaningful to patients (and their employers), and to organize patients to lobby for coverage. After a few years, Carticel was finally able to secure positive coverage decisions from most insurance plans. However, this was achieved through the extensive efforts Genzyme launched after the initial TEC ruling to foster patient activism and gain advocacy support from workers' compensation organizations.[8]

Developing a reimbursement strategy: balancing tactical and strategic considerations

An effective reimbursement strategy involves the complex coordination of at least four high-level categories of initiatives:
- **Clinical research** to generate the evidence necessary to demonstrate that a new device should be the standard of care.
- **Reimbursement initiatives** to determine existing codes, coverage, and payment levels, formulate the value proposition and pricing strategy, and prepare documentation that will be used to support coding and Medicare initiatives and private payer targeting.
- **Coding and Medicare initiatives** that focus on obtaining both new codes to supplement gaps in existing codes as well as positive coverage from government payers (primarily Medicare). This includes the engagement of specialty societies and physicians in supporting applications for new codes (if existing codes cannot be applied) and expansion in coverage by Medicare.
- **Private payer targeting** to leverage coding initiatives and expand coverage and payment by private payers.

Figure 5.6.1 provides a representative view of the individual tasks under each initiative and the associated timeline.

This chapter first provides a description of the mechanics underlying the key coding initiatives (establishing a new code, securing a positive coverage decision from Medicare, and establishing a payment level). Then, it focuses on strategic considerations as they relate to clinical research, reimbursement initiatives, and private payer targeting. Finally, a discussion on global reimbursement issues is presented at the end of the chapter.

Stage 5 Development Strategy and Planning

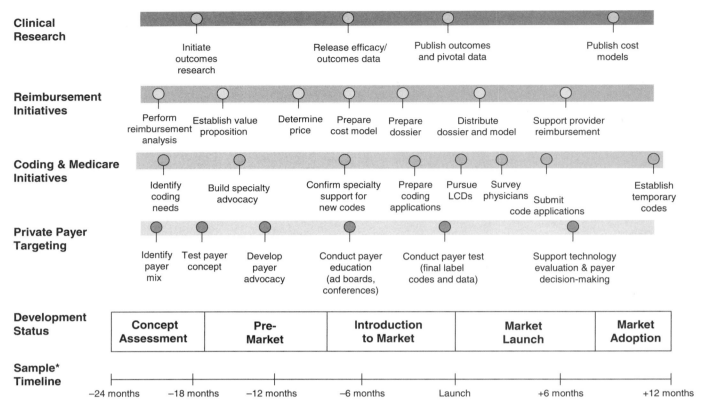

FIGURE 5.6.1
In this summary of key reimbursement strategy activities, each line represents a different initiative and the circles within each line show the individual activities to be pursued. The position of each activity reflects its general timing. The timeline at the bottom of the diagram maps these activities to the overall product development timeline. (Adapted, with permission, from Boston Healthcare Associates, Inc., "Hard Wiring EBM into Reimbursement Planning for Medical Devices," 2006.)

Revisiting reimbursement analysis

Before developing a reimbursement strategy, an innovator or company should revisit the reimbursement analysis performed during concept selection (see 4.3 Reimbursement Basics). Through the process of developing prototypes, choosing a final concept, and transitioning into development, important new information will have come to light that can potentially change the outcome of the initial reimbursement analysis. A re-examination of this information can help to verify whether existing reimbursement levels are adequate or if new codes and reimbursement levels will be needed. If the technology is completely new, this analysis should be performed using a currently marketed proxy device that most closely mimics the new treatment. Proxy devices should have a related function within the same or similar specialty, and should be used by the same type of provider in the same site of care. After revisiting the basic reimbursement analysis, the company should be able to answer the questions in Table 5.6.2.

If consultants, physicians, key opinion leaders (KOLs), and experienced specialty-specific billing managers have not yet been involved in the reimbursement analysis, the company should now approach these experts for input and advice. Such individuals are typically enlisted to help guide this process (and contribute to the creation of the reimbursement strategy), as real-time information on coding, reimbursement rates, and generally accepted practices often can be challenging to identify and interpret.

While payers will be the primary focus of this reimbursement analysis, the company should also consider

Table 5.6.2 *An effective reimbursement analysis allows the innovator to answer these basic questions.*

General reimbursement questions
Who will initially purchase the device?
Who will ultimately be responsible for payment?
Where will it be used, and who will use it?
What are the existing codes used to gain reimbursement of the device?
What type of coverage is associated with those codes?
What is the average payment level?
Are there important coverage and reimbursement issues associated with the device?
What was the process used to secure reimbursement for the proxy device?
Are reimbursement conditions stable or are future changes anticipated, such as restricted or reduced reimbursement?
Are there other warning signs in the reimbursement landscape (e.g., stories in the press, talk of class reviews, TEC assessments or demonstration projects) that may affect reimbursement?
Is the new technology sufficiently similar to existing devices and/or the proxy device to allow the company to capitalize on existing codes, coverage, and reimbursement payment levels?
Will it be necessary to obtain new codes, coverage determinations, and reimbursement levels? Or is it possible to work within existing reimbursement codes but expand coverage criteria and payment levels?

the perspectives of all stakeholders (see 5.7 Marketing and Stakeholder Strategy) that have the potential to influence reimbursement-related issues. Resistance to the new technology from such groups can delay or prevent reimbursement or lead to unfavorable coverage and reimbursement levels. However, important, influential stakeholders who become advocates for the technology can speed reimbursement and help minimize reimbursement-related risks to the company (see Advocacy section).

Coding, coverage, and payment
This section addresses the process for establishing new codes, coverage determinations, and payment rates within the Medicare system, as well as circumstances under which existing coding, coverage, and payment may be adequate. Medicare is the primary focus because its decisions and methods are followed by most private payers. The discussion that follows is organized to address coding, coverage, and payment. However, because the distinction between the three is not as crisp in real life, issues related specifically to payment levels also appear in the coding and coverage sections.

Coding
Within the United States, there are two ways to secure a code for reimbursement coverage: (1) to choose one from the codes that have already been established, or (2) to apply for the creation of a new code. Although it is sometimes necessary for a company to pursue new code(s), achieving reimbursement will almost always be faster, less complicated, and less expensive if a company can utilize existing codes. It is also significantly less risky – a factor that is increasingly important to investors and other stakeholders. As a result, existing codes for physician/facility billing and reimbursement should be used whenever the existing code descriptors match those of the new device and associated procedure.

Chapter 4.3 Reimbursement Basics provides a detailed treatment of the process for assessing and using existing codes for reimbursement (readers can also refer back to this chapter for information on the various types of codes and how and by whom they are used). When evaluating existing codes, the company should determine if the new device and its usage is mismatched in any way to the existing descriptor for the most relevant code or codes. If so, the language of the code will need to be expanded or a new code will be required. Importantly, minor differences in the indications or procedure associated with the new device relative to the established code descriptor can be grounds for a payer to reject the new device for reimbursement. Even if new codes are required, the presence of applicable existing codes may serve as a "bridge" for the company until new codes, coverage, and higher reimbursement have been secured.

If a new code is necessary, the company must choose a specific course of action based on the type of code needed and its purpose. Despite well-defined application processes, the creation of new codes is time consuming, resource intensive, and heavily dependent on clear differentiation and a strong value proposition for the product. For these reasons, potential investors in a company typically take a keen interest in whether or not a new technology can be adequately covered under existing reimbursement codes.

For a breakthrough technology, with strong stakeholder support and compelling clinical data, some companies will initiate code requests during the premarket phase, just after pivotal trials have been conducted and their results published, when they have a clear idea of the product's intended indications. For example, Johnson & Johnson (J&J) applied for a new code to cover its drug eluting coronary stent during its premarket period (see From the Field section). Yet, most code requests will not be considered until the device is in the market and there is a strong body of evidence to support their reimbursement. The process of applying for new codes depends on the type of code and its purpose.

CPT codes
CPT codes are used for physician and physician practice reimbursement. If a company finds that the existing CPT codes are inadequate for its purposes, it can seek to have a new code created or an existing code modified to apply to a new technology. To request a new (or modified) code, an application must be submitted to the American Medical Association (AMA). Rather than being submitted directly by the company, the application must be sponsored and managed by an appropriate professional society of physicians willing to advocate for the new code.[9] Though finding the right sponsor is essential, the complexities of society politics should not be underestimated. For example, societies may avoid lobbying for a specific device if the reimbursement environment for another device (manufactured by a different company that is financially supportive of the society) might suffer as a result.

Similarly, the time and effort associated with an application for a new code should not be taken lightly. The company, via the sponsor, is required to present and defend extensive evidence as to why a new code is necessary. Specifically, the application and supporting data for a new CPT code must demonstrate that:[10]

1. The service/procedure is [or will be] approved by FDA for the specific use(s) of the device.
2. The service/procedure is [or will be] performed by the many physicians and/or healthcare professionals in multiple locations across the country.
3. The clinical efficacy of the procedure/device has been well established and documented in peer-reviewed publications of clinical trials.

In the majority of cases, the AMA will not approve a new CPT code until pivotal trials have been performed and the results published, FDA approval/clearance has been granted, and there is evidence of widespread adoption. For most companies pursuing a new code, this means that their code application efforts cannot begin in earnest until the product is launched. Once the product is in the market, driving adoption and performing the studies required to justify code creation or revision can be resource intensive and time consuming. This process can take from three to six years in some cases, depending on the device, study design, follow-up period, and publication schedule of target medical journals. After evidence is collected and submitted to the AMA, the company might wait an additional one to two years while the AMA reviews the coding change request and makes a determination. For example, it took Metrika (the company with a point-of-care hemoglobin A1C (HbA1C) test for diabetes patients described in 4.3 Reimbursement Basics) seven years from the time of FDA approval to receive an adequate CPT code and reimbursement level. Importantly, companies must remember that, despite this lengthy process, there is

Table 5.6.3 *Categories of CPT codes.*

CPT code	Format	Application	RVUs
Category I	5-digit numeric code	FDA-approved procedures and technologies that are widely performed	Assigned RVUs by the AMA, then Medicare, such that payment is associated
Category II	5-digit alpha-numeric code	Procedures that are tracked for performance measurement	Never have RVUs assigned and never have payment associated
Category III	5-digit alpha-numeric code	Procedures or devices that are not yet approved by the FDA and/or have not yet demonstrated widespread usage and acceptance	Never have RVUs assigned and typically do not have payment associated

still no guarantee that the AMA will approve its coding request.

In addition to approving new codes, the AMA has a separate committee responsible for assigning relative value units (RVUs). As explained in 4.3 Reimbursement Basics, RVUs measure the resources required by a new procedure. They also provide the basis for establishing reimbursement levels. The AMA/Specialty Society Relative Value Scale Update Committee (**RUC**)[11] assigns RVUs to a new code based on evidence generated by the company and provided by the sponsor to demonstrate the time and effort involved using the device/performing the procedure. This evidence typically should include economic studies based on surveys or time and motion studies on physician's time, other practice expenses (such as labor and supplies), and malpractice risk.[12] This type of economic study requires physician society involvement to help survey facilities and physicians and compile the data into a rigorous financial model to be shared with payers. The RVU recommendations are forwarded to the Centers for Medicare and Medicaid Services (CMS), where a final RVU determination is made.

RVUs are used by CMS to calculate payment rates for CPT codes (see Payment section) – the number of RVUs assigned by the RUC is multiplied by a monetary conversion factor (and a geographic conversion factor, as well) to determine the amount Medicare will pay for the procedure or service performed. Interestingly, the assignment of payment rates based on RVUs is budget neutral, meaning that total Medicare funding for all CPT codes remains constant within any given budget year. When a new CPT code is created (or an existing code modified) and a value is assigned to that code, payments made through other CPT codes must be reduced to compensate for the change. Companies must recognize that this creates a disincentive for AMA to approve new codes, especially if they will increase total payment to one specialty at the expense of another. Budget neutrality can raise stakeholder issues, especially among specialty societies, as one sees its total payment rates go down and while another sees them go up. As a result, these issues need to be identified early and managed proactively.

There are three types of CPT codes assigned by the AMA, as shown in Table 5.6.3.[13] Most existing CPT codes that qualify for reimbursement are category I codes. Category II codes, applied for and used by health systems to document and monitor quality practices, are not applicable to device reimbursement. Category III codes are typically used as intermediate codes to track usage and establish the need for a CPT code. If a company applies for a category III code, it does so to build a history of widespread usage so that the category III code will graduate to category I.[14]

While the path to securing a category III code requires less supporting evidence and time, it does not always have payment linked to it[15,16] and can flag the procedure as experimental, potentially disqualifying it for coverage with some payers. Strategies that include category III CPT codes can pose risks, and may best be fully evaluated with the help of reimbursement consultants.[17]

Category III codes can take up to five years to become category I codes that trigger payment. For highly compelling technologies, however, this conversion can sometimes happen more quickly, or even immediately. For example, Conceptus Inc. applied for a category III code for its Essure device, which received FDA approval in late 2002 for the minimally invasive treatment of blockages in the fallopian tubes. The company's plan was to use the category III code while it worked to build adoption. However, based on the strong quality of life and cost savings data submitted with its application, the AMA instead granted a category I code only two years after FDA approval.

Another common strategy for billing devices and procedures that do not yet have a category I CPT code is the temporary use of a miscellaneous CPT code, with or without the listing of an assigned category III code. Miscellaneous CPT codes are grouped by anatomical system and procedure type for use by providers (physicians, hospitals, etc.). Often, a company will actively promote the use of a miscellaneous CPT code for the reimbursement of its device as it seeks to build adoption (see Miscellaneous Codes section below). Respironics, a manufacturer of devices for the assessment and treatment of sleep apnea, was granted a category III code 0089T for its Actiwatch device in January, 2005. However, all procedures billed in conjunction with the device are billed under 95999, the miscellaneous code for unlisted neurological and neuromuscular diagnostic procedures. In the company's sales literature, it urges providers to also include 0089T on their claim forms, so that evidence can be collected to justify the creation of a category I CPT code in the future.[18]

Obtaining codes for outpatient billing

In addition to the CPT code, a company may need to identify or establish an APC code if the service/procedure will be performed in an outpatient facility setting and the patient population includes the elderly (who are covered by Medicare). If an existing APC code or set of codes does not provide adequate reimbursement or the descriptors do not match that of the new device, then an application must be submitted to CMS for one or both of the following:

- A pass-through code (c-code) that covers the "pass-through" cost of the new device.[19]
- A New Technology APC code to reflect the added procedural and supply costs associated with a new technology and, in some cases, the bundled cost of everything including the new device.[20]

Typically, a transitional pass-through code or c-code will be issued to cover the cost of a device that is incremental to the services provided under an existing APC code, or set of codes. The cost of the device may be bundled into this transitional APC code, or may still be billed separately under a temporary pass-through code. A New Technology APC will only be issued in the case that the new device also warrants an entirely new procedure that cannot easily be described by an existing code or combination of codes.[21]

The Medtronic BRAVO capsule-based monitoring system for gastroesophageal reflux disease (GERD) provides an example of how this process works. Initially, this technology was assigned a temporary c-code C9712 and was mapped to a New Technology APC code 1506 in the outpatient setting. Existing CPT codes were used to bill for the professional services utilized (e.g., physician fees). Then, in January, 2005, a specific CPT code 91035 was issued for the procedure. In the physician office setting, this new CPT code covered the cost of the device (capsule) and the procedure. In the outpatient hospital setting, C9712 was discontinued, and the APC code 1506 became permanent and mapped to CPT 91035[22] (see 4.3 Reimbursement Basics for more information about how CPT and APC codes interact).

With rare exceptions, applications for pass-through codes and New Technology APCs cannot be submitted until FDA approval has been granted. Therefore, these processes must be initiated at launch, with the codes generally taking several months to be issued.

Obtaining codes for inpatient billing

If the service/procedure will be performed in an inpatient hospital setting, then a company may also need to identify or establish a prospective payment MS-DRG code. If an existing MS-DRG code does not provide adequate reimbursement or the description is not inclusive of the new device, then an application must be submitted to the CMS for the following:

- An ICD-9-CM procedure code (which is used for describing patient diagnoses and procedures, as well as justifying the MS-DRG codes used for non-physician related billing for a patient encounter).[23]
- A new medical services and technology add-on payment to reflect the added procedural and supply costs associated with a new technology and, in some cases, the bundled cost of everything including the new device.[24]

Though an existing ICD-9-CM code may be submitted as part of this application, typically, a new ICD-9-CM code is needed to identify the device and associated service on subsequent claims. The add-on payment is issued if the manufacturer demonstrates that the cost of the procedure with the new device exceeds a threshold specified by Medicare for the MS-DRG code that covers that procedure. If the data are convincing, the add-on payment is set at 50 percent of the average cost of the new device.

An example of a new technology that successfully obtained an add-on payment is Medtronic's Infuse bone graft. This product is used to treat a spinal condition called degenerative disc disease. The Infuse product, which contains recombinant human bone morphogenetic protein (rhBMP-2), provides an alternative form of treatment by eliminating the pain and complications associated with harvesting bone from a patient's hip.[25] Medtronic originally applied for a new add-on payment in 2002 for this device, but its application was rejected. It successfully reapplied in 2003, submitting new data that demonstrated that the cost of performing the procedure with the Infuse bone graft exceeded a threshold set by Medicare and, thus, qualified for the add-on. The add-on payment was set at 50 percent of the average cost of the bone graft (approximately $8,900) and it remained in place for two years.[26]

Miscellaneous codes

An option that is often available to companies is to use a miscellaneous code. CMS defines miscellaneous codes as codes that can be used when "there is no existing national code that adequately describes the item or service being billed."[27] Their advantage is that they can be used as soon as a new procedure is approved by the FDA and while the manufacturer is applying for a permanent code. The disadvantage is that claims with miscellaneous codes require more detailed documentation than usual claims and are manually reviewed by Medicare. Therefore, it is to the manufacturer's benefit to closely monitor and manage the process of submitting these claims so that they are reliably reimbursed at adequate levels.[28] Not managing the process may set a bad precedent that could affect the reputation of a new device if claims are routinely denied or payments are inadequate.

Coding and private payers

Codes set up by the AMA (e.g., CPT codes) or CMS (e.g., APC and DRG codes) are also used by private payers to bill services. If there is a long delay before unique codes are issued, private payers may ask providers to use an existing code to bill services in the interim, agreeing on expanded coverage and/or payment terms under that code. In many cases, private payers would rather have providers use some unique code, even if it is not the permanent code, so that they are able to track and monitor usage and more efficiently process claims.

Coverage

Even with codes in place, payments cannot occur without a coverage determination by the payers that the procedure/device is medically necessary and adds value relative to other available treatment options. A new coverage decision is always needed when a new code is developed, but may also be necessary even when an existing code is in place. This occurs when a device expands the indications addressed by earlier coverage decisions.

When seeking coverage from Medicare, there are two different approaches a company can pursue: (1) a national coverage determination (NCD) that is only relevant for new devices to be used in the inpatient setting, or (2) a local coverage determination (LCD) that applies in almost all other cases. National coverage determinations are made by CMS and apply to all US beneficiaries. Local coverage determinations are made by so-called local carriers and apply to beneficiaries in their geographic area (administration of Medicare benefits is handled by 28 local Medicare contractors throughout the United States).[29] Because claims adjudication takes place at the local level, providers and manufacturers must negotiate with local contractors, especially in the launch phase of a device (see below) and different LCDs may be in place for the same device under different contractors.

Local coverage determination

Requests for an LCD of a medical device are often submitted by local providers or professional societies acting on behalf of the company. Medical professionals among the Medicare contractors tend to have strong personal relationships with the local medical community. These relationships can be leveraged by the company as part of the LCD process. When making an LCD, a local contractor will consider a variety of inputs, including the results of a literature review, medical evidence submitted, physician testimony, and the outcome of existing systematic reviews (such as TEC assessments). Once a recommendation has been made, a final review by the contractor-based carrier advisory committees (**CAC**) is mandatory for all proposed local coverage decisions. Carrier advisory committees are made up of local medical experts in a particular specialty. Proposed LCDs are posted on the CAC website for 90 days so that public and CAC comments can be collected as input to the final decision. Once a decision is implemented, the coverage policies can be revised or expanded whenever new, sufficiently compelling evidence is presented to justify a change. Furthermore, LCDs are non-binding – so "one-off" rulings are often made at the contractor level. For example, local contractors may allow one-time access and/or deny or pay claims for interventions that preceded local coverage decisions.[30]

Typically, company representatives pursuing an LCD will visit the appropriate local Medicare contractors at launch in order to secure a coverage commitment for their new technology and associated services using miscellaneous and temporary codes. Until these efforts lead to favorable, published coverage decisions, confusion can exist among providers as claims are potentially denied or underpaid – a common occurrence during the early stages of a local coverage determination. Additional confusion can be caused by the fact that LCDs are not binding, can change at any time, or be highly inconsistent across the country. The rework associated with this uncertainty places a significant administrative and financial burden on the providers that attempt to use the new technology. Reimbursement strategies that include LCDs require the company to build and sustain strong local advocacy and relationships with providers, a relatively large reimbursement team to help manage issues, and persistence in working with local contractors. Success at the local level is most often achieved through a series of small wins, with all favorable policies requiring ongoing maintenance by the company.

National coverage determination

National coverage determinations are primarily used for inpatient procedures. However, companies may seek an NCD even when the LCD option is possible to avoid the inconsistencies and confusion that can be associated with multiple LCDs. National coverage determinations can be applied for through informal contact (usually through consultants or specialty representatives with personal connections to the CMS) or through a written application.

Although there are many benefits associated with an NCD, this is not a realistic strategy for all companies to pursue. Normally, the CMS will only consider NCD requests for a device or intervention that shows breakthrough potential in terms of its clinical benefits and/or cost effectiveness. The CMS may also make an NCD for highly controversial technologies that it deems to be unsuitable for individual rulings by the local contractors (in these cases, the CMS can trigger an NCD internally, without a request from the company).[31] Furthermore, the CMS stipulates that it will only consider an NCD for an innovation that affects a significant Medicare population. The process of pursuing an NCD can also be extremely time consuming resource intensive, and risky, making it infeasible for most small device start-ups.

Johnson & Johnson sought a coverage decision at the national level (in parallel with a new DRG code) before it launched the first drug eluting stent. However, to achieve such a ruling, the company had to invest heavily in garnering support for its device, even lobbying representatives of Congress to convince the CMS that there was widespread support for the technology. Johnson & Johnson's efforts were ultimately successful because the company had vast support and good timing, as well as strong efficacy and economic outcomes data to support a positive decision (see From the Field section for more information). Without the resources of a large company to devote to the pursuit of an NCD, this approach is considered highly risky. A much safer model is to use a series of LCDs to build a basis for national coverage.

National versus local coverage determination
When considering whether to pursue an LCD or an NCD, there are several other factors that a company should keep in mind. First, even with a published NCD, payment levels for devices and the associated procedures may still be set at the local carrier level, often with limited guidance from CMS. This means that a sizable effort is still required of the company at the local level to secure favorable payment terms. Second, NCDs can be problematic in that they are binding decisions that can result in unfavorable and often irreversible policies toward the company. For example, in July, 2005, the FDA approved a vagus nerve stimulation (VNS) device manufactured by Cyberonics for depression in patients who had failed other treatments. Though the device was previously approved and covered for epilepsy, the value of the procedure for depression remained controversial. At a cost of roughly $30,000[32] for the device and implantation procedure, payers were denying coverage for resistant depression. In July, 2006, Cyberonics requested that the NCD for VNS be expanded to include resistant depression. In May, 2007, an unfavorable NCD was issued, citing the lack of compelling evidence to justify such a move. The Cyberonics website now states that VNS for resistant depression can be approved for reimbursement on a case-by-case basis; however, following the unfavorable NCD, it is unlikely any payer will grant liberal coverage in the future – a factor that significantly (and permanently) limited the company's potential market opportunity.[33] Even though an NCD only formally applies to Medicare, private payers, while not obligated to follow this lead, are often influenced by decisions made by the CMS.

Coverage determinations for private payers
There are many similarities between the processes used by Medicare carriers and by private payers to evaluate a new technology and reach a coverage determination, but there are also important differences. Private payers have their own internal committees (a medical policy or technology assessment committee,) consisting of physicians, plan administrators, health economists, and statisticians. These committees evaluate the evidence supporting the clinical necessity of a new procedure, paying special attention to evidence obtained outside academic and well-controlled settings. Precedence coverage decisions by other major payers are also evaluated. For example, in reaching a coverage position for bariatric surgery, Cigna (a major private payer) based its decision on information provided in a CMS coverage decision, technology assessment reports by the BCBS TEC in the United States, the National Institute for Health and Clinical Excellence (NICE) in the UK, and a comprehensive review of the literature.[34]

The final outcome of a private payer coverage determination is either full coverage, no coverage, or coverage with restrictions. A significant difference between private payers and Medicare is that private payers may offer multiple plan designs, with certain plans specifically excluding certain procedures (even if they are medically necessary). Again, in Cigna's case, their coverage determination on bariatric surgery does not apply to health plans where such surgery is explicitly excluded.

Payment

For new procedure-based codes, CMS uses a formula that is based on resource utilization for the service and procedure to determine a payment level, or the rate at which a new, covered technology will be reimbursed. As previously explained, for most CPT codes, this calculation multiplies the RVU as determined by the AMA/Specialty Society Relative Value Scale Update Committee (RUC) by the conversion factor and local adjusters.

For CPT codes for diagnostic tests, Medicare establishes a payment level for a new code using one of two approaches: crosswalking or gap-filling at the local carrier level (unless an NCD has been issued). If an existing code is sufficiently similar to the new code in terms of cost, technology, and clinical use, CMS determines the payment amount by crosswalking the code – assigning a comparable reimbursement rate to the new code based on its similarities to one or more existing codes. If no comparable code exists, the gap-filling method will be used to determine a new reimbursement amount, which requires each local Medicare carrier to individually determine an appropriate payment amount for the new code. These recommendations are then shared with CMS, which analyzes the carrier-determined amounts and sets an appropriate payment level for the new code that will be used until the next annual schedule is published. Detailed economic studies by the sponsor can support these efforts.

Payment rates for APC and DRG codes are determined nationally and published on a standard schedule. Temporary pass-through and add-on payment rates will be equal to a fraction of the average cost of the new device and may be set either nationally or locally (the most common outcome). Medicare will use documentation provided by the company on the cost of the device, its own internal data collected using miscellaneous and temporary codes, as well as formal and informal testimony from key opinion leaders and representatives of specialty societies to determine the average cost of the new device and specify payment levels. After one to two years, final decisions will be made as to whether transitional codes will become permanent, and at what level of reimbursement, or whether they will be embedded within the APC and DRG codes. Again, economic studies by the sponsor can influence the payment level.

Payment by private payers

In general, the fundamentals for setting payment levels are similar for Medicare and private payers, but there are important differences in the processes they use. All payers place a strong emphasis on evidence-based medicine, cost, and the demonstration of value. However, Medicare is responsible for publishing codes and payment rates, so its decisions tend to serve as the pathfinder, heavily influencing the policies of other payers. As of 2008, Medicare also could not legally consider cost (for the time being) in the evaluation of new technologies. Furthermore, officials from the CMS, and to some extent local contractor representatives, may be subject to greater political pressures and lobbying. Private payers are less likely to face such pressures since they have no direct links to elected officials and because their policies and payment schedules are not in the public domain. However, private payers are beholden to their providers, and to the employers and individuals that subscribe to their plans. For innovators and companies, there is often a greater opportunity to develop business relationships with private payers, as public payers (and some non-profit plans, such as Kaiser) restrict their employees' participation in industry-sponsored events and activities. Regardless, whatever contacts and relationships a company has with any payers should be leveraged to help secure appropriate payment levels and help facilitate a smooth reimbursement process when new codes, coverage, and payment levels are being established.

When processing claims from miscellaneous and temporary codes, private payers have different formulas for calculating payment levels. Each private payer generally uses its own formula to reimburse for billed charges on manually submitted claims. With miscellaneous code claims, payment rates usually amount to a percentage of billed charges for services and/or devices or a flat predetermined allowable. Interestingly, some private payers routinely reject or delay all manually submitted claims until they are elevated for special review within the organization. Companies must understand and anticipate this potential outcome in their reimbursement strategies. For devices and procedures that have been recently approved, private payers will typically set their payment at some percentage (usually greater than 100 percent) of Medicare or at a rate that approximates similar procedures and devices.[35] In the hospital, private payers are more likely to pay for each device separately and not lump them into a per diem payment in order to keep hospitals from losing money. Payment rates vary substantially from one private payer to another and depend on heavy negotiation between providers, payers, and the manufacturer.

Table 5.6.4 provides a summary of all codes available for use by a company at launch and beyond, as well as related information about coverage, payment, and the application process for these codes.

Table 5.6.4 Different codes are used under different circumstances for seeking reimbursement for a new technology.

Service/device billed	Temporary codes	Coverage/payment information	Permanent codes	Payment	Application process
Device/procedure in physician office	Miscellaneous CPT (device plus procedure) CPT category III (device plus procedure)	Miscellaneous codes paid as billed charges or percentage of billed charges Coverage and payment determined by local carriers and private payers CPT category III may not yield payment	CPT category I (device plus procedure)	Published rate as printed in Medicare payment schedule Private payer pays negotiated rate Coverage and payment determined by local carriers and private payers (without NCD)	Miscellaneous codes can be used without application, but may lead to difficult claims processing Category I and III codes are applied for through the AMA with specialty society data
Device/procedure in Outpatient Prospective Payment System (OPPS) setting	Misc. c-code (device) Temporary c-code (device) New Technology APC (procedure or device plus procedure)	Coverage generally determined by local carriers Misc. code paid at billed charges, with high degree of risk Temporary c-code paid at or near cost of device New tech APC paid at crosswalked or gap-filled amount by local carriers	APC code (device plus procedure)	Coverage generally determined by local carriers C-code paid at or near negotiated cost of device (published nationally) APC paid at crosswalked or gap-filled amount determine by local carriers (without NCD)	C-code applied for through CMS New Technology APC applied for through CMS, but only when the cost of procedural component is significantly changed
Device/procedure in Health Insurance Prospective Payment System (HIPPS) setting	New ICD-9-CM procedure code (procedure) Existing DRG (device plus procedure)	Local fiscal intermediaries determine coverage No payment associated with ICD-9-CM code Inclusion of all codes and DRG qualifies for add-on payment	New ICD-9-CM procedure code Existing DRG (device plus procedure)	Local fiscal intermediaries determine coverage DRG rate is increased in presence of ICD-9-CM code that signals device use, based on crosswalked or gap-filled amount	ICD-9-CM procedure codes are applied for through CMS Add-on payment to DRG is applied for through CMS

Stage 5 Development Strategy and Planning

Strategic reimbursement considerations

To support effective, efficient coding, coverage, and payment decisions, a company must craft a compelling value proposition for payers, design clinical studies providing evidence to establish the value proposition, and marshal support from the physician and patient advocacy groups that can influence payer decision making.

Payer value propositions

As implied throughout the chapter, establishing clear, evidence-based value propositions that address key payer concerns is essential to gaining reimbursement in the current healthcare environment. Value propositions targeted at payers rationalize costs in terms of the benefits provided by the technology. These benefits may be qualitative or quantitative and can range from a presentation of cost-relevant efficacy endpoints to a complete cost/benefit model analysis (see Appendix 5.6.1 for more information on the types of financial models that can be used to capture the costs and benefits associated with a technology). Value propositions should be clear and simple, avoiding tenuous projections based on intermediate markers of efficacy. They should also be based on evidence from well-controlled clinical trials, whenever possible. Typically, data from randomized, controlled trials are used to support effective value messages. Using data collected from a patient population and provider type that resembles that of the targeted health plan is also important for maximizing a company's likelihood of success. The example below illustrates a strong, data-driven value proposition.

Working example: Sample value proposition for the cost of treatment with an ICD[36]

Research shows that implantable cardioverter defibrillators (ICDs) provide an invaluable form of life insurance for people most at risk, preventing sudden cardiac arrest (SCA) death 98 percent of the time. Evidence-based medicine has demonstrated that ICDs significantly reduce death among Americans at highest risk:
- 31 percent reduction in death among SCA survivors from a second event.
- 31 percent reduction in death among post-heart attack sufferers.

Despite these statistics, ICDs are underutilized:
- Fewer than 20 percent of currently indicated patients receive the benefits of an ICD despite being at high risk for sudden death.
- Although SCA is responsible for more deaths than breast cancer, lung cancer, stroke, and HIV/AIDS combined, spending on SCA prevention is modest when compared to other diseases (AIDS = $19.5 billion; stroke = $6 billion; lung cancer = $1.6 billion; breast cancer = $0.8 billion; SCA = $2.4 billion, including drug and device therapy).

The value of ICDs outweighs their cost to the system:
- An ICD costs approximately $25,000 (including implant costs), which equates to less than $10 per day over the average life of a device (four to five years).
- The cost per day of ICD protection has decreased by nearly 90 percent over the last ten years from more than $90 in 1990 to less than $10 today (equivalent to the cost of optimal medical therapy for these same patients).
- ICD Medicare expenditures are significantly less than for other cardiovascular procedures. In 2002, Medicare reimbursed $1.2 billion for ICD procedures versus $6.4 billion for stent implants and $7.8 billion for bypass surgery.
- The cost of ICD therapy per year is less than 0.2 percent of projected Medicare spending over the next ten years.

The impact of reimbursement on clinical trials

Just as company representatives should meet with the FDA early in the regulatory process to understand what evidence will be required to gain regulatory approval, they should do the same with payers to understand what data are necessary to convince payers about the value proposition and lead to reimbursement. In particular, payer feedback should be solicited on reimbursement-related endpoints for clinical trials. These endpoints, which are often incremental to endpoints included for

Table 5.6.5 *Data to support reimbursement must explicitly be collected as part of a clinical strategy.*

Concerns to be addressed by reimbursement-related clinical trial endpoints
What are future significant expenses for patients with this disease or condition?
What expenses could payers avoid as a result of funding the technology?
What inefficiencies exist with the current standard of care, and how does the new device address this?
What unacceptable risks are inherent in current treatment standards, and how does the device minimize these?
Do the outcomes that the device will deliver provide patients with a highly compelling reason to seek treatment?
Is the target patient population young, or of working age, and will the device affect their productivity or ability to work?

regulatory purposes, can be expensive to generate and they are not without risk, as failure to show a positive effect will work against efforts to support reimbursement. Gaining feedback from payers will help to determine that the resulting data will have a positive impact. Endpoints should be selected based on the likelihood of addressing important payer concerns (see Table 5.6.5 for a list of concerns to be addressed), for their probability of success, and their ability to support simple, clear value messages that resonate with payers, physicians, patients, and the public.

Outcomes data that address payer concerns are even more important if the company's goal is to market the product outside of the United States in a region or country that has a nationalized payer system to determine reimbursement, such as Great Britain, France, or Australia. Many countries outside of the United States require specific outcomes data and cost-effectiveness studies (see Appendix 5.6.1), and the company will have to translate these data to endpoints such as quality of life, disability, morbidity, and mortality, and will also need to provide payers with detailed models of cost and effectiveness.

Advocacy

Physician advocacy is another factor that is essential for reimbursement success. Value propositions targeted at physicians is one way to get physicians on board with a new technology (see 5.7 Marketing and Stakeholder Strategy for information about developing value propositions targeted at physicians and other stakeholders). They can also be used by physicians, when advocating for the product among their peers. Managed care plans are ever conscious of retaining good physicians and typically do not like to deny therapeutic resources that the doctors feel are integral to providing good care. Physicians often see new devices as important innovations that fill an unmet need or have the potential to increase safety and efficiency. Specialty societies are invested in helping physicians provide innovative care, gain prominence, and operate financially robust practices. Not only will physician specialty groups be called on to sponsor the creation of new billing codes, but they have the ability to generate pressure and influence health plan decision making. In building a case for reimbursement, it is imperative (more often than not) to develop enduring personal and professional relationships with physicians and officers of key specialty societies in order to secure strong support for the product.

Depending on the disease or condition in question, patient advocacy groups can also be mobilized to build a powerful case that demonstrates the medical necessity of the product, especially if the benefits of the technology are easy to understand and promise significant improvements in quality of life. Existing patient advocacy groups should be engaged early in the reimbursement process. Officers of these associations can be asked to help recruit patients for enrollment in clinical trials. Strong value propositions are important both to attract patients and to harmonize their advocacy message as they network with others who may have an interest in the new technology. Patient advocacy groups can have an impact on payers through public relations campaigns and direct appeals to health plans. The ability to demonstrate enhanced patient outcomes related to workplace productivity is highly valued by health plans (because they can use this information in their marketing messages to retain and acquire the employers that make up their customer base). Furthermore, private health plans do not want to risk losing members by denying patients access to a technology that is perceived as useful. The threat of negative publicity is a major consideration for payers when deciding whether or not to cover a particular technology.

To some extent, companies can leverage payer advocacy to extend the adoption of a new technology. Because many health plans look to other plans for consistency when establishing coverage policies, and dislike appearing more restrictive than their competitors, support for a device from a core group of payers can be used to influence other payers. This can be done informally, through personal and professional networking, or more formally, through publications and advisory boards.

Value propositions, trial data, and advocacy all work together to help facilitate a successful reimbursement strategy. The J&J drug eluting stent story demonstrates how these factors, in combination, can have a meaningful impact on reimbursement.

FROM THE FIELD | **JOHNSON & JOHNSON**

Taking a proactive approach to reimbursement strategy for a breakthrough technology

When J&J, through its subsidiary Cordis, helped pioneer the bare metal stent (BMS), a small mesh-like tubular scaffold to help hold open narrowed heart arteries, the company did little to proactively advocate for the reimbursement of its device. Assuming that payers would see the benefit that the BMS provided and, therefore, reimburse for the device accordingly, J&J did not start reimbursement discussions with CMS until it was nearing FDA approval for its stents. Moreover, J&J had not performed many of the economic studies that payers believed would best show the value created by the BMS. As a result, CMS did not grant full reimbursement for the device in the United States until three years after FDA approval (granted in 1994). During this interim period, hospitals lost profits in their catheterization laboratories when stents were utilized in patients because reimbursement was inadequate under the codes that had traditionally been used for angioplasty. Interventional cardiologists wanted to adopt the technology based on its clinical benefits to patients, but they were under pressure from their hospitals to avoid stenting based on the roughly $1,600 that each stent added to the cost of angioplasty. This was true even though the price of a stent in the United States was much lower than in Europe and Japan.

The tension over inadequate reimbursement created a fair amount of animosity between the medical community and J&J, which was the BMS market leader for several years with its Palmaz-Schatz coronary stent until Guidant captured sizable market share with its Multi-Link coronary stent in 1997. Importantly, the Multi-Link was launched at the same time reimbursement was corrected to reflect the full cost of stents. As a result, even though the Multi-Link was priced $100 to $200 more than the Palmaz-Schatz BMS, Guidant never faced the same pricing and reimbursement challenges that J&J had experienced in the market.

Johnson & Johnson intended to reclaim its leadership position in the stent market through the introduction of the first drug eluting stent (see the case example in 2.1 Disease State Fundamentals for more information). As the company worked on the development of this new technology, "We absolutely knew we had to have a better strategy on reimbursement than we did with our bare metal stents," said Bob Croce, J&J's group chairman for Cordis Corporation. "Frankly, there were no favors given on the bare metal stents. It really had a terrible impact on the hospital's budget, even though it was best for the patient. We didn't want to repeat that experience again."[37] Since the drug eluting stent (DES) would be more expensive than the BMS, Croce knew that a new DRG code and coverage rate would be needed (at that time, Medicare was still using the older DRG system, not the more recent MS-DRG). Obtaining a new DRG could be a difficult and time-consuming process, so the company committed itself to getting an early start on this challenge.

One of the first things J&J did was to put together a team that would focus on nothing but the company's reimbursement strategy. "We had a small staff that was totally dedicated to this issue," said Croce. "They were extremely competent. The group was headed up by a physician who was a cardiologist at one time and ran clinical trials for years before he joined the company. Plus, we had a couple of people who really understood CMS and another couple who understood private payers and insurance companies."

Because the company would need to have economic data to justify the creation of a new DRG, Croce insisted that the reimbursement team work directly with the clinical team to design studies that would satisfy the objectives of both groups. "We had our reimbursement people totally integrated with our clinical group," Croce said. The reimbursement team had a voice in trial design to make sure that whatever data was collected could be used to generate economic outcomes. "It has to be an integrated clinical/economic effort so that you end up with facts and don't have to just trust that the new treatment will be cheaper. It required that we do a few extra things in our studies, but it was well worth the extra effort," he recalled.

As J&J began to get the results from its early trials, it initiated a sweeping, proactive communication campaign. "We presented our preliminary data to CMS at the same time that we gave it to the FDA," explained Croce, speaking of a move that was unusual at the time. "We presented how much it would actually shift the cost of medicine from a higher cost bypass surgery to drug eluting stents," which would save CMS money in the long run, in addition to improving outcomes for patients. Every time new data was available, the company presented it to CMS and the FDA simultaneously. "You can't do this with every product," he pointed out, "but in this case, they were fascinated by it. CMS wanted to understand this because they knew the ball was going to come into their court fairly quickly."

In addition to establishing open, early, ongoing communication with CMS, J&J contacted members of the Bush administration and Congress to get a seat at the table and explain what made DES such a revolutionary technology. According to Croce, his group "tried to do the best job we could in informing everybody that would have a say in the process or could influence the process, and at least get our story across." Fortunately for J&J, the timing was right for increased government collaboration. As Croce described, "The administration was saying, 'We've got to work in concert to provide better healthcare more quickly to seniors, especially if it saves lives and saves the government money.' They were looking for something that the FDA and CMS could work together on to make something positive happen for patients. So this was an ideal situation."

Johnson & Johnson's efforts were also greatly aided by the highly promising clinical data for DES (with in-stent restenosis rates of zero to three percent and in-segment restenosis rates of nine percent, compared with 33 percent in the bare metal stent arm).[38] Furthermore, the company had the data it needed to develop a compelling economic model. "Even if five percent of patients avoided a bypass procedure by receiving a DES, it was a fairly significant saving. But we actually developed a model that allowed CMS to plug in whatever percent it wanted and we would show them the cost savings. It was a very clever system," said Croce. The model also took into account the potential savings from the reduction in follow-up procedures created by lower restenosis rates.

Johnson & Johnson's efforts ultimately convinced CMS to grant reimbursement for the DES nearly eight months *before* the technology had been approved by the FDA for sale in the United States. Private payers, such as Aetna and Blue Cross, quickly followed suit, paving the way for the rapid and widespread adoption of the new technology upon its US commercial release (April, 2003). Worldwide sales for J&J's Cypher® stent exceeded $1.5 billion in that year and became a phenomenal success for the company (see Figure 5.6.2).

Reflecting on the success of J&J's reimbursement strategy, Croce acknowledged that this approach would not work for every company. "This was so different and so revolutionary that it captivated people's attention," he noted. However, he advised all companies, large and small, to do medical economic studies as part of their clinical trials. "You have to make sure you produce economic data. Without it, it's too easy for payers to push you aside." Other factors working in J&J's favor included the facilities, knowledge, network, and money that it needed to get the data and gain an audience with the people who would ultimately make the reimbursement decision.

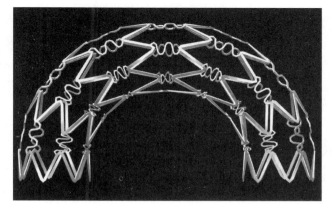

FIGURE 5.6.2
The Cypher drug eluting stent (courtesy of Cordis Corporation).

Pricing

Pricing decisions are a critical component of a company's reimbursement strategy. As described in 5.7 Marketing and Stakeholder Strategy, the company will need to set a price that would make the device affordable at the reimbursement levels it expects to secure from providers. To determine the appropriate price, many companies perform what is known as demand-curve analysis. This exercise involves assessing the price sensitivity of key payers to determine the range within which the company can secure adequate reimbursement, which in turn determines how to realistically price its product. The Genomic Health example provided in Chapter 5.7 demonstrates how value-based pricing works.

Reimbursement tactics

Throughout the reimbursement process, certain tactics can help a company assess payer segments, build important relationships, gather feedback from stakeholders, define product positioning to managed care organizations and payers, clearly present value messages and outcomes data, and provide reimbursement support to providers. More information about some of these tactics is provided below.

Payer segmentation

Because early-stage start-ups have limited resources, they typically need to prioritize which payers to approach. Approximately six months before launch, a company should initiate efforts to segment payers. Some payers will allow immediate use of an approved device (in accordance with product label) at launch, while others will mandate a six- to twelve-month waiting period. These more restrictive plans will allow exceptions to varying degrees. For example, some payers will allow device use after an appeal is made by a physician and found to be justified. Often, these waiting periods culminate in an intensive review of the technology (i.e., technology assessment), during which clinical and economic data are scrutinized and the new technology is compared holistically to others in its class or therapeutic area. The outcome of this analysis leads to a final decision about the reimbursement status of the device within that plan. Some payers will wait and follow the lead of TEC assessments, while other payers will be more likely to approve more quickly in a manner consistent with the label.

Information about how different payers respond to new technologies (based on primary payer research) can generally be purchased from a managed care consulting firm and then used to perform a preliminary payer segmentation. Common criteria used to segment payers include their openness to new technology and the leniency of their review process, as shown in Figure 5.6.3.

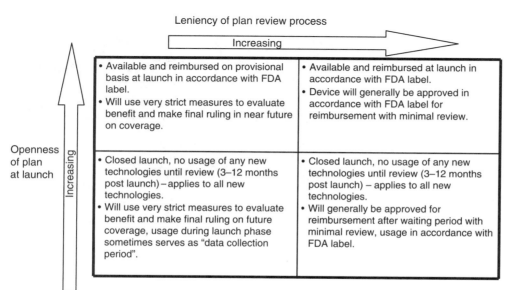

FIGURE 5.6.3
Issues to consider in segmenting payers.

Whenever possible, these segments should be confirmed by the company's reimbursement team. In addition, the number of patients covered by payers in each segments should be estimated to help the company more efficiently allocate the time and focus it will devote to each payer at launch. In general, a small device start-up should follow the 80/20 rule – targeting the top 20 percent of payers that cover 80 percent of patients. At a minimum, companies should include Medicare and one to three prominent private payers in their initial reimbursement strategy (for the private payers, identify the ones most likely to adopt/cover the new technology). Importantly, private payers are the largest single source of funding for healthcare in the United States, accounting for more than $1 out of every $3 spent on healthcare in 2005.[39] As a result, they must not be overlooked, especially if the company has a sizable non-Medicare population within its target audience.

Payer advisory boards

Establishing a **payer advisory board** (that consists of KOLs and medical directors from select payers in the target payer segment) provides an important way for a company to build relationships with and create awareness and support for the device among payers. Initially, smaller advisory board meetings can be held in order to gather feedback on the product. They can also be used to evaluate the technology's chances for reimbursement success at different prices and/or to validate clinical and economic endpoints. As clinical trials progress and the regulatory submission process begins, the nature of information gathered at an advisory board meeting will shift to the viability of reimbursement given possible outcomes, coding recommendations, and forward-looking strategies with payers. Just before launch, payers will be able to provide timely information about claims processing, payment amounts, patient prior authorization requirements, provider and facility requirements, and timing of technology assessments. They can also provide the company with feedback on the financial models, dossiers, marketing materials, and reimbursement literature that it has developed.

Successful advisory boards can be national, but are more often regional in nature. Their meetings are sometimes planned to coincide with managed care society conventions to make participation easy for payers. They can be held in person or through conference calls and/or webcasts. The typical agenda for a meeting begins with a physician KOL (who is an advocate for the technology and has participated in product testing or a clinical trial) presenting epidemiology data, evidence of an unmet need, clinical data, clinical vignettes, and proposed use to the other members of the payer advisory board. Ideally, this physician will have good relationships with managed care groups and will have presented to such an audience successfully in the past. A commercial representative from the company might next present product positioning, patient population information, progress with the FDA, the timeline for launch, proposed code use, value-added services, and the distribution plan. This may also be a good time to vet sections of the product dossier or the validity of economic models. In the event that there is a discrepancy between payer and patient understanding of unmet need, a patient advocate from the clinical trials can be a helpful addition to the meeting. Because an honorarium is offered to attendees, not all payer representatives will be allowed to participate in off-site industry-sponsored advisory boards (some health plans limit such employee involvement to avoid conflicts of interest). A list of questions to explore in the first advisory board meeting is presented in Table 5.6.6.

Publications

In addition to publishing clinical data that demonstrate the safety and efficacy of a device in peer-reviewed medical and specialty journals, it is important for the company to publish economic results and key value propositions in journals targeted to the managed care audience. Often an advocate medical director at a health plan will codevelop and coauthor an article with a physician KOL who was included in the clinical trial. There should also be a publication plan for favorable outcome endpoints from clinical trials as soon as they become available, not just in the specialty journals but in managed care journals. For devices that are likely to broaden their indications and usage over time, it is important to plan studies and publications well in advance to support such an expansion.

Table 5.6.6 *Questions such as these can be used to guide preliminary discussions with advisory board members.*

Preliminary advisory board questions
What is the decision maker's perception about the current standard of treatment in terms of efficacy, safety, and cost?
What is the prevalence of the disease within the plan population? What is the total cost of care?
What are the consequences of not treating a disease?
What are the current coding, coverage, and payment levels for the current standard of care? What are the exclusions and limitations of coverage? Copayments?
What is the reaction of the decision maker to the value proposition of the new treatment?
What would be desired endpoints from any studies? What would be desirable studies to support reimbursement?
What are the key considerations that will be taken into account when making a coverage decision?
What is the process for reaching a coverage decision and who will be involved?
What information/evidence does the decision maker expect to see? Will the company have the opportunity to orally present to the medical policy committee or any policy analysts?
What kind of economic models do they typically require from the company?
Are there any predicate products whose coverage decision or current reimbursement would affect the determination for the new product?
What price is acceptable for a payer and how would price and payment affect the coverage determination?

Presence at managed care conferences

In order to gain maximum reach with its messages, a company should submit data as abstracts at specialty society and managed care meetings to be published and presented by a KOL. Abstracts and oral presentations are good vehicles for generating positive press, and effective coordination can provide new coverage of the product at the industry, local, or national levels. Scientific symposia at national managed care meetings can be sponsored by the manufacturer to showcase clinical and economic data to payers. At some meetings, an exhibit booth is also available for sponsorship.

Reimbursement dossier

New technologies are increasingly being subjected to assessment by committees such as the BCBS TEC. Documentation is frequently distributed by the company and its sponsor (in the form of a **reimbursement dossier**) to support the product through this process. A product dossier can serve as the official source for all key information about the product. Typically, a dossier is prepared in advance by the company and distributed to payers just after final labeling becomes available from the FDA.

A dossier should have multiple sections that address the topics shown in Table 5.6.7. Because the content of a dossier is comprehensive, it can serve as a roadmap when presenting to payers. Compilation of the data and messages to complete the dossier is a lengthy and intensive process that requires the support of specialty medical personnel, as well as experts in cost-modeling, marketing, and managed care. Work should begin at least one year before approval to ensure completion before launch. Planning for dossier development should begin even earlier: during concept screening the company starts to anticipate the evidence that will be necessary to gain reimbursement. For reimbursement decisions outside the United States, the dossier (adapted to fit individual country standards) will be essential, as deliberative meetings with industry personnel and advocates are less likely to occur, and a well-documented need for the technology will be essential.

Provider reimbursement support

Providers will be less likely to use a new technology if they experience claim denials and lose money. As a result, a company must develop a comprehensive

Table 5.6.7 *A reimbursement dossier can be an important communication vehicle for information related to a company's reimbursement strategy.*

Topic	Description
Basic product information	Materials and mechanism of action, indications and product labeling, product cost, associated codes, access, and distribution
Place of product in therapy	Epidemiology of disease, approaches to treatment, alternative treatment options, product positioning, and expected outcomes of the technology
Key clinical and economic studies	Treatment populations, number of people studied, study designs, clinical and economic outcomes measured
Summary of results	All published and unpublished results of clinical studies
Disease management strategies	Overview of ancillary or disease care management to accompany treatment
Outcomes studies and economic evaluation supporting data	Results of outcome and economic studies
Modeling report	Information on cost model(s) and assumptions, results of inputs, parameter estimates, applicable time-horizon, and discounting
Product value and overall cost	Overall justification of cost given data presented
References	To support data provided

approach for helping providers gain reimbursement and be ready to deploy this approach on the day of the launch. This is especially important for technologies used in the physician office setting. The specific tasks that need to be completed prior to launch include the development of educational materials and claim submission packages that salespeople can distribute to providers to assist them with reimbursement. It can also be useful to set up a reimbursement hotline that providers can access with their billing problems. Companies are further encouraged to build on payer segmentation to create a list of payer-specific billing guidelines that the company can share with its first customers.

At launch, reimbursement support begins with educating providers on key reimbursement challenges and giving them access to the support systems that can help them confirm coverage and secure appropriate payment. The company should educate providers about specific payer requirements in their region, as well as identify payers that support reimbursement for the new technology. If health plans have limited coverage of the device to a subset of the indicated population, physicians and their staff should be trained on appropriate patient selection and prior authorization requirements.

Billing staff will require training on plan-specific billing, coding, and claim submission procedures and the claim submission package can be very helpful. The product sales force will be on the front line of reimbursement support and must also be trained to field reimbursement-related questions. Monitoring of payer actions should also take place. Furthermore, appropriate reimbursement expectations should be set with providers, internal stakeholders, and the investment community.

Throughout these pre-launch and launch reimbursement support activities, oversight by skilled legal counsel will be essential. Practice management support is one of the key areas related to reimbursement that gets companies into trouble. Because supportive information helps providers get paid, in some cases from government payers, great care must be taken to avoid activities that constitute practice building, kickbacks, inducements, or inadvertent counsel to break the law. A misstep in this area can mean huge fines, bad publicity, and added marketing restrictions from which it could be nearly impossible to recover.

For example, in 2008, Medtronic agreed to pay a $75 million settlement in a case alleging that one of its companies, Kyphon®, defrauded the Medicare system.[40]

Table 5.6.8 *This sample budget for a small- to mid-sized high-end therapeutic implant provides an order-of-magnitude estimate of the cost of developing and implementing a reimbursement strategy.*

Tactic	Market conception and premarket phase	Introduction to market phase	Launch and adoption phase
Reimbursement consultants	$50,000–$150,000	$100,000–$300,000	$100,000–$500,000
Value proposition and cost-model development	–	$50,000–$150,000	$50,000–$100,000
Advocacy and society sponsorships	$5,000–$100,000	$5,000–$100,000	$50,000–$150,000
Payer advisory boards	$40,000–$120,000	$200,000–$600,000	$200,000–$600,000
Publications	$30,000–$50,000	$50,000	$100,000
Presence at managed care meetings	$20,000	$50,000–$250,000	$50,000–$250,000
Dossier development	–	$100,000–$300,000	$50,000

According to government claims, Kyphon advised hospitals to bill Medicare for its spine operation procedure using more lucrative inpatient rates instead of less expensive rates for outpatient treatment. By keeping patients overnight (allegedly without cause), the facilities performing the *kyphoplasty* procedure were able to charge Medicare, over time, millions more than it potentially needed to pay. Kyphon, in turn, was able to charge more for its technology. While Medtronic maintained that Kyphon and its employees had not engaged in any wrongdoing,[41] the settlement was damaging to its reputation.

Maintaining a reimbursement strategy

Even after the device successfully launches and providers are being reimbursed for their services and recovering costs, a company must maintain an active reimbursement strategy. Early and active publication planning should yield outcomes data at regular intervals to support product expansion into new indications or patient populations. A company should proactively nurture relationships with managed care organizations and payers of all types, answering product questions as they arise, sharing new data, and refuting negative competitive claims. Reimbursement materials should also be continuously updated in response to code or label changes, price increases, entry of a new competitor to the market, or class reviews. Similarly, as codes tend to evolve for several years after launch, a company has to continually manage that process.

Table 5.6.8 provides an example of how a company's reimbursement strategy budget might break out across pre-launch, launch, and post-launch activities. In the post-launch phase, a company may have opportunities to develop mutually beneficial partnership arrangements with health plans. Companies may elect to work with private payers to track usage and performance outcome measures within their plan population. This type of partnership is often an extension of a contract in which the payer receives a discounted price or rebate by reaching predefined share or volume targets agreed upon with the manufacturer. This provides an incentive to both parties to co-promote coverage. For example, Conceptus Inc. issues a press release each time a new payer adds a favorable coverage policy for its minimally invasive procedure to address blockages in the fallopian tubes.[42] Companies can also use in-office pull-through activities, in the form of cobranded sales pieces promoting the device as "preferred" by this payer. Sales representatives can be used as well, to help advertise coverage policies in order to keep providers informed regarding which payers are reimbursing for the device. All of these activities must be carefully integrated and managed through an ongoing reimbursement strategy.

Global reimbursement issues

As described in 4.3 Reimbursement Basics, in most healthcare systems outside the United States, governments play a critical role in reimbursement and purchasing of new medical devices. In most advanced European nations, new devices need to undergo technology assessment before they are approved by use from nationalized healthcare systems. Therefore, a company should be prepared with economic and clinical studies to influence such evaluations. Most technology assessment groups follow policies similar to those discussed in Appendix 5.6.2. Pricing is usually set either through direct negotiation with the company or through **reference pricing** (i.e., setting a price relative to what is charged in another country).

When reference pricing is in place, the sequence in which a company releases a product in different markets can be important – it should first be released in countries where a higher price can be commanded. Developing nations, such as India and China, are increasingly relying on reference pricing. Given the complexities of the global healthcare marketplace, most small start-ups will focus their initial efforts in the United States. However, if for any reason it makes strategic sense to consider other markets, establishing a relationship with a local distributor who has local knowledge is the best way to penetrate these other markets.

GETTING STARTED

Creating a reimbursement strategy is challenging and complex. As noted, most companies will benefit from the services of an expert reimbursement consultant. However, before approaching a consultant, the innovator should perform a preliminary assessment of the reimbursement gaps that the strategy needs to fill so that a consultant with a proven track record for addressing such needs can be selected. Ideally, the company wants a consultant with experience in resolving coding, coverage, and payment gaps that are as close as possible to the specific ones it is facing. Large health economic consultancy firms can help with the development of reimbursement strategies, as well as individuals who previously worked in payer organizations before becoming independent consultants. In the United States, several such firms and consultants are members of the Medical Devices Manufacturers Association and their names are listed on its website.

See **ebiodesign.org** for active web links to the resources listed below, additional references, content updates, video FAQs, and other relevant information.

1. Assess the reimbursement landscape

1.1 What to cover – Repeat the basic reimbursement analysis performed as described in 4.3 Reimbursement Basics, using information gathered on the final solution concept under development and/or proxy devices. Be sure to understand the mechanics related to codes (existing versus new); coverage decisions (Medicare, large commercial payers, payers outside the United States); reimbursement levels; and the status of technology assessment in the given field. Identify critical gaps in codes, coverage, and reimbursement payments that should be addressed via a reimbursement strategy. Rank payers and technology assessment groups based on expected perception of the company's technology.

1.2 Where to look – Refer to 4.3 Reimbursement Basics.

2. Perform primary market research with payer decision makers

2.1 What to cover – Identify 10 to 15 target payers based on the number of lives they cover, the number of procedures they would cover, the leniency of the plan review process and the openness of the plan at launch, and the analysis in step 1. Research policies of any new payers identified. Identify key contacts within each payer organization (either through independent searches, by engaging a reimbursement consultant, or through referrals by KOLs, since

Stage 5 Development strategy and planning

medical directors in health plans can have links to KOLs). Interview medical directors and health policy analysts. Use the questions in Table 5.6.6 as the basis for each discussion, and the segmentation techniques in Figure 5.6.3 to identify the target payers.

2.2 **Where to look** – See Appendix 5.6.3 for a list of major US payers. Work with KOLs and reimbursement consultants to facilitate introductions to target payers and medical directors willing to be interviewed. In addition, refer to the following resources:
- **Technology assessment/evidence based reports**
  - Blue Cross/Blue Shield Technology Evaluation Center
  - NICE (in the UK)
  - ECRI Institute
  - California Technology Assessment Forum.
  - IOM Report on Technology Assessment Organizations – A comprehensive report of organizations that perform technology assessments.
- **Payer policies**
  - **America's Health Insurance Plans** (AHIP) – AHIP's website (www.ahip.org) provides a comprehensive list of private payers in the United States. Their websites can be used to research the policies of any new payers that have been identified.
- **Government policies**
  - Medicare coverage policies
  - AHRQ evidence-based analysis

3. Evaluate strategic options

3.1 **What to cover** – Determine whether existing code, coverage, and reimbursement levels are adequate or whether new codes and/or modifications in coverage and reimbursement are needed. Questions to consider include: Are existing codes, coverage, and reimbursement levels directly applicable? Is the current reimbursement level appropriate for the pricing strategy? Do the payers perceive a significant need that the product addresses and that justifies any required changes in coverage and reimbursement? Think about these questions from the perspective of both private and public payers. The same conclusion may not be reached for all payers. Determine whether new codes are needed and whether coverage determinations need to be modified for all or some payers.

3.2 **Where to look** – Utilize the results from the primary and secondary research performed in steps 1 and 2. Compare existing payment levels to pricing considerations from 5.7 Marketing and Stakeholder Strategy. Convene a payer advisory board, if appropriate.

4. Develop evidence

4.1 **What to cover** – Identify studies and publications used to support reimbursement for proxy devices. Determine studies, including specific clinical and economic endpoints, needed to support the reimbursement strategy. Consider primary and secondary data collection, clinical trial studies, database studies, registries, etc. Prioritize studies based on their costs and likelihood of influencing reimbursement decision makers. Develop a preliminary economic model and use it to identify gaps in available data. Share the model with consultants and KOLs to verify its strengths and weaknesses. Include studies in the clinical trial design plan.

4.2 **Where to look**
- Results from steps 1 and 2
- **Appendix 5.6.1** (for different types of economic models)
- Payer advisory board
- **Technology assessment reports and coverage policies for proxy devices**
- **PubMed** – For studies performed for proxy devices and used successfully to support reimbursement.
- **CPT background and categories of CPT codes** – The American Medical Association's website summarizes requirements for new CPT codes.

5.6 Reimbursement strategy

GETTING STARTED

- **Requirements for pass-through payments** – If the procedure will be performed as part of an inpatient hospital stay, then an application for a DRG add-on can be submitted – this is a supplemental sum to augment the standard DRG code when the new device is used. See CMS's "Application for New Medical Services and Technologies Seeking to Qualify for Add-On Payments Under the Hospital Inpatient Prospective Payment System for Federal Fiscal Year 2009."
- **National coverage determination requirements** – Medicare's website has documents on "Medicare Program; Revised Process for Making Medicare National Coverage Determinations" and on "Factors CMS Considers in Opening a National Coverage Determination."

5. Organize a timeline and develop a budget for reimbursement tactics at launch

5.1 **What to cover** – Prepare a reimbursement dossier and education material for payers and advocacy groups. Involve KOLs. Educate and communicate with specialty societies, the AMA, CMS, local Medicare carriers, private payers, and payers outside the United States. Initiate coding, coverage, and reimbursement initiatives. Identify contacts in each key constituency and map them to specific reimbursement plans. Determine priority order for targeting payers and the appropriate sequence and timing of all activities. Determine requirements in terms of consultants and in-house expertise necessary to execute the plan. Develop a preliminary plan for supporting reimbursement post launch.

5.2 **Where to look** – Take stock of decisions made in steps 3 and 4. Network with KOLs, clinical advisors, and reimbursement consultants. Review timeline for similar tactics for proxy device.

Credits

The authors would like to acknowledge Steve Fair and Sarah Garner for their help in developing this chapter. Many thanks also go to Robert Croce, Trena Depel, and Mitchell Sugarman for their contributions.

Appendix 5.6.1

More about cost models and their link to payer value propositions

Product value to payers is strictly defined by the costs and benefits associated with the technology. Some of the various economic models that can be used to construct compelling payer value propositions are described below.

Cost/benefit analysis

This is the strongest and most persuasive model to payers. It shows that the money spent on the device is lower than the total cost of the outcomes of the disease or of current standard therapy.

Example: Radical retropubic prostatectomy (RRP) is the most common therapy for patients with prostate cancer. Laparoscopic (LRP) and robot-assisted (RAP) prostatectomies have recently been introduced as minimally invasive alternatives to RRP. A study published in the *Journal of Urology* in 2004, attempted to quantify the costs and benefits of LRP and RAP compared to RRP.[43] The calculations involved both labor costs and supply costs for all three procedures, as well as the cost of purchasing and maintaining the *da Vinci®* robot from Intuitive Surgical Systems for RRP (see 5.8 Sales and distribution strategy for a description of Intuitive Surgical). Because the cost of the robot had to be spread over multiple procedures over multiple years, the analysis had to make assumptions on the number of procedures per year and the number of years. The results from the analysis are summarized in Table 5.6.9. This information demonstrates that while RAP reduces total operating room costs and hospital room and board compared to both RRP and LRP, the added cost of equipment and the robot cost (which is in addition to the equipment) make it significantly more costly than RRP. The analysis assumed a purchase price for the robot of $1.2 million, annual maintenance cost of $100,000, 300 cases per year with a seven-year period of analysis. Table 5.6.9 presents costs per case.

Cost-effectiveness model

In a **cost-effectiveness model**, cost is expressed per unit of meaningful efficacy, usually used comparatively across interventions.

Example: Initial hospital costs for a drug eluting stent (DES) are $2,881 higher than for a bare metal stent (BMS). Over the first year, follow-up costs for DES are lower by $2,571 primarily because revascularization events drop from 28.4 percent in the BMS group to 13.3 percent in the DES group. Therefore, over the first year the incremental cost of DES is $309 per patient but the revascularization rate is 15.1 percent lower. The incremental cost per revascularization avoided is $309/0.151 = $2,046.[44]

Cost-utility analysis

In **cost-utility models**, a cost is assigned for quality of life and years lived, based on clinical outcome measures related to quality of life and/or disability and mortality. Quality of life indexes and disability-adjusted life indexes are multiplied by the number of years gained. Costs are then expressed as a ratio. This model is often used by national health systems as the standard for evaluating reimbursement outside the United States. The threshold for allowing certain treatments is a dollar value per quality adjusted life year (QALY) or disability adjusted life year (DALY).

Example: Eight randomized trials evaluated the value of using implantable cardioverter defibrillators (ICDs) for primary prevention (see the InnerPulse story in 2.3 Stakeholder analysis). The studies showed that life expectancy increased by as much as 4.14 years (depending on the study) while costs increased by more than $100,000 in some instances (all these relative to an increase in life span by ten years, but with a quality of life ratio of 0.4 – a quality of life ratio equal to 1 is perfect health and 0 is death). In one specific study (SCD-HeFT), the cost increase related to ICD was $71,000 and

Table 5.6.9 *A per-case cost comparison.*

Cost component (per case in $)	RRP ($)	LRP ($)	RAP ($) With robot purchase cost	RAP ($) Without robot purchase cost
Total	5,554	6,041	7,280	6,709
Operating room	2,428	2,876	2,204	2,204
Standard equipment	75	533	1,705	1,705
Surgeon fees	1,594	1,688	1,688	1,688
Hospital room and board	988	514	474	474
Fluids and medications	150	78	72	72
Robot cost per case (purchase and maintenance)			857	286

the life expectancy increase was 1.40 years.[45] Assuming a quality of life of 0.75, this translates into a QALY increase of 0.75 × 1.40 = 1.05 and incremental cost per quality adjusted life year of $71,000/1.05 = $67,619.

Budget impact models

Budget impact models look at the cost and treatable population within the health plan, as well as the expected annual cost to the plan of covering the device. The results are normally evaluated in terms of per-member, per-month costs.

Example: A health plan has 50,000 members. The annual incidence of patients who expect to be treated by a new device is 0.1 percent. The total reimbursement for the device is $500 per patient. Therefore, the total budget impact of the device on the health plan is $25,000 per year, which equates to approximately $0.04 per member per month.

Cost models are particularly powerful when interacting with payers. For models shared pre-launch, costs may need to be estimated. Models should be sensitive to price increases and competitor discounts, as they can "live on forever" once distributed and work against the product in the future if the economics change. The value propositions derived from these models will become important tools to help build advocacy, and for use by advocates to favorably affect coverage and reimbursement policies.

The models described here are usually packaged with a user-friendly interface and can be modified by the payer to test a range of values to represent plan patient populations. Sometimes they are distributed by the company for use by the health plans. Other companies choose to present their findings to health plans and not "leave behind" the actual model (particularly in a highly competitive market). In addition to competitive considerations, there are market risks in the distribution of models, as they extrapolate clinical data over large populations and extended periods of time and these extrapolations may be proven wrong over time.

Appendix 5.6.2

Blue Cross/Blue Shield Technology Evaluation Center (TEC) assessment

For many private health plans, including Blue Cross/Blue Shield (BCBS), FDA approval is not sufficient to establish coverage for a particular intervention. Before drawing that conclusion, the plan will review data supporting the intervention and assess its benefits relative to comparable therapies, as well as its overall effect on health outcomes. While not all health plans conduct internal technology assessments, many follow the recommendations published by entities such as BCBS TEC. Blue Cross/Blue Shield and other plans are concerned that the limitations of clinical trials, used solely to gain FDA approval, fail to give the physician enough information to fully assess the strength of the new technology. Health plans under political pressure have been observed to cover technologies that have not been fully evaluated, only to find after a period of time that the interventions are definitively not helpful or even may cause harm.[46]

Tysabri, a groundbreaking new therapy for multiple sclerosis co-marketed by Biogen Idec and Élan, is one such example. This product was launched in the United States after accelerated FDA review in the fall of 2005. Many payers were asked to cover it before complete data from phase III clinical trials became available so that patients would not be denied access. However, in February, 2005, Tysabri was pulled from the market after causing death and severe disability in a handful of patients.[47]

Generally, interventions will be chosen for TEC evaluation if they have the potential to cause serious side effects, are indicated for a large treatment population, and/or the cost of the technology is high. The TEC committees make their final recommendations based on the following criteria:[48]

1. The technology must have final approval from the appropriate governmental regulatory bodies.
2. The scientific evidence must permit conclusions concerning the effect of the technology on health outcomes.
3. The technology must improve the net health outcome.
4. The technology must be as beneficial as any established alternatives.
5. The improvement must be attainable outside the investigational settings.

Conclusions are generally drawn on the basis of relevant evidence from clinical trials and analysis of cost models, and with only limited influence from expert opinion or current prevailing community medical practice. While BCBS allows comments and materials to be submitted and considered by BCBS officials involved in the decision, the final decision-making process is closed, and generally includes only internal or invited experts to testify. This is in contrast to the assessment process at the FDA, CMS, and certain other health plans. For this reason, it is important to reach BCBS early with materials and contacts to ensure that officials go into this meeting with enough information to rule favorably on the technology.

Appendix 5.6.3

Overview of US payers and integrated providers with more than $1 billion in sales (2008)

Company	Sales (millions)	Lives covered (millions)	Premiums (millions)	Description (from company annual reports)
UnitedHealth Group Inc.	75,431.00	75.40	68,800	A diversified health and well-being public company, serving approximately 70 million Americans. The company provides individuals with access to healthcare services and resources through more than 560,000 physicians and other healthcare providers and 4,800 hospitals across the United States. UnitedHealth Group conducts its operations through four operating divisions: Health Care Services, OptumHealth, Ingenix and Prescription Solutions.
WellPoint Inc.	61,134.30	34.80	55,865	A commercial health benefits public company serving approximately 34.8 million medical members as of December 31, 2007. An independent licensee of the Blue Cross Blue Shield Association (an association of independent health benefit plans), the company serves its members as the Blue Cross licensee for California and as the Blue Cross and Blue Shield licensee for Colorado, Connecticut, Georgia, Indiana, Kentucky, Maine, Missouri, Nevada, New Hampshire, New York, Ohio, Virginia and Wisconsin. The company also serves its members throughout various parts of the United States as UniCare.
Aetna Inc.	27,599.60	28.00	23,000	A diversified healthcare benefits public company, offering a range of traditional and consumer-directed health insurance products and related services, including medical, pharmacy, dental, behavioral health, group life, long-term care and disability plans, and medical management capabilities. Aetna's operations include three business segments: Health Care, Group Insurance and Large Case Pensions. Health Care consists of medical, pharmacy benefits management, dental and vision plans offered on both a risk basis and an employer-funded basis. Medical plans include point-of-service (POS), health maintenance organization (HMO), preferred provider organization (PPO), and indemnity benefit (Indemnity) plans. Group Insurance products include life, disability and long-term care insurance. Large Case Pensions manages a range of retirement products, including pension and annuity products for tax qualified pension plans.

Stage 5 Development Strategy and Planning
Continued

Company	Sales (millions)	Lives covered (millions)	Premiums (millions)	Description (from company annual reports)
Humana Inc.	25,290.00	18.30	Not reported	A full-service health and supplemental benefits public company, offering an array of health and supplemental benefit plans for employer groups, government benefit programs, and individuals. As of December 31, 2007, Humana had approximately 11.5 million members in its medical benefit plans, as well as approximately 6.8 million members in its specialty products. The company manages its business with two segments: Government and Commercial. The Government segment consists of beneficiaries of government benefit programs, and includes three lines of business: Medicare, Military, and Medicaid. The Commercial segment consists of members enrolled in its medical and specialty products marketed to employer groups and individuals.
Kaiser Foundation Health Plan Inc.	22,044.00	8.00	22,040	A non-profit managed healthcare company with more than 8 million members in California, Colorado, Georgia, Hawaii, Maryland, Ohio, Oregon, Virginia, Washington and Washington D.C. It is a public-benefit corporation that contracts with individuals and groups to arrange comprehensive medical and hospital services. The company also contracts with Kaiser Foundation Hospitals and Permanente Medical Groups, facilities with more than 10,000 doctors, to provide services to its patients. Kaiser Foundation Health Plan is part of Kaiser Permanente, an American integrated healthcare organization, and has headquarters in Oakland, California.
Cigna Inc.	16.50	9.30	9,800	A public company that provides healthcare and related benefits offered through the workplace. Key product lines include healthcare products and services (medical, pharmacy, behavioral health, clinical information management, dental and vision benefits, and case and disease management); and group disability, life and accident insurance. In addition, CIGNA also provides life, accident, health and expatriate employee benefits insurance coverage in selected international markets, primarily in Asia and Europe.
Coventry Health Care Inc.	9,879.50	5.00	8,690	A national managed healthcare public company operating health plans, insurance companies, network rental/managed care services companies and workers' compensation services companies. The company provides a range of risk and fee-based managed care products and services, including health maintenance organizations (HMO), preferred provider organizations (PPO), point-of-service (POS), Medicare Advantage (MA), Medicare Prescription Drug Plans, Medicaid, Workers' Compensation and Network Rental to a cross-section of individuals, employer and government-funded groups, government agencies, and other insurance carriers and administrators in all 50 states, as well as the District of Columbia and Puerto Rico. Coventry has two operating segments: Health Plans and First Health.

5.6 Reimbursement Strategy

Company	Sales (millions)	Lives covered (millions)	Premiums (millions)	Description (from company annual reports)
WellCare Health Plans Inc.	3,762.90	2.30	3,713	A public company that provides managed care services exclusively to government-sponsored healthcare programs, focusing on Medicaid and Medicare. It offers a variety of Medicare and Medicaid plans, including health plans for families, children, the aged, blind and disabled, and prescription drug plans. As of December 31, 2006 it served over 2,258,000 members nationwide.
Harvard Pilgrim Health Care Inc.	2,183.20	1.00	2,488	A non-profit health provider for the greater New England area. The company offers a variety of plan choices, including health maintenance organization, point-of-service and preferred provider organization plans. It also features a plan for Medicare beneficiaries called First Seniority. Harvard Pilgrim provides healthcare coverage in Massachusetts and Maine, while its subsidiaries provide coverage for other areas.
Advocate Health and Hospitals Corporation	2,080.00	Not Reported	Not Reported	The largest fully integrated not-for-profit healthcare delivery system in metropolitan Chicago, recognized as one of the top ten systems in the country.
Excellus Blue Cross Blue Shield	1,663.20	2.00	5,727	Part of a $4-billion family of private companies that finances and delivers healthcare services across upstate New York and long-term care insurance nationwide. Collectively, the enterprise provides health insurance to more than 2 million people and employs more than 6,000 New Yorkers.
HealthSpring Inc.	1,574.80	0.75	1,270	One of the largest public managed care organizations in the United States whose primary focus is the Medicare Advantage market. A concentration on Medicare programs allows them to understand the complexities of the Medicare regulations, design competitive products, manage medical costs, and offer high-quality healthcare benefits to Medicare beneficiaries in local service areas. Medicare Advantage experience also encourages collaborative and mutually beneficial relationships with healthcare providers, including comprehensive networks of hospitals and physicians experienced in managing Medicare populations.
Health Alliance Plan of Michigan	1,443.50	0.57	Not reported	A non-profit health plan based in Detroit, it provides quality healthcare coverage to 576,000 members. Over its 40-year history, HAP has embraced change and seized opportunities to provide its customers with innovative health coverage solutions. Today, HAP provides "one-stop" health benefits shopping to companies ranging from two to tens of thousands of employees.

Notes

1. "Report to the Congress: Medicare Payment Policy," Centers for Medicare and Medicaid Services, March 2006, http://www.medpac.gov/publications/congressional_reports/Mar06_Ch02a.pdf (August 20, 2008).
2. From remarks made by Steve Oesterle as part of the "From the Innovator's Workbench" speaker series hosted by Stanford's Program in Biodesign, April 27, 2007, http://biodesign.stanford.edu/bdn/networking/pastinnovators.jsp. Reprinted with permission.
3. "Medicare Program: Revised Process for Making Medicare National Coverage Determinations," Department of Health and Human Services, Centers for Medicare and Medicaid Services, September 26, 2003, www.cms.hhs.gov/DeterminationProcess/Downloads/FR09262003.pdf (March 4, 2008).
4. Shara Rosen, "Successful Commercialization Strategies for New Healthcare Products and Technologies," Kalorama Market Research, July, 2003, p. 90.
5. "Carticel: The Cost of This Alternative Therapy," http://biomed.brown.edu/Courses/BI108/BI108_1999_Groups/Cartilage_Team/matt/Carticel1.html (March 12, 2008).
6. Rosen, op. cit., p. 93.
7. Ibid.
8. Ibid., p. 94.
9. "How to Apply for a New CPT Code and How to Get It Valued," American Academy of Otolaryngology, http://www.entlink.net/practice/resources/upload/Final_CPT_RUC_Processes1.doc (March 6, 2008).
10. "CPT Background and Categories of CPT Codes," AMA, http://www.ama-assn.org/ama/pub/category/12886.html (March 6, 2008).
11. "The RVS Update Committee (RUC)," AMA, http://www.ama-assn.org/ama/pub/category/16401.html (March 6, 2008). The purpose of the RUC process is to provide recommendations to CMS for use in annual updates to the new Medicare RVUs. The RUC is a unique committee that involves the AMA and specialty societies, giving physicians a voice in shaping Medicare relative values. The AMA is responsible for staffing the RUC and providing logistical support for the RUC meetings.
12. "How to Apply for a New CPT Code and How to Get It Valued," op. cit.
13. HSS Inc. staff, "Understanding the Three CPT Code Categories," Advance, http://health-information.advanceweb.com/Editorial/Content/Editorial.aspx?CC=42599 (March 12, 2008).
14. "CAD Sciences Announces the Creation of the First Ever Category III CPT Code for Breast MRI CAD," DeviceSpace, January 10, 2006, www.devicespace.com/news_story.aspx?NewsEntityId=6631 (March 6, 2008).
15. "Coding Change Request Form Instructions," AMA, www.ama-assn.org/ama/pub/category/print/12888.html (March 6, 2008).
16. "New Category III CPT Code Created for NeoVista's Novel AMD Treatment," PR Newswire, January 22, 2008, www.prnewswire.com/cgi-bin/stories.pl?ACCT=109&STORY=/www/story/01-22-2008/0004740507&EDATE= (March 6, 2008).
17. E. R. Scerb and S. S. Kurlander, "Requirements for Medicare Coverage and Reimbursement for Medical Devices," *Clinical Evaluation of Medical Devices*, (Totowa, NJ: Humana Press, 2006), p. 74.
18. "Helpful Hints for Filing: Actigraphy Studies," Respironics, 2007, www.actiwatch.respironics.com/PDF/01-2107-ActigraphyHH.pdf (March 6, 2008).
19. "Process and Information Required to Apply for Additional Device Categories for Transitional Pass-Through Payment Status Under the Hospital Outpatient Prospective Payment System," Centers for Medicare and Medicaid Services, http://www.cms.hhs.gov/HospitalOutpatientPPS/Downloads/catapp.pdf (March 6, 2008).
20. "Process and Information Required for a New Technology Ambulatory Payment Classification (APC) Assignment Under the Hospital Outpatient Prospective Payment System (OPPS)," Centers for Medicare and Medicaid Services, http://www.cms.hhs.gov/HospitalOutpatientPPS/Downloads/newtechapc.pdf (March 6, 2008).
21. "Process and Information Required to Apply for Additional Device Categories for Transitional Pass-Through Payment Status Under the Hospital Outpatient Prospective Payment System," loc. cit.
22. "CPT Code for Capsule-Based pH Monitoring," Medtronic, http://www.medtronic.com/physician/gastro/c_bravo.html (March 7, 2008).
23. "Process for Requesting New/Revised ICD-9-CM Procedure Codes," Centers for Medicare and Medicaid Services, http://www.cms.hhs.gov/ICD9ProviderDiagnosticCodes/02_newrevisedcodes.asp#TopOfPage (March 6, 2008).
24. "Application for New Medical Services and Technologies Seeking to Qualify for Add-On Payments Under the Hospital Inpatient Prospective Payment System for Federal Fiscal Year (FY) 2009," Centers for Medicare and Medicaid Services, http://www.cms.hhs.gov/AcuteInpatientPPS/Downloads/FY2009_New%20Technology%20App.pdf (March 6, 2008).

25. "Infuse Bone Graft," Medtronic, https://www.infusebonegraft.com/ (November 2, 2008).
26. "In a Testament to Second Chances, InFUSE gets CMS approval for $8,900," Healthpoint Capital, August 4, 2003, http://www.healthpointcapital.com/research/2003/08/04/in_a_testament_to_second_chances_infuse_gets_cms_approval_for_8900/ (November 2, 2008).
27. Healthcare Common Procedure Coding System (HCPCS) Level II Coding Procedures http://www.cms.hhs.gov/MedHCPCSGenInfo/Downloads/LevelIICodingProcedures113005.pdf (March 11, 2008).
28. "Helpful Hints for Filing: Miscellaneous CPT Codes," Respironics, 2007, www.reimbursement.respironics.com/downloads/01-1879-CPT-HH%2007_12_07.pdf (March 6, 2008).
29. "Intermediary-Carrier Directory," U.S. Department of Health and Human Services, http://www.cms.hhs.gov/ContractingGeneralInformation/Downloads/02_ICdirectory.pdf (April 1, 2008).
30. M. I. Burken, "Case Studies on the Local Coverage Process," *Clinical Evaluation of Medical Devices*, (Totowa, NJ: Humana Press, 2006), p. 74.
31. "Medicare Program: Revised Process for Making Medicare National Coverage Determinations," loc. cit.
32. Lori Long, "Depression Treatments," MSA of King County, April 2006, http://www.msakc.org/Articles/DepressionTreatments.htm (March 7, 2008).
33. "CMS Issues Decision for VNS Therapy in Treatment-Resistant Depression," Cyberonics, May 7, 2007, http://www.cyberonics.com/pressroom/PressRelease_detail.asp?ID=9C4A3E6E-7EE9-448A-953F-CB7AA6802309 (March 7, 2008).
34. "Cigna HealthCare Coverage Position: Bariatric Surgery," May 15, 2007, http://www.cigna.com/customer_care/healthcare_professional/coverage_positions/medical/mm_0051_coveragepositioncriteria_bariatric_surgery.pdf (March 12, 2008).
35. Barbara Grenell, Debbie Brandel, and John F. X. Lovett, "Obtaining Reimbursement Coverage from Commercial Payers," *Medical Device Link*, http://www.devicelink.com/mddi/archive/06/01/009.html (March 12, 2008).
36. "Cost Effectiveness of ICD Therapy in SCD-HeFT," Sudden Cardiac Death in Heart Failure Trial, Medtronic, http://www.scdheft.com/reimbursement/index.html (November 25, 2008). Reprinted with permission.
37. All quotations are from interviews conducted by the authors, unless otherwise cited. Reprinted with permission.
38. Robert J. Applegate, "Drug-Eluting Stents: The Final Answer to Restenosis?" Wake Forest University Medical Center, http://www1.wfubmc.edu/articles/CME+Drug-eluting+Stents (January 4, 2006).
39. Barbara Grenell, Debbie Brandel, and John F. X. Lovett, "Obtaining Reimbursement Coverage from Commercial Payers," *Medical Device Link*, http://www.devicelink.com/mddi/archive/06/01/009.html (March 7, 2008).
40. "$75m Settles Allegation of Medicare Fraud," *The Boston Globe*, May 23, 2008, http://www.boston.com/business/articles/2008/05/23/75m_settles_allegation_of_medicare_fraud/ (November 2, 2008).
41. Ibid.
42. "Cigna Now Covering the Essure Procedure," Conceptus, February 15, 2006, http://www.conceptus.com/pr/2006/021506_1_press_release.html (March 7, 2008).
43. Lotan *et al.*, "The New Economics of Radical Prostatectomy," *Journal of Urology* (2004): 1431–1435.
44. Cohen *et al.*, "Cost Effectiveness of Sirolimus-Eluting Stents for the Treatment of Complex Coronary Stenosis," *Circulation* (2004): 508–514.
45. Sanders *et al.*, "Cost-Effectiveness of Implantable Cardioverter-Defibrillators, *New England Journal of Medicine* (2004): 1471–1480.
46. David M. Eddy, "Technology Assessment, Deployment, and Implementation in Prepaid Group Practice," *Toward a 21st Century Health System: The Contributions and Promise of Prepaid Group Practice* (Josey-Bass: San Francisco, 2004), pp. 85–107.
47. "Natalizumab," Wikipedia.org, http://en.wikipedia.org/wiki/Natalizumab (April 1, 2008).
48. "Technology Evaluation Center: Kaiser Collaboration," Blue Cross Blue Shield Association, http://www.bcbs.com/betterknowledge/tec/kaiser-collaboration.html (March 6, 2008).

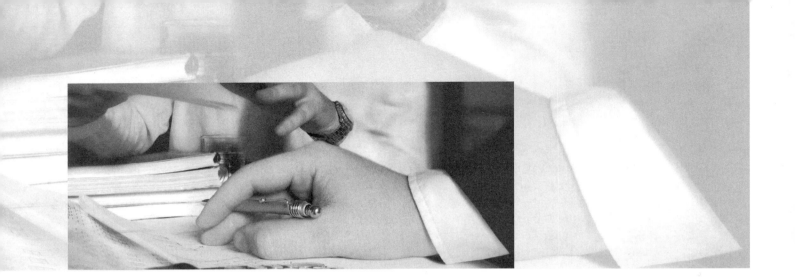

5.7 Marketing and Stakeholder Strategy

Introduction

Consider two new medical devices with similar clinical data establishing their safety and efficacy. One is hailed as an important advancement, receives broad professional society support, and is quickly adopted upon its commercial launch. Another is dubbed "experimental," blocked by key opinion leaders, and struggles in the market. Although there are many factors that can affect the adoption of a new technology, the difference in its success or failure often comes down to how well the company develops and executes a proactive, forward-looking, multi-faceted marketing and stakeholder strategy.

The greatest medtech innovations are disruptive. They deliver breakthrough improvements in patient outcomes and fundamentally change the practice of medicine. Yet, the "do no harm" principle of medical ethics supports a conservative culture that can be at odds with adoption of new technologies. This tension presents a challenge for innovators and requires them to establish and communicate the clinical value of new device technologies in such a way that the information serves as a catalyst for the desired change. A well-designed marketing and stakeholder strategy seeks to accomplish this by influencing the key decision makers who will drive the adoption of the technology. Because there are multiple stakeholders in the medical device field who can sway adoption decisions, an effective strategy must be multi-faceted and tailored to the unique perspectives of each primary stakeholder group. The key components of the strategy are the value proposition, the combination of products and services that will support the value proposition, the pricing strategy for the offering, and the mix of promotional activities that will be used to communicate the value proposition to the target audience.

OBJECTIVES

- Understand the basic functions of marketing as they relate to the commercialization of a new medical technology.
- Appreciate the importance of confirming the attitudes and perceptions of key stakeholders toward a specific need and/or new solution.
- Know how to define value propositions targeted toward the most important stakeholder groups to differentiate a device from existing treatments.
- Recognize how to a develop a marketing communication strategy to convey the product's value propositions to key stakeholders, and a public relations campaign to support product launch.
- Understand how to develop a pricing strategy that will enable a company to capture value.

The decisions made in developing a marketing and stakeholder strategy must be supported by those made as part of 5.3 Clinical Strategy, 5.4 Regulatory Strategy, and 5.6 Reimbursement Strategy (although these topics are referenced but not addressed in detail within this chapter). The marketing and stakeholder strategy also informs the activities described in 5.8 Sales and Distribution Strategy.

Marketing and stakeholder strategy fundamentals

The primary functions of marketing are to create, capture, and sustain value. These functions closely mirror steps in the biodesign innovation process, as shown in Figure 5.7.1. Early on, through the careful examination of the marketplace and efforts to develop a product that addresses an important unmet medical need, an innovator lays the foundations for *creating value*. In this chapter, with the creation of a marketing and stakeholder strategy, the focus shifts to the process of *capturing value*. The upcoming chapter on sales and distribution (see Chapter 5.8) is largely about *sustaining value*.

Capturing value relies on the company's marketing mix – the "4 Ps," which include positioning, product, price, and promotion. It starts with revisiting stakeholder and market analysis (Chapters 2.3 and 2.4) to better define the needs of the different audiences in order to articulate the product's value proposition, and position (differentiate) it in relation to emerging or existing products and services. In addition to defining a competitive advantage (see 5.9 Competitive Advantage and Business Strategy), this may involve refining the offering to improve its perceived value and positioning (e.g., adding a service component to a capital equipment product), determining the channel through which the product will reach the customer (direct, indirect, or hybrid), and deciding how product promotion (education, advocacy) will be managed. Ultimately, the innovator develops an appropriate price, as well as a reimbursement plan (see 5.6 Reimbursement Strategy) that enables the company to capture value in the market.

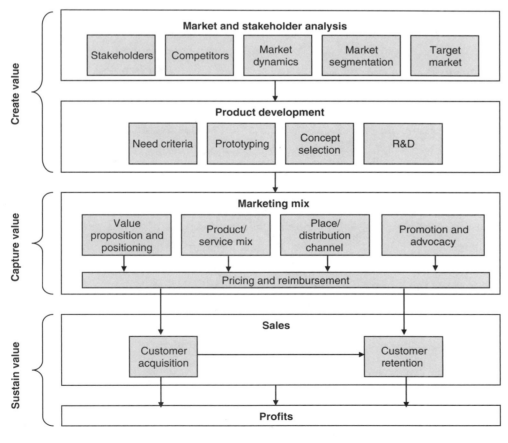

FIGURE 5.7.1
Sales and marketing functions, and their interaction with the biodesign innovation process (adapted from Alvin J. Silk, *What is Marketing?* (Harvard Business School Press, 2006); reprinted with permission).

Issues related to developing an effective marketing mix are addressed in more detail below.

Marketing mix

When an innovator begins to think about a marketing mix, one of the first priorities should be to confirm a solid understanding of the following three factors:
1. Who are the most important stakeholders.
2. What their opinions are toward the need *and* the new solution.
3. Where does alignment and/or conflict exist between different stakeholder groups.

As noted in 2.3 Stakeholder Analysis, one of the best ways to ensure that no important stakeholders are overlooked is to perform both *cycle of care* and *flow of money* analyses. Repeating these assessments is important because now much more is known about the specific solution concept, how it will be implemented, and by whom. More will also be understood about regulatory, intellectual property (IP), reimbursement, and clinical issues. As a result, the cycle of care and flow of money analyses may show different results from the original assessments (which were performed even before the need was clearly specified).

When revisiting these evaluations, physicians and other providers, patients, payers, and facilities all should be considered. It is also useful to think about subsegments within these major groupings to make sure that the interests of all important parties are addressed. For example, all physicians will not respond in the same way to a new innovation. One specialty may be enthusiastic about a new technology because it drives business to their field of medicine (as happened with angioplasty for interventional cardiologists), while another group may resist the innovation because it takes business away from their specialty (e.g., cardiothoracic surgeons). Companies preparing to launch a new innovation must think carefully about the factors that will drive the behaviors of each and every one of these key constituencies.

Value proposition and positioning

One way that a company can convince stakeholders to change their behavior is through the development of a clear value proposition (recall that value propositions directed at payers were introduced in Chapter 5.6). In marketing, a customer value proposition refers to the sum total of benefits that a company promises to a customer in exchange for payment (or other value-transfer).[1] Another way to think about a value proposition is as a marketing statement that summarizes why a customer should choose a particular product or service. This statement should convince a potential customer that one product or service will add more value or better solve a problem than other competitive offerings. Companies generally use value propositions to target the customers likely to benefit most from using their devices. An ideal value proposition will be concise and designed to appeal to the strongest drivers of a customer's decision-making behavior.

Developing value propositions

In the medical device field, there are a number of common value propositions that can be crafted to appeal to different stakeholder audiences. These value propositions are summarized in the stakeholder value proposition matrix shown in Table 5.7.1, along with their most common effect on different stakeholder groups and other important strategic considerations.

Choosing and proving a value proposition

Depending on the unique characteristics of a company's device, an innovator chooses the value proposition that is most compelling to each primary stakeholder group and that best differentiates the product from other alternatives. However, it is not enough to make empty claims about how a device will deliver value. To be effective, a value proposition must be backed by quantitative data that provides clear evidence of the value that will be delivered. Qualitative data can also be used to support a value proposition (e.g., testimonials and anecdotes from key opinion leaders–KOLs), but should ideally be augmented with quantitative information that supports those claims to achieve maximum impact. Table 5.7.2 shows the types of endpoints that can be studied to collect the evidence needed to support the common value propositions used in medical devices. For more information about these endpoints and the types of studies that support them, see 5.3 Clinical Strategy and 5.6 Reimbursement Strategy.

Table 5.7.1 *Innovators can choose from multiple types of value propositions in the medtech field.*

Value proposition	Stakeholder net effect +	Stakeholder net effect −	Examples	Strategic issues
Decreased procedure invasiveness (which translates into increased demand for the less invasive procedure)	Facility Patient	Payer	Laparoscopic (minimally invasive) procedures versus open surgery	• Can create stakeholder conflict if treatment shifts from one specialty to another • Some physicians prefer traditional though more invasive procedures • Spikes in demand create hardship for payers unless costs are reduced elsewhere
Reduced need for future treatment (e.g., monitoring or complications)	Payer Patient	Facility Physician	Drug eluting stents (reduce need for treatment of restenosis) Genomic Health's diagnostic test for predicting recurrence of early-stage breast cancer following surgery (see Chapter 2.4)	• Providers may resist treatment if it reduces their traditional reimbursement rate • Reduction in future treatment needs helps payers justify reimbursement for treatment
Increased need for future treatment or diagnostics (e.g., preventive screening, monitoring or treatment of complications)	Facility Physician Patient		Mammography screening to find malignancies early so that they can receive timely treatment	• Coverage and reimbursement rates for future treatment may be beneficial to physician and facility • Low-value future treatment should not interfere with a physician's ability to take on high-value new cases • Company should seek to have treatment embedded in professional society guidelines, if applicable • Clinical benefit should be significant to ensure patient compliance with increased treatment needs
Increased opportunity for bundling procedures	Facility Physician	Payer	3-D mapping and ablation for AF (two separate procedural steps are required to be done in same procedure)	• Company must establish mutually beneficial economics for providers and negotiate coverage by payers • Payers must manage additional complexity created by discounts in secondary procedures in order to provide sufficient reimbursement
Improved compliance with best practices/guidelines	Payer Patient		HbA1C point of care diagnostic (can be more easily performed, creating increased compliance)	• Cost of device may not fall within established reimbursement levels • May necessitate a new code if it adds on to providers' regular workload • Improved compliance may reduce long-term cost of care for payers and improve outcomes for patients
Reduced time per procedure	Facility Patient Physician		Percutaneous ICD – a less invasive ICD (see Chapter 2.3)	• May result in the downward revision of RVUs by Medicare and reduced payments by private payers • May create stakeholder resistance if traditional reimbursement levels are reduced • Payers may benefit from this if reimbursement can be lowered, but this could adversely affect facilities and patients

Table 5.7.1 *Continued*

Value proposition	Stakeholder net effect +	Stakeholder net effect −	Examples	Strategic issues
Reduced supplies per procedure	Facility	Physician	Improved navigation in colonoscopy (e.g., reduced need for sedation when using advanced colonoscopy – see EndoNav in Chapter 4.3)	• Company must establish that cost reduction from fewer supplies is greater than the additional cost of device • Payers may benefit from this if reimbursement can be lowered, but this could adversely affect facilities and patients

Table 5.7.2 *Specific endpoints can be studied to generate data that support common medtech value propositions.*

Value proposition	Economic and clinical endpoints to be studied by company to demonstrate value proposition
Decreased procedure invasiveness (which translates into increased demand for the less invasive procedure)	• Non-inferiority (or superiority) to established standard of care • Economic impact on payers: expected number of additional procedures, total cost • Financial impact on provider: cost per procedure, increased number of procedures
Reduced need for future treatment (e.g., monitoring or complications)	• Non-inferiority (or superiority) to established standard of care • Economic impact on payers: expected reduction in number of follow-up procedures, total cost • Financial impact on provider: loss of revenue from follow-up treatments, possible addition of other procedures from saved time, cost per procedure
Increased need for future treatment or diagnostics (e.g., preventive screening, monitoring or treatment of complications)	• Clinical benefit from additional follow-up • Economic impact on payers: incremental costs of additional treatment versus cost savings through clinical benefits • Economic impact on facilities and providers (balancing high-value and low-value activities)
Increased opportunity for bundling procedures	• Establish clinical benefit and economic impact for payer • Economic benefit for provider: increased throughput, amount of bundling opportunities, number of bundling procedures
Improved compliance with best practices/guidelines	• Clinical and economic benefits from increased compliance for payer • Cost of compliance for provider versus reimbursement level (i.e., how much does it cost the provider to comply with the guidelines and what is the reimbursement level – is it adequate for the provider to financially justify compliance?)
Reduced time per procedure	• Equivalence (or superiority) to gold standard • Economic impact on payers: expected number of additional procedures and total cost • Financial impact on provider: additional procedures, cost per procedure (e.g., supplies, capital equipment, labor), potential revenue
Reduced supplies per procedure	• Economic studies for providers comparing the cost of supplies with the current standard versus the new procedure

The essential issue to recognize is that a company must work cross-functionally to design clinical studies that will produce the evidence needed to satisfy the data requirements of the many different stakeholder groups it will seek to influence. Of course, these studies will need to be prioritized and/or sequenced based on their relative importance and potential impact on the company's ability to achieve its goals, especially in light of resource constraints.

As a company considers what studies to conduct, it is helpful to think carefully about who will be involved in those studies from a stakeholder perspective. Study design and management provides an effective way to get key opinion leaders involved with the company and familiar with its product. A KOL is a "thought leader" or individual widely respected in a field that can potentially influence the behavior of others through his/her professional status.[2] Physicians who are considered KOLs have deep experience that can be leveraged to help design and implement studies with a high probability of success. Their expertise and involvement also lends credibility and helps generate interest in the study results among the KOL's peer group. When deciding which KOLs to target, companies must think about factors such as whether they prefer academic or community physicians and the geographic location where the studies will be performed (see 5.3 Clinical Strategy).

In advance of study design and execution, KOLs can also be engaged to help a company develop and/or refine its critical value propositions. While KOLs are traditionally targeted within the physician population, innovators should remember that influential individuals can be found among facility administrators, biostatisticians, health economists, academics, allied professionals (e.g., nurses and technicians), and patient groups. All such experts can provide invaluable advice regarding which value proposition is most compelling to their associated target audience, and what data is required to effectively support that value proposition.

It is important to recognize that before a company can convince certain stakeholders of the value of its specific technology, it sometimes must persuade them that the need for a solution exists at all. Often, a company may have to perform or sponsor studies simply to establish the sense of a necessity in the minds of the target audience. For instance, consider the case of implantable cardioverter defibrillators (ICDs). Before the results of the MADIT-II (2001) and SCD-HeFT (2004) trials were released, ICDs were primarily used to treat a relatively small number of secondary prevention patients – those with a prior episode of sudden cardiac arrest. However, MADIT-II and SCD-HeFT demonstrated that ICDs dramatically reduced the mortality rates of primary prevention patients – those at high risk of sudden cardiac arrest who had not yet experienced it. As a result, the FDA expanded the indications for ICD implantation to include primary prevention patients. Similarly, Medicare and private payers modified their policies to provide coverage for ICDs used for primary prevention. This expanded the market for ICDs to an additional 1 million patients, since this group was much larger than the secondary prevention group. Physicians, payers, and patients just had to be shown in definitive terms that a legitimate need existed within this broader population.

A similar example exists within the treatment of patients with chronic kidney disease (CKD) before their condition progresses to end-stage renal disease (ESRD). While the need to manage heart disease in patients with ESRD was relatively well understood, heart disease was not proactively or consistently managed in most pre-ESRD patients. However, when a study was published in the *New England Journal of Medicine* in 2004,[3] which showed that pre-ESRD patients have a high risk of heart failure, it was enough to establish the need for a new treatment paradigm for this patient population. It also opened up a new field of treatment to innovators and companies pursuing opportunities related to CKD.

Note that the company's business strategy and competitive advantage can also support the value propositions it has defined. For each value proposition, it should carefully consider what strategies and potential company strengths will enable it to successfully deliver as promised (see 5.9 Competitive Advantage and Business Strategy for more information). The company's position in the healthcare value chain and its unique capabilities are an important source of leverage in bringing value propositions to fruition.

Product/service mix

Another mechanism for delivering on a value proposition is the product/service mix that a company defines. The idea is to create an offering with the greatest likelihood of delivering the company's value propositions. In general, the product/service mix is driven by the business model the company has adopted (see 4.4 Business Models). For example, many medtech companies with disposable or implantable devices have a pure product offering, which they may support through ancillary services (e.g., business development activities, education, and training). Capital equipment companies, on the other hand, generally have a more complex product/service mix that involves the bundling of multiple products (e.g., equipment, plus a disposable component), as well as a service (e.g., maintenance contracts and upgrade agreements). For instance, Accuray (profiled later in this chapter) includes future product upgrades as part of its basic product offering.

Place/distribution channel

With the total offering defined, an innovator can decide on the best way to get it into customers' hands. A marketing strategy without an effective way to reach the customer is simply an academic exercise. In some cases, placement is important (e.g., with physician sell and over-the-counter business models as described in Chapter 4.4). However, with most medical technology offerings, the channel is the most important consideration to address.

The company must decide how it will reach its customers (direct sales force, distributor, or a hybrid model), what channel management activities will be needed, and how much it will cost. At times, different channels are needed to reach different segments of the target audience. More details about various distribution approaches are provided in 5.8 Sales and Distribution Strategy.

Promotion and advocacy

Once key stakeholder value propositions have been articulated, the offering has been defined, and the channel has been decided on, then a strategy can be developed to begin communicating with the target audience. A company generally manages these communications through a marketing communication and/or public relations strategy.

Raising awareness through a marketing communication strategy

The purpose of a marketing communication strategy is to raise awareness among stakeholders regarding the need and/or solution before they are put in a position to make a decision or change their behavior (i.e., before the technology is launched).

Information that is disseminated by the company is considered a form of *direct awareness*. Information that comes from other sources – such as publications, conferences, KOLs, etc. – is considered *indirect awareness*. An effective marketing communication strategy typically leverages both direct and indirect awareness mechanisms in reaching its target audience.

Usually, a marketing communication strategy outlines which communication vehicles will be used for each key stakeholder group, what the key messages will be for each communication, and when each communication will be released. The idea is to create a sequence of communications for each stakeholder group that builds their awareness, introduces and then reinforces key messages (e.g., the value proposition), and prepares the target audience to make a decision or change its behavior at the time the device is launched. Table 5.7.3 presents the different communication vehicles that can be incorporated into a marketing communication strategy. Note that detailing – when sales representatives and clinical specialists visit physicians or hospitals – is omitted since this is considered part of the sales process, as discussed in 5.8 Sales and Distribution Strategy.

Importantly, companies should be careful not to create awareness of their technology among patients *before* physicians are fully informed. If a physician is blindsided by patient inquiries regarding a new product, it can cause resentment toward the company and negatively affect adoption or even drive adoption of competing products. This problem, while most common in the pharmaceutical business, is still something to watch out for in the medical device field with the emergence of direct-to-consumer (**DTC**) advertising. Typically, the best approach is to raise awareness first among KOLs, then expand communications to facilities and payers.

Table 5.7.3 *A combination of communication vehicles is the most effective way to reach target stakeholders.*

Vehicle	Description	Issues to consider
Peer-reviewed publications	Peer review is the process of subjecting an author's scholarly work or research to the scrutiny of others (peers) who are experts in the same field.[24] Peer-reviewed publications are considered more credible than other sources because they aggregate, filter, and validate author submissions independent of any outside influence or interested third party.[25]	• Submitted data must be based on clear evidence • Publication can take anywhere from three months to two years from the time of submission, depending on the publication, topic, and strength of the data; as a result, planning is difficult • Revisions/rewrites may be required prior to publication • Publication is not guaranteed • Any potential conflicts of interest between the company and the authors must be disclosed • May cost as little as $15,000 but can require hundreds of thousands of dollars to support the research that leads to the publication (conflicts of interest caused by the relationship between the research sponsor and the investigator that may undermine the scientific validity of any studies will need to be managed and disclosed properly)
Peer-reviewed (abstract) conference presentations	An abstract provides a concise statement of the major elements of a research project. Abstracts are reviewed against other submissions, with some subset being chosen for a brief presentation (or a poster) based on their fit with conference criteria.	• Submitted data must be based on clear evidence • Abstract requirements vary from conference to conference and should be well understood before a submission is made • The abstract, stating the purpose, methods, and findings of the research project, is submitted to conference organizers to inform them of work-in-progress or completed work that is available to be presented[26] • Abstracts must be submitted an average of 6 to 8 months prior to the conference
Technology/clinical talks	Technology talks are given by KOLs, either in the context of a meeting, as an evening session, or pre-conference session.	• Can be used to satisfy CME requirement (see below) • More extensive than the peer-reviewed presentation; provides opportunities for more in-depth coverage of a new technology
Continuing medical education (**CME**)	CME is required by physicians in most US states to maintain their licenses. It provides a way for physicians to stay informed and learn about new developments in their field. Content for CME programs is typically developed, reviewed, and delivered by faculty who are experts in their individual clinical areas.[27]	• CME programs must be certified by the Accreditation Council for Continuing Medical Education • Any potential conflicts of interest between the company and the authors must be disclosed • CME programs can be sponsored directly by companies or through professional societies • The costs are significant: a professional conference attended by 200 people for two days could be $100,000 or more; a one- to two-day academic conference for 50 people could be more than $50,000 (start with speaker honoraria of $5,999 to $15,000 for ten speakers, then add room, board, travel expenses, etc.)

Table 5.7.3 *Continued*

Vehicle	Description	Issues to consider
Reimbursement dossier	Reimbursement dossiers typically serve as the official source for all key information about the product, including the place of product in the diagnostic and therapeutic chain, results from key clinical and economic studies, disease management strategies, modeling report, product value and overall cost, and references (see 5.6 Reimbursement Strategy).	• Must be based on well-established and tested facts about the product • Can be time-consuming and resource-intensive to prepare, requiring specialized expertise and significant lead time • Must be customized to meet the needs and interest of each target audience • Cost can range from $50,000 (for a simple dossier) to as much as $250,000 (for a more complete deliverable)
Direct-to-consumer (DTC) advertising	DTC advertising refers to the promotion of medical devices to patients through newspapers, magazines, television, and the Internet. Companies also use brochures, videos, and other materials that are made available to patients in doctors' offices.[28] One recent trend is to focus on direct-to-patient (**DTP**) advertising (i.e., advertising in channels accessed more exclusively by patients, such as the *Diabetes Digest* for diabetic patients).	• DTC advertising is only legal in two developed countries: the United States and New Zealand • Not all medical technologies are well-suited to DTC advertising and this approach is rarely used in the industry (exceptions: elective Lasik surgery, Botox, drug eluting stents, Lap Band for bariatric surgery) • DTC advertising is considered somewhat controversial among regulators, some physicians, and some patient groups • DTC advertising is complicated and expensive

Once these stakeholders have a basic understanding of the technology and its value propositions, the company can begin targeting other physicians (who are not necessarily KOLs) and then patients (as appropriate) with its marketing messages.

Professional societies and thought leaders can be especially valuable advocates for a new technology. Societies can create awareness by incorporating new medical technologies in their practice guidelines, sponsoring symposia, and advocating payers for reimbursement (see 5.6 Reimbursement Strategy). Thought leaders not only influence other physicians, but can influence future generations of product development to create and sustain additional value. Ron Dollens, president and CEO of Guidant from 1994 to late 2005, credits relationships with thought leaders as one of the key factors beyond the success of Advanced Cardiovascular Systems (ACS). Advanced Cardiovascular Systems was acquired by Eli Lilly in 1984 and led by Dollens from 1988 until its spin-off in 1994 to form Guidant. According to Dollens:[4]

ACS had a sales organization that developed great relationships with the thought leaders. So it was associated with the kinds of people that were taking the therapy to another level.

Promoting through a public relations strategy

A public relations (PR) strategy can be an imperative element of a marketing/stakeholder strategy if the company anticipates strong resistance from one or more stakeholder groups. The concept of PR refers to the methods and activities employed by a company to establish and promote a favorable relationship with the public.[5] While PR companies are mostly accessible to large companies, focused and shrewd PR efforts can also work for smaller companies. For example, Athena Health, a company that provides web-based medical billing software and services for physician practices, published a ranking of insurers based on the timeliness of claim processing, which attracted considerable media attention and generated interest among physicians for its products.[6]

Not all device companies require an explicit PR strategy. However, if the company is concerned about certain potential barriers to the adoption of its technology, this may be necessary. For instance, payers are more likely to resist a new technology if it requires new reimbursement codes. Yet, if a company launches an effective PR campaign that engenders meaningful support from physicians (and their professional societies), payers will have a more difficult time delaying or denying coverage for the device. Similarly, if a physician turf war ensues upon the introduction of a new device, a PR campaign targeted at creating patient demand for the new technology can be an effective way to drive physician adoption, despite the conflict surrounding who will administer it.

Guidant employed a PR campaign to apply pressure to the Centers for Medicare and Medicaid Services (CMS) while seeking more widespread coverage for its ICDs. Although ICDs had been shown in several studies to prevent sudden death by detecting dangerous abnormal rhythms and shocking a patient back into normal rhythm, the CMS was willing to reimburse for the technology in only a fraction of the patients who could benefit from the device. Guidant sought to address this issue by funding a $7 million dollar prospective trial looking at the effect of ICDs on cardiac deaths among those who had suffered at least one heart attack. When the results became available, many private payers recognized the value of the device and agreed to reimburse for all patient types that had shown benefit in the study. However, Medicare took a more conservative approach, limiting its coverage to a high-risk subset of the patients studied in the trial (due largely to the potential billion-dollar price tag associated with expanding coverage to such a large group of patients). To help overcome this resistance, Guidant had advocates for the device place editorials in prominent mainstream newspapers and medical journals and took other steps over a two-year period to rally support among patient groups, specialty societies, and other ICD manufacturers (Medtronic and St. Jude). Eventually, at least partially in response to criticism that it was, to some extent, rationing medical care, CMS finally agreed to expand coverage.[7]

When a company is developing a PR strategy, some of the best resources to target include news organizations and journalists, patient advocacy groups, physician professional societies, hospital associations, and even members of Congress (if the value proposition associated with the device is compelling enough to capture the attention of this audience).

The Accuray story below highlights how marketing communication and PR techniques and activities can be brought into a comprehensive marketing and stakeholder strategy.

FROM THE FIELD

ACCURAY INCORPORATED

Using stakeholder analysis to develop an effective marketing strategy

Accuray Incorporated was founded in 1991 by John Adler, a Stanford neurosurgeon, to develop an image-guided radiosurgical device that could be used to ablate tumors without the need for traditional surgical resection. Adler first conceived of the idea while observing procedures performed with a predecessor device, called the GammaKnife, in Sweden in the 1980s. As Adler described, "Accuray really was developed in response to the shortcoming of the GammaKnife," which was able to treat brain tumors through non-invasive radiosurgery, but had to be anchored to the patient's body via bone screws. Accuray's product, a frameless stereotactic radiosurgical system, simplified radiosurgery by eliminating the need for these invasive, painful anchors while achieving comparable results. The Accuray system, called the CyberKnife®, also would enable radiosurgery in other less anchorable areas of the body, including the spine, chest, and abdomen. By expanding non-invasive surgical technology, Accuray would give surgeons the option to be more aggressive in treating cancer in multiple anatomic locations or near vital organs, while simultaneously lowering risk and patient recovery times (see Figure 5.7.2).

Starting with his connections at Stanford, Adler put together a team to develop the CyberKnife idea and bring it to the marketplace. Despite difficulties raising enough money to sustain the high, ongoing R&D-related expenses associated with a capital equipment device, Accuray managed to get the CyberKnife ready for human

FROM THE FIELD — ACCURAY INCORPORATED

Continued

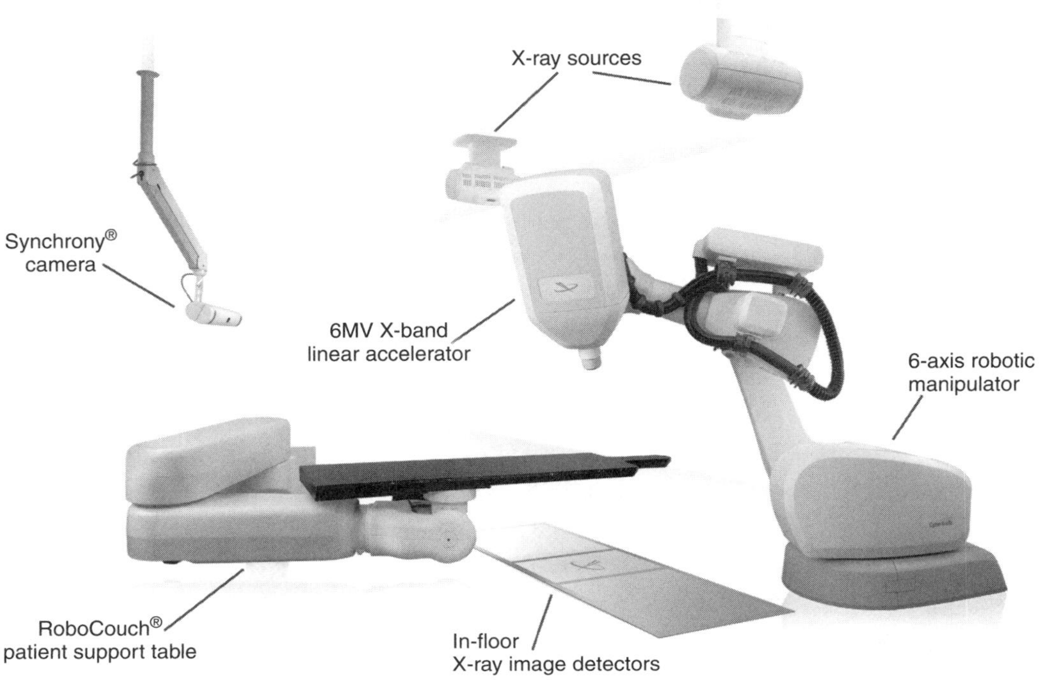

FIGURE 5.7.2 Accuray's CyberKnife system (courtesy of Accuray).

Labels: X-ray sources; Synchrony® camera; 6MV X-band linear accelerator; 6-axis robotic manipulator; RoboCouch® patient support table; In-floor X-ray image detectors.

testing under an investigational device exemption by 1994. Although the frameless design of the CyberKnife allowed Accuray to treat tumors in different parts of the body, Accuray originally focused on brain tumors, much like the GammaKnife, modifying existing procedures but not yet expanding outside the field of neurosurgery. "The first generation was dedicated to replacing many of the GammaKnife's procedures," said Adler, "because we could do it without the frame and modify what had been done to make it biologically a somewhat better procedure."[8]

The decision to begin by targeting neurosurgery was driven, in part, by a series of stakeholder issues. In this market, Accuray believed that it clearly understood the interests of key stakeholders and had the best chance of addressing their concerns. According to the company, the two most prominent stakeholder groups were neurosurgeons and radiation oncologists. Neurosurgeons had the primary relationships with brain tumor patients and had traditionally performed the open operations (to remove solid tumors) that the CyberKnife would replace. This population was receptive to new surgical procedures and innovations. Adler's personal connections with many neurosurgeons also helped generate excitement about CyberKnife within this community, paving the way for the acceptance of the device. With limited resources and funds to attract early adopters to the technology, Accuray became convinced that neurosurgeons would be a receptive audience.

Radiation oncologists, on the other hand, represented a stakeholder group that could be somewhat less receptive to the new device. Yet, Accuray had to earn their support because government regulations required radiation oncologists to participate in any form of radiation delivery, including radiosurgery. Because of various turf wars and financial issues, Accuray anticipated some resistance from radiation oncologists as physicians in different surgical specialties began using the CyberKnife. However, the company believed that it could mitigate these concerns in the neurosurgery field. Adler explained why: "Radiosurgery was developed by neurosurgeons and has been driven by this field for 35 or 40 years now. So, over time, radiation oncologists have come to accept neurosurgeons' role in this type of radiation delivery. Given such a history, the opposition from radiation oncologists could not be based simply on the fact that they control other forms of radiation." Having arrived at this conclusion, Accuray moved forward with efforts to attract both radiation oncologists and neurosurgeons to the use of CyberKnife in the field of neurosurgery.

When the company launched an improved CyberKnife in 2000, Accuray's technology was ready to begin expanding indications for the device beyond neurosurgery. However, with this move, the company recognized that it would face an even greater challenge convincing surgeons outside neurosurgery and radiation oncology to begin using the CyberKnife to treat patients with tumors in other regions of the body. For Accuray to have success in the long run, the company needed to fully understand the stakeholders involved in the broader radiosurgery arena and how to reach them.

Starting with physician stakeholders, Adler observed that surgeons in other specialties (non-neurosurgeons) tended to be less open to new innovations. Many of them were also threatened in other ways, having already seen their practices reduced by interventional pulmonologists, cardiologists, and gastroenterologists performing less invasive procedures for conditions that had traditionally been addressed through surgery. "What Accuray is doing is pretty heretical," said Adler. "It is anti-surgery in the eyes of true-blue surgeons because we're not cutting people open, there's no general anesthesia, and there's no blood loss – it's not nearly as dramatic as conventional surgery. So it strikes at the heart of what surgeons see themselves as being. It's not a trivial cultural shift to get surgeons to understand that they needn't be defined by blood loss and pain."

To help address concerns within this stakeholder group, Accuray had two primary arguments in its favor. The first, according to Adler, was that 90 percent of patients given the option would choose radiosurgery over traditional surgery. Additionally, surgeons reluctantly recognized that they needed to get involved in such procedures or risk losing this entire portion of their business to the radiation oncologists. Within the current healthcare delivery model, radiosurgery called for a surgeon and a radiation oncologist to be present during the procedure, with both specialists reimbursed separately for their time. Because of the manner in which it is used, CyberKnife procedures could possibly be performed by a single physician, and since most state regulations necessitated the participation of radiation oncologists, surgeons could potentially be seen as expendable participants. Recognizing this fact helped motivate surgeons to involve themselves in the adoption of the CyberKnife and carve out their role in the procedure, beyond just referring patients for treatment. "We achieved a sort of critical mass when surgeons started to realize that this was inevitable," recalled Adler. "They recognized they either had to get in the game, or they were going to be left behind."

Because the use of the CyberKnife in areas beyond neurosurgery would dramatically increase the involvement of radiation oncologists in treating a new class of patient, Accuray hoped that this stakeholder group might be receptive to the company's efforts to expand the use of the device. However, in reality, this turf battle turned out to be what Adler described as one of the company's "biggest problems." Instead of seeing opportunity to expand the use of radiation treatments to more patients "they worried about a new group of physicians playing with what was traditionally their toy. From a regulatory standpoint, it's pretty clear that radiation oncologists have been given complete authority by the state and federal governments to oversee all of this type of work, but they were still threatened." He continued, "They never looked and said, 'This is good because now we're going to be treating 80 or 90 percent more patients.' They don't think that way. Especially the more senior practitioners seemed more committed to creating their little fiefdom and keeping out any interlopers."

With both surgeons and radiation oncologists showing some reluctance to adopt the CyberKnife in areas outside the neurosurgery field, Accuray developed a robust marketing communication strategy to help drive adoption. The activities that the company undertook directly targeted the specific concerns of these surgeons and radiation oncologists, as well as patients, payers, and hospitals where the procedures would be performed.

When asked about the most successful marketing activity that Accuray employed to convince radiation oncologists and surgeons in other specialties of the benefits of the CyberKnife, Adler was adamant that it was the use of publications. "The single most effective strategy is to perform clinical outcomes studies and publish them in peer-reviewed journals," he said. When asked how many studies Accuray had performed, Adler answered, "Maybe a hundred. It never ends." The company's strategy was

> **FROM THE FIELD** — **ACCURAY INCORPORATED**

Continued

to perform relevant studies in specific specialty markets and then publish the results in the most prominent peer-reviewed journals in those fields to maximize the credibility of the results. Although it is more expensive to set up multiple studies, showing improved clinical outcome for each condition makes it much easier to market to surgeons, radiation oncologists, and payers. Accuray also presented its data to the targeted specialists at numerous conferences, as well as networking extensively with potential users.

Another interesting move by the company was the development of its own professional society: the CyberKnife Society. The purpose of this group was "bringing together diverse medical professionals affiliated with radiosurgery worldwide to foster scholarly exchange of clinical information, and to educate the general public with patient information on treating medical conditions, such as cancers, lesions, and tumors anywhere throughout the body using the CyberKnife, most of which are unreachable by other radiotherapy systems."[9] The CyberKnife Society had a strong online presence because, as Adler emphasized, "The Internet has proved to be a very powerful marketing tool." From a physician's perspective, he said, "The society is focused on providing ongoing professional training and provides a way to give [surgeons and radiation oncologists] updates and the medical tools they need to use this technology to its latest and greatest capacity." For patients, it is another source of information for learning about their conditions, creating demand for the technology, and networking with other CyberKnife patients.

According to Adler, once patients understand the CyberKnife procedure and how it compares to traditional, open surgery, most of them become intrigued and request more information about the procedure from their surgeons. Accuray seeks to encourage this behavior by sending "evangelists" (patients who have successfully undergone CyberKnife treatment) to patient advocacy and support groups. In their role as evangelists, patients also are vitally important to influencing the next groups of stakeholders: Medicare, private insurance companies, and other payers.

Accuray realized early on that its first adopters would be extremely important in helping the company convince the American Medical Association (AMA) and Medicare to create new codes and coverage rates that would allow physicians and facilities to be adequately reimbursed for CyberKnife procedures. The company leveraged its relationship with an early CyberKnife adopter, Georgetown University Hospital in Washington D.C., to help influence the Center for Medicare and Medicaid Services (CMS) regarding the value of its device during the decision-making process. Adler believed that CMS was the central stakeholder in this process because "private payers follow Medicare." "It's all a matter of persistence," he said, "and of applying enough pressure." To assist in swaying payers, Accuray assisted customers in the formation of the CyberKnife Coalition, a membership of the product's users, to work with lobbyists on issues related to reimbursement.

The last stakeholder group, which is among the most important for Accuray, includes the hospital administrators who make the decision to approve the multimillion dollar purchase of the CyberKnife. To justify such a sizable expenditure, Accuray needed physicians to actively lobby hospitals to buy the equipment. The decision to start with the neurosurgery field proved to be beneficial to the company, as neurosurgeons are often influential within hospitals based on the revenue and profits they generate. However, Accuray also discovered that it needed physicians who would be aggressive in their support for the product and willing to persevere through the lengthy, bureaucratic buying process typical in most hospitals. To help convince hospital administrators to invest in its systems, Accuray developed a customized, detailed business case that outlined how much income each hospital could generate, across specialties, through the CyberKnife product (see 5.8 Sales and Distribution Strategy for more information about Accuray's approach to sales). Although the company had experienced some success in driving adoption among hospitals, this stakeholder group remained difficult to satisfy and could be the primary impediment to more widespread adoption in the future.

Using what Adler referred to as "guerilla marketing techniques," Accuray had sales of more than $140 million in 2007, a 165 percent increase over the previous year.[10] The company went public in early 2007, causing its market capitalization to climb beyond $1 billion. Worldwide, the CyberKnife had been used to treat more than 40,000 patients.[11] As it continued to grow and become profitable, its success would be driven largely on whether it could meet the diverse, often conflicting needs of its stakeholders on an ongoing basis.

Pricing

Pricing begins with a deceptively simple question: what is the appropriate baseline price for an offering? The price of a product or service can play a critical role in encouraging or discouraging adoption, particularly in a third-party payer environment. Therefore, pricing not only involves establishing a baseline price for the customer, but recognizing the different ways through which prices (and the related issue of reimbursement) will influence adoption. Reimbursement-related issues are covered at length in 5.6 Reimbursement Strategy. However, they warrant some discussion in this chapter to the extent that they relate to pricing.

Before innovators think about setting a baseline price, they must first understand all of the costs associated with developing, manufacturing, and marketing the offering. They should also determine what sort of mark-up (profit) the company would ideally earn to support its overhead and ongoing development efforts. Once these factors are clearly understood, the next step is to perform an evaluation of real and perceived value associated with the offering. Cost/benefit analyses and the value propositions defined for the device can provide important inputs to this assessment.

Value-based pricing is typically the easiest type of pricing strategy for a company to support. If the price of a new technology can be directly linked to the value it will deliver (with the value exceeding the cost), buyers and payers are far more likely to support the adoption of the offering. The most persuasive value-based pricing argument is related to direct savings in healthcare costs. If a physician or hospital will save money with each device used (relative to the current standard of care), this gives the company a strong argument for justifying its price. Another common pricing argument is related to improved outcomes. If a device leads to improved results such that a payer saves money on follow-up care and/or treatment related to complications, this is frequently a compelling argument. Companies can sometimes encounter resistance, however, when the pricing argument is based on quality of life. Buyers and payers are generally willing to support the adoption of devices that lead to significant, measurable, evidence-based improvements in quality of life. Yet, their standards for demonstrating such a change are growing increasingly stringent, especially for high-end, high-cost devices that represent a sizable potential cost burden to the healthcare system.

Another way that companies can establish a baseline price is to perform a **comparables analysis.** By evaluating the pricing strategies (and associated reimbursement status) of similar offerings in the field, companies can gain valuable information to help them choose a price. In general, medical device pricing for established products should give the innovator a strong sense of what the market will bear. Comparables analysis can be accomplished through primary and secondary research. Performing this type of market research can also be helpful in terms of understanding the price sensitivity of key stakeholders (i.e., identifying the price point at which they will resist the technology).

As described in 2.4 Market Analysis, Genomic Health, a company addressing the need *for high-value, information-rich diagnostics based on patient-level genomic testing to predict the recurrence of early stage, N–, ER+ breast cancer and enable personalized treatment decisions*, faced an interesting pricing challenge with its product. Because diagnostic companies traditionally charged between $25 and $50 for their tests, commanding margins of just five to ten percent, the company had to pioneer an entirely different pricing paradigm to support the high cost of R&D and clinical studies necessary to bring its genomic-based test to market. To decide on a price, the company first performed a comparables analysis, focusing primarily on a genetic test that looked at the mutation of the BRCA-1 and -2 genes to assess a woman's hereditary risk of breast cancer (one of the few other genomic-based tests in the market at the time). This test was priced around $3,000 and was on its way to being reimbursed on a relatively broad scale. The company then performed a value-based pricing analysis, which evaluated the cost savings associated with the reduced need for chemotherapy among women with a low likelihood of recurrence. Through this analysis it determined that the total cost of chemotherapy for early-stage breast cancer patients was at least $15,000. If the test cut the number of patients undergoing chemotherapy by 50 percent (by predicting low recurrence risk), then the savings to the healthcare system would be roughly $7,000 per patient. This meant that Genomic Health could potentially command a price of up to $7,000 per test. Finally, the company performed additional market research, including a market survey with 30 or 40 medical directors. Through this effort, Genomic Health tested the price sensitivity of payers and discovered that they considered any test over $1,000 to be expensive. However, their reaction was not significantly different between price points of $1,500 and $4,500, assuming the test had clinical value and adequate validation to support high-value pricing. Evaluating all of these inputs, the company eventually decided on a price of roughly $3,500 for its product.[12]

Despite the relatively high costs of its test (by traditional diagnostics standards), Genomic Health anticipated that it might take the company approximately 18 to 24 months to gain consistent reimbursement for its product. However, when the company met substantial resistance from payers (driven, in part, by the concern that patients would take the test but then still pursue chemotherapy regardless of the test results), it took an innovative approach to driving adoption. Genomic Health agreed to enter into a **pay-for-performance** deal with a major commercial payer UnitedHealthcare. The insurer said it would reimburse for the test for 18 months while it would monitor the results with Genomic Health. If too many women still elected to receive chemotherapy, even if the test suggested they did not need it, then UnitedHealthcare would seek to negotiate a lower price on the grounds that the test was not having the intended impact on actual medical practice.[13] Although this type of arrangement has not typically been part of the pricing strategies of most medical device manufacturers, such deals have the potential to become increasingly common as payers, as a stakeholder group, become more powerful and demanding, and as innovators recognize that such deals can overcome payer resistance when the evidence base for their technology is evolving. In Genomic Health's case, its management team was confident that women would follow the course of treatment recommended by the test, and ultimately the pay-for-performance agreement could be used as a way to advertise the company's confidence in the test.

As this example suggests, choosing a baseline price has many subtleties. Additionally, the company must decide under what circumstances it might be convinced to deviate from its baseline price. Pricing strategies that require a company to adopt a more complex approach to pricing include **differential** and **bundled pricing**, **gainsharing**, and pay-for-performance. Details on each of these alternatives can be found below.

Differential pricing

Differential pricing refers to the basic concept of pricing the same product or service differently for different customer segments. For example, in some cases, medical device companies might negotiate discount pricing with large purchasers (e.g., group purchasing organization, integrated delivery facilities). While this strategy can be effective in driving volume, it has the potential to create conflict in the market among customers, as well as payers. It may also create legal challenges, if it is perceived as creating a financial inducement to physicians.

Bundled pricing

Bundled pricing refers to setting a single price for a combination of products and/or services. A medical device manufacturer might bundle service contracts or ancillary products and services with its primary offering to try to drive increased revenue. Bundled pricing can be a way of offering discounts to buyers while incentivizing them to buy a wider range of products and services than they would otherwise. While this works in some cases, it is not successful for all offerings. For example, Guidant offered bundled pricing on its catheter and guidewire products in an effort to drive more widespread adoption. However, because physician preference is so strong in this particular field, practitioners wanted to choose products "à la carte," despite the discounts that could be realized by purchasing bundled products.

Gainsharing

Gainsharing agreements between a hospital and its physicians represent another subtlety in pricing. Under these agreements, hospitals can negotiate reduced prices with certain manufacturers in exchange for increased volume. Gainsharing differs from differential pricing, however, in that physicians are given direct incentives to adopt certain devices. For instance, these plans often provide them with a percentage of the cost savings derived from reduction of waste and use of specific supplies during procedures.[14] While some observers view gainsharing as an effective cost-cutting tool, others perceive it to be laden with inherent conflicts of interest, an obstacle to innovation and proper patient care, and possibly even a violation of anti-kickback statutes.[15]

Pay-for-performance

As the Genomic Health example illustrates, companies may agree to set their prices contingent on the realization of specific results. If the company delivers on its value proposition (as measured by mutually agreed-upon performance metrics), a payer or customer will pay its baseline price. If not, certain discounts will be expected to justify the lower "payback" on the device. As with gainsharing, pay-for-performance arrangements are relatively new and are still somewhat controversial within the industry.

Any company should seek legal counsel when creating pricing strategies. The healthcare space is highly regulated and arrangements that may be perceived as creating an inducement for a physician to use a particular device or procedure can run afoul of the Stark Law. This law governs physician self-referrals, or the practice of a physician referring a patient to a medical facility (or form of treatment) in which s/he has a financial interest, be it ownership, investment, or a structured compensation arrangement.[16] While the law remains controversial, innovators should exercise appropriate caution to avoid a potential conflict of interest.

Medical marketing and ethics

The role of marketing in the promotion and adoption of new medical devices is not without controversies. When used properly, relationships between medical device companies and KOLs can be advantageous to a company and can lead to better products for patients and physicians. However, in some cases, they may be perceived as undermining scientific integrity and unduly influencing medical care. The Advanced Medical Technology Association (**AdvaMed**) has developed a code of ethics on its interactions with healthcare professionals that aims to address concerns about potential conflicts of interest. This code provides guidelines on product training and education, support of education conferences (such as those used for continuing medical education, CME), restrictions on gifts and meals provided as part of sales and marketing, and limitations on arrangements with consultants.[17]

However, there is considerable public concern that industry efforts on self-regulation, such as the AdvaMed code of ethics, may not be enough. US senators Chuck Grassley and Herb Kohl introduced legislation in 2007 that would require manufacturers of drugs, devices, and biologics to disclose the amount of money they give to doctors through research grants, payments, honoraria, and other mechanisms.[18] In an independent incident in 2007, five major orthopedic manufacturers were the subject of an investigation by the Department of Justice (DOJ) regarding their consulting practices. The DOJ accused the manufacturers of using these consulting agreements to induce surgeons to use their products. The companies agreed to pay $311 million in fines, accepted increased oversight by a federal monitor, and posted their consulting agreements on public websites.[19]

Marketing-related ethical concerns were further reflected in a recent article in the *Journal of the American Medical Association* (*JAMA*).[20] This article argued that there is a conflict of interests between the desire of companies to promote their product and physician commitment to patient care. Self-regulation of marketing activities by the industry is, according to the article, inadequate, with public disclosures ineffective in addressing the problem. The authors of the article proposed the elimination of common practices, including some that were considered acceptable according to the AdvaMed code of ethics (e.g., small gifts, CME participation, and even consulting and research contracts). In conclusion, the article asked academic centers to take the lead in eliminating such conflicts by restricting relationships between their physicians and device/pharmaceutical manufacturers.[21]

The innovator should recognize that some marketing activities are potentially controversial. The AdvaMed code of ethics provides a minimum standard to be met, but some feel that it may not go far enough. On the other hand, the prevailing view of the industry is that self-regulation and transparency is all that is needed and the restrictions proposed in the *JAMA* article or in legislation proposed by Senators Grassley and Kohl can inhibit both the development of new devices and physician training.[22] As a physician who was involved in founding, developing, or selling multiple medical technology companies, including Progressive Angioplasty Systems, Savacor and, most recently, the $1.4 billion acquisition of Conor Medsystems by Johnson & Johnson, Frank Litvack stated:[23]

> I really believe in disclosure. But disclosure alone is insufficient. The best way to avoid the pitfalls of any form of conflict of interest, in addition to complete disclosure, is independent research corroborated by multiple parties, independent core labs, independent clinical research organizations, and peer review. There has to be a system in place. In our efforts to assure the integrity of information, what we really don't want to do is to eliminate just about anybody who knows anything about developing medical products or technology. And it's unreasonable to think that people who are involved and who work very, very, very hard for many years should not be compensated. The world just doesn't work that way.

GETTING STARTED

While the help of a marketing consultant and key thought leaders will be invaluable at this point, the following steps can be used to help the innovator devise an effective marketing and stakeholder strategy.

See **ebiodesign.org** for active web links to the resources listed below, additional references, content updates, video FAQs, and other relevant information.

1. Revisit preliminary stakeholder analysis

1.1 **What to cover** – The company must be sure that it understands the attitudes and opinions of key stakeholders toward the clinical need and the solution, particularly in light of new information that became available about the technology after the preliminary stakeholder analysis was performed.

1.2 **Where to look** – Review 2.3 Stakeholder Analysis and the preliminary stakeholder assessment completed as part of needs screening. Revisit the resources listed in that chapter, including:
- **Primary research**
- **Professional societies**
- **Patient advocacy and support groups**
- **Online patient blogs**
- **Private insurance websites** – Sites such as The Regence Group (TRG) Medical Policy or Wellmark Blue Cross/Blue Shield Medical Policies outline under what circumstances medical services may be eligible for coverage under these private insurance programs and can provide insight into payer behaviors. The Blue Cross/Blue Shield Technology Evaluation Center (TEC) is another applicable resource.
- **Public insurance websites** – Sites such as Medicare's Coverage Center or the UK's National Institute for Health and Clinical Excellence (NICE) may provide relevant information.
- **Facility-related resources** – To locate facility-related perspectives, try conducting a Medline search using keywords that reflect the need and author affiliations/departments related to the people working in the facility (e.g., hospital management, pathology). Trade group websites, such as the American Hospital Association and the Medical Group Management Association are other possibilities (although the extent of information provided may be limited).

2. Develop value propositions and collect required evidence

2.1 What to cover – Create a stakeholder value proposition matrix (as shown in this chapter). Approach key stakeholders to obtain their input regarding which value propositions will resonate best with their affiliated target audience. Once the top value propositions are defined, review clinical trial designs to confirm that studies will support them. Seek input from experts on the proposed studies and what data will be required to make value propositions compelling to the target audience. Propose additional studies, as needed. Identify and approach potential study authors (e.g., community and academic clinicians, statisticians, economists). Monitor studies, performing ongoing data analysis to assess how trial results are progressing.

2.2 Where to look – Conduct primary research with stakeholders to collect the needed information. Perform detailed interviews with KOLs. Revisit 5.3 Clinical Strategy and 5.6 Reimbursement Strategy, as needed.

3. Define product/service mix

3.1 What to cover – Consider pure product versus product and service mix. Explore service offerings that can enhance the appeal of the underlying product (e.g., service agreements, business development activities, product bundles).

3.2 Where to look – Revisit 4.4 Business Models. Evaluate the offerings of competitors and substitutes. Perform additional interviews with key stakeholders, as required.

4. Make a channel decision

4.1 What to cover – Revisit customer segments and confirm their purchasing characteristics. Identify the appropriate channels to reach each segment. Consider a direct internal sales force, distribution agreement, and/or delegated sales force, based on what will be most effective in reaching the company's targets. Be sure to consider channels within and outside the United States, as appropriate.

4.2 Where to look – Refer back to 2.4 Market Analysis for segmentation data and 5.8 Sales and Distribution Strategy for more information about channels.

5. Develop an approach to promotion

5.1 What to cover – Create a marketing communication strategy that outlines vehicles, key messages, and timing for communications targeted at each key stakeholder group. This should include strategies for peer-reviewed publications, conferences, CME, reimbursement dossiers, and DTC advertising (as needed). Remember to sequence communications appropriately so that no stakeholder group is taken by surprise regarding inquiries about the device. Determine whether or not strong resistance is to be expected within or between any key stakeholder groups. If so, develop a PR strategy that engages appropriate professional societies, physician groups, patient groups, hospital groups, etc. that can help resolve the anticipated conflict and/or mitigate it so it does not negatively affect device adoption upon launch.

5.2 Where to look – Gather as much competitive intelligence as possible as inputs to developing these strategies, including:

- **Interviews with KOLs**
- **Competitive intelligence** – Marketing communication and/or PR strategies of competitors.
- **Press releases from competitors**
- **Advertising campaigns in medical journals, newspapers, and business journals for competitors and/or proxy devices**
- **Competitor presence at professional society meetings and conferences**

GETTING STARTED

- Competitor presence at patient advocacy and support group meetings
- Competitor presence on online patient blogs

6. Develop a pricing strategy

6.1 What to cover – Be sure that the company's costs are well understood relative to the value the offering provides. Perform comparables analysis and supporting market research (including the status of reimbursement and price sensitivity in the data collection effort). Set a baseline price, and then consider the different pricing strategy options available to the company (e.g., differentiated pricing, bundled pricing, gainsharing, pay-for-performance).

6.2 Where to look – See 5.6 Reimbursement Strategy for details on justifying the value of a device. Survey purchasers and perform secondary research to assess comparables.

Credits

The authors would like to acknowledge Steve Fair for his help in developing this chapter. Many thanks also go to John Adler for sharing the Accuray story.

Notes

1. *Marketing Management*, Kotler and Keller, 12th Edition, Prentice Hall, NJ.
2. "Pharmaceutical Marketing," Wikipedia.org, http://en.wikipedia.org/wiki/Pharmaceutical_marketing (February 21, 2008).
3. A. S. Go, G. M. Chertow, D. Fan, C. E. McCulloch, C. Y. Hsu, "Chronic Kidney Disease and the Risks of Death, Cardiovascular Events, and Hospitalization," *New England Journal of Medicine*, September 23, 2004, http://www.ncbi.nlm.nih.gov/pubmed/15385656?dopt=Abstract (March 11, 2008).
4. From remarks made by Ron Dollens as part of the "From the Innovator's Workbench" speaker series hosted by Stanford's Program in Biodesign, March 21, 2005, http://biodesign.stanford.edu/bdn/networking/pastinnovators.jsp. Reprinted with permission.
5. "Public Relations," Dictionary.com, http://dictionary.reference.com/browse/public%20relations (February 21, 2008).
6. "AthenaHealth Causing Trouble Again," The Health Care Blog, May 7, 2007, http://www.thehealthcareblog.com/the_health_care_blog/2007/05/cigna_ranks_no_.html (November 4, 2008).
7. Thomas M. Burton, "More Are Eligible for Heart Device – Accused of Rationing Care, Medicare Expands Use of Implantable Defibrillators," *The Wall Street Journal*, September 29, 2004, p. D4.
8. All quotations are from interviews conducted by the authors, unless otherwise cited. Reprinted with permission.
9. CyberKnife Society Home Page, http://www.cksociety.org/ (March 3, 2008).
10. "Accuray Incorporated," Hoovers, http://premium.hoovers.com/subscribe/co/factsheet.xhtml?ID=131014 (March 3, 2008).
11. "CyberKnife Frequently Asked Questions," Accuray, Inc., http://www.accuray.com/content.aspx?id=356 (March 3, 2008).
12. Supporting information drawn from "Genomic Health: Launching a Paradigm Shift … and an Innovative New Test," GSB No. OIT-49.
13. Brandon Keim, "Pay-For-Performance Pharmaceuticals: Satisfaction Guaranteed or Your Money Back," *Wired*, July 16, 2007, http://blog.wired.com/wiredscience/2007/07/pay-for-perform.html (March 11, 2008).
14. Lauren Uzdienski, "OIG Approves New Gainsharing Plan," HealthPoint Capital, November 28, 2006, http://www.healthpointcapital.com/research/2006/11/28/oig_approves_new_gainsharing_plan_advamed_critical/ (March 12, 2008).
15. Mark Prodger, "A Primer on Gainsharing," Hospital Buyer, http://www.hospitalbuyer.com/materials-management/cost-savings/a-primer-on-gainsharing-145/ (March 12, 2008).
16. "Stark Law," Wikipedia.org, http://en.wikipedia.org/wiki/Stark_Law (March 13, 2008).
17. "Code of Ethics," AdvaMed, http://www.advamed.org/MemberPortal/About/code/ (November 4, 2008).
18. " Grassley, Kohl Say Public Should Know When Pharmaceutical Makers Give Money to Doctors,"

Medical News Today, September 8, 2007, http://www.medicalnewstoday.com/articles/81822.php (November 4, 2008).
19. Michael Piechocki, "Five Orthopedic Device Companies Reach Settlement with DOJ for Investigation into Consulting Practices," Ortho Supersite, September 2007, http://www.orthosupersite.com/view.asp?rid=24187 (November 4, 2008).
20. Troyen A. Brennan, David J. Rothman, Linda Blank, David Blumenthal, Susan C. Chimonas, Jordan J. Cohen, Janlori Goldman, Jerome P. Kassirer, Harry Kimball, James Naughton, Neil Smelser, "Health Industry Practices That Create Conflicts of Interest: A Policy Proposal for Academic Medical Centers," *JAMA*, 2006, pp. 429–433.
21. Ibid.
22. Lisa Girion, "Medical Ethics Reform Urged," *Los Angeles Times*, January 25, 2006.
23. From remarks made by Frank Litvack as part of the "From the Innovator's Workbench" speaker series hosted by Stanford's Program in Biodesign, March 5, 2007, http://biodesign.stanford.edu/bdn/networking/pastinnovators.jsp. Reprinted with permission.
24. "Peer Review," Wikipedia.org, http://en.wikipedia.org/wiki/Peer_review (February 21, 2008).
25. "Strength in Numbers," Elsevier, http://www.elsevier.com/authored_subject_sections/P11/sin/value.html (February 21, 2008).
26. "How to Write an Abstract," UC Davis, http://urc.ucdavis.edu/howtowriteanabstract.html (February 21, 2008).
27. "Continuing Medical Education," Wikipedia.org, http://en.wikipedia.org/wiki/Continuing_medical_education (February 21, 2008).
28. "Direct-to-Consumer Advertising," SourceWatch, http://www.sourcewatch.org/index.php?title=Direct-to-consumer_advertising (February 21, 2008).

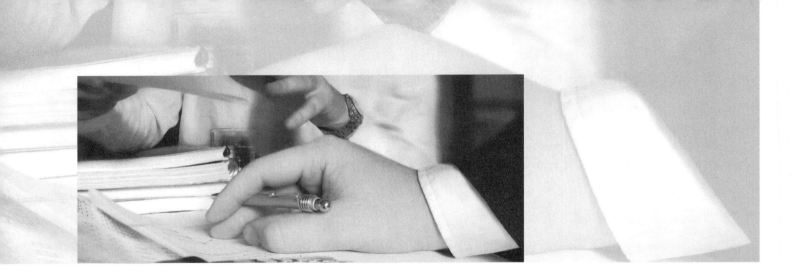

5.8 Sales and Distribution Strategy

Introduction

Talk to any experienced medtech innovator or executive about commercializing a new medical technology and they will inevitably ask: "How will you get it in front of the physician who will be using it?" It is not uncommon for a new device to target a specialty with thousands of physician-users, so it can be a formidable challenge to introduce the technology, educate them about its value, train them on its use, and then convince them to build their procedure volume. But all of these steps are necessary – usually in rapid succession – to create a sustainable business in the medtech field.

At this stage in the biodesign innovation process, a company defines the approach it will use to sell its device and deliver it to customers. This effort focuses on what is known as "the last mile," or the process of educating the key provider and physician purchasers about the device and its benefits, as well as working with them to make sure that the device is appropriately used as part of the patient care paradigm. The critical question to be addressed as part of this exercise is whether the offering is best promoted through a dedicated sales force (**direct sales model**) or if it can be sold through a partnership with some other entity (**indirect sales model**). Once this is decided, then specific strategies and tactics for executing the chosen approach can be put into place.

OBJECTIVES

- Appreciate the impact of the company's business model and other essential factors on its options for reaching customers (e.g., an indirect or direct sales and distribution model).
- Understand the most common medical device indirect sales model and distribution options and how they work.
- Understand how the direct sales model works in medtech.
- Learn how to determine the most appropriate sales and distribution model for a particular offering.

Sales and distribution fundamentals

Great companies are built based on their relationships with customers, and great managers maintain a relentless focus on creating these strong relationships. As John Abele, cofounder of Boston Scientific, put it:[1]

> It always drives me crazy when I walk into a company that posts its daily stock price in the lobby. In my mind, that's the wrong incentive. The incentive is to provide outstanding value to your customers. If you do that well, the stock price will eventually follow.

Abele's statement is a compelling reminder that the success of any medtech venture depends on its sales organization and on the choice of the right sales model to deliver a positive customer experience, working within market and financial constraints. As described in 4.4 Business Models, sales and distribution channel considerations form important factors in an innovator's decision regarding the selection of a business model for his/her company. Some business models lend themselves better to specific sales and distribution approaches than others. The differences arise primarily from the characteristics of the innovation or product offering, although customer factors also come into play.

Chapter 4.4 outlines a total of nine different business models, of which four are most common in the medtech field: disposables, reusables, implantables, and capital equipment. Although some notable exceptions exist, the typical product characteristics that correspond to these four business models are summarized below.

Disposables generally fall into two categories. First, there are low-value products (as measured by sales price in this context), such as lab supplies, syringes, or gloves. All of these rely on high sales, high volume, and low overhead to be profitable. They also tend to be low-complexity, non-differentiated products (meaning that physicians do not typically express a preference for which brand of the product they use). As a result, low-value disposables often require limited sales effort beyond the initial product launch, assuming that prices are kept low and quality remains consistent. Marketing and product-related information sharing is generally accomplished via supplier catalogs or online resources, with customer service representatives available by phone to handle more specific product inquiries and sales.

High-value disposables include products such as ablation catheters used in the treatment of atrial fibrillation or the automated anastomosis systems for cardiac artery bypass graft (CABG) surgery developed by Cardica, Inc. (described in 6.1 Operating Plan and Financial Model). These high-value disposables command relatively high prices, tend to be complex, and may be more appropriately handled through a direct sales model.

Many *reusable* products, such as surgical instruments, tend to be lower-value offerings that provide relatively low margins and depend on moderate to high volumes to be profitable. Like disposables, they can be marketed through supplier catalogs and/or Internet resources. However, some reusable products, if they are sufficiently complex, have a slightly higher value and/or command somewhat higher prices, such as ambulatory cardiac rhythm monitors (e.g., Holter monitors), which can be reused among many patients before requiring replacement. As a result, they may be sold and distributed by a third-party distributor that carries a complete line of complementary products.

When considering the characteristics of *implants*, it is useful to separate higher-value implants from those with moderate or lower relative value. High-value, complex implants, such as pacemakers and artificial joints, often require a knowledgeable and involved sales force in order to complete a sale and ensure proper device usage (through training, etc.). They usually have a somewhat longer sales cycle, as well. Depending on expected sales volume, high-value implants typically lend themselves to either a direct sales force or a specialized third-party distributor. Moderate and lower-value implants, on the other hand, may have characteristics that more closely resemble reusable products. As a result, it may be more appropriate to distribute them without the involvement of a direct sales force.

Capital equipment products, such as magnetic resonance imaging (MRI) machines and ultrasound equipment, are often highly complex. They provide high value to the user and, in turn, command high margins. These products tend to have a particularly long sales cycle that requires a prolonged, dedicated effort in order to make the sale. For this reason, the associated business model favors a direct sales force. Facilities generally act as buyers for capital equipment, with or without broad provider input. Because volume is low, the sales team is usually small, although auxiliary field personnel are required to provide servicing of the equipment.

These descriptions imply two fundamental approaches to sales and distribution: the indirect and the direct models. In an indirect model, one or more sales teams from distributors or third-party manufacturers serve as the primary point of contact with the end user to manage product sales and delivery. They typically are not dedicated exclusively to any one product from a single company, but instead represent a portfolio of complementary products from multiple companies. That said, while some distributors carry a broad range of products across multiple fields, others may be more specialized, targeting products in a particular field, specific types of physicians, and/or a narrowly defined geographic territory. In a direct model, the company builds its own internal sales force to handle all sales functions, and establishes a separate customer support division to manage distribution of the product to the end user. A third, hybrid model is becoming increasingly common in the medtech industry, in which a combination of direct sales force and distributors is employed. For example, a company may decide to use a direct sales force in the US market and local distributors in Asia and Europe. The business model, product characteristics, and preferred sales and distribution approach come together as shown in Figure 5.8.1.

While this view represents what is often found with medical devices, the decision regarding an approach is not always clear-cut. As previously emphasized, companies must determine the most effective sales and distribution model for capturing and sustaining value based on their chosen business model, their unique

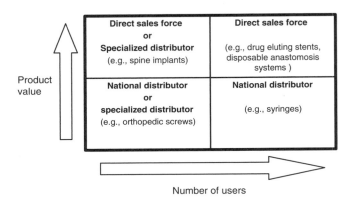

FIGURE 5.8.1
Sales force models are often linked to product attributes.

product offering, the customer being targeted, and the desired interaction with the customer. Customer accessibility and receptiveness are two important factors that play into the desired interaction that a company seeks to create with its target audience. For example, a direct sales model requires a high-touch relationship with targeted physicians. Such a model is only appropriate if a company can reliably gain access to those physicians through a direct channel, and they are generally willing to engage in the sales process based on how they have traditionally been contacted by other device manufacturers. As noted in 5.7 Marketing and Stakeholder Strategy, each medical specialty tends to have its own "personality" regarding how receptive physicians are to experimenting with new innovations, how curious they are to learn about new products, and how loyal they are to established treatments. Understanding the psyche of the doctors in the field where the company is trying to establish a foothold is essential to deciding on an optimal sales and distribution model. Examining the models used by competitors can be a source of invaluable insights.

The final choice of the sales model is also based on detailed financial analysis. At a high level, the innovator needs to determine the largest direct sales force that the potential revenue from the product can sustain and then decide whether such a sales force can deliver the assumed revenue. If not, an indirect model may be necessary (or it may even be possible that a stand-alone business may not be viable). The exact financial calculations that can be used to support the choice of the appropriate model are described in 6.1 Operating Plan and Financial Model, while 6.4 Licensing and Alternate Pathways explains the available options to the inventor when a stand-alone business is not viable. The section on Determining the Most Appropriate Sales and Distribution Model later in this chapter also provides more information.

On the surface, it may appear that only the indirect model is viable for new start-up companies. However, if the financial analysis performed as part of Chapter 6.1 demonstrates that a direct sales force is viable, then the innovator can seek to raise enough capital to build a direct sales force before the company has sufficient revenue (see 6.3 Funding Sources), as shown in the Kyphon® story later in this chapter.

Indirect models

The key advantage of an indirect sales model is that it enables a small or emerging company to access the market with minimal investment and without needing a dedicated sales force. The primary disadvantage is that the company gives up control of the sales process, which can mean that total sales are optimized and important customer feedback relevant to new product opportunities may be lost.

In the medtech field, there are several common approaches to adopting an indirect model for sales and distribution. These include entering into a sales and distribution agreement with a national or specialized distributor, or forming a third-party partnership with another manufacturer. The primary differences between these various approaches are outlined below (and summarized in Appendix 5.8.1).

National distributors

National distributors work especially well for disposables, reusables, and simple implants that have a relatively low unit price and low to moderate complexity. In these cases, the sales and distribution process is largely managed by a large national distributor or wholesaler.[2] The company generally sells to the distributor at a discount. The distributor, in turn, passes along the product at full price, keeping the difference as a form of compensation. The company maintains its own brand, while the distributor maintains the right to represent other products (i.e., the sales and distribution agreement is not exclusive). While this model is almost always adopted for commodity products, such as syringes, surgical gloves, and other such medical equipment, it can also be effective for somewhat higher-end disposables and reusables. For example, Acclarent, a company that manufactures and markets endoscopic, catheter-based tools to perform *balloon sinuplasty*™ (a procedure in which a balloon is used to dilate various areas of the sinuses) uses this kind of indirect model to sell and distribute its product to surgeons in much of Europe.

In most cases, the customer for the products sold and delivered through a national distributor is not the physician or other end user of the device. Instead, national distributors tend to interact with the purchasing departments of hospitals or clinics, or they interface with intermediaries, such as global purchasing organizations (**GPOs**) or buyers for integrated delivery networks (**IDNs**). (GPOs and IDNs are described in more detail below.) Once a contract is in place, buyers commonly place orders by telephone or online through systems managed by the distributor. Products are then delivered from the distributor's warehouse(s) via courier, with the distributor often managing inventory on behalf of the company.

The role of GPOs and IDNs in a national distributor model

Global purchasing organizations (GPOs) and integrated delivery networks both seek to leverage the purchasing power of a group of participants to negotiate for more favorable sales and distribution terms. Both groups have formed partly in response to rising healthcare costs. GPOs, such as Novation and Premier, organize multiple hospital groups, large clinics, and medical practices into buying cooperatives. Integrated delivery networks take the idea of consolidation one step further by aggregating hospitals, physicians, allied health professionals, clinics, outpatient facilities, home care providers, managed care, and suppliers into a single, closed network. Kaiser, Tenet, and Intermountain Health are examples of IDNs. Companies and/or their distributors may have the opportunity to negotiate sole source contracts with GPOs and large IDNs, giving greater pricing and delivery concessions in exchange for this preferred status.

Large GPOs may represent as many as 1,000 to 2,000 hospitals (making them greater in size than IDNs).[3] As a result, they have a high degree of negotiating power with companies. Approximately 72 percent of all medical system purchases go through a GPO. On average, GPOs are able to secure ten to fifteen percent discounts on supplies, devices, capital equipment, services, pharmaceuticals, and more, which they pass on to their members.[4] In return, GPOs earn a two to three percent contracting administration fee (**CAF**)[5] per contract. In the medical device field, this fee is typically paid for by the manufacturer. Many GPOs also provide value-added services to their members, such as product evaluation, training, and service and inventory management.

Integrated delivery networks are becoming sufficiently large to gain buying clout and act as their own GPO. A 2002 study from the American Hospital Association (AHA) reported that there are approximately 455 IDNs in the United States, averaging eight hospitals each. Approximately half of these IDNs are large enough to be able to negotiate independently and competitively with manufacturers.[6] In addition to size, IDNs also gain negotiating power as they include several healthcare venues (hospital, nursing home, etc.) across the continuum of care, offering an attractive place for companies (or their distributors) to contract a full line of products. Additionally, the integrated, single-ownership nature of IDNs greatly improves compliance with purchasing contracts, ensuring that the targeted sales volume is reached in exchange for the discount granted.

Specialized distributors

Another indirect model that is widely used in the medtech field is to work with specialized distributors. This approach is best suited to devices that are somewhat differentiated, relatively more complex, and offer at least moderate value to the user (as measured by physician preference for a brand name device, e.g., certain types of implants). Alternatively, this model works well for low-complexity products that complete a complex product line carried by the specialized distributor.

With a specialized distributor model, a company gains access to sales representatives who are specialists in a particular type of product or therapeutic area. The sales personnel that support this model may be employed by a central company. However, they often work as independent representatives, contracting directly to the company on a salary and/or commission basis without actually becoming employees. In either scenario, these representatives have strong customer relationships within a particular region, and are able to leverage these relationships to sell whatever product they are carrying at the time. Sole source distributors carry one company's products while multiple source distributors carry a few lines of products, ideally with clear differentiation along product characteristics and price. While it is in the best interest of distributors to carry as many brands as possible, manufacturers prefer exclusivity and can demand this if they are large enough.[7] Invivo Surgical Systems is an example of a multisource franchise distributor[8] that carries spinal fixation implant product lines and equipment. Invivo Surgical Systems limits its coverage to the Northeast and New England, and recruits its own sales personnel to sell its products.

Specialized distributors can be set up as master resellers or agents. The primary distinction lies in whether the specialized distributor takes ownership of the product. This distinction has important ramifications on the economics of the sale. Specialized distributors who are master resellers purchase the product from the company in anticipation of orders coming in from customers, and resell it to the end user. Because master resellers take title of the product, they have more power to negotiate contracts and discounts with the end user. Specialized distributors who act as agents, on the other hand, sell product on consignment and negotiate the sale on behalf of the manufacturer, earning a commission.[9] They have limited pricing flexibility and can only contract and discount at terms that are acceptable to the manufacturer. Although master resellers have more flexibility to negotiate the price than agents, they also take on more financial risk. They purchase goods from the manufacturer at a discount that is roughly equal to their target margin, but have no guarantee that the end user will buy at the retail price, or buy at all. Agents, on the other hand, receive a 20 to 30 percent commission[10] on each sale and are not responsible for product that does not sell. However, they do risk losing part of their commission if they accept a lower price from the end user.

Third-party partnerships

A third approach to indirect sales is to manage the process through a partnership with an established medical device manufacturer that has its own direct sales force (this form of partnership is different from partnerships focused on developing the product, which are described in 6.4 Licensing and Alternate Pathways). This model, while widely used in the pharmaceutical industry, is relatively new to medical devices. Lifescan's 2007 announcement that it planned to distribute its glucose monitor through Medtronic provides one example of this kind of partnership.[11] Another example is seen in Israel-based Medinol's decision to sell and distribute its drug eluting stent through Boston Scientific in the late 1990s. Sometimes worldwide sales and distribution rights are awarded as part of the deal. In other scenarios, the smaller company maintains local rights (e.g., in the United States), while essentially "outsourcing" sales and distribution in other markets (e.g., Europe).

Third-party sales and distribution partnerships are generally only available to companies whose products can be clearly differentiated, have at least moderate perceived value, and/or complete a complex system offered by the larger partner. The most common reason for small companies to enter into these partnerships is to avoid the large financial burden associated with building and maintaining a direct sales force. Companies may also consider such a deal if they lack the breadth in their product lines to sell to entities that wish to purchase a complete line of products. Established manufacturers also have more leverage with GPOs and other important buyers and provide smaller companies with access to these relationships.

There are two primary financial arrangements for forming these partnerships. The distinction is driven by whether or not the distributing company will manufacture the product. If the smaller company will be the manufacturer, a transfer fee is awarded each time a product is shipped to the distributing company. If the distributing company will be the manufacturer, then a royalty is paid (usually five to eight percent) to the smaller company based on net revenues.[12] In both scenarios, an upfront payment may be awarded to the smaller company to provide the financial means to bring the product to market.

While third-party sales and distribution partnerships can ensure the resources, infrastructure, relationships, and experience to quickly help get a product into practice, entering into such a relationship can be risky for a small device company, particularly if it is a single-product company. Because the larger company controls all of the customer relationships and manages pricing, training, and usage of the product, the smaller company is essentially putting its fate into the hands of its partner. Partnerships such as these are best employed only when some terminal event is spelled out in the partnership agreement toward which the company is working (e.g., an acquisition) or the company has another flagship product over which it exercises significantly greater control. Another scenario in which a third-party partnership might be appropriate is when the company is facing financial difficulties that leave it with few other alternatives.

Hybrid models

A hybrid distribution model that combines both a small direct sales forces and an indirect approach can be a good choice for some medtech companies, particularly in their early stages (a hybrid approach can also be called a two-and-a-half tiered model). The time, cost, and other resources associated with establishing a full direct sales force can be daunting to a young company. However, if a product is at least moderate in its complexity, value, and anticipated volume, the company could launch a small direct sales team whose efforts are complemented by an independent distributor. The direct sales force can target a narrow segment of the company's highest-value customers, while distributors or third-party partnerships with larger manufacturers are leveraged to give the company breadth in the total available market.

Often hybrid models are based on geography, with a company deploying a direct sales force in concentrated markets but relying on regional specialized distributors in diffuse, closed, or global markets. In these scenarios, the company may have the ability to gradually "take back" markets from distributors as its product gains momentum and revenues increase.[13] Osteotech, which makes bone implants that partially utilize donated tissue, is one example of a start-up company that utilized a hybrid model in order to rapidly gain a presence in a relatively large market. To address territories that extended beyond the capabilities of its small direct sales team, Osteotech worked with multisource agents to promote its product. However, over time, the company had plans to move toward a more direct model, investing $4 million in its sales and distribution channel in 2007 to make more "OsteoBiologic Specialists" available to drive the sales process.[14]

The benefits of a hybrid model are greatest when the sales team and the distributor work closely together. Micrel Medical Devices provides an example of the synergy that can be achieved.

FROM THE FIELD | MICREL MEDICAL DEVICES

The role of sales in understanding customer needs to help drive product expansion and growth

Micrel Medical Devices was founded in Greece in 1980 by Alexandre Tsoukalis and Nikitas Machlis to address a relatively rare condition known as thalassemia. This disease is a hereditary form of anemia (low red blood cell count) that is most prevalent among individuals of Mediterranean descent. Treatment requires regular blood transfusions, which often result in iron overload, as iron is a key constituent of red blood cells. Patients must take chemical chelators to bind and remove the iron from the system. While previously introduced intravenously, syringe style pumps, such as Micrel's Thalapump™, allowed chelators to be infused slowly under the skin of the stomach. As a result, patients could treat themselves at home, rather than receiving infusions at a hospital or other infusion center.

Based on the success of its early products, Micrel began expanding beyond this initial need, creating a full line of ambulatory volumetric and syringe infusion pumps for hospital and home care. The company's volumetric pumps are used for patient-controlled analgesia, parenteral nutrition, chemotherapy, and other therapies. Its syringe pumps are used to treat thalassemia, Parkinson's disease, immunodeficiencies with immunoglobulin infusion, and pain.

Micrel maintained a small sales and marketing force that worked closely with local distributors in its key markets.

In the home care market, the distributor focused on managing relationships with the home care providers (who typically bought the pumps and rented them to patients). Micrel's sales representatives were dedicated to interacting with patient groups and professional associations, as well as the home care providers to help drive demand for the company's products. "The sales people are very important to us because they translate market needs into requirements for R&D," said Tsoukalis, the company's managing and technical director.[15] As a company, Micrel prided itself on designing pumps that were easy to use, comfortable for patients, and uniquely suited to the specific needs within a particular patient or geographic segment. Its sales force was a key enabler of this process.

According to Tsoukalis, the sales team also was constantly on the lookout for new opportunities to help the company expand. For example, when competitor Baxter withdrew its parenteral pump from the European market, Micrel sales representatives rallied the company to seize the opening. Baxter had previously been dominant in parenteral pumps and its departure left a significant and sudden void, especially in the lucrative UK market. Micrel learned about the opportunity around April, 2006, but the home care providers "had a clear deadline to decide which pump they should go with by October, 2006," said Tsoukalis, a deadline set by the UK hospital system for the consideration of new entries for the coming year. With just six months to customize, seek approval for, and market a new device, the company aggressively went to work.

While the R&D team modified their pump design to meet the "many peculiarities" of the UK market (including greater output and specific patient setup and sterile environment requirements), Micrel's sales representatives established a continuous conversation with home care providers, patient groups, and regulatory contacts. They presented them with regular updates on Micrel's progress and diligently gathered feedback on patient needs and desired features. In the end, the results of a study, performed by a standing committee of the British Association for Parenteral and Enteral Nutrition (BAPEN) were highly favorable towards Micrel's new product, Rythmic™ (see Figure 5.8.2).

Tsoukalis described the feedback: "They said our pump was the best of all the pumps they checked ... [and] our offering was very, very well fitted to the needs in the UK." Micrel began shipping the Rythmic product to this market in early 2007, successfully replacing Baxter as the leader in parenteral pumps.

Tsoukalis gives due credit to Micrel's UK distributor in helping the company penetrate this market: "He knew exactly where to go and whom to talk to," as the company sought to make the necessary connections. The distributor and the sales representatives worked closely together through the design and development process to facilitate the desired end result. He also applauded the sales team. In addition to identifying key user requirements and maintaining ongoing communication with important stakeholders, "they also made the schedule, and managed the absolutely tight dates on which we had to deliver," said Tsoukalis. Sales' attention to suggestions from patient associations was cited as a major reason for Rythmic's top rating. In 2008, Micrel was seeking FDA approval for the Rythmic device (see Figure 5.8.3).

Leveraging a hybrid sales model that creates synergy between a direct sales force and local distributors, Micrel has replicated this expansion strategy in other markets, such as France, Sweden, Norway, and Germany. As of early 2008, the company was approaching 40 employees and $8 million in annual sales.

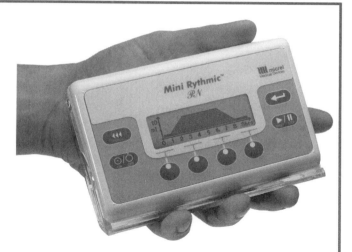

FIGURE 5.8.2
Micrel's Rythmic device (courtesy of Micrel Medical).

FIGURE 5.8.3
Micrel's sales representatives, Jean Christophe Cascailh and Jean Claude Rondelet, with Gordon Machray, the principal salesman from the company's UK distributor Eden Medical (courtesy of Micrel Medical).

Direct models

For more complex and expensive devices, such as capital equipment and high-value disposables or implants, a direct sales model is often required. The primary customer in these sales is generally the physician or the medical facility (hospital, clinic, ambulatory surgical center, etc.). Often, both the physician and the facility must be considered customers as they each have a role in the purchase process: physicians evaluate and endorse the performance of a product, while the facilities evaluate the economic ramifications of the purchase. More often than not, to convince these stakeholder groups to buy a product requires information, relationships, and interactions that only a direct sales force can provide. The decision to develop a direct sales force and the appropriate size of the force is complex and must be based on detailed bottom-up financial modeling, explained in Chapter 6.1.

With a direct sales force, a company hires its own sales people to engage providers (physicians and facilities) in a six-step process: (1) account prospecting, (2) relationship building, (3) pitching and closing the sale, (4) pricing and contract negotiations, (5) provider training, and (6) account management, as described below.

Step 1: Account prospecting

Focusing sales efforts on targets that will eventually lead to consummated business is a two-stage process. First, initial sales targets are set by the company based on the amount of calculated potential business a target can generate in the future (6.1 Operating Plan and Financial Model). Next, sales representatives visit or "call on" these targets. Over time, they identify and focus on the subset of providers who are most receptive. Sales representatives may also identify additional providers not included on the original list of targets through networking and other professional activities.

Step 2: Relationship building

Forming partnerships and developing trust with providers requires extensive formal and informal relationship-building activity. During this stage of the selling process, the sales representative is constantly evaluating providers' needs and seeking to convince them of the product's value in addressing those needs (see 5.7 Marketing and stakeholder strategy). The off-site social meetings popularized by the drug industry (such as dinners, sporting events, and company-sponsored weekend advisory board meetings) have become relatively commonplace in the medical device field as one vehicle for gaining access to providers. Educational events are also used as a mechanism for building relationships as part of the sales process. Providers often decide to enter into a relationship with a company based on their evaluation of a product's attributes, the sales representative's medical expertise and personal characteristics (such as honesty and integrity), and the commitment of the company to the account (demonstrated through time, special attention given to providers, and the future potential for research partnerships).

The process of building a relationship with a provider can be time consuming. It is not unusual for new sales representatives to be given months (or even years in some highly complex, specialized fields) to identify and develop productive relationships before they must generate sales.[16] Building relationships can be especially challenging in crowded therapeutic areas, where physicians may have deep loyalties to the products and sales personnel of established competitors. For example, in orthopedics, most representatives are expected to have convinced only ten physicians to try their products within the first two years of selling.[17]

Step 3: Pitching and closing

Ongoing product sales begin once the relationship has become adequately established and the product benefits are clear to the provider. Sales representatives are extensively trained to handle this stage of the sales process. With a particular patient or patient-type in mind, the representative uses product, competitor, and disease state information as well as an understanding of the provider's unmet needs in order to make a case for his/her company's product. Materials are often customized to each provider's specific circumstances and interests. Sales representatives must be prepared to handle objections from providers (such as doubts about performance, safety, or value). The actual "close" takes place when the representative asks for the product to be purchased for a specific patient or within a defined time frame, and the provider agrees.

Step 4: Pricing and contract negotiations

Pricing and contract negotiations may occur in parallel with pitching and closing the sale. Some providers want to address these issues before any final agreement is reached, while others prefer to address them after the product has been tested and found acceptable in a few patients. Within a single facility, price and contract negotiations for capital equipment and high-value implants are generally conducted between the sales representatives and contacts from purchasing and/or administration. While physicians are integral to making the recommendation to procure the product, they do not directly participate in the purchasing negotiations, which include agreeing on a price for the product and the negotiation of all relevant deal terms (delivery, training, usage, service, etc.). Outside a hospital or other larger facility setting (e.g., for products sold into doctors' offices), the physician is much more likely to serve as the primary point of contact for the end-to-end sales process, including pricing and contract negotiations.

As an integral part of the purchasing negotiation, especially for high-cost items, companies generally develop a business case model that can be customized by the sales representatives to provide a tailored demonstration to the buyer regarding how the product will improve outcomes and enable future profit in the buyer's environment (e.g., through billing the insurance company for the product and procedure, by expanding the provider's market, or through repeated use of capital equipment). In the case of strategic customers, such as large IDNs, major hospitals, or GPOs, contract negotiations are usually handled by the most experienced sales representatives or by higher-level company executives (the vice-president of sales and marketing or the chief medical officer). Such negotiations can be complicated and time consuming. Additionally, they are often strategic in nature and the buying decisions tend to involve large sums of money (e.g., in the form of pricing concessions).

Step 5: Provider training

Once a sale has been made or promised, sales representatives help the provider use and integrate the device into practice. For example, with an implant, they first train the provider on how to select the appropriate model and size of the device and then how to perform the procedure. If appropriate, sales representatives are commonly present in the first few procedures to assist the provider in selecting the model or size of device for each individual patient and help troubleshoot problems and issues as they arise (see Figure 5.8.4). For highly complex devices, it is not unusual for device companies to restrict the use of a device until physicians have completed appropriate training by a representative of the company (see the Kyphon story in this chapter). The quality of this interaction is critical in determining whether the provider will continue using the product.

Step 6: Account management

Post-sale account servicing includes everything that happens after the sale to support all servicing included in the contract (e.g., service for a piece of equipment, answering product questions, and assisting with reimbursement issues). Additionally, it is focused on customer retention and the development of future sales with the same account. By proactively addressing customer concerns, a sales force can understand what issues might cause a customer to try another product and, in many cases, make adjustments to the offering to circumvent the problem. Examples of common concerns in the medtech field that must be managed on an ongoing basis include difficulties with the device and/or procedure, lower than anticipated patient outcomes, and suboptimal customer service.

Importantly, account management is also intertwined with the development of new product ideas. As sales representatives assist providers with the use of the device and address product questions and concerns, they gather feedback that can be used by engineers and other technical experts within the company to improve and/or expand the company's offerings.

As this brief description of the direct sales process illustrates, there is a sizable investment of time and resources to make each sale. However, once physicians and facilities start reliably engaging the sales representatives and using the company's products, a strong bond is often formed that can be leveraged to make sales easier and less time-consuming in the future.

Direct sales teams are typically organized geographically, with each individual territory manager reporting

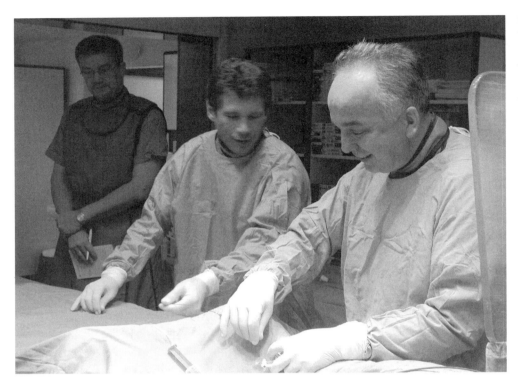

FIGURE 5.8.4
TransVascular, Inc. representatives Ludwik Firek and Ted Lamson train Dr. Tomasz Siminiak on a new procedure for injecting stem cells into the heart (courtesy of ExploraMed).

to a regional manager who, in turn, reports to a national sales director. The territory sales representatives have limited freedom in choosing which doctors and hospitals they target, as much of this direction comes from company management. Responsibility for the sale is shared somewhat between the regional manager and the territory manager. They take a team approach closing the sale, after which the regional manager handles the contract negotiations and the territory manager handles account maintenance and customer retention in collaboration with the sales representative. Any contracts and discounts are typically managed centrally by the company.[18]

In a direct model, the physical delivery of the device is typically made by the sales representatives or via a courier service (capital equipment requiring a major installation is, of course, one exception to this rule). The provider completes and submits an order form, calls the company's customer service department, or places the order directly through the sales representative. In turn, the product is shipped or hand-delivered. In some cases, multiple sizes or models of the product are left with the provider on consignment so that, during procedures, the device with the correct specifications is always available. Because the physical delivery of a device is not as complex as the underlying sales process, the manufacturer may engage a third-party distributor to take charge of deliveries, even as it maintains a direct sales force.

The role of GPOs and IDNs in the direct model

Although manufacturers of high-value devices traditionally have not entered into contracts with GPOs and IDNs,[19] groups that aggregate the purchasing power of multiple entities must still be viewed as important customers. Healthcare providers sometimes organize themselves into purchasing groups for buying complex, high-value medical devices. Similar to GPOs, these groups take individual physician input into account, but also rely heavily on pooled criteria for making buying decisions that satisfy the interests of all participants, such as the ability of the company to serve as a sole-source supplier of a device, customer service capabilities, pricing, and discounting. Companies faced with the challenge of selling high-value devices to such purchasing groups typically must work hard to sell them on the value of the products. Similarly, they must continue to innovate and improve to retain the business once a buying decision has been made.[20] However, it should be noted that, in most cases, these aggregate purchasers play a limited role in high-value devices because the purchasing decision is more directly driven by physicians (and aggregate purchasers exercise little control over physician choices).

The role of payers in the direct model

Payers should also be viewed as customers for more complex, high-value medical devices, as they are responsible for the incremental reimbursement of new and expensive products. As addressed in 5.6 Reimbursement Strategy, the reimbursement process is not unlike the sales process in this chapter: it involves identifying who will be the major payers, developing relationships early to understand what product information and value propositions will be required, lobbying for coverage and payment policy in anticipation of or in conjunction with launch, contracting and discounting, and payer account maintenance (i.e., fielding product questions or concerns as they arise). However, while "selling" to payers requires significant effort and focus, it typically only occurs once. After the product is accepted by the payer, less effort is needed to maintain product acceptance, although constant follow-up is necessary to keep the product in a favored position.

Sales force training

Any direct sales force requires extensive training to effectively represent and sell a complex product to the target audience. Such training should be delivered directly by the company that developed the device such that the sales people gain a deep understanding of the product's attributes, the disease state that it treats, how it is used in practice, how it compares to competitive products and/or the standard of care, and its core value proposition. In cases where the sales representatives will be asked to educate, train, and/or coach providers on the use of the device, even more in-depth training is required to prepare the sales force for the field. Representatives should be prepared to answer detailed and often technical questions related to clinical benefit, economic benefit, and billing and reimbursement. At sales meetings, representatives will often be asked to "role play" so they gain experience in addressing the many challenging questions that can arise in the sales process. At times, sales representatives are also given the opportunity to shadow doctors while learning about a new procedure or field of medicine.

It is important to bear in mind that what a company and its sales representatives can (and cannot) communicate to a buyer is regulated, to a large extent, by the FDA in the United States. For instance, when a device is cleared for use under a 510(k), the company is restricted from making claims above and beyond those of the substantially equivalent device. A company can claim new and/or different indications for devices approved under a PMA; however, these claims may not extend beyond the intended use for which the device has been approved (see 5.4 Regulatory Strategy).

Furthermore, sales representatives must be trained to adhere to the principles of medical ethics – in particular, the principle to "first, do no harm." As the representatives are closest to the patients on whom a device is used and to the physicians or users of the device, representatives have an obligation to raise any and all concerns to company executives and demand the timely resolution of any concerns. Recent stories surrounding the recall of major medical devices, such as implantable cardioverter defibrillators (ICDs) (see the Guidant example in 5.2 Research and Developement Strategy), underscore how important it is for a company to mount a timely response to any issues of safety and efficacy that arise. If sales representatives have been trained to identify and proactively communicate potential issues, a company may even be able to detect and correct problems early, before they pose a threat to patients, lead to recalls, or affect product and company reputation.

In addition, sales representatives should be aware of potential conflicts of interest that may arise as a result of their interactions with clinicians and proactively manage them. The AdvaMed code of ethics described in 5.7 Marketing and Stakeholder Strategy provides a minimum set of guidelines. As Steve MacMillan, president and CEO of Stryker Corporation, put it, "Our products are totally dependent on the surgeon that puts them in."[21] For this reason, close communications and relationships with these providers are critical and can reveal problems that ultimately lead to improved products. However, potential conflicts of interest that could arise may undermine the scientific credibility of any new product. Therefore, it is not only good ethics, but also good business practice, to manage conflicts of interest.

Interaction with marketing

Marketing and sales activities must be closely coordinated to maximize the company's efforts to capture and then sustain value. Well-defined marketing activities support the sales effort by providing sales representatives with all of the information necessary to effectively execute the sales process and also to help retain customers once the sale is made. Because of the highly scientific nature of complex medical device sales, promotion/advocacy is one of the most important elements of the marketing mix in terms of supporting sales. Medical education is especially important (based on data that has been published in peer-reviewed journals), particularly for devices that seek to change the practice of medicine. The Accuray story demonstrates how marketing and sales activities should work hand in hand.

FROM THE FIELD **ACCURAY, INCORPORATED**

A multifaceted approach to complex capital equipment sales

As described in 5.7 Marketing and Stakeholder Strategy, Accuray was founded by neurosurgeon John Adler in 1991 to develop a frameless stereotactic radiosurgical system called the CyberKnife®. According to Adler, the cost of a CyberKnife system can be upwards of $5 million, including the technology upgrades that were built into the initial deal. Each unit also required specialized facilities renovations to support the installation, which could add another $1 million to the cost of the system to a hospital. Given the level of required investment, the sales cycle for this product was typically lengthy, challenging, and bureaucratic, taking Accuray anywhere from 18 months to as much as seven or eight years to close a deal. See Figure 5.8.5.

To accomplish its sales goals, Accuray employed approximately ten full-time direct sales representatives. According to Adler, "We can't use entry-level representatives. These sales go right to the highest levels in a hospital and require sign-offs by VPs, purchasing committees, boards of directors, and all kinds of lawyers." In terms of their background, a few of Accuray's representatives had experience in radiation equipment, but most were seasoned medical device salespeople from heavyweight medtech companies such as Johnson & Johnson or GE.

Accuray's sales process was multifaceted and focused on both initial customer acquisition and customer retention. The primary customer was the hospital, and hospital administrators were the ultimate decision makers in the sales process. However, Accuray learned quickly that physicians, serving as medical champions, were crucial to almost every step in the sales process. "The medical champions are often interested surgeons who are early adopters, who want to have an advantage in the local medical marketplace, or who are just curious and like new toys and gadgets," said Adler. Initially, Adler leveraged his deep professional network to generate sales leads and acquire customers when Accuray was narrowly targeting the neurosurgery market. He explained, "I had a web of

FIGURE 5.8.5
Adler with the CyberKnife system (courtesy of John Adler).

relationships that I could use to seed different hospitals with champions, and then they would keep banging away," trying to convince the hospitals of the need for a CyberKnife system.

As Accuray grew, however, it needed to expand its community of medical champions beyond Adler's network. The company sought to persuade additional neurosurgeons of the benefits of the CyberKnife by presenting at neurosurgery conferences and other relevant professional society meetings. Moreover, the company involved key opinion leaders in managing clinical trials that compared the CyberKnife technology to conventional surgery. And, as noted previously, Adler underscored the importance of publishing the results of these studies in peer-reviewed medical journals. Once these physicians became convinced of the benefits of the CyberKnife, they would, in most cases, begin advocating the procedure (and the need for the device) to their hospitals. The more well-known or influential the physicians, the more effective they were in helping Accuray convert leads into sales based on their clout with hospital administrators.

After strong demand was fostered within the local physician community, the next challenge to Accuray's sales representatives was to convince multiple layers of hospital administrators to make the required investment. To accomplish this, the company devoted significant time and energy to creating a customized business case and financial models that demonstrated the economics of radiosurgery for each hospital considering the system. Adler explained, "The company now has well-oiled systems to show a hospital how many cancers of a specific type are in their particular referral network and what likely percentage they could capture, given current Medicare reimbursement rates, and how much money they could generate. One of the things that's been driving sales, while we've waited for our big blockbuster studies to mature, is that Medicare reimbursement has been good enough for hospitals to make a lot of money off of CyberKnife."

Adler further described how the company used revenue-sharing to drive initial sales. "One thing that we did early on to jump-start the business was to implement a shared-revenue model with the hospitals," Adler recalled. "The hospitals were required to build out a room to put the CyberKnife, but we put it in for no charge. And then we treated patients together and split the revenue. We also are linked in with various financing organizations that can provide loans at favorable rates for hospitals."

Another factor that enabled Accuray to persuade hospital administrators to purchase its product was the "halo effect" associated with the CyberKnife technology that could be used by hospitals to differentiate themselves in the marketplace. Adler explained, "To survive in the medical marketplace, hospitals need to stay ahead of the technology curve. Especially in America, they have to be perceived as really cutting edge, so that's one attribute that's appealing about our product. It's a very sexy technology. A lot of hospitals have used it as a marquee technology to market their institution as a whole. If you want some sex appeal in your hospital, this is probably, in my opinion, a better way to get it than by putting up a bunch of billboards and having a radio marketing campaign about how wonderful you are. There's more behind it. People look at it and say, 'This is pretty cool. This is Star Wars technology. It must be a good hospital.'"

As Accuray began to expand the indications for its product beyond the neurosurgery field, the company again engaged its network of medical champions to assist with physician education and expand the number of procedures performed in each hospital. Once a CyberKnife system was installed at a hospital, it could be used by multiple specialties as long as the practicing physicians had received the appropriate training. Expansion of the device to other specialties not only enabled such hospitals to better exploit the clinical value of Cyberknife, but helped them realize its potential as an investment sooner. In parallel, it could also help the company generate new sales leads as physicians networked in a viral fashion with their colleagues at other facilities (word of mouth among surgical specialists and radiation oncologists, especially within a given institution, remains a powerful marketing tool). As a result, it was in the best interest of both the hospital and the company for Accuray to manage the initial customer acquisition and simultaneously enhance customer value by expanding the clinical indications of the device. According to Adler, "I'm a big believer in guerilla marketing (see 5.7 Marketing and Stakeholder Strategy). The single best way we educated general and thoracic surgeons about radiosurgery was by getting neurosurgeons to have discussions with their colleagues at the scrub sink in the operating room. The best way to convert these guys is to have a neurosurgeon who works in the hospital and uses CyberKnife tell his colleague in thoracic surgery how the system works, how well it works, and why he should consider it."

Reflecting on Accuray's long and challenging sales history, Adler noted that the company's multifaceted approach had finally begun paying off. However, despite the sophistication of its sales strategy, Adler laughingly credited Accuray's success, especially in the company's early days, to "lots of chutzpah and just getting out there."

The link between sales and R&D

Just as the Accuray case demonstrates the linkage between sales and marketing, the Micrel example (presented earlier) illustrates the connection between sales and R&D. Companies with a direct sales model should rely on their sales representatives to play a key role in their ongoing research and business development efforts. By offering feedback from providers on the performance of a device during use, proactively gathering ideas for product improvements, and communicating this information to the development team, representatives can help fuel need-guided innovation (in terms of incremental improvements and new product directions). Companies must train their sales force to play this role for the company, setting up mechanisms and processes through which their feedback can be shared with, and acted upon, by the rest of the organization.

Determining the most appropriate sales and distribution model

As mentioned previously, companies must determine the most effective sales and distribution model for sustaining value based on their chosen business model; the complexity, value, and anticipated sales volume of the unique product offering; the customer being targeted; and the desired interaction between the customer, the company, and the device. In practice, however, there are other considerations that must be taken into account, including the size and financial resources of the company, degree of market complexity, and the extent of competition in the therapeutic area (see 6.1 Operating Plan and Financial Model for a more detailed discussion of the economics related to sales and distribution).

If a company lacks the resources and/or access to capital necessary to build a sales force, then it has no choice but to use a distributor or enter into a third-party partnership with another manufacturer, at least in the short term. When an indirect sales and distribution model is determined to be the best option, then the company should assess the characteristics of its product to determine which type of indirect approach makes the most sense. For low-complexity or undifferentiated products, distribution through large national distributors is typically the best strategy (either under a private label or the manufacturer's brand name). For a higher-complexity product with strong competitive attributes, the company can either sell through specialized distributors or via a partnership agreement with a larger, well-established manufacturer.

If the company has the resources to build a sales force, then a direct sales and distribution model becomes a possibility. A direct sales force is appropriate for complex, high-value products that can command high profit margins, but that will require significant support and training in return. When limited funding for a sales force exists, a hybrid model can be an effective strategy for moving a company from an indirect to direct sales and distribution model as shown in Figure 5.8.6.

Although many medtech companies fit the common approaches described within this chapter, there are notable exceptions. Companies are continuously finding innovative new ways to approach sales and distribution that give their products an edge. They also carefully consider dynamics within their specific medical fields that may cause an alternate approach to be appropriate. For example, St. Francis Medical, a company that developed the X-STOP® to alleviate the symptoms of lumbar spinal stenosis, took an approach to distributing its device that might have appeared unusual in any other medical specialty (see 4.4 Business Models for more information about St. Francis Medical). Traditionally, as a high-complexity, high-value implant, X-STOP would have been a good candidate for a direct sales and distribution model, especially since it would require a significant change in the established standard of care (shifting treatment of this condition from open surgery to a minimally invasive implant). However, the company instead decided to use an indirect model built around independent, specialty distributors in the spinal field. As referenced in Chapter 4.4, St. Francis anticipated physician adoption challenges based on the fact that the established surgical procedure was deeply entrenched in medical practice and well reimbursed. Spinal surgeons were also known for having strong relationships with the sales representatives of the major medical device companies. With hundreds of representatives on their sales teams and easy access to the most influential surgeons, companies like Medtronic, Johnson & Johnson, and Stryker had the potential to

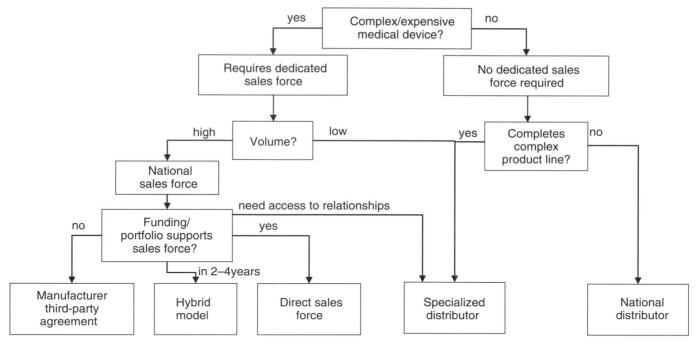

FIGURE 5.8.6
The innovator can follow this decision tree to help determine the appropriate sales and distribution model. (Note: for low-complexity devices, sales volume should be high because margins tend to be low.)

discredit St. Francis and its device by convincing them that the X-STOP was nothing more than a gimmick (despite its strong clinical data).

To proactively combat this possibility, St. Francis decided to tap into the highly experienced, well-connected network of independent specialty distributors that existed within the orthopedics field. These distributors, who worked as independent agents for both large and small medtech companies, were known for their strong relationships with spinal surgeons (and were potentially perceived to be more objective than the sales representative of the major device manufacturers). Kevin Sidow, St. Francis' CEO from 2004 until the time of the company's acquisition in 2007, explained the advantages of this approach: "If you built a direct sales force and had an unknown representative with no relationships come in with some gimmicky product, the surgeon wouldn't see him for six months to a year. On the other hand, if a close friend of his, whose opinion is respected, walks in and says, 'Look, I know this thing looks too simple to be true, but just go to training with me, review the data, and let's take a look,' the physician is far more likely to try the product."[22] Through these independent specialty distributors, St. Francis' product gained relatively rapid adoption, earning close to $58 million in worldwide sales in its first year following FDA approval.

While this approach was successful for St. Francis, Kyphon, another orthopedic start-up, flourished in the spinal market by building a direct sales force to market its implants to treat vertebral body compression fractures (see below). This apparent contradiction underscores the fact that there is no "right" answer when it comes to deciding on a sales and distribution model. In many cases, there can be distinct advantages related to developing a direct sales force. Often, the adoption rate of a new product is significantly faster when the company employs a direct model because the company's sales representatives typically only sell one product (product line) in each class, are more invested in advocating for the product, and benefit from greater synergy between the company's marketing and sales efforts. Building a direct sales force also can help a company establish itself as a credible player in a given market. Furthermore, information flows directly from the company to the end user (and back), positively affecting education and future innovation.

FROM THE FIELD: KYPHON, INC.

Investing in a high-touch direct sales model

When asked at what point in the biodesign innovation process she began to consider the sales and distribution model that would be most appropriate for Kyphon®, Karen Talmadge, company cofounder, director, and executive, said, "I began thinking about sales the second I heard about the product." She continued, "Once I met Mark Reiley, who conceived of the operation that became kyphoplasty, and went into the literature to understand the clinical need, I immediately had to think through every aspect of the business to see if we could build a company. That included the obvious things, like research and regulatory issues, but also how we would get this to the patient."

Talmadge, Reiley, and Arie Scholten, a custom medical device developer, cofounded Kyphon in 1994 to *correct the deformity and stabilize the spine in patients with a vertebral body compression fracture* (VCF). These fractures occur in the large blocks of bone in the front of the spine and most often are found in elderly patients whose bones have been weakened by osteoporosis or cancer. Within the US osteoporosis patient population alone, VCFs affect 750,000 people a year.[23] Treatment for this painful condition traditionally involves bed rest, over-the-counter pain medication, anti-inflammatory drugs, and back bracing.

In the mid-1980s, a surgical procedure was introduced in France, called vertebroplasty, which used X-ray guidance and spinal needles to inject specially formulated acrylic bone cement into the collapsed vertebral body.[24] The cement was intended to strengthen and stabilize the fracture, although the vertebral deformity was not repaired. This procedure, generally performed by radiologists, began to be used widely in the United States in the late 1990s.

The idea of kyphoplasty also began in the mid-1980s, when Mark Reiley, an orthopedic surgeon, thought of performing what became vertebroplasty, but rejected it due to concern about bone cement leaks. He set out to create a cavity in the bone, so the bone cement could be placed inside the bone under low pressure and fine control, with minimal invasiveness. Kyphoplasty was born when Arie Scholten proposed creating the cavity by adapting angioplasty balloon technology to function inside bone, and Mark Reiley instantly recognized that this would provide the additional benefit of restoring some or all of the lost vertebral anatomy. The surgeon would access the spine with two small cannulae and deliver orthopedic balloons to the center of the collapsed bone. The balloons were then inflated to compress the inner bone and push the outer bones apart to help restore the patient's spinal anatomy. The balloons were then removed and the filler (e.g., bone cement) was inserted to stabilize the vertebra and help alleviate the patient's symptoms.[25] Talmadge explained: "Beyond the fracture pain, spinal deformity is a profound problem in these patients. Studies link the vertebral deformity, independent of fracture pain, to loss of lung function, digestive problems, changes in gait that decrease independence and increase the risk of falls, loss of quality of life, increased future fracture risk, and increased risk of death. That is why we focus on the deformity along with the pain."[26] See Figure 5.8.7.

Talmadge, who was trained as a scientist, had more recently spent many years in the biotechnology sector working on business issues before joining the Kyphon team. One such role, in business development for a company called Scios, was instrumental in heightening her appreciation of the importance of sales, as well as aiding her understanding of the orthopedic market. As part of a Scios product development committee, Talmadge evaluated the market for a fibroblast growth factor (FGF) for aiding soft tissue repair in orthopedic uses. To support her research, she attended a meeting of the American Academy of Orthopedic Surgeons (AAOS).

"I went to some of the lectures to understand the clinical landscape, but I also spent a lot of time on the exhibit floor, going from booth to booth and asking people 'How do you sell your products?'" Talmadge recalled. "For the most part, people were unbelievably helpful. They

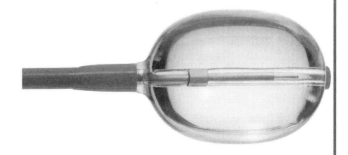

FIGURE 5.8.7
The KyphX Xpander® used to perform balloon kyphoplasty (courtesy of Medtronic Spinal & Biologics).

took me under their wing for the ten minutes we spoke and said, 'Well, here's how we've been marketing our products and here are some of the issues.' They described the pros and cons of the distributor networks and/or agents used by many orthopedic companies, noting that large companies had at least some direct sales people, requiring more expense and investment upfront, but enabling more attention and focus downstream."

Through her science and her introduction to the orthopedics field, Talmadge said, "I was fortuitously very prepared to understand the Kyphon concept, as well as the business issues around it." As such, she recognized, "For the product to succeed, we had to change the practice of medicine. Primary care physicians would have to refer patients for an operation they had never had up to that point. In order to do that, we had to do everything right. And getting everything right meant that the basic science, clinical studies, and professional education had to be rigorous." For this, the company worked with a small number of academic spine surgeons and experts in osteoporosis to be sure they were developing "all of the information that the physicians needed, as well as all the information that the future sales people would need," said Talmadge, "including the story of osteoporosis and the fractures, the anatomy, the technical aspects of the procedure, and the appropriate mechanical and clinical studies to document the outcomes. Then, the academic spine surgeons became our expert faculty, along with Mark Reiley, who had conceived of the operation."

But also, it was clear to Talmadge from the very beginning, "based on everything I had learned about orthopedics, we would need to have a direct sales force." Among other benefits, direct sales representatives would have a vested interest in helping to ensure that the product was used correctly in the field, to help prevent the consequences of product misuse, which could be devastating to spine patients. "Every decision we made as we got the company off the ground was informed by the fact that we were creating a new spine procedure and that we had a responsibility to patients to be careful," Talmadge noted.

To build its direct sales force, the Kyphon team started by hiring a few, select people with a combination of sales and professional education experience. This technical team of professional education managers began supporting the cases that were being performed as part of Kyphon's post-clearance clinical trial in early 1999 (the company's inflatable bone tamp technology was cleared for commercial use by the FDA under a 510(k) in 1998, and it launched a randomized controlled clinical study shortly thereafter to collect the clinical data necessary to support product adoption).

In an effort to be certain that only trained surgeons were using the products, and to help prevent the negative outcomes associated with product misuse, Kyphon requested that one of its technical experts be present for every kyphoplasty procedure that was performed, which, according to Talmadge, is "typical for the medical device industry." Other steps were unusual, she noted. The company restricted product use to surgeons participating in post-marketing clinical studies, and they did not sell the product to hospitals for inventory. "We did not sell products directly to the hospitals for about two years," Talmadge explained. Instead, the technical expert would bring the product with him/her to the procedure "and a purchase order would be created on the spot in the operating room," she said. As the technical experts observed these procedures, they "provided a resource to the surgeons for information about the product and about how other surgeons were performing the procedures they had observed. They also took notes on technical information about the product, such as how the balloons functioned at different pressures and volumes," data that Kyphon then used to refine its instruments and improve its education materials.

"Once we had built a core of knowledge, clinical outcomes data had been collected, and the randomized, controlled clinical trial comparing kyphoplasty to non-surgical management alone was underway, we were ready to develop a direct sales force," Talmadge recalled. In late 1999, the company hired a sales leader and began recruiting representatives. However, "The role of our sales force was different than a typical sales force because this was still in the pre-inventory days. We needed people who understood that this was market development, in addition to sales. Their role would be to help the physician educate local referring physicians about new treatment options, as well as to support the physician technically."

Talmadge described Kyphon's referral-based approach to market development in a field where most VCFs were diagnosed by primary care physicians: "The sales people would ask the surgeons, 'Who refers to you?' We also got market data on physicians in each area with elderly patients. And then we went and knocked on their doors and asked, 'Would you be interested in hearing about new treatment options for vertebral body compression fractures?' So, the sales people facilitated meetings where the surgeons who were doing the procedure would come and talk to a group of physicians in a lunch or dinner

> **FROM THE FIELD**
>
> **KYPHON, INC.**
>
> *Continued*
>
> setting. And they would talk to their colleagues about what they were doing, the outcomes they were having, and the clinical outcomes from studies as they became available."
>
> Kyphon also expected its sales people to work as a team, which Talmadge explained was "also unusual." This team-based approach was driven, in large part, by the company's ongoing commitment to having a Kyphon representative present at every procedure across its 20 sales territories, if this was requested by the surgeon. "If two cases were going on in the same territory on the same day and one person couldn't cover them both, we flew someone from another territory to bring the product in for the case and observe the procedure," said Talmadge.
>
> In terms of selecting representatives, she recalled, "The sales people that we hired didn't have to come out of spine. In fact, at times we felt that coming from a conventional spine hardware company might be a disadvantage because spine hadn't had many products that really involved a change in clinical practice." According to Talmadge, many of its first representatives came out of major medical device companies, such as Johnson & Johnson and Ethicon, "where the level of sales training is very, very high, and where they had experience selling devices for new procedures, such as the various laparoscopic techniques of the prior decade." Great emphasis was placed on finding individuals with the right traits. "We went for a series of characteristics that were most important. Intelligence, passion, commitment, and the highest ethical standards," she said. At the same time, representatives had to know how to sell in a hospital setting, including the approval process for buying new devices.
>
> In 2000, the early results of clinical outcomes studies documented the safety and effectiveness of kyphoplasty, while the randomized clinical trial became impossible to enroll. Patients did not want to participate in a study where they might not receive the operation (i.e., in the control group). At the same time, surgeons were frustrated with the policy that product use had to be restricted to clinical study sites. Recognizing this, the company decided to move the randomized clinical trial to Europe and begin a full market launch in the United States based on the rest of the clinical data. To support this shift, the company expanded the sales force to 20 representatives. What Kyphon found was that its sales grew proportionately with the number of sales people it added. As a result, it continued to increase the size of its sales force in advance of sales, along with its continuing investment in clinical studies. As surgeons began to gain more experience with kyphoplasty, the company became more comfortable with procedures being performed without a Kyphon representative present, increasing the leverage of its sales representatives in the process. However, even after this change, physicians continued to demand a high level of professional support from the company. "They love it," said Talmadge. "In fact, for us, it's a little frustrating. You would think after their hundredth kyphoplasty, they wouldn't really want us to be present, but they do. They really want it. Physicians love the technical expertise that the sales person brings. Because the physicians may have done a hundred kyphoplasties, but that sales person might have seen a thousand kyphoplasties. So there's a lot of knowledge in the room."
>
> While Kyphon's high-touch, direct sales model was instrumental to the company's success, Talmadge acknowledged that it was costly and resource intensive. "You have to decide early on that your market is large enough," she said, "because obviously this takes a significant investment." For Kyphon's shareholders, the investment yielded a tremendous return. In 2007, Medtronic acquired the company in a transaction with a total value of $4.2 billion.[27] By the end of that calendar year, approximately 310,000 patients had received kyphoplasty treatment.[28]

As illustrated in the Kyphon story, the time, effort, and expense of building and maintaining a direct sales force can be daunting to many medtech start-ups. While indirect sales and distribution models can be adopted with low upfront set-up costs and margin-sharing and/or commission compensation plans, the direct approach requires a large upfront investment and a significant payback period (see 6.1 Operating Plan and Financial Model for more information about the costs associated with sales and distribution and developing a sales model for a direct sales force). Again, each company must carefully consider what

5.8 Sales and Distribution Strategy

model makes the most sense based on its individual circumstances.

Global sales

Unlike the US market, where purchasing decisions tend to be decentralized and driven by physician treatment choices and/or hospital purchasing decisions, global markets purchasing tends to be more centralized, especially for more expensive devices. Often, it is driven by government procurement practices. Therefore, establishing relationships with local distributors can be a factor for success for any company seeking to penetrate a market outside the United States.

GETTING STARTED

Although the "last mile" is characterized by complex decisions, it is an exciting step in the biodesign innovation process because it signals how close the innovator is getting to the launch and commercialization of a product. The following steps provide an overview of the issues and questions that should be considered when choosing a sales and distribution model.

See **ebiodesign.org** for active web links to the resources listed below, additional references, content updates, video FAQs, and other relevant information.

1. Evaluate the impact of the business model on options for reaching the customer

1.1 **What to cover** – Assess the characteristics of the company's business model and product against the following dimensions (high/low): relative complexity, differentiation from competitors, price/value, and expected sales volume. Determine customers and end users.

1.2 **Where to look** – This assessment is subjective and can only be made relative to other products in the field. Therefore, research should be grounded in the evaluation of competitors and proxy products (see 2.4 Market analysis for information about assessing the competitive landscape and 6.1 Operating Plan and Financial Model for an approach to proxy analysis). Refer to 4.4 Business Models for key sales and distribution considerations related to each business model type. In addition, seek the counsel of other innovators, experienced sales representatives and distributors, and other advisors.

2. Assess the impact of intermediaries on sales and distribution

2.1 **What to cover** – Become familiar with the various types of GPOs in the United States, including large national GPOs and smaller specialized physician purchasing groups. Read through websites and review online bidding calendars to get a sense of products purchased by that organization. Consider contacting select GPOs for a phone interview to assess the contracting environment for a specific device. Learn more about IDNs and their associated purchasing teams. Read available literature on purchasing trends within the medical device field.

2.2 **Where to look**
- **Books and articles** – Refer to texts such as the following for a background on intermediaries and medical device purchasing:
  - L. R. Burns, *et al.*, *The Health Care Value Chain: Producers, Purchasers and Providers* (San Francisco: Jossey-Bass, 2002).
  - Richard Cohen, "The Narrowing Distribution Funnel: How to Get Your Medical Device to Market," *Medical Device Link*, 1999 (April 1, 2008).
- **GPO websites**
  - Novation
  - Premier
  - HealthTrust
  - Amerinet/Intermountain Health Care
  - Medassets
  - Consorta
- **IDN and hospital systems websites**
  - GroupHealth

- Kaiser Permanente
- Tenet

3. Choose a sales and distribution model

3.1 What to cover – Define how the sales process would ideally work, taking into account the unique characteristics of the medical field, business model, customer, and product, as well as the interaction the company hopes to create with the buyer. Take into account the effect of intermediaries and any special considerations for the product that mandate a particular type of distribution. Consider how devices will be shipped, stored, and serviced. Select the sales and distribution model that provides the best fit. Vet the model against the company's current financial position and projections (see 6.1 Operating plan and financial model).

3.2 Where to look – Refer back to figures within the chapter for high-level decision tools to help make this decision. To better understand the sales process with a direct model, try setting up a field visit with sales representatives in the selected therapeutic area, or attend trade shows/conferences and visit the exhibit halls to talk with representatives in that environment. To understand the process for an indirect model, it may be helpful to contact the following entities:

- **Large national medical distributors**
  - Baxter Healthcare
  - Henry Schein
  - Cardinal Health
  - McKesson
- **Independent specialized distributors**
  - Independent Medical Distributors Association
  - Health Industry Distributors Association
- **Analyst reports and press releases** – Review analyst reports to understand the portfolios of potential third-party partners and watch for recent press releases regarding activity in this space.

4. Coordinate marketing, training, and support activities

4.1 What to cover – Determine which resources and activities will be required to support the chosen sales and distribution model. Key issues to consider include sales training, customer service, and the linkage between sales and marketing activities. If an indirect model has been chosen, determine how to balance responsibilities and costs between the company and the distributor, taking into account internal resources as well as need for long-term control over the direction of product. Use the output of this assessment to inform the selection of one or more distributors/partners with complementary strengths and capabilities. Then create an internal marketing, sales training, and support plan designed to work in concert with the chosen distributor/partner(s). Develop a plan to ultimately recruit a vice-president of marketing and sales (this position is relevant in both the direct and indirect model).

4.2 Where to look – Refer to 5.7 Marketing and Stakeholder Strategy and 6.1 Operating Plan and Financial Model. Examine similar industry partnership examples for best practices on how to divide sales operations, marketing, and customer service functions. Work with consultants and colleagues to validate brand support requirements and the feasibility of outsourcing each component of these functions to distributors/partners. Look to partners, as well as consultants, to coordinate a cohesive plan.

Credits

The authors would like to acknowledge Sarah Garner and Jared Goor for their help in developing this chapter. Many thanks also go John Adler, Karen Talmadge, and Alexandre Tsoukalis for their assistance with the case examples. Nick Gaigh provided helpful insights into hospital purchasing decisions and GPOs.

Appendix 5.8.1

Summary of indirect sales and distribution models

	Sale process	Distribution process	Advantages	Disadvantages
National distributors (e.g., Owens Minor, Cardinal)	Distributors employ a sales force that personally calls on purchasing teams to sell products from multiple companies. Purchasers place orders through call centers or automated systems managed by the distributors.	Products are directly delivered by the distributors. Distributors often manage inventory on behalf of companies they represent, delivering product "just in time" from a network of warehouses.	• Provides companies with a relatively low-cost mechanism for selling and distributing a product. • May help providers manage inventory and provide other value-added services (e.g., data tracking) to manufacturers.	• Distributor sales representatives lack detailed knowledge about products and their relative merits. • Companies have limited control over how representatives present their products to the buyers. • Representatives do not have as strong an incentive to sell any particular product.
Specialized distributors (e.g., InVivo Surgical)	With a specialized distribution model, a company gains access to elite sales representatives with strong specialty knowledge and physician relationships. These reps manage all aspects of an intensive sales process. Distributors may take ownership of products (acting as resellers) or sell on commission (as agents).	Products are ordered by the buyer and hand-delivered by the sales representatives of the specialty distributor or sent via courier.	• Provides companies with access to high-quality sales representatives with strong, established physician relationships. • Reps have more specialized knowledge in the field. • Less financial risk than building an internal sales force. • Product may be packaged by specialty distributors as part of a complete line of products, making it easier for buyers to integrate something new into their practice.	• Company has slightly more (but still limited) control over how reps present their products to the buyer. • Because distributors "own" the customer, companies lose negotiating power over commissions and discounts. • If a change in distributor is necessary, company loses all relationships since they are owned by the third party. • Companies also have limited access to direct customer feedback and upward integration to drive future innovation (although useful feedback can sometimes be collected from the representatives).

Stage 5 Development Strategy and Planning

Continued

	Sale process	Distribution process	Advantages	Disadvantages
Third-party partners (e.g., Lifescan's agreement with Medtronic)	Original manufacturers license product concepts or manufactured products to third-party companies or manufacturers for marketing and sales. These third parties generally handle all aspects of sales operations and marketing, as though the products were their own. Original manufacturers may receive funding during development and ongoing fees or royalties post-launch.	Products are ordered via customer service departments or through sales representatives. Delivery is managed in accordance with the practices of the third-party manufacturers. (hand delivery, courier, and/or consignment). Products may be transferred to third parties before shipment to customer, or may ship directly from original manufacturers.	• Provides companies with access to high-quality, well-integrated, potentially large sales force with strong provider relationships. • Third-party manufacturers typically have strong, established presence in market and substantial leverage with buyers. • Companies avoid costs and risks of building internal sales force. • Companies can benefit from funding from agreements to aid product development.	• Companies have little or no control over how third-party reps present products to buyers and how products are prioritized in their overall product portfolios. • Long-term viability of companies is put at risk if third-party proves to be ineffective at selling or distributing its product. • Companies have almost no direct access to buyers to develop relationships and/or sales experience that can be leveraged for future product sales.

Notes

1. From remarks made by John Abele as part of the "From the Innovator's Workbench" speaker series hosted by Stanford's Program in Biodesign, February 2, 2004, http://biodesign.stanford.edu/bdn/networking/pastinnovators.jsp. Reprinted with permission.
2. The terms *wholesaler* and *distributor* are often used interchangeably, although there is a difference between the two: wholesalers sell to another intermediary (such as a pharmacy or a specialty distributor) while distributors sell directly to the end buyer. For simplicity, the focus of this chapter is on distributors since wholesalers are less frequently used in the medtech field.
3. L. R. Burns, *et al.*, *The Health Care Value Chain: Producers, Purchasers and Providers* (San Francisco: Jossey-Bass, 2002), p. 74.
4. "The Role of Group Purchasing in the Health Care System and the Impact on Public Health Care Expenditures if Additional Restrictions are Imposed on GPO Contracting Processes," Muse & Associates, September 2002, https://www.higpa.org/about/about_pubs.asp (April 23, 2008).
5. Burns, op. cit., p. 48.
6. Ibid., p. 67.
7. Healthpoint Capital, "Devices, Distribution and Dollars: Orthopedic Sales Backgrounder and Distribution Model Comparison," 2005, p. 9.
8. Invivo Surgical Systems, http://www.invivosurgical.com/product_lines2.htm (April 1, 2008).
9. Healthpoint Capital, op. cit., p. 8.
10. Ibid., p. 13.
11. "Medtronic Announces Alliance with LifeScan to Bring Leading Blood Glucose Meter to Its Diabetes Patients in United States," August 21, 2007, http://wwwp.medtronic.com/Newsroom/NewsReleaseDetails.do?itemId=1187707491527&lang=en_US (April 1, 2008).
12. Healthpoint Capital, op. cit., p. 10.
13. Ibid., pp. 9–10.
14. "2006 Annual Report," Osteotech, 2007, http://www.osteotech.com/financial/Ann_Rpt.pdf (April 1, 2008).
15. All quotations are from interviews conducted by the authors, unless otherwise cited. Reprinted with permission.
16. Healthpoint Capital, op. cit., p. 5.
17. Ibid.
18. Healthpoint Capital, op. cit., p. 9.
19. Burns, op. cit., p. 43.
20. Richard Cohen, "The Narrowing Distribution Funnel: How to Get Your Medical Device to Market," *Medical Device Link*, February 1999, http://www.devicelink.com/mddi/archive/99/02/003.html (April 1, 2008).
21. From remarks made by Steve MacMillan as part of the "From the Innovator's Workbench" speaker series hosted by Stanford's Program in Biodesign, May 21, 2007, http://biodesign.stanford.edu/bdn/networking/pastinnovators.jsp. Reprinted with permission.
22. From remarks made in an interview with the authors in Fall 2007. Reprinted with permission.
23. "Vertebral Compression Fractures," NeorosurgeryToday.org, March 2007, http://www.neurosurgerytoday.org/what/patient_e/vertebral_compression_fractures07%20.asp (April 3, 2008).
24. Ibid.
25. Steve Halasey, "Growing Up, Globally," November/December 2006, *Medical Device Link*, http://www.devicelink.com/mx/archive/06/11/cover.html (April 3, 2008).
26. Ibid.
27. "Medtronic and Kyphon Complete Merger," Medtronic, November 2, 2007, http://www.medtronic.com/Newsroom/NewsReleaseDetails.do?itemId=1194016103392&lang=en_US (April 7, 2008).
28. "Kyphon Corporate Fact Sheet," Kyphon, http://www.kyphon.com/pressRoom/0206/16000459-03_PR_Fact_Sheet.pdf (April 7, 2008).

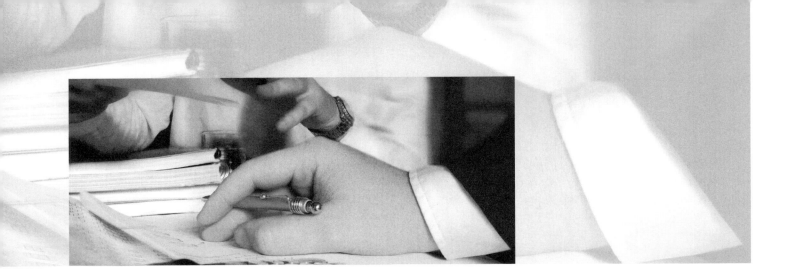

5.9 Competitive Advantage and Business Strategy

Introduction

As overwhelming as it may sound, figuring out an approach to intellectual property (IP), R&D, quality, clinical trials, regulatory approval, reimbursement, marketing, and sales is not enough. An innovator must also think about how all of these factors come together to create a compelling competitive advantage and, ultimately, a business strategy. Without an explicit point of view on how a company and its technology will be differentiated from the competition in the complex and dynamic medtech field, a product may never achieve or sustain its market potential.

Fundamentally, a company has a competitive advantage if its competitors cannot replicate its offering in some material way. The value of a competitive advantage is that it can prevent new competitors from entering a market based on the strength of the company's position and the barriers to entry that it has created through the development of key capabilities. It can also dilute efforts by established players to imitate what the company does well. Defining a competitive advantage is highly dependent on external factors within the market (e.g., IP, regulatory, customer and competitive dynamics), as well as factors internal to the firm (e.g., strengths, weaknesses, and organizational issues). Once defined, the company's business strategy should be built around the optimization of its competitive advantage.

OBJECTIVES

- Understand and know how to apply a fundamental framework for defining a competitive advantage.
- Recognize how to develop business strategies designed to capitalize on that advantage.

Competitive advantage and business strategy fundamentals

A company's competitive advantage is something special about the company that differentiates it from its competitors, creates value for its stakeholders, and prevents competitors from capturing the value it creates. By definition, a competitive advantage must be unique and/or difficult for others to replicate. No two companies will have exactly the same competitive advantage. In other words, there is no universal source of competitive advantage, even for similar companies in the same industry.[1] A value proposition is a description of how a customer will benefit from using a specific product, service or company (see 5.7 Marketing and Stakeholder Strategy). A competitive advantage, combined with an appropriate business strategy, enables a company to operationalize and deliver on its customer value proposition.

There are two primary types of competitive advantages: positional and those based on capabilities.[2]

Positional advantages

Positional advantages come from a company's ability to strategically position itself relative to its competition and its stakeholders. Positional advantages are, in general, quantifiable or measurable. A company enjoys a positional competitive advantage if it possesses an asset that is not easily obtainable, such as being first to market, obtaining key patents, or having the most widely recognized brand. For instance, companies that have direct access to the customer, through a strong direct sales forces or distributor network, achieve a positional competitive advantage by controlling key customer relationships (e.g., surgeons in the orthopedic market). A company that does not possess those relationships is at a distinct disadvantage when it tries to break into the market. Another way to achieve a positional advantage is to build a strong portfolio of products that offer comprehensive solutions that competitors cannot easily imitate due to technology limitations or other resource requirements. Broad IP portfolios can also be a source of positional competitive advantage. For example, Johnson & Johnson Interventional Systems achieved a strong positional advantage via its powerful Palmaz-Shatz patent portfolio surrounding its first coronary stent. Kyphon® achieved a similar advantage with its intravertebral balloon, as did ArthroCare with its coblation® technology.

Capabilities

Capability-based advantages are built on a company's know-how, or competencies, as well as its ability to leverage its capabilities to become better than its competitors in key areas. Capability-based advantages sometimes are harder to quantify, and they usually reflect value that is embedded within teams or organizations. For example, a medical device company that develops an exceedingly capable IP team that is able to continually evolve and improve the company's patent portfolio enjoys a competitive advantage based on this capability. Another real-world example is SciMed Lifesystems, which established a capability-based advantage by becoming an expert in rapidly evolving processes to produce catheter and stent products with improved performance characteristics. U.S. Surgical also established itself as a leader in minimally invasive surgical devices by developing a capability in advanced low-cost manufacturing techniques and rapid product iteration.

The interplay between positional and capability-based advantages

In general, positional competitive advantages are driven by a company's relationship to its external context. They are based on factors such as timing, size, location, specific assets, and access to important resources. These positional advantages are all relative to the other stakeholders in a field (customers, suppliers, or actual and potential competitors) and are constantly changing. Furthermore, positional advantages do not necessarily transfer effectively from one field to another. For example, Medtronic is the clear leader in cardiovascular solutions that utilize active implantable devices (e.g., pacemakers) and enjoys a strong positional advantage in terms of brand name and relationships in this field. However, the company has had more of a difficult challenge in trying to create a durable positional advantage in the field of catheter-based cardiovascular therapies.

In contrast, capability-based advantages are driven by factors internal to a company. They are based on a company's ability to do something better or less expensively than its competition and/or customers. These capabilities can be specific to a single process or generalizable across multiple processes. Capability-based advantages also may be transferable from one field to another. For instance, Pfizer's capability of effectively marketing and distributing drugs – frequently enabling the company to achieve the highest sales in a category within weeks or months of obtaining product approvals or acquiring those products – demonstrates how a capability can transfer from one medical field to another.

Sometimes the distinction between a positional advantage and capability-based advantage can be somewhat unclear. For instance, in the case of Pfizer, the company achieves a competitive advantage not only through its marketing capabilities, but also through the position it has achieved in terms of its relationships with physicians. Its broad product portfolio also makes it more appealing for physicians to spend time with Pfizer's sales representatives because they are able to address more products. Similarly, Medtronic has a positional advantage in implantable cardiovascular solutions, but has also developed a unique capability for continuously developing complex closed-loop systems that others cannot easily imitate. Innovators must recognize that multiple strengths often work in combination to give a company its advantage.

The link between competitive advantage and value propositions

Any number of strategies and competitive advantages can be combined to support a company's value proposition(s). For instance, if the value proposition of a company is to continually provide its customer with a constant flow of cutting-edge innovations, it might attempt not only to be first to market with any innovation (a **first mover strategy**), but also support that positional advantage with a strong, constantly-evolving IP development capability. Another company seeking to provide its customers with the value proposition of an improved user experience might alternatively choose to follow quickly behind the market leader, allowing them to learn from the mistakes and usability issues encountered by the first mover.

Value propositions must always be measured through the perspective of the customer. That is, they represent how a product, service, or company is viewed from the outside. The positional and capability advantages that are required to create and sustain a value proposition are what drive the creation of a competitive strategy for a business. For each value proposition a company defines for its stakeholders, it should carefully consider what strategies and potential advantages will enable it to successfully deliver on that vision and prevent imitation from its competitors.

Types of positional advantages in medtech

In the medical device field, many innovators and young companies initially seek to develop core assets that give them a positional advantage. Another way to think about a positional advantage is as an initial barrier to entry. There are seven main sources of positional advantage.

1. Intellectual property portfolio

Establishing an initial patent portfolio (e.g., through a picket-fence strategy – see 5.1 Intellectual Property Strategy) that is legally defensible can limit other companies' freedom to operate in a field and define a broad positional advantage for the company. Johnson & Johnson (J&J) and Medtronic are well known for establishing broad, highly-effective patent portfolios.

2. Key relationships

The strength of relationships with important customers and other stakeholders, particularly key physician opinion leaders, can help a company to more easily capture customers and/or more quickly gain regulatory approval. For example, direct sales representatives in orthopedics and pacemakers wield substantial power on behalf of their companies due to the strong relationships they hold with influential customers.

3. Strategic alliances

Partnerships with larger corporations and institutions can allow a new or smaller company to "borrow" reputation and keep other players from accessing the same limited resource (in the case of an exclusive alliance). Angiotech's relationship with Boston Scientific relating to drug eluting stent technologies provides an excellent example of an effective medtech partnership.

4. Geographic coverage/distribution channels

The ability to gain access to preferred suppliers for distributing products and/or providers for disseminating services to customers in key markets can lead to lower transaction costs and better terms than competitors can achieve. It can also be exceptionally difficult to replicate, as was the case for Stryker when it established strong, early distributor relationships outside the United States, most of which continue to hold to this day.

5. Brand

Premium brands often achieve premium pricing. Branding can also increase product awareness, which in turn can lead to higher sales. Companies that establish strong brand awareness within their customer base are at a significant positional advantage with respect to incoming competition. For instance, surgeons are often reluctant to switch brands, especially when the products are life-saving and highly successful, such as the heart valves sold by Edwards Lifesciences or the hip and knee implants sold by Zimmer.

6. Financial access

Capital is a finite resource. Locking in top-tier venture capitalists, banks, and other financiers can keep competitors from accessing the same capital resources, force them to deal with suboptimal terms, and/or leave them with less optimal alternatives. Companies such as Satiety, GI Dynamics, and Endogastric Solutions, which were early pioneers in the area of obesity, captured many of the desired medical device early-stage venture backers, making it much more difficult for later entrants to access capital.

7. Unique, exclusive, or low-cost access to a key resource

Key resources can include hard-to-find materials such as nitinol, proprietary gene sequencing databases, or limited tissue samples. Key resources can also include exclusive relationships with technology providers in specific fields, such as slippery coatings for catheters, polymer drug-elution platforms or silicone–polyethylene copolymer blends. A company that can gain access to key resources exclusively, more easily, or at a lower cost has a positional advantage over its competition.

Almost every successful medical technology company has, at least at some point in time, developed one or more positional advantages to fend off competition. For example, Fox Hollow, a company focused on the treatment of peripheral arterial disease (which was acquired by ev3), created an almost impenetrable barrier to competition by effectively developing strategic relationships with most, if not all, of the key thought leaders in the field. The use of key physician relationships has similarly been a fundamental basis of creating an advantage in the orthopedics industry (although, this practice recently has begun to fall out of favor as these relationships have become scrutinized for their potential to create conflicts of interest). Strategic corporate relationships offer another common approach towards achieving a positional advantage, such as J&J and Guidant's accord related to balloon catheters and Rapid Exchange (RX)™ technology.

Importantly, positional competitive advantages can be relatively perishable. For example, an army at the top of a hill enjoys a distinct positional advantage as it attacks an opposing force in the valley below. However, as soon as the army charges down the hill, its advantage deteriorates and it is suddenly on a level playing field with the opposition. In the medical device field, companies that enter the market with an advantageous position must continually seek to improve and/or defend their position to sustain the advantage. For instance, a company that successfully establishes meaningful relationships with key opinion leaders must continuously monitor and invest in those relationships or else a competitor may edge in and gradually appropriate them. Similarly, competitors will find ways to design around a company's initial IP position if the market is sufficiently compelling. And, with enough time and money, even new sales and distribution channels can be built to rival an established infrastructure.

Types of capability-based advantages in medtech

Innovators and young companies can also create a competitive advantage by developing a product or delivering a service that is better, cheaper, or more efficient than what is offered by the competition. While doing this creates a competitive advantage in its own right, capability-based advantages are often used to help sustain positional advantages. That is, a positional competitive advantage can be maintained by transforming it into a capability-based advantage and/or can be defended through the development of a strong capability in that area. In most cases, positional advantages can become capabilities if a company can learn to consistently perform better in the given area than the competition (just as capabilities can help create a positional advantage). The creation of specific systems and processes focused in one of five common areas can be the first step in developing a unique capability.

1. Intellectual property management

As noted, if a company seeks to recruit, develop, and retain a world-class IP team, it can transform a strong IP position into a best-in-class IP capability. This is accomplished through systems and processes designed to continually and aggressively monitor the IP landscape, anticipate competitive moves, and act more quickly than the competition to advance, strengthen, and defend the company's IP portfolio.

2. Relationship management

To create a capability-based advantage focused on key relationships, it is not enough to manage and protect the company's position with established thought leaders. Systems and process must be created for (and resources must be dedicated to) identifying subsequent generations of potential opinion leaders and developing relationships with them before the competition.

3. Alliance management

A company can develop a unique capability when it comes to partnering if it has the right processes in place and resources on-board to think about alliances in innovative new ways. Likewise, it can continually monitor the external landscape for mutually beneficial partnerships, and negotiate more exclusive (or otherwise more favorable) terms than the competition.

4. Human resource management

Human resources (HR) management refers to the creation of an optimal team of individuals with the essential competencies necessary to succeed in the market (e.g., technical, clinical, reimbursement). This is a capability-based advantage when it is achieved on an ongoing basis, despite the natural level of turnover inherent in any industry.

5. General management

Simply having good leadership and well-trained managers in key positions within a company could be a strong capability-based advantage. Installing management training and feedback programs can enhance and sustain these advantages.

Capability-based advantages can also be created in other areas where a company has unique process expertise, whether or not it is related to a positional advantage. For example, medtech companies may achieve capability-based advantages related to the following factors:

- **Time to market** – Having the people, processes, and systems in place to consistently get to market sooner than the competition.
- **R&D productivity** – Systematically generating more or better innovations than others in the field.
- **Low-cost manufacturing** – Achieving low-cost manufacturing through the acquisition of low-cost inputs has a positional advantage while achieving it through more efficient processes has a capability-based advantage.

Innovators must remember that even the most impressive capabilities are not a source of competitive advantage if one or more competing companies can match them. Capabilities can only become a source of sustainable competitive advantage if they are hard to imitate and/or the company can continuously improve upon them before others can catch up. When a competitive advantage resists competition, it is said to be sustainable.[3]

Often, capability-based competitive advantages are widely understood in an industry, but still difficult to replicate because no one is certain what causes them. Complex routines, structures, and individual attributes within an organization, combined with a high level of tacit knowledge, can make them difficult for competitors to imitate.[4] For example, Ethicon was widely considered to be the worldwide leader in surgical closure technology and, in particular, suture technology. However, in the 1970s, U.S. Surgical began making inroads unto Ethicon's markets with the introduction of its surgical staple technology as a substitute for traditional sutures in minimally invasive laparoscopic procedures. Buoyed by its success, U.S. Surgical decided to enter the suture market and directly compete with Ethicon in its core market by introducing its own line of sutures in 1991. However, U.S. Surgical dramatically underestimated the strength of Ethicon's suture manufacturing capabilities. The company did not realize how challenging it was to produce a high-quality suture that did not break, stayed attached to the suture needle, and performed consistently, package after package. In the end, the product released by the company failed to meet physician quality expectations and, as a result, never claimed the market share that the company expected. Ethicon's proprietary know-how (i.e., trade secrets) in suture manufacturing proved to be a formidable competitive advantage that U.S. Surgical could not overcome, even with its vast resources.

Fundamental business strategies

After a company has identified its potential sources of competitive advantage, it can effectively evaluate the best fundamental business strategies to pursue in order to optimize them. Importantly, no single strategy can be expected to work for a company indefinitely. Instead, an innovator should think about choosing the series of strategies that will enable the company to achieve long-term success.

In medtech, it is important to keep in mind that strategies, such as the ones listed below, are not mutually exclusive. Different strategies can be combined to capitalize on those advantages that best differentiate the company. While some of these strategies lend themselves more readily to positional rather than capability-based advantages, they can still be used in combination and/or in sequence to enable both types of competitive differentiation.

First mover

A first mover strategy refers to being the first company to offer a product or service in a market. Being first to market has significant advantages, including the ability to establish key IP that can serve as a barrier to entry, form relationships with important early customers and other resources, and define the most important product attributes and regulatory strategy that future competitors will have to consider. In some cases, a first mover can also create high switching costs for customers in order to keep future entrants from selling to the same customer base. This often happens with capital equipment, such as magnetic resonance imaging (MRI) and computed tomography (CT) machines, that require a large upfront investment by the hospital that purchases them.

However, first movers incur certain costs for the positional advantages this strategy typically affords them. These costs include the need to define the regulatory and reimbursement pathways, educate customers to drive adoption of a new product category, and train users on new techniques. These pathways can then be used by other companies to more quickly and less expensively create follow-on products. Typically, reimbursement and regulatory approval pathways paved by first movers are exploited by those that follow. But sometimes the first mover creates barriers by developing those pathways – they set the hurdles higher for those that wish to enter. For example, early developers of certain cardiac implants set a ten-year cycle testing standard that has required all that followed to abide by those criteria.

Importantly, a first mover advantage is only sustainable if the company in the lead continues to find creative ways of exploiting its first mover position. Otherwise, it will fall quickly into the ranks of the competition when other companies arrive in the market, forfeiting any benefits associated with being the first mover.

Another example of a successful first mover is Kyphon. Kyphon focused on treating vertebral compression fractures, a previously overlooked and undertreated market, by creating a new, minimally invasive procedure called kyphoplasty. By setting up a dominant IP portfolio, the company deterred many competitors from entering the market and remained the leader until it was acquired by the competitor that posed its greatest threat. Guidant, on the other hand, was unable to sustain its first mover advantage in the field of interventional cardiology after several years of being the market leader. Instead, it fell into more of a follower position for several years due to the company's early reluctance to explore new products, such as the area of coronary stents.

Fast follower

Instead of blazing the path to the market, **fast followers** often leverage other advantages to quickly capture market share from the first mover. Large companies in the fast follower position can often leverage their established distribution channels and customer relationships to this end. Both large and small companies may be able to introduce new features that create subtle differences between their technology and the first mover's product. This can allow the company to overcome barriers to entry created by the first mover's IP position (in that different IP can be pursued). New features can also be used to address known shortcomings of the first mover's technology, further eroding the strength of its position. Fast followers also benefit from the reimbursement, regulatory, and awareness/education campaigns that have already been conducted by the first mover.

Boston Scientific's drug eluting stent (DES) provides a good example of a successful fast follower strategy. Johnson & Johnson created a dominant first mover position in the DES market with its Cypher® stent, holding a monopoly position in the United States for nearly a year. During this time, however, J&J ignited tensions with many doctors by pricing its technology at levels that many considered to be unreasonable. The company also experienced significant supply problems, which further angered physicians when they could not obtain an adequate supply of the product to satisfy the patients on their waiting lists. When Boston Scientific launched its Taxus® stent, physicians already understood the benefits of a drug eluting device and were receptive to the new technology. The company intentionally took advantage of the relationships that had been damaged by J&J's supply and pricing issues by ensuring that it had plenty of supply on hand, and was working with hospitals to develop mutually beneficial pricing strategies. Furthermore, Boston Scientific promoted the product features of its DES that made it more flexible and easier to work with in direct response to criticisms of the Cypher stent, which was stiffer and more difficult to place. As a result, Boston Scientific captured nearly 70 percent of the DES market from J&J in its first seven weeks on the market.[5]

Me-too

Unlike the products of fast followers, **me-too products** are relatively undifferentiated from the products that are already on the market. Typically, me-too products seek to benefit from the technical development and market development of earlier players by creating a product that is cheaper to develop or manufacture and, thus, may be offered at a lower cost or with some very modest improvements. By watching earlier competitors, purveyors of me-too products often manage to avoid costly mistakes and sometimes can go through less arduous regulatory steps. Since me-too companies have lower development costs to recoup than the first movers and fast followers, they often compete based on price.

A prime example of a me-too product is found in SciMed's initial entry into the angioplasty catheter market. The company's first product was not tremendously differentiated from other competitors in the marketplace at the time. Interestingly, as noted previously in this chapter, SciMed ultimately developed a powerful iteration and rapid development capability that moved it away from this me-too strategy into a first-mover strategy.

Another example of a me-too product strategy is seen with generic versions of drugs. Generic drugs can save significant time in getting to market and can incur lower development expenses by using separate regulatory pathways from those used in new drug approvals. Once a generic enters the market, the original product must leverage its brand to retain some advantage. However, a positional brand advantage can be eroded quickly since the less expensive alternatives are favored by price-sensitive payers. This me-too strategy applies not only to generic drugs, but to any follower joining the market on the heels of another company's pioneering product, particularly once that product category has been well established. While fast followers often enter with differentiated products that they believe to be patentably distinct, generic followers enter only once key patents have begun to expire. In both cases, the followers have the distinct advantage of having watched the pioneer in the market make mistakes and adjust its strategies. If they can use these learnings to help them navigate whatever competitive barriers have been erected by the predecessor, they have the opportunity to gain market share much more rapidly.

Niche play

Rather than attempting to own a product category, **niche strategies** focus on owning the customer relationships in a specific, focused area of medicine. Companies pursue a niche strategy when a group of physicians, typically in a smaller subspecialty, remains underserved by the market. To meet the needs of a niche market, a company may tailor a selection of its products to the unique needs of the physician/customer group. The overarching goal is to gain a positional advantage through strong customer relationships as a method for blocking competitors. Another niche strategy is to target a particular geography (or demographic) that may not be compelling to the dominant, established player in the field, and then use this as a springboard into other areas as the company builds momentum.

American Medical Systems (AMS) executed a niche play to win the urologist market. While other companies called on urologists with the same sales representative that called on general surgeons, AMS positioned itself narrowly as a urological disorders business with products to treat conditions such as erectile dysfunction, incontinence, and prostate disease. By focusing exclusively on serving urologists and the diseases that urologists treat, the company sought to position itself "not only to benefit from growth in the urology market, but to drive it,"[6] as compared to some of its more broad-based competitors that participated in multiple market segments. This strategy was extremely effective, allowing AMS to build strong relationships with customers that others had trouble accessing.

Distribution play

Distribution play focuses on product breadth and channel relationships rather than on product superiority. By meeting the needs of customers through diverse product offerings, more flexible payment arrangements, or strategies to bundle products and/or services, a business may keep more focused entrants out of the market.

Stryker, a leading manufacturer of orthopedics products, offers a broad product portfolio and thus can be a single solution for all of a hospital's orthopedic products needs. Theoretically, it would be possible for Stryker to bundle the purchasing of hip and knee implants together and, thus, keep a new knee implant company out of an account. While the bundling of products is regulated by antitrust regulations like the Sherman Act, a company that offers a one-stop solution may win in the market – especially in commodity markets. Baxter and J&J are good examples of companies that use a distribution play strategy to maintain an advantage in hospital supplies (e.g., intravenous bags, gloves, bandages) and other lower-cost/commodity markets.

Original equipment manufacturer/licensing

Using an original equipment manufacturer (**OEM**) strategy, a company provides technology and/or components to another company that then assembles and sells the finished product. This strategy is commonplace in the computer industry where components, such as hard drives, processors, and video cards for a single computer, may all come from different manufacturers. Early-stage medical device start-ups looking to reduce their upfront capital commitments may contract with an OEM (e.g., for the manufacturing of machined or molded components). Or, they may become OEMs themselves by selling their physical products or licensing their technologies for use by another company. Some examples of OEM relationships include Gyrus' relationship with Ethicon (involving endosurgery equipment) or Surmodics' relationship with J&J (providing the polymer coating for the Cypher stent). A list of OEM providers is easily obtained from the Thomas Register as well as the Medical Device and Diagnostics Industry (MDDI) website.

The contracts involved in becoming an OEM and/or entering into a licensing agreement can range from simple to complex. One of the most strategically important terms is whether the contract is exclusive or non-exclusive. Exclusive agreements can limit the company with the technology from sharing it with any other firms (thereby giving the licensor a strong positional advantage). In turn, the licensor may or may not agree to use only this one technology and not seek alternative suppliers. In non-exclusive arrangements, both parties are free to make deals with other companies.

For example, Wilson Greatbatch is a large company whose primary strategy is to be the OEM of certain battery and electrical component technologies. Specifically, it brands itself as a leader in the development, design, and manufacture of components critical to implantable medical devices.[7] A typical arrangement for Wilson Greatbatch is to supply batteries and electrical components (technologies that are complicated and protected by strong IP positions) for the pacemakers produced by all the large implantable pacemaker manufacturers. As shown by this example, if a company owns a component or service that is considered rare but potentially can be widely used, an OEM strategy is a good way to capitalize on this competitive advantage. Other examples of products/services well suited to OEM strategies include high-volume, five-axis machining used to create orthopedic implants, which is almost completely controlled by the major orthopedic manufacturers, or guidewire technology, which is dominated by Lake Region Corporation.

Partnering

Sometimes two companies become partners so that each one can leverage the competitive advantage of the other (more information is included in 6.4 Licensing and Alternate Pathways). Typically, two companies form an exclusive, contractual agreement in order to shepherd a product into the marketplace. The most common **partnering strategies** occur in the pharma-biotech world and the pharma-drug delivery world. Usually the smaller entity makes a new technology available and the large pharmaceutical entity brings capital, expertise, and distribution to the partnership. Partnerships are often structured so that the large entity covers portions of the development costs and makes royalty payments. In return, the larger entity gains marketing and distribution rights for the product. An example of this arrangement was the partnership between Inhale Therapeutics, a smaller biotech company, and Pfizer, a large pharmaceutical company, to develop and bring inhalable insulin to the market (Exubera).

In the medical device field, partnering between small and large companies is often centered on a distribution, investment, or technology relationship. While obtaining a partner may result in fewer strategic options for a company, it also can reduce strategic options for competitors. Consequently, partnering can sometimes be a prudent step, especially when the useful strategic options generated from these relationships are considered relatively scarce. For example, consolidation has resulted in few acquirers of businesses or technology in some markets; thus, a key partnership could solidify one of those rare relationships and limit **exit** options for the competition. Such a scenario can be found in the interventional cardiology field where there are now just four primary players – Boston Scientific, J&J, Abbott Vascular, and Medtronic. If a small start-up is competing against a vast number of companies with similar products or services, a partnering relationship with one of these major players could provide it with a significant advantage against its peers, especially if only three others would potentially be capable of solidifying a similar relationship with the remaining major players. If one of the four major players has a more dominant position than the others, then a relationship with that leader could provide an even more powerful advantage.

Using competitive advantage to define the basis for competition

Beyond choosing the appropriate business strategy(ies) to capitalize on its competitive advantages, an innovator or company should seek to create a basis for competition in its field that plays to its own strengths and attacks the weaknesses of its competitors. The basis of competition refers to the features, benefits, or qualities that become central to the way businesses compete with each other. Price, quality, brand perception, customer support, training, or certain technological features could all serve as bases for competition. If a company is a first mover or early entrant to a relatively new field, it is usually able to exercise more control in this regard and essentially define the "playing field." One example is the way J&J, upon being the first entrant into the stent market, used "stent strut strength" as a basis of competition in the early days of coronary stenting. This forced all subsequent competitors to test their devices to a standard that was difficult to beat while addressing other customer needs. While this basis of competition was later eroded by competitors, it served as a significant barrier in the early days of stent development.

If a company is a later entrant to a well-established field, then it must try to change the basis of competition by developing capabilities and positional advantages in areas where leaders do not have them. For instance, if a small company intends to compete in a medical device market against Medtronic or J&J, it would be impractical to try to challenge these firms on the strength of their distribution positions and capabilities. However, by interacting with key stakeholders to understand these companies' areas of weakness, the smaller firm may be able to gain a foothold in the market. If physicians are frustrated by the ease of use of an established device, for example, the company might target innovation in this area. Or, if J&J or Medtronic has neglected customer service, a competitor might seek to develop deep capabilities in this area in an effort to win over customers.

In the field of active implantable devices, a completely new basis of competition has evolved to include web-based disease management systems (such as Medtronic's CareLink®) used by doctors and hosted by the companies that provide the implants to ensure improved patient care. Innerpulse, the developer of percutaneous implantable cardioverter defibrillators (PICDs) described in 2.3 Stakeholder Analysis, provides another example of how a new entrant can differentiate its offerings and redefine the competitive playing field. By choosing to focus on primary prevention of sudden cardiac death, Innerpulse has chosen to avoid head-to-head competition with the large, established ICD manufacturers that historically have targeted and successfully captured the secondary prevention market in contrast to the primary prevention market where they have been much more unsuccessful.

Importantly, choosing where to invest and what advantages to develop is essential. Otherwise, a company runs the risk of becoming a "jack of all trades, master of none," which can make it difficult to defend any competitive advantage that it may have. It is not realistic to expect that any company can develop all of the sources of competitive advantage discussed in this chapter. Often, as companies are getting started, they first identify gaps in the product offerings of their potential competitors. Then, they seek to determine which advantages to build in the near term and which ones to invest in or develop at subsequent stages of their strategic evolution in order to exploit these gaps. As with any strategy, decisions regarding competitive advantage should be reviewed frequently based on changes in the external and internal environments, to ensure they remain valid and attainable.

The ev3 story below demonstrates how one company explicitly defined and managed its competitive advantage.

FROM THE FIELD — ev3

Building a company on capabilities and business strategy

In late 2000, the formation of ev3, a company focused on delivering products for coronary, peripheral, and neurovascular applications, was somewhat unique in that Warburg Pincus, the private equity investors that backed the venture, funded a capability-based competitive advantage and a supporting business strategy rather than a traditional product. Jim Corbett, ev3's former CEO, described the advantage upon which the company was built. "We had assembled a team of executives experienced in the market who we thought could identify the right technology segments and then acquire or develop products to create a company," he said.[8]

The market ev3 would go after was the endovascular field. "The endovascular market had become very large over the preceding 10 or 15 years, and it had consolidated," Corbett explained. "That consolidation created some rather large global companies – Cordis/J&J, Boston Scientific, Medtronic, and Abbott Vascular. Those companies, in their success, had developed mega-product

FROM THE FIELD: ev3

Continued

categories that they were defending. So, if you're in CRM [cardiac rhythm management] or if you're in drug eluting coronary stents, you really cannot afford to lose your beachhead because there are billions of dollars of earnings at stake. The consequence of that was that the rate of innovation in the endovascular markets had dramatically decreased, because the big R&D centers were all focused on the defense or preservation of their market position rather than on creating new concepts, or new devices, or new market segments." ev3, with its experienced management capability, developed an explicit business strategy to target opportunities that existed within this "innovation gap."

The company's name refers to the three primary types of opportunities it would target in the profitable endovascular market: coronary, peripheral, and neurovascular devices and interventions. To establish its preliminary product position, ev3 rapidly acquired nine companies and/or technologies that its management team believed could be commercialized in a competitive time frame for a reasonable cost. Because physicians, particularly peripheral and neuro, had been relatively underserved by the major medical device players, ev3's strategy was somewhat of a niche play. For example, according to Corbett, the market leader in peripheral stent technology had not introduced an innovation for roughly eight years. In neurovascular, the market leader had been offering the same base product for nearly ten years. "So there were real opportunities for us to find the chinks in our competitors' armor, so to speak, and create platform technology, and innovate, and bring something new to market," he said.

ev3's idea was to use a niche strategy to establish a foothold and then expand its position. For instance, to become a dominant player in the neurovascular market, which was stronger in Europe than in the United States, ev3 added a geographic element to its niche strategy, which it reinforced through the development of new positional and capability-based competitive advantages. "Two-thirds of the global market for neurovascular products was (and is) outside the United States, which is very atypical for most medical device markets," explained Corbett. "Usually the US market is dominant, or is the largest segment between the two. We made it part of our early strategy to have direct selling operations in Europe. So we put a lot of effort into global distribution, which, again, is an uncommon choice for an early-stage medical device company. In fact, a lot of companies these days choose not to do it at all, and rely on improving their product in the United States because they actually plan for consolidation. But that was a very important distribution choice on our part and it paid a lot of dividends. In 2008, over $140 million of ev3's sales are projected to be outside the United States and it is the fastest-growing segment of the business."

Over time, ev3 recognized the need to further focus its business strategy and its approach to the market. By mid-2003, it was clear that the larger, established companies continued to enjoy a competitive advantage in the coronary field, where Corbett estimated they continued to spend as much as $1.5 billion per year on R&D for DES alone. Staying true to the company's desire to innovate around its major competitors, ev3 decided to drop the coronary market as a focus area. Looking back, Corbett said it was an easy decision to focus on the two fields with the greatest opportunity for innovation. At that time, many peripheral and neurovascular devices were nothing more than repurposed or scaled-down coronary devices, which left ev3 with plenty of room for entirely new devices and intervention innovations. In addition, development of these products was potentially much faster than coronary products and so offered a lower investment risk to ev3.

Another explicit decision made by the company was to invest in developing a strong capability in internal development and innovation. One of the primary reasons for doing so was to help give the company a more sustainable competitive advantage. "Product innovation is the core basis of competition in our industry," said Corbett. By developing a deep capability-based advantage in this area, ev3 would be able to "create a footprint of a company, not just a product line," he added, also noting proudly that, "In 2005 and 2006, we introduced more new products into the market than the four biggest endovascular competitors combined. And that was all internal development." One of the new products the company is most excited about is the EverFlex® peripheral stent. Fractures in peripheral stents, which occur in as many as 25 percent of all cases, cause restenosis, surgery, and amputations. ev3's EverFlex stent has been shown to have a fracture rate four to five times lower than the leading product in the category (see Figure 5.9.1).

When thinking about developing a competitive advantage, Corbett had this advice to entrepreneurs: "Resist *not* focusing. Both at the beginning and along the way, there is such an opportunity to take on new tasks. But it will take away from your ability to execute on your core strategy. Focus is the path to success, without question." He also underscored the importance of speed. "Speed is key. Obviously speed has to be conducted with excellence, and you should never do speed if it compromises quality, or ethics, or any of those types of matters. But the time you spend getting to market burns cash, and you need that cash. There is often a tinkering mentality that emerges in young businesses, especially when they're privately held. But speed is a core capability that you should try to develop from the start."

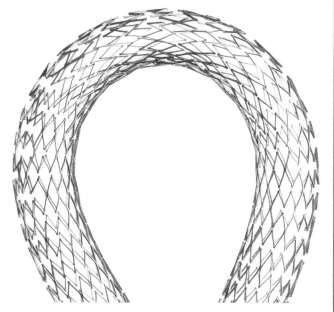

FIGURE 5.9.1
The EverFlex peripheral stent (courtesy of ev3).

Developing a statement of competitive advantage

When innovators or companies are working on defining their sources of competitive advantage and the business strategies needed to support them, they should clearly articulate them in a statement of competitive advantage. In developing such a statement, look at those created by other companies for ideas about how to effectively capture the key ideas (other recommended steps can be found in the Getting Started section below).

Unfortunately, detailed information is not always available regarding other companies' statements of competitive advantage because it is often considered proprietary. However, even in the absence of inside insights, the innovator can frequently discern the focus of a company's competitive advantage from the information it makes publicly available. For example, when a private company goes public:

- Evaluate its prospectus carefully. Usually there is a section entitled "Our strategy" or "Business strategy" that contains relevant information.
- Identify the aspects of this text that are indicative of the company's core assets and chosen fundamental business strategy.
- Use professional judgment and knowledge of the field to translate this information into a statement of competitive advantage.

Information gathered from public sources can be used to reconstruct a statement of competitive advantage, as show in the working example.

Working example: Kyphon, Inc.
Discerning a competitive advantage from public information

The following information comes from a Kyphon prospectus.[9] Kyphon, which specializes in devices that enable the minimally invasive treatment of spinal fractures caused by osteoporosis or cancer, went public in 2002. Within this public description of the company's strategy, the elements that correspond to Kyphon's core assets have been underlined, as follow.

Business strategy

Our goal is to establish treatments using our proprietary balloon technology as the standard of care in orthopedic applications. We are initially focusing our efforts on vertebral body compression fractures. The key elements of our strategy are to:

Penetrate the spinal market using a direct sales force

We believe that a *direct sales force* will allow us to most effectively educate and train physicians in the use of our products. Our products are sold directly to physicians by our experienced sales team, comprising 20 sales representatives, three managers and a vice-president of sales. By leveraging their *extensive spinal market experience*, our sales people are able to identify key physicians and provide effective case support to accelerate market adoption of our procedure. Our sales team is supported by two in-house coordinators and four field-based associates.

Educate referring physicians and patients

Patients with vertebral body compression fractures often are not referred to spine surgeons for treatment. Our objective is to *establish referrals from physicians who initially diagnose* vertebral body compression fractures to spine-focused surgeons who perform Kyphoplasty. As a result, we have implemented an *awareness marketing campaign* to educate internists, family physicians, gerontologists, and other primary care physicians about Kyphoplasty and its potential to be an effective therapy. As part of this campaign we provide educational materials to treating physicians, referring physicians and patients, and organize regional market seminars where surgeons trained in performing Kyphoplasty educate referring physicians.

Expand clinical support of the Kyphoplasty procedure

We are conducting *outcome studies* to increase awareness of the procedure within the medical community, to develop additional marketing claims and to support third party reimbursement. Through our own outcome studies and those of surgeons currently performing the Kyphoplasty procedure, we are gathering data for *peer reviewed journal articles* in support of reimbursement efforts.

Work with opinion leaders

We have obtained the advice and *support of nationally recognized spine surgeons* who are helping us to further develop our products and the procedure, to demonstrate the benefits of Kyphoplasty, and to obtain third-party reimbursement. Because these leading physicians help *set medical policy* in their respective areas of expertise and are experienced in outcome assessments, we believe they will help create patient referrals and advance third-party reimbursement. The reputations of these physicians and their leadership in professional societies help bring recognition and credibility to our products.

Expand surgeon adoption of Kyphoplasty through training

We have implemented specialized training programs and are *rapidly expanding the number of physicians trained in Kyphoplasty*. As of August 31, 2000, we had trained approximately 300 physicians in the United States and Europe and we plan to have trained more than 400 by the end of 2000. We support these physicians through professional development programs, which include funding local seminars, funding travel to national medical conferences and assisting in the preparation of scientific papers for publication.

Expand into additional orthopedic markets

We intend to leverage our *proprietary balloon technology platform* for other applications, including compression fractures of the wrist, knee and hip. These new applications involve refinements of our current products, and we intend to conduct outcome studies in these applications to support market adoption. We believe our *intellectual property position* and our *position within the orthopedic marketplace* will allow us to become the leading provider of minimally-invasive medical devices for the treatment of compression fractures.

From this overview, one can infer that Kyphon is developing a competitive advantage that depends on the core assets of:

IP – To protect its innovative technology in the area of spinal orthopedics, Kyphon defined its initial market narrowly and built a strong and extensive IP barrier to protect its desired position in the market. Through its IP strategy, the company also positioned itself for expansion into other markets.

Owning key relationships – To create additional barriers to entry for competitors, Kyphon has invested heavily in locking up the specialist market (including key opinion leaders in the field) through training and other professional development activities. The company also has gone after generalists (who diagnose the target condition and refer patients for treatment) to increase the strength of its relationships further and make it more difficult for competitors to gain a clinical customer base. To keep the clinical community engaged and convinced of its product's value, Kyphon is also performing ongoing outcome studies to be published in peer-reviewed journals.

Establishing core channels – To support the development and nurturing of its key physician relationships and further strengthen its barriers to entry, Kyphon developed a strong direct sales force. Given the specialized nature of the Kyphon product, this type of channel makes the most sense for sales and distribution. By investing in a team of highly educated sales representatives that can assist in training physicians, the company is positioning itself for easier, more rapid user adoption while also making it more difficult (and costly) for competitors to replicate its sales/distribution capability.

Based on the core assets indicated by the company and a basic knowledge of the industry, one might next infer that Kyphon is pursuing a niche play strategy. Rather than initially going after the general orthopedics market, it targeted a small subspecialty and has invested heavily in achieving a leadership position in this area. Kyphon also benefited from a first mover advantage that enabled it to leverage its core assets effectively (i.e., in locking up opinion leaders and creating an impenetrable IP barrier).

Given this positioning, the company's statement of competitive advantage might read something like this:

Kyphon will establish a leadership position in the minimally invasive treatment of spinal fractures caused by osteoporosis or cancer through a strong IP position (from which it can expand beyond this niche); deep relationships with specialists and key opinion leaders in the field developed and maintained by a highly educated, top-tier direct sales force; and widespread general awareness of the efficacy of Kyphoplasty among referring physicians.

GETTING STARTED

Identifying a competitive advantage and crafting an effective business strategy to support it is an iterative, ongoing process. The steps below highlight some of the issues and questions that should be considered when initiating this exercise.

See **ebiodesign.org** for active web links to the resources listed below, additional references, content updates, video FAQs, and other relevant information.

1. Understand the competitive advantages of competitors

1.1 What to cover – Carefully evaluate established or emerging competitors in the market to identify their strengths. Consider their weaknesses and what opportunities this creates for a new company to potentially establish a foothold in the market by addressing those gaps. Also evaluate companies that have established a leadership position in other markets for ideas and examples of how competitive advantages can effectively be developed. Review what is known about the competitive landscape to understand factors in the external environment that may be a source of competitive advantage. Then, name the competitive advantages that these companies have developed (i.e., what they do well that others cannot easily imitate), as well as the business strategies they have put into place to capitalize on them.

1.2 Where to look
- **Analyst reports** – If players in the competitive landscape are public, carefully review analyst reports for clues regarding a company's core assets, strengths, and sources of competitive advantage. These publications may also provide information about the chosen business strategies of public companies.
- **Personal networks** – Network with individuals in the field to gain insight into the competitive advantages and supporting business strategies that have been chosen by these companies, how well their strategies support their competitive advantages, and any difficulties they have encountered along the way (which might translate into opportunities for a new competitor).
- **2.4 Market Analysis** – Refer back to this chapter for information on the competitive landscape in the field in question.
- **6.1 Operating Plan and Financial Model** – Review this chapter for guidance on analyzing proxies.
- **6.3 Funding Sources** – See the types of research recommended in this chapter for other ideas for investigating companies of interest.

2. Identify the company's competitive advantage(s)

2.1 What to cover – Perform a detailed assessment of the company's own strengths, weaknesses, and assets. Think about which ones potentially can be developed into competitive advantages, taking into account important internal and external factors (such as the availability of time, money, and necessary connections). Evaluate how the company's competitive advantages align with the most urgent and important needs of customers in the market, as well as the strengths and weaknesses of competitors. Evaluate whether they are positional or capability-based advantages and how sustainable they are. Then, prioritize the list of potential advantages based on the likelihood that they could be achieved, sustained, and utilized to effectively differentiate the company.

2.2 Where to look – Involve key members of the company in this assessment to make sure nothing is overlooked and multiple perspectives are considered. Be honest about which competitive advantages are truly feasible, sustainable, and substantial enough to differentiate the company. Validate the assessment with an outside expert or advisor if additional objectivity is required.

3. Create a statement of competitive advantage

3.1 What to cover – Articulate the most promising competitive advantages that have been identified in a concise statement of competitive advantage. This statement should convey in one or two sentences what the company does, for whom it does this, how this uniquely solves an urgent need in the market, and why competitors cannot imitate it.[10] Many companies have competitive advantages, but few can clearly and explicitly articulate them. Beyond just being an important mental exercise to gain a deeper understanding of a company's advantages, a defined statement of competitive advantage helps the company more effectively persuade potential investors about its ability to capture the value created by its products and ultimately position itself in the market and convincingly demonstrate to customers why it is a more valuable alternative than the competition. Keep in mind that the statement must present a distinctive advantage that cannot easily be replicated by others and also can be demonstrated in concrete terms. Superlative claims will not stand up in the market. Continue honing the statement until it is clear, compelling, and "feels right" based on the fundamental strengths of the organization. It is also a good idea to test the statement with trusted advisors and a small group of customers for confirmation that it communicates a persuasive and differentiated reason for doing business with the company.[11] (The information included in the working example, Kyphon, Inc., also can be applied to developing a unique statement of competitive advantage.)

3.2 Where to look – As noted above, it is essential to involve all key members of the team in the development of the statement of competitive advantage.

4. Set a strategy

4.1 What to cover – Based on the definition of its competitive advantages, determine the fundamental business strategies that make sense to pursue. Start by eliminating strategies that are not applicable (e.g., first mover is not an option if another company is already in the market; me-too is impractical if the company's technology has differentiated features and benefits). Then consider which of the remaining strategies make the best match based on the business itself, the needs of the market, and the value propositions the company is trying to demonstrate to its stakeholders.

4.2 Where to look – Make this decision with the management team. Involve the board of directors for additional expertise and to validate the approach. Recognize that this is just one aspect of a holistic business strategy, and invest time and energy in making sure the rest of the strategy lines up around the fundamental business approach that has been chosen (and vice versa).

Credits

The authors would like to acknowledge Darin Buxbaum and Jared Goor for their assistance in developing this chapter. Many thanks also go to Jim Corbett and Julie Tracy for sharing the ev3 story.

Notes

1. The following overview is drawn from Garth Saloner, Andrea Shepard, Joel Podolny, *Strategic Management* (John Wiley & Sons: New York, 2001).
2. Ibid.
3. Ibid.
4. Ibid.
5. Shawn Tully, "Blood Feud," *Fortune*, May 31, 2004, p. 100.
6. American Medical Systems, S-1 Registration Document: Business Description, May 19, 2000.
7. Greatbatch.com, http://www.greatbatch.com/ (February 23, 2007).
8. All quotations are from interviews conducted by the authors, unless otherwise cited. Reprinted with permission.
9. Kyphon Inc. S-1 Form, 2000 http://www.secinfo.com/dut49.54We.htm (March 19, 2007). Reprinted with permission.
10. "Your Positioning Statement – The Single Most Important Sentence in Your Marketing Arsenal," Woodside Fund, http://www.woodsidefund.com/ent/articles/Your_Positioning_Statement.html (March 6, 2007).
11. Ibid.

Acclarent Case Study

Stage 5 Development strategy and planning

As the team began thinking about its next stage of work, it was time to find a name for the venture. After considering a series of different possibilities, they agreed on Acclar*ent*. The decision to embed ENT (ear, nose, and throat) within the new company name was intentional, as the group began to grow increasingly excited about the many opportunities to address unmet needs in the field.

Their next efforts were focused on transforming the *Balloon Sinuplasty*™ concept into a product and building a business capable of bringing that technology to market.

5.1 Intellectual property strategy

As described earlier, the team recognized intellectual property (IP) as an important element of its work early in the life of the company. At this point, it was decided that a full IP search would be conducted to evaluate the complete landscape of patents and study the prior art. While several relevant prior art references were found and added to the information disclosure statement made to the patent office, nothing was found that the team believed would stand in the way of proceeding with the chosen technology into commercialization. "At this point we were inventing at a very fast pace and we tried to capture as much as we could in our notebooks," Makower said. "It seemed that wherever we looked we saw more and more opportunities to expand the technology of Balloon Sinuplasty more broadly, along with other product extensions and other ENT needs that still needed innovation."

5.2 Research and development strategy

After the meetings with Sillers and Bolger, Chang, who would become the company's vice-president of engineering, remembered thinking: "Okay, we need to add to the team. I can't do all this by myself." His first move was to bring in an experienced engineer with a broad background. "I needed someone who was a jack-of-all trades, executes well, and who I worked well with under pressure and stress." To fill this role, Chang recruited a former colleague, Julia Vrany, with whom he knew he could partner to do everything from "clinical protocol writing, cadaver testing, development, working with vendors, quality, sterilization, validation and verification testing, and packaging – the full gamut of product development to get to first-in-man." Chang's next hire was a young engineer, John Morris, who was early in his career. "He was a Stanford grad – smart and hungry for experience," Chang said. As development progressed, Chang recognized the need for more balloon expertise so he identified an OEM vendor that was a specialist in this area. "So we had our super-broad generalist who was great at project management, our worker-bee engineer, and a great OEM partner who really knew balloons," he said. Together with Chang,

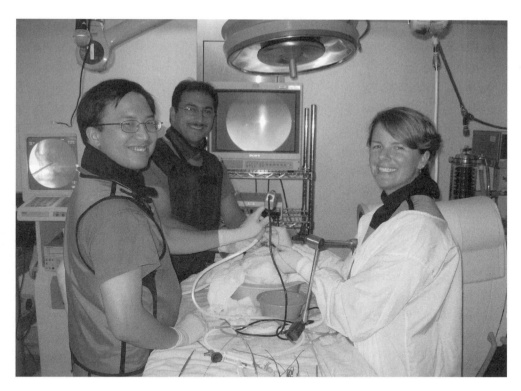

FIGURE C5.1
Members of the early team. From left to right: Chang, Makower, and Vrany (courtesy of ExploraMed).

these three pieces formed the team that would develop the working product used in the company's first-in-man study. "It's a balancing act," he added. "You need enough resources to get you there, but not so many that you sink the ship." See Figure C5.1.

At this point, they began working on the key technical challenges associated with the "family" of devices necessary to perform a balloon dilation procedure in the paranasal sinuses, which included not just the balloons, but the guides, inflation devices, and other key elements of the system. However, not surprisingly, many of the greatest issues to overcome had to do with the balloons. Chang elaborated:

> We needed something that was non-compliant, meaning it would keep its shape. We were fracturing bone, so we had to figure out what sort of pressures were required. And then, probably most significantly, we saw that we were puncturing balloons. We realized that we were placing more stress on the balloon than typically seen in a coronary or peripheral vascular procedure. The targeted bone was not only strong, but typically sharp. And since we recognized that doctors would likely be inclined to use the same balloon in multiple sinuses, the balloon would need to survive multiple cycles of placement, dilation, and removal.

This durability issue meant that the team could not use available balloon technology – Chang and his engineers agreed they would have to develop something unique. "It turned out that helping with development is exactly what our vendor did very well," he said. The next step was to move into a collaborative development loop, with Acclarent using input from its expert physicians (such as Sillers and Bolger) to refine the requirements it fed to the supplier. According to Chang, "Working with our physicians, we were really able to tune the catheter length, balloon dimensions, pressure requirements, and burst requirements – we quickly iterated a number of times between the doctors and the lab [where the testing occurred] and the balloon manufacturer. It was a great partnership."

Of course, noted Chang, "We were in super-stealth mode at the time. So we didn't really tell the supplier what we were doing, even though we had an NDA. After we had made several iterations and significant progress, they more frequently asked what we were doing. I'd say, 'Look, you've just got to understand that we're not going to tell you what we're doing. But I can tell you what the spec on the balloon needs to be.' It became a running joke. We would have a call about the next iteration and they would close with, 'And you're going to use this on … ?' And I would just answer 'Somewhere near, in, or around a body that may or may not be human.'"

Acclarent entered into an exclusive supply agreement with this vendor for balloons in the ENT field. "Our approach as a start-up company was that we didn't need to own every solution or every capability. There are a lot of people who know how to make balloons. Our goal was to find the one that would satisfy our needs. Once we identified the right supplier, we promptly transitioned from a project-based OEM relationship to more of a long-term partnership that we could scale with," Chang explained.

To get a working product, the team used a similar approach for developing the other devices it needed. Basic development, robust testing, and the detailed refinement of all specifications was performed in-house, but then external suppliers were entrusted to produce the actual devices. Under Facteau's leadership, prior to the commercial launch, Acclarent would make the decision to bring all development and manufacturing in-house. But, until then, this model served the company well.

Chang described the engineering environment at the time as requiring "all hands on deck." Everyone on the team was expected to contribute at full capacity. In terms of the philosophy, he said:

> Sometimes people fall into the trap of wanting a perfect solution, the perfect prototype. We weren't at all afraid of trying something a little crude and fast – in fact, the faster the better, because you can't replace time. You don't have to be perfect because you're going to learn so much from each prototype. Failure teaches you a ton, right? So just go, go, go, iterate as much as you can. We literally had labs two days apart where we'd go in and test the product to its limits. If something didn't work well, we'd go back, try to fix it, get another cadaver, and try again. It was all about speed. Otherwise, you could spend a month, or two, or three, or more trying to make the perfect prototype. And that just didn't make sense from our perspective.

Before long, the team got to the point where it had a system of devices that allowed physicians to get access to a sinus, demonstrate that they were truly inside it, put a balloon across, dilate the ostium, and show that they had enlarged it via fluoroscopy and endoscopy. See Figure C5.2.

At this point, Chang said, "We felt ready to prove it clinically – to demonstrate proof of concept, safety, and feasibility."

5.3 Clinical strategy

To prepare for first-in-man studies, Acclarent took a careful approach, working closely with its expert physicians to plan an appropriate course of action. When the team asked the physicians about the risks associated with performing the procedure in humans, they referenced skull-base fractures that could create serious complications. "We needed to determine if a high-pressure balloon dilation would cause a fracture to propagate in an uncontrolled manner or cause a 'tectonic plate' shift of bone and ultimately lead to a skull base fracture and cerebral spinal fluid leak, or worse," remarked Chang. If the company could demonstrate that this risk was minimal, Bolger, Sillers, and others advised, it would be ready to move into a first-in-man safety and feasibility trial.

To clear this hurdle, they brainstormed with the physicians to come up with a study that would demonstrate that Balloon Sinuplasty devices did not create these adverse events. In the end, they designed a cadaver study that would require a CT scan before and after the ostia were accessed and dilated. In order to test the "worst case scenario," the study would use the largest balloon Acclarent had developed and it would be inflated to its maximum pressure. A comparison of the CT scans would allow the team to assess any adverse events. Concerned about microfractures that might be missed in a CT scan, however, one physician suggested that a complete dissection might then be used

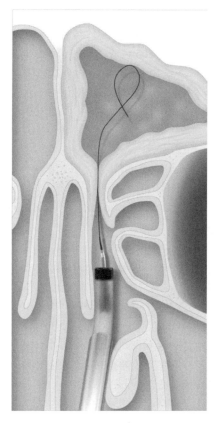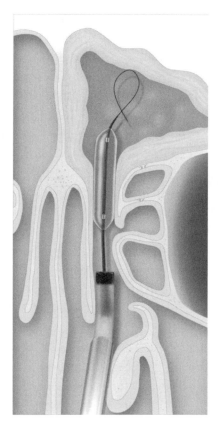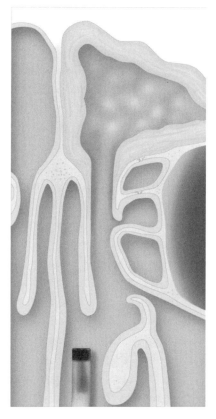

FIGURE C5.2
There are three primary stages of using the Balloon Sinuplasty technology. First, a flexible, atraumatic guidewire gains full access to the sinus cavity. Second, a balloon catheter smoothly tracks over the wire. Balloon dilation open up blocked ostia. Third, the devices are withdrawn, leaving open ostia with minimal tissue disruption (provided by Acclarent).

to examine the results of the procedure. According to Chang:

> So that's what we did. We took six cadaver heads, scanned them, put them in a cooler, jumped in the car to take them to the cadaver lab, dilated them, took them back and scanned them again, and then brought them back again to the lab where we cut them in half. The physician who was there [Bolger] spent three days with us. He flew out and dissected every single cadaver head and said, 'Okay, looks good. No fractures.' Every single one. We put the results together and the work resulted in a paper and the foundation for our argument to move into live humans.[1] This was the last key piece we needed to get the ethics committee comfortable with the idea that the technology was safe for human use.

Consistent with its stealth strategy, Acclarent decided to perform its first-in-man trial in Australia. This approach would allow the company to keep the study quiet while still generating results that would have a relatively high likelihood of being accepted in the United States by the FDA. In addition to having a population that enjoyed reasonably high standards of living and healthcare, Australia had a legitimate regulatory agency, the Therapeutics Goods Agency (TGA). "You can never be positive how your data will be received, but we were hopeful that it would be received favorably in the US," said Chang. Another advantage of performing the study abroad was to help Acclarent further delay any competitive threat and strengthen its IP position based on the results of the study before word of its approach got out in the domestic market.

In Australia, clinical trials were regulated by the TGA and required ethics committee approval prior to being launched. The process of gaining ethics committee approval was similar to being granted institutional review board (IRB) approval in the United States by the hospital or other facility where the procedures would be performed. The first step was to find an investigator willing to work with the company to perform the study. Fortunately, the expert physicians in the team's network were able to help them identify and secure the participation of an appropriate surgeon, Dr. Chris Brown in Melbourne, Australia. Acclarent then put together a package of information that included all the necessary background information, cadaver test data, and other relevant materials for an ethics committee review. "The ethics committee had a couple of questions for us, which we answered to their satisfaction," recalled Chang. "So they gave us a green light for doing the cases in Australia. But they limited us. They said, 'You can only do ten patients, and you can only treat two sinuses in each patient.' We said, 'Okay, that's good enough.'"

In total, it took the company about two months to get a green light to conduct the study. Then, with everything queued up in advance, it took another month or so to line up the patients, ship all required materials and supplies to Australia, and execute the trial. See Figure C5.3.

Chang described the outcome: "We went in and had unbelievable results – ten out of ten patients were successfully treated. It was beyond our wildest dreams. You just don't dare predict or hope for something that good. We were very fortunate." See Figure C5.4.

While an important clinical milestone had been achieved, Acclarent viewed this as just the beginning of its clinical strategy. The company's plan was to begin lining up its next human trials, which it would initiate in the United States shortly after receiving FDA clearance for its devices (the next study would become known as the CLEAR study). "We knew we needed more clinical experience and more data that we could share with people," said Chang. "So we started to line up investigators in the United States very carefully. The early physicians helped us identify more investigators and called ahead to make introductions. They would tell their colleagues, 'You're going to get a phone call from somebody. Talk to them. They're not going to tell you a whole lot until you sign an NDA.' This process was in anticipation of the need to increase our volume and get data from a mix of academic and private institutions."

There were also important lessons from the Australian study that required the company to change

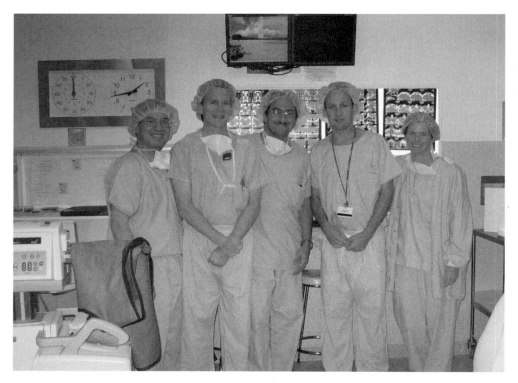

FIGURE C5.3
From left: Chang, Bolger, Makower, Brown, Vrany after completing the first clinical cases using Balloon Sinuplasty technology (courtesy of ExploraMed).

> **Title**: Safety and Feasibility of Balloon Catheter Dilation of Paranasal Sinus Ostia: A Preliminary Investigation
>
> **Authors**: Christopher L. Brown, MD, William E. Bolger, MD
>
> **Objectives**: Endoscopic sinus surgery (ESS) is an effective option for managing patients in whom medical therapy for rhinosinusitis fails. However, ESS is not always successful, and serious complications can occur. New techniques and instrumentation that improve outcomes and reduce complications would be seriously welcomed. Innovative catheter-based technology has improved treatment of several conditions such as coronary artery disease, peripheral vascular disease, and stroke. Recently, catheter devices have been developed for the paranasal sinuses. Cadaver studies confirm the potential use of these devices in rhinosinusitis. The objective of this investigation was to ascertain the feasibility and safety of these newly developed devices in performing catheter-based dilation of sinus ostia and recesses in patients with rhinosinusitis.
>
> **Methods**: A nonrandomized prospective cohort of 10 ESS candidates was offered treatment with a new technique of balloon catheter dilation of targeted sinus ostia. The frontal, maxillary, and sphenoid sinuses were considered appropriate for this innovative catheter-based technology. The primary study end points were intraoperative procedural success and absence of adverse events.
>
> **Results**: A total of 18 sinus ostial regions were successfully catheterized and dilated, including 10 maxillary, 5 sphenoid, and 3 frontal recesses. No adverse events occurred. Mucosal trauma and bleeding appeared to be less with catheter dilation than is typically observed with ESS techniques.
>
> **Conclusions**: Dilation of sinus ostial regions via balloon catheter-based technology appears to be relatively safe and feasible. Larger multicenter clinical trials are now warranted to further establish safety and to determine the role of this new technique.

FIGURE C5.4
An overview of the results of Acclarent's first-in-man clinical trial, as summarized in the abstract to the clinical journal article in which the results were eventually published (from *The Annals of Otology, Rhinology, and Laryngology*, April, 2006).

its clinical approach. For example, said Chang, "The outcome measure questionnaire that we used in Australia was much too complicated. People didn't understand it; they couldn't fill it out by themselves. We recognized that and switched to a simpler outcome measure instrument."

The results of the first-in-man trial also affected device development. Chang shared another example: "Based on our experience in Australia, we completely changed the design of our guides. We made significant modifications to get them ready for prime time. We spent a lot of time working on these types of changes before we went commercial."

5.4 Regulatory strategy

Acclarent needed a regulatory strategy for its entire family of devices. According to Chang, "There were three different balloon sizes, five different guides, an inflation device – just a ton of different things." The team had previously assessed which components of the system would likely fall into each FDA class, as part of its decision to pursue a 510(k) regulatory pathway. The next step was to identify appropriate predicates for each one. For simple components, such as tubing, guidewires, and needles, Acclarent readily found predicates. However, identifying devices upon which it could demonstrate substantial equivalence for the balloons took a little more searching. "From a balloon perspective, we looked hard for predicates in ENT," said Chang. "We actually found something in the eye, a lacrimal duct catheter, that was pretty close." This balloon device was developed for dilating constricted lacrimal ducts which connect the eye to the nasal cavity; thus, the location was similar to Acclarent's intended use. The company would make submissions, with good predicate devices, on all aspects of its system with the intent of having the FDA either exempt or clear them for commercial use.

A strategic decision would need to be made regarding the submission for the balloons. Acclarent had to determine whether or not to submit the data from its Australian study along with its 510(k) application for this device. It was unclear if the data would be required by the FDA as no other device in this category had all the required clinical data to receive clearance by the agency. For this reason, the decision was made to submit the 510(k) without clinical data, but to be prepared to provide it if the FDA made a direct request.

When Acclarent made its submissions for its class I devices, the FDA responded within ten days, indicating that all of them were exempt from premarket notification requirements. One week after the first submission, the company filed its 510(k) application for the balloons. Two months later, the FDA responded with a series of questions, including a request for clinical data to demonstrate that there would be no adverse events. "I think they probably expected us to file for an IDE," said Chang, "but we already had the study results in our back pocket." Within a week, the team had prepared a comprehensive response to the FDA's inquiry. "We scrambled that whole week to answer their questions in the most thorough and complete way we could. It was a true team effort. Within a week, we had a thorough and carefully reviewed clinical package with data submitted for the FDA's review. After a few additional phone calls and questions, we were able to provide the appropriate information that eventually led to our 510(k) clearance. The company was now one-step closer to its planned clinical trial in the United States, which it intended to launch immediately after clearance.

5.5 Quality and process management

In parallel with its clinical and regulatory efforts, the Acclarent team began thinking about a quality system and the need to bring some structure to the way the company managed important processes. According to Chang:

> When we started getting some traction and realized that we were marching toward a major clinical experience, we recognized that we needed to have a quality system and certain controls in place. Because we were running lean, we had a consultant come in and we worked with him to develop a fledgling, flexible quality system. We knew we needed design controls, processes for supplier audits, and other standard operating procedures so we could prepare for a commercial launch of the products.

The idea was to build a quality system that would initially allow the company to cover the FDA's requirements, but that had the capacity to effectively and efficiently scale with the company as it grew and faced the need for more structure and formality in its approach. "We started with standard operating procedures [SOPs] that were flexible and adaptable," said Chang. "We recognized as we grew, we could then tailor these SOPs and make them more specific. And that's really what has happened over the last four years."

With the plan in place, the team began executing it. Because Acclarent would be using contract manufacturers to produce its first family of products, Chang had particularly vivid memories of performing several supplier audits. "We went out to vendors and audited them with our check list. We examined their quality manual, how they received and inspected materials, as well as their manufacturing processes and methods of keeping lot history records, etc." Based on the outcome of these inspections, the team chose certain vendors to work with, ranging in size from large, established contract manufacturers to smaller "mom and pop" shops with specialized expertise in a certain area. "What we learned over time," said Chang, "was that some vendors were not as good at helping us develop processes, with appropriate quality systems, as others." Gradually, as Acclarent moved closer and closer to launch, it had to replace those suppliers with others that could better support the company's growing need.

5.6 Reimbursement strategy

As part of its preliminary reimbursement analysis, Acclarent received two favorable opinions that it would be able to use the existing FESS codes to obtain reimbursement for its Balloon Sinuplasty devices. However, as Lam Wang pointed out, "It's much more political than that." The company believed that it would be important to get the leading professional society in the field – the American Academy of Otolaryngology (AAO) – to endorse the position that the new balloon technology could be used with these codes. Lam Wang explained:

> In the end it's the third-party payers who have to pay for the technology. In general, when the volume of claims being submitted by surgeons adopting a new procedure is low, the claims can go under the radar. It's when the volume of

claims starts to get kind of high that the insurance companies start looking for reasons to not pay. When you hit this critical mass, they start questioning it. And it's really good if you have a professional society to support you, saying, 'We agree.' The third-party insurance companies don't pretend to practice medicine – they defer to the physicians. If the physicians say, 'We used these new products to perform the FESS procedure because of medical necessity and we agree that they should be covered by these codes,' they're more likely to agree.

To solicit AAO support, Acclarent's reimbursement consultant recommended that the company seek a letter from the society indicating that it agreed with the use of FESS codes to cover the Balloon Sinuplasty devices. Acclarent could then take such a letter to the American Medical Association (AMA) and work with that organization to make this recommendation part of the current addendum to its "coding Bible." Ideally, this letter would be created and provided by physicians familiar and experienced with the technology. Thus, the selection of investigators was an important step in developing this strategy. The good news was that it was clear from the feedback that the device's use fell squarely under the existing codes.

5.7 Marketing and stakeholder strategy

The company's efforts to secure reimbursement were closely coupled with its marketing and stakeholder strategy. While reimbursement looked at how physicians would be paid for performing the procedure with Acclarent's products, the larger marketing and stakeholder strategy sought to understand the other ways in which Balloon Sinuplasty devices would affect their practices so that Acclarent could promote the benefits and mitigate any downsides. Lam Wang described the in-depth interviews she conducted with a handful of physicians (who were all under NDAs): "My approach was to understand who they are and how their practices work. Fundamentally, I wanted to understand the product positioning of Balloon Sinuplasty within the physician's practice and within the treatment of this patient population."

According to Lam Wang, the responses she received varied significantly. Some physicians viewed the balloon as just another "club in their golf bag" that they would use when it made sense, or in hybrid procedures that used both traditional FESS and Balloon Sinuplasty tools. "One doctor took it on a sinus-by-sinus approach," she recalled. "He said, 'The maxillary sinuses are so fast and easy that I wouldn't bother to use the balloon. But, for the frontal sinuses, this is definitely an awesome solution.'" Other physicians viewed the balloons as a more fundamental leap in technology. "They saw it as a potential paradigm shift in how they would treat patients – no more sharp instruments, no more picking at tissue, but balloons, catheters, and wires." Some cited the fact that one of the four sinus types, the ethmoid sinuses, were not being treated as a reason why the device might have less utility. Comments such as these gave Acclarent a strong understanding of how Balloon Sinuplasty technology would potentially fit into these physicians' practices, how it would be used, and, by extension, how it would affect them financially.

Lam Wang did some additional research into the effect on facilities – primarily hospital outpatient facilities and ambulatory surgical centers. "I also looked at the ramifications of bringing endoscopic sinus intervention back into the office, but it was clear this was not going to be as easy." As the launch approached, Acclarent began revisiting the needs of patients and evaluating the interests of referring physicians (e.g., general practitioners who would refer patients to the sinus specialists).

When it was time to define clear value propositions for each of these stakeholder groups, Lam Wang sought to put these messages in context:

> I felt it was important to have a story. I believe that's the backbone of any marketing program, especially for a new company with a new technology. So I spent some time putting together the story for Acclarent. It's not enough to say 'new product, unmet need, done deal.' You really have to tell a story for why this technology makes sense. The value it provides is the core component of the messages you give out to all of your

> **Key Messages to ENT Surgeons**
> - There are over 600,000 chronic sinusitis or recurrent acute patients per year in the U.S. who have unmet clinical need. These patients are underserved by medical therapy; they are surgical candidates, but they are not getting FESS.
> - The goals of sinus surgery are to restore ventilation and normal sinus function by opening up blocked ostia while maintaining as much normal anatomy and mucosa as possible.
> - However, there are shortcomings associated with today's surgical instruments that make achieving those goals challenging. Straight, rigid tools are used in the tortuous sinus anatomy for a delicate, targeted procedure. As a result:
>   - More tissue removal than necessary often required just to gain access (e.g., ethmoidectomy to access sphenoid or frontal areas, uncinate regularly removed during maxillary antrostomy.
>   - Procedure hard to tolerate for patients – bloody, packing, general anesthesia, long recovery. Many patients hesitate to undergo surgery.
>   - Potential complications can be serious – penetration of eye or brain cavity with these straight, rigid steel instruments. FESS most common reason for litigation against ENT surgeons in the U.S.
>   - Frontal procedures are challenging and prone to iatrogenic scarring and stenosis – unrefined instruments make it easy to cut, tear or strip mucosal lining from the recess during ethmoid surgery.
> - Improvements to the field have been made – but still same basic problem of using straight, rigid tools.
>   - Microdebrider – powered instrument shaves away diseased tissue while sparing normal tissue.
>   - Image Guided Surgery – allows surgeon to see instrument location relative to historical CT landmarks.
> - There is a need for a new way to accomplish sinus surgery:
>   - Devices that conform to the tortuous sinus anatomy.
>   - Devices that stretch sinus tissue rather than tear or cut.
> - Now, there is a new, less invasive procedure alternative.
> - Balloon Sinuplasty uses soft, flexible balloon catheters to gently open up blocked ostia, allowing drainage and return to normal sinus function.
>   - Innovative platform change in technology from straight, rigid instruments to soft, flexible catheter-based devices.
>   - Devices conform to the tortuous sinus anatomy.
>   - Less need to remove sinus tissue in order to gain access.
>   - Allow improved access to more remote areas.
>   - Less likely to penetrate into the orbit or brain.
>   - Balloon catheters stretch sinus tissue rather than tear or cut.
>   - Less bleeding during and following surgery.
>   - Less likely to stimulate scar tissue.
>   - Lower potential of revision surgery.
>   - Adheres to physiologic principles and goals of sinus surgery.
>   - Safe, low rate of complications.
> - FDA cleared.
> - Balloon Sinuplasty works.
>   - Clinical data to date.
>   - Case studies.
> - We have thought leader support.

FIGURE C5.5
Acclarent developed a story that included their core value propositions for each key stakeholder group. The example shows some of the messages targeted at ENT surgeons (provided by Acclarent).

stakeholders, whether it's patients, surgeons, or facilities.

Once these value propositions were defined and embedded within the context of a story for each stakeholder group (see Figure C5.5), Lam Wang turned her attention to the development of a plan to support promotion and advocacy. "Our next focus was all about getting ready for the commercial launch," she said. The team hired an ad agency to help it establish a brand, as well as a look and feel for the product. This brand was used as the backbone of all marketing communications, including brochures and a website (note that the company's publication strategy was managed separately; also, see the Sales and Distribution Strategy section for information about professional training and education). Significant effort was also put into planning a major product launch at the American Academy of Otolaryngology – Head and Neck Surgery (AAO-HNS) conference, which would occur in September, 2005. This was the meeting at which Acclarent planned to transition out of stealth mode and announce its technology to the world. The launch plan for this event included a booth, where Acclarent staff would show the technology and answer questions throughout the conference, as well as a symposium at the show with presentation by thought leaders in the field (e.g., Bolger and Sillers) and a video demonstrating on a real patient how the procedure worked.

Pricing

Acclarent also needed to decide on a price for its products, an exercise that was closely linked to the company's business model, funding requirements, and its overall viability. "We had to figure out the average price we could sustain per procedure," said Lam Wang. She explained her approach:

> For the first pass, I revisited the average payment for FESS per sinus treated, and then by procedure. I understood after this analysis that, in a given FESS procedure, physicians could bill up to nine codes, and that the codes would stack, meaning that they would get paid the full amount for the first procedure done on a certain sinus and then receive 50 percent of each code after that. Next, I did this super in-depth analysis to determine what a typical case looked like at the time. That's when I went back to talk to our physician advisors. And I learned that there was a combination of sinuses that was typical. After that, I looked at an average case from a break-even perspective. If I was going to add cost to the system [by charging for reusable devices], I then had to figure out where I could take cost out of the system.

"I looked at it from a long-term perspective to make an economic justification for this new technology," said Lam Wang. Ultimately, the sum total per case benefit of what a physician or facility would gain from increasing patient flow and decreasing procedure time, taking into account what third-party insurance companies would pay, represented the high end of what Acclarent could charge. She continued, "I completed a twelve-page analysis and came up with a range – it was something like $1,200 to $2,000 per case. So not per balloon, not per device, not per sinus, but just per patient."

With this range in mind, the Acclarent team again approached its advisors to discuss pricing. For the purposes of discussion, the company used a figure of approximately $1,200 per procedure. Facteau explained what happened next:

> When we talked with some of our physician advisors about this, we got some initial push back. At that point, we decided to share some confidential information to help them gain an appreciation for what it really takes to build a medical device company from scratch. We explained to them how and why we came to our decision on price, we shared with them our P&L and cash flow assumptions, and pointed out that in our first year we would lose $16 million. We mentioned that we could do things a lot less expensively; however, we believed the right thing to do to build a great company is to invest in clinical research, physician training (which included a cadaveric experience), and innovative products. Couple those investments with a direct sales organization and we anticipated the need to raise $75 to 100 million before we would break even. Therefore, at that point, $1,200 of disposables per procedure really didn't seem to be as big of an issue for them, and they became supportive of the pricing strategy.

5.8 Sales and distribution strategy

In terms of sales and distribution, Facteau had a clear vision from the outset regarding the model he wanted to adopt:

> I had a strong bias that we should go direct, and that we needed to hire an experienced, clinically-savvy sales organization that could execute on our vision. We needed to train physicians really

well to preserve the safety of the device and the procedure, and to protect the Balloon Sinuplasty reputation as a safe tool. We thought about how we could leverage distributors, but the direct model made more sense. I was a strong proponent of this. When no one gave me any real resistance, internally or with our investors, we agreed to create a direct sales organization in the United States.

Lam Wang underscored the key advantage of this approach. While a direct sales force was more expensive and time consuming to build, she said, "It allows you to maintain more control."

Deciding on the type of sales representatives that would be best suited to represent Acclarent's technology turned out to be something of a challenge. "We asked ourselves, of all the people in our sales careers that we had interviewed, how many had come from ENT?" recalled Facteau. "And we couldn't point to one." For this reason, the company decided not to target sales people with ENT experience. Instead, they decided to look for people who had worked in interventional cardiology and had experience selling balloons, catheters, guidewires, and other equipment similar to the components of the Balloon Sinuplasty system. "That's where we started," Facteau said, "but that turned out not to be the right model for us." Ultimately, Facteau and team discovered that, while cardiology sales people understood the technology, most had not actually developed the skills to introduce a paradigm shifting technology into the operating room.

Ultimately, Acclarent determined that it needed sales people with deep experience training physicians and who had spent time in the operating room. According to Facteau, representatives who had been in a start-up before also tended to work out well, as did people who had helped develop new markets in the past. "Market development is a lot different than taking share in an established market," he noted. "And I think a lot of people don't really appreciate that. If you have a better mousetrap and there is an established market for it, that's a much easier, predictable sale than if you have to go out and create a whole new market from scratch."

Training

Through its direct sales force, Acclarent intended to invest substantially in the training it delivered to physicians on its technology. "Even though it had never been done before in ENT, we knew that's what we needed to do," said Facteau. The company believed firmly that Balloon Sinuplasty products would lead to improved results if they were used consistently by surgeons – the key was to get each one to use the products and perform the procedure in the same way to achieve comparable results to those achieved in the company's studies. "We had to standardize," Facteau recalled. "What was missing in sinus surgery was standardization. You could go to ten different hospitals and watch sinus surgery and you'd see ten different ways to do it." For this reason, the team worked diligently to define best practice processes and protocols that would be taught consistently to physicians and could easily be replicated post-training. Facteau elaborated:

> I started my career at U.S. Surgical, and I am of the opinion that they were the first to crack the code on physician training. They spent a significant amount of resources to ensure proper education for new medical procedures. One of the guiding principles to successful patient outcomes and adoption was centered on standardization. No matter what OR you were in, general surgeons placed trocars in the same location, they held the instruments the same way, and utilized the same retraction techniques to gain better visibility. This ultimately led to better outcomes. We set out to accomplish the same with Balloon Sinuplasty technology. 'If you do it this way, we believe you're going to get good results because we know the proper techniques resulted in good outcomes in our clinical studies.'

Led by experienced physicians and Acclarent's sales representatives, physicians would be trained on this standardized approach. For those surgeons open to the idea of training, the team was confident that this approach would protect patients and the reputation of the company while enabling desired results. However, as Facteau explained, "We had some concern that this level of standardization – our commitment to standardization – might not necessarily be well received, especially early on because it had never been done before in ENT by a manufacturer. Despite this potential resistance, the company believed strongly that this was an appropriate (and necessary) approach to take when introducing novel technology to a market that had not seen a great deal of innovation in decades. See Figure C5.6.

As far as building the sales organization, Facteau took a somewhat measured approach. For instance, he decided that Acclarent would not begin hiring any sales representatives until the company had received clearance from the FDA on its balloon technology. Other critical milestones were put into place to keep the rate at which the sales organization expanded in alignment with the commercial viability and adoption of the technology. While this made some members of the team a little nervous as the launch date rapidly approached with no sales force in place, "We felt like it was fiscally responsible – and the right thing to do," said Facteau.

5.9 Competitive advantage and business strategy

When the company began thinking about creating a competitive advantage, the Acclarent team relatively quickly focused on two core capabilities: developing a world-class direct sales force/training organization and establishing a pipeline of innovation. Facteau (shown in Figure C5.7) explained:

> The first was around the commercialization. We wanted to build the best ENT commercial organization in the world. There have been some good examples of companies in other specialties that have done that, but no one had tackled it in ENT. We benchmarked companies like Kyphon, Fox Hollow, and Perclose. We performed case studies in an attempt to understand what they did well and to try to learn from their mistakes. In doing this, one of the things we realized was that none of these companies developed a true pipeline of innovation that could drive organic growth for a long time. We decided that we wanted to be great at both, which is no easy feat.

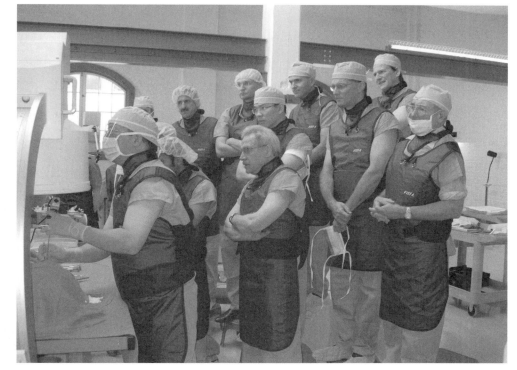

FIGURE C5.6
Surgeons being trained to use Balloon Sinuplasty products in Baltimore, MD (courtesy of Acclorent).

FIGURE C5.7
Facteau lays out his vision for Acclarent at a staff meeting in 2005 (courtesy of Josh Makower).

Developing a strong internal R&D capability was also consistent with Acclarent's business model and the company's desire to get ENT physicians "addicted to innovation." Lam Wang added, "Bill's vision was not just to focus on 'N,' but 'E' and 'T,' too. We want to be everything for the ENT surgeon." Sensing that untapped opportunities existed across the ENT field gave Acclarent a source of ideas that it could develop into an ongoing pipeline of innovative products. The decision to focus on developing this capability, dovetailed with Acclarent's decision to begin the process of bringing all R&D and manufacturing in-house.

Note
1. William E. Bolger and Winston C. Vaughan, "Catheter-Based Dilation of the Sinus Ostia: Initial Safety and Feasibility Analysis in a Cadaver Model," American Journal of Rhinology, May/June 2006, pp. 290–294.

IMPLEMENT Integration

STAGE 1 NEEDS FINDING
1.1 Strategic Focus
1.2 Observation and Problem Identification
1.3 Need Statement Development
Case Study: Stage 1

STAGE 2 NEEDS SCREENING
2.1 Disease State Fundamentals
2.2 Treatment Options
2.3 Stakeholder Analysis
2.4 Market Analysis
2.5 Needs Filtering
Case Study: Stage 2

STAGE 3 CONCEPT GENERATION
3.1 Ideation and Brainstorming
3.2 Concept Screening
Case Study: Stage 3

STAGE 4 CONCEPT SELECTION
4.1 Intellectual Property Basics
4.2 Regulatory Basics
4.3 Reimbursement Basics
4.4 Business Models
4.5 Prototyping
4.6 Final Concept Selection
Case Study: Stage 4

STAGE 5 DEVELOPMENT STRATEGY AND PLANNING
5.1 Intellectual Property Strategy
5.2 Research and Development Strategy
5.3 Clinical Strategy
5.4 Regulatory Strategy
5.5 Quality and Process Management
5.6 Reimbursement Strategy
5.7 Marketing and Stakeholder Strategy
5.8 Sales and Distribution Strategy
5.9 Competitive Advantage and Business Strategy
Case Study: Stage 5

STAGE 6 INTEGRATION
6.1 Operating Plan and Financial Model
6.2 Business Plan Development
6.3 Funding Sources
6.4 Licensing and Alternate Pathways
Case Study: Stage 6

Knowing what to do is not enough. One of our main recommendations is to engage more frequently in thoughtful action.

Jeffrey Pfeffer and R. Sutton[1]

Now this is not the end. It is not even the beginning of the end. But it is, perhaps, the end of the beginning.

Winston Churchill[2]

Integration

While this is the final stage of the biodesign innovation process, it is in fact the beginning – the beginning of your sustained effort to implement a product or business around the need you've identified and your invention or innovation. Integration enables execution on all of the plans you've developed in Stage 5.

A very specific focus of this stage, and indeed this book, is around the start-up process. Building and managing a small business, generating the business model, articulating a business plan, and navigating the complicated waters of fundraising are all essential components and are explored in depth. Remember, once outside sources are involved, the enterprise no longer belongs solely to the innovator.

The final chapter looks at alternate approaches to starting a business – that is, the options of partnerships, licenses, or the outright sale of an idea, completely relinquishing control to the new owner. If you have a great idea, but no time or inclination to devote to it, one of these pathways can provide a wonderful route to get a solution into practice.

Regardless of how you approach it, the journey should be fun. That's not to say that many lessons won't be learned the hard way, but the optimism and, indeed, idealism of the innovator can profoundly catalyze transformation in healthcare. Good luck.

Notes

1. Jeffrey Pfeffer and Robert I. Sutton, *The Knowing-Doing Gap: How Smart Companies Turn Knowledge into Action* (Perseus Distribution Services, 1999).
2. According to the Churchill Centre, this statement was made at the Lord Mayor's Luncheon, Mansion House, following the victory at El Alamein in North Africa, London, November 10, 1942 (see http://www.winstonchurchill.org/i4a/pages/index.cfm?pageid=388).

6.1 Operating Plan and Financial Model

Introduction

Engineers to design, prototype, develop, and manufacture a new product. Clinicians to run clinical trials. Statisticians to analyze the data. Professional management to develop marketing and sales strategies. Reimbursement consultants to secure codes, coverage, and payment. Sales people to get the technology into the hands of customers. Key opinion leaders to promote its adoption. All of these individuals are essential to the development and commercialization of a medical device. Yet, without a carefully integrated plan for managing and synchronizing their complex and interdependent efforts, the resources required to make a new treatment a reality can quickly spiral out of control, threatening the very existence of a new venture.

Through the operating plan, innovators specify precisely who will execute the strategies outlined through the development strategy and planning stage of the biodesign innovation process, when and in what order they will be performed, and with what resources. This information is then translated into costs and consolidated into an integrated **financial model** that also includes a detailed revenue plan for capturing a share of the potential market. By comparing the revenue projections to the cost estimates coming out of the operating plan, the innovator can confirm whether or not the potential market justifies the financial requirements for developing and commercializing the product. The operating plan and financial model further provide the blueprint for implementing the company's business strategy, realizing its vision, and monitoring the results it achieves.

OBJECTIVES

- Understand the process for developing an operating plan and cost projections.
- Create a revenue model and integrate with the cost projections into a financial model to support business planning.
- Know how to identify important strategic and tactical issues that should be reflected in the operating plan and financial model.
- Appreciate how to make appropriate medtech-specific assumptions when developing an operating plan and financial model.
- Learn to perform a proxy company analysis to validate all components of the operating plan and the financial model against a more established company with attributes similar to the new venture.

Operating plan and financial model fundamentals

Thomas Fogarty, innovator and founder of more than 30 medtech companies, highlighted the diversity of the skills required to bring a new device to the market:[1]

> Particularly in this day and age, you need people from different disciplines – you need intellectual property attorneys, you need corporate attorneys, you need regulatory experts, you need good engineers ... And, you need different types of engineers. You need the person who can conceptualize, the one who can prototype, the production engineers, and then what I call a 'finisher.' To bring this whole team together, you have to understand value allocation. All of these people create value. They bring something different to the table. If you think, just because you had the idea, you brought all the value, you're not going to be successful. Somebody has to implement your idea, and one individual can't do it.

A financial model for a new venture is a detailed, quantitative articulation of Fogarty's statement. It begins with the company's operating plan, tracking both the cost of developing the innovation and bringing it to the market, along with market revenue, over a period of five to seven years. For at least the first three years (and, ideally, the first five years), the model tracks costs and revenue on a quarterly basis and, after that, on an annual basis.

Costs include salaries, capital equipment, supplies, facility expenses, and cost of goods sold (**COGS**), which is the total material cost for manufacturing. Costs of goods sold are obtained by multiplying total sales by material cost per product sold.[2] Revenue is total sales multiplied by sales price. Figure 6.1.1 summarizes these basic drivers of the financial model.

The financial model has six components, which correspond to the six steps in its development. The *operating plan* provides an overview of the activities that must be performed to develop the technology, their timing, and the key milestones they support. The *staffing plan* specifies the personnel needed for executing the operating plan. The *market model* includes revenue

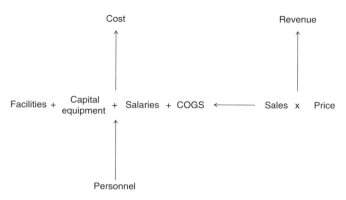

FIGURE 6.1.1
Cost and revenue are the fundamental drivers of any financial model.

and market share projections, as well as a preliminary assessment of the sales force required to achieve these forecasts. *Cost projections* then integrate all costs together: salaries, facilities, equipment, supplies, and COGS from product manufacturing. The *income statement* subtracts all costs from revenue to generate an accounting income statement. An important distinction is that, with accounting income, the cost of capital equipment is not subtracted in the year the equipment is purchased. Instead, it is spread across the lifetime of the equipment. However, the capital necessary to acquire the equipment is needed in the year of the purchase. *Cash flows* reflect this distinction and are calculated by subtracting from the revenue the actual expenses incurred in a given time period. The **cash flow analysis** then provides the basis for estimating a company's funding needs. Each of these components is explained in more detail in the sections that follow. Figure 6.1.2 provides for an overview of how these elements work together.

Throughout this chapter, an example will be referenced to illustrate each of the components of the financial model. The example will focus on a business that is developing an ablation catheter used for the treatment of atrial fibrillation (AF). Atrial fibrillation is a cardiac rhythm abnormality that causes the atria (the upper chambers of the heart) to contract irregularly. This condition creates a number of undesirable side effects ranging from palpitations and fatigue to the potential for a debilitating stroke. One way to treat AF is to destroy, or ablate, certain areas of the heart responsible for

Stage 6 Integration

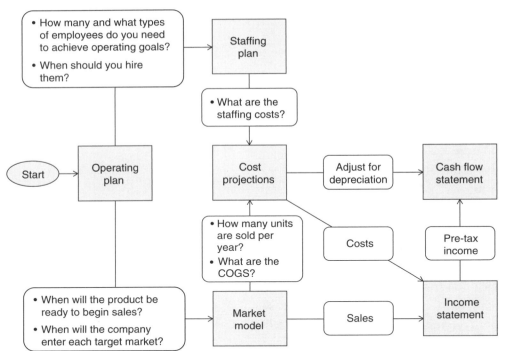

FIGURE 6.1.2
The six primary components of a financial model and how they work together.

the initiation and perpetuation of AF. This treatment is achieved through the use of an ablation catheter, which can be advanced into the heart via the blood vessels where it can be used to deliver energy to ablate the cardiac tissue (see 2.1 Disease State Fundamentals and 2.2 Treatment Options for more information).

Operating plan

The operating plan starts with the high-level milestones a company chooses to chart its progress. Many of these early milestones are related to R&D, as explained in 5.2 Research and Development Strategy. Overall, specific milestones vary from company to company. However, at a high level they would include:

1. Proof of concept
2. Product development progress
3. Manufacturing feasibility
4. Clinical trials and FDA submission
5. Reimbursement progress
6. Development of scalable manufacturing
7. Marketing and sales (United States, Europe, Asia)

A detailed operating plan would break down these high-level milestones into additional granularity (e.g., reimbursement initiatives could be divided by payer type). Figure 6.1.3 shows a sample operating plan by quarter for the hypothetical ablation catheter business (assuming a 510(k) regulatory pathway and reimbursement under existing codes).

Innovators may wish to think that the operating plan does not change. However, most ventures will face significant technical or clinical challenges that will force them to reconsider their plans and milestones. For example, the company featured in Appendix 6.1.1 (Cardica, Inc.) faced formidable challenges when the FDA changed the clinical trial requirements in the middle of its clinical studies as a result of safety problems with a competitor's product. The appendix explains how the company modified its strategy and operating plan in response to the unanticipated change.

Staffing plan

The staffing plan outlines the number of employees that need to be hired over time. The staffing plan is a direct outgrowth of the steps in the operating plan, since staffing requirements should be determined by the company's strategic and operating decisions. Essentially, an innovator needs to figure out the types of employees (in terms of skill sets or function) and

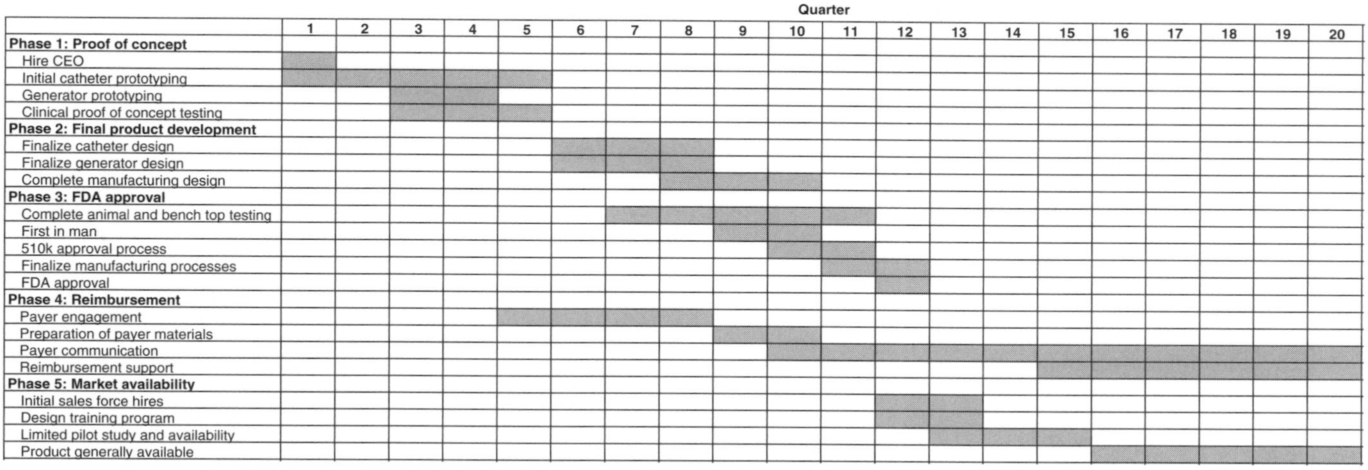

FIGURE 6.1.3
Sample operating plan for a hypothetical ablation catheter business.

	Quarter 1	2	3	4	5	6	7	8	9	10	11	12	13	14	15	16	17	18	19	20	Year 6	7
Manufacturing team																						
Engineers	0	0	0	0	0	0	0	0	1	2	3	4	6	6	6	6	9	9	9	9	9	12
Manufacturing assemblers	0	0	0	0	0	0	0	0	2	2	2	2	5	5	5	5	10	10	10	10	18	35
R&D																						
Catheter team																						
Engineers	1	1	1	1	1	1	1	1	1	2	2	2	2	2	2	2	2	2	2	2	3	5
Techs	2	2	2	4	4	4	4	4	4	4	4	6	6	6	6	6	6	6	6	6	7	10
Generator team																						
Engineers	1	1	1	1	1	1	1	1	1	1	1	1	1	1	1	1	1	1	1	1	3	5
Techs	2	2	2	2	4	4	4	4	4	4	4	4	4	4	4	4	4	4	4	4	7	10
SG&A																						
Salesforce	0	0	0	0	0	0	0	0	0	0	0	4	4	8	12	12	12	12	12	12	25	50
Marketing, business development, and reimbursement	0	0	0	0	0	1	1	1	1	1	2	2	3	3	3	3	3	4	4	4	5	5
Clinical advisor	1	1	1	1	1	1	1	1	1	1	1	1	1	1	1	1	1	1	2	2	2	2
Regulatory	0	0	0	0	0	0	0	1	1	1	1	1	2	4	4	4	4	5	5	5	5	5
Administrative asst.	0	0	0	0	0	1	1	1	1	1	1	1	1	1	2	2	2	3	3	3	4	5
Management	1	1	1	1	2	2	2	2	2	2	2	4	4	4	4	6	6	6	6	6	8	8

FIGURE 6.1.4
A sample staffing plan.

overall number of employees required to achieve the milestones set forth in the operating plan. The strategies developed in earlier chapters provide the basis for developing the desired operating plan.

Figure 6.1.4 shows a sample staffing plan that is driven by the operating plan above. The staffing plan is divided into three components: manufacturing, R&D, and **SG&A** (selling and general administrative) expenses. The levels in the staffing plan should directly correspond to the milestones in the operating plan. For instance:

- R&D staffing starts from the first quarter but it increases as product launch approaches in the twelfth quarter (Q12). The company hires a second catheter team R&D engineer in Q10 to support R&D activities related to the development of the manufacturing process. This should match the achievement of a specific milestone in the operating plan (e.g., completion of manufacturing design and initiation of the development of manufacturing process).
- The completion of manufacturing design (under Phase 2) is finalized in Q10 and manufacturing processes are finalized in Q11 and Q12. Accordingly, manufacturing employees are first hired in Q9, to participate in finalizing the manufacturing process before manufacturing actually begins.
- FDA approval is targeted for Q11, which is why the first members of the sales force are hired in Q12, since it would not make sense to hire them prior to having a product to sell. In this example, a focused

direct sales force strategy (selling directly to the physician customer) is assumed. The peak sales force size assumed (50) is relatively small for direct sales force models and should be rationalized based on a more detailed sales force model (the method used to develop such a model is discussed as part of the bottom-up market model later in this chapter). Other sales models, such as using distributors, would require a smaller in-house sales team and these numbers would be reflected in the staffing plan accordingly.

In general, staffing numbers will vary based on factors such as the regulatory pathway, reimbursement environment, and sales model that have been chosen. For example, a direct sales force can require as many as 100 to 200 sales representatives at a peak, while a two to three person sales team might be sufficient to support an indirect sales model. As well, the size of the target physician population (e.g., a smaller specialty as compared to the large numbers of primary care doctors) can significantly affect the required sales force. Reimbursement may require a team of up to ten people if an intense reimbursement effort is anticipated and a high level of support is desired. On the other hand, one or two consultants may be sufficient if codes and favorable coverage decisions are already in place.

One way to roughly check if the number of employees reflected in the staffing plan is reasonable is to use commonly accepted staffing ratios and formulas. For example:

- **SG&A to R&D** – Typically, the ratio of SG&A employees to R&D employees equals roughly three in a mature medical device company. (It is important to note that this ratio may be very different in the early phases of a business as a product is being developed.)
- **Product units per assembler** – If the company intends to assemble or build the product in-house, it must be sure to have enough assemblers to meet demand for the product. Calculate the total number of units produced by each assembler to make sure it is a reasonable match to anticipated demand and reasonable for an assembler to achieve in a given time period. If either of these are not the case, more or fewer assemblers may be needed.

The number of assemblers is linked to the market model, which defines the number of units that need to be produced for a given time period (see below).

Market model

There are two ways to approach market models: top-down or bottom-up. Fundamentally, the top-down model provides a high-level estimate of the market based on broad, embedded assumptions about a company's ability to access the market. In most cases, a top-down model is built based on the availability of strong comparable data from other companies in similar market spaces. A bottom-up model, on the other hand, is developed "from scratch" to provide the innovator with a granular understanding of the business and the market in which it operates.

While top-down models can be useful in developing "back of the envelope" market estimates (as described in 2.4 Market Analysis), innovators are encouraged to develop a market model from the bottom-up for inclusion in the financial model. This approach helps to ensure that the cascade of assumptions relied upon to develop the numbers are fundamentally sound and well understood. It lends credibility to the model and provides the innovator with confidence when asked to defend it. And, it provides the innovator with an opportunity to think through the sales process in depth and, thus, obtain a better understanding of the various challenges to capturing the market. These challenges cannot be revealed from a rough top-down model.

Top-down model

A top-down market model forecasts projected yearly revenue by outlining the segments of the market that will be addressed and by determining how many customers the company intends to service (the number of patients that will be treated, hospitals/offices that will be sold to, etc.), then multiplying the number of customers by the price of the innovation. This approach is useful for quickly confirming how attractive the market is to pursue. Figure 6.1.5 shows a sample market model; a detailed description of the calculations follow the figure.

Key assumptions		
Assumption	#	Source
# of paroxysmal AF patients	960,000	American Heart Association
Population growth rate	2%	US pop stats - baby boomers
% drug refractory	28%	New England Journal of Medicine
Average selling price (ASP) of ablation catheter ($)	2000	Company research
% annual decrease of ASP	1%	Estimate

US Market Model							
Year	Year 1	Year 2	Year 3	Year 4	Year 5	Year 6	Year 7
Key milestones			FDA Approval				
Paroxysmal AF patients	960,000	979,200	998,784	1,018,760	1,039,135	1,059,918	1,081,116
% drug refractory	28%	28%	28%	28%	28%	28%	28%
Refractory AF patients	268,800	274,176	279,660	285,253	290,958	296,777	302,712
% of refractory patients treated	2.5%	4.5%	7.5%	15.0%	25.0%	30.0%	35.0%
Procedures per patient	1.15	1.12	1.10	1.08	1.07	1.06	1.05
Total ablation procedures performed	7,728	13,818	23,072	46,211	77,831	94,375	111,247
Total market size ($)	15,456,000	27,498,756	45,683,536	91,042,470	152,572,428	184,078,320	215,901,728
% Market share of ablation patients	0.0%	0.0%	0.1%	1.0%	5.0%	12.0%	20.0%
Total units sold	-	-	23	462	3,892	11,325	22,249
Average selling price ($)	2,000	1,990	1,980	1,970	1,960	1,950	1,941
Revenue ($)	-	-	45,684	910,425	7,628,621	22,089,398	43,180,346

FIGURE 6.1.5
A sample bottom-up market model.

As noted, a top-down market model is based entirely on a number of statistical assumptions. Innovators can find many of these key statistics in scientific journals. Remember to cite all references to support the projections as this lends increased credibility. Also, draw on information developed as part of 2.4 Market Analysis.

To develop this particular top-down market model, the first step was to calculate the number of paroxysmal AF patients per year, starting with 960,000 in year 1 and increasing by two percent per year. The number of paroxysmal AF patients per year was then multiplied by the percentage of patients who are drug refractory (i.e., not responsive to pharmaceutical therapy) and, therefore, are candidates for ablation (28 percent). This represented the total number of patients who can be treated with ablation. That number was next multiplied by the percentage of drug refractory AF patients who are actually treated to get the number of patients treated per year. These percentages were taken from a Frost & Sullivan research report (it is always a good idea to incorporate documented statistics where applicable).

Because patients may need multiple procedures, the number of patients treated was then multiplied by the number of procedures per patient to produce the total number of procedures performed per year. The number of procedures performed per year was multiplied by the average selling price (ASP) per procedure ($2,000) to determine the total market size in dollars. To address the fact that the company will capture a fraction of this market, market share projections were multiplied by the total number of procedures per year to produce the target number of procedures per year, which corresponds to the total number of units sold per year. (Note that if more than one device was used per procedure, this would also need to be taken into account.) Finally, the total number of units sold per year was multiplied by the average selling price to produce the revenue target per year.

When determining market share, bear in mind that market share growth depends on a number of factors, including how innovative or unique the technology is, as well as how likely physicians are to adopt it. Factors such as how aggressive the company's sales plan is and how many sales representatives will be hired should also be taken into account. Finally, consider the number of competitors already in the market and the extent of their resources.

Experts caution that one should not be overly aggressive when modeling market penetration. In most cases, it is unlikely for a product to gain more than one percent market share in the first year unless there is substantial pent-up demand and the sales channel is already well established. Although there are some notable exceptions, the innovator should also assume that, in the long run, competition will join the market (if not already present) and that portions of the market will not necessarily be accessible. As a result, a long-term market share forecast of 15 to 30 percent is a realistic outcome, even in a successful scenario.

Bottom-up model

The bottom-up model takes a different approach to assessing the market as described in the seven steps listed below:

1. **Determine the fundamental unit of business** – In this first step, the key is to determine what drives business in the market. For a company selling capital equipment, the "unit of business" is the number of facilities that purchase the machine; the number of patients treated or number of physicians using the equipment is not the core driver of sales. For most medtech companies, however, physicians are often the key unit of business because they are the ones that drive the selection, sale, and use of a device. Unless a company is dealing with an over-the-counter or fee-per-use business model (see 4.4 Business Models), patients rarely serve as the fundamental business unit in the device arena.

2. **Consider the sales cycle** – The sales cycle involves all of the time and expense required to sell one unit of business. This includes the time and cost of raising the buyer's awareness of the technology, getting him/her interested, and closing the deal. If the buyer is a facility or hospital, standard purchasing processes and cycles (which can often be lengthy) must be taken into account. If the buyer is a physician, the time and cost of training him/her to use the device may also need to be calculated, as well as any follow-up training. For new sales reps, a learning curve must also be factored into these calculations (the time and cost of training the rep before s/he makes a first sale).

3. **Consider the adoption curve** – The adoption curve refers to the rate at which the buyer will utilize or consume the technology. For devices that present a relatively low risk and/or obvious benefits, the number of devices used over a given unit of time may grow relatively steadily. However, for higher-risk devices or technologies perceived as being more experimental, utilization may grow more gradually. For instance, a physician may perform a first procedure using the technology and then wait for several months to see how that preliminary patient responds before performing another. Certain facilities and medical specialties have a tendency to adopt new technologies faster than others, which should also be taken into account when developing the adoption curve. The key is to derive the anticipated utilization rate for a single unit of business on day one, day two, etc., to get a sense for how quickly the company will be able to build its sales volume. A typical metric that can be used to capture utilization is the percentage of the case volume of the business unit that utilizes the venture's products.

4. **Build the commercial effort** – At this point, the innovator stops thinking about what is required to make individual sales to a single unit of business and starts thinking about how many sales reps should be hired to build a reasonable business. From step 1, the innovator can determine how many units of business exist within the target market. From step 2, s/he understands how much total effort is required and expected to make each sale. From step 3, s/he has estimated a realistic adoption rate. Now, the challenge is to pull these factors together to determine how many sales reps are needed to grow the business at an appropriate, realistic pace. The key is to balance the innovator's desire for "reward" with his/her tolerance for risk. Hiring too many sales reps at once can be a costly mistake until key assumptions regarding the sales cycle and adoption curve have been tested in the market. On the other hand, the company needs to be able to drive enough sales within a realistic time frame to sustain its operations and keep its commitments to investors. Typically, it is advisable to run multiple scenarios (what would it look like to start with five sales reps? ten sales reps? fifteen sales reps?), then choose the one that strikes the best risk/reward ratio.

5. **Consider market development factors** – The next step is to consider other factors in the external environment that have the ability to affect the overall market model. For example, reimbursement coverage can have a major effect

on market adoption. If reimbursement has not yet been achieved, this could affect the size of the initial sales force that is appropriate. Similarly, if professional societies have not yet endorsed the technology or key data have not been published, it may be wise to start with a smaller sales force focused on converting early adopters until some of these other factors have been put into place to support more widespread sales. Importantly, the top-down market model does not explicitly take these types of factors into account, which is one of its inherent weaknesses.

6. **Factor in product evolution** – If the innovator anticipates that subsequent versions of the technology will become available within a one- to three-year time frame, this should be reflected in the market model. New versions of a technology have the potential to increase utilization based on the improvements made and/or features added that potentially make the technology relevant to a greater number of procedures.

7. **Consider other factors** – Finally, an innovator should look closely at the market, the buyer, and the technology to determine if there are any other factors that might affect sales. For example, some medical specialties are more seasonal than others (e.g., orthopedics is busiest in the winter and the summer when people participate in seasonal sporting activities). Buying behavior in certain medical specialties can also be influenced by major medical conferences that occur at a certain time of year (when physicians and hospital administrators convene to check out new technology in the field). These types of considerations should be factored into the overall market model and the timing of key market decisions.

The following example, on Cardica, Inc., demonstrates how a bottom-up market model is developed in practice within a medical device start-up. More details about Cardica and a more complete financial model for the company can be found in Appendix 6.1.1 on **proxy company analysis**.

FROM THE FIELD

CARDICA, INC.

Developing a bottom-up market model for a direct sales force focused on coronary artery bypass graft (CABG) surgery

Cardica, Inc. was cofounded by Bernard Hausen and Steve Yencho, PhD, in 1997 in order to design and manufacture proprietary products that automate the connection, or anastomosis, of blood vessels during CABG surgery. In CABG procedures, veins or arteries are used to construct "bypass" conduits to restore blood flow beyond closed or narrowed portions of coronary arteries. This is typically accomplished by suturing one end of the vein or arterial conduit to the aorta and the other end to the coronary artery at a site beyond the blockage. Cardica's product portfolio included two products by 2004: the C-Port® Distal Anastomosis System (referred to as C-Port) and the PAS-Port® Proximal Anastomosis System (referred to as PAS-Port). The C-Port product was used to connect a bypass graft to the coronary arteries while PAS-Port was to automate the connection of the graft to the aorta. In November, 2005, Cardica received 510(k) clearance for C-Port, but PAS-Port was deemed to require a PMA and a randomized clinical trial (more details about the difference are provided in the proxy example analysis in Appendix 6.1.1). As a result, management expected a potential three-year delay between the US commercial launch of its C-Port and PAS-Port systems. See Figures 6.1.6 and 6.1.8 for photos of the devices.

Early in the biodesign innovation process, Cardica's management team made the decision to develop a direct sales force. Bob Newell, Cardica's CFO, explained: "A key metric we use to measure progress in our business is the number of trained surgeons using the product. In the US there are about 3,000 surgeons performing approximately 250,000 CABG surgeries per year in 1,000 hospitals; but about 225 hospitals perform 50 percent of all procedures. So we can go after these higher volume facilities with a targeted sales force. Our sales force doesn't have to be really huge, like it would be if we were in interventional cardiology. In addition, we do

FROM THE FIELD — CARDICA, INC.

Continued

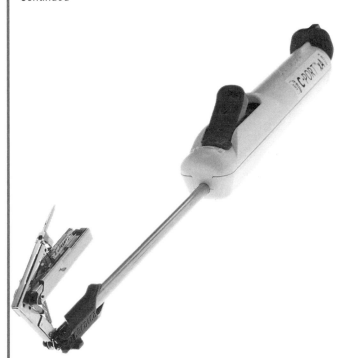

FIGURE 6.1.6
Cardica's C-Port xA distal anastomosis system (courtesy of Cardica, Inc.).

not need to target all surgeons in these hospitals. There are two primary segments: on-pump and off-pump CABG surgery. On-pump is the traditional way of doing bypass surgery where the heart is stopped and the patient is on a bypass machine, which filters their blood and keeps the blood flowing while the heart is stopped. About 75 percent of the market still does bypass surgery that way. A newer method of doing bypass surgery is called off-pump, or beating heart surgery, which means that the surgery is performed on a heart that has not stopped – it continues to beat and the blood is not bypassed out into a bypass machine. Beating heart surgeons tend to be early adopters, so most of our first customers are likely to be beating heart surgeons."[3] Cardica's management estimated that 225 surgeons performed the overwhelming majority of beating heart surgeries.

Cardica's basic sales strategy was to first train surgeons based on the use of the C-Port device. The company's goal was to have many high-volume beating heart surgeons trained by the time the PAS-Port product was approved and launched. This gave the company three years from the time of C-Port's launch to train its first adopters. Newell estimated that the main expense in educating a surgeon was the time it took the sales person to deliver the training. After a four- to six-hour initial instruction at the company's facilities, which included a brief excursion to the wet lab, the sales person would attend five to six cases performed by the surgeon. Once a surgeon was trained, it could take anywhere between two to nine months to reach steady-state sales for that customer.

Cardica management estimated that C-Port would have a theoretical total US market of 875,000 units based on an average of 3.5 distal anastomoses per procedure and a total number of 250,000 procedures. Of this amount, the beating heart segment would be about 25 percent or a total available theoretical market of just over 200,000 units. PAS-Port, on the other hand, would potentially have a total theoretical market of 375,000 units based on an average of 1.5 proximal anastomoses per procedure. (The difference between the average number of distal and proximal anastomosis is because some of the arterial bypass conduits, such as the internal mammary artery, are used with their proximal blood flow site intact and thus no proximal anastomosis is required.) The price for C-Port was set at $800 per anastomosis, while PAS-Port was $600 per anastomoses. Using these numbers, management estimated that the total theoretical annual US sales with 100 percent penetration would be $225 million for PAS-Port (250,000 procedures × 1.5 devices per procedure × $600 per device) and $700 million for C-Port (250,000 procedures × 3.5 devices per procedure × $800 per device). The corresponding numbers for the beating heart segment were: $56 million and $170 million (all numbers rounded to the nearest million).

The company then estimated that a salesperson would cost between $350,000 and $400,000 per year (fully loaded) and s/he could generate a maximum of $1.5 to $2.5 million in annual sales. Once a salesperson reached that threshold, a new salesperson would be needed in the territory to cover additional surgeons. It would take Cardica two to three months to train a salesperson, and the company anticipated that each salesperson could bring in and train three new surgeons per quarter (as long as s/he was not maxed out).

	C-Port® Launch					PAS-PORT® Launch			
Quarter	1	2	3	4	9	10	11	12
Direct Sales Force	3	3	3	10	10	10	10	10
Physicians Trained in Current Quarter	10	10	10	10	30	30	30	30
Physicians Trained (and retained) in Prior Q		7	7	7	20	20	20	20
Physicians Trained (and retained) Two Qs Ago			7	7	20	20	20	20
Fully Trained & Retained Physicians				7	80	100	150	140
Total Trained Physicians	10	20	30	40	190	220	250	280
Total C-Port Sales		37,037	111,111	222,222	1,666,667	2,000,000	2,333,333	2,666,667
Total Pas-Port Sales								
Total Sales		37,037	111,111	222,222	1,666,667	2,000,000	2,333,333	2,666,667
Quarterly Sales per Sales Rep		12,346	37,037	22,222	166,667	200,000	233,333	266,667
Total Sales Force Costs	300,000	300,000	300,000	1,000,000	1,000,000	1,000,000	1,000,000	1,000,000

FIGURE 6.1.7
The phase I market model (note: in this model, the number of total trained physicians reflects the cumulative number of physicians trained, ignoring the 2/3 retention rate; the delay in sales following the launch of both products reflects the need for physician training time).

To develop a bottom-up market model based on this information requires two phases of effort. Phase I covers the initial period between C-Port and PAS-Port launch. Phase II covers PAS-Port launch and beyond.

The goal of phase I is to train early adopters at the rate of 25 to 30 per quarter and to have as many as 250 to 300 or more surgeons trained and using C-Port by the time PAS-Port is approved. Assuming a two-third retention rate over a three-year period, this translates into roughly 180 to 200 retained surgeons by Q12. This calculation implies that about ten new surgeons can be trained each quarter in the first year. A reasonable target in Q5 to Q12 is 30 new trained surgeons per quarter (which translates into a direct sales force of ten reps). The model in Figure 6.1.7 outlines the details (each row in the model represents a quarter and the cost for a fully loaded salesperson is $400,000 per year or $100,000 per quarter).

A small sales force of three salespeople is in place for the first three quarters, until it increases to ten in the fourth quarter. The model assumes a quarter of training for new sales people; therefore, after training for one quarter, the additional salespeople allow Cardica to reach the target surgeon training level of 30 in Q5. The model keeps track of the number of surgeons trained in the current quarter, as well as in the previous two quarters in order to capture the three to nine months it takes for newly trained surgeons to reach steady-state sales. The calculation of total sales assumes surgeons trained in the current quarter do not perform any procedures, while surgeons trained in previous quarters perform increasingly more procedures until reaching ten percent of their case volume nine months after they complete training. The model indicates that the sales force will generate enough sales to cover its expenses by Q8 (not shown).

FIGURE 6.1.8
PAS-Port proximal anastomosis system (courtesy of Cardica, Inc.).

The next step is to consider phase II: PAS-Port launch and beyond. With an established base of 180 trained and retained off-pump surgeons (140 fully trained with more than two quarters of experience, plus 40 trained and retained for at least one or two quarters) by Q12, one would expect a successful launch and potentially a rapid, "hockey stick" pattern of adoption for PAS-Port. Cardica's management estimated that all retained C-Port surgeons could be trained on the PAS-Port within two quarters after the PAS-Port launch, given their familiarity and use of a Cardica device and relationship with a Cardica salesperson. New surgeons would continue to be

Stage 6 Integration

FROM THE FIELD — CARDICA, INC.

Continued

	PAS-PORT® Launch							
Quarter	11	12	13	14	15	16	17	18
Direct Sales Force	13	13	13	13	13	17	17	17
Physicians Trained in Current Quarter	30	30	30	30	30	30	30	30
Physicians Trained (and retained) in Prior Q	20	20	20	20	20	20	20	20
Physicians Trained (and retained) Two Qs Ago	20	20	20	20	20	20	20	20
Fully Trained & Retained Physicians	120	140	160	180	180	220	240	260
Total Trained Physicians	250	280	310	340	340	400	430	460
Total C-Port Sales	2,333,333	2,666,667	3,000,000	3,333,333	3,666,667	4,000,000	4,333,333	4,666,667
Total Pas-Port Sales				964,286	2,142,857	3,937,500	3,937,500	3,937,500
Total Sales	2,333,333	2,666,667	3,000,000	4,297,619	5,809,524	7,937,500	8,270,833	8,604,167
Quarterly Sales per Sales Rep	179,487	205,128	230,769	330,586	446,886	466,912	486,520	506,127
Total Sales Force Costs	1,300,000	1,300,000	1,300,000	1,300,000	1,300,000	1,700,000	1,700,000	1,700,000

FIGURE 6.1.9
The phase II market model.

trained at the rate of 30 per quarter on both PAS-Port and C-Port.

Further, management believed the penetration of PAS-Port into the surgeon's case volume would be more rapid than C-Port due to more accepted clinical benefit (one of the benefits of PAS-Port is that it avoids the clamping of the aorta that is required in manual procedures, which can lead to a stroke). They felt that it would be reasonable to expect a 25 to 50 percent penetration within three quarters from launch (the model shown assumes a penetration of 35 percent). Under these assumptions, Cardica's total quarterly US sales would be expected to exceed $8 million by Q17 as shown in Figure 6.1.9. This sales growth would strain the sales force of ten. Therefore, Cardica would need to build its sales force in anticipation of this growth. The model also assumes that recruitment and training for the sales force happens early in Q11 with a modest increase from 10 to 13 to anticipate the new, additional needs of the existing trained base. The sales force will then increase again to 17 by Q16. Other assumptions include: sales territories are split and new salespeople are added when the existing sales of a salesperson reaches the target $1.5 to $2.5 million per year and each new salesperson requires one quarter of training. The model shown in Figure 6.1.9 starts from Q10 and modifies the sales people row to capture the increased needs due to the product launch.

This model demonstrates that by Q18 a sales force of 17 will be able to cover 300 trained and retained surgeons, which represents about 17 surgeons per salesperson. Sales activities will include a mixture of supporting existing surgeons and bringing new surgeons on board. Total sales of $8.6 million per quarter will be reached in Q18. The reader will notice that despite the markedly increased sales force, the number of new trained physicians per quarter remains constant. This is because most of the salespeople are now spending more of their time maintaining and supporting existing surgeons and spend a smaller fraction of their time recruiting new surgeons. A summary of the assumptions made in the model are provided below:

- 3,000 surgeons performing CABG surgeries in the United States in 1,000 hospitals.
- 225 hospitals perform roughly 50 percent of all procedures.
- Of the 250,000 CABG surgeries annually in the United States, approximately 25 percent are "off pump" or beating heart procedures.
- Surgeons performing off-pump CABG surgeries will be most of the company's first adopters.
- Cardica's goal is to train 300 or more surgeons on the use of C-Port by the time PAS-Port is approved for commercial use.
- The company anticipates that it will take approximately three years (Q1 to Q12) from the time of the C-Port commercial launch to the approval of PAS-Port.
- Once a surgeon is trained, it takes two to nine months to achieve a steady-state sales volume.
- Approximately two-thirds of trained surgeons are "retained," or continue to use the product.

- C-Port would be used in approximately ten percent of the CABG procedures with an average of 3.5 anastomoses per procedure at $800 each.
- PAS-Port would be used in 25 to 50 percent of all beating heart procedures and 5 to 10 percent of on-pump procedures with an average of 1.5 anastomoses per procedure at $600 each.
- The cost of a Cardica direct sales rep would be approximately $350,000 to $400,000 per year, fully loaded.
- Each sales rep has the capacity to generate $1.5 to $2.5 million in sales per year.
- Once a sales person reaches that threshold, a new rep would have to be hired and the territory divided.

Discussion of the Cardica bottom-up model
This model is only a starting point. The firm can experiment further with its core assumptions, especially with respect to the ramp-up in Q12, training requirements for existing surgeons on PAS-Port, and the training of new surgeons once PAS-Port and C-Port are both in the market. Other considerations that could be included in the model are outlined below.

Reimbursement
The analysis shown above does not factor in the effect of reimbursement, if any, on sales. In reality, product adoption can be significantly influenced by the status of reimbursement coverage. As a result, this issue should be factored into the model.

Expansion in indications
While the model above focuses primarily on the adoption of PAS-Port and C-Port among surgeons performing beating heart surgeries, Cardica's management team believes that both PAS-Port and C-Port will eventually also be used in non-beating heart CABG surgeries, and also believes there may be potential applications in other vascular grafting procedures.

Changes in the business environment
The model can also be used to monitor changes in the external environment and progress within the business to determine if the company's targets remain attainable. Such a model should never be static. Management must actively monitor conditions and then factor them into the model such that its timing is optimal regarding issues such as when to expand the sales force or divide sales territories.

The sales organization
The current model does not provide details about how the initial sales territories will be organized. Typically, this kind of analysis should be complemented by a preliminary design of sales territories, since different regions are likely to drive different levels of sales. In the United States, it is common for companies to focus their preliminary sales effort in major, high-volume metropolitan areas such as New York, Boston, Chicago, San Francisco, Los Angeles, San Diego, Denver, and Miami. Utilization data from Medicare and the Dartmouth Atlas of Health Care[4] can be used to determine the highest-volume areas for purposes of designing the initial territories.

Reconciliation with the top-down model
As mentioned earlier, a quick back-of-the-envelope calculation shows that the total US markets for beating heart CABG surgeries in which PAS-Port and C-Port could be used were estimated to be $56 million and $170 million, respectively. Assuming a 35 percent penetration for PAS-Port and 10 percent penetration for C-Port into the total caseload of beating heart CABG surgeries (both are assumptions made in the bottom-up model), this brings the estimated annual US sales to $20 million and $17 million for total annual sales of $37 million, or slightly more than $9 million per quarter. It is reassuring that this quick, back-of-the-envelope, calculation is consistent with estimated quarterly sales of $8.6 million by Q18 in the bottom-up model. It should be noted that both of these sales estimates do not necessarily represent the final steady-state sales for Cardica but rather, steady-state sales in the first adopter segment of beating heart surgeries.

Table 6.1.1 *Medtech salaries can vary by geographic location and stage of the company's development, but generally fall within standard ranges.*

Typical medtech salary ranges
Vice-president – $150,000 to $200,000 per year
Scientist or engineer – $75,000 to $100,000 per year
Sales representative – $75,000 to $100,000 per year
Technician or lab worker – $40,000 to $70,000 per year
Assembler – $30,000 to $60,000 per year
Administrative employee – $30,000 to $55,000 per year
Consultants – $1,000 to $2,000 per day

Cost projections

Cost projections (also called the operating statement) calculate the estimated costs of the business, including manufacturing costs and operating expenses (**OpEx**). Manufacturing costs include both material costs (COGS), as well as manufacturing labor, facilities, and equipment. They capture how much it costs for the company to make the products it sells. Operating expenses captures all other costs not included in manufacturing costs, including R&D, sales staff, general and administrative functions, and non-production facilities costs.

A cost analysis can be performed at varying levels of detail. It is up to the innovator to decide how much detail is required to satisfy the target audience for the business plan.

Salary analysis

Using the ablation catheter example, the creation of cost projections begins with an employee salary analysis, which includes three main components:

1. Summarize the staffing plan by calculating the average number of hires per year by type.
2. Outline annual salary assumptions, or how much the company will have to pay in salary by employee type, for each position type (more information is provided below).
3. Calculate **fully burdened cost** per employee per year cost by position, taking into account employee healthcare and other benefits, insurance, computers, desk chairs, equipment, etc. for each employee. The fully burdened cost represents the total annual cost for each employee.

Benchmarks for average annual salaries are widely available. Table 6.1.1 provides some sample ranges that were developed based on estimates from Salary.com and then verified with an experienced innovator in the medical device field.[5]

Rather than figuring out the exact fully burdened cost, many experts use a back-of-the-envelope factor of two-times the annual salary for each employee (e.g., the fully burdened cost of a scientist would be $100,000 × 2 = $200,000). The 2× multiplier, used to account for benefits, insurance, etc., applies to all employees except manufacturing employees, for which a 1.5× multiplier should be used. Additionally, do not forget to take into account annual salary increases, as well as the need to offer stock to attract and retain high-caliber talent.

Figure 6.1.10 shows a sample salary analysis that builds on the staffing plan in Figure 6.1.4.

Manufacturing costs

Manufacturing costs have at least three primary components: manufacturing labor costs, manufacturing facilities costs, and raw materials costs (COGS). Figure 6.1.11 shows a sample model for capturing this data.

In this analysis, to calculate manufacturing labor costs, engineers per year (from the staffing model) were multiplied by their fully burdened salary cost to produce a total engineer cost. The total number of labor employees (e.g., assemblers, processors, testers) per year was similarly multiplied by their fully burdened salary cost to produce total manufacturing labor cost. These two figures were added together, providing the total manufacturing labor cost.

To calculate manufacturing facilities costs, the analysis started with an estimate for the cost of facilities space per square foot. This assumed $25 per square foot per year, inflated at 2.5 percent per year based on

	Year 1	Year 2	Year 3	Year 4	Year 5	Year 6	Year 7
Manufacturing team							
Engineers	0	0	3	6	9	9	12
Manufacturing assemblers	0	0	2	5	10	18	35
R&D							
Catheter team							
Engineers	1	1	2	2	2	3	5
Techs	3	4	5	6	6	7	10
Generator team							
Engineers	1	1	1	1	1	3	5
Techs	2	4	4	4	4	7	10
SG&A							
Salesforce	0	0	1	9	12	25	50
Marketing, business development, and reimbursement	0	1	2	3	4	5	5
Clinical advisor	1	1	1	1	2	2	2
Regulatory	0	0	1	4	5	5	5
Administrative asst.	0	1	1	2	3	4	5
Management	1	2	3	5	6	8	8

Salary assumptions	Factor	Salary ($)
Engineering	2	130,000
Technicians	2	50,000
Clinical, regulatory & QA	2	130,000
Sales, marketing, bus. development	2	100,000
Administration	2	35,000
Manufacturing	1.5	35,000
Management	2	200,000
Yearly salary increase	2.5%	

Fully burdened salary - per employee ($)	Year 1	Year 2	Year 3	Year 4	Year 5	Year 6	Year 7
Engineering	260,000	266,500	273,163	279,992	286,991	294,166	301,520
Technicians	100,000	102,500	105,063	107,689	110,381	113,141	115,969
Clinical, regulatory & QA	260,000	266,500	273,163	279,992	286,991	294,166	301,520
Sales, marketing, bus. development	200,000	205,000	210,125	215,378	220,763	226,282	231,939
Administration	70,000	71,750	73,544	75,382	77,267	79,199	81,179
Manufacturing	52,500	53,813	55,158	56,537	57,950	59,399	60,884
Management	400,000	410,000	420,250	430,756	441,525	452,563	463,877

FIGURE 6.1.10
A sample salary analysis.

top tier and second tier office space in Silicon Valley.[6] Next, space per employee was calculated by assuming 250 square feet per employee. This estimate can be reduced over time to 210 square feet per employee as manufacturing processes become more efficient.

Next, the total space required was determined by multiplying the space per employee by the number of manufacturing employees.

Then, assumptions were made about projected facilities space, for instance that the company can buy facilities space in blocks and always has more facilities space than needed, but employee moves are limited to a reasonable number. In this example, the first two years of manufacturing (years 3 and 4) are satisfied by 3,000 square feet. A move is made in year 5 to a 15,000 square foot facility. Finally, the projected facilities space was multiplied by the cost of facilities space to produce the manufacturing facilities cost.

To calculate raw material costs, the total units forecast to be sold (taken from the market model) was multiplied by the cost of raw materials per product. In the model, it is assumed that the cost of raw materials drops by ten percent per year as production volume increases to reflect learning and volume discounts ($600 in year 3 down to $394 in year 7).

Finally, these three cost elements were summed and that figure was taken as the total manufacturing cost. It should be emphasized here that the process of developing

Stage 6 Integration

Manufacturing labor	Year 1	Year 2	Year 3	Year 4	Year 5	Year 6	Year 7
# of engineers	-	-	2.5	6.0	9.0	9.0	12.0
Fully loaded employee cost ($)	260,000	266,500	273,163	279,992	286,991	294,166	301,520
# of direct labor	-	-	2.0	5.0	10.0	18.0	35.0
Fully loaded employee cost ($)	52,500	53,813	55,158	56,537	57,950	59,399	60,884
Manufacturing labor cost ($)	-	-	793,222	1,962,633	3,162,424	3,716,676	5,749,180
Manufacturing facilities	Year 1	Year 2	Year 3	Year 4	Year 5	Year 6	Year 7
Cost per sq. foot ($)	25.00	25.63	26.27	26.92	27.60	28.29	28.99
Inflation	n/a	2.5%	2.5%	2.5%	2.5%	2.5%	2.5%
Sq. footage / employee	250	250	250	240	230	220	210
# of manufacturing employees	-	-	4.5	11.0	19.0	27.0	47.0
Sq. footage required	-	-	1,125	2,640	4,370	5,940	9,870
Projected sq. footage	-	-	3,000	3,000	15,000	15,000	15,000
Manufacturing facilities cost ($)	-	-	78,797	80,767	413,930	424,278	434,885
Raw materials	Year 1	Year 2	Year 3	Year 4	Year 5	Year 6	Year 7
Units sold	-	-	23	462	3,892	11,325	22,249
Raw material costs per unit ($)	-	-	600	540	486	437	394
% improvement	n/a	n/a	n/a	10%	10%	10%	10%
Raw material & packaging costs (COGS) ($)	-	-	13,843	249,539	1,891,298	4,953,558	8,758,685
Manufacturing Costs	Year 1	Year 2	Year 3	Year 4	Year 5	Year 6	Year 7
Manufacturing labor cost ($)	-	-	793,222	1,962,633	3,162,424	3,716,676	5,749,180
Manufacturing facilities cost ($)	-	-	78,797	80,767	413,930	424,278	434,885
Cost of Goods Sold ($)	-	-	13,843	249,539	1,891,298	4,953,558	8,758,685
Total ($)	-	-	885,862	2,292,939	5,467,652	9,094,512	14,942,750

FIGURE 6.1.11
A sample model for estimating manufacturing costs.

a manufacturing cost model is tightly coupled with the company's R&D and manufacturing strategy. Answers to the following questions must be available before an accurate model can be developed. What will the manufacturing process be? Where will manufacturing be done (in-house, outsourced in the United States or outsourced outside the United States)? What components can be made in-house and what components can be purchased or outsourced? Who will be the suppliers of the outsourced components and who will supply the raw material(s)? What kind(s) of equipment will be needed? What are the cost projections for these components?

Operating expenses

Operating expenses (OpEx) refers to the following components in the model: R&D staff costs, clinical trials costs, SG&A staff costs, and non-manufacturing facilities costs. Figure 6.1.12 shows a sample of an OpEx analysis.

To calculate OpEx, estimate R&D staff spending by multiplying the number of R&D-related employees per year by their fully burdened salary for each type of employee for a given year. In this example, engineers and technicians are considered R&D employees. All employee data is taken from the staffing plan and salary analysis.

Next, estimate clinical trials costs as follows. First, review the operating plan and 5.3 Clinical Strategy to determine the length of the clinical trials and the number of patients. The trial length and strategy should be determined early, as part of the overall operating strategy. In this example, the trials will be completed in six quarters: quarters 7 through 12. Therefore, year 2 will include two quarters or 0.5 years of clinical trials (quarters 7 and 8) and year 3 will include all four quarters or a full year of trials. The number of patients participating in the clinical trials is assumed to be 100. Since year 2 includes 0.5 years of clinical trials, it includes 50 patient-years. Similarly, year 3 includes 100 patient-years. Then, make another assumption about the cost per patient-year, $15,000, which is a reasonable (but somewhat optimistic) estimate for invasive medical device clinical trials. (Cost per patient-year is the cost of one patient

R&D	Year 1	Year 2	Year 3	Year 4	Year 5	Year 6	Year 7
Staff costs							
# of engineers	2.0	2.0	2.8	3.0	3.0	6.0	10.0
Fully loaded employee cost ($)	260,000	266,500	273,163	279,992	286,991	294,166	301,520
# of techs	4.5	8.0	8.5	10.0	10.0	14.0	20.0
Fully loaded employee cost ($)	100,000	102,500	105,063	107,689	110,381	113,141	115,969
Staff costs	970,000	1,353,000	1,644,228	1,916,865	1,964,787	3,348,968	5,334,590
Clinical trials costs							
Est. # of patient-years in each year	0	50	100	0	0	0	0
Cost per patient-year	15,000	15,000	15,000	15,000	15,000	15,000	15,000
Total cost of trials	0	750,000	1,500,000	0	0	0	0
R&D costs ($)	970,000	2,103,000	3,144,228	1,916,865	1,964,787	3,348,968	5,334,590

SG&A	Year 1	Year 2	Year 3	Year 4	Year 5	Year 6	Year 7
Staff costs							
# of sales reps	0.0	0.0	1.0	9.0	12.0	25.0	50.0
Fully loaded employee cost ($)	200,000	205,000	210,125	215,378	220,763	226,282	231,939
# of clinical, regulatory & QA	1.0	1.3	2.0	4.5	6.3	7.0	7.0
Fully loaded employee cost ($)	260,000	266,500	273,163	279,992	286,991	294,166	301,520
# of marketing & bus. development	0.0	0.8	1.5	3.0	3.8	5.0	5.0
Fully loaded employee cost ($)	200,000	205,000	210,125	215,378	220,763	226,282	231,939
# of administrative assistants	0.0	0.8	1.0	1.5	2.8	4.0	5.0
Fully loaded employee cost ($)	70,000	71,750	73,544	75,382	77,267	79,199	81,179
# of management	1.0	2.0	2.5	4.5	6.0	8.0	8.0
Fully loaded employee cost ($)	400,000	410,000	420,250	430,756	441,525	452,563	463,877
Staff costs	660,000	1,360,688	2,195,806	5,895,976	8,132,341	12,784,913	18,984,181
Facilities costs							
Cost per sq. foot ($)	25.00	25.63	26.27	26.92	27.60	28.29	28.99
Inflation	n/a	2.5%	2.5%	2.5%	2.5%	2.5%	2.5%
Sq. footage / employee	200	200	200	200	200	200	200
# of employees	8.5	14.0	17.8	32.5	40.0	64.0	100.0
Sq. footage required	1,700	2,800	3,550	6,500	8,000	12,800	20,000
Projected sq. footage	3,000	3,000	10,000	10,000	10,000	25,000	25,000
Facilities cost ($)	75,000	76,875	262,656	269,223	275,953	707,130	724,808
SG&A costs ($)	735,000	1,437,563	2,458,463	6,165,199	8,408,295	13,492,043	19,708,990

Operating costs	Year 1	Year 2	Year 3	Year 4	Year 5	Year 6	Year 7
R&D costs ($)	970,000	2,103,000	3,144,228	1,916,865	1,964,787	3,348,968	5,334,590
SG&A costs ($)	735,000	1,437,563	2,458,463	6,165,199	8,408,295	13,492,043	19,708,990
Total operating costs ($)	1,705,000	3,540,563	5,602,691	8,082,064	10,373,082	16,841,011	25,043,579

FIGURE 6.1.12
A sample OpEx model.

participating in a clinical trial with one-year follow-up). Multiply the patient-years by the cost per patient-year and sum over the years involved, in this case years 2 and 3, to yield the total clinical trials cost.

Calculate SG&A staff spending in the same way as R&D staff spending. In this example, SG&A employees include sales reps, marketing and business development, clinical and regulatory employees, administrative assistants, and managers. Determine non-manufacturing or SG&A facilities costs using a method similar to that applied to manufacturing facility costs but using a higher initial cost per square foot. Then, take the R&D, SG&A, and facilities expenses and sum them to calculate the company's total OpEx.

It is worth noting that clinical and regulatory staff is sometimes categorized under R&D staff and sometimes under SG&A. In this example they are allocated under SG&A.

Be aware that clinical trials costs can vary widely, depending on the invasiveness of the device and length

Table 6.1.2 *The elements of an income statement.*

Item	Description
Revenue (Sales)	Total sales for the year.
Manufacturing costs	Total cost of the products *actually* sold by the company. In the case of ablation catheters, this includes cost of raw materials and labor to assemble the device and any other component that went directly into the production of the device. Does *not* include expenses such as marketing, sales costs, management salaries, etc.
Operating (gross) margin	Revenue minus manufacturing costs.
OpEx	All the other expenses associated with running the business that were not incorporated into manufacturing costs. Includes items such as R&D, facilities rentals, SG&A, company functions, etc.
Operating income	Operating margin minus operating expenses.

of study follow-up. Trials for non-implantable therapeutic devices and diagnostics with a short follow-up period can cost as little as $2,000 per patient-year (or $3,500 per patient on average). However, these costs can climb to as much as $100,000 per patient for an implantable or therapeutic device that typically requires a lengthy follow-up period. Trial expenses may also depend on whether some of the costs for treating the patient (e.g., physician and facility reimbursement) will be covered by Medicare or private payers.[7] Additionally, the number of patients in the trial should be based on the number of patients statistically needed to establish a particular clinical result. Clinical trial strategy and planning is discussed in more depth in 5.3 Clinical Strategy.

Income statement

The income statement brings together all of the elements of the financial model into a unified view of the company's expected financials. It is also known as an earnings statement, statement of operations, or profit and loss (P&L) statement, and includes the line items shown in Table 6.1.2.

Constructing the income statement is quite simple, since it only requires the innovator to pull together his/her previous calculations. Figure 6.1.13 shows a sample of an income statement.

When investors examine an income statement, they typically apply a series of guidelines to check and see if the financial plan is realistic. The following are principles that apply to mature medical device companies:

- A typical gross margin at maturity should be around 70 percent (gross margin equals revenue less manufacturing costs). Many companies target a 60 percent gross margin in their initial business plans.
- R&D is roughly 10 to 15 percent of sales at maturity.
- SG&A expenses are roughly 30 percent of sales.
- While not usually highlighted as a separate line on the income statement, SG&A facilities expenses are roughly 1 percent of sales.

Double check the income statement against these ratios to help anticipate how it will be received by potential investors, and/or make adjustments. In this example, the only expense category that deviates from medical device norms is SG&A. Therefore, innovators should examine carefully that component of the income statement and either rationalize it or determine ways to reduce it.

In the ablation catheter business example, the business loses money for the first six years due to important early investments in R&D, clinical trials, etc. In general, it takes longer for device companies to become profitable than it did in the past due to the changing, more competitive environment. The operating margin for the sample company reaches seven percent in year 7, which is modest for a medical device company.

6.1 Operating Plan and Financial Model

	Year 1	Year 2	Year 3	Year 4	Year 5	Year 6	Year 7
Revenue ($)	-	-	45,684	910,425	7,628,621	22,089,398	43,180,346
Manufacturing costs ($)	-	-	885,862	2,292,939	5,467,652	9,094,512	14,942,750
Gross margin ($)	-	-	(840,178)	(1,382,514)	2,160,969	12,994,886	28,237,595
Gross margin (% of sales)	N/A	N/A	N/A	-152%	28%	59%	65%
Operating expenses							
R&D costs ($)	970,000	2,103,000	3,144,228	1,916,865	1,964,787	3,348,968	5,334,590
% of sales	N/A	N/A	6883%	211%	26%	15%	12%
SG&A costs ($)	735,000	1,437,563	2,458,463	6,165,199	8,408,295	13,492,043	19,708,990
% of sales	N/A	N/A	5382%	677%	110%	61%	46%
SG&A facility costs ($)	75,000	76,875	262,656	269,223	275,953	707,130	724,808
% of sales	N/A	N/A	575%	30%	3.6%	3.2%	1.7%
Total operating expenses	1,705,000	3,540,563	5,602,691	8,082,064	10,373,082	16,841,011	25,043,579
% of sales	N/A	N/A	12264%	888%	136%	76%	58%
Pre-tax operating profit ($)	(1,705,000)	(3,540,563)	(6,442,869)	(9,464,578)	(8,212,112)	(3,846,125)	3,194,016
Operating margin	N/A	N/A	N/A	N/A	N/A	-17%	7%

FIGURE 6.1.13
A sample income statement for the hypothetical catheter ablation business.

Cash flow statement

At this point, the innovator is ready to determine the exact cash needs of the business. These are not the same as the net result of the income statement. The discrepancy is due to an accounting concept called depreciation. Sometimes when a business spends cash, it does not record it as a cost on the income statement right away. For example, the company may purchase a computer for $1,000, which has a useful life of three years. On the income statement, the company may record a cost of $333 per year, representing the expended value of the computer each year. However, it still requires $1,000 cash upfront to make the purchase. Similarly, the company will need to spend cash on raw materials to build its product, but it will take time before the products are sold (and therefore recognized as sales on the income statement). As a result, cash will move out of the hands of the company before the company recognizes the corresponding revenue.

When a company's cash is equal to zero, the business is essentially bankrupt. For this reason, having a cash requirements plan is extremely important. Such a plan can also help the company balance the need to be frugal with the need to allow appropriate spending to support the growth of the business – a delicate but essential balance that every company must strike.

Figure 6.1.14 shows a **cash flow statement**, incorporating financing.

To complete a cash requirements plan, the innovator needs to begin with an actual pre-tax **operating profit** (loss) from an income statement. Then, elements representing cash flow "out the door" can be added in, which are not immediately deducted from the income statement. This includes (as shown in the model used here) capital equipment purchased in three categories:

- Cost of capital equipment per employee. This includes cost of computers, phones, desks, etc. In the model, it is estimated to be $7,500 per employee.
- Cost of clean rooms in the years that manufacturing facilities are developed.
- Cost of manufacturing equipment.

Next, consider raw materials costs (taken from manufacturing analysis when developing cost projections) needed to build up inventory. In this example, this is taken to be 35 percent of the total raw material costs.

Subtract these elements from the pre-tax operating profit (loss) to produce total cash flow per year. Next, add cumulative cash flow from the prior year (presumably $0 prior to first year) to total cash flow from the current year to produce cumulative cash flow for the

Stage 6 Integration

	Year 1	Year 2	Year 3	Year 4	Year 5	Year 6	Year 7
Pre-tax operating profit (loss) ($)	(1,705,000)	(3,540,563)	(6,442,869)	(9,464,578)	(8,212,112)	(3,846,125)	3,194,016
Capital equipment purchased							
# of non-mfg employees	8.5	14.8	21.8	41.5	52.8	78.0	117.0
Cost of computers, phones, desks / employee ($)	7,500	7,500	7,500	7,500	7,500	7,500	7,500
Total cost of computers, phones, desks ($)	63,750	46,875	116,250	195,000	200,625	384,375	493,125
Clean room ($)	0	0	300,000	0	0	600,000	0
Manufacturing equipment ($)	0	150,000	150,000	300,000	150,000	300,000	200,000
Total capital equipment purchase ($)	63,750	196,875	566,250	495,000	350,625	1,284,375	693,125
Raw material costs ($)	0	0	13,843	249,539	1,891,298	4,953,558	8,758,685
Factor	0.35	0.35	0.35	0.35	0.35	0.35	0.35
Inventory build ($)	0	0	4,845	87,339	661,954	1,733,745	3,065,540
Total cash flow from the year ($)	(1,768,750)	(3,737,438)	(7,013,964)	(10,046,917)	(9,224,692)	(6,864,245)	(564,649)
Cumulative cash flow ($)	(1,768,750)	(5,506,188)	(12,520,152)	(22,567,069)	(31,791,760)	(38,656,006)	(39,220,655)
Cash needs	Year 1	Year 2	Year 3	Year 4	Year 5	Year 6	Year 7
Cash balance ($)	(1,768,750)	(2,506,188)	2,479,848	(7,567,069)	10,208,240	3,343,994	2,779,345
Suggested amount financed ($)	3,000,000	12,000,000		27,000,000			
Post-financing cash in the bank ($)	1,231,250	9,493,813	2,479,848	19,432,931	10,208,240	3,343,994	2,779,345

FIGURE 6.1.14
A sample cash flow statement.

current year. Prior to any financing, the cash balance listed under cash needs will be the same as cumulative cash flow.

Cash balance and financing needs

Once the cash flows are calculated, the innovator can determine the company's financing needs (i.e., funds it needs to raise) and cash balances over time. The cash balance will always need to be positive so financing will need to occur before the cash balance becomes zero. By looking at the cash requirements based on cumulative cash flow (i.e., what is the minimum cash infusion that would make the cumulative cash flow positive), the innovator can input in various amounts representing the financing to be raised. The choice of how much money to raise and at what milestones (or points in time) are key strategic decisions (see 6.3 Funding sources). Implicit in this decision is also how long each financing round will allow a company to have a positive cash flow.

The spreadsheet in Figure 6.1.14 shows how, starting with the desired financing, the innovator can calculate the post-financing cash in the bank by adding the (pre-financing) cash balance to the current year financing. All subsequent year (pre-financing) cash balances must be adjusted by adding the cash flows to the cash balance. This will need to be repeated for each year until a company has sufficient pre-tax operating profits (based upon revenue) which can independently keep cash flow positive.

The financing milestones (i.e., when to raise money from investors) are a critical output from the financial model. In some cases, it becomes evident that not enough important milestones occur within a certain time frame to align investor interest with the funding needs of the business. As a result, it may be necessary to either raise more money upfront to cover the company through these periods, or reconsider the operating plan to allow for a more continuous flow of value-building milestones (milestones that demonstrate measurable progress toward the market). See 6.3 Funding Sources for more information.

When developing the cash flow model, the innovator will face the reality that the process of building a company involves working on an extremely tight budget. This can be a challenge, but Mir Imran, founder and CEO of InCube Labs, pointed out the silver lining:[8]

I really believe that the good part of not having sufficient capital is that it really forces you to think through your expenses more clearly and spend the money less frivolously.

A note on profitability

In developing a financial model, it is important to determine a realistic profit goal. This profit goal will directly affect the company valuation (i.e., how much the company will be worth) since the valuation is a function of earnings (the higher the earnings, the higher the valuation – see Chapter **6.3**). Other considerations that impact the business' valuation include:

* **Revenue ramp** – How quickly the company expects to grow its revenue.
* **Time to profitability** – How long it will take the company to break even, and then turn a consistent operating profit.
* **Operating profit percent** – What operating profit the company expects to achieve as a percentage of revenue.
* **Competitive benchmarks** – Benchmarks and comparable data related to the type of business model the company has chosen (e.g., disposable, reusable, capital equipment; see 4.4 Business Models) and the sector/industry within which the company operates (e.g., devices, genomics, pharmaceuticals).

Typically, a reasonable operating profit goal should be 30 percent of pre-tax income in the long term. However, how the company achieves this goal will depend upon its go-to-market strategy. A company might choose a direct sales strategy, in which case it would not have to share revenue with distributors, resulting in high gross margins; but it would have to hire a large sales force, resulting in high sales costs. Alternatively, the company could choose to use distribution partners, resulting in lower gross margins but lower sales costs. Another option is to blend the two strategies, employing distributors in some markets and direct sales in others. To determine a likely profit model/goal, examine comparable companies (see below) to gain insights into reasonable revenue and cost projections. In the example presented here, the operating profit margin in year 7 was relatively low, reflecting the high cost of building a direct sales force. As part of developing the financial model, the innovators should consider the implications of different strategic choices (e.g., direct sales force versus hybrid).

Proxy companies

Proxy companies, also known as comparable companies, are those whose operations resemble what is required to commercialize an innovator's innovation. The concept of comparables is that historical precedents can be used as a basis of validating a financial model. In the broadest sense, comparables will include other medical device companies. Further refinement would narrow comparable companies to ones that have, or went through, similar development challenges, clinical and regulatory pathways, are focused on the same disease state, and ideally have similar products.

In analyzing a proxy company, there are three primary objectives:

* Determine the milestones that the company selected and/or achieved in developing its business.
* Identify the main risk factors that the company has faced in developing its business and develop a plan to mitigate similar risk factors.
* Reconstruct the company's operating plan and financial model from the details that have been uncovered (so that they can be used as a benchmark).

Investors, as well as innovators, use proxy companies to determine appropriate valuations for a new venture using the valuation models outlined in 6.3 Funding Sources. Therefore, an added benefit of proxy company analysis is that it can help innovators be better prepared in their discussions with investors since proxy companies play an important role in reaching a fair, market-driven valuation for the company.

This analysis can be performed at a high level or in great detail, depending on the needs of the innovator. A thorough analysis will enable a new enterprise to benchmark and modify its plans against multiple comparable companies, learning from their successes and failures. A high-level or more narrowly focused analysis will provide an innovator with directional information or the answer to a specific question (e.g., How long did

Table 6.1.3 *Typical markers for proxy company identification.*

Timing	Company focus	Valuation
Years 1 and 2	• Product development • Intellectual property (IP) filing • Pre-clinical discovery	Value less than $25 million since clinical impact and ability to franchise has not yet been determined
Years 3 and 4	• Clinical value established • Product defined after multiple iterations • Regulatory approval complete • IP issued	Value of approximately $100 million because there is a vision of sustainability, synergy, and franchisability
Years 5 and 6	• Revenues ramping up • Franchise established • Pipeline of product iterations • Clinical utility established	Value between $100 million and $2 billion

comparable companies spend in clinical trials? What are reasonable fund-raising goals and milestones?).

While innovators can learn from the assessment of almost any proxy company, it is most helpful to look for those companies that have been considered a success in terms of their time to market and eventual valuation. When the innovators can select from multiple proxy companies they can use the markers in Table 6.1.3 to determine which of these companies will more closely resemble their proposed venture in terms of milestones and realized valuations (note that these values apply to implantable/therapeutic devices but may vary for diagnostics, pharmaceuticals, and other types of products).

See Appendix 6.1.1 for a sample proxy analysis.

Note on intrapreneurship

Financial models for a new technology that will be developed and commercialized by an established company (this type of development is often referred to as "intrapreneurship") have the same basic components as the model described in this chapter for start-up companies. However, there are at least two key differences to consider.

Leveraging existing resources

Unlike innovators starting a new stand-alone venture, **intrapreneurs** may be able to leverage existing resources within the organization to manage some of the costs. For example, existing R&D engineers may be engaged in the projects, existing capital equipment may be used, and the existing sales force may be leveraged. In that respect, one of the first questions to be resolved is which of the existing resources can the intrapreneur access for the project? However, even if existing resources *can* be used, they are not cost free and they should not necessarily always be employed. The finance department of the organization should provide guidance on the cost of existing resources. And because multiple projects may be competing for the attention of limited internal resources, this can create time completion risks. Therefore, the intrapreneur should consider developing multiple financial models involving different degrees of reliance on internal resources, and use these different models in making a recommendation to senior management in terms of the extent of using internal resources.

Complementarities with existing products and company strategy

The financial model should explicitly account for complementarities with existing product portfolios and company strategy and should also provide an assessment of the impact of the project on the company's financial statement. Questions to address include whether the new project will defend existing market share or be used to increase market share. In both cases, the financial benefits to the organization should be articulated. In the defensive case, the market model should explain how market share may decrease without the new innovation.

In the offensive case, the market model should highlight how the new innovation will increase revenue without cannibalizing existing sales, and whether the increase in revenue will be significant enough to make a difference. Consider for example a company like Medtronic. In November, 2006, Medtronic's quarterly revenue grew 11 percent to $3.08 billion.[9] To sustain this growth trend, Medtronic would need an additional $1.36 billion in new revenue per year. This means that new product development projects that senior management will sponsor will need to have a significant market potential in the hundreds of millions if not billions of dollars. Therefore, the intrapreneur's financial model should demonstrate that the project will have a significant impact on the company's earnings.

GETTING STARTED

Much of the step-by-step information for completing the six primary elements of the financial model has been provided in the body of the chapter. However, the following section puts those six components into context and demonstrates how they are part of a larger approach to developing and validating a meaningful and reasonable financial model.

See **ebiodesign.org** for active web links to the resources listed below, additional references, content updates, video FAQs, and other relevant information.

1. Develop detailed bottom-up financial model

1.1 What to cover – Once the innovator has a quick, high-level understanding of the company's financials, it is necessary to create a more detailed financial model to ensure the accuracy and credibility of the directional information. In a "bottom-up" analysis, the individual parts of the financial model are developed in detail. Then the parts are linked together to form the final plan. Following the guidance and samples provided in the background section, the innovator can create the:

- **Operating plan** – Key inputs to developing the operating plan include a solid understanding of the steps required to build the innovation, a detailed understanding of how the project will be regulated by the FDA (including the estimated length of the clinical trials), and knowledge about how the product will be sold. Important outputs from the operating plan are an understanding of when the first sale will occur and the estimated timing for other critical milestones that affect the overall financial model.
- **Staffing plan** – Key inputs to developing the staffing plan include the operating plan and an understanding of the complexities associated with building the innovation. Pay close attention to the milestones and determine how many staff will be required to achieve each one. Additionally, think about the impact that each milestone will have on the company and its staffing levels as it moves forward.
- **Market model** – Develop both a bottom-up and a top-down market model and then reconcile the two. In the bottom-up market model, identify: (1) the key metrics used to measure progress – trained physicians, installed hospitals, etc.; (2) training times and requirements per trained physician (or other appropriate metrics); (3) time to ramp up and steady-state per trained physician; (4) target performance metrics per sales rep; (5) number of sales representatives that will achieve and maintain the target metrics.
- **Cost projections** – To prepare cost projections, the staffing plan is an important input, along with competitive benchmarks for employee salary costs. The company must also consider how much manufacturing and non-manufacturing facilities space it will need, and should then get estimates about pricing from the local market. Equally important is developing a strong understanding of the raw materials costs associated with the product so that these can accurately be taken into account when developing the cost projections.

GETTING STARTED

- **Income statement** – The income statement provides a summary of the company's anticipated financial condition. Be sure that the ratios (R&D/sales, SG&A/sales, etc.) are reasonable or it may be an immediate warning sign to investors that the entire financial model is faulty.
- **Cash flow statement** – Important inputs to the cash flow statement include the company's most recent income statement, the number of projected employees, estimates of cash spent per employee, cash likely to be spent on equipment, and estimates of inventory required. The output of this statement is the cumulative cash the business will spend, which can be translated into the amount of money the company needs to raise.

1.2 Where to look
- **Other innovators** – These individuals may be able to provide valuable anecdotal information about appropriate staffing levels, ratios, and costs, having previously faced similar financial decisions.
- **Industry consultants** – Consultants in the field may also possess valuable information about common industry figures, such as salary benchmarks, facilities costs, and raw materials costs.
- **Other resources for top-down and bottom-up market modeling**
  - 2.4 Market Analysis and 5.7 Marketing and Stakeholder Strategy chapters.
  - **Hospital Compare** – Includes Medicare data regarding the volume of procedures performed per hospital in the United States.
  - **Dartmouth Atlas of Health Care** – Database that documents glaring variations in how medical resources are distributed and used in the United States.
  - **Procedure files from Medicare**

2. Identify proxy companies

2.1 What to cover
Direct competitors typically make for the strongest comparables. However, sometimes meaningful historical comparables exist that are no longer in business. When direct comparables are not available, gradually relax the constraints to broaden the sample set of companies that can act as proxy companies. Start by identifying approximately ten potential proxy companies and then narrow down this list to the one to four companies with the most relevant challenges and products. Remember to focus on companies that have demonstrated success in execution.

2.2 Where to look
- **Yahoo! Finance** – Yahoo! Finance aggregates the financial data of public companies. By looking up a known public competitor, a hypertext link in the left-hand column leads to a web-page of "Competitors."
- **VentureXpert** – Provides data on mergers and acquisitions, initial public offerings (IPOs), and venture capital funding based on data from the Securities Data Corporation. Users can search the database to find detailed information about private start-ups including funding, investors, and executives.
- **VentureSource** – Another database that contains detailed information about private start-ups including funding, investors, and executives.
- **Hoovers Online** – A database of product profiles for companies, which includes competitors.

3. Develop proxy company analysis (referred to as top-down analysis)

3.1 What to cover
Once appropriate proxy companies are identified, the next step is to analyze each one. Find operational information, including the company's key milestones, and try to reconstruct the operating plan (see Appendix 6.1.1 for a sample proxy company analysis) via a top-down analysis. In a top-down analysis, an overview of the model is developed based on available information. Then, each part is refined (with additional

information and assumptions added) until the entire model can be validated. The data to search for are outlined below. Public companies can be excellent proxy companies for two reasons. First, becoming public is an important milestone in itself that indicates some degree of success and suggests that the company's operations potentially are worth analyzing (see Chapter 6.3 for more information). The second reason is that a company prospectus can be a rich source for operational details. A prospectus (also known as an S-1 filing made to the Securities and Exchange Commission) is a document released by a company prior to becoming public in which it seeks to tell a meaningful story about its history and future prospects to prospective investors. In doing so, the company must describe its operating history in some detail. The prospectus will not lay out the company's entire operating plan and will likely omit data regarding the company's earlier years. However, it provides valuable information that may be used to piece together the company's operating history.

- **High-level operating plan** – Timeline of key operating milestones.
- **Financing milestones** – Timeline of major financing events.
- **Clinical trials history** – Timeline of clinical trials including duration, location, and number of patients enrolled.
- **Other milestones** – Additional milestones, such as sales and distribution agreements signed.
- **Operating costs and staffing levels** – Historical operating expenses and staffing levels by employee type.
- **Clinical trials costs** – Estimate of cost per person of the clinical trials process.
- **Cash flow** – Cumulative cash flows.
- **Information extracted from the Prospectus of Public Companies**
  - **Management's discussion and analysis of financial condition and results of operations** – Among other information, this section provides a qualitative discussion of the company's operational progress and key milestones achieved. It also comments on the changes in year-to-year financial results and provides some historical cash flow data. This financial information can be useful in determining how much a company had to spend in order to achieve the milestones it achieved.
  - **Business** – This section provides a fairly rich description of the company's business, including information about the market, targeted disease, products, strategy, clinical trials sales and marketing, competitors, manufacturing, reimbursement, IP, government regulation, employees, facilities, and litigation. The clinical trial data in this section can help to determine the length and complexity of the clinical trials as well as any specific regulatory hurdles the company has faced. It may also detail the current number of employees, employee types, and facilities sizes.
  - **Financial section** – This section, at the end of the document, provides detailed financial information, including the company's balance sheet, income statement, and statement of cash flows. The number of years of historical data available in the prospectus will vary by company.
3.2. **Where to look** – Information about private companies is more difficult to find than public ones. Venture capital databases (such as those listed below) can be useful for locating certain details, including the company founding date and information about funding rounds (e.g., number of rounds, money raised per round, selected **post-money valuations**). They will also provide non-specific indicators of a company's stage at each round of funding (selected from among "early stage," "late stage," "expansion stage," etc.). However, they are unlikely to offer detailed information about a company's operating history, requiring the innovator to have to locate this data from other sources.
- **VentureXpert**

- **VentureSource**
- **SEC Edgar Online** – Provides online access to business and financial information, including SEC filings, for more than 30,000 global companies.
- **Personal networks** – Contact individuals and experts in the field who may be able to share relevant information.
- **Cold contact** – Consider contacting executives at the proxy companies to request information that has not yet been made publicly available. Recognize that there is certain information that these individuals will be more/less likely to share and prepare a set of appropriate questions that will result in additional knowledge, without overstepping the bounds of professionalism and privacy. Also consider what reasons these executives would have to help (or not) and prepare your request accordingly. At all times, it is critical to take an ethical, honest approach to this kind of data collection, particularly when dealing with potential competitors.

4. **Compare and rationalize bottom-up and top-down approach**

4.1 **What to cover** – Once the proxy company analysis is complete, compare the results to the bottom-up financial model created in step 1 of the process. Identify where significant variance exists. Look for divergence in the following areas:
- Key milestones
- Clinical trial process
- Staffing levels
- Underlying cost assumptions
- Costs to reach each milestone
- Time to begin sales
- Time to profitability

Also consider if the financial model should be adjusted based on the learnings and observations from the proxy company analysis. Were additional risks or potential concerns identified that should be addressed in other parts of the business plan? Review all aspects of the business plan again and make adjustments (or seek additional input) before finalizing the document. Remember that this is an iterative process and the financial model will probably need to be refined several times before it is ready for review by potential investors.

4.2 **Where to look** – Refer back to the working version of the business plan and financial model and data gathered from the proxy company analysis. The innovator should also reach out to members of his/her personal network with questions or concerns.

Credits

The editors would like to acknowledge Steve Fair, David Lowsky, and John Cavallaro for their help in developing this chapter, as well as John White for contributing many of the financial models. Many thanks also go to Bernard Hausen, Bob Newell, and William Younger for their assistance with the Cardica story. Special recognition goes to Larry Tannenbaum, who passed away in 2008. Larry was a long-time supporter of Stanford's Program in Biodesign and this chapter is based, in part, on his lectures.

Appendix 6.1.1

Proxy company analysis: Cardica, Inc.

This example presents an overview of how to analyze and reconstruct an operating plan using only publicly available information about a proxy company. The analysis reflects information available as of March, 2008. The insights of the proxy company's management team are shared throughout the document, but were not used to perform the analysis.

- **Company name** – Cardica, Inc.
- **Founded** – 1997
- **Description** – Cardica designs and manufactures proprietary automated anastomotic systems used by surgeons to perform coronary artery bypass surgery. In coronary artery bypass grafting, or CABG, procedures, veins or arteries are used to construct "bypass" conduits to restore blood flow beyond closed or narrowed portions of coronary arteries. Cardica's first two products, the C-Port® Distal Anastomosis System, referred to as the C-Port system, and the PAS-Port® Proximal Anastomosis System, referred to as the PAS-Port system, provide cardiovascular surgeons with easy-to-use, automated systems to perform consistent, rapid and reliable connections, or anastomoses, of the vessels, which surgeons generally view as the most critical aspect of the CABG procedure.[10]

Performing proxy company analysis

Table 6.1.4 presents the main components of proxy company analysis and how they can be used to inform a financial model and plan.

The end goal of this analysis is to refine and validate the company's own business plan, and in particular its operating expenses model, financial model, and funding requirements. The analysis of proxy companies should enable the company to:

1. Confidently state that its plan is reasonable (i.e., expenses and timelines are reasonable and appropriate, and the clinical strategy is relevant).
2. Identify reasonable financing milestones and potential valuation points.
3. Identify potential risk factors (based on challenges faced by proxies) and integrate risk mitigation strategies into the business plan.

To achieve these objectives, the company may wish to examine between one and four proxy companies. Try to include companies with various levels of success in this analysis for a diversity of perspectives and lessons learned.

High-level operating plan

In preparing the Cardica proxy company analysis, the first step was to identify the key operating milestones achieved by the company. The primary source of information for this step was the company prospectus. The section entitled "Management's discussion and analysis of financial condition and results of operations" summarized the key milestones achieved by the company to date.

As stated in the prospectus, 1997 to 2002 "consisted primarily of start-up activities, including developing the C-Port and PAS-Port systems, recruiting personnel and raising capital."[11] The prospectus did not contain any details about the intermediate milestones between the company's founding in 1997 and the start of clinical trials in 2003, so this period was designated as the product development phase. During this time frame, the team was probably quite small and focused on core R&D activities to develop the company's initial products and prepare for clinical trials. The period from 2003 to 2005 included all clinical trials, start of sales, and development of R&D and sales distribution partnerships.

Figure 6.1.15 contains an operating timeline of key company milestones, including major financing events, clinical trials history and outcomes, and sales and distribution milestones. As discussed in the main body of this chapter and also in 6.3 Funding Sources, key operating milestones and financing milestones (which are

Stage 6 Integration

Table 6.1.4 *These components work together to inform an effective proxy company analysis.*

Component of proxy analysis	How to	Purpose
High-level operating plan – Timeline of key operating milestones	Evaluate benchmarks from: Prospectus VentureXpert VentureSource	To validate the company's own operating plan in terms of timeline and staffing requirements.
Financing milestones – Timeline of major financing events and valuations (see Chapter 6.3 for a discussion of valuations)	Evaluate benchmarks from: Prospectus VentureXpert VentureSource	To validate the company's own financing needs and potential valuations.
Clinical trials history – Timeline of clinical trials, including duration, location, number of patients enrolled, and important regulatory events	Evaluate benchmarks from: Prospectus Medline Search Clinicaltrial.gov	To validate the company's clinical strategy and match the clinical and regulatory strategies to financing milestones and the operating plan.
Other milestones – Additional milestones, such as sales and signed distribution agreements	Evaluate benchmarks from: Prospectus	To identify other relevant milestones and embed them into an operating plan.
Operating costs and staffing levels – Historical operating expenses and staffing levels by employee type	Use staffing levels and operating cost information from prospectus. Fit these levels into a more refined operating cost model. Divide broader categories of employees reported in the prospectus into specific subcategories using relevant ratios. Iterate to develop a complete model, including reasonable assumptions about staffing costs.	To help the company develop its own operating expenses analysis, including detailed staffing levels and salary information.
Clinical trials costs – Estimate of cost per person of the clinical trials process	Obtain total clinical trial costs and divide by number of patients to get rough estimate of clinical trial cost per enrolled patient.	To help the company estimate the cost of its own clinical trials.
Cash flow – Cumulative cash flows	Examine financial statements of proxy companies and make adjustments for unusual items such as grants of stock options.	To help the company develop its own cash flow models.

discussed in more detail below) often go hand in hand. For example, Cardica's first clinical trial, the PAS-Port European clinical trial, began in quarter 2 (Q2) 2002, which is the same quarter that the company completed its 4th round of financing ($18.6 million). The C-Port and PAS-Port II trials both began in Q3 2003, the same quarter that the company closed its 5th and 6th rounds of financing (together worth $14 million). In each case, the cash infusion may have been a necessary prerequisite for the company to launch each clinical trial.

Financing milestones

VentureXpert and VentureSource were used to identify Cardica's funding rounds. VentureXpert reported three funding rounds along with their post-money valuations (the value assigned to a company after each successful round of funding; see Chapter 6.3). However, the earliest reported round occurred in 2001, four years after the company's founding. Since Cardica's income statement revealed that the company had spent several million dollars to get to this point, this data seemed incomplete.

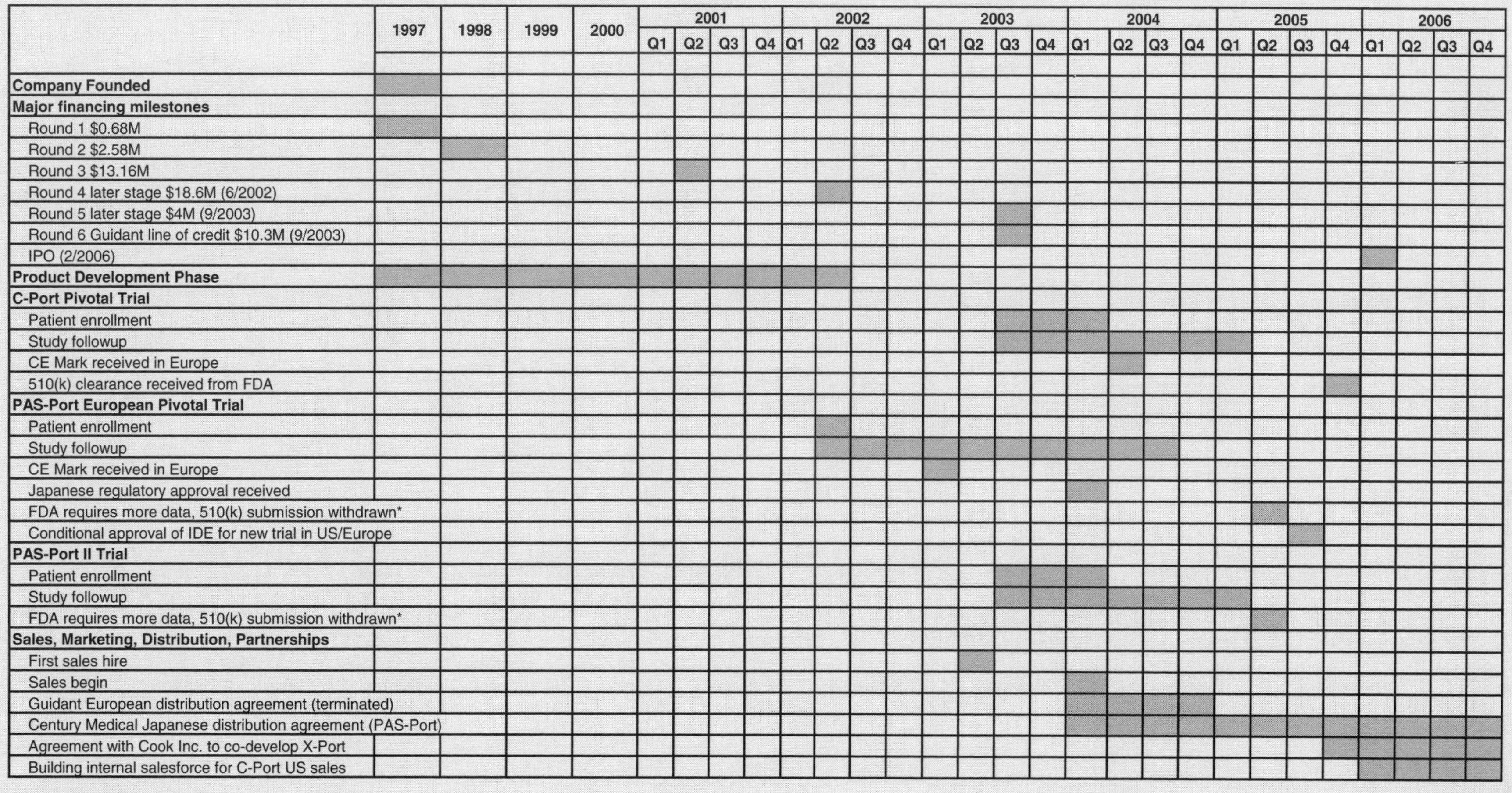

6.1 Operating Plan and Financial Model

	1997	1998	1999	2000	2001 Q1 Q2 Q3 Q4	2002 Q1 Q2 Q3 Q4	2003 Q1 Q2 Q3 Q4	2004 Q1 Q2 Q3 Q4	2005 Q1 Q2 Q3 Q4	2006 Q1 Q2 Q3 Q4
Company Founded										
Major financing milestones										
Round 1 $0.68M										
Round 2 $2.58M										
Round 3 $13.16M										
Round 4 later stage $18.6M (6/2002)										
Round 5 later stage $4M (9/2003)										
Round 6 Guidant line of credit $10.3M (9/2003)										
IPO (2/2006)										
Product Development Phase										
C-Port Pivotal Trial										
Patient enrollment										
Study followup										
CE Mark received in Europe										
510(k) clearance received from FDA										
PAS-Port European Pivotal Trial										
Patient enrollment										
Study followup										
CE Mark received in Europe										
Japanese regulatory approval received										
FDA requires more data, 510(k) submission withdrawn*										
Conditional approval of IDE for new trial in US/Europe										
PAS-Port II Trial										
Patient enrollment										
Study followup										
FDA requires more data, 510(k) submission withdrawn*										
Sales, Marketing, Distribution, Partnerships										
First sales hire										
Sales begin										
Guidant European distribution agreement (terminated)										
Century Medical Japanese distribution agreement (PAS-Port)										
Agreement with Cook Inc. to co-develop X-Port										
Building internal salesforce for C-Port US sales										

FIGURE 6.1.15
Cardica's estimated operating timeline.

Table 6.1.5 *Cardica's financing activity.*

Round	Date	Amount raised	Post-money valuation	Stage	Investors
1st	1997	$0.68 million	$4.94 million	Early stage	Individual investors
2nd	1998	$2.58 million	$10.64 million	Early stage	Individual investors
3rd	June 14, 2001	$13.16 million	$31.43 million	Early stage	Sutter Hill Ventures Allen & Co.
4th	June 13, 2002	$18.6 million	$62 million	Later stage	Sutter Hill Ventures Allen & Co. Guidant Corp.
5th	September 30, 2003	$4 million	$85 million	Later stage	Guidant Corp.
6th	August, 2003	$10 million	N/A	Line of credit	Guidant Corp.
IPO	February, 2006	$35 million	$110 million (post-IPO) $75 million (pre-IPO)	IPO	N/A

The VentureSource database revealed a total of five venture funding rounds. This case illustrates the importance of double-checking information when possible.

In August, 2003, Guidant extended a $10.3 million line of credit to Cardica, as described in the company's prospectus:

> In August 2003 Guidant extended a line of credit to Cardica for $10.3 million. The company has drawn down this line of credit and currently has a long-term loan of $10.3 million outstanding from Guidant, due in August 2008. Interest of 8.75 percent per year accrues during the life of the loan and is due at maturity.[12]

Combined with Guidant's participation in Cardica's 4th and 5th venture rounds, the company has received a cumulative total of $14 million from Guidant. Table 6.1.5 shows a summary of all of the company's financing events.

Clinical trials history

The clinical trial history, shown in Table 6.1.6, was obtained from the company's initial public offering (IPO) prospectus.[13] The prospectus also showed regulatory results to date (see Table 6.1.7).

As noted, Cardica has had difficulty obtaining US regulatory approval for its PAS-Port product. Initially expecting a one-year trial process, the company submitted the results of two years' worth of trial data. Ultimately, the FDA required more robust data and Cardica withdrew its 510(k) submission. This seems to have been a blow to Cardica and may have contributed to lowered expectations and a reduced valuation heading toward their IPO.

Working example: Cardica management perspective: FDA approval difficulties

Bernard Hausen, cofounder and CEO of Cardica, Inc., explained how changes in the FDA requirements stalled product approval after Cardica's clinical trials had been completed: "At the time, there was a competitive product to the PAS-Port already approved called the Symmetry device from St. Jude Medical. It was approved by the FDA with hardly any clinical data and was selling very well. Suddenly, some patients started coming back six to nine months after surgery with symptoms of angina, which was indicative of occlusions or narrowing of the blood vessels in the connections that this device created. The FDA realized that it had approved the Symmetry device with far too little clinical data and that any similar products, such as ours, would have to pass through a very different hurdle than the one St. Jude had to pass." To help determine specifically what manufacturers with a product in this space would have to prove to receive approval, the FDA convened

6.1 Operating Plan and Financial Model

Table 6.1.6 *Cardica's clinical trial history.*

Study	Number and location of sites	Enrollment start date	Enrollment completion date	Number of patients	Objective	Length of follow-up
C-Port pivotal trial	5 European sites	July, 2003	February, 2004	133	Determine safety and efficacy of distal anastomotic device	12 months
PAS-Port European pivotal trial	3 European sites	June, 2002	September, 2002	55	Determine safety and efficacy of proximal anastomotic device	24 months
PAS-Port II trial	4 European sites	July, 2003	February, 2004	54	Increase data pool for study and efficacy with an improved PAS-Port device	12 months

Table 6.1.7 *Cardica's regulatory history.*

Product	Date	Result
C-Port	April, 2004	CE mark received in Europe
	November, 2005	510(k) clearance received in United States
PAS-Port	March, 2003	CE mark received in Europe
	January, 2004	Japanese regulatory approval received
	Ongoing	510(k) clearance has proved elusive. Submitted results of 3/6-month follow-up data in application for 510(k) clearance in April, 2003. After FDA redefined objective performance criteria for safety/efficacy of anastomosis products, company resubmitted pooled data from two PAS-Port trials. In April, 2005, FDA panel decided it required more robust data, and company withdrew 510(k) submission. Received IDE for a new randomized prospective clinical trial to be conducted in the United States and Europe.

a panel of experts who redefined the standards. The impact of the new requirements on Cardica was that "the clinical data we had accumulated at that point was not sufficient to convince the FDA to approve us, which meant we had to start all over again with that product." This development, which Cardica saw as largely outside of its control, was a central factor in reducing Cardica's valuation for its IPO, as well as forcing the company to raise more capital in order to fund these additional clinical trials.

Other milestones

Cardica entered into the following sales and distribution partnerships:

- **Cook license agreement** – In December, 2005, Cardica entered into a license, development, and commercialization agreement with Cook Incorporated, related to development of the X-Port Vascular Access Closure Device, a product candidate in pre-clinical animal model studies as of early 2007.

Stage 6 Integration

- **Guidant European distribution agreement** – Cardica entered into an agreement with Guidant Corporation for European distribution of the PAS-Port and C-Port systems. The agreement was signed in May, 2003, amended in January, 2004, and ultimately terminated in September, 2004. This agreement accounted for 48 percent of total revenue in fiscal year 2004 and 50 percent of total revenue in fiscal year 2005.
- **Guidant development/supply agreement** – Cardica entered into a development/supply agreement with Guidant in December, 2003 to develop an aortic cutter for Guidant's Heartstring product, and manufactured the first 10,000 aortic cutters. Guidant subsequently outsourced future production of the aortic cutter to a third-party contract manufacturer. Cardica would receive a modest royalty for each aortic cutter sold beyond 2005, but does not expect these royalties to contribute significantly to revenue.
- **Century Medical distribution agreement** – Cardica distributes the PAS-Port system in Japan through an exclusive distributor, Century Medical Inc. Sales to Century produced 33 percent of fiscal 2005 net revenue. Century is responsible for the development of the anastomotic device market in Japan, possessing a direct sales organization of 16 reps and providing clinical training and support for end users in Japan. Cardica provides promotional support and clinical training to Century. Agreement expired in June, 2008.

Working example: Cardica management perspective: partnership agreements

Unlike most medical device start-ups, which try to get one product into the market before developing another, Cardica decided early on to diversify its product portfolio. In addition to developing both the C-Port and the PAS-Port systems in parallel, the company sustained a number of other smaller projects (e.g., the company's aortic cutter). Hausen saw this strategy as more of a biotech approach that involved focusing the company on developing a core competency from which it could create many potential products. He explained: "It made sense for us to take our core competency, which is in developing these miniature stapling devices, and see where else in the human body they could be used and how they could be developed into additional products. This would give us not just two legs to stand on, but three, four, or five legs in case one of them was to get cut away."

While this diversification strategy had some advantages (e.g., allowing the company to begin generating revenues from the sale of its C-Port product when it was forced to regroup around the FDA approval requirements for the PAS-Port), it proved to be expensive. When Cardica did not have the cash to continue developing the various products in its project pipeline, William Younger, a managing director of Sutter Hill Ventures and a member of the Cardica board of directors, saw partnerships as a potential solution to this problem. "As a young company, Cardica could not afford to sustain development on all these products and still afford the distribution when we eventually got a product approval," he said. "So there had to be sponsorship and a distribution arrangement that got the company royalties." It was this necessity, driven by Cardica's approach to diversification, that led to the formation of the partnerships listed above.

Cumulative sales

As of September, 2005, Cardica sold over 250 C-Port systems and 2,400 PAS-Port systems worldwide. Sales began in fiscal year 2004. Lacking information about individual 2004 and 2005 unit sales, for the purposes of financial modeling, the assumption was made that Cardica sold 100 C-Port and 700 PAS-Port systems in 2004 and 150 C-Port and 1,700 PAS-Port systems in 2005, for a total of 800 systems in 2004 and 1,850 systems in 2005 (a 2005 to 2004 ratio of 2.3). This assumption was guided by the ratio of historical total 2005 revenue ($2,056,000) to total 2004 revenue ($836,000).

Operating costs and staffing levels

One of the key objectives in analyzing Cardica as a proxy company was to infer its historical operating

Table 6.1.8 *Excerpt from Cardica's income statement (2006 Cardica prospectus; reprinted with permission).*

Operating costs and expenses (in 000s)	2001	2002	2003	2004	2005
Cost of product revenue (includes related-party costs of $1,377, $1,180, $306, and $0 in fiscal 2004, fiscal 2005, and [three] months ended September 30, 2004 and 2005, respectively)				2,105	2,478
Research and development	5,058	5,765	6,698	5,826	6,289
Selling, general and administrative	1,166	1,635	1,936	1,809	3,753
Total operating costs and expenses	6,224	7,400	8,634	9,740	12,520

costs and staffing levels so that these figures could be used to validate a bottom-up financial model for a new enterprise. Two key pieces of available data were especially useful in this regard: (1) 2005 staffing levels and (2) 2001 to 2005 operating cost information (both retrieved from the prospectus).

Note: when presenting snapshots from the company's financial model or financial data from Cardica's prospectus, fiscal years are used throughout. Fiscal years for this company run from July of a given calendar year to June of the next calendar year. For example, fiscal 2003 is July 2002 through June 2003.

Staffing levels for 2005 were outlined accordingly:

> As of November 30, 2005, we had 42 employees: including 16 employees in manufacturing; one employee in sales and marketing; four employees in clinical, regulatory, and quality assurance; five employees in general and administrative; and 16 employees in research and development.[14]

Operating cost data for 2001 to 2005 (each fiscal year ending in June) were taken from Cardica's income statement and shown in Table 6.1.8.

The goal was to use these pieces of available data, along with Cardica's operating plan, to estimate the company's staffing levels and overall costs on an annual basis, from 1997 through 2005. This was done in two steps.

Step 1: Structure the operating cost model

The objective of this step was to define the specific elements of the operating cost model so that they accurately reflect the company being analyzed. These elements include cost line items and employee types.

Define costs

The line items underlying Cardica's cost model included: (1) R&D, (2) SG&A, and (3) COGS. In most medical device company financial models, employee costs account for most of the first two categories. However, upon reading the Cardica prospectus, it seemed appropriate to include additional cost line items. Under R&D, "cost of clinical trials" and "capital expenditure" were added. Under SG&A, "other SG&A expenses" were added. A section for "facilities" also was included under SG&A. All of these additions are explained in the section below entitled Cost Line Items Added. The operating cost model used here was based on the general model described in the main part of the chapter.

Bear in mind that different companies have different costs and the cost model should reflect the specific business being analyzed. Also note that the analysis can get as detailed as needed in this area, depending on how carefully the prospectus is reviewed. Just be careful not to spend too much time modeling costs that are unique to the company being analyzed and will not generalize to another venture.

Define employee types and how they are categorized within the income statement

There were five general employee types mentioned in the prospectus: (1) manufacturing, (2) sales and marketing, (3) clinical, regulatory, and quality assurance (QA), (4) general and administrative, and (5) R&D. To begin the analysis, manufacturing was divided into the

Stage 6 Integration

Table 6.1.9 *Cardica staffing estimates.*

Employee type (from prospectus excerpt)	Cardica income statement category
Manufacturing engineer	Cost of product revenue
Manufacturing technician	Cost of product revenue
R&D engineering manager (VP R&D)	Research and development
R&D engineer	Research and development
R&D technician	Research and development
Clinical management (VP Clinical)	Research and development
Clinical, regulatory and quality assurance	Research and development
General management (CEO, CFO)	Selling, general and administrative
Sales and marketing (VP)	Selling, general and administrative
Administrative assistant	Selling, general and administrative

employee types of manufacturing engineer and manufacturing technician. Engineering management and clinical management were also added as types within R&D, reflecting the fact that two senior managers (the vice-president (VP) of clinical and regulatory affairs and the VP of research and development), had been categorized in the income statement within the R&D line item and not within a more general management line item (senior management positions were identified from the "Management" section of the prospectus). This allowed for the definition of a higher salary category for these two positions. Table 6.1.9 captures how each employee type was mapped to an income statement category.

Step 2: Assign staffing and costs to the model

The objective of step two was to arrive at a reasonable estimate of annual staffing levels and costs. The general strategy here was to start from actual available data, assign estimates to each of the other unknown items, and then iterate until all items seemed reasonable alongside one another and the costs seemed to reflect the known operating milestones on an annual basis. The analysis should begin in a year where both staffing levels and total costs are known. In the case of Cardica, only 2005 staffing levels were available from the prospectus, so the analysis began with 2005 and was performed backwards.

Define number and type of employees in 2005

Since the prospectus only stated the number of employees within each of five broad employee types, it was necessary to make judgments about how these numbers should be divided across the more specific employee types:

- The 16 manufacturing employees mentioned in the prospectus were divided into groups of 5 engineers and 11 technicians. The general trend is that as a company is defining its manufacturing processes, there is a roughly equal number of technicians and engineers, or there are slightly more technicians. As manufacturing processes become streamlined, the ratio of technicians to engineers increases substantially, up to 3 or more manufacturing technicians per engineer.
- In this case, the company's manufacturing labor and materials costs were nearly the same in 2004 ($2,105,000) and 2005 ($2,478,000), despite a significant increase in unit sales from 2004 to 2005. (Note that while the common accounting

convention used in text is to include only material costs in the calculation of COGS, the Cardica prospectus includes both manufacturing staff and materials costs as part of COGS.) This suggests that the company made a large investment in manufacturing staff and capabilities when it began manufacturing in 2004. Nevertheless the company was only about a year into manufacturing, and therefore past the initial stage but still ramping up sales and refining its manufacturing processes. In 2004, the company had reallocated some of the engineering costs from the R&D category to manufacturing, suggesting that the manufacturing team was still being fully developed. Therefore, an allocation of 11 manufacturing technicians to 5 engineers seemed reasonable.

- The 16 employees referred to as R&D employees in the prospectus were allocated as follows: 1 engineering manager (VP R&D), 4 engineers, and 11 technicians. This engineer to technician ratio is roughly equivalent to the ratios described in the main part of this chapter, or roughly 1 to 3 R&D technicians per R&D engineer.
- The 4 employees referred to as clinical and regulatory staff in the prospectus were divided into a VP position and 3 clinical staff.
- The 6 employees referred to as SG&A in the prospectus were allocated into 1 VP position for sales and marketing, 2 administrative assistants, and 3 general managers (CEO, CFO, and one additional manager).

Assign 2005 employee salaries by employee type
First, salary levels were defined for each employee type. The initial values chosen were based on standard industry values (provided in Table 6.1.1). Factors for the fully burdened salaries (to account for benefits, equipment, etc.) were defined using standard medical device industry values: 1.5× for manufacturing employees and 2× for all other employee types. Finally, the salary levels and factors were used to produce fully burdened salaries for each employee type as well as totals for 2005 R&D staff costs, SG&A staff costs, and manufacturing staff costs.

Use 2005 known operating costs to estimate additional cost line items
Now that estimates for the total 2005 R&D, SG&A, and manufacturing staffing costs were understood, the only unknown cost items remaining were the additional cost line items defined in step one. Estimates could be assigned to these additional line items within each category such that each of the projected totals nearly matched the actual 2005 historical total operating cost values.

Tweak the 2005 estimates
Some experimentation was required with the employee salaries by type and the cost line items to make the projected 2005 operating costs seem realistic while remaining in alignment with the actual 2005 historical costs.

Assign 2001 to 2004 staffing levels and cost items
With 2005 staffing levels and costs estimated, 2001 to 2004 estimates were assigned next. The 2005 staffing levels and costs served as a guidepost, since 2001 to 2004 values should have increased reasonably towards 2005.

One of the most challenging aspects of creating this model was assigning staffing levels by employee type to years 2001 to 2004. The staffing levels for each year had to make sense relative to the key operating milestones (clinical trials progress in 2002 to 2004, start of sales in 2004, etc.), as well as to 2005 staffing levels. Once staffing levels were assigned and corresponding fully burdened costs calculated, values could be assigned to the additional cost line items for each year so that the total projected costs aligned with the historical costs from 2001 to 2004.

Iterate
The steps outlined above were repeated and the model was revised until the figures made sense alongside each other. Examples of some of the key considerations used to tweak the cost model included:

Stage 6 Integration

- The 2001 to 2005 costs projected by the model had to approach the actual 2001 to 2005 historical costs. In the spreadsheet, there is a line under each section called "discrepancy," which indicates the variance between the two sets of figures.
- The year-to-year increase in staff needed to reflect key company milestones. These milestones included the closing of the first major round of financing, as well as the start of sales and manufacturing.
- The company could not go bankrupt in any year. In other words, post-financing cash in the bank (under cash flow) had to be positive in each year. Because the company fundraising was minimal until its first major cash infusion in 2001, the projected staffing levels until 2001 were also quite low.

Some additional considerations are discussed in the section Additional Cost and Staffing Considerations. This process required several iterations until all considerations were met and the model was determined to be sound.

Figure 6.1.16 contains the actual version of the cost model. Some of the key considerations in selecting reasonable values for staffing levels and costs are outlined in Figures 6.1.17 through 6.1.25.

Actual Financials ($)	1997	1998	1999	2000	2001	2002	2003	2004	2005
(from prospectus)	Year 1	Year 2	Year 3	Year 4	Year 5	Year 6	Year 7	Year 8	Year 9
Revenue								836,000	2,056,000
Manufacturing costs	-	-	-	-	-	-	-	2,105,000	2,478,000
Total capital expenditures							757,000	914,000	882,000
R&D					5,058,000	5,765,000	6,698,000	5,826,000	6,289,000
Selling, general & administrative					1,166,000	1,635,000	1,936,000	1,809,000	3,753,000
Total operating expenses (excl. manufacturing costs)					6,224,000	7,400,000	8,634,000	7,635,000	10,042,000
Total operating expenses					6,224,000	7,400,000	8,634,000	9,740,000	12,520,000
Total operating loss							8,634,000	8,904,000	10,464,000

FIGURE 6.1.16
Cardica's actual financials (2006 Cardica prospectus; reprinted with permission).

Staffing	Year 1	Year 2	Year 3	Year 4	Year 5	Year 6	Year 7	Year 8	Year 9
Manufacturing									16
Engineers	0	0	0	0	0	0	0	5	5
Assemblers	0	0	0	0	0	0	0	11	11
Research & development									16
VP, research & development	0	0	0	0	1	1	1	1	1
Engineers	1	1	1	1	8	8	8	4	4
Technicians	0	0	1	1	11	11	11	11	11
VP, clinical and regulatory affairs	0	0	0	0	1	1	1	1	1
Clinical, regulatory and QA	0	0	0	0	3	3	3	3	3
Sales, general & administrative									6
VP, sales & marketing	0	0	0	0	0	0	1	1	1
Administrative assistant	0	1	1	1	1	1	1	1	2
General mgmt (CEO, CFO, other)	1	1	1	1	2.1	3	3	3	3
Total	2	3	4	4	27	28	29	41	42

FIGURE 6.1.17
Estimated staffing levels.

6.1 Operating Plan and Financial Model

Salary assumptions (2005)	Factor	Salary ($)
Engineering	2	130,000
Technicians	2	50,000
Clinical, regulatory & QA	2	130,000
Sales & marketing	2	100,000
Administration	2	35,000
Manufacturing	1.5	35,000
Management	2	200,000
Yearly salary increase	2.5%	

Fully burdened salary ($)	Year 1	Year 2	Year 3	Year 4	Year 5	Year 6	Year 7	Year 8	Year 9
Engineering	213,394	218,729	224,197	229,802	235,547	241,436	247,472	253,659	260,000
Technicians	82,075	84,127	86,230	88,385	90,595	92,860	95,181	97,561	100,000
Clinical, regulatory & QA	213,394	218,729	224,197	229,802	235,547	241,436	247,472	253,659	260,000
Sales & marketing	164,149	168,253	172,459	176,771	181,190	185,720	190,363	195,122	200,000
Administration	57,452	58,889	60,361	61,870	63,417	65,002	66,627	68,293	70,000
Manufacturing	43,089	44,166	45,271	46,402	47,562	48,751	49,970	51,220	52,500
Management	328,299	336,506	344,919	353,542	362,380	371,440	380,726	390,244	400,000

FIGURE 6.1.18
Salary assumptions.

Operating Expenses (Op Ex)	1997	1998	1999	2000	2001	2002	2003	2004	2005
R&D spending	Year 1	Year 2	Year 3	Year 4	Year 5	Year 6	Year 7	Year 8	Year 9
Staff costs									
# of engineering management	0	0	0	0	1	1	1	1	1
Fully loaded employee cost ($)	328,299	336,506	344,919	353,542	362,380	371,440	380,726	390,244	400,000
# of engineers	1	1	1	1	8	8	8	4	4
Fully loaded employee cost ($)	213,394	218,729	224,197	229,802	235,547	241,436	247,472	253,659	260,000
# of technicians	0	0	1	1	11	11	11	11	11
Fully loaded employee cost ($)	82,075	84,127	86,230	88,385	90,595	92,860	95,181	97,561	100,000
# of clinical management	0	0	0	0	1	1	1	1	1
Fully loaded employee cost ($)	328,299	336,506	344,919	353,542	362,380	371,440	380,726	390,244	400,000
# of clinical, regulatory, QA	0	0	0	0	3	3	3	3	3
Fully loaded employee cost ($)	213,394	218,729	224,197	229,802	235,547	241,436	247,472	253,659	260,000
Total R&D staff costs	213,394	218,729	310,427	318,188	4,312,325	4,420,133	4,530,637	3,629,268	3,720,000
Cost of clinical trials ($)	0	0	0	0	400,000	950,000	1,700,000	1,600,000	2,000,000
Capital expenditure (0.6 of total in 03-05) ($)	50,000	65,000	80,000	100,000	350,000	400,000	469,340	566,680	546,840
Proj. annual R&D expenses ($)	263,394	283,729	390,427	418,188	5,062,325	5,770,133	6,699,977	5,795,948	6,266,840
Discrepancy ($)					-4,325	-5,133	-1,977	30,052	22,160

FIGURE 6.1.19
Research and development expenses.

	1997	1998	1999	2000	2001	2002	2003	2004	2005
SG&A spending	Year 1	Year 2	Year 3	Year 4	Year 5	Year 6	Year 7	Year 8	Year 9
Staff costs									
# of sales & marketing	0	0	0	0	0	0	1	1	1
Fully loaded employee cost ($)	164,149	168,253	172,459	176,771	181,190	185,720	190,363	195,122	200,000
# of admin assistants	0	1	1	1	1	1	1	1	2
Fully loaded employee cost ($)	57,452	58,889	60,361	61,870	63,417	65,002	66,627	68,293	70,000
# of management	1	1	1	1	2	3	3	3	3
Fully loaded employee cost ($)	328,299	336,506	344,919	353,542	362,380	371,440	380,726	390,244	400,000
Total SG&A staff costs	328,299	395,395	405,280	415,412	824,415	1,179,321	1,399,167	1,434,146	1,540,000
Facilities costs									
Cost per sq. foot ($)	25.00	25.63	26.27	26.92	27.60	28.29	28.99	29.72	30.46
Inflation	n/a	2.5%	2.5%	2.5%	2.5%	2.5%	2.5%	2.5%	2.5%
Sq. footage / employee	250	250	250	250	250	250	250	250	250
# of employees	2	3	4	4	27	28	29	41	42
Sq. footage required	500	750	1,000	1,000	6,775	7,000	7,250	10,250	10,500
Projected sq. footage	3,000	3,000	3,000	3,000	12,000	12,000	12,000	12,000	18,000
Total facilities cost ($)	75,000	76,875	78,797	80,767	331,144	339,422	347,908	356,606	548,281
Other SG&A spending ($)						130,000	200,000	0	1,700,000
Proj. total SG&A spending ($)	403,299	472,270	484,076	496,178	1,155,559	1,648,744	1,947,075	1,790,752	3,788,281
Discrepancy ($)					10,441	-13,744	-11,075	18,248	-35,281

FIGURE 6.1.20
Selling and general administrative expenses.

Stage 6 Integration

Operating expenses (excl. manufacturing costs)	Year 1	Year 2	Year 3	Year 4	Year 5	Year 6	Year 7	Year 8	Year 9
Proj. annual R&D expenses ($)	263,394	283,729	390,427	418,188	5,062,325	5,770,133	6,699,977	5,795,948	6,266,840
Proj. total SG&A expenses ($)	403,299	472,270	484,076	496,178	1,155,559	1,648,744	1,947,075	1,790,752	3,788,281
Total operating expenses (excl. manufacturing costs) ($)	**666,693**	**755,999**	**874,503**	**914,366**	**6,217,884**	**7,418,877**	**8,647,052**	**7,586,700**	**10,055,121**
Discrepancy ($)					6,116	(18,877)	(13,052)	48,300	(13,121)

FIGURE 6.1.21
Total operating expenses.

Total Manufacturing Costs	1997	1998	1999	2000	2001	2002	2003	2004	2005
Manufacturing labor	Year 1	Year 2	Year 3	Year 4	Year 5	Year 6	Year 7	Year 8	Year 9
# of engineers	-	-	-	-	-	-	-	5	5
Fully loaded employee cost ($)	213,394	218,729	224,197	229,802	235,547	241,436	247,472	253,659	260,000
# of direct labor	-	-	-	-	-	-	-	11	11
Fully loaded employee cost ($)	43,089	44,166	45,271	46,402	47,562	48,751	49,970	51,220	52,500
Manufacturing labor cost ($)	-	-	-	-	-	-	-	**1,831,707**	**1,877,500**
	1997	1998	1999	2000	2001	2002	2003	2004	2005
COGS	Year 1	Year 2	Year 3	Year 4	Year 5	Year 6	Year 7	Year 8	Year 9
Units sold	-	-	-	-	-	-	-	800	1,850
Raw material costs per unit ($)	-	-	-	-	-	-	-	340	340
% improvement	0%	0%	0%	0%	0%	0%	0%	0%	0%
Raw material & packaging costs ($)	-	-	-	-	-	-	-	**272,000**	**629,000**
	1997	1998	1999	2000	2001	2002	2003	2004	2005
Manufacturing Costs	Year 1	Year 2	Year 3	Year 4	Year 5	Year 6	Year 7	Year 8	Year 9
Manufacturing labor cost ($)	-	-	-	-	-	-	-	1,831,707	1,877,500
Raw material & packaging costs ($)	-	-	-	-	-	-	-	272,000	629,000
Total manufacturing costs ($)	-	-	-	-	-	-	-	**2,103,707**	**2,506,500**
Discrepancy ($)	-	-	-	-	-	-	-	1,293	(28,500)

FIGURE 6.1.22
Total manufacturing costs.

Sales	1997	1998	1999	2000	2001	2002	2003	2004	2005
	Year 1	Year 2	Year 3	Year 4	Year 5	Year 6	Year 7	Year 8	Year 9
C-Port units	-	-	-	-	-	-	-	100	150
PAS-Port units	-	-	-	-	-	-	-	700	1,700
C-Port average selling price (ASP) ($)	-	-	-	-	-	-	-	800	800
PAS-Port average selling price (ASP) ($)	-	-	-	-	-	-	-	800	800
C-Port revenue ($)	-	-	-	-	-	-	-	80,000	120,000
PAS-Port revenue ($)	-	-	-	-	-	-	-	560,000	1,360,000
Total revenue ($)	-	-	-	-	-	-	-	**640,000**	**1,480,000**

FIGURE 6.1.23
Total sales.

Clinical Trials	1997	1998	1999	2000	2001	2002	2003	2004	2005	Total
	Year 1	Year 2	Year 3	Year 4	Year 5	Year 6	Year 7	Year 8	Year 9	
C-Port pivotal trial patient-years	0	0	0	0	0	0	66.5	133.0	33.2	
PAS-Port European trial patient-years	0	0	0	0	0	0	55.0	55.0	13.8	
PAS-Port II trial patient-years	0	0	0	0	0	0	0.0	40.5	13.5	
Total # of patient-years	-	-	-	-	-	-	121.50	228.50	60.49	410
Total clinical trial cost ($)										6,650,000
Cost per patient-year ($)										**16,200**
Total # of patients enrolled										242
Cost per patient ($)										**27,479**

FIGURE 6.1.24
Clinical trial expenses.

6.1 Operating Plan and Financial Model

Income Statement	1997	1998	1999	2000	2001	2002	2003	2004	2005
	Year 1	Year 2	Year 3	Year 4	Year 5	Year 6	Year 7	Year 8	Year 9
Revenue ($)	-	-	-	-	-	-	-	836,000	2,056,000
Total manufacturing costs ($)	-	-	-	-	-	-	-	2,103,707	2,506,500
Gross margin ($)	-	-	-	-	-	-	-	(1,267,707)	(450,500)
Total operating expenses (excl. manufacturing costs) ($)	666,693	755,999	874,503	914,366	6,217,884	7,418,877	8,647,052	7,586,700	10,055,121
Pre-tax operating profit ($)	(666,693)	(755,999)	(874,503)	(914,366)	(6,217,884)	(7,418,877)	(8,647,052)	(8,854,408)	(10,505,621)
Discrepancy ($)							(13,052)	49,592	(41,621)

Cash Flow	1997	1998	1999	2000	2001	2002	2003	2004	2005
	Year 1	Year 2	Year 3	Year 4	Year 5	Year 6	Year 7	Year 8	Year 9
Cash balance (= operating profit) ($)	(666,693)	(742,691)	962,805	48,440	(6,169,445)	(428,321)	9,524,627	670,219	4,164,598
Amount financed ($)	680,000	2,580,000	-	-	13,160,000	18,600,000	-	14,000,000	-
Post-financing cash in the bank ($)	13,307	1,837,309	962,805	48,440	6,990,555	18,171,679	9,524,627	14,670,219	4,164,598

FIGURE 6.1.25
Income statement and cash flow.

Cost line items added

The following information explains the cost line items added to the operating cost model.

Clinical trials expenses

While employee costs generate most R&D expenses, an additional line-item was needed under R&D to reflect the actual clinical trials cost. This excerpt from the prospectus indicated that the clinical trial process accounted for a significant portion of R&D costs:

> Research and development expenses fluctuate with the stage of development of, the timing of clinical trials related to, and the status of regulatory approval of our products.[15]

Once this line-item was added, 2001 to 2005 clinical trial costs were assigned: $400,000 in 2001, $950,000 in fiscal 2002, $1,700,000 in 2003, $1,600,000 in 2004, and $2,000,000 in 2005. It was assumed that the company incurred some pre-trial preparatory costs in 2001, and that 2002 trials costs were lower than the following years since the trials began toward the end of 2002. From 2003 to 2005, relatively constant clinical trial costs were assumed, with a slight increase in 2005. There is no way to deduce from the prospectus a precise timeline of how the costs were incurred. The costs may have been highest at the start of the trials reflecting registration fees or upfront payment to trial service providers, or may have been spread evenly over time. In this model, it is assumed that the bulk of the trial costs are incurred during 2003 to 2005 when most of the trial activity occurs.

> **Working example:** Cardica management perspective: additional clinical trial expenditures
>
> In 2005, after the FDA rejected Cardica's existing clinical trial data and required the company to complete an additional study, Cardica found itself facing clinical trial expenses that it had not anticipated in its operating plan, nor built into its financial model. This setback required a two-pronged approach. First, Cardica negotiated with the FDA on a shorter time frame for its new trial. "Our original trial had six-month data," Hausen explained. "The FDA wanted one-year data, and we were able to convince them to compromise on nine-month data for the new trial." This compromise has been estimated by the authors to lead to direct clinical trial savings of about half a million dollars, as well as enabling Cardica to come to market in the United States three months sooner than it would have otherwise. Second, this change forced Cardica to use more of the proceeds from its IPO to fund clinical trials, rather than committing them to marketing expenses. While this change would obviously negatively impact IPO pricing, the ability to change strategies quickly enabled Cardica to make the best of a bad situation. In the medical device world, which is ever changing, it is critical to be able to change directions when necessary and refrain from lamenting over lost opportunities.

Table 6.1.10 *Excerpt from prospectus detailing company equipment asset values in 2004 to 2005 (2006 Cardica prospectus; reprinted with permission).*

Equipment	2004 (in 000s)	2005 (in 000s)
Computer hardware and software	$362	$375
Office furniture and equipment	$144	$154
Machinery and equipment	$1,899	$2,832
Leasehold improvements	$461	$461
Construction in process	$174	$11
	$3,040	**$3,833**

Capital expenditures

This line item was added to R&D to reflect the costs of capital equipment necessary to perform the R&D process. The company statement of cash flows indicated that the company spent $757,000 in fiscal 2003, $914,000 in 2004, and $882,000 in 2005 on total purchases of property and equipment. To estimate what portion of these values were allocated to R&D equipment, Note 4 in the company's prospectus was referenced (see Table 6.1.10), which provided an estimate of the proportion of company capital equipment value represented by "machinery and equipment," namely $1,899,000 out of $3,040,000 (62 percent) in 2004, or $2,832,000 out of $3,833,000 (74 percent) in 2005.[16] The 62 percent figure was used to determine the estimates for R&D capital expenditures in 2003, 2004, and 2005. Estimates for 1997 to 2000 were ramped up gradually over time, followed by a sizable jump from 2000 to 2001 to reflect a substantial increase in staffing and financing in 2001.

Facilities

A facilities section was added under SG&A costs. The following excerpt from the prospectus detailed the current facilities situation:

> We currently lease approximately 29,000 square feet in Redwood City, California, containing approximately 19,000 square feet of manufacturing space, 7,000 square feet used for research and development and 3,000 square feet devoted to administrative offices. Our facility is leased through August, 2008. We believe that our existing facility should meet our needs for at least the next 24 months.[17]

All facilities expenses were allocated under SG&A, rather than broken up across SG&A, R&D, and manufacturing, for the following reasons: (1) Cardica only possessed one facility, which the various departments shared, and the amount of space used by any one of the three departments was small, and (2) the company had only recently begun sales, and cumulative sales figures were low, so the cost of goods attributable to facilities costs would probably be quite low.

Other SG&A expenses

This line item was added under SG&A costs to reflect additional non-staff costs. This was originally motivated by a sizable increase in SG&A expenses of $2 million, from $1.8 million in 2004 to $3.8 million in 2005, due to a non-cash stock-based compensation expense related to loans made to three directors to purchase shares in Cardica's **common stock**. This was not an operational expense but had to be accounted for in order to derive an accurate estimate of operational SG&A expenses in 2005. This line item also captured non-cash stock-based compensation, travel, and professional services expenses.

Additional cost and staffing considerations

The following were additional considerations in setting costs and staffing levels.

Research and development

As noted above, in assigning values to the cost of clinical trials line item, every effort was made to match the progress of the clinical trials from 2003 through 2005. This line item included filing and regulatory costs, consulting costs, as well as the cost of producing the units of the products used in the trial. An analysis of clinical trials is provided in a separate section below.

According to the prospectus (p. 40), R&D expenses decreased by $872,000 between fiscal years 2003 and 2004, from $6.7 million to $5.8 million, "primarily attributable to the reallocation to cost of product revenue of $817,000 of personnel-related costs of manufacturing overhead included in research and development expenses in 2003."[18] In other words, Cardica reallocated some of its engineers from R&D costs to manufacturing in 2004. Staffing levels in the model reflect this change. From 2003 to 2004, R&D engineers decreased from eight to four and manufacturing engineers increased from zero to five (reflecting that four engineers were reallocated from R&D to manufacturing and one additional manufacturing engineer was hired in preparation for start of sales).

It was assumed that R&D engineering and technician staff remained small from 1997 to 2000 and jumped significantly in 2001, reflecting the significant round 3 financing it received and acceleration of product development efforts.

It was assumed that the VPs of R&D and clinical and regulatory were hired in year 5, upon completing round 3 financing and having the resources to put in place a more senior management team. Clinical staff increased from zero in 2000 to three in 2001, giving this staff a two-year head start to coordinate with the product team prior to beginning clinical trials.

Note that a fractional staffing level (e.g., 2.1) signifies that one of the hires was made midway through the year. Sometimes fractional staffing levels were assumed in the model in order to make the total costs approach the historical costs.

> **Working example:** Cardica management perspective: R&D costs
>
> As companies develop products and begin to focus on manufacturing and commercialization, many decide to decrease their R&D costs until they can develop solid revenues and prove that a market exists for their product. Cardica, however, continued to maintain high R&D expenditures as it began to roll out its C-Port and PAS-Port systems. Younger cited the company's physician stakeholders as one of the primary reasons Cardica needed to sustain its R&D expenditures: "We face a difficult market development problem because we have a very conservative customer base that is hard to please and will only use the best products." Hausen added, "Our R&D budget remains big because we need to keep making the product better and better, so that whenever physicians come up with a reason not to use our products, we overcome that resistance with a better offering." While each company must set an R&D spending level that is appropriate for its business (and stakeholders), it may be dangerous to automatically assume that such expenses costs will decrease dramatically over time.

Selling and general administrative expenses

Since sales began in 2004, the VP of sales and marketing was not hired until 2003.

Manufacturing costs

The manufacturing team, including one new engineer and 11 technicians, was not hired until 2004, the year sales began.

In setting raw materials costs, the number of units sold in 2004 and 2005 was estimated to be 800 and 1,850, respectively. The company prospectus states that it had cumulative sales through September 30, 2005 of 250 C-Port Systems and 2,400 PAS-Port systems worldwide, for a total of 2,650 systems. Lacking better information, the assumption was made that the two systems have the same raw materials costs. It was also assumed that of the 2,650 systems sold, 800 were sold in fiscal year 2004 and 1,850 in 2005.

No improvement in raw materials costs was assumed, since sales had only recently begun and sales

Stage 6 Integration

levels were relatively low at the time of the IPO. Such improvements in costs are usually generated by high-volume production.

Revenue

Revenue for Cardica was difficult to model because the company had only two years of revenue at the time of their public offering and a significant portion was attributable to sales and development partnerships that had recently been terminated or downgraded. The Guidant European distribution agreement, which accounted for 48 percent of total 2004 revenue and 50 percent of fiscal 2005 revenue, had been terminated, and Cardica's role in supplying Guidant with aortic cutters had been downgraded from a manufacturing to a pure royalty relationship. Therefore, Cardica's revenue outlook heading into 2006 was unclear. A simple model of C-Port and PAS-Port sales could be created, but no attempt was made to model the other revenue sources because they did not seem to reflect future sales. Because the revenue model did not account for all of 2004 and 2005 revenues, the discrepancy between historical sales and the model's projected sales was not calculated. When filling in the revenue line in the income statement, historical 2004 and 2005 revenues were used.

> **Working example:** Cardica management perspective: redefining the potential market
>
> While the problems with FDA approval delayed Cardica's launch of the PAS-Port in the United States and led to increasing clinical trial costs, Hausen viewed the FDA's change as a potential "blessing in disguise" with regard to potential, long-term revenue. He explained: "When we were initially in the middle of the FDA approval process, there were nine other companies competing for FDA approvals of similar products. This raising of the bar by the FDA and additional public scrutiny for our devices got rid of all of our competition virtually overnight. We suddenly became a company in a $1 billion-plus potential market, completely by ourselves." While Medtronic later began clinical trials for a competing product, this shift in market dynamics allowed Cardica to reevaluate its US sales potential and model higher revenues in later years than was previously realistic.

Clinical trial costs

One of the objectives in analyzing Cardica was to develop a reasonable model for the cost of clinical trials in this sector. Previously, the cost of clinical trials per year was chosen so that the model's total R&D costs per year closely approximated the actual values in years 2001 through 2005. The next question to answer was how to use those clinical trial costs to calculate a reasonable cost per patient in the clinical trial.

First, the sum of the clinical trial costs was calculated in the model from years 2002 through 2005. This is the model estimate of the total spent on clinical trials ($6,650,000).

The start date and duration of each clinical trial was also examined to determine the number of patients actively involved in each clinical trial within each fiscal year. For example, in the case of the C-Port pivotal trial, patient enrollment began in fiscal Q3 2003 and ended in Q1 2005 and there were 133 patients involved in that trial; 133 patients were assigned to each quarter between Q3 2003 and Q1 2005. This date range spanned half of fiscal year 2003 (Q3, Q4), all of fiscal year 2004, and a quarter (Q1) of fiscal year 2005. Therefore, 66.5 patient-years were assigned to 2003, 133 patient-years to 2004, and 33.2 patient-years to 2005 for the C-Port pivotal trial. This process was repeated for the other clinical trials to produce a total number of patient-years per fiscal year and per trial.

Finally, the model's estimate of the total spent on clinical trials was divided by the total number of patient-years to produce an estimate of the cost per clinical patient-year. The end result was $16,200 per patient-year.

Another metric for cost of clinical trials is the cost per patient, rather than the cost per patient-year. To calculate this, the total estimate for clinical trial costs ($6,650,000) was divided by the total number of patients enrolled in the trials (242), yielding a result of $27,479 per patient. See Table 6.1.11.

It is worth noting that because information was lacking about the timing of clinical trial costs, as well as the rate of patient enrollment, it did not seem reasonable to try to analyze the clinical trial costs on an annual basis. That is, the total clinical trial costs were analyzed as they related to the total number of

Table 6.1.11 *Clinical trial analysis.*

	2003	2004	2005	
	Year 7	Year 8	Year 9	Total
C-Port pivotal trial patient-years	66.5	133	33.2	
PAS-Port European trial patient-years	55	55	13.8	
PAS-Port II trial patient-years	0	40.5	13.5	
PAS-Port II trial patient-years	121.5	228.5	60.49	410
Total clinical trial cost				$6,650,000
Cost per patient-year				**$16,200**
Total # of patients enrolled				242
Cost per patient				**$27,479**

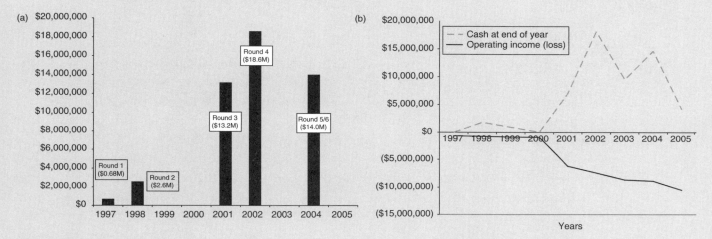

FIGURE 6.1.26
Cumulative cash on hand, operating income, and financing events (financing data from VentureSource).

clinical patient-years or the total number of patients. Also, bear in mind that the approach chosen here is quite rough.

Cumulative cash flows

In calculating cumulative cash flows, net income value was used without accounting for depreciation and other adjustments that would give a more precise number for cash flow. Examining the Cardica financial statements, it was clear that many of their adjustment items were details specific to Cardica that were not instructive to model. For example, in fiscal 2004, the net operating loss was $10,950,000 and net cash used was $7,417,000, an adjustment of $3,533,000. Yet $2,599,000 out of the $3,533,000 adjustment was due to stock-based compensation on grants of stock options to employees. Because the size and timing of this adjustment was specific to Cardica, it was not worth including in the model. Figure 6.1.26 illustrates cumulative cash on hand, operating income (loss), and financing events over time.

Lessons learned

Cardica faced significant hurdles at the time of its IPO:
- The company had so far failed to achieve FDA approval for its PAS-Port system, and only recently (November, 2005) received FDA approval for its C-Port system.

- Sales had only begun in 2004, and 2005 revenues were limited.
- The termination of its European distribution agreement with Guidant in September, 2004 was a setback, as the agreement accounted for 48 percent of total revenue in fiscal 2004 and 50 percent of 2005 revenue. At the time of its IPO, the company lacked a European distribution strategy and expected limited sales in Europe going forward.
- All of these factors probably contributed to Cardica's pre-IPO decrease in valuation. Between its 5th venture financing round in September 2003 and their prospectus filing on February 1, 2006, the company's valuation decreased from $85 million to $75 million.

Conventional wisdom in the field states that the proxy companies should have already reached profitability and achieved significant market penetration. In that respect, Cardica is not necessarily an ideal proxy company. On the other hand, the company appears to have persevered within a difficult and changing environment. Its approach and financial model can provide valuable lessons to innovators. Most ventures ultimately face challenges that may appear insurmountable. The ability of their management to work through the challenges in their path to the market is a critical success factor.

With that in mind, here are some lessons to take away from the proxy analysis:

- **FDA regulatory hurdles** – Cardica faced significant hurdles during the PAS-Port regulatory process. The FDA redefined the objective performance criteria in the middle of the process, requiring Cardica to resubmit data. Ultimately, the FDA required more robust data that required Cardica to withdraw its 501(k) application and embark on an entirely new clinical trial. If one were to develop a product similar to Cardica's, it would be important to further investigate these hurdles to understand what went wrong in the regulatory process.
- **Extraordinary cost items** – There may be some extraordinary cost items present in the income statement, such as stock options expenses or other non-operating expenses. For example in 2005, Cardica's SG&A expenses contained a $2.0 million stock compensation expense. Be aware of these expenses and account for them appropriately so that they are not improperly included in the operating expense estimates.
- **Focus on relevant costs** – When reconstructing the operating plan, try to focus on the cost items that will also likely apply to the company and spend less time on the items that are unique to the proxy company.
- **Identify successful companies in the space** – Before picking apart a particular proxy company, identify proxy companies that have excelled and whose operating plans might be worth emulating.

Cardica post-IPO

Cardica has had mixed success since going public in February, 2006. The company reported increases in revenues and was making progress with clinical trials. However, it is still without a distribution partner in Europe and is a long way from profitability.

FDA approval for PAS-Port

In September, 2008, Cardica's PAS-Port device gained its long-awaited FDA approval in the United States, following results from a 220-patient randomized clinical trial performed at multiple sites in the United States and Europe. The study met its primary endpoint of non-inferiority to hand-sewn anastomosis. Approval was granted five years after the company filed its original FDA application.

European strategy

Cardica is still without a European distribution partner after its Guidant agreement was terminated. This has significantly hurt sales in Europe. The company's February, 2008 10Q states, "We expect to rely on third-party distributors for substantially all of our international sales. If we are unable to establish adequate sales and marketing capabilities, independently or with others, we may not be able to generate significant revenue." Cardica is licensed and approved to sell its products in Europe but, without a distribution partner, it has not seen the uptick in usage that it expected to achieve. As Hausen noted, "Europe has lots of surgery, but no money," hypothesizing that the low reimbursement in

Europe for heart surgery will not enable Cardica to ever generate the kind of revenue that it expects to achieve in the United States after PAS-Port approval.

Revenues and profits

Revenues have increased since 2005, but still remain far below levels that Cardica needs to achieve profitability. Product revenues for fiscal year 2007 (ending June 30) were $2.1 million, while development revenues were $1.3 million. This compares to operating costs in 2007 of $19 million, leading to continued large losses for the company. Cardica had to recall 55 C-Port devices in the second quarter of 2007 due to a supplier manufacturing defect, which negatively impacted sales and resulted in lower than hoped for revenue growth. Cardica has had successes, such as the $2.0 million it has received in milestone payments from Cook Incorporated for work done to develop the Cook vascular closure device (formerly called the X-Port vascular closure device). Large increases in near-term revenue growth and achieving profitability, however, seem to rely more and more on FDA approval of the PAS-Port device and Cardica's subsequent ability to convince physicians that their device provides an advance over traditional cardiac surgery.

Overall, Cardica's mixed record post-IPO has led to a stagnating stock price, which on December 8, 2008 stood at $4.36 per share after opening in February, 2008 at $10.00 per share. The Cardica example, in many ways, illustrates many of the difficulties inherent to the medical device industry, as profitability has been elusive and FDA approval a long and often painful process. However, continued persistence and a belief in the ultimate success of the PAS-Port product in fundamentally revolutionizing the way physicians perform heart surgery has sustained Hausen and Younger's commitment to Cardica, and may ultimately lead to future market success.

Reconciliation with the bottom-up market model

The bottom-up market model for Cardica, which is included in the main text of this chapter, is meant to reflect the situation faced by the company around the time of the C-Port launch with the added insight that it would take several years before the PAS-Port product would be launched. One of the assumptions made within that model is that the major breakthrough for the company would occur when PAS-Port is launched in the United States. This assumption was partly driven by Cardica's experience with PAS-Port in Japan, where the company captured 25 percent of the total CABG graft market (both off-pump and on-pump) within four years. In 2005, approximately 250,000 CABG surgeries were performed in the United States.[19] While that number was stagnant for some time, it was starting to grow again in 2006 as physicians began to question the efficacy of the less invasive alternative of percutaneous coronary interventions (PCI) and particularly drug eluting stents. The growth in 2006 was modest, only 0.8 percent.[20] This implies that if Cardica's PAS-Port is launched sometime in late 2008, and that the company replicates its success in Japan in the US market, it could have annual PAS-Port sales of about $78 million by 2011. This figure is higher than the bottom-up estimate of $34 million shown within the chapter. The estimate in the bottom-up market model represents total sales in the off-pump market, whereas the higher estimate of $78 million (extrapolated based on the Japan experience) represents expansion into the on-pump segment.

Notes

1. From remarks made by Thomas Fogarty as part of the "From the Innovator's Workbench" speaker series hosted by Stanford's Program in Biodesign, January 27, 2003, http://biodesign.stanford.edu/bdn/networking/pastinnovators.jsp. Reprinted with permission.
2. Walter T. Harrison and Charles T. Horngren, *Financial Accounting* (Prentice Hall, 2007).
3. All quotations are from interviews conducted by the authors, unless otherwise cited. Reprinted with permission.
4. The Dartmouth Atlas of Healthcare, http://www.dartmouthatlas.org/ (April 10, 2008).
5. "Salary Wizard," Salary.com, http://www.salary.com/personal/layoutscripts/psnl_default.asp (February 22, 2007). Verified with Larry Tannenbaum, lecturer in Stanford University's Program in Biodesign.
6. Cushman and Wakefield, "Silicon Valley Office MarketBeat," Mid-Year 2006, http://www.siliconvalleyonline.org/PDF/Trends-Office.pdf (February 22, 2007).
7. "MedicareCoverage: Clinical Trials, Centers for Medicare and Medicaid Services, http://www.cms.hhs.gov/clinicalTrialPolicies/Downloads/finalnationalcoverage.pdf (April 10, 2008).
8. From remarks made by Mir Imran as part of the "From the Innovator's Workbench" speaker series hosted by Stanford's Program in Biodesign, April 28, 2004, http://biodesign.stanford.edu/bdn/networking/pastinnovators.jsp. Reprinted with permission.
9. Matthew Kirdahy, "Medtronic Growth Still Foggy," *Forbes*, November 21, 2006, http://www.forbes.com/markets/2006/11/21/medtronic-earnings-biotechnology-markets-equity-cx_mk_1121markets03.html (November 14, 2008).
10. Drawn from Cardica Inc. S-1 Form, 2006, http://www.sec.gov/Archives/edgar/data/1178104/000116923205005200/d65698_s-1.htm (March 8, 2007). Reprinted with permission.
11. Ibid.
12. Ibid.
13. Ibid.
14. Ibid.
15. Ibid.
16. Ibid.
17. Ibid.
18. Ibid.
19. "Heart Disease and Stroke Statistics," American Heart Association, 2008 Update, http://www.americanheart.org/downloadable/heart/1200082005246HS_Stats%202008.final.pdf (November 14, 2008).
20. "CABG Procedures to Rise in 2007," Millennium Research Group, June 28, 2007, http://www.mrg.net/news_newwin.php?news_id=236 (November 14, 2007).

6.2 Business Plan Development

Introduction

In today's medtech environment, having a formal business plan is not nearly as valuable as the process a company goes through to create one. Whether or not the plan is a 40-page Word document, a 20-slide PowerPoint presentation, or a 3-page executive summary, there is a tremendous benefit that comes from developing a holistic view of the business and a common vision for what it is trying to achieve.

The business plan provides a summary of all of the in-depth strategy and planning work that has been completed to this point in the biodesign innovation process. Among other information, it describes the medical need, the offering, the market potential, and the development and commercialization pathway. By putting this material in writing, as part of one cohesive overview, innovators are forced to crystallize their thinking about how they will communicate with internal and external audiences and the "story" they will tell about their business. Having a clear, compelling story helps them prepare for important interactions with potential employees, investors, and customers (whether or not a copy of the plan is ever shared with these audiences). It can also be used as a mechanism for aligning the goals of everyone who is involved in making the business a reality. While the development of a business plan should not take on a life of its own and eclipse the importance of staying focused on the execution of the operating plan, the disciplined thinking and integrated view of the business that it imposes on the company can be invaluable.

OBJECTIVES

- Learn to identify and develop the key components of a business plan.

- Understand the strategic value of a business plan, both internally in managing the new venture and externally in communicating with potential investors, partners, and employees.

- Recognize that a business plan does not directly lead to funding, but helps innovators rigorously prepare to present and defend the opportunity to potential investors.

Business plan fundamentals

A business plan serves three primary purposes: planning, management, and communication. As a planning tool, it outlines the steps that a company will take in order to bring its product to market and become financially viable. It also captures the overarching strategy behind the venture and the detailed steps that support the execution of that strategy, along with the required resources. As a management tool, the plan sets important goals and is then used to monitor progress toward them. Finally, the plan can be a tool for framing and guiding communications, both internally and externally. Internally, it facilitates cross-functional interactions and helps maintain the alignment of goals and activities across the different functions within the company. Externally, the plan can be used for recruiting purposes, to convince other businesses to enter into a strategic partnership, and to persuade investors to provide capital. With increasing regularity, however, the business plan itself is not directly shared externally, but rather serves as a vehicle for organizing management's thinking in support of important communications.

At an overarching level, the business plan also puts forth a cohesive argument for why the company is viable. Ron Dollens, president and CEO of Guidant from 1994 to late 2005, highlights several key ingredients of such an argument:[1]

> To me, there are three things that dictate whether a business is going to be successful or not. First is its operational capabilities – its product development, sales and marketing, cost structure management, and quality management systems. The second thing is the market structure – is it in a marketplace that rewards doing things well and has sufficient growth opportunities? Finally, there are environmental effects to consider. These are things that, even if the company does all of those other things well, can cause a disastrous outcome. In our field, one of these environmental effects is called health policy.

Many innovators mistakenly believe that the primary driver of business success is the underlying technology, especially its unique aspects. However, as Dollens' quote reflects, there are other considerations that are equally important to establish in the business plan:

- **Strong execution capabilities** – A great team with a proven track record for solving problems with limited capital.
- **Compelling market** – Large and growing revenue potential, a significant clinical need, as well as clearly defined financing needs that can generate an attractive return for investors.
- **Accessible market** – Well-defined reimbursement pathways and a community of stakeholders that is likely to be supportive of the innovation.

As an example, when a business plan is used to guide conversations with investors, the innovator should recognize that investors are more willing to put their money into market- or customer-driven companies than those primarily driven by technology (assuming, of course, that the underlying technology is viable). As a result, it is even more important for the entrepreneur to make a compelling case regarding the size and growth of target markets, potential sales, and expected profitability than it is to educate stakeholders on the detailed *features* of the innovation. Market strength can be shown by demonstrating unique *benefits* that will make the innovation compelling to target customers (e.g., a short pay-back period). The ability to demonstrate existing market interest is essential and can be accomplished through testimonials and/or survey data from customers (or potential customers) in the target market. Using statistics to size a market can further support the market's strength, as long as the market claims are realistic and well documented. It is essential to carefully analyze all available market data to be certain that the market estimates and expected sales figures included in the business plan are feasible, as well as defensible if/when investors question them[2] (see 2.4 Market Analysis and 2.5 Needs Filtering).

Once an innovator develops a convincing case for the market potential, s/he can create the financial projections needed to internally determine whether the opportunity is worth pursuing and, ultimately, to develop a financial case for potential investors. Like market estimates, financial projections must be realistically attainable given the industry, competitive landscape, competitive advantage of the innovation, and other factors that will affect the adoption of a new product or service. Similarly, financial costs must be accurately reflected in the projections. If an innovator fails to include or understates important costs (or overstates market estimates), not only will the projections be inaccurate, but s/he will also be likely to lose credibility with prospective investors (see 6.1 Operating plan and financial model). Studying how other companies of the same size or in the same field have expanded, how their sales figures have grown, and how these figures compare over time to similar entities serves as an effective check on whether the figures outlined in the financial projections of a business plan are truly realistic (refer to Appendix 6.1.1 Proxy Company Analysis). The innovator should also recognize that despite his/her efforts to provide realistic estimates, most investors will discount sales projection by at least 50 percent and may also inflate costs. Therefore, the innovator should "stress test" the financial model to make sure that it remains compelling if sales and costs are off by as much as a factor of two.[3]

The business plan should also be action oriented. It must include the tactical and operational activities that outline how the innovation will be developed and commercialized. Ultimately, the business plan is a template for executing a strategy that addresses an unmet medical opportunity. As Mir Imran, serial inventor, entrepreneur, and founder of InCube Labs, put it, "I have learned that good ideas are a dime a dozen. It's the execution that matters."[4] In fact, one guideline encourages innovators to develop plans that are ten parts implementation for every one part strategy.[5]

Structuring the business plan
Business plans typically include the following sections.[6]

Cover and title page
On the cover sheet, include the name of the company, address, phone, e-mail address, and month/year the plan was created. Innovators frequently submit business plans with too little information about how the company can be contacted for questions and additional information. The title page repeats this information with "Copy number xx" in an upper or lower corner. Tracking the number of copies will help innovators manage how many times the plan has been produced and distributed. It will also reassure investors that the plan has not been shared with too wide an audience.[7]

Executive summary
Create a summary of the most important aspects of the business plan. The executive summary is widely considered the most critical component of the business plan. Investors, partners, and other stakeholders always read it first and often make the decision whether or not to pursue further conversations, or even read the rest of the business plan, based on the caliber of this overview. Some experts suggest writing a draft of the executive summary first to help guide the development of the overall business plan. Others suggest saving the executive summary for last, after all necessary supporting information has been developed. In either scenario, it is essential to make sure the executive summary is concise, well-written, and compelling. (See below for more information about creating the executive summary and Appendices 6.2.1 and 6.2.2 for samples.)

Table of contents
Provide a well-designed table of contents to help readers navigate the document. At a minimum, headings and page numbers should be provided for each of the major sections.

The business
This section should include a brief history of the company, its strategies, and its plans for the future, including its timeline and critical milestones. Profile the management team briefly, highlighting those experiences that give members credibility in the relevant field and demonstrate established patterns of success. Since recruiting qualified talent is such a concern for any new venture, share any information that will address the company's ability to attract new hires in key roles as the business expands (e.g., personal networks).

The clinical need
Provide an overview of the disease state and the medical need that the innovation is intended to treat. Include evidence that the need is real and significant, including opinions of key clinical advisors. Make a strong case

indicating how the innovation addresses an important gap in the existing treatment landscape.

The product/service
Describe the innovation in detail from raw materials to end result. Include a detailed description of how the product will be used and who will use it. Provide evidence that potential users find the product compelling. This could include preliminary studies with users or quotes from user interviews. Address primary and secondary suppliers, production processes and costs, capacity, quality assurance, and after-sale service. If the innovation is still under development, outline precisely how much additional research and development is needed, technical risks, and risk mitigation mechanisms.

The market
Define the initial target market for the innovation. Outline the size of the market, as well as its projected growth and other important trends. Explain key characteristics of the market, including how buying decisions are made, how the market is segmented, important aspects of the competitive landscape, and what kind of market position is most appropriate for the innovation. Describe mitigation strategies for addressing the most critical market risks (e.g., how to differentiate the innovation from existing alternatives or how to defend the innovation against second-generation products if it is the first to market) and the competitive landscape. Discuss potential market expansion strategies.

Intellectual property (IP)
Describe the proprietary features of the innovation that will enable the company to practice the invention, referencing the status of all IP filings. Outline any IP risks and planned mitigation strategies.

Regulatory strategy
Describe the regulatory pathway for the invention and other key aspects of the regulatory strategy. Provide the current status of all regulatory efforts, including benchtop validation, animal studies, or clinical trials. Outline any regulatory and/or clinical risks and planned mitigation strategies.

Clinical studies
Describe the clinical studies the company proposes to sponsor. Include a detailed description of the endpoints, sample sizes, duration, and location, as well as key investigators. Provide evidence to support integration of studies with regulatory, reimbursement, and marketing efforts, and information demonstrating that the company has (or can access) the necessary expertise to successfully conduct these studies.

Reimbursement
Describe the reimbursement strategy for the product. Outline the likelihood of (and justification for) receiving reimbursement. Provide examples of the reimbursement codes to be used if they exist, or outline the status of efforts to secure reimbursement to date. Outline reimbursement risks and planned mitigation strategies.

Sales and marketing
Clarify the company's competitive advantage and its product's value proposition. Outline specific plans for the company's sales process, including how sales people will be hired, trained, and supported as well as how leads will be identified, targeted, converted into customers, and then retained. Describe the company's methods of clinical education, sales, distribution, and pricing, as well as its timing for market entry. Cover plans for advertising and public relations, as appropriate. Be certain to address the costs associated with sales and promotional activities and provide a convincing explanation of how the chosen methods will result in the greatest return for the most reasonable investment.

Financial information
Include detailed financial forecasts, as well as a statement of funds needed and how they will be used. Refer to 6.1 Operating Plan and Financial model for more detailed information on what specific financial information should be provided.

Management team
Profile key members of the company's management team in greater depth. It can also be helpful to highlight advisors or board members if they are well known and respected in the field. Include details about the

company's organizational structure and staffing plans, if they are relevant to its ability to execute on the strategy outlined in the business plan.

While the sections of the business plan do not need to be presented in the specific order listed above, this organization represents a logical flow that is used by many innovators. Table 6.2.1 provides an overview of how the various aspects of the business plan correspond to the various core activities in the biodesign innovation process. It also summarizes some of the key questions that each section should seek to address.

Creating an effective executive summary

The executive summary should be brief – the maximum length is ten pages in cases where no formal business plan will be presented – and self-contained. It should provide a succinct and compelling argument that the proposed business has the key ingredients of success. As the innovator begins to prepare an executive summary, the following template may prove useful:

- **Opening** – Start with a compelling statement about the clinical need, proposed solution, and market opportunity.
- **Body** – Provide the most important high-level information about each of the most relevant topics outlined in Table 6.2.1, focusing on actions that have been taken and results that have been achieved to date. Include evidence to support claims about the extent of the need and the likelihood that the solution will address the need. Describe key risks and propose mitigation strategies. Use headings to make the organization of the summary intuitive and clear. Make sure the text flows smoothly from one section of the overview to the next, using transitions as needed.
- **Conclusion** – Conclude the summary with a clear statement of the purpose of the business plan (e.g., "The purpose of this plan is to raise $2 million in funds to establish the technical feasibility of the product and perform animal testing …"). If the company clearly articulates its capabilities and needs in the executive summary, it will have a significantly greater chance of engaging the reader.

Always develop the executive summary with the intended reader in mind. For example, if the business plan is to be used to raise funds from investors, then it is important to highlight the market potential of the idea, along with key financial projections and requirements in the overview. If the plan is to be used to seek input on the idea from expert clinicians, then more emphasis might be placed on the need, regulatory path, reimbursement issues, and other related subjects. In either scenario, explain or avoid terminology or in-depth concepts that might be unfamiliar or off-putting to the target audience. The most effective executive summaries, along with the business plans themselves, demonstrate an understanding of the reader's priorities, issues, and questions, and address them in a straightforward and comprehensive manner.

When developing an executive summary, the innovator should thoughtfully consider the following issues:[8]

- **The executive summary is more than a brief description of the business and its products.** The executive summary should provide a holistic overview of the entire business plan, focusing on the most important and compelling information about the opportunity. Be careful not to let information from any one section of the plan dominate the executive summary at the expense of providing a comprehensive summary of the overall plan.
- **The executive summary is not an extended table of contents.** For example, do not tell the reader that the plan includes market projections. Instead, share the highlights of that forecast and their implications.
- **Do not exaggerate.** While the innovators want to excite readers enough to seek more information, exaggerated claims must be avoided. Provide a positive yet realistic perspective on the business and its opportunities, as well as its risks. Experienced investors and savvy business people will recognize hype, which will only serve to undermine the plan's credibility.
- **Do not cut-and-paste.** Given its importance, the executive summary should summarize, not duplicate, content in the larger plan. Write it with a fresh perspective, taking care to weave together the key points from the major sections of the plan into a cohesive narrative from beginning to end.

Stage 6 Integration

Table 6.2.1 *The major sections of the business plan can be mapped to specific steps in the biodesign innovation processes.*

Major business plan sections	Questions to address	Relevant activities in the biodesign innovation processes (from which to pull information)
The business	• What is the focus of the business and why? • How can the business and its strategy best be described? • Who is the business intended to serve? • What is the business' competitive advantage?	1.1 Strategic Focus 2.3 Stakeholder Analysis 4.4 Business Models 5.9 Competitive Advantage and Business Strategy
The clinical need	• What is the need being addressed? • Why is this need important? • How is the need currently being addressed (if at all)? • In what ways are current solutions inadequate?	1.2 Observation and Problem Identification 1.3 Need Statement Development 2.1 Disease State Fundamentals 2.2 Treatment Options 2.5 Needs Filtering
The product/service	• What is the proposed solution to the need? • How will it be used (and by whom)? • How does it better address the need than what is currently available? • What is the value proposition? • How will the company prove that the solution is technically feasible? • What are the most critical technical risks and how will they be mitigated? • How will the product be manufactured and where?	3.2 Concept Screening 4.6 Final Concept Selection 4.5 Prototyping 5.2 Research and Development Strategy 5.5 Quality and Process Management 5.7 Marketing and Stakeholder Strategy
The market	• Who is the target customer? • What is the market size? • How fast is it growing? • Who are the primary competitors? • How will the company differentiate itself from the competition? • What are the barriers to entry?	2.4 Market Analysis
Intellectual property	• How will the company protect its IP? • How strong is its IP position?	4.1 Intellectual Property Basics 5.1 Intellectual Property Strategy
Regulatory strategy	• How will the company get its product cleared/approved for the market? • What will be required to demonstrate safety and efficacy?	4.2 Regulatory Basics 5.3 Clinical Strategy 5.4 Regulatory Strategy
Clinical studies	• How will the company collect safety and efficacy data? • Where will the studies be performed and who will be the key investigators? • What other endpoints will be studied (and why)?	5.3 Clinical Strategy
Reimbursement	• What is the reimbursement pathway? • What codes will be used? • How much resistance is anticipated?	4.3 Reimbursement Basics 5.6 Reimbursement Strategy

Major business plan sections	Questions to address	Relevant activities in the biodesign innovation processes (from which to pull information)
Sales and marketing	• Why will customers be compelled to use the product? • How will the company reach customers and generate a profit?	5.7 Marketing and Stakeholder Strategy 5.8 Sales and Distribution Strategy
Financial information	• What are the company's financing needs? • What does it intend to do with the money raised (according to what timeline)? • How (and when) will it generate a return for investors?	6.1 Operating Plan and Financial Model 6.3 Funding Sources
Management team	• Who are the key individuals that make up the company? • What specific qualifications do they bring to bear?	6.1 Operating Plan and Financial Model

Writing the business plan

Before starting to write the business plan, the innovator should answer the questions captured in Table 6.2.1 and keep them visible while writing. The key points that best address each question should be summarized in a single paragraph at the beginning of each section of the plan. Then, the innovator can expand on each point by providing additional information and more detail in the text that follows. In terms of the way the plan is written, experts advise that it should be:[9]

- **Succinct** – Emphasize quality and clarity over length.
- **Detailed** – Address important issues directly and with sufficient information to mitigate major concerns. Be specific.
- **Professional** – Be sure the plan is well written (e.g., error free and appropriate in tone and style). It should also be professional in appearance (but not excessive in the cost to prepare and produce it).
- **Clear** – Use simple and readily understandable language that is appropriate for the plan's target audience.

In total, a business plan should not exceed 40 pages (including the executive summary). Important supporting information can be provided in appendices, as needed. Whenever possible, obtain and look at other business plans that were used for both successful and unsuccessful businesses to gain an appreciation of the nuances and details of the different sections and how they can be organized.

Other helpful hints when developing a business plan include the following:[10]

- **Do not put it off** – Too many companies wait until they absolutely have to before developing a business plan. One benefit of having a business plan in place before it is formally required is that it can help the company save time and money, if it is used as a guidepost for prioritizing efforts and resources. It can also expedite the process of raising funds because it will enable the innovator to anticipate and respond to questions raised by potential investors in a rapid manner.
- **Address key risks** – An innovator should always be honest about the main challenges a company faces, because these almost always become apparent as stakeholders evaluate the business plan. Building a company is often about identifying trouble spots early and then mitigating them. Make sure the plan identifies at least two key risks and outlines how they will be lessened.
- **Focus on what is most important** – Be careful to present only that information which is most critical to the success of the business. A company that presents its top 15 or 20 priorities could be perceived as being unfocused rather than strategic.

The following case example demonstrates how the various aspects of a business plan come together to support planning, management, and communication.

FROM THE FIELD

A TEAM IN STANFORD'S BIODESIGN INNOVATION PROGRAM

The role of the business plan in facilitating internal control and external communication

As described in 3.1 Ideation and Brainstorming, Darin Buxbaum and his team collaborated to address the need for *a less invasive way to help morbidly obese people lose weight* as part of Stanford's interdisciplinary Biodesign Innovation Program (see Figure 6.2.1). Upon graduating from the program, the team founded a company called HourGlass to develop a device that would provide the 9 million morbidly obese patients in the United States with a less invasive alternative to bariatric surgery. Buxbaum described the company's approach this way: "Over the next few years there are a series of first-generation transoral obesity devices coming to market that are much less invasive than surgery, but they are going to be large, bulky, and difficult for physicians to use. Our company, HourGlass, has an IP-protected device that is so easy to use that it has the potential to cut procedure times in half. And it's this attribute of being easier to use and also longer lasting, with a lower reoperation rate, that we think will drive a tremendous shift in the market to our next-generation device."[11]

As Buxbaum and his teammates progressed through the biodesign innovation process, they prepared to develop a business plan for two primary purposes: internal control and external communication. Internal to the team, Buxbaum noted that, "A business plan for our team served the purpose of getting everyone on the same page. In many ways, it provided a unifying vision that helped drive the team to achieve the same goals. And it helped make sure everyone had the same understanding of how long the project may take and just how much funding it might require." He added, "It is also a repository for new ideas and information. For instance, we're always on the lookout for new competitive information or new information about public companies that are in the obesity space that we can use to tweak the business plan."

Of course, the team also used the business plan to begin communicating with potential investors and experts in the field about HourGlass and the company's device. Specifically, they prepared three types of documents: (1) a high-level executive summary, (2) the detailed business plan, and (3) the "pitch," which ended up being multiple versions of a slide-based presentation that captured key elements of their plan. Regarding the executive summary and the main business plan document, Buxbaum said, "You don't want to disclose anything that you wouldn't be comfortable sharing with a competitor," emphasizing that innovators never know into whose hands their information may fall. For the pitch, he advised, "There are umpteen versions of the pitch. There are the ones you feel comfortable leaving behind and the ones you don't. Everyone has varying degrees of how comfortable they feel about disclosures. In general, it depends on the IP space and just how crowded it is and what the risks are if other people find out about what you're doing." He also emphasized that the pitch should be tailored to the specific needs and interests of each unique audience,

FIGURE 6.2.1
The HourGlass team (by Anne Knudsen, courtesy of the Stanford Graduate School of Business, 2006).

based on what the team is trying to accomplish. "The pitch should be customized based on the perspectives of each audience ... even for different partners within the same firm. It's a constantly changing document," he said.

When asked what a good business plan should include, Buxbaum said, "Investors are always going to think about questions like: 'Is the market large enough for them to make an adequate return? How is the competition in the space? How is the intellectual property in the space and what other barriers to entry exist? And, is this really the right team to pull it off?' I think those are really the core of what has to be addressed in the business plan." The challenge is to identify the key points and focus research efforts there. "You don't have the time or resources to delve into the minutiae on every single topic," commented Buxbaum, "so you have to figure out which topics are going to be the most salient or troublesome." These are the areas that should be addressed in greater detail.

Buxbaum noted that the team's business plan evolved over time. "At first, it was more product focused. I think the way things changed is that we did a lot more primary market research." After developing their early plan, the team talked with patients, scientific advisors, CEOs of other medical device companies, key technical people in medical devices in general (and also specifically in the field of gastrointestinal systems), former regulators from the FDA, and competitors. These diverse perspectives helped them enhance the richness of their analysis. For example, said Buxbaum, "It's not just understanding how big the market is, but whether or not it is possible to capture the market."

In terms of potential pitfalls that the team encountered, Buxbaum explained, "We didn't spend enough time on the operating plan and setting milestones. We initially thought about raising a round of funding as a key milestone. But it's not. It's a good internal milestone to help motivate a team. But the milestones on your operating plan, and in your business plan, should be all about taking risk out of the business. Financial risk is part of it, but at the very beginning there are so many other risks, like technical risks, biologic risks, and market risks. Raising cash is just a means to eliminate those other risks." As these risks are retired, the company becomes more valuable to investors. Buxbaum continued: "Our original milestones were focused on prototyping the device. Then, we knew we had to do some animal work before going into humans. After that, it would be phase I human trials and pivotal trials. Then, we would go for regulatory approval and sales. That was a good high-level structure to work off, but there's so much to do before you ever start moving into animals."

As the team members collected more information about what they needed to do to effectively move forward, they were able to think about more specific milestones to gate their progress. "It helped to talk to experts in the field to understand what the key risks were," said Buxbaum, and then build milestones around eliminating them. "We also talked to people who had failed in the field. They told us what the hardest thing was for them, and that was valuable to hear. In general, people are willing to discuss why they failed without divulging information that's proprietary or confidential." This data, too, was used as an input into the development of more detailed milestones. "Once we built the operational plan around smaller, intra-company milestones, it became much more manageable." It also helped the team communicate a more compelling plan to investors.

Reflecting on the first time the team presented their plan to groups of investors, Buxbaum recalled, "We did get feedback that, 'What you guys are doing is interesting. This is a different idea than what we're doing now, but we like it.' But, they told us, 'We would like to see some animal data.' We recognized the significance of pre-clinical data to certain firms, even at this early stage." However, he noted that not every group responded the same way. As a caution to other teams, Buxbaum said, "Even though some people liked our pitch, it didn't speak to other people. Investors come into a pitch with their own preconceptions. And there's a lot of groupthink [or consensus thinking]."

He also advised innovators to "Be very clear when you are presenting your business plan about how much you're going to disclose. If you go into a presentation with mismatched expectations, it's just going to frustrate whoever you're talking to. If you are not willing to disclose certain things, you should tell them that as you're making the meeting. And, even then, they're still going to push. It's their job." In terms of maintaining confidentiality, Buxbaum also explained that certain audiences may be more trustworthy than others. "I would love to just be able to tell people what we're working on because, when you share, you get more feedback. But you have to protect the company. In the end, you have to use your own judgment."

Summarizing the importance of a business plan, Buxbaum emphasized, "The business plan ideally addresses all of the reasons why other people think your idea might fail. There is a great deal of skepticism surrounding new ideas. The business plan outlines how you plan to mitigate key risks." In early 2008, he described HourGlass as an early-stage company raising funding with IP and animal data.

Business plans and funding

Some medtech innovators erroneously believe that the primary goal of writing a business plan is to raise funds. In reality, the focus should be on how they will develop a business. A carefully crafted business plan will not lead to funding on its own. However, the exercise of rigorously thinking through the development and commercialization plan for an innovation (and then effectively presenting it to potential investors) can be instrumental in the fundraising process (see 6.3 Funding sources).

Importantly, funding from external investors will not sustain a new venture indefinitely. A company needs profitable, revenue-generating products to become sustainable. As investors are required in the current investment environment to commit increasingly large sums of money over a longer time horizon to reach an exit (i.e., until the company is acquired or completes an initial public offering (IPO)), they have higher expectations (and perform more stringent **due diligence**) related to a company's operational capabilities. A compelling business plan demonstrates how a company will become profitable, as well as the management team's ability to run the business for as long as necessary for investors to generate a return. The InterWest Partners case example explains the role of the business plan in external communications with investors.

FROM THE FIELD | **INTERWEST PARTNERS**

The role of investors in screening new business opportunities

As a partner specializing in life sciences investments with venture capital firm InterWest Partners, Ellen Koskinas hears about hundreds of new medical device business opportunities each year. However, she said, "It's been a long time since I've seen a formal business plan with tabbed sections." Koskinas continued, "These days, full business plans are somewhat of a historical remnant. That said, I actually look at the process of creating a business plan as an important opportunity for innovators to really organize and synthesize their own thoughts on the various dimensions of the business."

To give innovators a sense for the process venture capitalists (VCs) go through in evaluating a business opportunity, Koskinas described her firm's approach: "Entrepreneurs typically initiate contact by sending an executive summary. An executive summary could be anywhere from three to ten pages, outlining the very high-level aspects of the business that they're planning to pursue. It's actually just a very tightly written business plan. Even though it's concise, it hits upon most of the major business questions." Then, if the firm is interested, she said, "We'll schedule a fairly detailed, 90-minute presentation where the entrepreneur covers each of those different areas in depth. That gives us the foundational layer of information." This "pitch" is often supported by a PowerPoint presentation that expands on the information included in the executive summary.

Based on the outcome of the preliminary meeting, the InterWest team decides whether or not to initiate due diligence. For those opportunities that cross the preliminary bar, key information is requested from the innovator in specific categories. Among other areas, InterWest makes queries related to product development, regulatory, clinical, reimbursement, and the market opportunity. Koskinas elaborated on the types of questions that might be asked: "Generally, we want to go into depth on product development. What is the current development status, what does the product development plan look like, and what other products are behind it? For regulatory, we might look at the likely regulatory pathway, where they are with the FDA, and are there predicates that suggest that this actually could follow a 510(k) path as opposed to a PMA? In the reimbursement category, is there existing reimbursement coding? If not, what's the step-wise progression to establish that, and which consultants are helping them navigate that landscape?" Importantly, Koskinas pointed out that the due diligence requirements were likely to vary from investor to investor: "They are typically tailored to the specific interests that each venture group might have." For example, Koskinas has a background in product development so tends to probe deeply in this area, requesting a detailed product specification, bench-top data, and other documentation that allowed her to understand the technical requirements, device performance characteristics, and product manufacturability. Innovators are expected to gather the information outlined on the due diligence list and submit it to the VC relatively quickly. "If entrepreneurs have gone through the exercise of creating a business plan, it enables them to more effectively respond to questions and be prepared to provide the information that VCs are likely to request during due diligence," she noted.

In parallel with the review of due diligence information provided by the company, the VC firm conducts some of its own independent research. "We do a lot of phone calls to the entrepreneur's advisors to get their input on various aspects of the plan," Koskinas explained. They also independently contact physicians and other subject matter experts to gather specific feedback on certain aspects of the business, if that expertise does not exist in-house. Recently, many VCs had begun using clinician networks, such as those offered through companies like the Gerson Lehrman Group and Leerink Swann. These networks includ thousands of physicians who are available to venture and hedge fund investors to help them assess the merits and potential shortcomings of different clinical opportunities. Input from impartial third parties could be helpful. "Many of the clinicians that are advising a company are inherently biased toward the technology's potential," said Koskinas. "So we really want to talk to clinicians who are skeptical about a new technology or may even be naysayers so that we fully appreciate the risks and understand the clinical issues we need to consider as we enter the market." In addition, the company could benefit from access to this specialized expertise, as diligence learnings frequently can be valuable in shaping the company's future plans. "More specific knowledge is required for making smart investments today," she added.

Innovators who progress through the due diligence process and gain both the life science sector and partnership approval are issued a **term sheet** (a non-binding document outlining the key terms of a funding agreement – see 6.3 Funding Sources). "After that, we do a much more in-depth IP review," Koskinas said. "Almost everybody does that after the term sheet has been accepted, because it's a very expensive process. It can run anywhere from $10,000 to $100,000. We typically try to keep ours to a more reasonable amount, since there is some IP risk with any device technology we fund. We need to be able to assess the degree of that risk, but cannot really predict what might unfold. In some cases, the IP landscape is particularly complex and deeper patent diligence is essential."

If no major IP hurdles are uncovered and the other term sheet requirements are met, then InterWest typically funds the venture at this point. Over the past few years, Koskinas commented, the process of evaluating business opportunities has become more stringent, primarily because the VCs are increasingly required to put up more money, over a longer period of time, to reach an exit. With a higher bar to achieve a successful IPO or acquisition transaction, venture capitalists need to more fully assess the organizational and capital requirements to commercialize the technology and gain market traction. As a result, said Koskinas, "I would still recommend that entrepreneurs write a business plan in order to really think through the various aspects of the company's development before they seek funding, so that they can address these topics."

In terms of getting an executive summary to a VC, Koskinas encouraged innovators to tap into their networks. "When a deal comes in from someone we know – another entrepreneur, a lawyer, an accountant, consultants, or another credible source – we tend to look at it more closely. While we generally review all of the deals that are sent in, proposals that come in over the transom typically get lower priority. So, for the most part, it really pays to use your network." For those without an expansive established network, she recommended attending industry conferences, such as Medtech Insight, or academic presentations as a way to meet prospective investors.

When asked how InterWest screens the hundreds of executive summaries it received down to a more manageable number, Koskinas said, "In general, there's a very quick filtering process at most firms based on whether or not an opportunity falls within our strategic focus. There are just so many deals out there, that we've got to stay focused in terms of how we spend our time." For example, the InterWest team members focus more closely on opportunities in cardiovascular, spine, obesity, aesthetics, orthopedics, and ophthalmology, and tend to be less enthusiastic about the executive summaries they see in other clinical areas. This point underscors the importance of careful research when innovators began targeting potential investors. "The other quick calculation we do in our head," Koskinas continued, "is whether or not it's a big enough opportunity given the development, regulatory, and reimbursement path. For the most part, we're targeting technologies that address at least a $500 million market. If it's a smaller market than that, the opportunity would have to be something really innovative and capital efficient to get our attention."

> **FROM THE FIELD — INTERWEST PARTNERS**
>
> *Continued*
>
> Finally, she commented on the importance of preparing high-quality materials to share with potential investors. Any proposal seriously considered by the firm has to at least preliminarily address the key business questions and be well constructed. "The executive summary or PowerPoint presentation is the first interaction we have with an entrepreneur. We see it as an important reflection of the entrepreneurial team's thought process and credibility," she said.
>
> When developing a business plan, Koskinas had this advice: "Be as comprehensive as you can in thinking through the whole business evolution. When we see a technology that is intriguing, but the team hasn't thought much about the downstream business issues, it suggests a lack of experience. We recognize that we'll need to be more actively involved with the company formation and early development. While we enjoy taking on that role, there are bandwidth constraints that prevent us from doing that too often." She also encouraged innovators to carefully think through their messaging. "Be really clear about the value proposition, why it's superior to existing or emerging competitors' technologies, and how the different constituencies [physicians, patients, payers] are impacted by adoption of the technology."

Business plans and teams

Most innovators and investors alike acknowledge that it is not new technologies that create new businesses, but rather teams and people. An effective business plan articulates the key team and management structure behind the venture. Even more importantly, it helps the members of the founding and management teams better articulate their goals. Every individual involved in a new venture has his/her own aspirations and objectives. The exercise of building the business plan helps reconcile and potentially find synergies among those expectations. For example, some founders place significant emphasis on ownership and control. If, in the process of developing the business plan, it becomes clear that necessary external funding will dilute management control and ownership, this discovery creates the opportunity for an open discussion about how the issue will be handled.

When it comes to team-related issues, there are several difficult questions that should be addressed as part of the business planning process to ensure that the management structure put in place lays the foundation for a successful venture:

- What will be the equity ownership of each of the founders?
- How will equity and stock options be distributed to key employees?
- What is the process through which these decisions will be made?
- What will be the responsibilities of the founders and early employees?
- What roles will the founder and early employees play as the company evolves and grows?
- Who will be the CEO (and will there be an interim CEO to be replaced by a permanent CEO)?
- What are the desired attributes of early employees?
- How and from where will important hires be recruited?

A thoughtful discussion of these questions should occur early in the business plan development phase because it can help highlight differences in expectations. Furthermore, the answers may affect the firm's strategic directions and funding options, as well as its overall culture and performance. However, it should be recognized that while there is some flexibility in how to address these questions, market forces may constrain the range of viable alternatives. Consider, for example, the question of how equity and stock options will be distributed to key employees. The answer depends, to a large extent, on what other companies offer their key employees. For instance, if the current trend is to offer two percent of the company to a key senior executive, deciding to offer less may make the company's offer non-competitive. In contrast, offering more may not be considered acceptable by investors. More details on the question of equity ownership for key employees are presented in 6.3 Funding Sources.

Business plans and intrapreneurship

Even though most people tend to think of business plans in the context of entrepreneurship, these plans also play a critical role in intrapreneurship (entrepreneurship within the context of an established company or organization). Importantly, when intrapreneurs develop business plans to support an opportunity, they need to consider how they leverage existing resources within the firm and what complementarities exist with existing product portfolios, as described in 6.1 Operating Plan and Financial Model. They should also consider how the opportunity fits within the existing organizational structure and the extent to which it meets defined financial hurdles. Large firms, within which intrapreneurship often occurs, tend to have well-defined organizational structures that may inhibit or support important relationships with internal and external constituencies that are needed to bring an opportunity to fruition. They also have clearly defined capital budgeting processes and requirements for the return on investment (**ROI**) of any project that is undertaken. So, the financial model should be adapted to reflect internal processes and requirements. Factors such as these can serve as enablers or barriers to intrapreneurs as they craft a business plan.

GETTING STARTED

As soon as a company initiates development strategy and planning activities, the time is right to start thinking about the creation of a business plan. The steps below can be used to guide the creation of an effective plan.

See **ebiodesign.org** for active web links to the resources listed below, additional references, content updates, video FAQs, and other relevant information.

1. Define the purpose and audience for the business plan

1.1 What to cover – Depending on its stage of development, a company may have different reasons for creating a business plan. Early stage ventures may use a business plan to clarify the key opportunity and funding requirements in order to seek capital from venture capitalists (equity investments). More established companies may use a business plan to map out a change in strategy or a plan for accelerated growth. As a result, it is important to clarify the purpose of the business plan before a company begins developing a draft. Similarly, the company must define its audience and understand the unique needs and interests of those audience members.[12]

1.2 Where to look – Once the purpose of the business plan has been clearly defined and the target audience has been identified, innovators should tap into their networks to clearly understand what the target audience looks for in a business plan. For example, if the objective is to seek funding, talk with retired VCs or other innovators who have recently made a "pitch" to VCs for advice and guidance. If the purpose is to solicit clinical input from expert physicians, informally talk to trusted clinicians about their preferences and requirements.

2. Confirm what is unique about the company's innovation and/or industry

2.1 What to cover – Before developing a draft of the business plan, it is important to think about what is different or unique about the company's innovation and/or industry. Audience members experienced in a particular industry or field will know what critical success factors, potential pitfalls, unusual characteristics, and atypical dynamics the company is likely to face. An effective business plan not only acknowledges these unique factors, but includes information about how the company will address them. Make a list of

GETTING STARTED

such factors and refer back to it throughout the business plan development process to ensure that each issue is addressed.[13]

2.2 Where to look – In getting started, it may be helpful to revisit with the product's value proposition(s) as defined in 5.7 Marketing and Stakeholder Strategy and the company's competitive advantage as defined in 5.9 Competitive Advantage and Business Strategy. Colleagues within the company should be able to help brainstorm on what other issues to include on this list. External advisors (e.g., board members) can also provide useful feedback.

3. Develop an outline

3.1 What to cover – The next step is to develop an outline for the business plan. Before developing detailed content in any one section, think about the business plan as a whole. Start by following the conventional organizational structure for business plans in the medtech field (see the Structuring the Business Plan section of this chapter) and then make modifications based on the most logical and compelling flow of information for communicating the company's innovation and needs. Within each section, make notes about the information that should be included, data that needs to be collected, supporting documentation that needs to be compiled, and ideas for effectively transitioning between or linking the various components of the plan. During this process, it will be helpful to read as many example business plans as possible (see Appendices 6.2.1 and 6.2.2 for samples and some high-level analysis of the plans).

3.2 Where to look
- **U.S. Small Business Administration** – Planning tools to help in the development of a business plan.
- **Local Small Business Development Centers (SBDCs)** – Available across the United States to provide assistance to small businesses with business planning and other start-up activities.
- **Bplans.com** – Resources and tools for developing business plans, including some business plan examples in the medical device field.
- **Entrepreneur.com** – A website with step-by-step instructions on creating a business plan.
- **BusinessPlans.org** – Sample business plans from the MootCorp® Competition at the University of Texas.
- **Carnegie Mellon University Business Plan Index** – An index of business plans by business types available online at the CMU library website.
- **The nuts and bolts of business plans** – An MIT course on Business Plans with extensive online resources.
- **Business planning software** – Software packages designed to help lead innovators through the process of writing a business plan, including: Planware Business Plan Pro, BRS Business Plan.

4. Conduct research and compile supporting documentation

4.1 What to cover – Once the general structure of the business plan has been mapped out, an innovator can begin to conduct research in an organized manner. Refer back to the various activities in the biodesign innovation process for the resources that can be leveraged to locate important data. Remember that an effective business plan will be fact based and specific. In addition to performing secondary research, it may be helpful to consider primary research with key stakeholders. In addition to referencing this information within the text of the business plan, results from primary research can be included as supporting documentation. Other supporting documentation to compile may include copies of patents, licenses, and other legal documents, letters of intent from customers and/or suppliers, testimonials, résumés of the management team, and other documentation that demonstrates the credibility of the venture.

4.2 Where to look – See Table 6.2.1 for a mapping of the biodesign innovation process to the major sections of the business plan. Refer back to these documents for resources that can be used to conduct additional research in each area.

5. Write the plan, seek input, and iterate

5.1 What to cover – Once all the necessary data have been collected and organized into the various sections of the outline, it is time to begin writing the plan. As mentioned, some experts suggest drafting the executive summary first and then using it to guide the development of the overall business plan. Others suggest saving the executive summary for last and then developing it by using the most compelling information from each section. All innovators have to decide on the system that works best for them. Regardless of the approach taken, remember that business plan development is iterative. Once a first draft has been developed, seek the input of colleagues and advisors (see 4.1 Intellectual property basics for addressing issues of confidentiality, and be explicit about asking people if they have a potential conflict of interest *before* sharing a draft with them), and then use their input to iterate and improve the document. Continue this process until the business plan is truly clear, informative, and compelling. However, it is important to recognize that each reviewer will have his/her own preferences and that all feedback should not be considered equal. Once feedback has been collected from a handful of individuals, compile the questions, comments, and issues, and then carefully evaluate the data. Correct obvious problems and address common concerns. For issues about which there is no common agreement, probe further to be sure that a change is warranted. A company must be careful not to dilute its position by responding to every reviewer's individual preferences.

5.2 Where to look – Leverage the innovator's network to find appropriate advisors and contacts to review the business plan.

Credits

The authors would like to acknowledge Darin Buxbaum and Ellen Koskinas for their assistance in developing the case examples. We would like to thank Todd Alamin, Tom Goff, John MacMahon, and David Miller for their contributions.

Appendix 6.2.1

Sample executive summaries[14]

Executive summaries take many different forms. The two that follow are actual examples from teams in Stanford University's Program in Biodesign (which have been modified only to disguise the identity of the companies and their management teams). As such, they do not necessarily represent an "ideal" executive summary; instead, they are meant to provide the innovator with a perspective on how others have addressed this challenge. By identifying which aspects of the two examples are appealing (and which are not), the innovator can use these insights to inform and guide the creation of his/her own executive summary.

Example 1: Stanford Biodesign Systems

Stanford Biodesign Systems (SBS) has developed a proprietary platform for the diagnosis and treatment of cervical spine pain that vastly improves on existing procedures. The founders are Dr. Fred Terman, a Stanford spine surgeon, and Leland Stanford, Jr., chief product designer. The company has raised approximately $400,000 to date.

About the products

The ***diagnostic device*** is a single-use disposable device with no capital equipment requirements. The device anticipates FDA clearance in Q3 of 2007 and is projected to generate sales revenue in 2008. Stanford Biodesign Systems has strong clinical evidence validating its procedure from a 25-patient study at Stanford University. The device uses a proprietary catheter system to deliver "unobtainium"[15] to remove pain sensations in a new procedure.

The device also has minimal technology development risk and can be easily manufactured. Existing reimbursement codes can be used for the procedure; there is a strong possibility for increased reimbursement with more clinical data. Stanford Biodesign Systems believes that this device will become the standard method of evaluating cervical spine pain.

The ***therapeutic device*** builds off the technology of the diagnostic device and replaces existing invasive surgeries such as fusion or disc replacement with an effective, specific, and minimally invasive treatment. This device uses a catheter system suitable for long-term implantation to deliver unobtainium when patients feel pain sensations. The therapeutic device would initially use a generic unobtainium, but could later be used as a platform to deliver a wide range of substances in a focused and controlled method to the specific site of pain.

The market

We estimate that 200,000 patients with persistent, severe cervical spine pain would be indicated for our diagnostic procedure per year. Assuming a diagnostic device usage of 1.5 per procedure and an average selling price of $500 per device, we have estimated the diagnostic market size to be $150 million in the United States and $250 million worldwide. The market for the therapeutic is much larger, on the order of $800 million to $1.5 billion, based partly on a higher average selling price and on a larger patient population seeking help prior to surgery – 300,000 to 400,000 per year.

The team

In addition to the founders, many experienced executives, venture capitalists, and entrepreneurs advise the company, including Dr. Jane Stanford (Stanford), Dr. Leland Stanford, Sr. (UCSF), William Hewlett (Sequoia Hospital), and David Packard (SPARCmed) on the clinical side, and Eddie Money and Johnny Cash on the business side.

The company anticipates signing an A-round of funding in Q4 2007.

For more information, contact Leland Stanford, Jr. at (650) 555–1212 or lstanford@stanford.edu.

Example 2: ABC Medical Design

ABC Medical Design (ABC) has developed proprietary technology in the field of embolic protection devices. ABC's protective devices benefit patients, clinicians, and payers of the healthcare system by enabling safer and more cost-effective modalities for the treatment and prevention of neurological strokes. The primary purpose of ABC's technology is to improve the performance outcomes of two related procedures: surgical carotid endarterectomy (CEA) and minimally invasive carotid angioplasty and stenting (CAS). These protection devices are targeted for use in the carotid arteries and other anatomical sites wherein embolization is increasingly recognized as a severe threat to the health of an expanding patient group.

The need: stroke prevention

The carotid arteries are among several common sites of atherosclerotic disease. In 2000, approximately 760,000 symptomatic strokes occurred in the United States. With over 158,000 fatalities, stroke is the third leading cause of death nationwide. Twenty-five percent of stroke survivors die within one year and 50 percent of strokes occur within the hospital setting. Particles, referred to as emboli, dislodge from atherosclerotic plaques, travel in the bloodstream and block important arterial vessels, causing 24 percent of all strokes. The direct and indirect costs of stroke in the United States were $45.4 billion in 2001.

Unfortunately, the current procedures to treat carotid disease include an inherent risk of dislodging emboli that could lead to a stroke. Such a complication is the very outcome these procedures are intended to prevent. In light of this risk, certain treatments of these vessels are currently performed only on people who demonstrate neurological symptoms associated with the narrowing of the carotids.

Present standards for treatment

Currently, patients visit a physician after experiencing conditions such as a minor stroke, dizziness, fainting, or a transient ischemic attack. On average, these patients' carotid arteries have narrowed (stenosed) such that 85 percent of their cross-sectional area is blocked. The current gold standard for addressing carotid stenosis is a surgical approach. Carotid endarterectomy (CEA) is performed by a vascular surgeon who dissects a portion of the neck to access the carotid artery and then removes the plaque within. More than 150,000 CEAs are performed each year in the United States, with a reported neurological complication rate of five to seven percent.

Carotid angioplasty and stenting (CAS) is a less invasive approach that leverages minimally invasive technologies and is used pervasively in the treatment of coronary artery disease. However, the widespread adoption of CAS over CEA has been delayed due to a high prevalence of neurological complications:

- Carotid endarterectomy (CAE) procedures: five to seven percent complication rate.
- Carotid stenting and angioplasty (CAS): nine to eleven percent complication rate.

These complications relate primarily to the release of emboli, and have been recently addressed in part by first-generation embolic protection devices. ABC's immediate goal is to further develop its technology and provide markedly superior protection and performance over existing protection methods. Such advances are expected to accelerate the development of the potentially lucrative CAS market.

Growing population with carotid plaques

Despite its current risks, roughly 150,000 people seek appropriate treatment for carotid disease each year, while more than 2.6 million go untreated. This number of CEA procedures has been growing at an annual rate of four percent. Interventionalists state that if the risk of embolization were to be removed from CAS and CEA procedures, a fourfold increase (600,000 treatments per year) would occur immediately. Asymptomatic carotid stenosis of greater than 50 percent is present in 4.5 percent of the US population over 65, and 10 percent of those over 85. In 2020, nearly 5 million Americans are predicted to have an annual risk greater than 4.5 percent of suffering a stroke. The public is increasingly aware of the risks presented by carotid artery disease. Long's Drug Stores recently began offering ultrasound screening for carotid artery disease at selected locations for a nominal fee, in response to this awareness.

Products

ABC has developed specific devices for both the surgeon and the interventionalist. For the vascular surgeon, a solution amenable to 510(k) approval has been developed. This device will offer an early product roll-out and early revenue stream potential. Currently, no embolic protection device other than a traditional clamp is used for these surgical procedures; thus, this particular market offers no new competition.

For the inteventionalist's approach, ABC offers a proximal protection device to facilitate carotid angioplasty and stenting. FDA regulatory approval for this product will likely be a fast-tracked premarket approval (PMA). Furthermore, the core technology behind ABC's proximal protection device is likely to find application in other regions of the vascular anatomy wherein embolization presents a potential complication during procedures.

Proximal versus distal protection for carotid artery stenting

All but one of the competing CAS embolic protection technologies involves placing a protection device distal (downstream) of the lesion. The mere delivery of distal protection has been estimated to release 12 percent of the procedure's emboli *prior to the establishment of protection*. This is an inherent risk of distal devices that ABC's proximally placed protection technology overcomes. Only one other company, Acme Medical Science, is presently proposing a proximally placed solution. When compared to ABC's proprietary technology, Acme's approach has several shortcomings that will allow our solution to be established as the preferred mode of protection.

Strategy

By improving the success rates of both the CEA and CAS procedures, ABC aims to raise the expectations for their outcomes and lower the threshold of the severity of the disease at which a clinician and patient will elect to undergo treatment.

Reimbursement

Medicare recently decided to reimburse carotid angioplasty and stenting when associated with IDE class B devices, when used in a clinical trial. As the clinical trials further demonstrate the need for embolic protection, CMS will be under pressure to establish more definitive reimbursement for these devices.

Financing

To date, the founders have funded ABC to develop intellectual property, proof-of-concept prototypes, and perform bench-top testing. ABC is now seeking $2.5 million to further develop these products with the specific milestone of initial use in humans.

Management team

Bob Smith*, chief technical officer – Smith is presently committed full-time to ABC. Smith holds a Masters in electrical engineering and an undergraduate engineering degree in mechanical engineering, both from Stanford University.

Mark Jones*, Design Engineer – Jones is presently committed full-time to this project. Jones holds a BS in product design from Stanford, with a focus on medical device and rehabilitation design.

Tom Taylor*, Director of Medical Technology – Taylor is working part-time with this project. He is currently an MD candidate at Stanford Medical School and is also a research associate at Stanford's Center for Research in Cardiovascular Interventions.

Jane Doe, Chief Financial Officer – Doe will continue to work as the part-time consulting CFO. Doe has an extensive background in accounting and financial management. She is the former finance director for the Lucille Salter Packard Children's Hospital at Stanford, providing financial oversight of a $200 million budget.

* Denotes coinventors of ABC's core technologies.

Appendix 6.2.2

High-level analysis of executive summary examples

Note that this analysis does not seek to evaluate the caliber of the business opportunity being presented, but rather the quality of the executive summaries themselves.

	Example 1 SBS	Example 2 ABC
+	• Concise. • Clearly written and easy to understand for any reader.	• Provides a relatively comprehensive overview of business plan. • Clearly addresses status of regulatory, reimbursement, and clinical activities. • Clearly articulates the market opportunity. • Clearly written and easy to understand for any reader.
−	• Does not provide a comprehensive summary of the business plan. • Fails to address key issues relevant to all medical device opportunities (e.g., status of IP, regulatory, clinical trials, and reimbursement). • Does not conclude with clear statement of purpose. • Does not cite data sources.	• A bit too lengthy – summary of the need and current standard of treatment could be made more concise to reduce total length to two pages. • Does not address status of IP activities. • Does not cite data sources.

Notes

1. From remarks made by Ron Dollens as part of the "From the Innovator's Workbench" speaker series hosted by Stanford's Program in Biodesign, March 21, 2005, http://biodesign.stanford.edu/bdn/networking/pastinnovators.jsp. Reprinted with permission.
2. Stanley E. Rich and David E. Gumpert, "How to Write a Winning Business Plan," *Harvard Business Review*, May 1, 1985, p. 130.
3. Ibid., p. 133.
4. From remarks made by Mir Imran as part of the "From the Innovator's Workbench" speaker series hosted by Stanford's Program in Biodesign, April 28, 2004, http://biodesign.stanford.edu/bdn/networking/pastinnovators.jsp. Reprinted with permission.
5. Tim Berry, "Keys to Better Business Plans," Bplans.com, June 15, 2004, http://www.bplans.com/dp/article.cfm/198 (May 15, 2007).
6. "Writing an Effective Business Plan," Deloitte Touche Tohmatsu International, 1993.
7. Rich and Gumpert, op. cit., p. 136.
8. Based on information provided in the "eBusiness Plan Tutorial," Pearson Prentice Hall, http://myphliputil.pearsoncmg.com/student/bp_turban_introec_1/ExecSumm.html (March 21, 2007).
9. Rich and Gumpert, op. cit., p. 136.
10. Tim Berry, "Common Business Plan Mistakes?" Bplans.com, June 15, 2004, http://articles.bplans.com/index.php/business-articles/writing-a-business-plan/common-business-plan-mistakes/ (November 4, 2008).
11. All quotations are from interviews conducted by the authors, unless otherwise cited. Reprinted with permission.
12. Darrel Zahorsky, "Create a Money Winning Business Plan," www.About.com, http://sbinformation.about.com/cs/businessplans/a/bpoutline.htm (May 15, 2007).
13. Ibid.
14. These executive summaries have been modified to disguise the identity of the companies and their management teams; however, they were prepared by actual medical device innovators. Reprinted with permission.
15. A fictional substance used to disguise the example.

6.3 Funding Sources

Introduction

In 2007, US venture capitalists invested $3.9 billion in 365 medical device start-up deals. This corresponded to a 40 percent increase in investment in the medtech sector over the previous year.[1] More venture funds are available than ever before for turning great ideas into products. The key for innovators is to enter into the right, mutually beneficial funding relationships that maximize their chances for successful product development and commercialization. While this may involve venture capital, there are other sources of funding that can also be advantageous and should not be overlooked.

When evaluating funding sources, the goal of the innovator is to develop a process for identifying potential investors, understanding their requirements, and determining the best way to approach them and present the opportunity to invest. The most effective funding strategies typically align the need for capital with the retirement of key risks so that the company receives the best possible valuation relative to its stage of development. Another objective of this process is to recognize the advantages and disadvantages of different funding sources, particularly in light of the fact that investors will become long-term partners in the process of establishing and expanding the business.

This chapter focuses specifically on funding sources for companies that, with enough capital, can generate adequate cash flows to become self-sustaining entities. However, if the need addressed by a new solution does not support a stand-alone company, different funding options are presented in 6.4 Licensing and Alternate Pathways.

OBJECTIVES

- Identify the different sources and types of funds potentially available to innovators.
- Understand the criteria used by investors to evaluate an investment opportunity.
- Learn simple valuation models used by investors.
- Understand term sheets and how their key aspects can affect a new venture.
- Appreciate the strategic value investors can bring to a venture.

Funding fundamentals

Many innovators find the process of raising funds for a new venture frustrating and disappointing. They are often perplexed about why investors fail to grasp the potential of their ideas and decline to fund their companies. Mir Imran, an experienced device entrepreneur who has experienced funding from the perspective of both the innovator and the investor at different times during his career, summarized the situation this way:[2]

> I used to really ponder why these venture capitalists didn't invest in all my companies and give me big checks. I finally got the answer when I started writing checks. When you put on the investor hat, you ask a different set of questions. You're looking at risk – how to measure and gauge risk. So I have sympathy for both sides.

In the medtech industry, the costs (and related risks) of forming a new venture can be particularly high due, in large part, to the clinical and regulatory requirements associated with products in this field. It is not uncommon for a start-up that intends to bring a new device to the market through the postmarket approval (PMA) regulatory pathway to require $50 to $100 million in funds and five or more years of development. Table 6.3.1 provides a small sample of medtech companies and their funding requirements.

Before seeking funding, a company must have a clear execution strategy and business plan that explains why the innovation will be successful and elaborates on its market potential, funding requirements, and the team that will implement the plan (see 6.1 Operating Plan and Financial Model and 6.2 Business Plan Development).

The innovator representing the company must be prepared to make a succinct, compelling presentation to potential investors, which convinces them of the value of the business and lays out an approach for meeting the company's funding requirements as it progresses. By carefully understanding the company's funding needs and the types of investors best suited to meet these needs before making a "pitch," an innovator can increase the likelihood of a fair valuation if (or when) funding is offered. An effective funding strategy will also help the innovator understand when and how to enter into investor relationships that will be long-lasting and mutually beneficial.

Sources of funding

Company funding can come from a number of different sources. Common types of medtech investors are described below.

Friends and family

Friends and family investors refer to members of an innovator's personal network with adequate means to make an investment in the company. Generally, this is one of the least expensive forms of financing since friends and family are usually more flexible than professional investors in terms of their timeline and expected level of return. However, innovators sometimes worry that "goodwill" investors may not truly understand the level of risk involved in a new enterprise and fear negatively affecting these personal relationships.

Angels

Angel investors are experienced investors using their own wealth. Typically, angel funding is less expensive than other funding alternatives, with investors sometimes willing to take on more risk. However, angels with a good understanding of the company's specific medical field can be difficult to find. Since money may need to be raised from several angels to meet the capital needs of the company, managing the expectations of numerous angels may also be daunting.

Venture capital (VC)

Venture capitalists (VCs) are professional investment managers who specialize in investing in companies with the potential for high returns.[3] Venture capitalists typically raise money from institutional investors (e.g., pension-funds, charitable foundations), wealthy individuals, or private and public entities. The money raised is put into a fund. Funds vary in size, but can range from $50 million to $2 billion and beyond. Each fund has a specific investment profile (e.g., medical devices, biotech, information technology) and an investment horizon (e.g., from three to seven years). The capital in the fund is invested in start-ups that fit the fund's profile with the

Table 6.3.1 *Device start-up funding requirements vary widely, but are not insignificant (compiled from Datastream).*

Company	First funding date	Date of initial public offering (IPO)	Total funding	IPO market cap
Cardica Inc.	6/14/2001	2/3/2006	$28,140,000	$97,800,000
Conor Medsystems Inc.	2/1/2000	12/14/2004	$78,925,000	$406,920,000
Micrus Endovascular Corporation	11/1/1996	6/1/6/2005	$58,779,000	$150,740,000
SenoRx Inc.	4/14/1998	3/29/2007	$54,570,000	$122,540,000
Xtent Inc.	7/18/2002	2/1/2007	$45,975,000	$375,680,000

expectation of reaching a liquidity event – that is, the company will be acquired or will go public – within the fund's defined investment horizon.[4] Venture capitalists are known for having "deep pockets" when it comes to qualified investments. They usually collaborate with the companies they invest in such that the companies benefit from substantial start-up and industry experience. Venture capitalists also have the ability to provide multiple rounds of funding and have extensive networks of contacts (including other potential investors) that can be leveraged to assist the company. However, VC funding is expensive since innovators must be willing to give away significant equity to attract investment (see section on Equity later in this chapter), and the due-diligence associated with the funding process can be time and labor intensive. Venture capitalists also have a clear objective for their involvement in a company – a strong financial return on investment – which may be based upon an exit strategy that may or may not be aligned with the company's long-term objectives.

Corporate investment

There are two ways that corporations can provide funds to start-ups: (1) through the purchase of equity in support of a research and development (R&D) or a licensing agreement, or (2) traditional venture investments. **Corporate investments** are often made by large companies (e.g., Johnson & Johnson) for strategic as well as financial reasons, since corporations seek to exploit synergies between projects in their internal portfolios and innovation occurring in the external environment.[5] This type of funding can be less expensive to an innovator and can bring with it unique forms of leverage (e.g., access to established distribution channels and new technology). The association with a major corporation can also lend credibility to a young company. However, an innovator involved in this type of relationship may receive limited value in return for building the business. Conflicting agendas may arise as the corporate investor looks out for the corporation's best interests. There also may be issues surrounding the ownership of new intellectual property (IP) that is generated, which may be beneficial to both the start-up and the corporation. Corporate investments also may be susceptible to changes in the economy and provide innovators with limited opportunities for follow-on funding. Exiting from a corporate investment also may become complicated if there is more than one bidder but the corporate investor has been granted first rights to an acquisition. On the other hand, in many cases a strong, mutually beneficial relationship with a corporate investor provides an innovator with a smooth and valuable exit strategy.

Customers

At times, customers may make direct investment in the development of new products or services. For instance, physicians may invest in devices in their field of expertise (e.g., when a cardiologist invests in the development of a cardiac device that meets a specific clinical need that s/he has helped to identify). Other times, a manufacturing company may strategically invest in a technology company that supplies a key future component. Customer funding transforms the traditional customer value chain from *design → manufacture → sell* to *sell → design → manufacture*.[6] This funding approach is usually one of the least expensive options. Upfront investment by customers can also dramatically increase an innovation's credibility. However, customer funding can sometimes lead to conflicts when the company seeks to sell the innovation to other companies or promote the innovation in ways that may not be what the customer, who invested in the innovation's development, had in mind.

Government grants

The US Department of Defense (DoD) funds a billion dollars each year in early-stage R&D projects at small technology companies. These projects must serve a DoD need and have commercial applications. Through its Small Business Innovation Research (SBIR) program, small technology companies (or individual innovators who form a company) can receive up to $850,000 in early-stage R&D funding. The Small Business Technology Transfer (STTR) program provides up to $850,000 in early-stage R&D funding directly to small companies working cooperatively with researchers at universities and other research institutions. Recipients of these grants retain the IP rights to the technologies they develop. Funding is awarded competitively, but the process is relatively user-friendly.[7] The primary advantage of government funding is that the financial interests of the company's founders do not get diluted. Being the recipient of such a grant can also be a source of credibility for a start-up. On the other hand, the DoD expects grant recipients to perform diligent research (comparable to what would be required by the most demanding academic institution). In addition, competition for funds can be fierce, and the review cycles for awarding funds can be lengthy. With a cap at $850,000 per project, the amount of funding granted through the programs also may not be sufficient to meet a company's needs. Other sources for SBIR and STTR funds include the National Science Foundation and National Institutes of Health.

Banks

Banks are commercial or state institutions that provide financial services, including loans, to individuals and businesses. Debt is the most typical form of financing that banks provide to companies (see Types of Funding section).

Public investors

For more established companies, the general public can also invest by purchasing shares of the company traded in public stock markets such as the New York Stock Exchange or the London Stock Exchange. For this to happen, a company needs to be listed on the stock exchange through a process known as an initial public offering (IPO). More information about IPOs is provided under the section on Liquidity Events.

Choosing the best investor for the company

When considering funding sources and different types of investors, it is imperative for innovators to be explicit about what is most important to them. For example, is it essential to find an investor who can make a long-term commitment (and potentially contribute to multiple funding rounds)? Or, is it a higher priority to find an investor that will not restrict the company's strategic options (e.g., by requiring a large ownership stake or by taking majority **voting rights** that would make the company dependent on the investor's direct support on critical decisions)? The amount of relevant experience and expertise required (or desired) in an investor will vary from innovator to innovator and company to company, depending on the strength of its advisor base and/or board of directors.

Carefully considering the best investor for a company is especially important because medtech capital often comes with "faces" attached to it (individuals with specific expectations and requirements about how the company should conduct business). This is generally true, regardless of whether the young company is funded by angels, VCs, corporate money, or customers, but not if it is funded by government grants. Whereas a bank would traditionally give a loan to an established business contingent on the use of its business assets as collateral, investors in a medical device start-up are handing over millions of dollars in return for owning a fraction of the company and the innovator's promise to build a successful company. For taking on the increased risk, investors in start-ups typically anticipate higher returns, seek to exercise more control, and expect to own more of the company. Not only can the capital carry with it an expected return many times greater than a bank loan, but it may be granted on the condition that the representatives of large investors become board members of the company and have input and voting control over the company's future. For this reason, it is critical to select investors that can add value to the company and ensure that their goals are aligned with those of the company's management team.

The relative advantages and disadvantages associated with various investor types are summarized in Appendix 6.3.1.

Stages/uses of funding

During the earliest stages of its existence, a company's worth is low because there is a great deal of risk and the company has not yet proven itself. This means that the cost of raising money is expensive – a large percentage of ownership in the company will be given up for relatively small amounts of money. As the company accomplishes its major milestones (see 6.1 Operating plan and financial model), it is able to raise increasing amounts of money to support hiring, production, marketing, and other value-building activities. As progressively more milestones are met, resulting in lower risk for the investor, acquiring funds becomes less expensive to the start-up.[8]

As a company moves through the stages of its development, it may benefit from using different types of investors to provide different types of funding to meet its evolving needs, as explained below.

Seed funding

Defining early-stage prototypes and proof of concept can require extensive mentoring, depending on the experience of the team. Typically, this round relies on **seed funding** from friends and families and/or angels due to low capital requirements (i.e., $50,000 to $100,000). Government grants are another option. Many innovators also turn to bootstrapping (financing a small firm without raising equity from traditional sources or borrowing money from a bank; e.g., using internally generated retained earnings, credit cards, second mortgages, and customer advances).[9]

Start-up funding

In this stage, start-ups often look for sources of capital that can make investments in future rounds, as well as add value to completing the management team (e.g., angels, VCs). Common milestones during **start-up funding** are related to product development, animal testing, potential registration with the FDA, and the initiation of early-stage (small) clinical trials. This type of funding often requires capital in the millions to tens of millions of dollars.

Expansion funding

When the innovation is in early-stage clinical trials, **expansion funding** is needed to ensure the completion of clinical trials, the initiation of additional trials, and an initial product launch in limited markets. Such funding can often be provided by VCs and/or corporate investors.

Mezzanine funding

Mezzanine funding is typically used when some of the most significant risks have been resolved but the company has yet to generate sufficient revenue to be self-sustaining. This form of funding is usually available once the product is approved and has been launched. It is used to build distribution channels, fund sales and marketing campaigns, and expand/develop product lines.

These four common stages of funding are visually represented in Figure 6.3.1.

Types of funding

The two primary ways for an innovator to finance a company are equity and debt. Equity is generally used earlier in the biodesign innovation process (sometimes in conjunction with debt). Debt typically comes later, when a company is generating revenue and has tradable assets.

Equity

Equity refers to a share of ownership in the business received by an investor in exchange for money. This is the most common way that friends and family, angels, VCs, and corporations invest in medtech start-ups. The primary advantages of **equity funding** are that equity contributions do not have to be paid back, even if the company goes bankrupt. The company's assets do not have to be used as collateral. In addition, the company

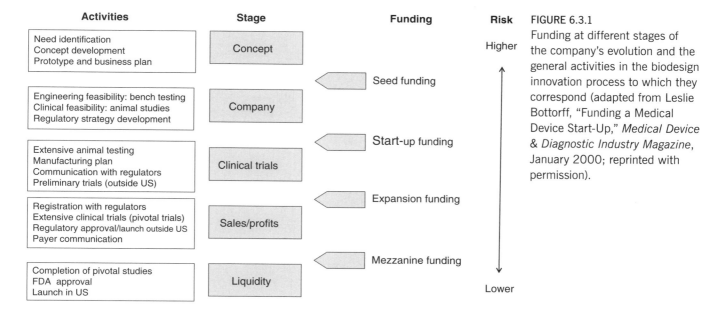

FIGURE 6.3.1
Funding at different stages of the company's evolution and the general activities in the biodesign innovation process to which they correspond (adapted from Leslie Bottorff, "Funding a Medical Device Start-Up," *Medical Device & Diagnostic Industry Magazine*, January 2000; reprinted with permission).

benefits from more cash since no monthly payments are due. On the other hand, equity investments require the innovator to relinquish some ownership of the business. Equity investors may, in turn, assert their ownership rights by seeking input into how the business should be run. The company also may be expected to share its profits with its equity investors through the payment of dividends (profits can be shared as dividends or reinvested in the business as retained earnings). Importantly, dividend payments to investors are not tax deductible.[10]

Equity investments take two primary forms: common and **preferred stock**. Both common and preferred shareholders own a portion of the company, but they are granted different rights in exchange for their investments. Common stock confers on shareholders voting and pre-emptive rights (the right to keep a proportionate ownership of the company by buying additional shares when new stock is issued). Preferred stock often gives shareholders a liquidation preference, meaning that preferred stock holders are paid before common stock holders if a company is sold or its assets are liquidated. Holders of preferred stock are also given priority regarding the payment of dividends. Based on the rights associated with preferred stock, it is often considered less risky and generally favored by institutional investors, such as VCs and corporate investors.

Individual and angel investors (as well as founders and early employees), on the other hand, more often are issued common stock. Founders own common stock, while key employees may be given preferred stock as part of an incentive program.

Debt

Debt refers to money that is borrowed by a business. Debt must be paid back by borrowers (usually in monthly payments of principal and interest over a fixed period of time, similar to a mortgage). It is generally obtained from individuals, banks, or other traditional lenders. One of the main advantages of **debt funding** is that the innovator usually does not have to turn over any ownership in the company or future profits to the lender, and the lender does not exercise any control over how the business is managed. Interest on loans can also usually be deducted on the company's taxes. On the other hand, debt financing requires that a company have an adequate cash flow to repay the loan. Loans to start-ups are generally considered risky. As a result, they are likely to have relatively high interest rates, they may require a co-signer or guarantor on the loan, and they may specify that the company's assets be used as collateral (which means they can potentially be seized if the company fails to make its payments). Too much debt also may negatively affect a company's credit rating and impair its ability to raise money in the future.[11]

Beyond traditional bank loans, different types of debt financing are available to innovators. Leveraged finance refers to funding a company with more debt than would be considered normal for that company or industry. The most common form of debt financing is a bond (typically high-yield bonds or "junk" bonds rated below investment grade – i.e., less than triple-B).[12] A bond is issued by the company and sold through an investment bank to investors. The proceeds from the sale of the bond are provided to the company (this is the principal) and the company promises to pay the bond owners periodic interest payments and finally the principal at the bond's expiration date. Because these products are costly to the borrower (i.e., they carry a high interest rate) based on the additional risk to the lender, they are most often used to achieve a specific, often temporary, objective. For example, a medtech start-up might employ leveraged finance to help it expand its sales force or invest in cash-generating assets. This option is generally available for companies with existing revenue.[13]

A **bridge loan** is another interim debt financing option available to individuals and companies. Usually bridge loans can be arranged relatively quickly (with limited documentation) and are used to span a period of time before additional financing can be obtained (e.g., before the next milestone is achieved and a new round of funding can be initiated). As with leverage finance products, bridge loans are typically more expensive than conventional debt financing because they command higher interest rates. Points and other costs of the loan also must be amortized over a shorter period of time. Some lenders may even ask for equity participation as a condition for granting a bridge loan or for an option to convert the loan into equity.[14]

Venture-backed loans can serve the purpose of a bridge loan to help the company reach positive cash flows and achieve more attractive future valuations. The typical size of such financing ranges between $1 million and $15 million, and they mature in two to four years. The loan is secured by the assets of the company (IP may be one of these assets). The lenders are compensated for the risk they take through a mix of high interest rates and **warrants**. Warrants give the lenders the right to purchase a number of shares of the company's stock at a price per share paid by other investors purchasing the stock. The terms of the loan specify the maximum number of shares the warrants will cover as a percentage of the principal. From a practical perspective, venture-backed debt financing may be an attractive option for a company when its prospects appear to be the most promising: shortly after it achieves a major product milestone and closes a round of equity financing.[15]

Convertible bonds offer a hybrid debt-equity alternative to companies seeking financing. These products are a type of bond that can be converted into shares of stock of the issuing company, usually at some pre-announced ratio.[16] Convertible bonds usually have a low coupon rate (the predetermined payment promised to the bond-holder in return for his or her loan of money to the bond-issuer), which means that the cash interest payments are lower than other forms of debt financing. However, in exchange for lower payments, the investor gains the ability to convert the bond to common or even preferred stock, usually at a substantial discount to the company's market value. This, in turn, can dilute the innovator's ownership stake in the company. Convertible bonds are issued by the company and purchased by the investor(s) providing the funding.[17]

Equity versus debt

There are significant differences between equity and debt both from the investor's perspective and from the innovator's perspective. For investors, lenders always have the first claim on the company's assets, followed by preferred shareholders, and then common shareholders. What this means is that, if the company is sold, the proceeds will first be used to pay any outstanding loans and then distributed to the preferred shareholders. Whatever is left over will then be distributed to the common shareholders. For innovators, debt does not involve loss of management control or **dilution** of ownership for the existing shareholders. However, it is only available in later stages in the start-up, as the company gets closer to achieving positive cash flows.

For all practical purposes, equity is the most common type of funding available to all but the most advanced medtech start-ups (in terms of the milestone they have satisfied).

Approaching investors

When seeking funding from any source other than a bank, it is preferable to use one's connections to gain access to investors. Cold calling, blind mailings, or online submissions of a business plan rarely lead to positive results, and may actually hurt a company's chances of being funded in some cases. Finding direct referral sources to investors is essential, particularly for innovators and companies seeking VC investment. See Figure 6.3.2.

Referrals to investors can be gained through any number of different sources. Other innovators, as well as service providers (attorneys, consultants), can be rich sources of leads. If innovators do not have direct contacts in the industry, they should network to find them. For example, if an innovator knows a VC or angel investor who works in the high technology sector, s/he might be able to make a referral to another VC or angel

Table 6.3.2 *Innovators should spend nearly as much time screening investors as investors spend screening investment opportunities.*

Questions for screening investors
At what stage does the individual or firm usually invest?
What is the typical investment level over the life of a company?
What is the typical amount invested in each round?
For VCs, when was the current fund started? (If more than five years ago, then there may not be adequate funds remaining for additional financing rounds.)
For individuals, how did the investor earn his/her money? How many other companies is s/he involved in?
How much time will be spent with the company? What expectations does the investor or firm have regarding involvement?
What prior deals has the individual or firm done in the industry?
Does the company have references (other innovators with whom it has worked) who can be contacted?

who works in the healthcare or medical device field. Be assertive, but respectful in networking and pursuing referrals.

If an innovator has no choice other than to make a cold contact, s/he should be sure to focus on investors or firms whose criteria seem appropriate to the business. Names and relevant information about potential investors can be found through online directories of VCs and angel investors. Once potential targets are identified, conduct thorough research to understand their existing portfolio, track record, investment criteria, management style, and other factors that can help determine if there is a good fit between the business and the investor. The questions listed in Table 6.3.2 can help in this assessment.

Before making contact with prospective investors, an innovator should at least have an "elevator pitch"[18] ready and a sense of the company's funding requirements. Having an executive summary of a well-developed business plan available to share is even better (see 6.2 Business plan development).

FIGURE 6.3.2
Cold calling, blind mailings, or online submissions of a business plan rarely lead to positive results (reprinted with permission of Ted Goff).

Table 6.3.3 *Common questions asked by professional investors (based on Ross Jaffe, "Introduction to Venture Capital," October 6, 2004; reprinted with permission).*

Area of inquiry	Questions asked by professional investors
Technology or service concept	• Is the product/service concept clear? Does it make sense? • Is there sufficient proof of principle or evidence of feasibility? • Are there adequate proprietary aspects – patents, trade secrets, or other barriers to entry? • Can it be manufactured at a reasonable expense? • Are there regulatory issues?
Market size and dynamics	• How large is the market, realistically? What is the actual addressable population? • Does the company have realistic potential to obtain substantial revenues in the market? • Is the decision making of purchasers and users well understood? • Are there reasonable marketing and sales costs? Sales cycles? Distribution systems? • Are the business relationships between referral sources, purchasers, providers, and consumers well understood? • Does the company have a strong competitive position? • Is the technology/service consistent with market, regulation, and reimbursement trends?
Management team	• Is the management team smart? Are they knowledgeable about this business? • Does the management team have a proven record, particularly in this business? • Do managers have high levels of honesty and integrity? Can they be trusted? • Do they have reasonable expectations for the business, particularly for the difficulties of product/service development, rate of company growth, capital requirements, ultimate business size and profitability?
Business model and financial requirements	• What are realistic revenue and expense projections for the company? • How much capital will be required to reach positive cash flow? • What are realistic expectations for the timing and sources of this cash? • What are the realistic exit opportunities for investors in this deal?
Valuation and deal structure	• Will the valuation of the investment in this deal afford a high probability of a substantial return (40 percent or greater annual rate of return)? • Will the deal allow for enough capital to be put to work to make the investment worth the time and effort? • Who will be the co-investors? Are these parties good to work with? • How can this investment be structured to minimize the technological and financial risk?

Understanding what investors want

In an ideal world, what most investors are seeking is an attractively valued company with a strong, experienced management team that can implement a well thought-out business model to sell a proprietary technology or service in a large, worthwhile market, and grow a substantial, profitable company over a short time frame with a clear exit strategy for the investors. In other words, they are seeking huge rewards with little risk.[19]

Rodney Perkins, founder of multiple companies including ReSound Corporation, Laserscope, Collagen Corporation, Novacept, and Pulmonx, offered this advice to innovators as they prepare to address investor questions:[20]

> When you're talking with VCs, any complexity makes them nervous. You want to describe a very straightforward, single product; you're going to sell it to a known market, this is the development risk, this is the regulatory risk, and make sure that these are clearly understood.

Most investors tend to make preliminary inquiries in five primary areas: (1) technology or service concept,

Table 6.3.4 *Different investors have investment criteria that differ in certain areas.*

	Angels	VCs	Corporate investors
Market size	Smaller, emerging markets	Large, established markets with $500 million or more in sales	Same as VCs but with an emphasis on markets in which they already operate
Investment size	$50,000 to $250,000	$5 to $20 million (or even more)	$1 to $20 million
Expected return	May accept lower returns if investment is aligned with mission	4 to 10 times initial capital invested	Same as VCs
Capital intensity	Smaller markets with lower requirements; often look for markets untapped by VCs	Willing to enter market with intense competition if potential reward is large enough	Same as VCs
Strategic fit	More likely to be mission driven	Seeking next blockbuster device, often regardless of specific field	Looking for opportunities that complement their existing portfolio
Investment timeline	Sometimes flexible; could be longer than eight years	Four to eight years; tend to prefer devices on 510(k) versus PMA regulatory pathway	Same as VCs but sometimes may be shorter for corporate investors
Ownership target	Small but will eventually want the start-up to seek funding from VCs or corporate investors	30 to 80 percent in early rounds, rising as high as 95 percent of company by the time of exit	Ultimately may seek to acquire technology
Board representation	Often	Almost always	Rarely

(2) market size and dynamics, (3) management team, (4) business model and financial requirements, and (5) valuation and deal structure. Table 6.3.3 provides examples of the types of specific questions VCs may ask in each of these areas when assessing an opportunity. Answers to these questions should have been prepared by the innovator as part of the business plan development process described in Chapter 6.2.

It is important for innovators to recognize that, even though all investors may be interested in the answers to these kinds of questions, different types of investors may have different expectations about what the answers should be to attract their investment. Table 6.3.4 compares some general guidelines for what angel investors, VCs, and corporate investors are often looking for when they evaluate a business opportunity.

Investor due diligence

The process investors go through to find out the answers to the questions listed in Table 6.3.3 (and how opportunities measure up to their investment criteria) is referred to as due diligence. Due diligence is an iterative process that requires ongoing investigation and discovery. Each due diligence exercise is different, with the approach and required time varying from situation to situation and company to company. However, the steps outlined in Figure 6.3.3 are generally followed.[21]

Information collected via this process is continually reviewed. If negative information comes to light at any point, an investor may decide to abandon the due diligence process (and abstain from investing in the company). When dealing with VCs, this process can take from as little as four weeks in rare cases to as long as three years. The typical range is three to nine months. Other types of investors will also conduct due diligence, but they may not spend quite as much time completing the process. However, fundraising, regardless of investor type, usually takes much longer and requires more effort than most innovators estimate (see Table 6.3.5).[22]

Stage 6 Integration

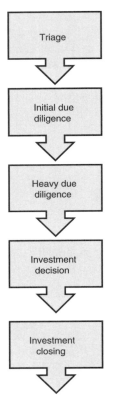

- Direct or indirect contact between the investor and innovator, often resulting in the business plan being shared.
- Review of the business plan and/or initial meeting between the investor and the innovator or management team.

- Initial discussion of concept by investor with some of his/her partners (if appropriate); investor makes initial calls to contacts in the technology/service area to get a general opinion on the concept and identify issues that need to be addressed.
- One or more meetings with innovator/management team to discuss technology/service concept, market, management experience, business/financial model, and valuation expectations.

- Calls or meetings with company references on technology/service concept, market, management experience, etc.
- Calls or meetings with independent experts or knowledgeable individuals among the technology/service area's actual or potential customers, and independent references on management. Investors may hire independent consultants to evaluate aspects of the business, particularly technical, market, regulatory, reimbursement, or IP issues.
- Independent analysis of financial projections and valuation scenarios.
- Further discussion with innovator/management team around specific business issues and company valuation.

- Based on due diligence. If positive, a term sheet is generated (see section on Term Sheets later in this chapter).

- If agreement is achieved on the investment terms, legal and patent due diligence is done during the process of closing the investment to ensure that there are no "hidden" issues about which the investors are unaware (e.g., lawsuits, problems in company capitalization, problems in contracts with employees, suppliers, or others that may impede the growth of the company or cause problems at exit).

FIGURE 6.3.3
The due diligence process becomes increasingly rigorous and thorough as the innovators and investors approach an agreement (based on Ross Jaffe, "Introduction to Venture Capital," October 6, 2004; reprinted with permission).

Table 6.3.5 *The average time between funding rounds takes longer than many innovators anticipate (compiled from VentureXpert; data represent a sample of 130 medtech firms that received one or more rounds of venture funding between November 1, 1992 and December 31, 2007).*

Funding round	Number of companies	Average time to next funding event (months)	Average amount of funding (in 000s)	Average number of investors (in current round)
1st	130	15.60	$6,829	2.3
2nd	90	15.66	$8,076	3.2
3rd	68	17.33	$10,009	4.4
4th	50	15.35	$11,230	4.7
5th	33	12.69	$10,236	5.4
6th	20	14.53	$11,657	4.5
7th	9	10.75	$6,138	4.1

Milestones

Funding is typically provided to a new venture in a staged manner. Rather than committing too much funding upfront, investors provide financing as the business demonstrates its ability to accomplish the milestones laid out in its business plan. This process enables investors to periodically update their information about the firm, monitor its progress, review its prospects, and evaluate whether to provide additional funding or abandon the project. It also enables them to exercise greater control over the direction of the company.[23] It is important for innovators to seek funding in stages, since the attainment of significant milestones in between funding rounds may strengthen their position from a valuation and ownership standpoint during the next round of financing negotiations (see below).

Funding milestones represent significant events in the life of a start-up and should be selected with great care. The operating milestones chosen when creating the operating plan in 6.1 Operating Plan and Financial Model can serve as the starting point for selecting funding milestones. From an investor perspective, funding milestones represent points when a sizable amount of technical, clinical, or market risk has been eliminated from the company. Often these points will coincide with milestones in the operating plan, such as proof of concept, clinical trial initiation, regulatory approval, etc. Ultimately, the final funding milestone for investors is the exit event, be it an initial public offering (IPO) or acquisition (which is commonly known as the "harvesting event"). See Figure 6.3.1 for sample milestones that investors will expect to be completed between various stages of funding.

As each of these milestones is achieved, the company has the potential to reduce its cost of capital. Initially, at the earliest stage of investment, the company's worth is low and the cost of raising money is high, because the business has no proven track record and thus poses significant risk for the investors. Yet, as the company begins to meet its milestones, it is able to raise increasingly large amounts of money at more competitive rates as the risks to investors decline.[24] Three of the most important hurdles for a company to overcome in demonstrating increased value include:

- **Technical feasibility** – Relies heavily on engineering, science, and clinical interactions and is accomplished when the company has proven data regarding in vitro (seed funding), animal use (some seed funding and start-up funding), and human use (start-up funding and expansion funding).
- **Product feasibility** – Depends primarily on R&D and clinical and regulatory expertise. Product feasibility demonstrates commercial viability and is proven through the completion of pivotal trials, regulatory approvals, support from key opinion leaders, and first commercial sales (some start-up funding and expansion funding).
- **Company feasibility** – Relies on continued R&D, sales and marketing, and manufacturing to demonstrate sustainable profitability. Company feasibility can be shown through revenue and profit growth, a full product/technology pipeline with multiple generations of devices being developed, strong brand identification in the market, and defect-free quality (expansion funding, mezzanine funding, and IPO).

As each of these milestones is achieved, the company essentially increases its value. To get from one milestone to the next, the innovator must evaluate how much capital is required. Much of this information will be laid out in the company's operating plan. Plans should be revisited and reviewed with a funding perspective in mind to ensure that the link between capital needs and funding rounds makes sense given the current investment environment. Specifically, there are times when capital resources are ubiquitous, and other times when funding is virtually unavailable due to macro-economic trends outside the control of a start-up. When deciding on funding milestones, innovators should select milestones that the market views as significant barriers to the success of a company. This will yield the greatest increase in company value and, thus, the lowest dilution for innovators and previous-round investors when each subsequent round of funding is sought.

Valuation

A company's valuation, or the worth assigned to the business, is directly affected by the following factors:[25]

- The current and expected future valuations of comparable companies in the public and, when available, private marketplace: the higher the valuation of comparables and the more optimistic their outlook, the higher the valuation.
- The supply and demand for capital at the time of financing: shortage of alternative investment options for investors increases the valuation; abundance of alternative options decreases the valuation.

- Intangibles unique to a specific company, including the quality of the management team, a company's competitive advantage, and its likely pace of revenue growth and profitability: more experienced teams with proven track records can negotiate higher valuations.
- The nature and timing of an expected exit for the investor: the closer the time to exit and the higher the certainty of the exit, the higher the valuation.
- The implications of future capital raises, as well as needs to expand a company's **option pool** on the company's capital structure going forward: the more future rounds needed, the lower the valuation.

Investors often refer to pre-money and post-money valuations. Pre-money refers to a company's value *before* it receives outside financing (or the latest round of financing), while post-money refers to its value *right after* it gets outside funds.[26] The pre-money valuation reflects the value assigned by investors to the assets the company has developed to date and the promise that the company holds. The post-money valuation is always equal to the pre-money valuation plus the capital raised. In general, as the company makes progress, its value increases. This is reflected in the valuation when the pre-money valuation in the next round of funding exceeds the post-money valuation at the end of the previous round of funding.

Working example: Pre- and post-money valuation for Conor Medsystems

Conor Medsystems was founded in 1999 to develop a new generation of drug eluting stents (DES). Over the next five years it raised more than $78 million before going public in December, 2004. In February, 2007, it became a wholly-owned subsidiary of Johnson & Johnson. Table 6.3.6 presents the history of Conor Medsystems' funding rounds, valuations, and percentage ownership structure until its initial public offering in 2004, as well as a summary of the returns realized by the investors.

The post-money valuation is always calculated by adding the amount of funding raised to the pre-money valuation. For example, in the first round of funding on February 1, 2000, the pre-money value assigned by investors to the company based on its progress and plans to date was $2.675 million. The funds raised, $0.325 million, are then added to that amount to give the post-money valuation ($3 million), which is the value of the company, including its assets, the promise of its plans, and the cash just raised.

Taking the example one step further helps demonstrate the effect of dilution – as more investors provide money in exchange for shares in the company, the increased number of shares outstanding reduces the percentage ownership of existing shareholders.[27] To calculate the percentage of the company owned by investors in the first round of funding, simply divide the amount of funding raised by the post-money valuation. For example, after the completion of the first round of funding, the investors provided $0.325 million to acquire shares in Conor Medsystems. With a post-money valuation of $3 million, this means that the investors acquired approximately 11 percent of the company.

$$\text{Ownership} = \frac{\text{Investment in current round}}{\text{Post-money value}} \quad (6.3.1)$$

The remaining percentage (89 percent) stays with the founders/management and employee option pool. In each round, the same initial calculation is performed (e.g., $1,500,000/$10,000,000 = 15 percent in the second round) to determine the percentage of the company sold to the new investors. However, an additional calculation is needed to compute the effect of each subsequent round of funding on the ownership percentages from previous rounds. For example, in the second round, the ownership of the original investors and founders is reduced by 15 percent (the ownership share of the new investors). The new investors' share comes from the fraction owned by the original investors and founders. For instance, if the founders originally owned 89 percent of the company, after the new round of funding they will own 89 percent of the fraction of the company retained by the original

Table 6.3.6 Valuation for Conor Medsystems (compiled from VentureXpert and Conor Medsystems prospectus).

Date	Funding raised (in 000s)	Pre-money valuation (in 000s)	Post-money valuation (in 000s)	Founder/mgmt ownership	Investor ownership Series A	Series B	Series C	Series D	Series E	IPO
02/01/00	$325	$2,675	$3,000	89%	11%					
11/01/00	$1,500	$8,500	$10,000	75.65%	9.35%	15%				
06/27/02	$10,200	$17,800	$28,000	48.11%	5.95%	9.54%	36.4%			
10/22/03	$28,000	$29,000	$57,000	24.88%	3.08%	4.93%	18.82%	48.28%		
08/01/04	$38,900	$111,100	$150,000	18.42%	2.28%	3.65%	13.94%	35.76%	25.93%	
12/14/04 (IPO)	$78,000	$409,000	$487,000	15.47%	1.91%	3.06%	11.71%	30.03%	21.78%	16.02%
Summary										
Initial investment					$325 K	$1.5 M	$10.2 M	$28.0 M	$38.9 M	
Terminal value				$75,339	$9,302	$14,902	$57,028	$14,624	$10,606	
ROI					28.6 x	9.93 x	5.59 x	5.22 x	2.72 x	
CAGR					110%	75%	93%	356%	630%	

Numbers may be subject to rounding errors.
ROI = return on investment (see below for more information).
CAGR = compounded annual growth rate (see below for more information).

> **Working example:** *Continued*
>
> investors and founder (which is 85 percent). So the founders' diluted share of the company is 89 percent multiplied by 85 percent, which leads to the diluted share of 75.65 percent. Similar calculations apply for investors in series A (their new ownership is 11 percent of 85 percent, which is 9.54 percent).
>
> When Conor Medsystems went public in 2004, the terminal value for each investor is simply the value of their shares at the time of IPO (calculated as the number of shares multiplied by the price per share). Then, the ROI for each shareholder is given by:
>
> $$\text{ROI} = \frac{\text{Terminal value}}{\text{Initial value}} \quad (6.3.2)$$
>
> For example, for A round investors, the ROI is $9,301/$325 = 28.6$. That is, the original investment made by round A investors grew 28.6 times.
>
> The compound annual growth rate (CAGR) is the annual growth that the initial investment of each investor experienced. This is calculated using the following formula (which assumes compounding occurs continuously over time):
>
> $$\text{CAGR} = \frac{1}{\text{Time between investment and exit}} \ln\left(\frac{\text{Terminal value}}{\text{Initial value}}\right) \quad (6.3.3)$$
>
> For the funds provided by round A investors to grow 28.6 times from February 1, 2000 to December 14, 2004, this means that the continuously compounded annual growth rate was 110 percent. In other words, the investment grew at an annual rate with continuous compounding of 110 percent per year.[28]
>
> This example provides the funding requirements and valuations for Conor Medsystems, as reported after the fact. However, to derive a company's funding requirements, the innovator would look to the operating plan and financial model. Funding requirements should be set such that the company has enough money to reach the next major milestone in the operating plan, enabling it to demonstrate risk reduction and secure the next round of funding (see below).

Strategic considerations: how much funding between rounds and at what valuation?

As previously explained, the amount of funding a company needs to raise can be determined based on how much is required to complete critical milestones between rounds of fundraising. The achievement of each major milestone should be linked to the retirement of a major risk, such that it becomes easier to attract funding in each subsequent round. For example, as a company prepares to move from bench tests into animals, it should calculate how much money is needed to complete successful animal trials before it has to fundraise again.

The capital and time required to reach each funding milestone should come from two sources. First, they come from the operating plan (and should be validated relative to a proxy company or companies). Second, incremental capital may be needed due to deviations from the plan. Each round must include a "cushion" to address these deviations, since running out of cash between valuation points can be incredibly costly and potentially jeopardize the business. However, keep in mind that raising too much capital needlessly dilutes the ownership of the innovator and the previous investors.

Determining the value of the company at each round of funding is both an art and a science. Importantly, these calculations must be market based (i.e., what other investors are willing to get in?). A company can get started by doing simple modeling to develop "back of the envelope" valuations based on expected returns and potential exit valuations.

Two common methods of valuing the start-up at each round are (1) discounting terminal value and (2) a comparables analysis. The premise behind discounting the terminal value is that investors require a certain return on their invested capital. While not a definitive valuation method, this approach represents a good exercise to understand the drivers of valuation. During earlier rounds of funding when the venture is

Table 6.3.7 *Common discount rates for new medtech projects.*

Risk level	Example	Expected return
Risk-free project	Build a new plant to make more of an existing product when there is a surge in demand	10 to 15 percent
Low-risk project	Make incremental improvement in existing products	15 to 20 percent (above corporation's goals for return to shareholder)
Low- to medium-risk project	Develop next generation of existing product	20 to 30 percent
Medium-risk project	Develop new product using existing technology to address markets served by other products of the corporation	25 to 35 percent
Medium- to high-risk project	Build new product using existing technology to address new markets	30 to 40 percent
High-risk project	Build new product using new technology to address a new market	35 to 45 percent
Extremely high-risk project	Build new product using new technology to address a new market when there is an unusually high level of risk associated with one or more of these factors	50 to 70 percent

more risky, investors expect higher returns than in later-stage rounds when risk has decreased. The critical components in determining the company value at each round are listed below:

- **Terminal value** – With the current exit strategy, what amount can investors expect the company to be worth? As an example, a start-up may determine that it could be acquired by a large medical device company for $400 million. The terminal value is often based on what comparable companies received at their exit event but can also be based on the future cash flows that the product may generate after the exit event. For instance, in the Conor example the terminal value was the IPO pre-money value of $407 million. There could be an alternative terminal value that would represent an acquisition.
- **Duration** – The time frame between the specified round and the exit event.
- **Discount rate** – The discount rate is the return that investors expect to be compensated for putting their capital at risk. Table 6.3.7 illustrates typical discount rates for different types of projects. As a rule of thumb, the discount rates may get smaller with each subsequent round of funding.
- **Calculation** – For each round of funding, discount the terminal value back by the expected duration between that round and the exit using the discount rate. The general form of this equation is:

$$\text{Post-money valuation} = \frac{\text{Terminal value}}{(1+\text{Discount rate})^{\text{Duration}}} \quad (6.3.4)$$

For example, a rough calculation for Conor's valuation on December 1, 2000 could be as follows: assume a terminal valuation of $100 million, a discount rate of 70 percent, and a duration of five years (time to IPO). $100/(1 + 0.70)^5$ = $7 million. The actual realized valuation at that point was $3 million, which implies that the investors assumed either a longer duration, a lower exit, or required a discount rate in excess of 70 percent.

In addition to using the terminal value, a comparables analysis will help a company target a realistic valuation. This analysis starts by selecting a comparable company based on at least the following criteria: stage of funding, field and application, and founders' experience. The average pre-money and post-money valuation for each round of funding and stage of those companies is assessed next. Post-money valuations for the innovator's company should be based on pre-money valuations of comparable companies, plus the estimated operating expenses to achieve the next major funding milestone from the company's financial model.

The best way to secure a favorable valuation is to have multiple interested investors with multiple term sheets. While these analyses can help a company target a reasonable valuation, many other factors can influence the final numbers. The experience of the team, competitive threats, investor interest in the specific space, and macroeconomic market conditions can all play large roles in determining the final valuation. Appendix 6.3.2 provides another valuation example, using a slightly different approach, for the company analyzed in 6.1 Operating Plan and Financial Model.

Valuation, ownership, and dilution

Innovators should keep in mind that the company's valuation will determine their ownership percentage. As the company progresses and more capital is raised, owners should expect their ownership percentage to shrink. However, in parallel, the valuation of the company is expected to increase, which can lead to a higher total value for the owners.

In some cases, it is not uncommon for the valuation of a company to decrease between rounds of funding (a so-called "down round"). While this may be disappointing, it usually reflects a temporary setback that may be reversed in the future. Yet, start-ups are a risky investment, with some of them never generating a return on the capital and time invested. Investors are aware of these risks and seek to mitigate them in two ways: (1) by requiring high ownership stakes in the companies they fund, such that the returns on successful ventures help counterbalance investments in failed start-ups, and (2) by incorporating antidilution measures in any deal that prevent their equity investments from losing value (see section on Term sheets). Warrants are a vehicle used to protect investors from dilution. These give the investor the option to purchase additional shares of the company's stock at a pre-specified price or else face the dilution of their ownership percentage.[29] Innovators can help retain ownership in the company by carefully and proactively evaluating and managing key risks.

Sizing the option pool

Another important decision facing the founders and early investors is how much of the company's stock to reserve for key employees. Stock ownership, typically in the form of stock options, is a crucial incentive that can be used to recruit and retain valuable team members. According to analysis performed by the Silicon Valley-based law firm of Wilson, Sonsini, Goodrich & Rosati, roughly 20 percent of the company's shares should be reserved for the option pool after the first round of funding.[30] Following series A, the amount of stock options reserved for employees (as a percentage of the company's total stock) should follow the rough rules of thumb presented in Table 6.3.8.

Term sheets

The funding process culminates in a signed contract between the investor and the company. The important details of the final contract are typically worked out through a term sheet, which outlines the terms for a deal and serves as a letter of intent between the investor providing funds and the company receiving them.[31] The two most important functions of the term sheet are to summarize all of the important financial and legal terms related to a contemplated transaction, and to quantify the value of the transaction.[32] Term sheets are commonly used by VCs, corporate investors, and **syndicates** of angel investors, among other investor types.

Importantly, term sheets are not legally binding documents because they are put in place before the investors complete legal and patent due diligence. However, both parties involved in the deal are expected to interact in good faith, preserving the essence of the agreed-upon terms until the closing of financing. Term sheets are usually prepared by the lead investor and presented to the company's CEO. The term sheet becomes an expression of the investor's interest in a company and outlines the term by which the investor is interested in investing. Following the presentation

Table 6.3.8 *Maintaining competitive stock incentive program levels for essential personnel is central to a company's hiring and retention strategy (from Doug Collom, "Starting Up: Sizing the Stock Option Pool," The Entrepreneurs Report, Wilson, Sonsini, Goodrich & Rosati, Summer 2008).*

Employees (by position)	Post-series A preferred stock
CEO	5 to 10 percent
Vice-presidents	2 to 3 percent
CFO	1 to 2 percent
Director level	<0.5 percent

of a term sheet, a series of discussions between the company and the investor ensue, with the expectation by the investor that the company will accept the proposed terms. If a company is being courted by multiple investors, then the contents of a term sheet are potentially subject to negotiation. Once agreed to, proposing changes to the term sheet can severely undermine the working relationship between the company and the investor. As a result, any potential changes should be considered carefully and initiated only under rare circumstances.[33]

Not surprisingly, term sheets vary significantly from deal to deal – not only in substance, but in style and structure. Because the term sheets set forth all of the details surrounding a funding agreement and may have long-term implications for the business, innovators should confer with top lawyers and trusted advisors when reviewing them. There are typically anywhere from 8 to 18 sections within a term sheet. The most common sections are described in Appendix 6.3.3.

Managing the funding process

Whenever possible, companies should proactively manage the funding process and try to create competition among investors. During presentations to investors, anticipate their concerns and be prepared to address them in detail. Also, be prepared to talk "off script" as unanticipated questions arise. Listen carefully before providing answers, and be clear and concise in all responses. The importance of paying attention to what investors ask cannot be overemphasized, since it is likely that if one investor asks a question, another may be interested in the same issue. Thus, gathering answers to typical questions prior to each meeting is extremely beneficial.

Once an innovator receives a term sheet, s/he may be tempted to seek other investors if the proposed deal does not meet expectations. However, even if there is no exclusivity clause in the term sheet, approaching one investor with the details from the term sheet of another may backfire. Especially in the relatively small and interconnected world of VC financing, "shopping a deal" can potentially undermine an innovator's reputation and credibility. It is more professional and effective to do this "shopping" in advance – approaching two or three potential investors simultaneously to try to get multiple term sheets at once and then deciding who to work with to best meet the company's financing needs and expectations.[34]

The NeoGuide story below demonstrates how one company managed the funding process.

FROM THE FIELD: NEOGUIDE SYSTEMS

The challenges of fundraising in good times and in bad

As described in 4.4 Business Models, Amir Belson cofounded NeoGuide Systems in 2000 to develop a better way to perform colonoscopy (see Figure 6.3.4). Belson's idea for a computer-controlled endoscope would allow a colonoscopist to guide the scope more precisely and avoid the use of force. As a result, the technology would significantly reduce the patient discomfort that occurred in the vast majority of colonoscopy procedures. It would also dramatically improve the physician's ability to consistently achieve complete navigation of the colon.

Belson joined forces with two lecturers from Stanford University to cofound NeoGuide. Belson contributed the solution concept and IP, while the other cofounders had access to enough funding to enable the company to get started. Once the idea was a little more developed, they decided to initiate a series A round of funding. However, as they began meeting with investors, it became clear that many were not excited about opportunities in the gastroenterology (GI) market. "The potential of the market is very big," Belson said.[35] "But I think what scares investors is that there is no real history of success in this market for start-ups. The entrenched players are big Japanese companies [Olympus, Fuji, and Pentax] that are difficult to compete against. GI is not a popular area." As a result, he continued, "It was hard to get VCs to listen to our story. But once they heard it, they liked it." Specifically, many of the potential investors who agreed to meet with the NeoGuide team were enthusiastic about the technology. Eventually, Belson and his colleagues persuaded Versant Ventures and The Angel's Forum to provide $3.5 million in funding. "We convinced them how we would be different and what value we would bring to the table in the GI market," Belson recalled.

The primary milestone that NeoGuide set out to achieve before it would raise additional funds was to build a working prototype. Much of the $3.5 million raised went to hiring a team of ten engineers to achieve this goal. "We also hired an entrepreneur-in-residence from Versant and he became the president of NeoGuide," said Belson.

By December, 2002, the company had a working prototype and was ready to raise an additional $5 million to get the device into animals and then into humans. "To make a long story short," Belson said, "it was a very bad time for fundraising at the beginning of 2003. Sixty-five VCs told us 'no' between January and June, 2003. At that point in time, we ran out of money and had to let everybody go." Faced with the inability to get funding, Belson considered taking full ownership of the idea and going back to Israel to develop the solution himself. However, Versant (again with The Angel's Forum) offered him bridge funding in the amount of $125,000, encouraging him either to continue trying to raise the necessary funds or to license the technology for someone else to develop. "So I started doing more meetings," Belson recalled. "I tried VCs that said 'no' early in the process and VCs that had never heard the story, although there were very few investors that hadn't heard the story other than angels, corporations, and investors from Israel." Despite this challenge, Belson continued to pound the pavement for several months seeking an interested party. This time, however, Belson did the funding presentations himself, in an effort to bring credibility, enthusiasm, and persuasiveness to the pitch.

FIGURE 6.3.4
Amir Belson of NeoGuide (courtesy of Amir Belson).

The break NeoGuide needed to raise its series B funding finally ended up coming from one of the first VCs to tell the company "no." An investor from Utah Ventures initially declined to fund the company based on advice from his own gastroenterologist. When he described the NeoGuide device to his physician, highlighting the fact that it was designed to reduce patient discomfort and increase colonoscopy completion rates, "the gastroenterologist said he didn't have those problems. He didn't need the device," Belson recalled. When this investor met with Belson for the second time, he shared this story with him. At this point, Belson asked the investor, who was a cardiologist himself, "How many physicians do you know who will happily tell you they have a big problem?" After reviewing data that showed as many as 20 to 30 percent of colonoscopy procedures failed to reach completion (as measured by the scope reaching the cecum), the investor went back to his gastroenterologist. According to Belson, "He said, 'I know *you* don't have any problem. But do me a favor. Tell me the truth.' To which he responded, 'Oh, you want the truth?' and proceeded to validate the need NeoGuide's technology was developed to address."

At the same time that Utah Ventures came around, 3i Ventures also got interested, expressing a desire to lead the round. From there, NeoGuide again began to build momentum. "We had Kaiser, Versant, 3i, Utah Ventures, and Arboretum rushing in to do the deal," said Belson. "It was as though they suddenly thought, 'Oh, I'm going to miss out on something.' They recognized it was a good technology, that the need was big, and they believed we could do it right." While Belson had still hoped to raise the original $5 million that the company needed to get the technology into humans, he was prepared to accept $1.8 million. Instead, "All of a sudden, I had $15 million in funding and the round was oversubscribed." Funding closed in November, 2003, but at a pre-money valuation that was $1 million less than the post-money valuation after round A.

Using the $15 million, Belson began the process of rebuilding NeoGuide. He hired 50 people over the next three years, completed human clinical trials, and received 510(k) FDA approval for the device before looking to raise the company's next round of funding. NeoGuide had no trouble attracting investors to its series C round of fundraising. It closed at $25 million in November, 2006. The company's plan was to spend this money preparing the technology for manufacturing. It would also invest a sizable portion in commercialization activities, including the development of a direct sales force to reach "the high caliber centers and gastroenterologists who are already waiting for the device," said Belson.

Although NeoGuide relied primarily on VC funding, Belson had several other ventures that had tapped different funding sources. For example, when Belson developed the idea for Thermocure, a non-invasive therapeutic hypothermia device to cool a patient's body core temperature during stroke and myocardial infarction, he raised his first $125,000 from VCs but then switched to angel investors when the project failed to progress under VC management. "When I raised the first $125,000 from the VC, I told them they would have to manage the project because I really don't have time. And so they managed it, but I found that they did not succeed in taking the technology forward despite the funds that were used. So I was really unhappy with it, and they, too, were concerned with the fact that there was no progress." When the VC offered to loan Thermocure the next $125,000 that this project needed, Belson declined. Instead, "I called two groups of friends who might want to invest. And this started a horse race. They both wanted to put in the money." For another company, called ZipLine Medical, Belson also sought angel funding. He contacted an acquaintance about the company. When that individual asked his lawyers to structure the deal, "the lawyers came back and said, 'We'll do this for you, but we also want to invest in it,'" recalled Belson. "It actually became bigger than I wanted." Reflecting on his experiences in seeking funding from different sources, Belson said, "There are many different models that can work. The important thing is to be sure you understand the pros and cons of each one."

Regardless of what type of funding is being sought, Belson also advised innovators and companies to get comfortable hearing "no." "Be prepared to hear 'no,' " he said, "and be creative about how you deal with those issues." Yet, more than anything, Belson emphasized the importance of being committed to an idea before pursuing it. "You have to ask yourself, how much you really believe in what you're doing. Many people will tell you you're wrong, so you really have to believe in your idea to move forward. And once you decide, you really have to go for it and be willing to take personal risks. Decide how much time to give it, and then do everything you can to make it happen."

Stage 6 Integration

Liquidity events

Early-stage investments in any medtech company are **illiquid** – the investors cannot easily sell their ownership stake. But eventually, most investors will want to make their investment liquid to allow them to realize a return. This typically occurs through an acquisition or by entering the public markets through an IPO. Acquisitions and IPOs are best undertaken when the innovator, his/her management team, and the board of directors believe that the company's value has been maximized (meaning that they have done all they can with the resources at their disposal) such that the only way to continue to grow and mature is to be acquired or raise additional money by going public. However, the liquidity events (also referred to as exit events because they allow investors to extract themselves from the investment) are often influenced by the external market environment and may not be available to the company at a desired time. Investors should be willing to provide capital to sustain the company until the external environment improves. Another alternative available to companies is to enter into a licensing agreement as a means of generating revenue and providing a payback to investors. More information about licensing is provided in 6.4 Licensing and Alternate Pathways.

Acquisition

According to Ellen Koskinas, a partner specializing in medical devices at InterWest Partners, mergers and acquisitions account for as many as 80 to 90 percent of device company exits.[36] The sale of a company is usually driven by the strategic fit between the assets or technology of the acquiree and the strategy of the acquirer. Acquisition can be an attractive exit strategy to investors because they receive cash and tradable stock that can be acted on immediately, while also avoiding much of the volatility and risk that can be associated with an IPO (see below). It can also provide investors with a reasonable way to exit a troubled company.[37]

An acquisition occurs *outright* when the acquiring company takes full and complete ownership of the business. The primary advantage of this approach is that it can provide a clean, and potentially desired, break for the founders and current management team, who may be replaced. The major disadvantage is that an outright acquisition typically results in less value (lower price) and an immediate loss of control of business to the founders and current management team. Another approach to an acquisition is to do an earn-out – staged milestones culminating in a purchase. The founders and employees usually continue to be involved under the aegis of the acquiring company to help the acquiring company meet these milestones. The benefits of this approach are that it often results in greater value while the founders retain some control of the business through the transition. On the other hand, there may be less certainty regarding the level of value ultimately realized (as this may depend on how much the acquirer invests in developing and promoting the product), the payout milestones may not be under the founder's direct control, and the founders may need to remain with the company, allowing themselves to be directed by the acquiring organization. More information about acquisitions is provided in 6.4 Licensing and Alternative Pathways.

Initial public offering

An IPO refers to the first sale of a company's common shares to public (versus private) investors. The main purpose of an IPO is to raise capital for the company. It also provides private investors with a potential exit strategy because they can then sell their shares to public investors. However, there are typically restrictions on when the original investors can sell their shares after the company goes public and it is not uncommon for the original investors to have to keep their shares for several years before they can start divesting them. Initial public offerings also impose heavy regulatory compliance and reporting requirements on the business.[38] Due to these increased requirements, specifically the Sarbanes-Oxley Act of 2002, the frequency of IPOs by start-ups has decreased significantly. The rash of IPOs in the late 1990s also died down after the decade's economic "bubble burst," with many of the medical device companies that went public during this period failing to deliver on their high valuations. In late 2006 and for some time in 2007, the number of

IPOs appeared again to be on the upswing, but that window was short lived and was all but closed by 2008.[39,40]

With acquisitions and IPOs, innovators need to be cautious not to sell too soon. Innovators often act too quickly and, as a result, limit the potential future value of the company. However, there may be good personal reasons for an innovator to sell early, such as a desire to try other industries or medical fields or to move back to an early-stage company if the current company has grown too large for the innovator's interest. Also, there is always the pure financial motive of wanting to make at least some return for one's time and effort rather than continue with some degree of uncertainty and risk. In reality, however, the right time to sell will be determined by the market.

GETTING STARTED

Medtech funding can be challenging and complex. However, the following steps can be used to begin the process of analyzing funding requirements and investors from the viewpoint of the innovator.

See **ebiodesign.org** for active web links to the resources listed below, additional references, content updates, video FAQs, and other relevant information.

1. Identify comparable companies

1.1 What to cover – Investors look at historical information, in conjunction with current information, to predict financial trends for a business. Since start-ups lack historical financial data, the next best source of information comes from comparable companies. As mentioned in 6.1 Operating Plan and Financial Model, comparable companies, or proxy companies, are those whose operations resemble what is required to commercialize the innovator's idea. After reviewing proxy company selection criteria and the corresponding analysis that has been completed in Chapter 6.1, put on an investor's hat to scrutinize these examples from a funding perspective. Consider what lessons can be learned from the funding strategies taken by the proxy companies, as well as what risks can be avoided. Anticipate what questions investors might ask based on the experiences of these companies. When approaching individual investors, it is also important to understand which companies they think are the most appropriate comparables based on their personal experience and thus may require additional analysis.

1.2 Where to look
- **Other chapters** – Review 6.1 Operating Plan and Financial Model and 6.2 Business Plan Development.
- **Yahoo! Finance** – This website aggregates the financial data of public companies. By looking up a known public competitor, a hypertext link in the left-hand column leads to a web-page of "Competitors."
- **VentureXpert** – Provides data on mergers and acquisitions, IPOs, and VC funding based on data from the Securities Data Corporation. Users can search the database to find detailed information about private start-ups including funding, investors, and executives.
- **VentureSource** – Another database that contains detailed information about private start-ups including funding, investors, and executives.
- **Hoovers Online** – A database of product profiles for companies, which includes competitors.

2. Confirm funding milestones and capital needs

2.1 What to cover – To select funding milestones, start by reviewing the operating plan developed as part of the business planning process. Compare the operating milestones in that document with the operating and

Stage 6 Integration

GETTING STARTED

financing milestones analyzed from proxy companies. At this point in the process, similarities should exist between the business plan milestones and those of comparable companies. The most critical part of selecting financing milestones is to identify significant risks for the start-up and determine how these risks may differ from those of other comparable start-ups. Once funding milestones are selected, characterize the overall process by describing the number of rounds of funding needed, the average amount of each round, and the average duration between milestones. When determining the amount of capital and time needed to reach each funding milestone, again consider the plan laid out in the operating plan. Also take into account potential deviations from the plan and build in an appropriate cushion. Use the comparable company that has already been studied to check these estimates. As a final check of the funding milestones and capital needs, talk to seasoned innovators and investors.

2.1 Where to look
- VentureXpert
- VentureSource
- **Other chapters** – Review 6.1 Operating Plan and Financial Model and 6.2 Business Plan Development.

3. Determine a company valuation

- **What to cover** – Assessing valuation is critical to splitting the value, ownership, and control of the company between the founders, previous-round investors, and the next round of investors. Begin by discounting the terminal value for the company and then perform a comparables analysis as described in the Valuation section of this chapter. Once the pre- and post-money valuations for each round are calculated, determine the percentage ownership of each stakeholder. Then, take into account the many other factors that can influence the final numbers (e.g., experience of the team, competitive threats, investor interest in the specific space, macroeconomic market conditions).

3.1 Where to look
- VentureXpert
- VentureSource
- LexisNexis

4. Research and select investors

4.1 What to cover – Based on the company's stage of development, consider the different types of investors, as well as the advantages and disadvantages of each type of funding. Then, determine from which types of investor(s) to seek funding. Research specific firms (or individuals) to assess the potential fit between the business and the investor's interests and requirements. Carefully assess the track record and the reputation of these investors to avoid entering into discussions with those who may inadvertently disclose confidential information to potential competitors.

4.2 Where to look
- **Online directories** – Online investment directories can provide basic information to those seeking funding, for example vFinance.com, Venture Capital Directory.net, VCgate, and Inc.com's Angel Investor Director are examples of online directories that may be useful. Some are available at no cost; others require a subscription.
- VentureXpert
- Venture Capital websites

5. Approach investors

5.1 What to cover – Many experts recommend starting with a group of ten potential investors and keeping seven to eight active at any given time to help ensure the funding needs of the business are met. Whenever possible, leverage personal and professional connections to make introductions to potential investors. Follow up with a well-articulated elevator pitch. Be prepared to present a business plan and prototype shortly thereafter. Decide ahead of time how much you are comfortable disclosing and be open about disclosure limitations. Most investors may refuse to sign an NDA so you need to have your provisional patents submitted and all your trade secrets well protected. In general, raising funds is more challenging than many investors anticipate, taking up to 12 months or longer for each round.

5.2 Where to look
- **Personal network** – Be diligent, assertive, creative, and professional in seeking referrals to appropriate investors.
- **Service providers** – Lawyers, accountants, consultants and other professional service providers with which an innovator has worked can be an excellent source of referrals.
- **Other chapters** – See 6.2 Business Plan Development.

Credits

The editors would like to acknowledge Darin Buxbaum, Jared Goor, and Asha Nayak for their help developing this chapter. Many thanks also go to Amir Belson, of NeoGuide, for the case example and Ross Jaffe, of Versant Ventures, for contributing material. We have also benefitted from relevant presentations made in Stanford University's Program in Biodesign by Allan May, Ken Kelley, Richard Lin, Sami Hamadi, and Leslie Bottorff.

Appendix 6.3.1

Advantages and disadvantages of various investor types

Source of funds	Advantages	Disadvantages
Friends and family	• Least expensive funding source • Flexible alternative for early-stage funding	• Funding typically comes with limited expertise that can benefit the young company • Investors may not understand level of inherent risk • Investors may not be able to participate in subsequent rounds of funding
Angels	• Moderately priced, early-stage funding source • May be willing to take on considerable risk • Investors will typically seek to exercise less control than VCs and corporate money, but may want a board position • May be able to act more quickly than VCs	• Angels may or may not have expertise that benefits the company • Angels may or may not be able to invest in subsequent rounds • Increasing sophistication of angels is improving value-add, but negatively affecting terms and level of control from the company's perspective • May need multiple angels to meet funding needs – managing many individual investors can become complex
VCs	• Typically sophisticated investors with valuable experience to share with innovators and young companies • Can provide access to vast network of contacts • Often willing (and able) to take on considerable risk • Able to participate in multiple rounds of funding	• May seek to exercise considerable control over venture's direction, management, and exit; will almost always insist on a board position • Often require considerable share of ownership in exchange for investment • Expect high returns and other terms that may not be perceived as flexible or as favorable to company
Corporate investment	• Can lead to meaningful product/project synergies • Can provide access to valuable resources (e.g., technologies, sales and distribution channels, etc.) • May be less expensive than VC funding • May lend to young company's credibility • May provide company with a "built-in" exit strategy	• Inventors and founders may receive limited value in return for building the business • May encounter conflicts as corporate investors look out for their own best interests • May encounter issues related to IP ownership • Follow-on funding may be put at risk by changes in corporate investor's financial position • May complicate or limit value realized from exit strategy if corporate investor has "**right of first refusal**"
Customers	• Usually one of the least expensive funding options • May lend to a young company's credibility • Can provide valuable market-based insights that are relevant to product development	• May lead to conflicts if the company seeks to sell the innovation to the customer's competitor • Customer may seek to limit the ways in which the innovation is marketed to be consistent with its best interests

6.3 Funding Sources

Source of funds	Advantages	Disadvantages
Government grants	• A truly inexpensive funding option that does not require the founders to part with any equity in the company • Government does not exercise influence over any business decisions • May strengthen a young company's credibility	• Highly competitive • High expectations regarding the rigorousness of research performed • Funding review cycles can be lengthy • Funding capped at $850,000 per project
Banks	• Access to funds generally does not require the company to share ownership • Bank does not exercise influence over any business decisions • Interest payments are tax deductible • Can be secured relatively quickly and can be used to help bridge short-term financing gaps	• Business assets must be used as collateral • Affect cash flow as regular payments of principal and interest must be made • Start-ups pay a premium on their loans due to the associated risk • Start-ups may have difficulty getting loans if they do not have revenue or any tradable assets • Too much debt may affect a company's credit rating

Appendix 6.3.2

Company valuation example

This appendix uses data from the fictitious company in 6.1 Operating Plan and Financial Model. A summary of important company financial characteristics is shown in Table 6.3.9.

Steps for valuing the company

1. **Determine an appropriate exit value** – Assuming that the likely exit strategy is acquisition by a larger company, a common way to value a company is by revenue and earnings multiples. Revenue multiples for high-growth medical device companies in the recent marketplace have been between 4× and 9× of sales. The SDC Platinum database by Thomson Financial is the authoritative source on merger and acquisition transactions and these multiples can be calculated from it.[41] Using these revenue multiples yields an exit value between $172 million and $388 million (calculated by multiplying year 7 revenue by the exit multiple of 4 to 9). The remainder of this example assumes an exit value of $259 million.

2. **Understand necessary returns for investors and market norms** – An early-stage investor's expected returns will depend on the risk s/he associates with the project. If this project is considered higher-risk than even most medical device start-ups, investors may look for a 70 percent annual rate of return. For a seven-year investment, this would be a 41× return (calculated by taking 1.7, the annual return, and raising it to the 7th power, the number of years, to get $1.7^7 = 41$). Later rounds of funding will dilute the ownership of earlier stage investors, so investors will attempt to calculate the dilution from future financings when determining what percentage of the company they should own. Investors will also gauge similar transactions to find realistic pre-money valuations for the company. All of these factors can be taken into account to develop a potential capitalization table (see Table 6.3.10).

Investors in series A may realize that later investors may not value the pre-money worth of the company as high as $10.75 million and $40.44 million in series B and series C, respectively. For example, if series B investors use a 60 percent discount, the series B pre-money valuation will be $3.44, representing a significant loss of value. Alternative valuation models can be developed to stress test the effect of such assumptions.

Table 6.3.11 shows how the ownership shares of the various investors, as well as the founders and management team, changes over time.

Table 6.3.9 *Company financial highlights ($ in millions).*

	Year 1	Year 2	Year 3	Year 4	Year 5	Year 6	Year 7
Financing Required	$3.00	$12.00		$27.00			
Revenue			$0.05	$0.91	$7.63	$22.09	$43.18
Earnings Before Taxes	($1.71)	($3.54)	($6.44)	($9.47)	($8.21)	($3.84)	$3.19

6.3 Funding Sources

Table 6.3.10 A summary of potential returns ($ in millions).

	Year	Investment	Pre-money valuation	Post-money valuation	% of shares acquired by investors each round	Target IRR for investors	Target return multiple	Years to exit
Series A	1	$3.00	$3.31	$6.31	48%	70%	41×	7
Series B	2	$12.00	$10.75	$22.75	53%	50%	11×	6
Series C	4	$27.00	$40.44	$67.44	40%	40%	4×	4
Acquisition	7			$259				

Table 6.3.11 Ownership projections over time ($ in millions).

	Pre-money	Postmoney	Investor One investment	Investor One ownership	Investor Two investment	Investor Two ownership	Investor Three investment	Investor Three ownership	Founder & mgmt ownership
Start-up									100%
Series A	$3.00	$6.31	$3.00	48%					52%
Series B	$10.75	$22.75		22%	$12.00	53%			25%
Series C	$40.44	$67.44		13%		32%	$27	40%	15%
Acquisition	$259	$259		13%		32%		40%	15%
Value after acquisition			$34.87		$81.96		$103.72		
Multiple on investment			11.6		6.8		3.84		

Note: This is a fairly realistic example for a successful venture in terms of returns for investors and the percentage of ownership retained by the founders and management team. With a typical stock option pool size ranging between 5 to 10 percent of the company's equity, in this example the founders will have retained 5 to 10 percent of the company at the end.

Appendix 6.3.3

Common sections of terms sheets[42]

Section	Contents
Summary of financing	This opening section summarizes the contents of the term sheet and provides an overview of the transaction being proposed. It usually includes: • The name of the investors • The name of the company • The amount of financing being offered • The number of newly issued shares • The purchase price per share • The post-financing capitalization structure (which enables the reader to calculate the pre-money and post-money valuation). Any milestones that must be met for the release of funds will also be outlined in this section. Linking funding to milestones set by investors allows investors to more closely monitor company progress and manage their downside risk. However, such milestones may be perceived as a red flag to a company wary of relinquishing too much control and/or forfeiting its flexibility to adjust the operating plan as new information becomes available.
Dividend provisions	**Dividend provisions** outline the conditions for when/if dividends will be paid. They also address issues such as whether dividends will be cumulative or non-cumulative, and the priority order in which dividends will be paid to shareholders. As with most sections of the term sheet, dividend provisions can be considered investor favorable, neutral, or company favorable, depending on how they address these issues.
Liquidation preference	The liquidation preference outlines the terms governing the transaction if a company is closed down. While preferred shareholders will be given priority over common shareholders, the term sheet often takes this one step further by defining a multiple on the value of their initial investment that preferred and common shareholders will receive. According to these terms, the multiple promised to the preferred shareholders would be paid before any proceeds would be given to other shareholders.
Redemption	A **redemption** clause is used to associate a finite number of years (usually five to eight) with an investment. In essence, it forces an exit for investors at a defined point in time if a liquidity event has yet to occur. The terms outlined within a redemption clause specify the time frame associated with the exit, as well as the specific terms governing the rate and price at which shares will be repurchased from investors.
Registration rights	**Registration rights** outline the rights of preferred investors in forcing the company to register its stock for public offering under SEC rules. Essentially, they give investors another mechanism for creating an exit, if market conditions are perceived as favorable.[43] The issues addressed in this section include the number of registrations that the company is obligated to complete, the period of time in which these registrations are required, and the economic cost of going through the registration.

Section	Contents
Conversion and automatic conversion	When a **conversion** happens at the time of an IPO or acquisition, it is generally accepted that preferred shares will convert to common shares on a 1:1 basis. The reasons for including this clause in the term sheet is to give preferred shareholders a mechanism for converting to common shares if the IPO is likely to generate a return higher than the one stipulated in the liquidity preference section. The automatic conversion clause describes other variables in the event of an IPO, such as the amount of money that will qualify the IPO as acceptable to the preferred shareholders. Its purpose is to ensure that any automatic conversion price (or amount-raised hurdle) is set high enough to make forced conversion attractive.
Dilution	One of the single most important issues to investors, as a company grows, is how new rounds of financing will affect the value of their investment on a per share basis. Dilution clauses stipulate how conversion prices will be calculated if future rounds of financing are dilutive to preferred shareholders. If future stock is issued at a price lower than the current round, an antidilution clause helps ensure that preferred investors continue to hold an equal (or near equal) percentage of ownership in a company without committing more capital.[44] There are two primary approaches for structuring antidilution clauses: full-ratchet and weighted-average. Full-ratchet provisions are the most favorable antidilution clauses to investors. Under this approach, the conversion price is adjusted to ensure that the new price factors in the total amount of capital invested, and preserves the full percentage ownership of the preferred shareholders, typically at the expense of common shareholders and/or individuals holding stock options in the company.[45] With the weighted-average approach, the conversion price is calculated on a weighted-average basis, therefore only partially offsetting the effect of dilution.[46] As a result, it is considered more favorable to the company and common stock holders.
Right of first refusal	The right of first refusal gives preferred investors the right to exercise influence over the sale of any shares (preferred or common) within the company. Typically, investors use this clause to retain the ability to purchase shares or restrict their sale.
Voting rights	Voting rights are included in term sheets to ensure that all shares are treated equally in the event that a shareholder vote is called. Typically, one vote per share is granted for both preferred and common stock.
Protective provisions	**Protective provisions** outline the percentage of preferred shareholder approval needed for a company to take certain actions. The purpose of this clause is to help prevent common shareholders from diluting the power of preferred shareholders or "selling the company out from under them."[47] Actions potentially covered under protective provisions include the issuance of new stock, major financial transactions (merger, sale, payment of dividend), or other events affecting the course of a company's business (reorganization, amendments to the company's certificates of incorporation, increase or decrease in the size of the board).
Information rights	**Information rights** define what and how much information about the company is shared with investors. Under the terms most favorable to investors, companies must provide access to almost any document or system (audited or unaudited, financial or otherwise) as part of the standard inspection and visitation rights afforded to preferred shareholders. In more neutral scenarios, investors are regularly provided with, for example, audited annual financial statements and unaudited quarterly financial statements. Access to monthly financials and the company's operating plan are other items that may or may not be included in the term sheet.

Continued

Section	Contents
Board composition	The board composition clause outlines the number of board seats and how they will be filled. Typically, companies seek to build a board with representation by both common and preferred stock holders. Investor-favorable composition would give the preferred shareholders majority control. A more neutral arrangement might give preferred shareholders (investors) and common shareholders (company management) an equal number of seats, with one additional seat granted to a mutually agreed-upon independent participant.
Employee matters	From an investor's perspective, the conditions outlined in this section are designed to help retain important managers and employees while limiting the amount of power they exercise over the stock of the company. Issues typically addressed under employee matters include the number of shares of common stock reserved for option pools and other employee programs, vesting periods for common shares, restrictions on the sales of common stock by employees, the determination of whether life insurance policies will be taken out for key executives, and directors' insurance.
Conditions precedent	The conditions precedent outlines what steps must be taken before the financing deal proposed in the term sheet can be finalized. For example, it may specify that due diligence must be completed, including a review of the company's operating plan, IP strategy, regulatory strategy, clinical trials plan, and other relevant documentation. It may also require the completion of IP, confidentiality, and non-compete agreements with employees; review of the company's compensation, stock allocation, and vesting programs; and board approval. This section may also contain an exclusivity (or "no shop") provision, which bans discussions between the company and other investors for a defined period of time.
Expenses	This section outlines which expenses, for due diligence and legal counsel, are to be paid for by the company upon closing.

Notes

1. "2007 Venture Capital Investing Hits Six Year High At $29.4 Billion," PWCMoneytree.com, January 21, 2008, https://www.pwcmoneytree.com/MTPublic/ns/moneytree/filesource/exhibits/07Q4MT_Rel_FINAL.pdf (July 22, 2008).
2. From remarks made by Mir Imran as part of the "From the Innovator's Workbench" speaker series hosted by Stanford's Program in Biodesign, April 28, 2004, http://biodesign.stanford.edu/bdn/networking/pastinnovators.jsp. Reprinted with permission.
3. Leslie Bottorff, "Funding a Medical Device Start-up," *Medical Device & Diagnostic Industry Magazine*, January 2000, http://www.devicelink.com/mddi/archive/00/01/004.html (January 23, 2007).
4. Bob Zider, "How Venture Capital Works?" *Harvard Business Review*, November 1, 1998.
5. "How Corporate Venture Capital Investing Increases Innovation," Knowledge@Wharton, October 19, 2005, http://knowledge.wharton.upenn.edu/article.cfm?articleid=1299&CFID=7207756&CFTOKEN=25757895 (January 24, 2007).
6. "Customer Funding," SBIR Resource Center, http://sbir.us/library/keynote/slide4.html (January 24, 2007).
7. More information about these programs can be found at the DoD SBIR & STTR Programs website at http://www.acq.osd.mil/osbp/sbir/.
8. Bottorff, op. cit.
9. Valdim Kotelnikov, "Bootstrapping," Venture Finance Step-by-Step Guide, http://www.1000ventures.com/venture_financing/bootstrapping_methods_fsw.html (January 23, 2007).
10. Minassian, op. cit.
11. Mark Minassian, "Debt Versus Equity Financing," About.com, http://biztaxlaw.about.com/od/financingyourbusiness/a/debtvsequity.htm (September 14, 2007).
12. Ian Giddy, "What Is Leveraged Finance?" Giddy.org, http://giddy.org/dbs/structured/LevFinarticle.htm (September 14, 2007).
13. Ibid.
14. "Bridge Loan," www.wikipedia.org, http://en.wikipedia.org/wiki/Bridge_loan (September 14, 2007).
15. John Mao, Andrew Hirsch, Christine Foster, "Does My Startup Qualify for Venture Debt Financing?" The

15. (continued) Entrepreneur's Report, Wilson, Sonsini, Goodrich & Rosati, Summer 2008, http://www.wsgr.com/publications/PDFSearch/entreport/Summer2008/private-company-financing-trends.htm#5 (November 4, 2008).
16. "Convertible Bonds," www.wikipedia.org, http://en.wikipedia.org/wiki/Convertible_bond (September 17, 2007).
17. Ibid.
18. The elevator pitch is a brief description of the opportunity (and the associated product) that is concise enough that it can be delivered to an investor during an elevator ride (30 to 90 seconds).
19. Ross Jaffe, "Introduction to Venture Capital," October 6, 2004.
20. From remarks made by Rodney Perkins as part of the "From the Innovator's Workbench" speaker series hosted by Stanford's Program in Biodesign, April 14, 2003, http://biodesign.stanford.edu/bdn/networking/pastinnovators.jsp. Reprinted with permission.
21. Ibid.
22. Ibid.
23. Antonio Davila, George Foster, Mahendra Gupta, "Staging Venture Capital: Differential Roles of Early Versus Late Rounds," February 2003, http://www.olin.wustl.edu/workingpapers/pdf/2003-07-003.pdf (January 24, 2007).
24. Bottorff, op. cit.
25. Alex Wilmerding, *Deal Terms* (Aspatore Books: Boston, 2003), p. 18.
26. "What's the Difference Between Pre-Money and Post-Money," Investopedia.com, http://www.investopedia.com/ask/answers/114.asp (January 24, 2007).
27. "Dilution," Investopedia.com, http://www.investopedia.com/terms/d/dilution.asp (January 24, 2007).
28. Ln = the natural logarithm.
29. According to Investopedia, a warrant, like an option, gives the holder the right but not the obligation to buy an underlying security at a certain price, quantity, and future time. However, unlike an option, which is an instrument of the stock exchange, a warrant is issued by a company. Companies will often include warrants as part of a new-issue offering to entice investors into buying the new security. A warrant can also increase a shareholder's confidence in a stock, if the underlying value of the security actually does increase over time. See http://www.investopedia.com/articles/04/021704.asp (March 20, 2007).
30. Doug Collom, "Starting Up: Sizing the Stock Option Pool," The Entrepreneur's Report, Wilson, Sonsini, Goodrich & Rosati, Summer 2008, http://www.wsgr.com/publications/PDFSearch/entreport/Summer2008/private-company-financing-trends.htm#4 (November 4, 2008).
31. Alex Wilmerding, *Term Sheets & Valuations* (Aspatore Books: Boston, 2006), p. 9.
32. Ibid.
33. Ibid.
34. Wilmerding, *Term Sheets & Valuations*, op. cit., pp. 19–20.
35. All quotations are from interviews conducted by the authors, unless otherwise cited. Reprinted with permission.
36. See "Drug Eluting Stents: A Paradigm Shift in the Medical Device Industry," GSB No. OIT-50, p. 12. Reprinted with permission of Ellen Koskinas.
37. Jaffe, op. cit.
38. "Initial Public Offering," Wikipedia.org, http://en.wikipedia.org/wiki/Initial_public_offering (January 24, 2007).
39. "Medical Device IPOS Are Back ... For Now," MedGadget, November 28, 2006, http://medgadget.com/archives/2006/11/medical_device_2.html (January 30, 2007).
40. "IPO Round Up: Is the Window Slamming for Life Sciences?" Venture Beat, February 2007, http://venturebeat.com/2008/02/07/ipo-roundup-is-the-window-slamming-shut-for-life-sciences/ (November 4, 2008).
41. Learn more about SDC Platinum at http://www.thomson.com/content/financial/brand_overviews/SDC_Platinum (April 14, 2008).
42. Derived in part from Chapter 3 of *Term Sheets & Valuations*, by Alex Wilmerding, and used with permission. Copyright © 2006 Thomson Reuters/Aspatore.
43. Jaffe, op. cit.
44. Wilmerding, *Deal Terms*, op. cit., p. 63.
45. Ibid., p. 64.
46. Ibid., p. 65.
47. Jaffe, op. cit.

6.4 Licensing and Alternate Pathways

Introduction

A doctor who is dedicated to his practice comes up with an innovative concept but has no aspirations to become an entrepreneur. An engineer is advised that her concept potentially represents an interesting product feature but will not support a stand-alone company. In the course of pursuing one device idea, a development team uncovers another compelling opportunity but does not have the bandwidth or resources to pursue both. Under these kinds of circumstances, how can these innovative ideas still be used to change medicine?

While some innovators become entrepreneurs and start their own medical device companies, many more have an impact on their fields by licensing or selling their ideas, or partnering with another company to bring a solution to fruition in the market. Just as not every innovator is suited (or desires) to run a company, not every idea will best be realized through the creation of a stand-alone business. Sometimes the involvement of an outside entity allows an idea or technology to more quickly, successfully, and/or cost-effectively achieve its potential.

OBJECTIVES

- Recognize that not every medtech invention must (or should) be developed into a stand-alone business.

- Understand a process for deciding to take an alternate development pathway and an approach to managing the preparation and execution of a successful transaction.

- Know the key aspects of partnering, licensing, and sale/acquisition deals.

Licensing and alternate pathways fundamentals

The decision to pursue an alternate pathway is often dictated by the personal situation of the person or team developing the idea. Some innovators may have other priorities and skills that make the entrepreneurial track unappealing or unlikely – such as the inability (or lack of desire) to stay focused on a single area for too long, or another role that prevents them from diverting their attention to the pursuit of the idea. Others may have the appropriate qualifications and interests, but not the tolerance for risk that is often required in founding a medtech business. Still others may be part of a team or company with insufficient resources to pursue a new opportunity or an idea that falls outside of their currently defined scope and focus. In cases such as these, the innovator's circumstances make the creation of a stand-alone business undesirable. As described in 1.1 Strategic Focus, the individual or team with the idea should undergo a meaningful self-assessment before deciding whether or not to launch a start-up company.

Factors and constraints related to the idea itself can also affect this decision. To warrant the creation of a business, the innovator must ask him/herself if a stand-alone company would be able to create and capture adequate value from the idea, or if another entity would be in a better position to do so. Preliminary financial modeling (as outlined in 6.1 Operating Plan and Financial Model) can expose at least two common obstacles: lack of market access issues or issues related to return, timing, and cost.

Lack of market access

The bottom-up market analysis technique described in Chapter 6.1 may reveal that the cost of sales for a particular product is greater than the potential margin offered by that product. If so, the new product cannot reasonably be brought to the market in a profitable way. In this scenario, there is lack of market access and the idea cannot justify the creation of a dedicated company.

In well-established markets, competitors can "lock up" customers by making significant investments in sales and distribution infrastructures and long-term buyer/supplier relationships that cannot easily be replicated by small, start-up companies. For example, an innovator with a new idea for a syringe product would find it almost impossible to compete against Johnson & Johnson, Baxter, or Becton Dickinson, with their hundreds of sales representatives, deep catalogs of related products, and established connections with the appropriate buyers. While any one of these companies could quickly and inexpensively introduce a new syringe, a start-up company would almost certainly collapse under the strain of the time and effort required to access the buyers, get their attention, and convince them to purchase such a product from a small, unknown entity. In other words, the cost to acquire new customers would be prohibitive to this new market entrant.

Issues of market access can also be problematic when the product in question is generally bundled and sold with other products. For instance, an innovator who has developed a new introducer for a guidewire would have difficulty marketing his/her innovation because introducers (and guidewires) are usually sold in a set, along with the other elements of a complete catheter system. On its own, buyers see an introducer as a low-value item. As a result, they are unlikely to purchase it separately, even if it performs better than the introducers they have used in the past. The convenience of purchasing the introducer as an integrated part of an overall system usually outweighs any incremental performance benefit, particularly when the bundle is offered at a discount to buying each product individually. These are just a few of the ways market access can support a rationale to license or sell a product concept rather than develop it oneself.

Returns, timing, and cost issues

Product development costs can be another obstacle to the creation of a stand-alone business. The financial burden of developing a new technology *de novo* can be staggering. It is important to complete an assessment of these costs and the timing of expected returns to determine if the potential profitability of the technology will be sufficient to generate a significant payback on the required investment within a reasonable timeframe.

Importantly, not all investors have the same expectations regarding the timing and level of return on their investments. The innovator should consider this before selecting investors to fund a given idea. As described in 6.3 Funding Sources, venture capitalists may be primarily interested in funding ideas with billion-dollar markets and the potential for a 5X to 10X return on their invested capital. On the other hand, angels, as well as friends and family, may be willing to explore much smaller markets that yield significantly lower returns. However, if there is a poor fit between investor expectations and the innovation, this may be another powerful reason to pursue another pathway for development.

Types of alternate development pathways

Alternate development pathways leverage other entities to more effectively develop or commercialize an invention. There are three primary alternatives available to innovators who decide not to pursue a stand-alone business: partnering, licensing, and sale/acquisition. While these options differ in many ways, they are particularly distinct in terms of the degree of innovator involvement and certainty of the payoff upon the completion of a transaction. Regardless of which path is chosen, it is essential to obtain legal advice when pursuing agreements of this type.

Partnering

In a partnership, the two entities share responsibility for the development and/or commercialization of an idea or invention. In turn, certain costs and returns are distributed between them. In a partnership, each partner provides something that the other needs. For example, partnerships are a potentially favorable alternative when an innovator possesses intellectual property (IP) and technical capabilities whose value extends beyond a single idea. By partnering, this innovator might be able to explore the development and commercialization of multiple products by parsing each product to a different partner. In cases such as this, the innovator is typically a development partner and may choose to retain core IP right and development responsibilities, while his/her partner handles sales and marketing.

Partnership agreements can take many forms and the flow of cash and responsibility between each organization can be tuned to the unique arrangement between the two companies. In this way, partnerships are an effective way to manage risk when necessary technologies, IP, expertise, and resources are spread across more than one organization. Moreover, they can provide a mechanism for large companies to evaluate smaller companies as an intermediate step towards an outright acquisition. For example, in November, 2004, AstraZeneca entered into a five-year research and development alliance with Cambridge Antibody Technology (CAT) to perform monoclonal antibody research.[1] Less than two years later, in May, 2006, the company made a bid to acquire CAT. This approach allowed AstraZeneca to gain greater insight into the strengths and weaknesses of its target than it would have gleaned during any standard due diligence process prior to an acquisition.[2]

In medtech, some skepticism exists regarding the value of partnerships. According to Steve MacMillan, president and CEO of Stryker:[3]

> I'm not a big fan of joint ventures because, at the end of the day, you often have two very strong companies, both of whom are used to getting their way. Therefore, you end up making a lot of compromises, and neither side feels great. For all of us who've been involved with partners, there is a time and a place to come together for technologies, but generally with a shorter-term time frame, or a clearer endpoint, the 50–50s are hard for the long run.

This opinion is typical of many senior executives who have had the challenging experience of trying to make partnerships work. Because responsibilities are shared in a partnership, the innovator and his/her team play a critical role in keeping the project moving, working closely with representatives from the other entity. Having a well-defined management structure is important to sustaining a successful partnership. Choosing partners carefully, based on shared cultural values and work styles, can also contribute to desired results. Without alignment in these areas, partnerships can quickly run into difficulties.

For instance, in September, 2002, Eli Lilly announced a partnership with a small biotechnology company, Amylin Pharmaceuticals Inc., to develop and market a new diabetes drug. In the absence of a clearly defined partnership structure, the two companies argued over everything, including study design, the presentation of clinical results, manufacturing, and positioning of the drug. These conflicts were exacerbated by a clash of cultures between the two companies, with Amylin employees described as impulsive and Lilly employees afraid to make mistakes.[4] The ongoing management struggle nearly derailed the project until senior executives intervened, giving both sides a mandate to cooperate. In addition to becoming more sensitive to their different work styles, the two sides evaluated every task to be accomplished before the project's impending milestones. One partner was assigned primary responsibility for each activity, while the other was given a mechanism for providing input (e.g., Lilly took the lead on marketing to doctors and consumers since it had more marketing experience; Amylin took responsibility for physician and patient education since it was more familiar with the technology). Such a clear delineation of ownership allowed the project to get back on track. Ultimately, as each side performed, more trust was also created in the partnership.[5]

These kinds of challenging experiences are perhaps one of the reasons why true partnerships (as opposed to more "supplier/service provider" relationships) are relatively rare in the medtech industry. The lack of medtech partnerships provides a stark contrast to the biotech and pharmaceutical industries where partnerships are prevalent and often highly successful. Many different factors contribute to this disparity. In biotech and pharmaceuticals, partners often come together to share the tremendous cost and risk associated with developing and commercializing their products. Considering that the margins offered by many pharmaceutical products approach 90 percent, there is usually adequate value in most product launches to be shared between one or more partners. In these cases, the decision to partner may be part of a company's fundamental business strategy, as the story of Inhale Therapeutic Systems and its partnership with Pfizer demonstrates.

FROM THE FIELD | **INHALE THERAPEUTIC SYSTEMS**

Building a business around a partnership strategy

Inhale Therapeutic Systems was founded in Palo Alto, California in July, 1990 by John Patton and Robert Platz. The company's goal was to build on the innovative work that Patton and Platz had initiated in the area of inhalation systems for therapeutic macromolecules. Robert Chess, who held a series of executive positions within Inhale since 1992, including president, CEO, and chairman, described the company's focus: "The original concept was pulmonary delivery – essentially providing an alternate delivery mechanism for peptides, proteins, and other macromolecules using the lung as opposed to injection. Inhale developed a family of four different technologies: powder processing, formulation, packaging, and a device to enable the drug to get down to the bottom of the lungs where nature could do its work and it would be absorbed into the blood stream."[6]

Because pulmonary drug delivery was a concept that could potentially apply to many different products across a wide variety of medical specialties, Inhale recognized the value of partnering and built this into its business strategy early in the biodesign innovation process. Without partners to help with development and commercialization, there was no way Inhale's technology would reach its full potential. Beyond this, Chess elaborated, "We needed partners for a few reasons. One of them was because, at that time, getting access to the actual manufactured drug was quite difficult, both for patent reasons and also because there were very few companies that had the expertise and investment for doing large-scale protein manufacturing. In addition, bringing these products to market was very capital intensive and required large-scale clinical trials. And so we thought that the best thing for us to do would

FROM THE FIELD: INHALE THERAPEUTIC SYSTEMS

Continued

be to focus on our competency, which was the broad-based technology platform, and let the partners provide the capital and also the clinical expertise." The Inhale team further recognized that there were significant risks associated with the reformulation of emerging and existing drugs for pulmonary delivery. "So, partnering was a diversification strategy for us," added Chess.

One of Inhale's most well-known partnerships was formed relatively early in the company's history. In January, 1995, it announced an agreement with Pfizer Inc. to collaborate on the development of insulin products using Inhale's pulmonary drug delivery system. Chess, who was CEO at the time the deal was signed, described the work the company did to get ready to initiate the partnering process: "As a new company with a new technology, we knew we needed to have some proof, in man, that would show how patients responded to insulin using this delivery mechanism. The nice thing about insulin is that you can measure its level in the blood, so a phase 1 trial actually shows you something useful. I'm sure we would not have been able to get any significant deal without this clinical data."

As they completed the proof of concept, Chess and team began scouting appropriate partners in three categories. "In the first group, we had people who were currently in the insulin business. At that time, there were three major players: Eli Lilly, Novo Nordisk, and Hoechst," recalled Chess. "The second group included companies that were in the diabetes business, but not in insulin. So this would be a logical extension of their product line. In the third group were companies that were just looking for a big hit. Our preference early on was to find a partner in the second category – someone who was in diabetes, but not insulin – because inhaled insulin wouldn't be cannibalizing a current product. This would give them more incentive to bring it to market because it would be all incremental sales to them."

Once a list of prospective partners had been identified, Inhale contacted them one by one. According to Chess, all of the major insulin players were interested, along with several other companies. Through a process of progressive disclosure, Inhale eventually entered into later-stage due diligence with several companies before negotiating a contract with Pfizer via the company's business development team.

In these negotiations, Inhale focused, in part, on securing favorable financial terms. "What's the **royalty**, what are the milestones, is there an equity investment and, if so, at what premium?" summarized Chess. However, the company also invested significant energy into understanding Pfizer's commitment to the project. Chess explained: "We paid lots of attention to who is going to be implementing the project. It's important for you to determine if those are really strong, competent people, and whether or not they are dedicated to the success of the project. The business development people may tell you that they're very enthusiastic, but once the deal gets signed, they move on to the next thing. So you don't want to go into a partnership without meeting the people who will do the work and looking them in the eye. You also want to try, during the process, to get them enthusiastic about what you're doing. If they feel like the project is being pushed upon them, this is a concern. And the other thing is to figure out if this is a project that's strategically important to the company, because if so, it's less likely to get cut later on."

Inhale also devoted substantial effort to clarifying the details of the contract related to issues such as IP, diligence provisions, terminations, and cost sharing. "Small things, like what FTE [full time equivalent] rates will we use and what's the definition of product sales, can make a large difference in the final outcome," he noted.

According to Chess, beyond the economic terms, Pfizer seemed primarily concerned with issues of control. For instance, he said, "Manufacturing control was important to them. They didn't really trust a small company to do the manufacturing. Control over decision making and the clinical process was also a priority."

Eventually (after roughly 18 months from the first contact to a signed deal), Inhale and Pfizer finalized their agreement. Certain financial terms, including royalty rates, were not disclosed (although Chess suggested they were set as a percentage of sales between the high single-digits and low double-digits). Providing an overview of the deal, he said, "We made money through both the royalties and the manufacturing of the devices and powders. Pfizer paid for our costs related to their project. We were responsible for the base technology, although they paid for a portion of the powder filling development. They made an equity investment in the company of $5 million upon signing at a 25 percent premium to market price.

And then they had to put in a second investment of $5 million at a 25 percent premium to market when we entered phase 2b of clinical testing. There were milestone payments that happened at key production points during the development process. At that time, these were in the low $10s of millions." In addition, Pfizer was responsible for providing the insulin, and managing the clinical trials, regulatory pathway, and marketing for the product. Once development was completed, Inhale's only role beyond the product's launch would be to make the device and to supply Pfizer with the powdered formulation of the drug.

Reflecting on the negotiations, Chess commented, "You have to make the other side believe you have alternatives. Otherwise, you have no leverage. And you negotiate much better if you believe you have alternatives."

In terms of managing this partnership, Inhale and Pfizer both named program managers to the effort. They also created a joint development committee that reviewed progress on a quarterly basis. Executives from both companies met "a few times a year" to stay abreast of key issues. "Those were the formal processes," said Chess. "And then, given the nature of the work, there were conversations going back and forth between various Pfizer and Inhale people every week."

Since entering into the partnership with Pfizer, Inhale (now called Nektar) had completed more than 30 partnership deals. "One of the lessons is that a partner, like Pfizer, who does a great job of diligence is more likely to be a good partner going forward, even though it takes longer to get the deal done," offered Chess. "This is because they understand where the risks are. So when things go wrong, they say, 'I thought this might be a problem,' and they don't panic. Our worst partners are the ones who sign the deals more quickly, without really doing their homework. Then, when something goes wrong, it's somewhat of a surprise to them."

For innovators considering partnership, Chess offered this advice: "Take a look at the cost of bringing the product to market and the expertise that's required. Is that something that your company can do, or do you need a large company in order to decrease the risk of bringing it forward? Think about the dilution associated with doing it yourself. Also, is the validation of a partnership deal important for your company? How broad is your technology base, and how many different applications could it have? Because, if you have a very broad technology base, it's unlikely you can really exploit all opportunities yourself."

While Inhale/Nektar had successfully pursued a partnership strategy, Chess acknowledged that this

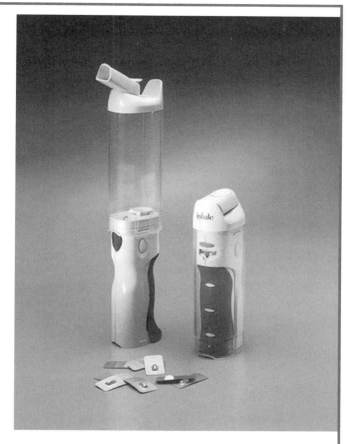

FIGURE 6.4.1
The Exubera product (courtesy of Rob Chess).

approach was not without its challenges. Of particular concern was the fact that the company relinquished a significant amount of control over the products it developed. The company faced one example of this when Pfizer discontinued sales of the inhaled insulin product, called Exubera (see Figure 6.4.1), in October, 2007 after a series of different factors hindered the product's adoption in the United States and abroad.

When considering a partnership strategy, Chess advised innovators to be cautious: "Partners can drop products for many reasons other than technical, and you need to keep that in mind. There are a lot of reasons to keep control of a product. But if you decide to partner, look for a company that's financially strong, not just so they give you good financial terms but so they don't end up having to cut your project when they get acquired or when they're not doing so well."

Based on the experience gleaned through its many partnerships, Nektar had expanded its strategy in 2008 to focus on developing its own products in addition to sustaining its codevelopment approach.

Licensing

A license involves the transfer of an idea or invention from the innovator to a licensee in exchange for ongoing royalties and/or other payments.[7] Licensing deals are sometimes used when the potential payoff from the invention is highly uncertain and/or the innovator does not possess the technical capabilities, experience, time, or money that will be needed to develop the product. While the prospect of royalties and payments may seem attractive to an innovator, it comes with a loss of control.

Depending on the other agreements obtained as part of the license, the innovator may or may not have a contractual obligation to participate in an ongoing manner. Often, the innovator can add significant value by consulting to the licensee as a technology expert or product advocate within the targeted community. However, even in these scenarios, his/her ongoing involvement is more limited compared to partnering. Rarely will the licensor be in the position to drive the ongoing management of a technology once it has been licensed. Innovators must be prepared to relinquish control when they enter into a licensing deal. For this reason, the selection of the appropriate licensee may depend more on the licensee's ability to deliver on development and commercialization than the potential economics of the transaction.

Beyond financial terms, licenses include a description of exactly what rights are being licensed,[8] the field of use and territory for which the rights are being granted, and whether or not the license is to be **exclusive**, **non-exclusive**, or sole (see Figure 6.4.2).[9]

An exclusive license grants only the licensee (and not even the licensor) the right to use a technology. A non-exclusive license allows the licensee to use the rights within the field and within whatever other limitations are provided by the license, but reserves other rights for the licensor or other parties. A sole license grants the same rights to the licensee and licensor, but prevents the licensor from granting rights to anyone else. In general, exclusive licenses tend to engender greater commitment from the licensee and lead to higher payments to the licensor. Exclusive licenses, however, can be higher risk for the licensee because the achievement of milestones (and the flow of royalties) rests in the hands of one party. Non-exclusive and partially exclusive licenses allow

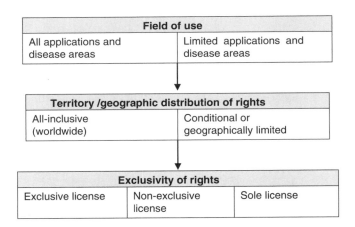

FIGURE 6.4.2
There are different types of licenses that grant different rights to the licensee and licensor.

for multiple paths and partners to bring a technology to market. On the downside, though, these structures limit a licensee's ability to use the license to differentiate their product. For this reason, exclusive licenses tend to have more value than non-exclusive licenses.

Nothing is more important to the success of a license (or a partnership) than choosing the right entity with which to do business. Although licensing agreements include provisions under which the licensor may terminate the license or license back the technology if the licensee does not commercialize it, there is often a substantial opportunity cost associated with the time lost in pursuing those actions.

Sale/acquisition

In some cases, an innovator may choose to sell an idea outright, completely relinquishing control to the acquirer. Acquisitions can happen at any stage in the biodesign innovation process. If the transaction happens early, the acquirer usually obtains only the rights to the technology and IP. If the acquisition occurs later, the acquirer may obtain all of the assets of the nascent company. The reasons why companies acquire IP, technology, and/or companies vary, but include the desire to obtain a broad portfolio of patents protecting current or anticipated products. They also may seek to expand into new areas or to block areas from the competition. Or, acquirers may choose to make an acquisition when they do not want to develop the product themselves or to track and pay royalties. Acquiring smaller companies with interesting technologies can also be one way that larger companies extend and enhance their R&D efforts. As Richard Gonzalez, former president and COO of Abbott, described: "We have a strategy that basically says we aggressively invest in internal R&D, but we supplement that investment with opportunities on the outside."[10]

In certain acquisitions, where there is still substantial development or adoption risk, an acquirer may not wish to pay the complete value of the deal until certain milestones have been reached. Such an arrangement is generally referred to as an earn-out. An earn-out is a contractual provision stating that the seller is to obtain additional future compensation when certain predefined goals are met (otherwise the earn-out will be forfeited). For example, an earn-out might be linked to the achievement of regulatory clearance, positive results from a particular experiment or test, the achievement of predefined gross sales or earnings targets, or the performance of certain individuals for a specified period of time. As described in 5.3 Clinical Strategy, the founding team of DVI faced a sizable earn-out linked to the successful completion of a critical clinical trial when the company was acquired by Eli Lilly in 1989. An earn-out can provide the innovator with a higher total value from the acquisition, but the total value is completely dependent on the achievement of milestones that may or may not be assured or under the innovator's control. Overall, although earn-outs carry considerable risk for the seller and are often the source of disputes and litigation, they can provide more value to the innovator if the triggers for the earn-out are achieved.

An equity loan paired with a put/call option is another structure that is often deployed when substantial risks remain to be resolved in the future, but the buyer believes that the price desired by the seller is not consistent with the current value. In this setting, the parties agree on some level of funding, via a loan or equity investment, and a prearranged acquisition price based on certain milestones that are believed to be achievable within the funding period. This structure allows the seller to potentially eliminate the issues standing in the way of achieving full value, and also allows the buyer to rationalize the price upon completion of the milestones. This structure has its downside, however, as things usually do not go precisely as planned and the two parties can often be left with a tough situation when the money runs out, but the milestones have not been achieved. Further, venture shareholders tend to be negative about such structures because they view them as limiting their potential return and their ability to drive the value to a maximum; thus, a "capped upside." Regardless, this structure is a viable alternative to consider when there is important strategic value in securing a corporate relationship, and there is also a need for more predefined outcomes.

In an acquisition, both parties must accept the tenet that the price in such a transaction is rarely a perfect approximation of the value of the asset (i.e., the buyer either overpaid or got a bargain, just as the seller was either paid at a premium or a discount). Value to the innovator from a licensing deal more closely tracks the true value of the innovation since royalties are usually linked directly to revenue. In contrast, with an acquisition, the agreed-upon value of the transaction represents a "best guess" made at a fixed point in time. Once negotiated, the outcome remains the same, regardless of how the asset actually performs in the market (unless there is an earn-out, which makes it possible for the value from the deal to better track the true value of the innovation). Despite this uncertainty, companies often try to acquire assets rather than license them in an effort to gain direct control over the asset, and also to shift the cost of the asset off the income statement (where it would be if license payments were to be made).

Building value toward a partnership, license, or sale

Once an innovator decides to investigate an alternate development pathway, it may not be clear at the onset which of these paths will be followed. In many cases, the actual outcome is determined by the negotiations that lead to the deal. For instance, an innovator may approach a company to propose a licensing agreement, but the final deal may take the form of a sale/acquisition. Innovators are advised to stay flexible and consider the advantages and disadvantages of the different financial, ownership, and continued-involvement features of any deal.

Regardless of the development pathway, the innovator must determine the optimal timing for involving another entity. It can be challenging to know how far to develop an idea into a technology or product before licensing it, selling it, or entering into a partnership. Typically, the further one proceeds through the biodesign innovation process, the more risks are eliminated and, thus, the more valuable the asset should be. This, in turn, should lead to more favorable terms in any transaction. On the other hand, each stage of the biodesign innovation process requires the innovator to invest increasing levels of capital and time. Some innovators may prefer to involve another entity earlier in the process to avoid the need to raise significant funding. Depending on the nature of the invention, licensors, acquirers, and partners may also prefer to get involved sooner rather than later, so that they can help shape the direction of product development. For example, a company seeking to acquire new catheter navigation technology may want to purchase it before the design is finalized so that it can be optimized to work with the rest of its own catheter system.

There is no proven formula for calculating the best time to embark on an alternate development pathway; however, innovators can rely on the evaluation of a number of factors to help them determine the optimal timing for their unique technology. These factors, outlined below, represent value in any licensing, acquisition, or partnership deal. While it may not be desirable (or even necessary) for an innovator to address each one, s/he should understand how these factors contribute to retiring risk and, therefore, driving value in a transaction.

People

An innovator should start out by assessing the value that s/he would personally bring to any deal. For example, if an innovator with an idea to improve the treatment of kidney disease is a nephrologist with 20 years of experience, s/he will have less difficulty convincing someone about the clinical merits of a solution than an individual with a general business or engineering background. However, if the technology is further along in its development, licensees, acquirers, and partners may also be interested in issues related to manufacturability or issues that someone besides a physician might be better suited to address. Rarely can one individual speak to all of these concerns. As a result, innovators should perform an honest self-assessment and then surround themselves with others who can help increase their credibility in other areas of importance.

Because individuals who complement the innovator's strengths and weaknesses can add tremendous value to a transaction, the quality of one's employees, advisors, consultants, and board members is important. Advisors and consultants lend credibility and expertise to the innovator. It can be particularly powerful for physician advisors (and customers) to provide testimonials regarding how they have used a technology in their practice and/or its anticipated impact on treatment in the field. The selection of board members is also important and should receive special attention as this is a common area where innovators make mistakes. Just because someone is a leading physician in the field or a trusted advisor does not mean that s/he would make a good board member. Board positions should be reserved for only those individuals to whom the innovator would be willing, at times, to turn over control of the company – for example, when a board vote is required for a critical decision. For this reason, it is important for board members to possess a strong general business perspective, as well as specialized expertise in an appropriate field. Ideally, they will also have previous experience serving on a board and will understand the full scope of the associated financial responsibilities.

When the technology being developed is in a highly specialized field that requires hard-to-find skill sets, an outside entity may be enthusiastic about a large, knowledgeable base of employees and an established set of advisors and board members. This is particularly true for partnerships that require the innovator and his/her team to stay engaged in the ongoing development of a technology. With licenses and acquisitions, however, keep in mind that sometimes outside entities may prefer to pursue technology with limited employee overhead.

Intellectual property

It is essential for an innovator to protect his/her IP position before initiating discussions regarding a partnership, license, or sale/acquisition. An idea or invention has little value if it has not been protected. Without a patent filing and a confidentiality agreement, the idea can be seen as nothing more than a suggestion, which carries with it no IP rights. In contrast, with an advanced IP position, the innovator can realize significant value from a transaction (see 4.1 Intellectual Property Basics and 5.1 Intellectual Property Strategy for more information about patent filings, prior art searches, and freedom to operate analysis).

Market assessment
The extent to which a market assessment increases the value of a deal depends, in part, on the target audience that the innovator approaches. If the innovator hopes to license or sell a technology to the world leader in a given field, that entity surely has more knowledge about that field than the innovator could ever muster. In these cases, it may simply make more sense to complete a survey of 20 or 30 potential customers to demonstrate that there is demand for the innovation. On the other hand, if the innovator has a technology that could take an outside entity in a new direction or require it to look at a market in a different way, then a more extensive market assessment may be useful. The key is to determine what potential partners, licensees, and acquirers know about a market and then tailor the market assessment to complement and expand that understanding (see 2.4 Market Analysis for specific information about what to include in a market assessment).

Prototype/proof of concept
Prototypes and proofs of concept play a central role in eliminating risk, and therefore adding value. In this context, a prototype refers to any non-clinical model that demonstrates the basic feasibility of the technology. Proof of concept generally refers to a more developed version of the device that can be used in a pre-clinical test to demonstrate its benefits (e.g., in an animal or bench-top study). As noted in 4.5 Prototyping, there is no substitute for being able to *show* someone how an idea will actually work. For this reason, having a working prototype is highly recommended for any innovator seeking to license, sell, or partner. Pre-clinical and clinical data is obviously beneficial, but is not always needed depending on the type of deal and the interests of the outside entity.

Plan for reimbursement, regulatory, and clinical advancement
As noted in the Inhale case example, having clinical data can be an important factor in getting an outside entity interested in an idea. However, if an innovator will not have any pre-clinical or clinical data to use as leverage in a transaction, s/he should at least have a well thought-out clinical plan. Similarly, s/he should have a strong understanding of the regulatory and reimbursement pathways associated with the device, potentially including an opinion from one or more regulatory and reimbursement consultants. These assessments give the licensee, acquirer, or partner a realistic evaluation of a technology's risk and, thus, can better allow them to assess its potential value. More information about the types of information to cover in clinical, regulatory, and reimbursement plans can be found in 5.3 Clinical Strategy, 5.4 Regulatory Strategy, and 5.6 Reimbursement Strategy.

Financial model
Like the market assessment, the value of an extensive financial model varies, depending on the nature of the deal. For a well-developed business with employees, other resources, and infrastructure in place, a financial model can be important (and likely would have been developed before such investments were made). For an early idea that has not progressed as far in the biodesign innovation process (e.g., an innovator and one employee working on early-stage prototypes), it may not make sense to spend significant time creating a detailed financial forecast, especially if the innovator is seeking to license or sell the idea to a company that already has an established infrastructure. In these circumstances, the more accurate financial model is determined by the employees, resources, and infrastructure of the buyer itself (following the transaction) than the seller; thus, a model created by the seller may be viewed as relatively useless. For example, an innovator could spend dozens of hours working on a detailed revenue model based on a small team of 20 sales people, but if the acquirer has an established sales force of 150 people, that 20-person model is of little value. When creating a model purely for selling purposes, an innovator should take this issue into account.

Ultimately, the financial model has to fit with the business of the acquirer (or licensee), not the innovator's vision of what s/he would do. Often, the innovator is better served to develop a high-level view of the finances – revenue and cost forecasts – to help justify the opportunity rather than a detailed financial model as described in 6.1 Operating Plan and Financial Model.

Preparation for the process

Before initiating contact with any outside entity, the innovator must be properly prepared. The more prepared one is going into negotiations, the stronger his/her negotiating position will be and the better the result. Innovators who casually enter into negotiations often find themselves feeling "ambushed" by outside entities more experienced at controlling these kinds of interactions.

The first step in preparing is to perform some due diligence on the entities (sometimes called "targets") that are most likely to be interested in entering into a deal. Questions to ask include:

- *Who are they* – What companies have similar or complementary interests that might be helped by the innovator's technology? What is their current position in the market and how will the technology help them (e.g., increase market share, boost revenue, introduce innovation in a stagnant product area, etc.)?
- *What is their reputation* – Which ones are most likely to be open to doing a deal? Which would be best to work with?
- *Who will handle negotiations* – Who within the company has responsibility for licensing, acquisitions, and/or partnerships? What can be learned about this person (or people)?
- *What is their track record* – What kind of deals have they done in the past? How successful have they been?
- *What is the best way to approach them* – Does the innovator have people within his/her network that could potentially make an introduction? If not, what are the other options for making contact?
- *What opportunities are there to create competition among targets* – Are some of the targets direct competitors? Who is chasing whom in the market?

Once a handful of appropriate targets are identified, the innovator should next begin thinking about his/her team – specifically, who the innovator needs to help represent him/her in the process. It is critical for an innovator to have a legal advisor who is closely involved in each step of the process. A business advisor can also be helpful, particularly someone with experience doing licensing, acquisitions, or partnership deals.

Carefully putting together an information packet is also important. Jamey Jacobs, vice-president of research and development for Abbott Vascular in Santa Clara, California advises innovators to think about the information packet as a tool that could be used by a potential target to "jump start" an attractive R&D project.[11] The packet should be complete, detailed, and customized to the unique needs and interests of each target. The more the innovator can do to inspire confidence that the technology/idea can be used by the target from the first day to accelerate a project, the higher the odds of a successful transaction.

When deciding what information to include, confidentiality is, of course, a critical issue. As described in 6.2 Business Plan Development, one effective strategy is to think about what information should be considered confidential and what can be shared more openly. The information that is shared openly should not reveal anything proprietary about the technology. However, it must be compelling enough to get a target interested in doing a deal. Most companies will not sign a non-disclosure agreement (NDA) until they are convinced that an idea is attractive to them, so the innovator must strike a careful balance when deciding what to share. For example, the executive summary of one's business plan (in slide format or another written form) might be what is disclosed without an NDA, while the details that make up sections of the business plan itself might be what can be shared on a confidential basis. This information, too, should be summarized in a slide presentation (also known as the "pitch"). Even more detailed information should be organized and pulled together to support due diligence (actual patents, case-by-case trial results, lists of vendors, copies of contracts signed, etc.), but should not be presented to the target(s) before there has been a formal expression of interest by the company. A letter of intent, which is a written (although non-binding) statement of the partner, licensor, or acquirer's intent to enter into a formal agreement, would be one appropriate way for an outside entity to express interest prior to the initiation of the due diligence process.

Defining a best alternative to a negotiated agreement

With all of this information in hand, innovators can begin thinking about their negotiating position. Specifically, they should invest considerable time determining their best alternative to a negotiated agreement (**BATNA**).[12] In negotiation theory, one's BATNA is the course of action that will be taken if a negotiation fails to lead to an agreement. Once defined, an innovator should never accept a worse outcome than his/her BATNA.[13] In this way, a BATNA is a sort of safety net and can also serve as a point of leverage in negotiations.[14] While the Inhale partnership story spoke to value of understanding one's alternatives, a simplified scenario provides an example of a BATNA in action. An innovator has an offer to license his/her technology for royalties that will equal approximately $1,000,000 per year through the remaining 15 years on the patent. With this as a BATNA, the innovator would likely not accept a price that is less than the net present value of all those payments to sell the company. If an innovator has no other options available at the time of a negotiation (and would not consider developing the idea him/herself), then the BATNA is zero, meaning that almost any prospective deal would be worth considering.

To effectively evaluate alternatives against one's BATNA, innovators are cautioned to take into account all considerations, such as relationship value, time value of money, and the likelihood that the other party will live up to its side of the bargain. Considerations such as these are often hard to assess since they are qualitative and uncertain, rather than easily measurable and quantifiable.[15] However, they can make a significant difference when choosing between an offer and a BATNA.

Importantly, in addition to defining one's own BATNA, an innovator should think carefully about the BATNA of the target. By understanding the outside entity's own best alternative to a deal, the innovator can gain insight into the company's behavior and anticipate how it might react to different demands. Think about what the target's alternatives are, who else could the company work with, and if there is someone else offering a substitute for the innovator's technology that could be pursued. Also consider what would happen if the outside entity simply decided to do nothing. Many times, this provides a target with an easier alternative than those available to the innovator.

Managing an effective process

Once an innovator has completed adequate preparation, s/he can initiate and then proactively manage an effective process. As one innovator put it, "The goal is to drive the process (see Figure 6.4.3), not to let the process drive you."

The first step is to lay out a timeline for all negotiations. This may be driven by external factors (e.g., there is only enough cash to last another six months). But, ideally, it will be initiated when there is an optimal level of value associated with an idea or invention (e.g., positive results from a small animal study are now available, but it would be good to get a deal done before it is time to launch first-in-man testing). In general, partnership, licensing, or acquisition deals can take anywhere from 6 to 18 months, with smaller companies tending to move faster than larger ones.

Within this broad timeline, it is useful to assign dates to important interim milestones. For example, an innovator might seek to make contact with all appropriate targets within 30 days, set aside 45 days for presentations, meetings, and arriving at a letter of intent, allocate another 45 days for due diligence, and then allow 60 days for negotiating terms and closing the deal. Establishing this sort of a schedule allows the innovator to engage multiple targets and keep them moving forward at roughly the same pace to maximize his/her chances of a favorable outcome (assuming the

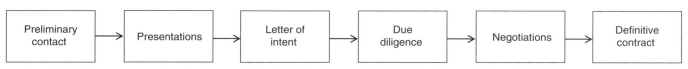

FIGURE 6.4.3
A carefully managed process leads to improved results.

Table 6.4.1 *The financial terms of licensing and partnership deals can be similar – greater variation between these deals exists in the level of control and ongoing responsibility formally maintained by the innovator (compiled from Shreefal S. Mehta,* Commercializing Successful Biomedical Technologies *(Cambridge University Press, 2008): pp. 93–95).*

Financial terms	Descriptions
Upfront payment	A lump sum paid by the licensee/partner to the innovator at signing.
Patent prosecution and maintenance fees	The licensee/partner may be asked to pay legal and USPTO fees for maintaining the patent.
Milestone payments	If the technology succeeds at further stages of development, the licensee/partner may be required to make additional payments to the innovator as risk is reduced. Milestone payments may be variable (e.g., if the outcome is good, the innovator receives $100,000; if the outcome is bad, the payment is reduced to $25,000).
Royalties	A percentage of commercial sales paid by the licensee/partner to the innovator. Royalties can scale upwards or downwards based on the volume of sales (e.g., if sales reach 500,000 units per year, the innovator receives a rate of three percent; if they reach 1,000,000, s/he received five percent). They can also have minimums and caps (e.g., regardless of sales, royalties will not fall below $500,000 per year or exceed $1,000,000 per year).
Equity considerations	Licensees/partners may purchase equity in the innovator's company as part of its payments and/or the innovator may accept equity in a licensee/partner's company in lieu of cash payments.
Sublicensee and sublicense fees	If a patent is sublicensed to another party by the licensee, then a portion of the sublicense payments made to the licensee may be passed through to the innovator as the original licensor. These payments can be structured as a flat fee or a percentage of sales. (Note that this particular term would not be relevant to a partnership agreement.)
Royalty anti-stacking provisions	If a licensee has to license other patents to get a product to market (e.g., a stent that requires the licensing of both a drug and a delivery mechanism), it may be economically infeasible for the licensee to pay full royalties to both the innovator (as the original licensor) and a separate, secondary licensor. Anti-stacking provisions provide a mechanism for the licensee to reduce the burden of dual (or stacked) licensing fees. (Note that this particular term would not be relevant to a partnership agreement.)

timeline is realistic). Without a timeline, innovators too often find themselves having to decide on one offer before the other targets have had enough time to determine if they are interested. In a best-case scenario, the innovator will receive multiple offers within days of one another, with the targets aware of the competition they are facing in trying to partner, license, or acquire the idea.

With a realistic timeline defined, the innovator can begin accessing the targets to determine which ones are interested in engaging in the process. Introductions are usually followed by lengthier, more detailed conversations. Due diligence is initiated once interest has been formally expressed by one or more targets. Then, for those who are most serious, formal negotiations ensue, a partner, licensee, or acquirer is eventually selected, and the deal is closed.

Components of a transaction

While acquisitions are treated somewhat differently, there is significant overlap in the common financial deal terms for licenses and partnerships. A typical medtech license or partnership agreement may include some (but not necessarily all) of the terms outlined in Table 6.4.1.

The following case example on Rapid Exchange highlights the way the concepts presented in this chapter work together to result in a successful licensing deal.

FROM THE FIELD

RAPID EXCHANGE

Licensing a new technology in an established field

Paul Yock was a fellow in angioplasty at Sequoia Hospital in Redwood City, California during the early days of this ground-breaking procedure. In angioplasty, a balloon catheter is inserted through an artery in the patient's groin or arm, carefully moved up the aorta using a long metal guidewire, and then dilated to open a blood vessel that is blocked by plaque. "There was 4 ½ feet of balloon catheter and, in order to move the balloon in and out of the heart without losing track of its position, the guidewire had to be over two times as long to allow the wire to stay in and the catheter to come out," explained Yock, making this a two-person procedure. "The more skilled physician was in charge of moving the balloon up and down on the wire, and the less skilled person, positioned down at the patient's feet, had to keep the wire stable. The junior member was supposed to compensate for the movement of the senior person to keep the wire in the same position," he continued. "The trouble was that this was a ten-foot long wire and moving the tip even half an inch could be a problem. So it required a dance, a coordination that was very awkward." As a trainee at the time, Yock recalled spending considerable time on what he referred to as "the wrong end of the procedure."

"The senior physicians didn't really perceive it a problem area because they had graduated beyond thinking about it," he said.[16] "But, in retrospect, it was kind of a glaring need." After inevitably losing the guidewire position in a couple of patients during his fellowship, Yock started thinking about better device designs. He made drawings and then developed a prototype rapid exchange catheter. Ultimately, this led him to the development of a technology that became know as Rapid Exchange (RX)™ balloon angioplasty system (see Figure 6.4.4). RX allowed one physician (using two hands) to perform the entire procedure.[17]

Yock was fortunate to be training under John Simpson, a renowned cardiologist, prolific inventor, and founder of the leading producer of catheter systems at the time, Advanced Cardiovascular Systems (ACS). He shared the RX idea with Simpson who, in turn, referred him to a patent attorney. "At that time," Yock noted, thinking back on his days as a young physician, "paying for the

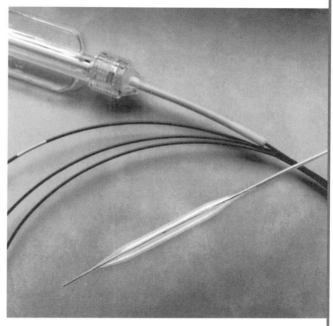

FIGURE 6.4.4
Guidant's ACS RX Comet VP™ Coronary Dilatation Catheter is one example of a balloon angioplasty system that leveraged Rapid Exchange technology through a licensing deal (courtesy of Abbott).

patent was an issue for me. But I decided to go ahead and do it." Yock concluded that the best approach for commercialization was to license the technology, and targeted Simpson's company ACS. "To tell you the truth, I never thought seriously about starting up a new company around RX. I was intent on becoming an interventional cardiologist and had no interest in taking a different pathway." Instead, Yock saw ACS as an exciting company that was emerging as a powerful player in the angioplasty field. He also viewed it as the ideal candidate to license the RX technology. "With the old style, the over-the-wire systems, you introduced both the balloon catheter and the guidewire together. With my system, the guidewire could go in first, without the catheter. That required really good guidewire performance." Yock had paid careful attention to the other companies in the field and felt that ACS had the best guidewire performance and overall positioning to benefit from RX. And, "The fact

Stage 6 Integration

FROM THE FIELD — **RAPID EXCHANGE**

Continued

that they were a local company was a nice coincidence," he added.

Yock had previously met Carl Simpson, the vice-president of R&D at ACS (no relation to John Simpson) and arranged a meeting with him to introduce the RX concept. Prior to his initial meeting, Yock made arrangements to put a non-disclosure agreement (NDA) into place. "I didn't have my own NDA," Yock said. "I talked with Carl ahead of time about what the non-disclosure would be and he provided the company's standard two-way version." Yock also took an extra precaution regarding his disclosure. "I was aware that what they were working on could potentially be influenced based on the discussion. So I came in with a written document that outlined my ideas of examples for how this catheter could be used," he recalled. Carl Simpson signed this document, acknowledging the contents as belonging to Yock in advance of the meeting.

With the disclosures taken care of, Yock demonstrated the RX system using a heart model and his prototype. Simpson was immediately intrigued. According to Yock, "Carl was excited about the system so he called in the VP of sales, who just happened to be down the hall. As soon as the VP of sales saw how it worked he said something like, 'Yep, I can sell that.'"

Just a few weeks later, Yock was invited back to meet with Ron Dollens, ACS's president and CEO at the time. ACS had done some of its own due diligence (including a thorough IP assessment) and was ready to negotiate licensing terms. The company proposed an agreement that would tie Yock's royalty payments to the system's sales volume. As sales increased, the royalty rate would be ratcheted downward, with the increase in volume compensating for the lower royalty percentage at each predefined sales threshold. He asked the company to cover the patent expenses, which it agreed to do. "There was also an indemnification of me, as an individual, for product-related malfunctions. That was extremely reassuring. I didn't think to bring that up, but I was happy to see it in the contract," he commented. In exchange, Yock exclusively licensed his patent to ACS and agreed to cooperate in any RX patent litigation incurred by the company. "There was also language that if they sublicensed the patent, I would get a royalty stream from it, which turned out to be important," he said, as RX was eventually sublicensed by multiple companies.

Yock described the fact that the licensing process went so quickly and so smoothly as "remarkable" and admitted that it was "somewhat unrepresentative" of what most innovators face. "These days, companies are really sophisticated in dealing with inventors," he said, which makes it much more difficult for those interested in licensing their technology to enter into mutually favorable agreements. "A lot of power sits with the companies, so the inventor has to create an incentive for them to want to move quickly and aggressively. That will happen spontaneously if the invention is good enough," noted Yock. However, in some cases, it helps if the inventor can orchestrate a competitive situation by approaching more than one company at a time with his/her technology.

Another difference in the current medtech environment is that companies typically want to have more risks eliminated before licensing an idea. "I was extremely lucky to be able to license RX just off a prototype," noted Yock. "Now you have to be prepared to take your invention further, typically into animal testing at a minimum." Beyond clinical and device feasibility issues, he underscored the value of addressing other workstreams in the biodesign innovation process – such as regulatory, reimbursement, sales – as a mechanism for eliminating risk for the licensee.

Of course, he pointed out, there are also potential risks to the licensor: "The biggest pitfall, I think, is that a company licenses your idea and does nothing about it. At the time they license it, they may have every good intention of developing the technology. But if it doesn't stay on the priority list, nothing is going to happen. So having some form of milestone – a certain stage reached at a certain time – in the contract is really important, if you can get it." Examples of milestones include FDA approval or first commercial sale. Even in the case of RX, Yock had to stay closely involved through the first year or so of the licensing contract to urge the development

> process forward. He also acted as an evangelist of sorts to help garner stakeholder interest in the technology. According to Yock, "Initially, it just wasn't clear whether people were going to like it or use it. It seems strange in retrospect, but it was a major change of practice. We had push-back from unexpected places. For example, people were worried that if we took the trainees out of the picture, because it was a single operator system, it would be harder to develop the next generation of interventionalists."
>
> Fortunately, Yock and ACS were able to overcome these challenges and RX went on to become the dominant balloon angioplasty system used on a worldwide basis.

For partnership deals, there are other terms that must be considered in the development of the agreement, including those that provide a precise description of the near-term development activities to be undertaken, specific responsibilities and cash flows for each party, a list of milestone events and proposed timing, resources or activities to be provided by both parties, and a management/oversight structure. Usually upon the completion of a phase (e.g., development) or at the termination of the agreement, there may be a decision point for one or both parties regarding whether or not they will continue the partnership and, if so, what the next steps will be. It is possible for partnership activities to continue from one phase to another (e.g., development into commercialization). However, in most cases, one of the parties assumes responsibility for taking the product into commercialization in predefined markets. Similarly, one of the parties assumes responsibility for manufacturing. While arrangements such as these are relatively common, each agreement is unique, with the terms defined based on the interests of the partners and the reasons why a partnership makes sense to them.

A note on licensing when development occurs in an academic context

Innovators working in an academic environment should keep in mind that they may need to enter into a licensing deal, regardless of the development pathway they choose. This is because the university with which they are affiliated is likely to have ownership of their ideas and inventions. Typically, the title to any patentable inventions conceived or reduced to practice (in whole or in part) by faculty, staff, or students of the university is assigned to the university, regardless of the source of funding.[18] What this means is that innovators may be required to license their inventions from the university if they intend to make, use, or sell the technology.

Every major university has an office of technology licensing (OTL), which is responsible for managing its IP assets. Typically, the goals of an OTL are to promote the transfer of technology from the university into practice, as well as to generate income to support ongoing research and education at the institution. The important thing for innovators in a university setting to know is that they must disclose their inventions to the OTL at their academic institution and work collaboratively with the office to reach a licensing agreement *before* developing a stand-alone business or entering into a sublicense, acquisition, or partnership based on the technology in question. The Spiracur example below demonstrates how innovators can reach mutually beneficial agreements with an OTL.

FROM THE FIELD — SPIRACUR

Negotiating technology licensing with a university

As referenced in 4.1 Intellectual Property Basics and 5.1 Intellectual Property Strategy, Moshe Pinto, Dean Hu, and Kenton Fong joined forces to address the need *to promote the healing of chronic wounds*. Because the invention was developed while they were still students at Stanford, they needed to work through the university's OTL to submit their preliminary provisional patent application and negotiate terms surrounding the licensing of the invention.

In approaching these negotiations, Pinto emphasized the importance of "preparation, preparation, preparation." The Spiracur team believed that it achieved a positive outcome by carefully understanding the motivations and needs of the OTL and being prepared to offer mutually beneficial solutions. "To give you an example," Pinto said, "we had to negotiate a royalty rate, which the OTL told us usually ranged from one to five percent. Obviously Stanford wanted to start at the high end of this range, but we ended up closer to the one percent. What we did to justify that was build a *proforma* model with financial statements that showed the typical margins in this marketplace. We used competitive analysis from established players selling such technologies as comps and made a compelling case that we wouldn't be able to have a viable business if we paid such a high royalty rate. The people at OTL were reasonable. They looked at what we did and felt comfortable based on the data we presented to them that they would be justified in agreeing to a lower royalty rate."

Pinto also underscored the point that a university such as Stanford is not "an economic animal." "It has a broader interest in the community that it factors into its decisions," he said. "One thing that was clear to us was that the folks at the OTL were not interested in maximizing profits and making money for Stanford. They wanted to see the technology disseminated to basically make the world a better place. When we figured this out, we started thinking about how we could accommodate this interest while still maintaining the exclusivity of the IP." Ultimately, Spiracur proposed a creative solution under which it would make the technology available at cost to charitable organizations, such as the Gates Foundation, the Ashoka Foundation, and the United Nations Health Program, working in developing countries. It would also offer the same arrangement to five African governments. This agreement would take effect in 2012, when the product was fully developed. "We know that every 30 seconds someone has a limb amputated because of a diabetic foot ulcer somewhere in the world, primarily in Africa," Pinto said. "But this is not our natural target market. These folks can't afford to buy the solution. So this agreement would satisfy the altruistic needs of Stanford and feel very good for us, without affecting our anticipated revenue stream for investors."

The key, underscored Pinto, is to recognize that the interests of the university and the interests of the team are not necessarily at odds. If you take a more collaborative approach, he noted, "You can find a middle ground."

GETTING STARTED

While deciding whether or not to take an alternate development pathway is a major strategic decision, it does not have to be a scary one. The following process, along with some honest evaluation, can help the innovator work through this important yet potentially exciting determination. It also lays out a straightforward course of action for making a partnership, license, or acquisition happen.

See **ebiodesign.org** for active web links to the resources listed below, additional references, content updates, video FAQs, and other relevant information.

1. Determine whether or not to pursue an alternate development pathway

1.1 **What to cover** – Carefully consider whether or not it makes sense to pursue a stand-alone business given the innovator's personal interests, skills, and circumstances. From a product perspective, evaluate if the market cannot reasonably be accessed, the product can stand on its own, and the alignment between the required investment to develop a product and the expected return of the most likely investors. Based on the

6.4 Licensing and Alternate Pathways

results of this analysis, determine if an alternate development pathway is warranted and, if so, which pathway appears most promising.

1.2 Where to look – Revisit 1.1 Strategic Focus and refer back to the results of the proxy company analysis performed as a part of 6.1 Operating Plan and Financial Model and 6.3 Funding Sources.

2. Assess the best timing for a transaction and build value toward that goal

2.1 What to cover – Determine the optimal timing for involving another entity. Evaluate the value that the innovator and his/her idea bring to a potential transaction, including an assessment of people, IP, the market, the technology (prototypes and proof of concept), plans for clinical trials, regulatory, reimbursement, and the financial model. Begin building value in the areas of greatest interest to prospective targets.

2.2 Where to look
- Refer to competitive analysis performed as part of 2.4 Market Analysis.
- Review the annual reports of publicly traded companies in the field for details on deals that closed.
- Perform web searches for information about the timing of comparable deals.

3. Identify and evaluate targets

3.1 What to cover – Identify the most promising targets and perform detailed due diligence on them. Research products, their experience in doing deals, their reputations, gaps in their product offerings that may make a deal viable, key contacts, how to gain an introduction, and how to create competition when it is time for the transaction.

3.2 Where to look
- Review company profiles via databases such as Hoover's Online, OneSource's U.S. Business Browser, Million Dollar Database, and Reference USA.
- Read analysts' reports for publicly traded companies that could be potential targets.
- Seek input and ideas from members of the personal networks belonging to the innovator, board members, employees, and other advisors.
- Attend trade shows and conferences and visit booths of potential targets.

4. Prepare to initiate contact

4.1 What to cover – Put together the right team to manage the process and perform negotiations. Prepare an information packet with confidential and non-confidential disclosures that can be shared in a progressive manner. Put together an NDA and have it ready to go. Think carefully about an appropriate BATNA.

4.2 Where to look
- *Getting to Yes: Negotiating Agreements Without Giving In* by Roger Fisher, William Ury, and Bruce Patton, (Houghton Mifflin Books, 1991).
- If getting a personal introduction is not possible, make arrangements to attend trade shows or other professional networking events where targeted companies will have managers and executives in attendance.

5. Manage the process

5.1 What to cover – Lay out a timeline for all critical activities and set milestone dates. Make contact with targets and communicate the timeline. Proactively manage targets through the process, including introductions, presentations and meetings, expressions of interest, detailed due diligence, negotiating terms, and closing the deal.

5.2 Where to look – Make every effort to keep the process and timeline in synch with critical milestones in the company's operating plan as described in 6.1 Operating Plan and Financial Model. Obtain advice from trusted advisors and legal counsel.

GETTING STARTED

Credits

The editors would like to acknowledge Robert Chess of Nektar Therapeutics and Moshe Pinto of Spiracur for their assistance with the case examples. Many thanks also go to Fred Khosravi of Incept LLC for his contributions in reviewing and editing the chapter.

Notes

1. "Astrazeneca and Cambridge Antibody Technology Announce Major Strategic Alliance to Discover and Develop Human Antibody Therapeutics in Inflammatory Disorders," AstraZeneca International, November 22, 2004, http://www.astrazeneca.com/pressrelease/3878.aspx (September 16, 2008).
2. "AstraZeneca: CAT's in the Cradle," *Pharmaceutical Business Review*, May 16, 2006, http://www.pharmaceutical-business-review.com/article_feature.asp?guid=B4A56667-EBAA-45BD-AE84-12A7CABB362E (September 16, 2006).
3. From remarks made by Steve MacMillan as part of the "From the Innovator's Workbench" speaker series hosted by Stanford's Program in Biodesign, May 21, 2007, http://biodesign.stanford.edu/bdn/networking/pastinnovators.jsp. Reprinted with permission.
4. Leila Abboud, "Trial and Error: How Eli Lilly's Monster Deal Faced Extinction – But Survived," *Wall Street Journal*, April 27, 2005, p. A1.
5. Ibid.
6. All quotations are from interviews conducted by the authors, unless otherwise cited. Reprinted with permission.
7. Shreefal S. Mehta, *Commercializing Successful Biomedical Technologies* (New York: Cambridge University Press, 2008), p. 88.
8. While many licensing deals focus on patents and patent rights, it is important to recognize that technology development rights, product distribution rights, and brand names can also be licensed.
9. Ibid., p. 89.
10. From remarks made by Richard Gonzalez as part of the "From the Innovator's Workbench" speaker series hosted by Stanford's Program in Biodesign, June 9, 2006, http://biodesign.stanford.edu/bdn/networking/pastinnovators.jsp. Reprinted with permission.
11. From in-class remarks made by Jamey Jacobs as part of Stanford's Program in Biodesign, Winter 2008. Reprinted with permission.
12. Roger Fisher, William Ury, Bruce Patton, *Getting to Yes: Negotiating Agreements Without Giving In* (Houghton Mifflin Books, 1991).
13. Ibid.
14. "Best Alternative to a Negotiated Agreement," Wikipedia.org, http://en.wikipedia.org/wiki/Best_alternative_to_a_negotiated_agreement (August 21, 2008).
15. Ibid.
16. Jeffrey S. Grossman, *Innovative Doctoring* (2006).
17. Burt Cohen, "Four Minus Two Equals Two – Hands, That Is," The Voice in the Ear, October 5, 2005, http://www.ptca.org/voice/archives/2005_10.html (August 5, 2008).
18. "Our Policies," Stanford University Office of Technology Licensing, http://otl.stanford.edu/inventors/policies.html#patent (August 21, 2008).

Acclarent Case Study

Stage 6 Integration

With strategies defined to address the many different aspects of Balloon Sinuplasty's development and readiness for commercial launch, Acclarent faced the challenge of integrating all of these plans into a unified vision of the business.

6.1 Operating plan and financial model

According to Lam Wang, a company's operating plan and financial model "has to do with the culmination of all of your assumptions, and then how you intend to go about executing your plans." In developing Acclarent's model, "we based a lot of it on prior experience," she said. However, the team also tried to benchmark proxy companies to validate its approach. "When you're trying to do something that's never been done before, like what we have set out to do in ENT, there really aren't a lot of models to look at," explained Facteau. "We had no idea what the adoption of the technology would be. In the early days, we struggled with how fast we should be training doctors, how fast we should expect them to turn into customers, and at what rate they should be reordering. We really had no peer data to look at."

Investigating companies outside ENT turned out to be more fruitful. "We probably looked at ten companies: Fox Hollow, Venus, Kyphon, Perclose, and others. Our CFO did a phenomenal job of laying out all their operating expenses and revenues," Facteau remembered. The team assessed these benchmarks from a time perspective (e.g., zero to three years), but then also in terms of revenue thresholds ($20 million and under, $21 to 50 million, etc.). These revenue bands allowed Acclarent to determine where (and when) the companies made major infrastructure investments and incurred certain costs. "Then we drilled down one more layer and looked at their business model. Is it similar? Is it disposables? Is it a market development type situation or is it a market share stealer?" Facteau added. "It wasn't perfect, but it gave us a way to plan and also to keep a report card of how we were doing as we moved forward. We also used this information to justify our operating plan and financial requirements to the board. We said, 'If this is the type of growth you're expecting from our company, these are the types of companies that have done it, and this is the investment that it took in order for them to get there. So if you believed those companies were a success, then you should be prepared to fund us at this level and we'll execute.'"

In terms of developing the model, Lam Wang described key aspects of the approach:

> We assumed it would be a surgeon training model, meaning that the way the product would be introduced to the field and generate revenue was through surgeon training. So I modeled that there would be national, regional, and local courses, meaning large, medium, small, and at each of these courses there would be so

many doctors per course, per month. Once they were trained, we assumed their business would increase at a certain case ramp rate. The first month they were trained, we assumed they would do no cases, but then it would start to build until they reached some sort of steady state. To calculate a ramp rate, I looked at what were they doing at the time with FESS and then made some more assumptions, like they would be partly converting cases from conventional FESS and partly drawing from the pool of nearly 600,000 new patients to get to some steady state. After 12 months, they would reach a steady state. So that's how the model would grow. But then I added some reality factors, like surgeon productivity. Even in the best case scenario, maybe only half of the surgeons would be actually performing the way they should. There was also some seasonality in chronic sinusitis that had to be taken into account. Then there was the competition factor. So for all the cases we thought we were going to get, we had to consider that we might lose some portion to new technology in the future. Taking all of these factors into account got us to a monthly number of cases, then we multiplied that by the $1,200 price to get to revenue per month. From there, we could estimate the expenses associated with growing that top line and the staffing that would be required.

Among other things, the financial model was used to determine how much money Acclarent needed to raise to get through the product's preliminary commercialization (approximately $14.5 million). Key near-term milestones in the company's operating plan included receiving FDA clearance for the balloon technology, launching a post-marketing clinical study, and building up its commercial sales force once it verified safe and efficacious results from this study.

Reflecting on the experience, Facteau commented that, "The model is only as good as the information that goes into it. Therefore, you really have to hone in and focus on the assumptions you're making. Challenge yourself to benchmark other companies. And be honest with yourself and your team's ability to execute."

6.2 Business plan development

Once the operating plan and financial mode were developed, Acclarent was able to assemble it with the other information it had developed regarding the clinical need, the market opportunity, its technology, and its chosen IP, regulatory, reimbursement, and sales/training pathways. These materials were then used as the basis for important communications with investors, stakeholders, and other contacts. "These presentations included the main guts of a business plan," Facteau said. "But we never formally developed a plan or sent it out. Remember that we were in stealth mode this entire time, so all of our external communications were really targeted."

6.3 Funding sources

When it came to funding, Acclarent was in an enviable, although rare, position. As part of ExploraMed II, the entity had received all of its seed funding from ExploraMed to get its efforts off the ground. When it was time to raise additional capital, Makower was able to tap into his established relationship with venture capital firm NEA relatively quickly. He and Facteau easily closed a Series A round of funding ($14.5 million) in January, 2005. This was the only fundraising the company completed prior to the commercial launch of its technology. Investors Versant Ventures and NEA shared the round.

Importantly, the company's fundraising efforts corresponded to the completion of its first-in-man clinical study. The positive outcome of that trial was used to help pique investor interest and build confidence in the viability of the technology. Despite this careful timing, Facteau emphasized how fortunate the company was in so readily gaining access to the funds it required: "The credit has to go to Josh on this one." He continued, "At the end of the day, you need to find people who know what they're doing in the fundraising world. Whether you put them on your board, bring them into your company, or have some other type of mentor relationship with them, it's invaluable. You need someone to help you understand what it's going to take to raise money and what investors are going to look for. Josh played this role for us and it was really beneficial."

Into commercialization: what happened?

Approximately one month after speaking with representatives from the FDA by phone, Acclarent received notification that its balloon technology had received 501(k) clearance (see Figure C6.1).

Achievement of this important regulatory milestone set into motion a whirlwind of activities as the team began aggressively pressing forward toward a September, 2005 commercial launch at the upcoming AAO-HNS conference.

One of the first things Acclarent did was formally launch its next clinical trial, the CLEAR study. Based on the legwork it initiated immediately following its first-in-man results, the company already had several IRB approvals lined up and could immediately begin enrolling patients. According to Facteau, by the time of the commercial launch, "We already had phenomenal results, having treated 75 patients at nine sites with zero complications and a significant improvement on quality of life."

Facteau also gave the green light for hiring a sales team. His first move was to bring on a vice-president of sales and marketing, who joined in July, 2005. "I really put him in a very difficult situation," Facteau recalled. "I said, 'Alright, go hire a training organization, a marketing organization, a sales management team, and a sales organization by September.'" By the launch, the company had added nine sales representatives, but was still in the process of assembling a complete team. To help these new members rapidly come up to speed, said Facteau, "We challenged everyone in the organization to observe cases [being performed as part of the CLEAR trial], to learn from each other, and to commit to the standardized approach so we could train the same way, every single time."

FIGURE C6.1
In April, 2005, Acclarent received FDA clearance on the core component of its Balloon Sinuplasty system (provided by Acclarent).

```
DEPARTMENT OF HEALTH & HUMAN SERVICES            Public Health Service
                                                 _____
                                                 Food and Drug Administration
                         APR 5  2005             9200 Corporate Boulevard
                                                 Rockville MD 20850

ExploraMed II, Inc.
c/o William M. Facteau
President & CEO
2570 W. El Camino Real
Suite 310
Mountain View, CA 94040

Re: K043527
    Trade/Device Name: Relieva Sinus Balloon Catheter
    Regulation Number: 21 CFR 874.4420
    Regulation Name: ENT manual surgical instrument
    Regulatory Class: Class I
    Product Code: LRC
    Dated:    March 24, 2005
    Received: March 25, 2005

Dear Mr. Facteau:

We have reviewed your Section 510(k) premarket notification of intent to market the device
referenced above and have determined the device is substantially equivalent (for the indications
for use stated in the enclosure) to legally marketed predicate devices marketed in interstate
commerce prior to May 28, 1976, the enactment date of the Medical Device Amendments, or to
devices that have been reclassified in accordance with the provisions of the Federal Food, Drug,
and Cosmetic Act (Act) that do not require approval of a premarket approval application (PMA).
You may, therefore, market the device, subject to the general controls provisions of the Act. The
general controls provisions of the Act include requirements for annual registration, listing of
devices, good manufacturing practice, labeling, and prohibitions against misbranding and
adulteration.
```

As the AAO-HNS conference approached and the team's launch plans began to come together, Facteau, Makower, and the team had to evaluate their stealth strategy and decide when (or if) the time was right to break their silence. Ultimately, they decided to contact a targeted list of key opinion leaders in the field to contact prior to the launch. Makower explained:

> We identified an important group of physicians in the specialty that we didn't want to surprise by our launch. We wanted them to be able to say to their peers that they knew about what we were doing if they were asked and ensure they wouldn't be caught off guard on the floor at the conference. We approached them a month in advance to explain the technology and share our clinical experience. Unfortunately, many of them were insulted by the fact that they hadn't been involved sooner and somewhat in disbelief that we had moved so quickly to be in a position to show up with clinical and cadaver data.

By the time the conference arrived, Acclarent was hit with "a tidal wave of interest." Chang described the first morning of the show:

> The exhibit floor opened at 9:30 so we met around 8:30 at the booth to give a pep talk and to make sure we were all prepared. Our booth was in the very back of the hall, literally by the bathrooms and the concessions. We were as far away from the entrance as possible. At 9:30, we finished our huddle and I picked up my briefcase to move it out of the way. By the time I put it down and looked up there were at least 40 people at the booth already. It just hit me. Clearly these people had sought us out. They had not meandered through the exhibit hall. They wanted to come and find out exactly what was going on and made a bee-line for our booth.

The team faced crowds of interested physicians throughout the entire conference. It was not until the company was formally approached by representatives of the AAO-HNS that the team's excitement turned to other emotions. "There were people from the Academy and certain committees that came over to our booth and basically said 'Your products are not covered by existing codes. This is different, and we will publish a position statement recommending the use of miscellaneous codes,'" recounted Facteau. Under the use of miscellaneous codes, physicians and facilities would not receive any reimbursement for performing sinus surgery that involved the use of the Balloon Sinuplasty products.

Facteau described what happened next:

> After that, the process just took on a life of its own. Within 60 days there was a dedicated task force that was formed by the president of the AAO-HNS at the time to evaluate our technology. Dr. David Kennedy – who was very influential and well connected in the field – headed up this task force, which included four or five other people. Dr. Kennedy had been one of the individuals responsible for establishing the FESS codes and their values. In parallel, members of the Academy took a very aggressive approach and began calling all of the trained doctors on our website and telling them that they did not recommend the use of existing codes, which was unprecedented in ENT at the time. In response, we did everything that we thought we could do. We really tried to communicate with some of the leaders in the Academy to better understand their concerns and explain our position. We immediately set up a meeting with CMS to get its opinion on the matter. Fortunately, CMS said 'We agree with you. This is a device. It's a tool that's used in sinus surgeries. This technology has been around in other specialties. We don't create new codes for devices and tools, per se, we do for new procedures. Whether you create an opening using a shaver, or back biter, or your balloon catheter, it doesn't matter. The codes are broad enough to interpret the use of any of those tools.'

Acclarent also invested significant time in formulating responses to refute the concerns and educate its strategic advisors and trained physicians that their position on coding was sound. "We spent a lot of money and resources on experts reviewing the language of the codes and the vignettes to make sure we weren't

FIGURE C6.2
Facteau (left), Bolger (center) and Dr. Fred Kuhn (right) review documents and discuss ideas on how to respond to the coding debate that ensued at the AAO-HNS in Los Angeles, 2005 (courtesy of Acclarent).

overlooking anything, after multiple confirmations that the existing codes apply we were able to respond with conviction and not take 'no' for an answer," said Facteau. We could not allow a few politically motivated individuals to dictate our future, because it could have fundamentally killed the company. So we persevered and did what we believed was the right thing to do for our physicians, patients, and the company. See Figure C6.2.

At the crux of the Academy's argument was the concern that using FESS codes to reimburse sinus surgery performed using Balloon Sinuplasty products would potentially lead to the devaluation of FESS reimbursement. Some individuals perceived that FESS would be significantly easier to perform with the new tools, which would affect reimbursement work values. Opponents also speculated that the new technology would cause surgeons to operate on patients sooner than they really should. This could, in turn, affect utilization and cause the codes to be revalued.

Another common concern was that radiologists and cardiologists would begin performing sinus surgery. "Could that happen?" asked Facteau. "Maybe. But the likelihood is very, very small." He elaborated:

Our counter position was, 'Listen, guys, this is great technology that's going to benefit a lot of patients. If you don't embrace it, others will.' There are many examples where that has happened in medicine before. Cardiology is one. Cardiovascular surgeons could have been the ones performing angioplasty in the early days, but they made a conscious decision that they were surgeons, they cut tissue, they do surgery, and they opted not to embrace interventional techniques. As a result, there was an opportunistic group that developed – called interventional cardiologists – who were ready to take advantage of their decision. So we thought it was a valid argument, and so did our advisors. They said, 'Sinus surgeons deserve to use this technology. We're the ones that need to embrace it. We need to research it. We need to study it. And we need to determine where it fits, not someone else.' So, this coupled with the fact that we were trying to build a dedicated ENT company eventually got us through that concern.

"We had to respond to every one of these arguments," Facteau continued. "We had to organize our thoughts so that our answers were consistent, well thought out, and well articulated as to why we didn't think these things weren't going to happen. We also spent a lot of time saying, 'This is technology that is going to make a positive difference in patient lives. We have to figure out a way to make it stay here.'"

Through this process, Acclarent's advisors and trained physicians became important evangelists – a role that Acclarent appreciated but had not initially expected them to play. "Our advisors probably found themselves defending Balloon Sinuplasty and Acclarent more than they ever wanted to. But they were just trying to do the right thing for the patient and the specialty as a whole."

According to Facteau, the task force to investigate Balloon Sinuplasty was formed in November, 2005. Its position statement, dubbing the procedure as experimental and recommending the use of miscellaneous codes, was released in the Academy's January, 2006 bulletin. "It had some pretty disparaging comments in there, as well, about the company from a marketing perspective," Facteau said. "It was pretty alarming and obviously very, very difficult to deal with at the time." Circumstances worsened for the company when the *New York Times* published a critical article in May, 2006. In October, 2006, the American Rhinologic Society (ARS – a subdivision of AAO) published its own unfavorable position statement on the technology.

The struggle and debate raged on, in both open and closed forums, at every sinus surgery meeting and committee meeting from that point onward. Despite the controversy, Facteau, Acclarent, and the doctors who believed in the technology and its place in sinus surgery did not give up. It was not until March, 2007 that the physician users of Balloon Sinuplasty (numbering in the hundreds at that point) were finally successful in getting the AAO-HNS to overturn its position. See Figure C6.3.

The ARS followed in May, 2007 with a positive statement of its own. These reversals in opinion were primarily driven by several thought-leading physicians who took it upon themselves to fight for the technology, as well as the persistence of Acclarent and the strength of the clinical data that it continued to amass. In September, 2006, the company presented the results of its CLEAR study at the AAO-HNS conference. It also hosted a symposium at that meeting, which included a live case demonstration (more than 200 doctors attended). Through ongoing clinical work, a focused publication strategy, and the sharing of information between physicians, Acclarent slowly began changing surgeons' minds.

However, the increasing acceptance of Balloon Sinuplasty among sinus surgeons had little impact on the attitude of insurance companies. While the professional societies initially had failed to support the balloon as medically necessary, most private payers flagged the technology as investigational, which essentially meant it would not be covered on grounds that there was insufficient clinical data to support it. Eventually, as the clinical evidence mounted many began reimbursing sinus surgery performed with Acclarent's tools, but this was not without tremendous hours of effort, physician lobbying, and more painful persistence. But, said Facteau in 2008, "To this day we are still fighting to get certain private insurance companies, like the Blue Cross/Blue Shield National Association and Wellpoint, to eliminate their investigational status. We've done over 50,000 procedures. 150,000 sinuses have been treated with our technology with virtually no complications associated with the technology, yet they still say it's investigational."

Reflections on Acclarent's past and future

Acclarent progressed through the biodesign innovation process and into the market at breakneck speed.

Sinus balloon catheterization is a surgical technique for the treatment of sinusitis, during which a wire-guided balloon catheter is inserted into the paranasal sinus and inflated in order to dilate the targeted ostium.

The board-approved policy declares that the evidence regarding the safety of sinus balloon catheterization has been supportive, and that balloon catheterization is a promising technique for the treatment of selected cases of rhinosinusitis. These include those without polyposis involving the frontal, sphenoid or maxillary sinuses either in conjunction with or in place of conventional instrumentation.

FIGURE C6.3
From the AAO-HNS position statement on Balloon Sinuplasty issued March 19, 2006 (from the American Academy of Otolaryngology – Head and Neck Surgery).

According to Facteau, "We moved from concept to commercialization in 18 months, which is pretty unusual in the device industry." Reflecting on the company's experience, he said:

> Being in the medical device business, I had anticipated that we would face challenges albeit I thought they would come from clinical, regulatory, or other technical workstreams – would the devices work, could we make them, and other issues that typically come with getting a device to market. But these things we sailed through surprisingly easily. The challenges came from other places – politics, adoption, reimbursement, and the societies. I just didn't anticipate those issues. The key was to respond to them with facts and data and stay focused on overcoming them. These issues definitely slowed us down. As a result, we had to raise more money than we initially anticipated, and make investments in areas that we hadn't contemplated.

When asked how the company's decision to adopt stealth mode may have affected the company's launch, he replied:

> None of us had been in ENT, we had no relationships, we had really no idea what process we should follow to work with the American Academy of Otolaryngology and who the key players were. So, I think a lesson learned here is to understand the politics, understand the Academy and society's positions and process, and communicate. Get them involved, even if you're in stealth mode. Put some NDAs in place and be able to engage the key players. It could make your life a lot easier.

Approximately three years after its commercial launch, Acclarent had trained more than 4,000 physicians to use its technology and was rapidly building sales momentum. Fifteen peer-reviewed clinical publications had validated that Acclarent's technology was safe, effective, and offered significant quality of life benefits over common alternatives. Both AAO-HNS and ARS had endorsed reimbursement of the technology under existing FESS codes, and CMS as well as many private insurance companies were consistently reimbursing the tools when used in sinus surgery. Beyond the United States, the company was selling into 45 other countries around the world. With a team of nearly 300 employees, Acclarent also had developed a significant pipeline of products to support its organic growth for years to come.

Despite these promising initial results, "We still have a long way to go to get this to be the standard of care," said Facteau. Balloon Sinuplasty had so far achieved less than ten percent penetration into the existing surgical market. "There are still a lot of physicians that we need to train and that we need to move along the adoption continuum to get them to use this more frequently and think about this as the standard of care for treating chronic sinusitis." The company was also still working to become profitable.

Yet, Facteau, Makower, and Chang remained confident about Acclarent's ability to overcome these challenges and become a leader in the ENT field. Looking back across the arduous journey of the biodesign innovation process, Facteau commented:

> Do not underestimate the importance of people. At the end of the day, it is probably the most important thing that will make or break a company. My father used to tell me, 'You can't coach speed. You can either run fast or you can't.' I think you can apply a similar principle with people and business. You can't teach people to get out of bed in the morning, they either have the drive or they don't. You have to find the ones with the right work ethic and the right mind set. If you get people on your team that have conviction, believe in the product, and have the desire, the work ethic, and the heart, to do extraordinary things – that's the piece that makes great companies.

With a strong team intact, Acclarent was moving boldly toward its goals.

Credits

Many thanks go to William Facteau, John Chang, Sharon Lam Wang, and Earl "Eb" Bright for their substantial contributions to the development of this case.

Image credits

Front cover: 'Egg shell broken in two' courtesy of Eric Bean/Getty images; SONATA$_{TI}$^{100TM} Cochlear Implant. Courtesy of MED-EL. MED-EL Cochlear implants are the result of an innovative research tradition spanning over 30 years. The SONATA$_{TI}$^{100TM} cochlear implant represents the latest development in implantable hearing technology.

Stage 1: 'To the hospital' © iStockphoto.com/Rebecca Grabill

Stage 2: 'Stethoscope' © iStockphoto.com/jgfoto

Stage 3: 'The team' © iStockphoto.com/David Newton

Stage 4: 'Surgeon' © iStockphoto.com/atbaei

Stage 5: 'Business people': © iStockphoto.com/Mikael Damkier

Stage 6: 'Strength in numbers' © iStockphoto.com/Jacob Wackerhausen

Index

510(k), 278
  abbreviated, 280
  clinical data for, 463-64
  cost, 284
  criticism of, 280
  de novo clearance, 280
  marketing considerations, 462-63
  meeting with the FDA, 464
  predicate device, 279, 461
  review process and timing, 280-81
  special, 280
  strategies, 461-63
  submission, 280
  substantial equivalence, 279, 461
Abbott Vascular, 125, 399, 462
Abele, John, 118, 275, 557
acceptance criteria, 5, 7-9, 12, 146, 202
Acclarent, 51-55, 165-71, 205-6, 320, 378-84, 596-608, 727-33
Accuray, Inc., 542, 545-48, 568-69
acquisition, 696, 714-15
Adler, John, 545-48, 568-69
Advanced Cardiovascular Systems (ACS), 373-75, 414, 544, 721
Advanced Medical Technology Association (AdvaMed), 551, 567
advocacy, 101, 503, 512, 516, 517-18, 542-45
Alamin, Todd, 200-202
Allen, Michael, 308-9
alternate pathways
  acquisition, 696, 714-15
  building value toward, 715-18
  financial terms, 720, 723
  lack of market access, 709
  licensing, 713-14, 723
  managing, 719-20
  partnering, 710-11
  preparing for, 718-19
  returns, timing, and cost, 709-10
  sale, 696, 714-15

American Medical Association (AMA)
  code of ethics, 16
  CPT codes, 302, 508, 509-10
  Relative Value Scale Update Committee, 509, 513
American Medical Systems (AMS), 8, 587
American Society of Quality (ASQ), 475
Amylin Pharmaceuticals Inc., 711
anatomy, 64-65
Angiotech, 582
animal studies, 418, 428, 429
ArthroCare Corporation, 411-13, 581
AstraZeneca, 710
Atritech, 372

Bakken, Earl, 6
Bariatric Partners, 330-32
barrier to entry, 361, 388, 395, 580, 582
baseline, 64, 369, 371
BATNA, 718-19
Baxter, 587
Belson, Amir, 328-29, 693-95
bench testing, 348-49, 417, 418
Bennett, Ian, 200-202
Blue Cross/Blue Shield, 306, 504
Blue Cross/Blue Shield Technology Evaluation Center (TEC), 307, 504, 513, 530
Boston Scientific, 475, 561, 582, 586
brainstorming, 176, 178
  capturing ideas, 180-81, 186, 194-95
  choosing a facilitator, 183
  choosing a topic, 182
  IP ownership, 186
  managing the session, 185-86
  preparing for, 183-84
  rules, 178-82, 191
  selecting participants, 182-83

  use of artifacts, 181, 184
  use of computers, 181-82
breadboard, 364-66
bridge loan, 682
Bujalski, Edmund, 330-32
business model, 319
  choosing, 320-21, 337-38, 557
  factors affecting, 320
  operationalizing, 336
  validating, 335-36
business model types
  capital equipment, 327-28, 557-58
  disposable, 321-23, 557
  fee per use, 332
  implantable, 325-26, 557
  over-the-counter (OTC), 334, 542
  overview, 321
  physician-sell, 335, 542
  prescription, 334-35
  reusable, 324-25, 557
  service, 329-30
business plan
  considerations, 658-59
  executive summary, 659, 661, 672-75, 683
  funding, 666
  intrapreneurship, 669
  purpose, 658
  structure, 659-61
  teams, 668-69
  writing, 661-63
business strategies
  distribution play, 587
  fast follower, 586
  first mover, 585-86
  me-too, 586-87
  niche play, 587
  OEM/licensing, 587-88
  overview, 585
  partnering, 588
Buxbaum, Darin, 187-89, 663-65

Index

C.R. Bard Inc., 483
cadaver testing, 350, 418, 428
Cahill, Colin, 200–202
Cambridge Antibody Technology (CAT), 710
Cardica, Inc., 557, 619–23, 623, 637–55
Cardiovascular Imaging Systems, 279
Cardiva Medical, 462
cash flow statement, 629–31
cash requirements, 630, 677
CE mark, 287, 468–69
Center for Biologics Evaluation & Research (CBER), 275, 459
Center for Devices & Radiological Health (CDRH), 275, 276, 459, 460
Center for Drug Evaluation & Research (CDER), 275, 459
Centers for Medicare and Medicaid Services (CMS), 100, 275, 459, 509, 511
certificate of conformity, 287
Chang, John, 53–55, 165–71, 205–6, 378–84, 596–608, 727–33
Chess, Robert, 711–13
Churchill, Winston, 610
Cierra, Inc., 152–55
Cigna, 513
circuit diagram, 364
clinical strategy, 425
    considerations, 426–27
    purpose, 426
clinical studies
    case control, 431–32
    cost, 432–33, 626, 627–28
    ethics, 426, 440–41, 443–44
    FIM, 418, 429, 430–31
    GCP, 442–43
    geographic location, 290, 433–34
    GLP, 430
    IACUC, 429–30
    non-clinical, 428–29
    pre-clinical, 429
    randomized, controlled, 432
    registries, 431
    trial nomenclature, 432
clinical study goals
    marketing, 427
    regulatory, 427
    reimbursement, 427

clinical trial design
    core laboratories, 441
    data entry/monitoring, 441
    DSMB, 441
    endpoints, 429, 434–35, 538
    hypothesis, 434
    IDE, 436–37, 455
    informed consent, 440–41
    investigators and center, 435–36
    IRB, 437
    patient enrollment, 439–40
    size, 435
    statistical design, 452–54
    statistical power, 432, 435, 453
    tips, 444
clinical trial management
    data management, 442
    research site staff, 441–42
    sponsor personnel, 442
coding. See *reimbursement*
common stock, 681
comparables analysis, 128, 549, 631–32, 637–55, 690
competitive advantage, 581
    as basis of competition, 588–89
    capabilities, 581, 583–85
    link to value proposition, 581, 582
    positional, 581, 582–83
    statement of, 591
Computer Motion, 400–403
Concentric Medical, 323–24
concept maps, 197, 199
concept screening, 194
    clustering, 195–97
    concept mapping, 197–99
    concept vs. approach, 195, 202
    eliminating ideas, 199–200, 202
    organizing ideas, 195–99
    seeking expert advice, 200
concept selection, 367
    alternative to Pugh method, 370
    Pugh method, 368–73
    taking a structured approach, 368
conflict of interest, 13–14, 443–44, 551
conformity assessment, 287
Conor Medsystems, 688–90
contract research organization (CRO), 442
contracting administration fee (CAF), 560

convertible bonds, 682
copyright, 211
Corbett, Jim, 589–91
Cordis. See *Johnson & Johnson*
cost of goods sold (COGS), 613, 624
cost projections
    manufacturing costs, 624–26
    operating expenses, 626–28
    salary analysis, 624
Croce, Bob, 61–63, 518–19
Curtis, Gary, 323–24
Cyberonics, 513
cycle of care, 22, 97–99, 538

Davila, Liz, 333–34
debt funding, 681–83
Deem, Mark, 77–78, 152–55
defensive patent strategies
    continuations, 396–97
    layered protection, 397
    picket fence, 396, 397
depreciation, 629
design controls, 477, 478, 483, 485
design drawing
    computer-aided design (CAD), 359
    cross-sectional, 358–59
    isometric, 358
    layout, 358–59
    orthographic projection, 359
design patent, 211
design validation, 478
design verification, 478
Devices for Vascular Intervention (DVI), 444–47, 715
Diasonics, 489–91
dilution, 682, 692
direct sales models
    interaction with marketing, 568
    link to R&D, 570
    process, 564–66
    role of payers, 567
    sales training, 567–68
discount rate, 691
discounting terminal value, 690
disease function. See *mechanism of action*
disease progression, 66
disease state research, 60
    anatomy, 64–65
    clinical outcomes, 68

Index

clinical presentation, 67
economic impact, 69
epidemiology, 69
pathophysiology, 66
performing, 61, 63–64, 70
physiology, 64–65
summarizing, 70
Dollens, Ron, 544, 658, 722
Drucker, Peter, 58
due diligence, 685

earnings statement. *See income statement*
Edison, Thomas, 58, 177
Edwards Lifsciences, 583
elevator pitch, 683
Eli Lilly, 710, 715
Emphasys Medical, 77–78
Endogastric Solutions, 583
EndoNav, 300–301
Engelson, Eric, 152–55
engineering. *See R&D*
epidemiology, 69
equity funding, 680–81
equity vs. debt, 682–83
Ethicon, 585, 587
ethics
    AdvaMed code of ethics, 551, 567
    AMA code of ethics, 16
    clinical studies, 426, 443–44
    conflict of interest, 13–14, 443–44, 551
    fiduciary duty, 14–15
    IACUC, 429–30
    in marketing, 551–52
    in observations, 28
    informed consent, 440–41
    overview, 13–16
    principles, 15–16
    reimbursement support, 522–24
    stakeholder analysis, 111–13
    Tuskegee syphilis experiment, 14
ethnographic research, 25
European Patent Office (EPO), 227
ev3, 589–91
Evalve, Inc., 351–54
evergreening, 399
exclusion criteria, 439
exempt devices, 278
exit event. *See liquidity event*

ExploraMed, 8, 11–12

Facteau, William, 378–84, 596–608, 727–33
Federal Food and Drugs Act, 274
Fielding, Louie, 200–202
financial model, 613
    cash balance and financing needs, 613, 630
    cash flows, 613, 628–31
    cost projections, 613, 623–28
    income statement, 613, 628
    market model, 613, 616–23
    operating plan, 613, 614
    profitability, 631
    staffing plan, 613, 614–16
financing milestones, 630, 685–87
first-in-man (FIM) testing, 418, 429, 430–31
Fischer, Frank, 437–39
flow of money, 97, 99–101, 538
Fogarty, Thomas, 22, 177, 300, 613
Food & Drug Administration (FDA)
    approval letter, 293
    background, 274–75
    definition of medical device, 275
    device classification, 276–78
    fees, 284
    GCP, 442–43
    general controls, 278
    GLP, 430
    GMP, 483
    HDE, 283–84
    IDE, 282–83, 436–37, 460–61
    MDUFMA, 284
    QSR, 278, 282, 475, 483–85, 499–500
    quality inspections, 485–89
    regulatory pathways, 278–82, 461
    role in reimbursement, 504
    working with, 464, 465, 472
Food, Drug, and Cosmetic Act, 274
Ford, Henry, 2
Foundry, The, 9, 77–78, 152–55
Fox Hollow, 583
freedom to operate (FTO), 213–14, 390–92
From the Field
    Accuray, Inc., 545–48, 568–69
    ACS, 373–75
    ArthroCare Corporation, 411–13

Bariatric Partners, 330–32
Cardica, Inc., 619–23
Cierra, Inc., 152–55
Concentric Medical, 323–24
Diasonics, 489–91
DVI, 444–47
Emphasys Medical, 77–78
ev3, 589–91
Evalve, Inc., 351–54
ExploraMed, 11–12
Genomic Health, 135–37
Heartport, 466–67
HourGlass, 187–89, 663–65
Inhale Therapeutic Systems, 711–13
InnerPulse, 108–11
InterWest Partners, 666–68
Intuitive Surgical, 400–403
iRhythm, 231–34
Johnson & Johnson, 61–63, 518–19
Kyphon, 572–74
Medtronic, 6
Metrika, 308–9
Micrel Medical Devices, 562–63
NeoGuide, 328–29, 693–95
NeoTract, Inc., 419–22
Northwestern University, 43–44
OEC Medical Systems, 489–91
Perclose, 284–86
Rapid Exchange, 721–23
Respironics, 30–32
Simpirica Spine, 200–202
Spiracur, 392–93, 723–24
St. Francis Medical, 326–27
Stanford University, 187–89
The Foundry, 77–78, 152–55
University of Cincinnati, 30–32
Ventritex, Inc., 437–39
VISX, 333–34
functional blocks, 341, 344–45, 346, 411
funding
    dilution, 682, 692
    due diligence, 685
    managing the process, 693
    milestones, 686–87
    preparing for, 677
    stages/uses, 680
    strategic considerations, 690–92
    term sheet, 692–93
    valuation, 687–92, 702

Index

funding needs, 630, 677
funding sources
   approaching investors, 683
   choosing investors, 679–80
   types of investors, 677–79, 700–701
   understanding investors, 683–85
funding types
   debt, 681–83
   equity, 680–81
   equity vs. debt, 682–83

Gelsinger, Jesse, 229, 437
General Agreement on Tariffs and Trade (GATT), 219
Genomic Health, 106–7, 135–37, 549–50
Genzyme, 504–5
Getting Started
   alternate pathways, 724–25
   brainstorming, 189–91
   business model, 337
   business plan, 669–71
   business strategy, 593–95
   clinical strategy, 447–50
   competitive advantage, 593–95
   concept screening, 203–4
   concept selection, 375–77
   disease state research, 70–72
   financial model, 633–36
   funding, 697–99
   intellectual property, 234–37
   IP strategy, 403–5
   licensing, 724–25
   market analysis, 137–41
   marketing strategy, 552–54
   need statement, 47–49
   needs filtering, 157–59
   observation, 33–35
   operating plan, 633–36
   prototyping, 355–57
   quality, 492–95
   R&D strategy, 422–23
   regulatory, 290–92
   regulatory strategy, 469–71
   reimbursement analysis, 312–15
   reimbursement strategy, 525–27
   sales and distribution strategy, 575–76
   stakeholder analysis, 113–15
   stakeholder strategy, 552–54
   strategic focus, 16–17
   treatment research, 89–90
GI Dynamics, 583
Gifford, Hanson, 77–78, 152–55
Given Imaging, 280
Glenn, Ben, 231–34
Global Harmonization Task Force (GHTF), 287
global purchasing organization (GPO), 559–60
Gonzalez, Richard, 715
Good Clinical Practices (GCP), 442–43
Good Laboratory Practices (GLP), 430
Good Manufacturing Practices (GMP), 483
group purchasing organization (GPO), 565, 566–67
Guidant, 465, 488, 545, 583, 585
Gyrus, 587

Hawkins, William, 6
Health Care Financing Administration, 304
Health Care Professionals Advisory Committee Review Board (HCPAC), 304
health hazard evaluation, 488
Health Insurance Portability and Accountability Act (HIPAA), 28
health maintenance organizations (HMOs), 306
healthcare value chain, 541
Heartport, 130, 466–67
hospital inpatient reimbursement
   obtaining new codes, 510–11
   using existing codes, 304–5
hospital outpatient reimbursement
   obtaining new codes, 510
   using existing codes, 305
HourGlass, 187–89, 663–65
humanitarian device exemption (HDE), 283–84

ideation, 176
   approaches besides brainstorming, 189
   guidelines, 177–78
IDEO, 178–82, 191
Imran, Mir, 5, 45, 630, 659, 677
incidence, 69
inclusion criteria, 439
income statement, 628
Indian Patent Office (IPO), 227
indications for use (IFU), 274, 280, 462, 465
indirect sales and distribution models
   national distributors, 559–60
   specialized distributors, 560–61
   third-party partnerships, 561
informed consent, 440–41
Inhale Therapeutic Systems, 588, 711–13
initial public offering (IPO), 696–97
InnerPulse, 108–11, 126–27, 589
innovation notebook, 28–29, 146, 186, 228, 355
Institutional Animal Care and Use Committees (IACUC), 429–30
Institutional Review Board (IRB), 283, 437, 460
integrated delivery network (IDN), 559–60, 565, 566–67
intellectual property (IP), 211
   documentation, 217–18, 228
   freedom to operate (FTO), 213–14, 390–92
   interference proceeding, 213
   inventorship, 228–30
   law, 211
   patentability, 212–13
   prior art, 214–19, 390
   public disclosure, 212, 218, 230–31
Intermountain Health, 559
International Organization for Standardization (ISO), 483
international patenting
   European Patent Office (EPO), 227
   in China, 227–28
   in India, 227–28
   Indian Patent Office (IPO), 227
   International Searching Authority (ISA), 227
   Patent Cooperation Treaty (PCT), 226
   PCT process, 226–27
   State Intellectual Property Office (SIPO), 228
InterWest Partners, 666–68
intrapreneurship, 632–33, 669
Intuitive Surgical, 400–403, 462
investigational device exemption (IDE), 282–83, 436–37, 460–61

Index

investors. *See funding sources*
Invivo Surgical Systems, 560
IP strategy
   defensive, 396–97
   freedom to operate (FTO), 390–92
   hiring legal help, 393–94
   international, 394–95
   licensing, 391–92
   managing a patent portfolio, 395–96
   offensive, 397–99
   patent coverage, 389–90
   risk, 391–92
   timing, 399
iRhythm, 231–34
ISO quality system, 483–85, 497–98

Johnson & Johnson, 61–63, 508, 512, 518–19, 581, 582, 583, 586, 587, 588

Kaiser Permanente, 307, 559
Kelley, David, 208
Kelley, Tom, 174, 347
Kelso, David, 43–44
key opinion leader (KOL), 427, 506, 521, 541, 551, 583
Koskinas, Ellen, 666–68, 696
Kumar, Uday, 231–34
Kyphon, 469, 523, 572–74, 581, 585, 591–93

Lake Region Corporation, 588
Lakein, Alan, 386
Lam Wang, Sharon, 165–71, 378–84, 596–608, 727–33
Lamson, Ted, 11–12, 419–22
Lebeau, Guy, 103, 408
leveraged finance, 682
lexicographer, 391
licensing, 229, 713–14, 723
Lifescan, 561
liquidity event, 687, 695–97
Litvack, Frank, 426, 552
local coverage determination (LCD), 302, 511, 512

MacMillan, Steve, 567, 710
Makower, Josh, 11–12, 51–55, 165–71, 205–6, 378–84, 596–608, 727–33
market analysis, 117
   dynamics, 123–27

global considerations, 137
   needs, 127–28
   performing, 118
   pitfalls, 134–35
   segmentation, 118–21
   size, 121–23
   summarizing, 131–34
   target, 128–31
   willingness to pay, 128
market development, 320
market model
   bottom-up, 122, 618–23, 709
   top-down, 121–23, 616–18
market segmentation, 118–21
marketing
   capturing value, 537
   communication strategy, 542–44
   direct-to-consumer (DTC), 101, 335, 542
   ethics, 551–52
   PR strategy, 544–45
   primary functions, 537
marketing mix
   place/distribution channel, 542
   positioning, 538–41
   price, 549–51
   product/service mix, 541–42
   promotion and advocacy, 542–45
master resellers, 560
May, Allan, 489–91
mechanism of action, 61, 66, 79, 81, 88, 195
Medicaid, 100
Medical Device Amendments Act, 274
medical device reporting (MDR), 480, 496
Medical Devices Directive (MDD), 287, 288
medical necessity, 504
Medicare, 100, 107, 120, 300, 301–2, 302–3, 507–11, 511–13, 513–14
Medinol, 561
Medtronic, 6, 464, 465, 510, 511, 523, 561, 582, 589
Metrika, 105, 308–9, 508
Micrel Medical Devices, 562–63
mission statement, 5–7
Moll, Fred, 400–403

national coverage determination (NCD), 302, 511, 512–13

National Health Service (NHS), 99, 312
National Institute for Health and Clinical Excellence (NICE), 312, 513
need, 37
   blue-sky, 46–47, 145, 152
   categorizing, 46–47
   criteria, 43, 156–57, 199, 346, 368
   incremental, 46, 145, 152
   managing risk, 44–46
   mixed, 47, 152
   specification, 155–57, 160–63, 199
need statement, 38
   definition, 21
   developing, 38–39
   embedding a solution, 40
   outcome measures, 42–43
   scope, 40–41
   validation, 41–42
needs filtering, 143
   approach, 144
   assigning ratings, 146
   documenting, 146, 150
   eliminating options, 150
   producing scores, 146–50
   screening criteria, 144–46
   screening thresholds, 150–52
Nektar Therapeutics. *See Inhale Therapeutic Systems*
NeoGuide, 134, 328–29, 693–95
NeoTract, Inc., 12, 419–22
Newell, Bob, 619–23
non-disclosure agreement (NDA), 230–31, 269
non-provisional patent. *See utility patent*
Northwestern University, 43–44
notice of allowance, 225
notified bodies, 287
null hypothesis, 452

observation
   conducting, 25–27
   documenting, 28–29
   ethics, 28
   knowing when to stop, 29–30
   perspectives, 22–24, 27
   preparing for, 24–25
   statement, 21
observer effect, 26
OEC Medical Systems, 489–91

Index

offensive patent strategies
    blocking, 398
    bridging, 398
office action (OA), 225, 389
Office for Human Research
    Protections (OHRP), 437
Office of Combination Products (OCP),
    275, 276, 460
Office of Device Evaluation (ODE), 275
Office of In Vitro Diagnostic Device
    Evaluation and Safety, 275
Office of Initial Patent Examination
    (OIPE), 224
office of technology licensing (OTL),
    394, 723
off-label use, 82, 85, 284, 436, 464,
    467–68
operating expenses (OpEx), 626
operating plan, 614
option pool, 692
Osteotech, 562

Palmaz, Julio, 177, 228
partnering, 710–11
patent, 211, See also utility or
    provisional patent
    agent, 393–94
    attorney, 393–94
    claims, 221–22
    criteria for obtaining, 212–13
    first to file, 213
    first to invent, 213
    international, 226–28
    searching, 214–19
patent classification index, 216
Patent Cooperation Treaty (PCT), 226
patent litigation, 212, 389, 399–400
pathophysiology, 66
Pauling, Linus, 2
Perclose, 284–86, 350, 462
Percusurge, Inc., 468
Perkins, Rodney, 230, 684
personal health information (PHI), 28
personal inventory
    acceptance criteria, 7–9
    mission, 5–7
    relevance to alternate pathways,
        709
    strengths and weaknesses, 7
Pfeffer, Jeffrey, 610
Pfizer, 582, 588

physician reimbursement
    obtaining new codes, 508–10
    using existing codes, 303–4
physiology, 64–65
pilot study, 432, 451
Pinto, Moshe, 122–23, 392–93,
    723–24
pivotal study, 432, 451
Plain, Hank, 284–86
Popovits, Kim, 136
Porter's five forces, 123–25
postmarketing study, 432, 451
pre-amendment devices, 279
predicate device, 279, 461
preferred provider organizations
    (PPOs), 306
preferred stock, 681
premarket approval (PMA), 281
    clinical data requirements, 281
    cost, 284
    meeting with the FDA, 465
    modular, 282
    post-approval requirements,
        465–66
    product development protocol
        (PDP), 282
    review process and timeline, 282
    strategies, 464–65
    submission, 282
    traditional, 282
premarket notification, 278
prevalence, 69
pricing
    bundled, 550
    differential, 550
    establishing a baseline, 549
    gainsharing, 550–51
    link to reimbursement, 549
    pay-for-performance, 550, 551
    value-based, 549
prior art, 212, 214–19, 221, 225, 390
Privitera, Mary Beth, 30–32
problem, 21
    identifying, 21–22, 27–28
    statement, 21
process management
    applications beyond quality, 491–92
    in product development, 474
    principles, 474
professional associations
    as advocates, 503, 544

    as stakeholders, 103
    listing, 18
    role in reimbursement, 517
profit and loss (P&L) statement. See
    income statement
profitability, 631
proof-of-concept, 408, 614, 687
prototypes
    electrical, 363–66
    materials science, 361–62
    mechanical, 358–60
prototyping
    combination products, 345
    functional blocks, 341
    goal, 340
    in the testing continuum, 348–50
    inhouse vs. outsourcing, 350
    iterative nature of, 340–41, 348, 354
    models, 342–44
    timing, 347–48
    tips, 354–55
    user involvement, 346–47
prototyping guidelines
    choose model, 342–44
    define requirements, 346–47
    identify functional blocks, 344–45
    identify questions, 342
    understand what is known,
        345–46
provisional patent, 211
    considerations, 220
    converting to utility patent, 219–20
    cost, 219
    definition, 219
    filing, 220
    sample, 238
provisional protection, 399
proxy analysis, 631–32, 637–55
public relations strategy, 544–45
Pugh method for concept selection
    assign scores and rank concepts,
        371–72
    confirm concepts and choose a
        baseline, 371
    identify user and design
        requirements, 369–70
    iterative nature of, 369–70
    overview, 368–69
    selection matrix, 372
    weight user and design
        requirements, 370–71

quality assurance (QA), 476, 483
quality control (QC), 476, 483
quality management system (QMS), 474
   CAPA, 477, 480–81
   definition, 476–77
   design controls, 477, 478
   equipment and facility controls, 477, 481
   implementing, 481–83
   ISO, 497–98
   management controls, 477–78
   materials control, 477, 481
   objectives, 477
   P&PC, 477, 478–80
   QSR vs. ISO, 483–85
   records, documents, and change controls, 477, 481
quality system audits
   compliance/enforcement, 486–87
   recalls, 487–89
   what to expect, 485–86
Quality System Regulation (QSR), 278, 282, 475, 483–85, 499–500
Quality Systems Inspection Technique (QSIT), 485

R&D (research and development), 407, 408
R&D lab, 424
R&D plan
   engineering resources, 416–17, 424
   link to strategy, 407, 409
   staffing, 414–15, 626
   testing methods, 417–19
   timelines, 419
R&D strategy, 407, 408
   approach, 409–11
   link to manufacturing costs, 626
   link to quality, 409
   milestones, 409–10, 614
   purpose, 408
   technical challenges, 410–11
reduction to practice, 213
referral patterns, 99
regulation outside the U.S.
   Australia, 288–89
   Brazil, 289
   Canada, 288
   China, 290
   Europe, 287–88

Global Harmonization Task Force (GHTF), 287
   India, 290
   Japan, 289
   Russia, 289–90
regulatory strategy
   510(k) clearance, 461–64
   environment, 459
   IDE, 460–61
   link to reimbursement, 459
   off-label use, 467–68
   outside the U.S., 468–69
   premarket approval, 464–66
   targeting within the FDA, 459–60
   working with the FDA, 464, 465, 472
reimbursement analysis, 299
   importance of, 300–301
   performing, 307, 309, 505–7
reimbursement in the U.S.
   background, 301
   coverage determination, 511–13
   decision criteria by stakeholder, 504
   Medicare coding, 301–2, 507–11
   Medicare coverage, 302, 511–13
   Medicare payment, 302–3, 513–14
   miscellaneous codes, 511
   new technology add-ons, 305
   private payers, 306–7, 511, 513, 514
   role of the FDA, 504
   self-pay, 307
   summary of codes, 316
reimbursement outside the United States
   considerations, 309–11
   France, 311
   Germany, 311–12
   India, 312
   strategy, 524–25
   United Kingdom, 312
reimbursement strategy
   advocacy, 517–18
   budget, 524
   developing, 505
   goal, 504
   impact on clinical trials, 516–17
   maintaining, 524
   payer value proposition, 516, 528–29
   pricing, 519–20
reimbursement tactics
   conferences, 521–22
   dossier, 522

payer advisory board, 521
payer segmentation, 520–21
publications, 521
reimbursement support, 522–24
Relative Value Scale Update Committee (RUC), 509, 513
relative value units (RVUs), 304, 509
research and development. *See* R&D
Resource Based Relative Value System (RBRVS), 304
Respironics, 30–32, 510

Safe Medical Devices Act, 274
sales agent, 560
sales and distribution models
   choosing, 558–59, 570
   direct, 558, 563–68
   global considerations, 575
   hybrid, 558, 561–62
   indirect, 558, 559–61
   summary, 577–78
Satiety, 583
SciMed, 586
Scott, Randy, 135–37
SG&A expenses, 627–28
Shelhigh Inc., 483
Sherman Act, 587
Sidow, Kevin, 326–27, 570–71
Simpirica Spine, 200–202
Simpson, Carl, 373–75, 722
Simpson, John, 177, 373–75, 444–47, 721
simulated use testing, 348–49, 418
solution bias, 35, 39–40, 146, 195
Spiracur, 87, 122–23, 392–93, 723–24
St. Francis Medical, 326–27, 570–71
St. Goar, Fred, 351–54
Stack, Richard, 97, 108–11, 118
staffing plan, 614–16
staffing ratios, 616
stakeholder analysis, 95
   cycle of care, 97–99, 538
   ethics, 111–13
   flow of money, 97, 99–101, 538
   global considerations, 107–8
   output, 111
stakeholder interests
   facilities, 104–6
   overview, 101
   patients, 101–2
   payers, 106–7
   physicians, 102–4

Index

standard of care, 9, 504
standard operating procedure (SOP), 430, 481
Stark Law, 551
Starling, Bill, 108–11
State Intellectual Property Office (SIPO), 228
statistical power, 435
statistical trial design, 452–54
Stevens, John, 466–67
strategic focus
  articulating, 9–10
  ethics, 13–16
  global considerations, 12–13
  statistics to consider, 9–10
Stryker, 583, 587
substantial equivalence, 279, 461
substantially equivalent (SE) determination, 281
superseding need, 45–46
Surmodics, 587
Sutton, R., 610
SWOT analysis, 125–26

Talmadge, Karen, 572–74
target market
  accessibility, 129, 709
  adoption of new technology, 129–30
  investor funding, 129
technical specifications, 347, 408
Tenet, 559
term sheet, 692–93, 704–6
terminal value, 691
testing continuum
  bench testing, 348–49
  in clinical strategy, 427–32
  in R&D, 417–19
  overview, 348
  simulated use testing, 348–49
  tissue testing, 349–50
Thapliyal, Hira, 411–13
Thornton, Troy, 351–54
tissue testing, 349–50, 418
trade secret, 211
  instead of patent, 211, 389–90
trademark, 211
treatment analysis
  clinical profile, 81
  economic profile, 84–85
  emerging profile, 85–86
  overview, 79
  summary, 87–88
  utilization profile, 85
treatment gaps, 75, 76–77, 92
treatment landscape, 87–88
treatment research, 74
  approach, 78–79
  considerations, 75–76
  gap analysis, 76–77, 92
  goal, 75
treatment types, 75
Tsoukalis, Alexandre, 562–63
turf war, 102, 545

US healthcare financing, 100
US Patent and Trademark Office (USPTO), 213, 216, 224–25
US Surgical, 585
US third-party payers, 28, 531–33
UnitedHealthcare, 550
University of Cincinnati, 30–32
user and design requirements, 346–47, 368, 408
utility patent, 211
  claims, 221–22, 390–91
  cost, 225–26
  drawing, 221
  filing and review, 224–25
  multiple filings, 389
  preparing, 222–24
  priority date, 219, 225
  sample, 257
  separating into parts, 389
  specification, 220–21, 391

valuation, 687–92, 702
value analysis, 128, 549
value proposition, 538
  choosing/proving, 538–41
  developing, 538
  link to competitive advantage, 581, 582
  payer-specific, 516, 528–29
  physician-specific, 517
Ventritex, Inc., 437–39
venture-backed loan, 682
VISX, 333–34

warrant, 682, 692
Will, Allan, 444–47
Wilson Greatbatch, 588
World Intellectual Property Organization (WIPO), 226

Yock, Paul, 721–23

Zimmer, 583